FATIGUE

Neural and Muscular Mechanisms

ADVANCES IN EXPERIMENTAL MEDICINE AND BIOLOGY

Recent Volumes in this Series

Volume 377
TISSUE RENIN–ANGIOTENSIN SYSTEMS: Current Concepts of Local Regulators in Reproductive and Endocrine Organs
Edited by Amal K. Mukhopadhyay and Mohan K. Raizada

Volume 378
DENDRITIC CELLS IN FUNDAMENTAL AND CLINICAL IMMUNOLOGY, Volume 2
Edited by Jacques Banchereau and Daniel Schmitt

Volume 379
SUBTILISIN ENZYMES: Practical Protein Engineering
Edited by Richard Bott and Christian Betzel

Volume 380
CORONA- AND RELATED VIRUSES: Current Concepts in Molecular Biology and Pathogenesis
Edited by Pierre J. Talbot and Gary A. Levy

Volume 381
CONTROL OF THE CARDIOVASCULAR AND RESPIRATORY SYSTEMS IN HEALTH AND DISEASE
Edited by C. Tissa Kappagoda and Marc P. Kaufman

Volume 382
MOLECULAR AND SUBCELLULAR CARDIOLOGY: Effects of Structure and Function
Edited by Samuel Sideman and Rafael Beyar

Volume 383
IMMUNOBIOLOGY OF PROTEINS AND PEPTIDES VIII: Manipulation or Modulation of the Immune Response
Edited by M. Zouhair Atassi and Garvin S. Bixler, Jr.

Volume 384
FATIGUE: Neural and Muscular Mechanisms
Edited by Simon C. Gandevia, Roger M. Enoka, Alan J. McComas, Douglas G. Stuart, and Christine K. Thomas

Volume 385
MUSCLE, MATRIX, AND BLADDER FUNCTION
Edited by Stephen A. Zderic

A Continuation Order Plan is available for this series. A continuation order will bring delivery of each new volume immediately upon publication. Volumes are billed only upon actual shipment. For further information please contact the publisher.

FATIGUE

Neural and Muscular Mechanisms

Edited by

Simon C. Gandevia
Prince of Wales Medical Research Institute
Sydney, New South Wales, Australia

Roger M. Enoka
The Cleveland Clinic Foundation
Cleveland, Ohio

Alan J. McComas
McMaster University
Hamilton, Ontario, Canada

Douglas G. Stuart
The University of Arizona
Tucson, Arizona

and

Christine K. Thomas
University of Miami
Miami, Florida

Technical Editor

Patricia A. Pierce
The University of Arizona
Tucson, Arizona

SPRINGER SCIENCE+BUSINESS MEDIA, LLC

Library of Congress Cataloging-in-Publication Data

On file

Proceedings based in part on the symposium on Neural and Neuromuscular Aspects of Muscular Fatigue, held November 10–13, 1994, in Miami, Florida

ISBN 978-1-4899-1018-9 ISBN 978-1-4899-1016-5 (eBook)
DOI 10.1007/978-1-4899-1016-5

Originally published by Plenum Press, New York in 1995
Softcover reprint of the hardcover 1st edition 1995

10 9 8 7 6 5 4 3 2 1

The challenge is to understand fatigue from the level of the single muscle fiber to the whole organism. [Photograph from Abbott, Bigland, and Ritchie, *Journal of Physiology* (London) **117**:382, 1952. Reprinted with permission.]

PREFACE

This volume describes the current state of our knowledge on the neurobiology of muscle fatigue, with consideration also given to selected integrative cardiorespiratory mechanisms. Our charge to the authors of the various chapters was twofold: to provide a systematic review of the topic that could serve as a balanced reference text for practicing health-care professionals, teaching faculty, and pre- and postdoctoral trainees in the biomedical sciences; and to stimulate further experimental and theoretical work on neurobiology.

Key issues are addressed in nine interrelated areas: fatigue of single muscle fibers, fatigue at the neuromuscular junction, fatigue of single motor units, metabolic fatigue studied with nuclear magnetic resonance, fatigue of the segmental motor system, fatigue involving suprasegmental mechanisms, the task dependency of fatigue mechanisms, integrative (largely cardiorespiratory) systems issues, and fatigue of adapted systems (due to aging, under- and overuse, and pathophysiology). The product is a volume that provides comprehensive coverage of processes that operate from the forebrain to the contractile proteins.

The neurobiology of muscle fatigue received minimal attention throughout the first six decades of the 20th century. Then, in the mid- to late 1970s, a series of reports appeared on neuromuscular aspects of fatigue that were featured in influential international symposia held in 1980 (London, UK), 1990 (Paris, France), 1992 (Amsterdam, The Netherlands), and 1994 (Miami, Florida). This sudden acceleration of interest in the study of neural and muscular aspects of fatigue is attributable to at least three factors: the practical importance and relevance of the topic; its long-neglected fundamental and clinical significance; and the type of integrated biological thought and experimentation needed to advance the field, extending from the cellular/molecular level to the behaving organism. This latter integrative emphasis is a feature of the present volume in all nine subsections. The volume also stresses the need for neuroscientists to expand our understanding of the role of the CNS in fatigue processes to the same extent as our current understanding of peripheral neuromuscular mechanisms. To this end, several chapters in the present volume emphasize how the full experimental armamentarium of the field of neuroscience can be applied to the neural aspects of fatigue.

While the volume forms a cohesive whole, we have respected the individuality of our authors, including innumerable nuances in their definitions of muscle fatigue. Similarly, in selected areas, there is a degree of overlap between chapters such that the reader can contemplate how the same issue is approached by different laboratories. These features of the volume should prove helpful when reviewed in pre- and postdoctoral training programs. To facilitate this possibility, the nine subsections of the volume are provided with a brief introduction prepared by the editors.

Since the mid-1970s, Brenda Bigland-Ritchie has been in the forefront of the study of the neuromuscular aspects of muscle fatigue. Using human subjects, she has shown to what extent the fatigue findings from animals or isolated tissues apply to real-life situations. More than this, she has stimulated inquiry into the search for reflex mechanisms which may serve to match the activity of the spinal cord with that of the fatiguing muscles. The present generation of scientists who study muscle fatigue owe much to her contributions and enthusiasm for the field. The chapters that follow document her impact. They also provide the rationale for why the editors and contributing authors have enthusiastically dedicated this volume to Dr. Bigland-Ritchie. Her efforts have been exemplary.

Simon C. Gandevia
Roger M. Enoka
Alan J. McComas
Douglas G. Stuart
Christine K. Thomas

ACKNOWLEDGMENTS

We greatly appreciate the dedication, effort, and talent that our technical editor, Patricia Pierce, has displayed in bringing this research monograph to successful fruition. We would also like to thank several University of Arizona colleagues: Dr. Edwin Gilliam, Dr. Jennifer McDonagh, Robert Gorman, and George Hornby for their help with the review and executive processing of the chapters in this volume.

Much of the work in this volume was presented by the authors of each chapter at an International Symposium, *Neural and Neuromuscular Aspects of Muscle Fatigue*, held in Miami, FL, November 10-13, 1994. Designated as a Satellite Symposium of the Annual Meeting of the Society for Neuroscience, the symposium honored Professor Brenda Bigland-Ritchie and was organized by Drs. Roger Enoka, Simon Gandevia, Alan McComas, Douglas Stuart, and Christine Thomas. The presentations in Miami by 33 invited scientists from Canada and the United States were supported largely by their own research funds, with some assistance from the University of Miami (Department of Neurological Surgery, Miami Project to Cure Paralysis) and the University of Arizona (Research Office of the College of Medicine, and Regents' Professor funds of Douglas Stuart). Travel awards were provided to 37 scientists from other countries (Australia, Belgium, Denmark, Finland, France, Germany, Israel, Norway, Slovenia, Switzerland, The Netherlands, UK, Yugoslavia) by the American Physical Therapy Association (Research and Analysis Division, and Section on Research), the Muscular Dystrophy Association (Conference grant no. 46168), the National Institutes of Health (National Center for Medical Rehabilitation Research, NICHD) and the Universities of Miami and Arizona. Ten trainee awards were provided by The Miami Project to Cure Paralysis. We would also like to thank Bette Mas (Miami Project to Cure Paralysis, University of Miami) and Lela Aldrich, Lura Hanekamp and Joan Lavoie (Department of Physiology, University of Arizona) for their considerable efforts to ensure the success of the symposium. The abstracts and discussion summaries from the symposium will be published by *Muscle & Nerve* in the Fall of 1995, as sponsored by the Henry L. and Kathryn Mills Charity Foundation and the Cleveland Clinic Foundation.

The Editors

ACKNOWLEDGMENTS

[illegible]

[illegible]

CONTENTS

LOOKING BACK

Brenda R. Bigland-Ritchie

Department of Pediatrics
Yale University School of Medicine
New Haven, Connecticut 06510

I do not know why I was selected for this honor; a book and a symposium both dedicated to me. Why me when so many others seem more worthy? But having been selected, I want first to express my deep gratitude and appreciation to those who decided to pay this quite extraordinary tribute to my work - the organizers, Roger Enoka, Simon Gandevia, Alan McComas, Douglas Stuart, and Christine Thomas. I do not know how they first got this idea, but I thank them for it more than I can say.

When asked to write this chapter, I was encouraged to look back over my career, and to describe the early events that influenced my path. It was all an adventure; mainly, of course, the adventure we all share when we keep nibbling away at some scientific problem until one of nature's well-kept secrets is at least partially revealed. I also found great satisfaction in exploring the limits of my own abilities at each stage in my career. In looking back, I realize now just how greatly I was influenced by the standards set by my early teachers, and also by chance encounters with people who just happened to be in the right place at the right time. We seldom know where fate may lead us. However, many incidents are recalled simply because they were amusing and came from an era now long gone.

From time to time, over the years, I have been told that I have played some kind of role model for other, younger women who faced problems similar to mine. If that is so, then it may also be of interest to hear how it felt to be a woman during a period in which public attitudes about the role of women in the workplace have changed so much, and how this has affected me. And what was the impact on my research of having chosen to interrupt my career for many years to raise my children, of changing countries; and what it took to get up and running again in physiology, a task I could never have achieved without the constant support, generosity and insistence of others. And finally, why was the effort all so very well worth while?

EARLY YEARS

I was an only child who spent her most formative adolescent years in British war-time (1939-45), mostly secluded in a country boarding school safely away from London bombing. The experience of those hard and anxious times was probably important for all of us who

Fatigue, Edited by Simon C. Gandevia et al.
Plenum Press, New York, 1995

later engaged in research; for the shortages of almost everything fostered a certain ingenuity in improvising. By luck, I was ready to start my undergraduate years at University College London in 1945, when the war in Europe was just over. What could be more exciting for a raw, overly protected, and inexperienced school girl than to find that the majority of her classmates were older ex-service men and women returning from a wide range of exciting and highly responsible war-time experiences, ranging from sea captains to prisoners of war assigned to such jobs as building the Burma Road. They opened our eyes as no ordinary freshman class could do to an awareness of what real life, responsibility and suffering could be. In those early years, I learned far more from them than from any of my formal classes (which I took fairly lightly at that time).

In 1945, the College had just returned from 4 years evacuation to Bangor, Wales, where large numbers of Welshman had enrolled. One of my fondest memories is of glorious male choruses which frequently erupted to fill the cloisters with harmony. This could happen at any time, as when sitting in the cafeteria someone would start humming, then the refrain was quickly taken up from different sides until it engulfed everyone.

I entered college not knowing which branch of science most attracted me. At one stage, I applied to take a degree in chemistry. Thank goodness they turned me down! Then someone said that a new degree course called *Physiology* was to be started. None of us had heard of that subject or what it would include; but we all rushed off to register because it meant taking courses with medical students whom every one knew were the most fun! So started a career which has brought such reward throughout my life, but which I might so easily have missed. I cannot now imagine being equally intrigued by any other topic. Most of the directions my life has taken since then have also been shaped by just such unpredictable chance happenings.

By going to University when I did, I was in time to catch the coat-tails of many of the outstanding scientists of those times. Sir Henry Dale, Gaddum, and other equally famous figures were all invited frequently to speak at our Medical Student Society evening meetings. Neither they nor we ever considered that this should entail an honorarium. How times have changed! I was taught by such people as Lovatt-Evans (colleague of Starling), Leonard Bayliss, Bernard Katz, to name a few. But the most profound early influence, then and later, came from A.V. Hill. My introduction to him was quite amusing. In those days, an important feature of science education, unfortunately lost today, was to emphasize the concept that all scientists stand on the shoulders of their predecessors, and that one can learn a lot from both the achievements and mistakes of those who went before. Thus, each lecture started with a brief history of how it had developed and by whom. They might start with such ancients as Socrates, or Aristotle, or more recent figures including Galvani, Cajal, Claude Bernard, Helmholz, Sherrington, etc. – all revered figures, but usually long dead. When the lectures on muscle began, reference was made at once to A.V. Hill, with the lecturer usually glancing upwards. I thus got the image of the great man playing a harp on some cloud in heaven. It was only much later I discovered that he was quite alive and well, but just happened to work on the floor above.

A.V. HILL

My first actual meeting with *A.V.*, as everyone called him, came through Murdoch Ritchie (later my husband). Murdoch was one of three young scientists (Bernard Abbott and Eric Denton were the others) whom A.V. recruited when he returned to science after spending the war years as a leading science advisor to the British and American governments. A.V. decided then that for modern biology to progress satisfactorily, it must, henceforth, engage the skills of engineers and physicists. So he proceeded to found what became the first officially named department of Biophysics. Murdoch, like the others, already had a degree in physics and mathematics, and had spent the final year of the war assigned to work on radar research as a *boffin boy*. They joined our classes when A.V. told them to go learn some biology.

Eventually I graduated and started thinking about a job. Things were so different in those days, and so much better than for young people today. I never worried as to whether there would be a job for me, or what that job would be, or whether there would be funds to let me do it. Something would turn up, and it did. I was lucky enough to be given a fellowship to stay on working in the same Physiology Department. I was assigned to work with another new graduate, Barbara Jehring, and we were given a large empty room to work in. After what seemed like many weeks, the professor (G. L. Brown) walked in one day and asked us whether we had yet decided what we were going to work on, how we would do it, and what equipment we would need. In those days, we were all expected to make our equipment, which we constructed in the excellent workshops provided by each department. Our first simple experiments were designed to answering a question posed by the biochemists as to whether hypertrophy from administration of growth hormone did make muscles stronger – the first time, as far as I know, that this question had been addressed. What frustrating experiments! We were lent a string galvanometer from which to measure force from the deflection of a beam of light focused onto a small mirror, with the magnitude of the deflection registered at the other side of the room. Hence the room had to be totally dark. Our final conclusion was that growth hormone did not affect muscle strength, but at that time I had little faith in our data. Curiously, this conclusion seems to have survived (although I have seldom seen our paper referred to). By that time, Barbara and I were both about to marry, and we consoled ourselves with the thought that, if necessary, we could use our new names in future to dissociate ourselves from those first bumbling efforts.

My next adventure, suggested directly by A. V. Hill, was clearly planned and executed much more seriously, and also much more fun. At that time, A.V. was measuring the heat produced in isolated frog muscles when they were allowed to shorten or were stretched. He suggested that Bud Abbott, Murdoch and I make similar comparisons of the relative metabolic costs in man by measuring oxygen consumption in each case. Despite his reputation for mathematical exactitude, and the esoteric and highly theoretical appearance of his work, he maintained a strong interest throughout his life in applying his conclusions to explain whole animal or human performance and athletic achievements measured in the field. One essential requirement for our proposed experiments was to be sure that the shortening and lengthening contractions were both made by identical muscle groups, moving at the same velocity, while exerting equal tensions. Only the direction of movement should be reversed. To do this, we borrowed a primitive bicycle adapted as an ergometer and hitched it back-to-back with the one the department already had (Fig 1). One cyclist pedaled forward (his quadriceps muscles shortening) while the muscles of the other cyclist were stretched by resisting the motion with enough force to prevent any change in speed. This assured that the forces in both directions were identical. Expired air was collected from each subject and its oxygen content measured. Enormous differences were revealed between the energy requirements in each case. Despite his obvious awareness as a physicist, A.V. insisted that the two movements be called *positive* and *negative* work. He had no problem with these terms since, when a muscles is stretched, work is done on it.

At A.V.'s urging, we twice demonstrated these experiments at the Royal Society where they aroused enormous interest and amusement. I have fond memories of the Lord Mayor of London, complete with all his ceremonial regalia, asking to *have a go*. Our fame quickly spread through London where the equipment came to be named the *push-me, pull-you*, after Dr. Doolittle's two-headed animal that never knew which way it was going. Many years later, a well-known scientist wrote about how a svelte young woman could resist

Figure 1. Apparatus used in 1952 to compare the metabolic costs of positive and negative work (from Abbott BC, Bigland B & Ritchie JM (1952). The Physiological cost of negative work. *Journal of Physiology (London)* **117**, 380-390; their Fig. 1).

and quickly exhaust the efforts of a husky young man, but had forgotten that these two were Murdoch and I! In later studies, we replaced the forward *positive* cyclist by a motor!

A.V. was a painfully shy man with no small talk. One of the things we all dreaded was to find ourselves with him at some departmental social function. But all was well if one could get him on to one of his favorite scientific theories, such as why dimensionally similar animals – whales and dolphins, or race-horses and greyhounds – can run or swim at the same top speed; or why the heart rates of humming birds and elephants must be so different. He would work it all out mathematically from their corresponding masses, the relative stresses on their bones and tendons, and their metabolism. In 1927, at a muscle meeting at Rockefeller University, someone asked how long it would take to run up the Empire State building. The next day the answers were to be compared with the best performance of two top athletes. Others made wild guesses which all turned out to be far too long. A.V., however, made a quick calculation, based on force-velocity, length-tension, and power-velocity relations to come up with an answer that was correct to within a fraction of a minute. For him, the role played by art, music or other forms of culture was best satisfied by the beauty he saw in mathematics applied to every aspect of life. His detachment from popular culture was well illustrated when I jokingly suggested to him one day that, with all the amusement we had caused, I should resign from our bicycle experiments so that they could be authored by *Abbott & Costello*. (Jose del Castillo occupied the next lab.) He just said "Why?". Clearly he had never heard of these comedians!

However, these experiments and their notoriety were important at a more serious level, in that they spawned a series of future studies by Asmussen, ourselves and many others who came to recognize that muscles resist stretch as often as they shorten, and that to coordinate their actions accurately in each mode requires quite different control strategies by the CNS. Our experiments also helped to promote A.V.'s famous challenge to biochemists in which he asked them, among other things, to show whether the reactions of the Krebs cycle may be reversible and if work done on a muscle (negative work) might result in the re-synthesis of ATP. I remember many years later being told by a now famous scientist that when he first read our paper as a graduate student, he went to the gym to see if he could exhale oxygen while climbing down a rope!

OLOF LIPPOLD

The next person to have a major impact on formulating my research interests was Olof Lippold. What a creative mind and imaginative teacher! It was during the next two years which I spent in his laboratory that I became convinced that the questions I found most fascinating concern how the CNS controls muscular performance, rather than about the properties of the muscles themselves. He has always been a strong advocate of human physiology, and working with him made me realize that many questions I wanted to address could only be answered by work on human volunteers; a view I have maintained ever since. When I joined him, Olof had demonstrated a linear relation between integrated EMG and force during isometric contractions of human gastrocnemious. Together we showed a similar linear relation during movement, with differences in slope between those for isotonic shortening and lengthening similar to those predicted from the oxygen measurements during cycling. We also obtained quite respectable force-velocity curves based on EMG.

Obviously these experiments raised questions as to what integrated EMG measures mean in terms of motor drive. Did different levels reflect changes in motor unit recruitment, or was rate coding the dominant feature – questions that continue to engage many of us today. While there were already a few reports on this topic, Olof and I were probably the first to make a systematic study relating the frequency of action potentials recorded from single motor units to the force exerted during voluntary contractions of human muscles. We recorded spike-trains on miles of bromide paper which we subsequently unrolled down the length of the main corridor so that we could measure inter-spike intervals using a meter ruler on our hands and knees.

FIRST RETIREMENT

I gave up my job in 1953 when my first child was born, as most women did then. I could have continued to work but I decided, somewhat piously, that having children was my choice and that they, therefore, deserved my full attention. I was so sure, but now I believe this decision may have been a mistake. I might have been a better mother if I had retained some intellectual interests with which to balance the more tedious aspects of domestic life and the inevitable sense of isolation. I certainly wanted to spend a lot of time with the children in the first few years; but had it not been for other unanticipated events, I believe I would have worked again much sooner than I did. At that stage in our lives we were relatively poor, always wondering whether we could make it to the end of each month.

MOVE TO AMERICA

In 1956 we were invited to spend a year in the United States, just as visitors. That was one of the best years ever. We traveled madly every weekend, thinking we might never have another chance again. But again fate stepped in. My husband was offered a permanent position at The Albert Einstein College of Medicine, then brand new, and so we were faced with the hardest decision of our life. It is not easy to abandon a land that has been good to you, or to abandon family and friends; but the job and prospects in the U. S. were so much better than we were likely to find in Britain then. Besides, we liked it in the U.S. So we stayed.

RETURN TO PHYSIOLOGY

For several years I tried to fill my intellectual void by volunteer work (the League of Women Voters, attempts to desegregate the local schools, low-income housing, etc.). But I always became disenchanted when the people who appeared to be most dedicated failed to act on their convictions if, in practice, this might threaten the quality of their own lives. By this time, my children were in high school, and my husband was so deeply engrossed in his own science that he had little time for me. At aged 40, with perhaps another 40 years to go, I felt useless and lonely. I had to find an absorbing occupation of my own. But I was also convinced I could never catch up with the science I had missed. How could I compete after all these years?

At this stage my husband, who always had more faith in my abilities than I did, persuaded me to work part-time in his laboratory, just to see how it felt. I remember reluctantly agreeing to try this on the conditions that I was regarded as only *a useful pair of hands, not required to think*! Of course this venture was only partially successful. He was always so much smarter and quicker at each job than I. Nevertheless, I am proud of the three papers we published on the mode of action of local anesthetics. Part of our problem was that we were unable to leave the lab behind when we went home. I remember several times one of us waking in the night to say "perhaps if we tried it this way it would work." Indeed, I am filled with admiration for any couple who work well professionally together.

I still did not know what I wanted to do more permanently. I kept telling myself that it would not be possible for someone of my age to compete with today's students. Then fate played its hand again. A friend told me that Hunter College was seeking someone to teach just one section of a physiology course. Although I had always been quite certain that I did not want to teach, I realized that this would be one way to find out what I was capable of doing. I would no longer have excuses to put that off. If one had a class to teach at 8 a.m., one better have something relevant to say. To my amazement, I discovered several striking things about myself. First, I could indeed learn much faster and more comfortably than the students. Certainly I could no longer memorize as they could – but who needs to memorize? Maturity had taught me to discriminate between what was important and what was not. Without my noticing it, my powers to reason had become so much better. Second, I found that teaching at that level was enormously rewarding. Few things gave me a greater thrill than to ignite a gleam of excitement and a flash of understanding in the dull eye of a student who has previously asked only *What do we need to know for the exam?* I also found that learning the techniques of effective teaching, particularly to less able students, could be as challenging as designing the right protocol for a good experiment. Most difficult of all was how to keep challenging the really able students, while not swamping those with more limited abilities. Finally, it taught me once again how wrong I can be about those things that I am most certain I know about already. I had found a most satisfying role for myself where I least expected.

COMPLETING MY Ph.D

However, I soon realized that I must face a major hurdle if I hoped that university level teaching might provide a reasonable basis for a late-breaking, alternative career. At that time, I did not yet have a Ph.D.! I had been registered to take one, and could easily have done so during the four years I worked at University College London, where I had published more than enough papers to meet all Ph.D. requirements. Indeed, my supervisors strongly encouraged me to complete one when they heard I would be leaving to care for children.

They even offered to give me the oral exam in the maternity hospital! But I refused. In those days a Ph.D. was not considered a necessary requirement for a successful academic career in Britain; and I was so sure that I would be much too old to need one later.

At that point, I was all for giving up, but I had to live with a husband who kept nagging me to write to London University to see whether they would, in fact, still allow me to complete a thesis, even fifteen years after the original deadline had expired! Eventually I agreed to write just to shut him up, although it seemed utterly impossible that permission could be granted. Imagine my chagrin when they finally replied some three months later that my application to complete my Ph.D. had been accepted. By that time, my supervisor had left the University and the external examiner they had originally appointed had died, so I had no one to advise me. But eventually these details were settled. Now I was left with no alternative but to prove myself by writing a real thesis, in contrast to just mugging up facts from some textbook before a class. The next summer I sent one child away to camp and the other home to grandparents in England and got to work.

My job was to re-write my original experiments as though they had been done only about six months ago instead of sixteen years; a daunting task when the field had changed so much. At the time when I was a student, ideas about the nature of muscular contraction were so different. Then we talked about dash-pots and visco-elastic elements, heat measurements, and mathematical equations to describe different types of muscle behavior, but we had absolutely no knowledge of sliding filaments or cross-bridge structures. These new concepts had all burst upon us during the years when I had been scientifically dormant, so I was faced with lots of reading. Once again, I was lucky in that, during that time, rather little detailed attention had been given to my particular corners of interest, a comparison of the cost of shortening and lengthening contractions executed at different velocities, integrated EMG as a quantitative measure of CNS muscle activation, and the motor unit firing patterns that underlie them.

Initially, my aim in writing a thesis was simply to acquire the paper qualification to apply for a faculty position – a piece of paper with three letters on it which I viewed as a bureaucratic chore. I did not think it would make me one iota better at my job. But, in fact, it turned out to be a task I enjoyed and one which largely restored my confidence in my abilities. If I could master one small section of physiology, I could surely do it in other areas should the need arise. More importantly, I found the niche for myself in science which I had been seeking. I saw my old experiments with new eyes, finding in them new concepts that had not occurred to us initially. I became excited once again about how these ideas might be developed. I was eager to try again to add some new pieces to the puzzle; and I came to believe that I had found a different way to look at things. And these ideas were my own; no one else had put them in my head. Clearly the most difficult step for mature women who wish to re-enter the work place after raising children is to overcome their lack of confidence in their ability to compete. Thus, in my case, not writing my thesis at the proper time turned out to be blessing in disguise.

Although I got my Ph.D. in 1969, it was still some years before I did any more experiments to explore those new ideas. First, I had to choose once more between family needs and my own. By that time, I had spent three years teaching happily at Marymount College in Tarrytown, N.Y. Then my husband was offered the chairmanship of Pharmacology at Yale. Of course we discussed the impact of this move on my fledgling new career; but in reality, I knew that my female pre-conditioning gave me no choice. Whatever he said, I could never be happy unless I put his needs first. Nowadays, major academic institutions often overcome this problem for professional couples by offering jobs to both partners if they want someone badly enough. But this was unheard of then. Nor would I have considered my own status at that time worthy of making that demand. However, after some searching, I was lucky enough to find a new home at Quinnipiac College and eventually I got tenure there.

QUINNIPIAC COLLEGE

Quinnipiac College is a liberal arts college with teaching as its primary function. Nevertheless, in order to advance in rank, we had to prove we had contributed successfully to basic research. Of course, at that time, such colleges provided neither space nor funds for these activities. Clearly, *research* in that context meant something one could keep in a drawer and pull out between classes. Furthermore, the teaching load was such that anything more than token research was quite impossible. Nevertheless, with the help of another new faculty member (Joe Woods), I decided to try my hand at writing a full NIH research grant application. I was not sure that we could put something together that would justify submission, but I certainly enjoyed trying. As I remember, my main aim was to get some impartial assessment of the ideas I had expressed in my thesis. I feared that the encouragement I had received previously might be mainly out of kindness. I certainly did not expect our application to be funded. Nor do I believe that it would have been, except for another piece of extraordinary good luck, the presence on the study section of Doug Stuart, Milic-Emili and others who happened to remember our early papers and decided to give us another chance.

GETTING STARTED AGAIN

What could we do! We had no place to work. The only space the college could find for us to work was a large storage closet in the gym, next to the men's weight room and right over the commercial dryers. It was not surprising that we spent most of our time battling mechanical and electrical interference problems. Despite this, however, Joe Woods and I eventually produced three papers designed to validate some of the ideas I had expressed in my thesis. In doing this, we were helped enormously by friends at the John B. Pierce Laboratory. They built our motorized bicycle ergometer in their workshop and Hans Graichen often rescued us from electronic problems. (Quinnipiac had no technical backup facilities.) Thus, when our grant came up for renewal in 1976, we asked the site visitors to arrange for us to rent permanent lab space at Pierce, which they did. So started a long and fruitful time. From that point, my subsequent adventures in human physiology is largely on the published record. However, there is one more tale I have to tell. That is how and why I first became interested in fatigue.

HOW FATIGUE SET IN

Until 1974 my research, like that of most others at the time, was directed towards neuromuscular responses evident in short-term exercise. I had not given much thought to how these responses might change as exercise continued. Then chance stepped in again. My husband and I went to Cambridge to attend a meeting of the British Physiological Society, and then have lunch with A.V. Hill. By that time, he was living there in retirement. His son David Hill was also there, together with a young and energetic Welshman, Richard Edwards, whom I had not met before. When I mentioned that I was looking for somewhere to work the following spring when Murdoch would be in Cambridge on sabbatical leave, Richard invited me to join him at the Royal Postgraduate Medical School at Hammersmith. There I also met and worked with David Jones. We had a fabulous time that year. I became fascinated with the changes in muscle properties induced by prolonged activity, and more particularly, by the strategies that the nervous system must adopt if effective control is to be maintained.

Thus my visit to Richard's laboratory completely changed the direction of all my future research, and all because of one lucky lunch-time encounter.

MY MORE RECENT COLLEAGUES

I have talked about a few of the many people who set me on my way in my early days, and whose examples I have tried to emulate in later life. But it is the folk with whom I have worked more recently who should, of course, get whatever credit the passage of time and my peers may choose to attribute to the work on fatigue performed in my laboratory. It is they who made everything possible and exciting.

First, Joe Woods, my stalwart ally from the beginning. I met him on my first day at Quinnipiac. From then on, we instinctively turned to each other for advice and support in every aspect of our professional lives – teaching, student problems, grant writing and all our earlier experiments, both at Quinnipiac and later at the John B. Pierce Laboratory. Without his help and encouragement, we could never have got started. It was a most unhappy day when Quinnipiac chose to make him Dean and our collaboration had to end.

Then my thanks to my various post-doctoral students, each with me sequentially for 2-3 years: Carl Kukulka, François Bellemare, Christine Thomas, Andy Fuglevand, and now Michael Walsh. They all came from such different backgrounds, and brought with them such different skills and personal philosophies. I have enjoyed knowing each one and have learned so much from them. I also appreciate, more than I can say, the contributions of the many scientists from other laboratories with whom I have collaborated, either when they visited us or we went to them. There are too many to name individually, but among them I want to mention, with particularly deep gratitude and affection, Olof Lippold, Roland Johansson, Nina Vøllestad and Simon Gandevia who have collaborated with us repeatedly. In addition to their well-known scientific talents, they also gave to me a very special friendship. And, of course, my husband, Murdoch Ritchie. Without his constant support, encouragement and often bullying I would never have had the courage to think that I could overcome the many obstacles in the way of a mature, immigrant woman wishing to return to academia after being absent from the action for so long.

THE SCIENTIFIC CONTRIBUTIONS OF BRENDA BIGLAND-RITCHIE

C. K. Thomas,[1] R. M. Enoka,[2] S. C. Gandevia,[3] A. J. McComas,[4] and D. G. Stuart[5]

[1] The Miami Project to Cure Paralysis
University of Miami School of Medicine
Miami, FL 33136
[2] Department of Biomedical Engineering, Cleveland Clinic Foundation
Cleveland, Ohio 44195
[3] Prince of Wales Medical Research Institute
Randwick, New SouthWales 2031, Australia
[4] Department of Biomedical Sciences, McMaster University
Hamilton, Ontario, L8N 3Z5, Canada
[5] Department of Physiology, University of Arizona College of Medicine
Tucson, Arizona 85724

ABSTRACT

Brenda Bigland-Ritchie has made seminal contributions to our understanding of skeletal muscle physiology - the energy cost of muscle when it shortens or is forcibly stretched, the relationship between EMG and force, the behavior of single motor units, and above all, the processes underlying neuromuscular fatigue. More than this, she has stimulated inquiry into the search for reflex mechanisms which may serve to balance the activity of the spinal cord with that of the fatiguing muscles. Her use of human volunteers for much of this work is extraordinary, and represents a major strength. Equally important are her well known and widely cited manuscripts. Not only are her findings clearly described and depicted, but every attempt is made to relate her results to the fatigue processes measured in animal or isolated tissue preparations.

INTRODUCTION

Fatigue is common, complex, and controversial (Bigland-Ritchie, 1981b). Common in that it confronts us in our daily tasks. Complex because the processes which fail vary with the exercise performed. Controversial due to discrepancies in its definition and interpretation. Despite this complexity, Brenda Bigland-Ritchie has heightened our awareness of neuromuscular fatigue as a subject for scientific study and has shown it to be approachable through physiological experimentation. It has taken a careful mix of inquiry, use of effective

Fatigue, Edited by Simon C. Gandevia et al.
Plenum Press, New York, 1995

techniques, and collaboration. In this chapter, five themes are addressed. These include: understanding muscle function from measures of oxygen uptake, examination of the sites of fatigue, exploration of the difficulties in interpreting EMG signals, the reciprocity of influences between spinal cord and muscle, and examples of how effective methods and collaboration can enhance our understanding of muscle fatigue. In viewing these areas, we highlight some of the findings of Brenda Bigland-Ritchie and propose why we think they are so valuable.

PARTNERSHIPS BETWEEN MUSCLE, HEART AND LUNG

Brenda Bigland began her research when there was a systemic view towards understanding muscle function. At the time, work done was typically correlated to oxygen uptake. But with the elegant demonstrations of the force-velocity relationships of striated muscle (Hill, 1938), it became possible to relate work in terms of the number of active muscle fibers and their activation rate. Since each fiber exerted a larger force during stretch than during shortening (Hill, 1938; Katz, 1939), varying the numbers of active fibers could explain any differences in energy cost for these two activities. Brenda Bigland's first scientific challenge was to determine the oxygen requirements of shortening and lengthening contractions (positive and negative work, respectively). To do this, it was necessary to find movements which were mirror images of each other. Two bicycles were linked together back to back (see Chapter, Looking back). To ensure that the placement and timing of activity in each limb was identical, one person pedaled forward while the other pedaled in reverse. Not only was the energy cost of positive work much greater than that of negative work (Abbott et al., 1952), but for the same rate of work, oxygen uptake was many times greater for work involving great force and low speed compared to work involving small force and high speed (Abbott & Bigland, 1953). All these data were explained in terms of fewer fibers being needed to perform negative compared to positive work. In terms of muscle energetics, they also substantiated the concept that muscle shortening was not the converse of muscle lengthening (Hill, 1938; Abbott et al., 1951).

These same concepts were explored further some twenty years later when Brenda Bigland-Ritchie re-entered science. The energy requirements of the fibers (rate of metabolic heat production and rate of ATP and phosphocreatine breakdown) had been shown to fall substantially during stretch compared to during shortening (Wilkie, 1968; Curtin & Davies, 1970). Thus, during positive and negative work, were there differences in the number of active fibers and in the energy requirements per fiber? With a new and improved motor-driven ergometer which better confined the pedaling work to the specific muscles involved (Bigland-Ritchie et al., 1973), oxygen uptakes were compared to measures of integrated surface EMG (IEMG) (Bigland-Ritchie & Woods, 1974). If the IEMG could provide a reliable measure of muscle activity during voluntary contractions, it was argued that it would be possible to examine changes in oxygen uptake per active muscle fiber. Interestingly, the IEMG and particularly the oxygen uptake were less during muscle stretching versus shortening (Fig. 1, open and closed symbols respectively). Thus, those muscle fibers which were active required less oxygen when they were stretched compared to when they were shortened (Bigland-Ritchie & Woods, 1976).

These manuscripts also signaled the reappearance of a well defined scientific style which has continued in ever-sharpening focus in her work on neuromuscular fatigue. However, in examining detail at the whole muscle and motor unit levels, Brenda Bigland-Ritchie is unusual in that she never assumes that the particular test muscle works in isolation. Rather, the entire organism may actually determine how well a particular part functions. For example, co-ordination between different body systems was the mechanism proposed to

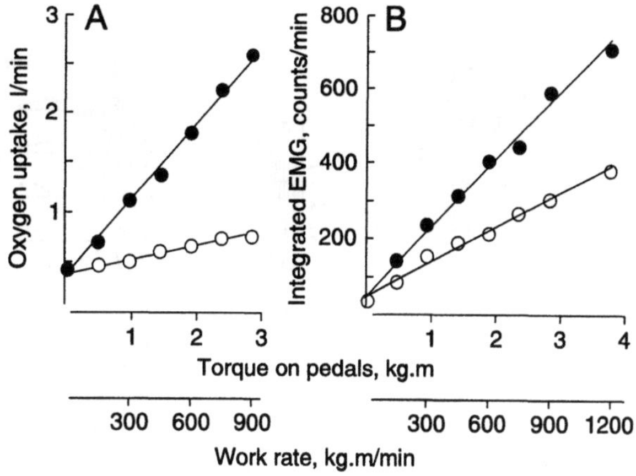

Figure 1. Mean oxygen uptake (*A*) and integrated EMG (*B*) versus mean torque on the pedals (or work rate) for positive (filled symbols) and negative work (open symbols). All data are from one subject (n=75 experiments conducted over several months) and only data from the left leg are shown in *B*. Pedaling was at 50 rev/min. (adapted from Bigland-Ritchie & Woods, 1976).

explain why dynamic but not isometric exercise performance was disproportionately impaired under hypoxic conditions (Bigland-Ritchie & Vøllestad, 1988). These kinds of hypotheses are invaluable in that they pose novel ways of interpreting data and they serve to inspire new experiments.

FINDING THE FACTORS THAT ARE RATE LIMITING IS AS IMPORTANT AS FINDING THOSE THAT ARE NOT

One hallmark of Brenda Bigland-Ritchie is her strong advocacy for evaluating each possible fatigue site rather than assuming only one is at fault and others are intact. Most discussions of fatigue sites focus on three possible areas: central fatigue, failure at the neuromuscular junction, and deficits in the muscle. Each area can be further subdivided to include other processes as shown in Fig 2*A*. But which of these processes fail, when, and how? Furthermore, is this rather unidirectional depiction of events, starting with the excitatory input to higher motor centers and ending with the energy-dependent interaction of actin and myosin, much more complex and interactive?

Central Fatigue

Early on, there were objective methods to evaluate whether a contraction was the strongest possible, and whether voluntary drive could be maintained during sustained efforts. For example, when a contraction was produced by voluntary effort or by tetanic nerve stimulation, similar forces were generated (Fig. 2*B*; Mosso 1915; Bigland & Lippold, 1954b; Merton 1954). Such findings were taken to indicate that fatigue was a peripheral phenomenon. The method of twitch occlusion originally suggested by Denny-Brown (1928) was adapted by Merton (1954) to assess contraction maximality. As voluntary force increased within a muscle, that force which could be evoked artificially by stimulation was reduced, then occluded completely during a maximal voluntary contraction. Could these initial findings be reproduced for different muscles, and during other types of contractions?

It is usually possible to activate a muscle maximally during brief contractions but during fatiguing contractions this is much more difficult (Bigland-Ritchie et al., 1978; Belanger & McComas, 1981; Gandevia & McKenzie 1988b; Thomas et al., 1989), particularly for the diaphragm during expulsive maneuvers (Bellemare & Bigland-Ritchie, 1987). While the latter difficulty may arise because of the rather unphysiological nature of this

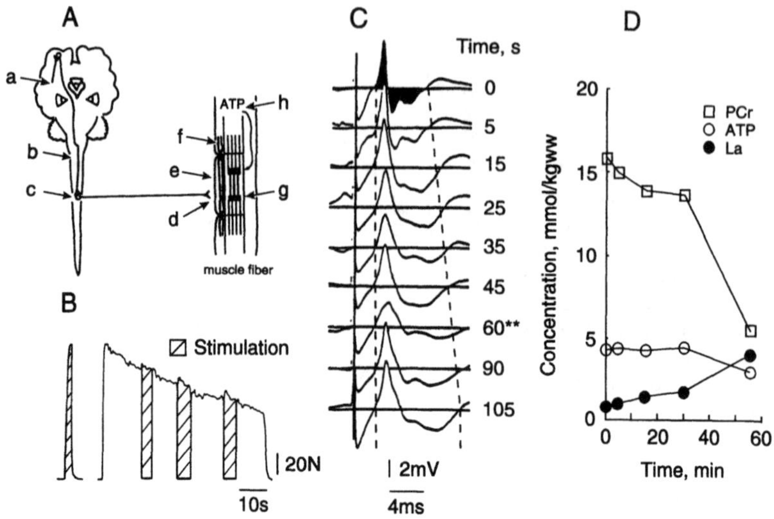

Figure 2. *A*, potential fatigue sites: a, excitatory input to the motor cortex; b, excitatory drive to the lower motoneuron; c, motoneuron excitability; d, neuromuscular transmission; e, sarcolemma excitability; f, excitation-contraction coupling; g, contractile mechanism; h, metabolic energy supply. *B*, 60 s maximum voluntary contraction of adductor pollicis which was interrupted periodically with 50 Hz stimulation (cross-hatching). Similar forces were always generated by stimulation and voluntary effort indicating an absence of central fatigue. *C*, intramuscular M-waves recorded during a fatiguing co-contraction of both adductor pollicis and first dorsal interosseous. At 60s (**) the M-wave declined. It was restored by repositioning the stimulating electrode (90s). *D*, mean metabolite concentrations (lactate, ATP, PCr) changes during repeated target force (30% MVC) contractions performed for 6s followed by 4s rest. Exhaustion was reached at 55 minutes. Data are from 5 subjects. (*A-D* adapted from Bigland-Ritchie, 1981a, 1983b, 1982, and 1986a, respectively).

behavior (McKenzie et al., 1992), it may also relate to the need for the diaphragm to continue functioning without contractile failure. As such, these data raise the intriguing possibility that there are substantial interactions between the potential failure sites, a phenomenon which needs our attention.

In this type of research there was a slow evolution of the methods needed for the assessment of central fatigue. To avoid injury, supramaximal nerve stimulation was replaced by percutaneous stimulation of a constant fraction of the muscle (Bigland-Ritchie et al., 1978). When bilateral tetanic phrenic nerve stimulation was too painful and inconsistent, the already established method of twitch occlusion was extended to evaluate muscle activation during sustained contractions (Bellemare & Bigland-Ritchie, 1987). Resolution of the small evoked forces was also a concern. Thus, voltage clamp circuits (Belanger & McComas 1981) or those which amplified only the AC force components were developed (Hales & Gandevia 1988; Westling, Johansson & Bigland-Ritchie, unpublished). In some cases, two briefly separated stimulation pulses were delivered to enhance the force output (Gandevia & McKenzie, 1985, 1988a). Another innovation was electrical stimulation of the motor cortex during fatiguing contractions (Merton et al., 1981). With no sign of electrical decrement during these protocols, both central and neuromuscular junction excitability must have remained intact, confirming support for the peripheral nature of fatigue. Such methods are also critical for evaluating central fatigue in neurological disorders which can partially disconnect the central and peripheral nervous systems (for example, spinal cord injury). Then, the same intact innervation which can be driven by voluntary effort is also stimulated (Thomas, 1993).

Neuromuscular Junction and Fiber Plasmalemma

Demonstration of junction and/or plasmalemma failure requires comparisons between the responses evoked by nerve stimulation and direct muscle stimulation. Since the latter is difficult in human experiments due to the large voltages required, measurements of M-waves have largely been used to evaluate integrity of the neuromuscular junction and plasmalemma (Merton, 1954). This in itself has brought controversy, primarily because investigators have measured and interpreted the waveforms differently (Bigland-Ritchie, 1987). Thus, while many studies have concluded that there is little evidence to support junction or plasmalemma failure as sites for fatigue (for example, Merton, 1954; Bigland-Ritchie et al., 1982; Duchateau & Hainaut, 1985; Thomas et al., 1989), other studies conclude the contrary (for example, Stephens & Taylor, 1972; Aldrich et al., 1986; Milner-Brown & Miller, 1986; Bellemare & Garzantiti, 1988). On several occasions Bigland-Ritchie has attempted to resolve these differences by repeating the experiments and eliminating methodological differences. Two important factors seem to have arisen from this. First, the M-wave amplitude can often be restored by repositioning of the stimulating electrode so depressed M-waves do not always imply failure (Fig. 2*C*). Second, any M-wave measurements must take into account the slowing of impulse conduction velocity which occurs with fatigue.

An alternative way of evaluating junction integrity has involved examining the regularity of motor unit interspike intervals. While continuity of impulses suggests junction integrity, the possibility of central fatigue must be precluded. That motor unit firing rates during maximal voluntary contractions rarely seem to exceed those stimulation rates where failure has been shown to occur (Krnjevic & Miledi, 1958; Bellemare et al., 1983), also lends plausibility to the idea that neuromuscular junction block may indeed be rare during voluntary contractions.

Muscle

Use of muscle biopsies, NMR spectroscopy and skinned fiber preparations have added much to our understanding of muscle metabolism during human fatigue. However, study of other processes beyond the neuromuscular junction has come largely from animal studies (reviewed by Fitts, 1994). In this area, Bigland-Ritchie has constantly urged others to proceed, but with standardized protocols. Then direct comparisons between human and animal data are possible, and human data can be more fully interpreted.

Perhaps the most difficult issue to address has been the potential failure of excitation-contraction coupling. It is often singled out as a contributing factor because of the disproportionately low force in response to low versus high frequencies of activation (Edwards et al., 1977). Alternatively, it has been identified as a failure site by exclusion. During the first 30 min of repeated low intensity submaximal intermittent contractions of quadriceps, little change was seen in muscle metabolites in the presence of intact central drive and neuromuscular junction function (Fig. 2*D*). Only failure of excitation-contraction coupling was left to explain the data (Bigland-Ritchie et al., 1986a; Vøllestad et al., 1988). However, as the exercise proceeded, metabolic changes did occur and revealed the novel finding that exhaustion was closely related to depletion of phosphagen stores (Vøllestad et al., 1988).

However, no matter what the conclusion, Brenda Bigland-Ritchie has always left room to question the importance of the findings. For example, are the proposed fatigue sites linked directly to the force loss, or are the observed changes by-products of other processes? Similarly, only a few studies have examined fatigue during dynamic contractions (for example, de Haan et al., 1989) but the relative importance of central and peripheral factors in these conditions deserves the same systematic investigation that has been given to isometric contractions.

UNSCRAMBLING THE ELECTROMYOGRAM

Surface EMG

Throughout the years Bigland-Ritchie has made careful and informative campaigns to clarify our understanding of EMG-force relationships. During brief voluntary isometric or constant velocity, shortening or lengthening contractions, the relationship between the integrated surface EMG and force can be linear (Lippold, 1952; Bigland & Lippold, 1954a; Milner-Brown & Stein, 1975) or non-linear depending on the muscle examined (Fuglsang-Fredriksen, 1981; Woods & Bigland-Ritchie, 1983). Cautionary notes suggest that Bigland-Ritchie, as well as others, thought some often overlooked factors which could explain the discrepant results. For example, the surface record may not reflect the activity of the muscle in general because fiber distributions are uneven for different units and the signals from deep units are attenuated at the surface (Clamann, 1970). Potentials may also sum in a non-linear way yet produce linear results (Milner-Brown & Stein, 1975). Similarly, if some motor units are already firing at optimal rates for force development when the whole muscle force is submaximal, further effort may increase the impulse frequencies of these units without adding tension (Tanji & Kato, 1973a,b). Alternatively, force and/or EMG signals may be contaminated from synergists or antagonists. Different roles of recruitment and rate coding in various muscles or during different tasks may also change the overall EMG-force relationship (Bigland-Ritchie, 1981a; Woods & Bigland-Ritchie, 1983).

Surface EMG changes with fatigue are equally complex. While motor unit synchronization (Lippold et al., 1960; Lippold, 1973) and muscle conduction velocity slowing with fatigue may lead to increases in EMG, reductions in either motor unit rate or in recruitment may lead to decreases. How these factors influence the surface EMG, and change for different muscles, types of exercise, degrees of ischemia and recovery is still unclear . Thus surface EMG recordings alone leave too much room for speculation on the possible fatigue site (Bigland-Ritchie et al., 1978; Bigland-Ritchie, 1981b, 1987; Bigland-Ritchie & Woods, 1984).

Careful evaluation of much published data from evoked or voluntary contractions has also brought the whole EMG-force relationship into question. For example, how can the EMG potentiate while the force fails? Alternatively, how can the EMG almost disappear with minor decrements in force? What mechanisms underlie dramatic changes in EMG-force relations which accompany minor increases in use? (see Kernell et al., 1987; Hicks et al., 1989; Thomas et al., 1991a). These kinds of data have prompted a thought-provoking conclusion: "no fixed relation is to be expected between force and EMG" (Bigland-Ritchie, 1988).

Intramuscular EMG

Surface EMG recordings cannot distinguish the importance of the number of active muscle fibers versus the frequency of excitation in force generation (Bigland & Lippold, 1954b). But that kind of information was necessary to decipher the pattern of motor unit recruitment during a brief voluntary contraction, as well as the rate at which units responded. Early motor unit research documented an orderly sequence of motor unit recruitment, showed that it was dependent on the task being performed, and characterized unit discharge rates during weak contractions (Adrian, 1925; Adrian & Bronk, 1929; Smith, 1934; Lindsley,

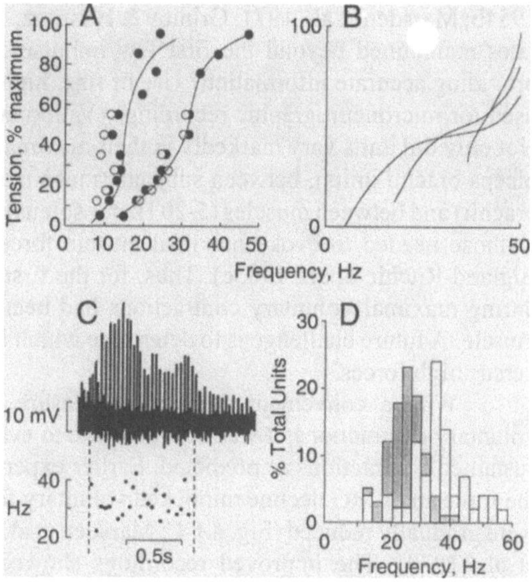

Figure 3. *A*, abductor digiti minimi tension (relative to maximum) compared to minimum and maximum unit discharge frequencies (left and right curves respectively) during voluntary contractions (closed symbols) and after almost complete pressure block of the ulnar nerve near the elbow (open symbols). *B*, typical force-frequency behavior proposed for single motor units during voluntary contractions. *C*, potentials from one motor unit that change in amplitude as the tungsten electrode is moved slowly through the muscle. The instantaneous unit rate is below. *D*, distributions of unit firing rates during maximum voluntary contractions of biceps brachii for two subjects. *(A,B* adapted from Bigland & Lippold, 1954b; *C,D* adapted from Bellemare et al., 1983).

1935; Denny-Brown & Pennybacker, 1938; Gilson & Mills, 1941). But, how did unit discharge frequencies vary for different units, and how were these frequencies altered with greater increases in muscle force?

Two tacks were taken to explore what was going on during stronger voluntary contractions. The first was to develop selective recording electrodes from fine wire. The other was to block activity in some axons by local nerve pressure (Bigland & Lippold , 1954b). These strategies taught us that units of higher threshold were recruited at higher firing rates and that each unit followed a sigmoid-shaped force-frequency relation. Lower threshold units reached the same maximal firing frequencies as high threshold units but at lower whole muscle forces (Fig. 3*A,B*). Although the maximal firing rates were not measured consistently, they suggested that only units recruited at the highest forces fired above 35 Hz. All these data were important in providing the first detailed analysis of unit rate coding in human muscle. Moreover, they gave a framework from which to examine further the role of frequency modulation and recruitment in force generation.

Substantial contributions have subsequently been made by others. For example, how recruitment and rate coding varies for different muscles (Kukulka & Clamann, 1981), additional detail on how unit frequency changes with force (Monster & Chan, 1977), the finding that human units are activated in order of increasing force output during isometric contractions which are slow (Milner-Brown et al., 1973) or fast (Desmedt & Godaux, 1977), and the observation that this pattern can be altered during eccentric contractions (Nardone et al., 1989; Howell et al., 1994). However, despite the current vast literature on these topics, it is interesting to note that our current models of motor unit behavior (Heckman & Binder 1993) still rely greatly on the data described by Bigland and Lippold (1954b).

Extracting motor unit potentials from the complex interference pattern of the electromyogram (EMG), particularly at maximal forces, still represents a major challenge. Although comparisons between voluntary and stimulated forces had shown that human voluntary drive can maximally activate all motor units in a muscle (Mosso, 1915), it was unclear what the unit firing rates accompanying such efforts actually were. Experiments involving partial nerve block, aberrant motor innervation, and partial denervation suggested that unit firing rates may range from 35 to 60 Hz, and go up to 200 Hz (Bigland & Lippold

1954b; Marsden et al., 1971, Grimby & Hannerz, 1977; 1983). But, were these extraordinary rates maintained beyond the first few impulses, and were these compromised situations providing accurate information? Use of fine, high impedance, tungsten electrodes typically used for microneurographic recordings (Vallbo et al., 1979) gave the answers (Fig. 3*C,D*). Not only did units vary markedly in their maximal rates (12-60 Hz for adductor pollicis and biceps brachii units), between subjects (mean maximal rates between 22-35 Hz for biceps brachii) and between muscles (5-20 Hz for soleus), but these rates were rather slow compared to those needed to evoke maximal muscle force by stimulation (Bellemare et al., 1983; Bigland-Ritchie et al., 1983c). Thus, for the first time, the limits of frequency modulation during maximal voluntary contractions had been defined for the population of units in a muscle. A future challenge is to determine which firing rates belong to units recruited at low versus high forces.

With a convenient method to measure motor unit firing rates during maximal voluntary contractions, it was now possible to evaluate whether these rates declined during sustained contractions as predicted. Earlier experiments involving stimulation showed that the force and EMG decline mimicked voluntary force and EMG loss if the stimulation rates were gradually reduced (Fig. 4*A-C*; Marsden et al., 1969; Bigland-Ritchie et al., 1979; Jones et al., 1979). The improved recordings showed that unit discharge did decline during maximal voluntary contractions such that the time courses of the force, contractile speed and rate reductions closely followed each other (Fig. 4*D*). As there was no evidence for central fatigue or neuromuscular junction failure, the force loss must have resulted from contractile failure (Bigland-Ritchie et al., 1983a,b). How then could muscle be activated optimally with lower and lower motor unit firing rates? The obvious hypothesis was that contractile slowing meant lower frequencies were adequate for full tetanic fusion (Fig. 4*C*). Maintenance of higher rates would actually be disadvantageous, since energy would be wasted in additional ion pumping and excitation failure would be more likely to occur. However, were these changes in firing rate mere coincidence, or were they under regulatory control?

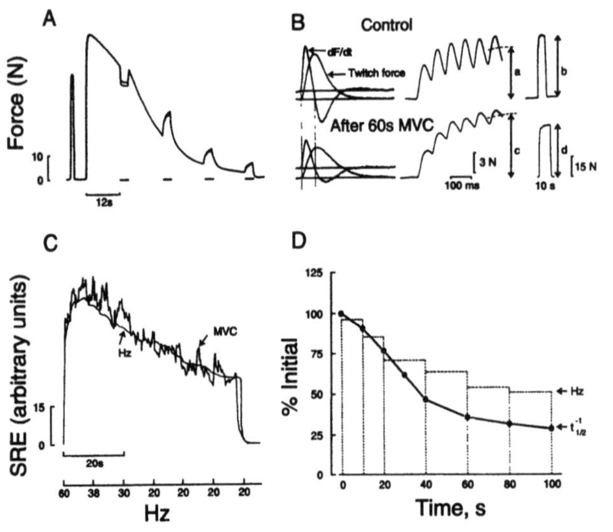

Figure 4. *A*, force from adductor pollicis during 80 Hz and 20 Hz (solid lines) stimulation of the ulnar nerve. Muscle ischemia was maintained throughout the protocol. Note that 20 Hz stimulation produces more force than 80 Hz towards the end of the contraction. *B*, mechanical responses to ulnar nerve stimulation after a 5s MVC (upper records) and after a 60s MVC (lower records) with blood occlusion. Stimulation was at 1, 7 and 50 Hz in each case (left to right records respectively). After fatigue the 7 Hz tetanus/twitch ratio increased from 1.67 to 2.50. The force in response to 7 Hz stimulation increased (a/c) by 15%, while the fraction of the maximum available force (a/b) increased from 0.17 to 0.24%. *C*. smoothed rectified EMG record from adductor pollicis during a maximum voluntary contraction and a stimulated contraction (smoother line) in which the frequency was decreased as shown (the amplification was four times greater for the MVC). *D*, changes (% initial) in mean motoneuron firing rates and relaxation during a 60s MVC. (*A-D* adapted from Jones et al., 1979, Bigland-Ritchie et al., 1983b, 1979 and 1983a respectively).

MUSCLE CONTROL, CNS CONTROL OR BOTH?

Fatigue has long been known to involve force loss and contractile slowing (Hill, 1913; Abbott, 1951). Yet what regulates and limits these processes? Moreover, how do these muscle responses integrate with other changes, and particularly those in the central nervous system? For example, what factors underlie the decline in motoneuron firing rates during sustained maximal voluntary contractions when there is adequate central drive and neuromuscular transmission? On some level, the possibility of reflex control from the muscle was evident to many. However, it needed to be shown. With the extraordinary dedication of the research volunteers, it was shown that the force and motoneuron discharge reductions which occur in response to sustained maximal voluntary contractions fail to recover if the fatigued muscle is kept ischemic (Fig. 5). Thus, it was proposed that factors from the exercised muscle must signal its fatigued character, and that discharge of motoneurons at the initial activation rates would confer no functional advantage to the slowed muscle (Bigland-Ritchie et al., 1986b). Viewed as such, these data provide a way to understand how a fatiguing muscle can continue to work optimally despite a decline in the firing rates of its constituent motor units. The argument for this reflex was then substantiated further by eliminating the possibility that transmission fails at the neuromuscular junction (Woods et al., 1987). These results perhaps provide the most persuasive evidence we have that dynamic factors in the muscle can play critical roles in motoneuron regulation.

One gauge of the importance of this work was the impetus it gave for new experiments. A flurry of new questions were immediately apparent. For example, which afferents contribute to these effects and under what circumstances? How widespread are these reflex effects in the spinal cord and higher brain centers? What triggers induce these reflex effects? Do independent changes in motoneuron excitability (Kernell & Monster, 1982) also play a role? Furthermore, how important are these reflexes when the exercise is submaximal or induced by stimulation? We soon learned that the reflex effects were not specific to the fatigued muscle but also carried over to synergistic muscles (Hayward et al., 1988; McComas et al., 1994), though the effect may vary in magnitude for different muscles or tasks (Dacko & Cope, 1994). Similarly, muscle slowing induced by cooling or length changes failed to reproduce the changes in motor unit firing rates typically measured during fatigue (Bigland-Ritchie et al., 1992a,b). Whether ischemia or metabolic factors are of more importance still needs to be evaluated. Similarly, amid the speculation that group III and IV afferents were

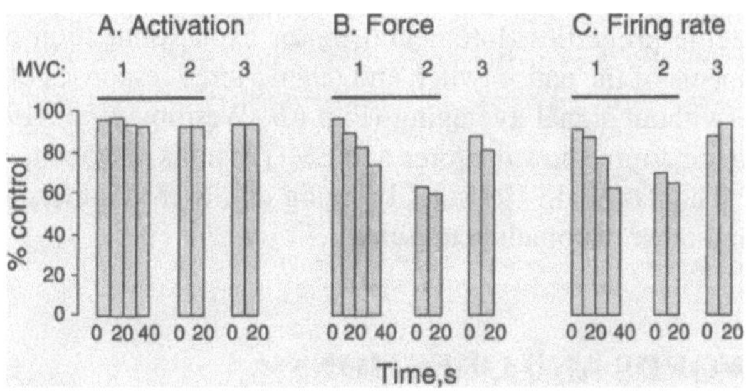

Figure 5. Mean quadriceps muscle activation (*A*), force (*B*) and unit firing rate (*C*) expressed relative to initial control values. Subjects performed a 40s maximum voluntary contraction (MVC1) then two 20s maximum voluntary contractions (MVC2, MVC3) with a 3 minute rest between each contraction. The muscle was kept ischemic during MVC1, the 3 min rest and during MVC2 as indicated by the solid line (adapted from Woods et al., 1987).

involved in the reflex (Bigland-Ritchie et al., 1986b; Woods et al., 1987), came evidence that group Ia spindle afferents may also contribute (Hagbarth et al., 1986) and that both excitatory and inhibitory influences seem to play a role (Gandevia et al., 1990). Clearly the entire significance of fatigue-induced control of motoneuron firing rates will only be appreciated after the completion of many more experiments.

METHODS AND COLLABORATIONS YIELD REWARDS

Apart from brief forays into experiments which examined the effects of growth hormone on muscle performance (Bigland & Jehring, 1952), how temperature influences the effects of neuromuscular blocking drugs (Bigland & Zaimis, 1958; Bigland et al., 1958), and the functional characteristics of local anesthetics (Ritchie et al., 1965a,b; Ritchie & Ritchie, 1968), Brenda Bigland-Ritchie's scientific contributions have all come from human experiments. Thus, what some perceived as limiting, that is, data from human experiments, she used to great advantage. Some of the value lay in viewing whole organism behavior without anesthesia and surgery, and some in examining the actual limits of performance and how it was controlled. The experimental strategy included the willingness to experience these limits first hand, the ability to recognize the talents of other people and to pull them into collaborations, and effective methodology. For example, when challenged to find a task involving the same work, movements and speed to measure the oxygen cost of positive and negative work, what more elegant solution could there have been than the two bicycles depicted on the frontispiece of this book (Abbott et al., 1952)? Similarly, when individual motor unit potentials could not be discerned from the interference pattern during high force voluntary contractions, a steady stream of innovations brought her closer to the eventual goal: the development of more selective electrodes inserted intramuscularly, partial nerve block from pressure (Bigland & Lippold, 1954b), counting unit potentials over a fixed amplitude (Bigland-Ritchie & Lippold, 1979), and the use of tungsten microelectrodes to record the firing rates of populations of motor units during maximal voluntary contractions (Bellemare et al., 1983; Bigland-Ritchie et al., 1983a,c). When bilateral phrenic nerve stimulation was needed to describe the contractile properties of the diaphragm, and tetanic stimulation was too painful and inconsistent, the method of twitch occlusion (Denny-Brown, 1928; Merton, 1954; Belanger & McComas, 1981) was extended to evaluate muscle activation during brief and fatiguing voluntary contractions (Fig. 6*A*; Bellemare & Bigland-Ritchie, 1984, 1987; Bellemare et al., 1986). More recently, Bigland-Ritchie initiated and saw to fruition the successful development of an alternative technique for measuring human motor unit contractile properties. Intraneural motor axon stimulation permitted the first detailed measurements of the entire twitch and tetanic force responses of the population of units in a muscle without signal averaging (Fig. 6*B*; Westling et al., 1990). Moreover, it became possible to determine how the force and EMG of units respond to defined activation patterns (Fig. 6*C*; Thomas et al., 1991a,b), bringing us closer to understanding differences between human and other mammalian muscles.

PICTURES, WORDS AND QUESTIONS

It is a special privilege to know Brenda Bigland-Ritchie. With both word and graph, her papers address well defined issues, are characterized by cautious interpretation and, at the same time, show her willingness to share ideas and to inspire new experiments. These manuscripts are worth investigating. And, if we ask who has advanced our knowledge of

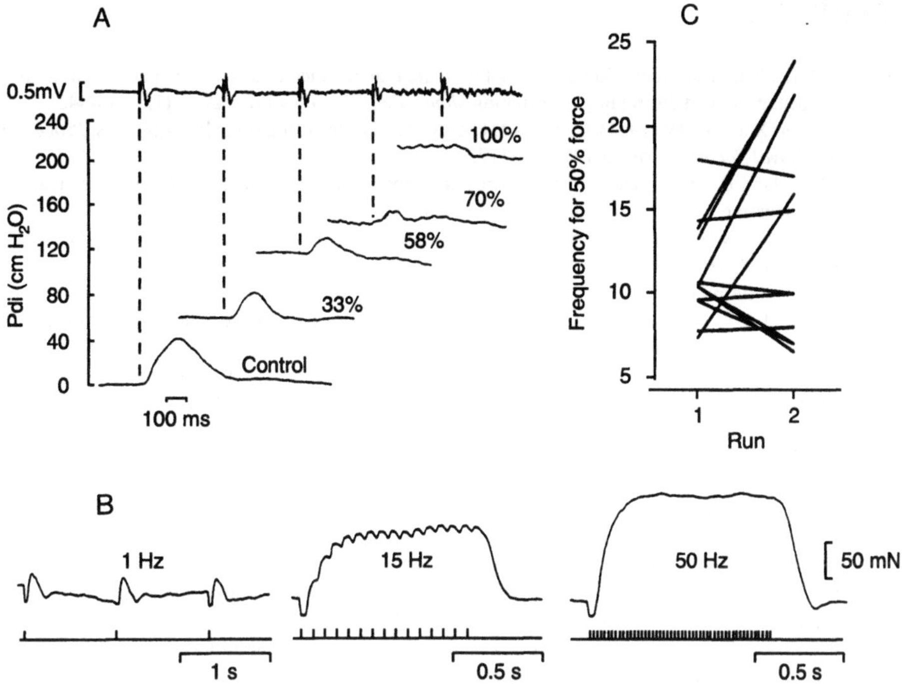

Figure 6. *A*, left EMG (upper) and Pdi (lower) records when maximal shocks were delivered to the phrenic nerve bilaterally during combined Mueller-expulsive maneuvers at different $\%Pdi_{max}$ in the sitting position. The vertical dashed lines mark the time of stimulation delivery. Note the magnitude of the response reduces as Pdi level increases. *B*, resultant force records from one thenar motor unit when its axon was stimulated with trains of pulses at 1, 15 and 50 Hz. *C*, frequencies that produced 50% of initial maximum tetanic force for 12 different thenar motor units before (run 1) and after (run 2) a Burke fatigue test. (*A* adapted from Bellemare & Bigland-Ritchie, 1984, *B-C* adapted from Thomas et al., 1991a).

neuromuscular fatigue, Dr. Bigland-Ritchie is not the only scientist who has. But clearly, she has made significant and imaginative contributions, perhaps more than anyone.

ACKNOWLEDGMENTS

The authors thank Ms. Bette Mas and Dr. Mary Pocock for editorial assistance and much library work, and Drs. Vladimir Esipenko and James Broton for help with the figures. The author's research is currently supported by: United States Public Health Service (USPHS) grant NS 30226, the Department of Defense, and The Miami Project to Cure Paralysis (*C.K.T.*); USPHS grants AG 09000 and NS 20544 (*R.M.E.*); the National Health & Medical Research Council of Australia and the Asthma Foundation of New South Wales, Australia (*S.C.G.*); the Medical Research Council and the Natural Sciences and Engineering Research Council (NSERC) of Canada (*A.J.McC.*), and USPHS grants GM 08400, NS 20577, NS 01686 and NS 07309 (*D.G.S.*). The authors' attendance at the 1994 Bigland-Ritchie conference was supported, in part, by NSERC (*A.J.McC.*), NS 30226 (*C.K.T.*), the Cleveland Clinic Foundation (*R.M.E.*), the Muscular Dystrophy Association (USA; *S.C.G.*), University of Arizona Regents' Professor funds (*D.G.S.*), and the University of Miami (*S.C.G.* and *A.J.McC.*).

REFERENCES

Abbott BC (1951). The heat production associated with the maintenance of a prolonged contraction and the extra heat produced during large shortening. *Journal of Physiology (London)* **112**, 438-445.

Abbott BC, Aubert X & Hill AV (1951). Changes of energy in a muscle during very slow stretches. *Proceedings of the Royal Society, Series B* **139**, 104-117.

Abbott BC & Bigland B (1953). The effects of force and speed changes on the rate of oxygen consumption during negative work. *Journal of Physiology (London)* **120**, 319-325.

Abbott BC, Bigland B & Ritchie JM (1952). The physiological cost of negative work. *Journal of Physiology (London)* **117**, 380-390.

Adrian ED (1925). Interpretation of the electromyogram. *Lancet* **208**, 1229-1233.

Adrian ED & Bronk DW (1929). The discharge of impulses in motor nerve fibres. Part II. The frequency of discharge in reflex and voluntary contractions. *Journal of Physiology (London)* **67**, 119-151.

Aldrich TK, Shander A, Chaudhry I & Nagashima H (1986). Fatigue of isolated rat diaphragm: role of impaired neuromuscular transmission. *Journal of Applied Physiology* **61**, 1077- 1083.

Belanger AY & McComas AJ (1981). Extent of motor unit activation during effort. *Journal of Applied Physiology* **51**, 1131-1135.

Bellemare F & Bigland-Ritchie B (1984). Assessment of human diaphragm strength and activation using phrenic nerve stimulation. *Respiration Physiology* **58**, 263-277.

Bellemare F & Bigland-Ritchie B (1987). Central components of diaphragmatic fatigue assessed by phrenic nerve stimulation. *Journal of Applied Physiology* **62**, 1307-1316.

Bellemare F, Bigland-Ritchie B & Woods JJ (1986). Contractile properties of the human diaphragm *in vivo*. *Journal of Applied Physiology* **61**, 1153-1161.

Bellemare F & Garzaniti N (1988). Failure of neuromuscular propagation during human maximal voluntary contraction. *Journal of Applied Physiology* **64**, 1084-1093.

Bellemare F, Woods JJ, Johansson R & Bigland-Ritchie B (1983). Motor-unit discharge rates in maximal voluntary contractions of three human muscles. *Journal of Neurophysiology* **50**, 1380-1392.

Bigland B, Goetzee B, MacLagan J & Zaimis E (1958). The effect of lowered muscle temperature on the action of neuromuscular blocking drugs. *Journal of Physiology (London)* **141**, 425-434.

Bigland B & Jehring B (1952). Muscle performance in rats, normal and treated with growth hormone. *Journal of Physiology (London)* **116**, 129-136.

Bigland B & Lippold OCJ (1954a). The relation between force, velocity and integrated electrical activity in human muscles. *Journal of Physiology (London)* **123**, 214-224.

Bigland B & Lippold OCJ (1954b). Motor unit activity in the voluntary contraction of human muscle. *Journal of Physiology (London)* **125**, 322-335.

Bigland B & Zaimis E (1958). Factors influencing limb temperature during experiments on skeletal muscle. *Journal of Physiology (London)* **141**, 420-424.

Bigland-Ritchie B (1981a). EMG and fatigue of human voluntary and stimulated contractions. In: Porter R, Whelan J (eds.), *Human Muscle Fatigue: Physiological Mechanisms*, pp. 130-156. London: Pitman Medical.

Bigland-Ritchie B (1981b). EMG/force relations and fatigue of human voluntary contractions. In: Miller D (ed.), *Exercise and Sports Sciences Reviews (Vol. 9)*, pp. 75-117. Franklin Institute Press.

Bigland-Ritchie B (1987). Respiratory muscle fatigue: posters, methods, and more. In: Sieck GC, Gandevia SC, Cameron WE (eds.), *Respiratory Muscles and Their Neuromotor Control*, pp. 379-389. New York: Alan R. Liss.

Bigland-Ritchie B (1988). Nervous system and sensory adaptation. In: Bouchard C, Shephard RJ, Stephens T, Sutton JR, McPherson BD (eds.), *Exercise, Fitness and Health*, pp. 377- 383. Champaign, IL: Human Kinetics.

Bigland-Ritchie B, Cafarelli E & Vøllestad NK (1986a). Fatigue of submaximal static contractions. Lars Hermansen Memorial Symposium: Exercise in Human Physiology. *Acta Physiologica Scandinavica* **128** (suppl. 556), 137-148.

Bigland-Ritchie B, Dawson NJ, Johansson RS & Lippold OCJ (1986b). Reflex origin for the slowing of motoneurone firing rates in fatigue of human voluntary contractions. *Journal of Physiology (London)* **379**, 451-459.

Bigland-Ritchie B, Furbush FH, Gandevia SC & Thomas CK (1992a). Voluntary discharge frequencies of human motoneurons at different muscle lengths. *Muscle & Nerve* **15**, 130- 137.

Bigland-Ritchie B, Graichen H & Woods JJ (1973). A variable-speed motorized bicycle ergometer for positive and negative work exercise. *Journal of Applied Physiology* **35**, 739- 740.

Bigland-Ritchie B, Johansson R, Lippold OCJ, Smith S & Woods JJ (1983a). Changes in motoneurone firing rates during sustained maximal voluntary contractions. *Journal of Physiology (London)* **340**, 335-346.

Bigland-Ritchie B, Johansson R, Lippold OCJ & Woods JJ (1983b). Contractile speed and EMG changes during fatigue of sustained maximal voluntary contractions. *Journal of Neurophysiology* **50**, 313-324.

Bigland-Ritchie B, Johansson R & Woods JJ (1983c). Does a reduction in motor drive necessarily result in force loss during fatigue? In: Knuttgen HG, Vogel JA, Poortmans, J (eds.), *Biochemistry of Exercise (International Series on Sport Sciences Vol. 13)*, pp. 864-870. Champaign, IL: Human Kinetics.

Bigland-Ritchie B, Jones DA, Hosking GP & Edwards RHT (1978). Central and peripheral fatigue in sustained maximum voluntary contractions of human quadriceps muscle. *Clinical Science and Molecular Medicine* **54**, 609-614.

Bigland-Ritchie B, Jones DA & Woods JJ (1979). Excitation frequency and muscle fatigue: Electrical responses during human voluntary and stimulated contractions. *Experimental Neurology* **64**, 414-427.

Bigland-Ritchie B, Kukulka CG, Lippold OCJ & Woods JJ (1982). The absence of neuromuscular transmission failure in sustained maximal voluntary contractions. *Journal of Physiology (London)* **330**, 265-278.

Bigland-Ritchie B & Lippold OCJ (1979). Changes in muscle activation during prolonged maximal voluntary contractions. *Journal of Physiology (London)* **292**, 14P-15P.

Bigland-Ritchie B, Thomas CK, Rice CL, Howarth JV & Woods JJ (1992b). Muscle temperature, contractile speed, and motoneuron firing rates during human voluntary contractions. *Journal of Applied Physiology* **73**, 2457-2461.

Bigland-Ritchie B & Vøllestad NK (1988). Hypoxia and fatigue: How are they related? In: Sutton JR, Houston CS, Coates G (eds.), *Hypoxia: The Tolerable Limits*, pp. 315-328. Indianapolis, IN: Benchmark Press.

Bigland-Ritchie B & Woods JJ (1974). Integrated EMG and oxygen uptake during dynamic contractions of human muscles. *Journal of Applied Physiology* **36**, 475-479.

Bigland-Ritchie B & Woods JJ (1976). Integrated electromyogram and oxygen uptake during positive and negative work. *Journal of Physiology (London)* **260**, 267-277.

Bigland-Ritchie B & Woods JJ (1984). Changes in muscle contractile properties and neural control during human muscular fatigue. *Muscle & Nerve* **7**, 691-699.

Clamann HP (1970). Activity of single motor units during isometric tension. *Neurology* **20**, 254-260.

Curtin NA & Davies RE (1970). Chemical and mechanical changes during stretching of activated frog skeletal muscle. *Cold Spring Harbor Symposia on Quantitative Biology* **37**, 619-626

Dacko SM & Cope TC (1994). Patterns of MG motor unit activity following fatigue of a close synergist in the decerebrate cat. *Abstracts, International Symposium on Neural and Neuromuscular Aspects of Muscle Fatigue* (Miami, FL, November 10-13, 1994, pp. 35. Miami, FL: Miami Project to Cure Paralysis, University of Miami.

de Haan A, Jones DA & Sargeant AJ (1989). Changes in velocity of shortening, power output and relaxation rate during fatigue of rat medial gastrocnemius muscle. *European Journal of Physiology* **413**, 422-428.

Denny-Brown D (1928). On inhibition as a reflex accompaniment of the tendon jerk and of other forms of active muscular response. *Proceedings of the Royal Society Series B* **103**, 321-336.

Denny-Brown D & Pennybacker JB (1938). Fibrillation and fasciculation in voluntary muscle. *Brain* **61**, 311-333.

Desmedt JE & Godaux E (1977). Fast motor units are not preferentially activated in rapid voluntary contractions in man. *Nature* **267**, 717-719.

Duchateau J & Hainaut K (1985). Electrical and mechanical failures during sustained and intermittent contractions in humans. *Journal of Applied Physiology* **58**, 942-947.

Edwards RHT, Hill DK, Jones DA & Merton PA (1977). Fatigue of long duration in human skeletal muscle after exercise. *Journal of Physiology (London)* **272**, 769-778.

Fitts RH (1994). Cellular Mechanisms of Muscle Fatigue. *Physiological Reviews* **74**, 49-94.

Fuglsang-Frederiksen A (1981). Electrical activity and force during voluntary contraction of normal and diseased muscle. *Acta Neurologica Scandinavica* **63** (suppl. 83), 1-60.

Gandevia SC & McKenzie DK (1985). Activation of the human diaphragm during maximal static efforts. *Journal of Physiology (London)* **367**, 45-56.

Gandevia SC & McKenzie DK (1988a). Activation of human muscles at short muscle lengths during maximal static efforts. *Journal of Physiology (London)* **407**, 599-613.

Gandevia SC & McKenzie DK (1988b). Human diaphragmatic endurance during different maximal respiratory efforts. *Journal of Physiology (London)* **395**, 625-638.

Gandevia SC, Macefield G, Burke D & McKenzie DK (1990). Voluntary activation of human motor axons in the absence of muscle afferent feedback. The control of the deafferented hand. *Brain* **113**, 1563-1581.

Gilson AS Jr & Mills WB (1941). Activities of single motor units in man during slight voluntary efforts. *American Journal of Physiology* **133**, 658-669.

Grimby L & Hannerz J (1977). Firing rate and recruitment order of toe extensor motor units in different modes of voluntary contraction. *Journal of Physiology (London)* **264** , 865-879.

Hagbarth K, Kunesch EJ, Nordin M, Schmidt R & Wallin EU (1986). Gamma loop contributing to maximal voluntary contractions in man. *Journal of Physiology (London)* **380**, 575-591.

Hales JP & Gandevia SC (1988). Assessment of maximal voluntary contraction with twitch interpolation: an instrument to measure twitch responses. *Journal of Neuroscience Methods* **25**, 97-102.

Hayward L, Breitbach D & Rymer WZ (1988). Increased inhibitory effects on close synergists during muscle fatigue in the decerebrate cat. *Brain Research* **440**, 199-203.

Heckman CJ & Binder MC (1993). Computer simulations of motoneuron firing rate modulation. *Journal of Neurophysiology* **69**, 1005-1008.

Hicks A, Fenton J, Garner S & McComas AJ (1989). M wave potentiation during and after muscle activity. *Journal of Applied Physiology* **66**, 2606-2610.

Hill AV (1913). The heat production in prolonged contractions of an isolated frog's muscle. *Journal of Physiology (London)* **47**, 305-324.

Hill AV (1938). The heat of shortening and the dynamic constants of muscle. *Proceedings of the Royal Society Series B* **126**, 136-195.

Howell JN, Fuglevand AJ, Walsh ML & Bigland-Ritchie B (1994). Motor unit firing during shortening and lengthening contractions. *Society for Neuroscience Abstract* **20**, 1759.

Jones DA, Bigland-Ritchie B & Edwards RHT (1979). Excitation frequency and muscle fatigue: Mechanical responses during voluntary and stimulated contractions. *Experimental Neurology* **64**, 401-413.

Katz B (1939). The relation between force and speed in muscular contraction. *Journal of Physiology (London)* **96**, 45-64.

Kernell D, Donselaar Y & Eerbeek O (1987). Effects of physiological amounts of high- and low-rate chronic stimulation on fast-twitch muscle of the cat hindlimb. II. Endurance-related properties. *Journal of Neurophysiology* **58**, 614-627.

Kernell D & Monster AW (1982). Motoneurone properties and motor fatigue. An intracellular study of gastrocnemius motoneurones of the cat. *Experimental Brain Research* **46**, 197- 204.

Krnjevic K & Miledi R (1958). Failure of neuromuscular propagation in rats. *Journal of Physiology (London)* **140**, 440-461.

Kukulka CG & Clamann HP (1981). Comparison of the recruitment and discharge properties of motor units in human brachial biceps and adductor pollicis during isometric contractions. *Brain Research* **219**, 45-55.

Lindsley DB (1935). Electrical activity of human motor units during voluntary contraction. *American Journal of Physiology* **114**, 90-99.

Lippold OCJ (1952). The relation between integrated action potentials in a human muscle and its isometric tension. *Journal of Physiology (London)* **117**, 492-499.

Lippold OCJ (1973). *The Origin of the Alpha Rhythm*. London: Churchill Livingston.

Lippold OCJ, Redfearn JWT & Vuo J (1960). The electromyography of fatigue. *Ergonomics* **3**, 121-131.

Marsden CD, Meadows JC & Merton PA (1969). Muscular wisdom. *Journal of Physiology (London)* **200**, 15P.

Marsden CD, Meadows JC & Merton PA (1971). Isolated single motor units in human muscle and their rate of discharge during maximal voluntary effort. *Journal of Physiology (London)* **217**, 12-13P.

Marsden CD, Meadows JC & Merton PA (1983). "Muscular wisdom" that minimizes fatigue during prolonged effort in man: Peak rates of motoneuron discharge and slowing of discharge during fatigue. In: Desmedt JE (ed.), *Motor Control Mechanisms in Health and Disease*, pp. 169-211. New York: Raven Press.

McComas AJ, McFadden L, Newberry R & Sacco P (1994). *Abstracts, International Symposium on Neural and Neuromuscular Aspects of Muscle Fatigue* (Miami, FL, November 10-13, 1994, pp. 35. Miami, FL: Miami Project to Cure Paralysis, University of Miami.

McKenzie DK, Bigland-Ritchie B, Gorman RB & Gandevia SC (1992). Central and peripheral fatigue of human diaphragm and limb muscles assessed by twitch interpolation. *Journal of Physiology (London)* **454**, 643-656.

Merton PA (1954). Voluntary strength and fatigue. *Journal of Physiology (London)* **123**, 553-564.

Merton PA, Hill DK & Morton HB (1981). Indirect and direct stimulation of fatigued human muscle. In: Porter R & Whelan J (eds.), *Human Muscle Fatigue: Physiological Mechanisms*. London: Pitman Medical.

Milner-Brown HS & Miller RG (1986). Muscle membrane excitation and impulse propagation velocity are reduced during muscle fatigue. *Muscle & Nerve* **9**, 367-374.

Milner-Brown HS & Stein RB (1975). The relation between the surface electromyogram and muscular force. *Journal of Physiology (London)* **246**, 549-569.

Milner-Brown HS, Stein RB & Yemm R (1973). The orderly recruitment of human motor units during voluntary isometric contractions. *Journal of Physiology (London)* **230**, 359-370.

Monster AW & Chan H (1977). Isometric force production by motor units of extensor digitorum communis muscle in man. *Journal of Neurophysiology* **40**, 1432-1443.

Mosso A (1915). *Fatigue, 3rd Ed.* Translated by Drummond M & Drummond WD. London: Allen and Unwin.

Nardone A, Romano C & Schieppati M (1989). Selective recruitment of high-threshold human motor units during voluntary isotonic lengthening of active muscles. *Journal of Physiology (London)* **409**, 451-471.

Ritchie JM & Ritchie BR (1968). Local anesthetics: Effect of pH on activity. *Science* **162**, 1394-1395.

Ritchie JM, Ritchie B & Greengard P (1965a). The active structure of local anesthetics. *Journal of Pharmacology & Experimental Therapeutics* **150**, 152-159.

Ritchie JM, Ritchie B & Greengard P (1965b). The effect of the nerve sheath on the action of local anesthetics. *Journal of Pharmacology & Experimental Therapeutics* **150**, 160-164.

Smith OC (1934). Action potentials from single motor units in voluntary contraction. *American Journal of Physiology* **108**, 629-638.

Stephens JA & Taylor A (1972). Fatigue of maintained voluntary muscle contractions in man. *Journal of Physiology (London)* **220**, 1-18.

Tanji J & Kato M (1973a). Recruitment of motor units in voluntary contraction of a finger muscle in man. *Experimental Neurology* **40**, 759-770.

Tanji J & Kato M (1973b). Firing rate of individual motor units in voluntary contraction of abductor digiti minimi muscle in man. *Experimental Neurology* **40**, 771-783.

Thomas CK (1993). Muscle fatigue after incomplete human cervical spinal cord injury. *Journal of the American Paraplegia Society* **16**, 87.

Thomas CK, Bigland-Ritchie B & Johansson RS (1991a). Force-frequency relationships of human thenar motor units. *Journal of Neurophysiology* **65**, 1509-1516.

Thomas CK, Johansson RS & Bigland-Ritchie B (1991b). Attempts to physiologically classify human thenar motor units. *Journal of Neurophysiology* **65**, 1501-1508.

Thomas CK, Woods JJ & Bigland-Ritchie B (1989). Impulse propagation and muscle activation in long maximal voluntary contractions. *Journal of Applied Physiology* **67**, 1835-1842.

Vallbo AB, Hagbarth KE, Torebjork HE & Wallin BG (1979). Somatosensory, proprioceptive, and sympathetic activity in human peripheral nerves. *Physiological Reviews* **59**, 919-956.

Vøllestad NK, Sejersted OM, Bahr R, Woods JJ & Bigland-Ritchie B (1988). Motor drive and metabolic responses during repeated submaximal voluntary contractions in humans. *Journal of Applied Physiology* **64**, 1421-1427.

Westling G, Johansson RS, Thomas CK & Bigland-Ritchie B (1990). Measurement of contractile and electrical properties of single human thenar motor units in response to intraneural motor-axon stimulation. *Journal of Neurophysiology* **64**, 1331-1338.

Wilkie DR (1968). Heat work and phosphorylcreatine break-down in muscle. *Journal of Physiology (London)* **195**, 157-183.

Woods JJ & Bigland-Ritchie B (1983). Linear and non-linear surface EMG/force relationships in human muscles: An anatomical/functional argument for the existence of both. *American Journal of Physical Medicine* **62**, 287-299.

Woods JJ, Furbush F & Bigland-Ritchie B (1987). Evidence for a fatigue-induced reflex inhibition of motoneuron firing rates. *Journal of Neurophysiology* **58**, 125-137.

SECTION I

Fatigue of Single Muscle Fibers

While the bulk of muscle fatigue can be attributed to processes beyond the neuromuscular junction, it is the central nervous system (CNS) that actually decides when the process of muscle fatigue will occur. Chapters in this section examine the crucial processes that take place beyond the neuromuscular junction. A common definition of fatigue is used in these chapters, namely a contraction-induced reduction in the ability of the muscle to produce force and/or to shorten. The experimental work relies primarily on the use of single muscle-fiber preparations to study the key events that impair force generation.

Chapter 1 (Edman) assesses the extent to which force is impaired due to events that occur at the crossbridges, or to more proximal events within the cell which impair crossbridge activation. It is argued that calcium-induced activation is less likely to be impaired than events at the myofibrils, at least during *moderate* fatigue. The relative importance of these events depends on the exact protocol and stimulus regimen that is used to induce fatigue.

In **Chapter 2** (Stephenson and colleagues), the processes that transpose the sarcolemmal action potential to the actomyosin motor are investigated. Many of the studies here use a preparation in which a single muscle fiber is skinned so that the *intracellular* environment can be manipulated to produce effects on force. This permits many crucial steps in excitation-contraction coupling to be examined. As the processes involved in excitation-contraction coupling are not all-or-none unitary events, emphasis must be placed on the molecular events involved in voltage sensing and initiation of signal transmission from the t-tubular system to the sarcoplasmic reticulum (via dihydropyridine receptors) and on the variety of factors that modify calcium release from the sarcoplasmic reticulum. An elevation in intracellular Mg^{2+}, which occurs in fatigue, may significantly impair calcium release and ultimately crossbridge activation.

Many examples of muscle fatigue are associated with intracellular acidosis. In **Chapter 3** (Allen and colleagues), the processes impaired by a fatigue-induced rise in hydrogen ion concentration are considered. While many features of fatigue can be mimicked by intracellular acidosis (such as an impaired maximal shortening velocity and slowed rate of relaxation), it is not the only factor that can produce them. This view is supported by observations on patients with myophosphorylase deficiency. They do not generate lactate during exercise, yet they fatigue faster than normal subjects. This chapter also reiterates the biochemical reality that hydrogen ions generated by exercise will affect many critical steps in force generation, including the turnover of crossbridges and the force produced by each attached crossbridge.

Ion exchange mechanisms have long been known to change in exercising muscle. In **Chapter 4** (Sjøgaard & McComas), the local and distant effects produced by the initial rise in interstitial potassium are reviewed. Local increases in extracellular potassium will be most marked in isometric exercise. While they can impair sarcolemmal and t-tubular function, regulatory mechanisms exist that range from local activation of the Na^+/K^+ pump in both active and adjacent quiescent fibers to the reflex adjustment of blood flow and ventilation. This discussion introduces the theme that specific mechanisms exist to inform the CNS about the progress of muscle fatigue.

The processes reviewed in this section define those crucial biochemical events that must underlie fatigue of single fibers acting not only *in vitro* but also *in vivo*. However, particular challenges, taken up in later sections, are to assess the relative importance of these events and to decide which are more important when force is produced by whole motor units *in vivo*.

1

MYOFIBRILLAR FATIGUE *VERSUS* FAILURE OF ACTIVATION

K. A. P. Edman

Department of Pharmacology
University of Lund
Sölvegatan 10
S-223 62 Lund
Sweden

ABSTRACT

Two principal mechanisms underlying fatigue of isolated muscle fibers are described: failure of activation of the contractile system and reduced performance of the myofibrils due to altered kinetics of crossbridge function. The relative importance of these two mechanisms during development of fatigue is discussed.

It is common knowledge in muscle physiology that the mechanical performance of an isolated muscle preparation is greatly dependent on the time that the muscle is allowed to rest between successive contractions. For example, a single twitch interposed in a series of isometric tetani can affect both the time course and the amplitude of one or more of the subsequent tetani. This indicates that the extra contraction causes some *disturbance* within the contractile machinery or in the excitation-contraction coupling, which takes the muscle a relatively long time to overcome. Thus, the mechanical performance of striated muscle is quite dependent on the stimulation history of the muscle fibers, and this likely holds true for muscles operating *in situ* in the body as well as muscle preparations studied in an organ bath. Progressive shortening of the resting intervals between contractions may greatly reduce the force generating capability of the muscle. This reversible state of force depression, referred to as *fatigue*, also includes a lower rate of rise of force and a slower relaxation (e.g., Edwards et al., 1975; Fitts & Holloszy, 1978; Edman & Mattiazzi, 1981). Two main causes are thought to underlie fatigue in an isolated muscle preparation: 1) failure of activation of the contractile system; and 2) reduced capacity of the myofibrils to produce force at any given state of activation (*myofibrillar fatigue*). The relative importance of these two factors during development of fatigue in isolated muscle will be discussed below. Myofibrillar fatigue will be discussed first.

Fatigue, Edited by Simon C. Gandevia et al.
Plenum Press, New York, 1995

MYOFIBRILLAR FATIGUE

An obvious difficulty in simulating human or animal fatigue in an isolated muscle preparation is that the motor control mechanisms normally operating *in vivo* are lacking in the organ bath. Presently, there is no clear information on the extent a muscle fiber can be stimulated by its motor nerve *in situ* in the body. Needless to say, an isolated muscle fiber can easily be stimulated until some aspect of the excitation-contraction mechanism fails and irreversible changes appear, but there is still no evidence that such a situation will ever appear *in vivo*. As will be discussed, it is feasible to produce a moderate, and fully reversible state of *fatigue* in an isolated muscle fiber by using a stimulation routine that does not lead to failure of activation of the contractile system. Single fibers isolated from amphibian leg muscles are most frequently used in these studies. The remarkable viability of isolated amphibian muscle fibers makes these preparations particularly well suited for the long-lasting and demanding experiments involved in the study of muscle fatigue.

Isometric Force

Fig. 1 illustrates the characteristic changes of the isometric tetanus that are produced during moderate fatigue in a frog muscle fiber. In the control series (Fig. 1*A*), the fiber is stimulated to produce a 1-s isometric tetanus at regular 5- or 15-min intervals. Fatigue is produced by reducing the intervals between the tetani to 15 s (Fig. 1*B*). Isometric force, using this stimulation regimen, declines to a steady level at approximately 75% of the original value after twenty-five to thirty contractions. The *shape* of the tetanus myogram is also changed in a characteristic way during fatigue; there is a lower rate of rise of force during the onset of the tetanus and a slowing of the relaxation phase (note the prolonged time to the shoulder of the relaxation phase). It is of interest to note, however, that these changes are all fully reversed after return to the control stimulation frequency (Fig. 1*C*).

There is now good evidence that moderate fatigue does not involve failure of activation of the contractile system. In studying this problem, caffeine has served as a useful tool based on its ability to potentiate the release of activator calcium from the sarcoplasmic reticulum. Fig. 2 illustrates the effects of 0.5 mM caffeine on the isometric twitch and tetanus after the tetanic force has been reduced by approximately 25% by frequent tetanization (15 s intervals between tetani). Caffeine can be seen to increase the rate of rise of force in both twitch and tetanus and to potentiate the isometric twitch. However, caffeine does not affect the amplitude of the tetanus. These results suggest that even in the fatigued state there is a capacity for further release of calcium from the sarcoplasmic reticulum. The lack of effect of caffeine on the tetanus amplitude clearly shows, however, that the contractile system is maximally activated during a tetanus in moderate fatigue.

Caffeine, added to the external medium in higher than twitch-potentiating concentrations, induces graded contracture of intact muscle fibers by releasing calcium from the sarcoplasmic reticulum.* This effect of caffeine is independent of membrane excitation and provides a means of activating the contractile system to the desired level by side-stepping the excitation-contraction coupling (e.g., Delay et al., 1986; Klein et al., 1990). Fig. 3 shows an experiment in which contractures induced by 3 and 15 mM caffeine have been interposed between 1-s tetani under control (rested state) conditions, and after development of moderate fatigue in a single muscle fiber. As can be seen in the inset diagram of Fig. 4, 3 mM caffeine induces submaximal contracture whereas 15 mM caffeine is more than sufficient to mechani-

* Contracture is the term used for shortening of muscle that does not involve membrane excitation. It should not be confused with a permanent shortening of soft tissues.

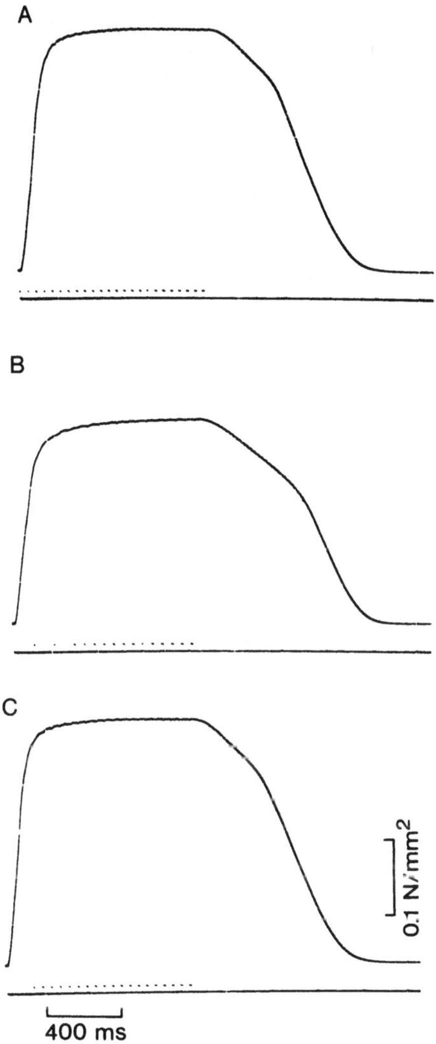

Figure 1. Isometric tetani of a single muscle fiber recorded after attainment of steady state at two contraction frequencies. Interval between contractions: *A*, 15 min; *B*, 15 s; *C*, return to 15 min intervals. Temperature, 2.2 °C (reproduced with permission from Edman & Mattiazzi, 1981).

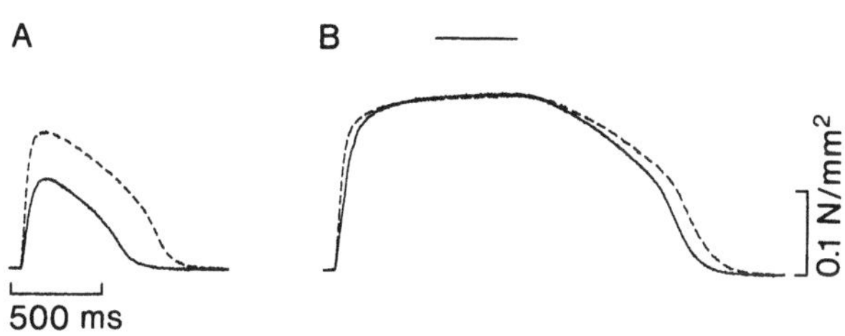

Figure 2. Effects of caffeine on isometric twitch (left) and tetanus (right) after development of fatigue. Continuous lines, no caffeine. Dashed lines, presence of 0.5 mM caffeine. Horizontal bar above tetanus myogram: maximum tetanic force recorded before fatigue (reproduced with permission from Edman & Lou, 1990).

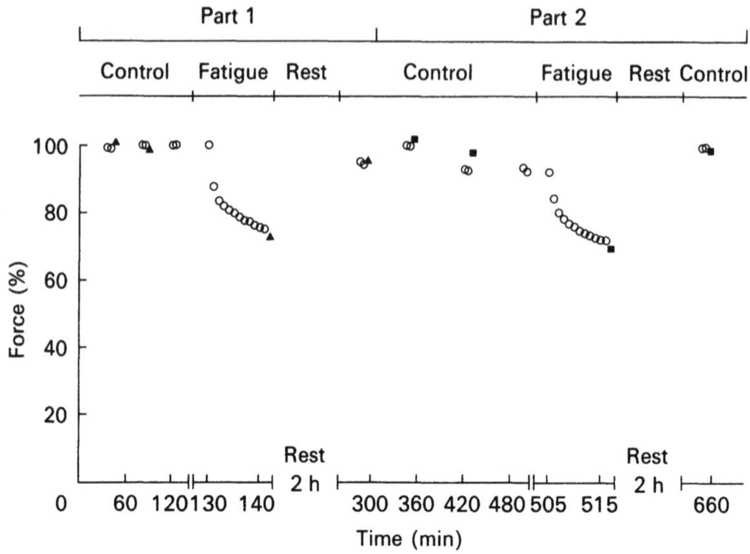

Figure 3. Relative changes of tetanic force and contracture responses to caffeine during development of moderate fatigue in a frog single muscle fiber. ○, isometric tetanus; ▲ and ■, contracture responses to 15 and 3 mM caffeine, respectively. The sequence of control, fatigue and rest periods is indicated by marked intervals above the data points. The time scale is expanded during the fatigue period. Only some of the tetanic responses are plotted in the diagram for clarity. Tetanic tensions throughout the experiment are normalized to the mean value of tetanic force derived during the control period of part 1. Contracture tensions being induced by two different caffeine concentrations are normalized to values derived during the control period in each respective part of the experiment. Note that tetani and caffeine contractures are depressed to approximately the same degree (from Edman & Lou, 1992).

cally saturate the contractile system of the intact frog muscle fiber. However, independent of the caffeine concentration used, the amplitude of the caffeine-induced contracture is reduced to nearly the same degree as is the isometric tetanus by fatiguing stimulation (Figs. 3 and 4). Similar to the situation during the tetanus, the rate of rise of force during the caffeine contracture is also reduced in the fatigued state (Edman & Lou, 1992). These results strongly suggest that the decrease in mechanical performance during moderate fatigue is not attributable to failure of activation of the contractile system. The fatigue process apparently makes the myofibrils less capable of producing force even though they are fully activated by calcium.

Vmax, Force-Velocity Relation and Power

Fatigue does not merely involve the fiber's capacity to produce force; the maximum speed of shortening (V_0) and the power output are affected as well. V_0 is preferably measured by the slack-test (Edman, 1979) in an isolated muscle preparation as this ensures that shortening occurs at virtually zero load and, furthermore, that the measured velocity is not affected by recoil of elastic components acting in series with the contractile units. The first detailed study of the relative changes of V_0 and isometric force during development of fatigue was performed by Edman and Mattiazzi (1981). In a more recent publication (Curtin & Edman, 1994) the effects of moderate fatigue on the entire force-velocity relationship, including measurements at loads exceeding maximum tetanic force (P_0), have been described. V_0 is not significantly affected during the initial stage of muscle fatigue but starts

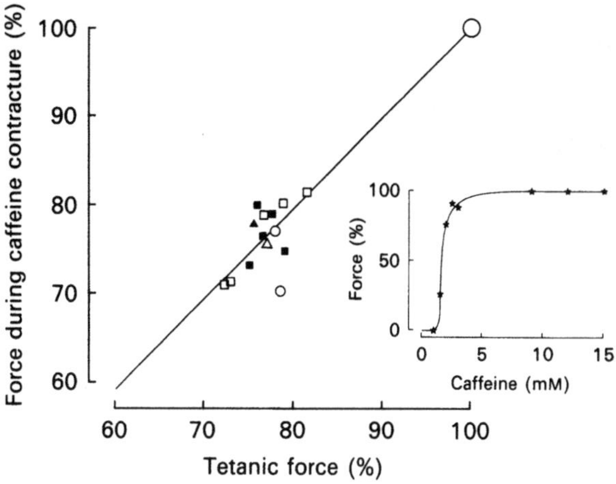

Figure 4. Relation between maximal tetanic force and maximal contracture response to caffeine during development of moderate fatigue. Data expressed as percentage of tetanus and contracture tensions recorded under control (prefatigue) conditions in respective fibers. The control value (indicated by large open circle) is the mean of two to four separate recordings in each fiber. Caffeine concentrations (mM): ❑, 3; ○, 6; Δ, 9; ▲, 12; ■, 15. Continuous line, linear regression based on all data points. Inset: relation between peak contracture tension and caffeine concentration; each data point is the mean of two to nine contractures performed in nine fibers (reproduced with permission from Edman & Lou, 1992).

to decline at a point where the tetanic force has been reduced by approximately 10% of its rested-state value (Fig. 5). As the fatiguing process goes on, however, V_0 becomes increasingly affected and finally reaches approximately the same degree of depression as P_0 does during moderate fatigue.

The fact that V_0 is reduced during fatigue provides a relevant piece of information to the evaluation of the cellular mechanisms involved in muscle fatigue. The maximal speed

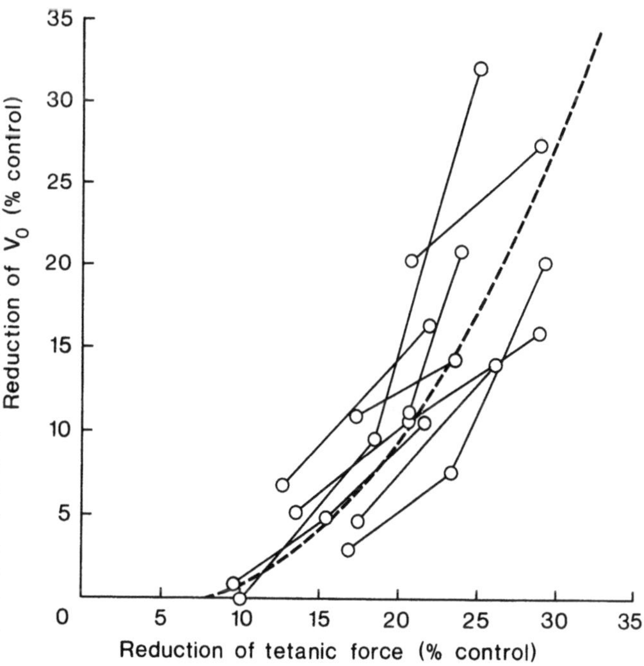

Figure 5. Decrease in velocity of unloaded shortening (ΔV_0) in relation to force depression (ΔP_0) during development of moderate fatigue in single muscle fibers. Each set of data connected by a solid line is from a single fiber. Dashed line drawn from the equation $\Delta V_0 = 0.006\ \Delta P_0^{2.48} - 1.0$ derived by regression analysis from all data points. Co-ordinates expressed as percent of V_0 and tetanic force recorded under control (prefatigue) conditions in respective fibers (reproduced with permission from Edman & Mattiazzi, 1981).

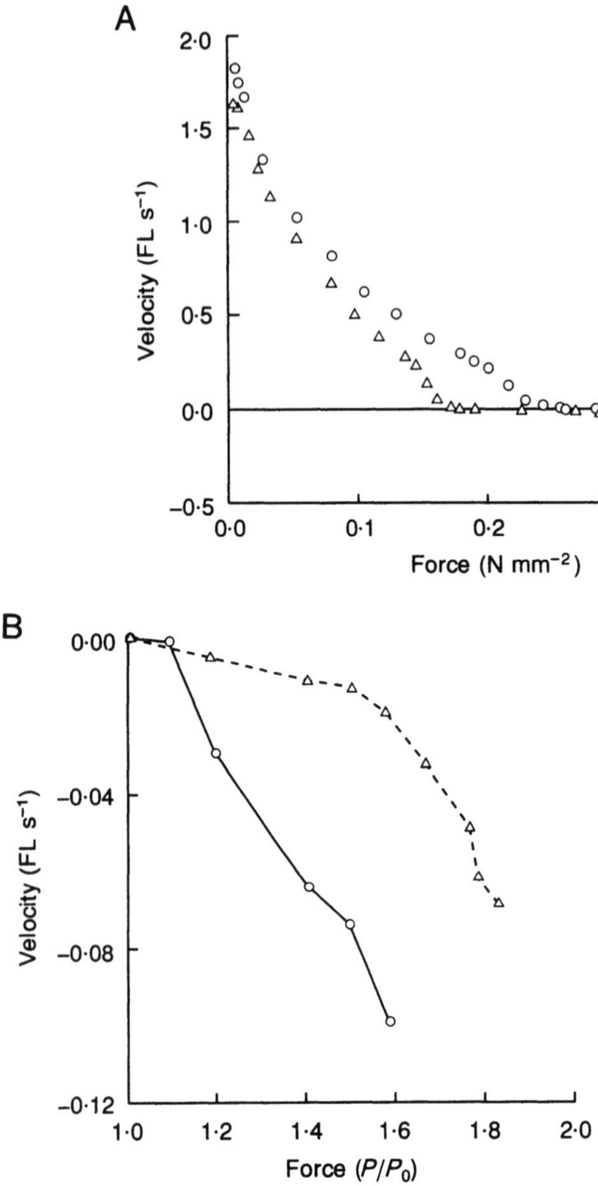

Figure 6. *A*, force-velocity relation of frog muscle fiber at rested state and moderate fatigue. FL is the fiber length at resting sarcomere length of 2.1 μm. Positive velocities for shortening, negative velocities for lengthening. Control (O), fatigue (Δ); *B*, normalized form of the force-velocity relation for lengthening. Force expressed as a fraction of the isometric force measured in control and fatigue, respectively (from Curtin & Edman, 1994).

of shortening in vertebrate skeletal muscle is, unlike the isometric force, independent of the degree of activation of the contractile system. This has been demonstrated in both frog intact muscle fibers (Edman, 1979) and skinned fiber preparations of amphibian and mammalian muscles (Podolsky & Teichholz, 1970; Brenner, 1980; Ferenczi et al., 1984. V_0 is furthermore insensitive to the amount of overlap between the thick and thin filaments (Edman, 1979) supporting the view (Huxley, 1957) that the speed of shortening at zero load is independent of the actual number of myosin crossbridges interacting with the thin filaments. The finding that V_0 does change during the course of fatigue thus strongly suggests that fatigue involves a specific change of the kinetic properties of the myosin crossbridge that is not attributable to altered metabolism of calcium in the excitation-contraction coupling.

V_0 may be presumed to reflect the maximal cycling rate of the crossbridges, and this property is now generally thought to be determined by the rate constant of detachment of

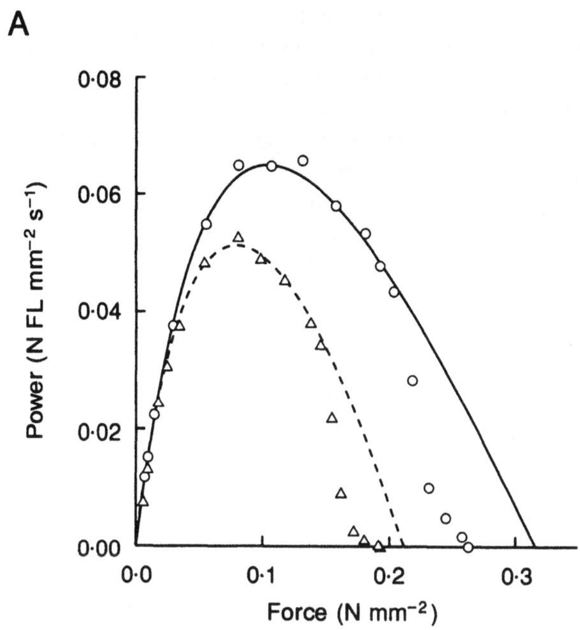

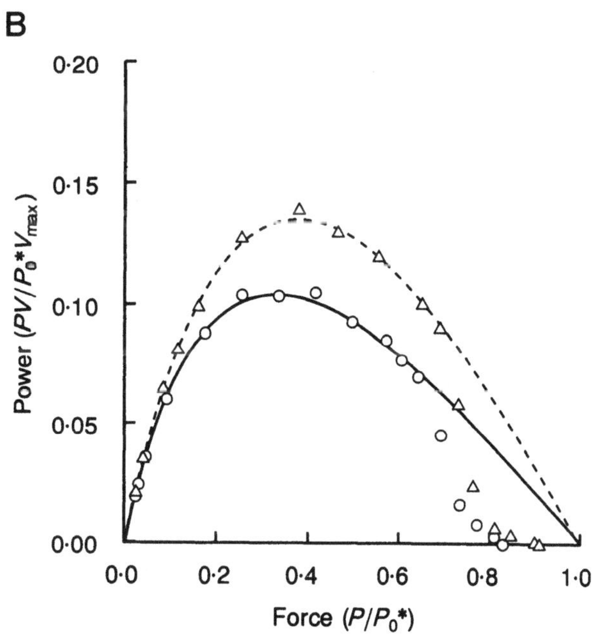

Figure 7. Power-force relations for shortening in the control state and in the moderately fatigued state. *A*, power calculated as the product of force and velocity expressed in N mm^{-2} and fiber lengths (FL) s^{-1}, respectively. *B*, power calculated after normalizing force (P) and vlocity (V) data with respect to maximum force (P_o^*) and maximum speed of shortening (Vmax), respectively. Control (○), fatigue (Δ) (from Curtin & Edman, 1994).

the crossbridges at negative strain, that is when the crossbridges are in a position where they brake the sliding motion of the myofilaments (Huxley, 1957).

Another expression of the altered crossbridge function during muscle fatigue is the change in shape of the force-velocity relation during fatiguing stimulation. The force-velocity relation for shortening has been shown to consist of two distinct curvatures, both having an upwards concave shape, located on either side of a break-point near 78% of P_0 (Edman, 1988). Moderate fatigue reduces the main curvature of the force-velocity relation but does not affect the biphasic nature of the force-velocity curve or shift the location of the breakpoint relative to P_0 (Fig. 6). As is evident from Fig. 7*A*, there is a true reduction of the power output during fatigue at all force levels. It is essential to point out, however, that since the force-velocity relation becomes straighter during fatigue, the fiber is able to maintain a higher power output than would otherwise be the case.

This is demonstrated in Fig. 7*B* where the mechanical power at rested state and during moderate fatigue has been normalized with respect to isometric force and maximum speed of shortening. Note that the relative power points for fatigue lie above the controls and that maximal power is attained at a higher relative force in the fatigued state. The decrease in curvature of the force-velocity relation can be presumed to make the muscle less efficient (Woledge, 1968). The muscle may thus have to spend somewhat more energy for a certain amount of work. This may be the price the muscle has to pay to better maintain its power output during fatigue.

The force-velocity relation at loads exceeding P_0 expresses the muscle's ability to resist stretching and forms a smooth continuation of the force-velocity relation during shortening (Edman, 1988). The flat force-velocity relationship around P_0 confers a high degree of stability upon the contractile system. Any tendency of a fiber segment to be stretched during contractile activity (due to pull by stronger segments in series) is thus counteracted by the build-up of a resistive force in the weaker (yielding) part. The recruitment of extra force during stretching is only to a minor part attributable to an increased number of attached crossbridges judging from the modest increase in fiber stiffness (Lombardi & Piazzesi, 1990; K.A.P. Edman, unpublished observations). Apparently the crossbridges are able to hold a greater force when the muscle is stretched and the bridges are dragged along the thin filaments against their normal course than when the bridges operate under isometric conditions. As is evident from the following, fatigue does not reduce the bridges' ability to resist stretch.

Fig. 6*A* shows the effects of fatigue on the force-velocity relation during stretching, that is for loads higher than P_0, along with results for shortening. Force is here expressed in absolute values. While fatigue reduces P_0 and lowers the speed of shortening at each load less than the isometric force, the force-velocity relations for stretching nearly overlap in the control and fatigued state. Thus, for a given speed of elongation, the fiber produces a proportionately higher force after fatigue than in the control state. This is further illustrated in Fig. 6*B* in which the force-velocity relation for lengthening is shown with force normalized to the corresponding isometric value, P_0, in control and fatigue. This change in contractile behavior during fatigue is likely to be of significance under *in vivo* conditions as a muscle, weakened by fatiguing exercise, may yet be able to resist stretch nearly as well as in the non-fatigued state.

Fiber Stiffness

The instantaneous stiffness of active muscle provides information on the number of myosin crossbridges that are interacting with the thin filaments (Huxley & Simmons, 1971). For this measurement a fast, low-amplitude length perturbation is applied to the muscle either in the form of a single, very fast length step (Huxley & Simmons, 1971) or a high-frequency sinusoidal length oscillation (e.g., Julian & Morgan, 1981; Cecchi et al., 1986; Edman & Lou, 1990). Simultaneous measurements of force and instantaneous stiffness have recently been carried out during development of fatigue in frog single muscle fibers (Edman & Lou, 1990, 1992; Curtin & Edman, 1994). Results from such measurements are illustrated in Fig. 8 for fibers subjected to moderate fatigue by repetitive tetanization at 15-s intervals. It is clear that stiffness is much less depressed than force is during fatiguing stimulation. This is readily seen from the inset myograms of Fig. 8 and from the pooled data in the main diagram, which shows maximum tetanic force plotted against maximum tetanic stiffness during development of fatigue in six different fibers. A regression analysis based on these results

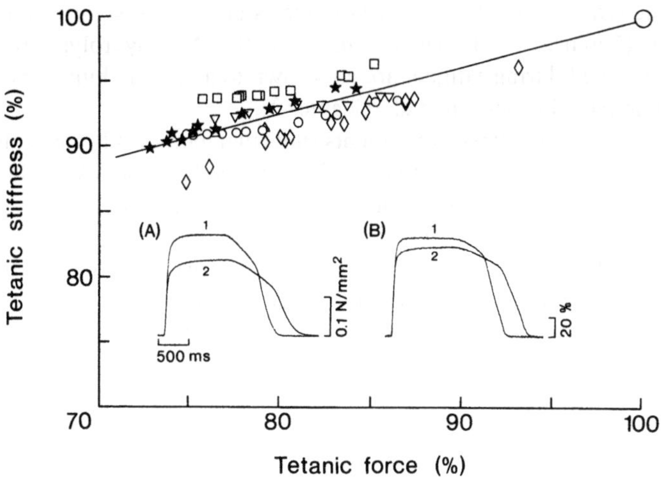

Figure 8. Relation between maximum tetanic stiffness and maximum tetanic force during development of moderate fatigue in six single muscle fibers of the frog. Data normalized with respect to maximum force and maximum stiffness recorded under control conditions in respective fibers. The control value (indicated by large open circle) is the mean of five separate recordings in each fiber. Data from a given fiber are denoted by the same symbol. Line, least squares regression of stiffness upon force. Insets: superimposed records of tetanic force (*A*) and tetanic stiffness (*B*) under control conditions (traces 1) and after fatiguing stimulation (traces 2) (from Edman & Lou, 1990).

shows that for a 25% depression of the maximum tetanic force during fatigue there is merely 9% reduction in fiber stiffness. Another discrepancy between force and stiffness concerns the rising phase of contraction. As pointed out before, the initial rate of rise of force during tetanus is reduced during moderate fatigue. By contrast, the rate of rise of stiffness was found to remain virtually unchanged (Edman & Lou, 1990).

These observations would seem to make clear that the decrease in force during moderate fatigue is only partly due to fewer crossbridges being attached to the thin filaments. Fatigue evidently reduces the force output of the individual crossbridges and this accounts for the major part of the force decline during moderate fatigue. As already discussed, fatigue also reduces V_0, the maximum speed of shortening, which is yet another expression of the modified performance of the crossbridge during development of fatigue. It should be noted that these changes occur while the contractile system is fully activated (see earlier).

The above changes in crossbridge performance during fatigue, including the slight decrease in number of attached crossbridges, therefore seem to be unrelated to the myoplasmic free calcium concentration. There is reason to believe, on the other hand, that the altered kinetics of crossbridge function is attributable to the intracellular metabolic changes that are known to arise during muscle fatigue.

Metabolic Changes during Fatigue and Their Influence on Muscle Function

Fatiguing exercise leads to, among other things, reduction of ATP and accumulation of products of the ATP hydrolysis, i.e., ADP, orthophosphate (P_i) and H^+ (Spande & Schottelius, 1970; Edwards et al., 1975; Dawson et al., 1978, 1980; Nassar-Gentina et al., 1978; Kawano et al., 1988). The change in MgATP concentration is quite small during moderate fatigue (Dawson et al., 1978, 1980) and is unlikely to significantly affect cross-bridge function (Ferenczi et al., 1984). However, the ATP hydrolysis products, in the concentrations reached during fatigue, are all known to affect the kinetic properties of the contractile system (see Stephenson et al., Chapter 2).

Results from skinned fiber experiments show that increased P_i concentration reduces isometric force (Altringham & Johnston, 1985; Chase & Kushmerick, 1988; Cooke et al., 1988; Godt & Nosek, 1989; Stienen et al., 1990, 1992; Dantzig et al., 1992. It is believed the power stroke of the crossbridge is initiated as P_i dissociates from the myosin head. An increased P_i concentration would thus be expected to increase the fraction of crossbridges that are in a low-force producing state. The influence of P_i on maximum speed of shortening, however, is less well understood. There is agreement that P_i does not affect V_0 at a pH of 7.0-7.4 (Altringham & Johnston, 1985; Chase & Kushmerick, 1988; Cooke et al., 1988). At pH 6.00, on the other hand, P_i was found to have no effect in one study (Cooke et al., 1988) but to have a marked depressant effect in another (Chase & Kushmerick, 1988). Mg-ADP has been stated to slightly increase the isometric force of skinned fibers, in the presence of millimolar concentrations of Mg-ATP, and to exert a small inhibitory effect on V_0 (Cooke et al., 1988). Finally, lowered pH has been convincingly shown to reduce both V_0 and isometric force in skinned muscle fibers (Chase & Kushmerick, 1988; Cooke et al., 1988; Godt & Nosek, 1989). A valuable study that has yet to be performed on the skinned fiber preparation would be to explore the *combined* action of the hydrolysis products on the force-velocity relation using concentrations corresponding to those recorded at various degrees of fatigue. The relative changes in V_0 and isometric force that are known to occur during fatigue in the intact fiber (see further Edman & Mattiazzi, 1981) would serve as a useful guide in such a study.

The effects of lowered intracellular pH (pH_i) are testable in intact single muscle fibers by varying the concentration of the permeant acid CO_2 in the bathing fluid. Based on this approach the contractile effects of intracellular acidification have been compared with the mechanical changes that are induced by fatiguing stimulation in frog muscle fibers (Edman & Mattiazzi, 1981; Edman & Lou, 1990, 1992; Curtin & Edman, 1989, 1994). Overall there is a striking similarity between the changes induced by the two interventions. Lowering of the intracellular pH reduces both V_0 and maximal tetanic force (Edman & Mattiazzi, 1981). Intracellular acidification also induces the characteristic changes of the isometric myogram that are seen in fatigue, that is reduced rate of rise of force during the onset of tetanus and slowing of the linear phase of force relaxation (Edman & Mattiazzi, 1981; Edman & Lou, 1990). Similar to the situation in moderate fatigue, the decrease in active force during intracellular acidification is associated with a relatively small reduction in fiber stiffness suggesting that lowered pH does reduce the force output of the individual bridge (Edman & Lou, 1990). Furthermore, reduced pH_i, like fatigue, increases the fiber's ability to resist stretch presumably by making the myosin crossbridges adhere more firmly to the actin filaments during activity (Curtin & Edman, 1989, 1994).

Although the changes induced by fatigue are, in large measure, reproducible by intracellular acidification, there is evidence that lowered pH_i does not fully account for the contractile change during moderate fatigue (see Allen et al., Chapter 3). This is evident when the relative changes in force and stiffness are considered for the two interventions. For example, a given decline in tension during intracellular acidification is associated with an even smaller drop in stiffness than during fatigue (Edman & Lou, 1990). Other examples of such differences between the two interventions are given by Edman and Lou (1990) and Curtin and Edman (1994). One or more of the other metabolic factors discussed above, notably P_i, are therefore also likely to play a part, in addition to lowered pH_i, during muscle fatigue.

Myoplasmic Free Calcium Concentration

Measurements of the intracellular free calcium concentration, $[Ca^{2+}]_i$, have been performed during fatiguing stimulation using various calcium indicators such as aequorin, Fura-2, indo-1 (Allen et al., 1989; Westerblad & Allen, 1991, 1993) and Fluo-3 (Edman et al., unpublished data). The results of these studies show that the tetanic $[Ca^{2+}]_i$ undergoes only slight changes during moderate fatigue, in line with the above conclusion that the contractile system remains maximally activated under these conditions. In a recent study, using Fluo-3 and accounting for the on- and off-rate constants for the calcium-dye complex (see further Caputo et al., 1994), the tetanic $[Ca^{2+}]_i$ during moderate fatigue did not drop below the level required for maximum tetanic force at rest (Edman et al., unpublished results). Westerblad and Allen (1991), using Fura-2 in a study of mouse muscle fibers, observed no decline in tetanic free calcium concentration as the force was reduced to approximately 80% of the rested-state value by fatiguing stimulation. At lower degrees of fatigue, tetanic $[Ca^{2+}]_i$ was actually found to rise above the control level (Lee et al., 1991; Westerblad & Allen, 1991). The situation is quite different, however, when the fiber is brought into a more profound state of fatigue by frequent stimulation. Under such conditions, as described in the following section, there is a marked reduction of the $[Ca^{2+}]_i$ transient during tetanus (Lee et al., 1991).

FAILURE OF ACTIVATION

Most previous studies of fatigue in isolated muscle preparations have employed a stimulation program which produces a more profound depression of force than that described in the previous section. The fatiguing protocols employed left very little time, generally no more than 1-3 s, for recovery of the muscle between the stimuli. As a result of such an intense treatment, the force produced during a 1-s tetanus is reduced within 15 min to approximately 40% of the rested-state value; over the same period of time the isometric twitch response declines to merely 5-10% of the control value (Eberstein & Sandow, 1963; Edman & Lou, 1992). This state of force depression will be referred to in the following as *excessive fatigue* (Edman & Lou, 1992) in order to distinguish it from the moderate, myofibrillar form of fatigue (approximately 70% of rested-state force), discussed in the preceding section.

The first convincing evidence that excessive fatigue involves failure of activation of the contractile system was presented by Eberstein and Sandow (1963). They demonstrated that a single muscle fiber, nearly exhausted by frequent stimulation and whose twitch response was virtually abolished, was nevertheless capable of producing an almost maximal contracture response to caffeine. This finding has subsequently been confirmed in several studies on amphibian and mammalian muscles (Kanaya et al., 1983; Jones & Sacco, 1989; Lännergren & Westerblad, 1989; Edman & Lou, 1992). High-frequency stimulation apparently leads to some failure of the excitation-contraction coupling that is overcome when the fiber is activated by caffeine, which releases calcium from its storage site in the sarcoplasmic reticulum (see further Delay et al., 1986; Klein et al., 1990). Another conclusion that may be drawn from this finding is that the intracellular calcium store is not depleted during high-frequency stimulation. In fact, the above results would seem to make clear that there is enough calcium available in the store during excessive fatigue to fully activate the contractile system.

There is now evidence that fast repetitive stimulation leads to failure of the inward spread of activation. This was first reported by Gonzalez-Serratos and colleagues (1981) and later, in a more detailed form, by Garcia and colleagues (1991). These authors observed that myofibrils in the center of the fiber were unable to actively shorten below slack length during tetanus in the fatigued state indicating that they were not properly activated by the electrical stimulus.

Using an experimental approach similar to that employed by Garcia and colleagues (1991), combined with rapid freeze-fixation, Edman and Lou (1992) did not find any sign of defective activation of the central myofibrils during *moderate* fatigue, that is under conditions when the tetanic force was depressed by less than 30% of the control value. However, as the tetanic force was depressed further, central myofibrils exhibited a wavy appearance when the fiber shortened below slack length at zero, indicating inactivity. In fact, at an advanced state of fatigue (tetanic force reduced below 50% of the control) only a relatively thin layer of myofibrils in the periphery of the fiber remained active while the rest of the fiber showed clear signs of being inactive or submaximally activated (Edman & Lou, 1992; their Fig. 8).

The mechanism underlying the failure of inward spread of activation during excessive fatigue is not yet fully understood. There is good reason to presume, however, that repetitive electrical stimulation will cause accumulation of potassium in the T-tubules (Hnik et al., 1976; Juel, 1986) when too short a time is available to restore the potassium concentration between the stimulation volleys. The increased potassium concentration will gradually depolarize the tubular membrane and this will impair the inward conduction of the electrical impulse. At high stimulation frequencies only the outer layers of the fiber may be accessible for activation by the action potential while the interior of the fiber stays totally inactive. Although this mechanism would seem to adequately explain the experimental results, the possibility cannot yet be excluded that fatiguing stimulation leads to reduced calcium sensitivity and that this effect somehow becomes more pronounced towards the center of the fiber (Allen et al., 1993). It is known that P_i and pH_i both reduce the myofibrillar calcium sensitivity (Fabiato & Fabiato, 1978; Kentish, 1986; Godt & Nosek, 1989; Metzger & Moss, 1990; Lynch & Williams, 1994), but it would require that a concentration gradient for these two products develops across the fiber during fatiguing stimulation. At present there is no experimental evidence to suggest that such a gradient arises.

As described earlier, the instantaneous stiffness of the muscle fiber is only slightly reduced during moderate fatigue when the decrease in mechanical performance is mainly attributable to altered kinetics of the myosin crossbridges (see earlier). However, there is a drastic change of the force-stiffness relationship at a point where the inward spread of activation begins to fail during fatiguing stimulation. Thus, as the tetanic force is reduced below approximately 70% of the rested-state value (interval between contractions 1-2 s), any further decrease in force is associated with a nearly proportional reduction in fiber stiffness (Edman & Lou, 1992). The progressive decrease in force at this stage of fatiguing stimulation is therefore largely due to fewer bridge attachments being formed. This supports the conclusion (see above) that failure of activation finally becomes the main cause of the force decline in a fiber subjected to frequent stimulation.

Further evidence in support of this view comes from intracellular calcium measurements. Thus, as observed in both frog and mammalian muscle fibers (Allen et al., 1989; Lee et al., 1991; Westerblad & Allen, 1991), the tetanic $[Ca^{2+}]_i$ is progressively reduced as the isometric force is depressed below the level (near 70% of the control value) at which stiffness starts to decline steeply. In one study, Allen and coworkers have indeed been able to demonstrate a *transient* of tetanic $[Ca^{2+}]_i$ across the fiber during intense fatiguing stimulation. Using a stimulation program in which relatively long tetani were produced at short intervals, these authors recorded a lower concentration of $Ca^{2+}{}_i$ in the center than in the periphery of the fiber (Westerblad et al., 1990). This observation accords with the idea that frequent stimulation eventually leads to failure of the inward spread of activation. However, in a later study the same laboratory, using a somewhat different stimulation routine, found no indication of a $[Ca^{2+}]_i$ gradient across the fiber during excessive fatigue (Westerblad et al., 1993). More experimental evidence is apparently needed to settle the problem of whether failure of inward activation during fatiguing stimulation is associated with an uneven distribution of $[Ca^{2+}]_i$ within the fiber.

SUMMARY

While *myofibrillar fatigue* is likely to play a relevant part during muscular exercise *in vivo* (see earlier), the physiological significance of *failure of activation* is less clear. For activation failure to arise the muscle fiber has to undergo an intense stimulation program that may not be *permitted* in a muscle operating *in situ* in the body. There is experimental evidence suggesting that the motor input to the muscle is progressively reduced during development of fatigue *in vivo* (Marsden et al., 1971; Bigland-Ritchie et al., 1986; Garland & McComas, 1990; Gandevia et al., Chapter 20). This implies that the altered contractile state of the muscle is somehow sensed and that this information is utilized in a feed-back loop to modulate the frequency of motor stimuli to the muscle. Failure of activation may therefore play a less significant role during development of fatigue *in vivo* than it does when a muscle is fatigued in an organ bath where no restrictions are placed on the stimulation program.

ACKNOWLEDGMENT

The author wishes to thank Dr. Edwin E. Gilliam for his helpful comments on the manuscript. The author's laboratory has been supported by grants from the Swedish Medical Research Council (project No. 14X-184), the Crafoord Foundation, and the Medical Faculty, University of Lund, Sweden. Attendance of the author at the 1994 Bigland-Ritchie conference was supported, in part, by the Muscular Dystrophy Association (USA) and the American Physical Therapy Association (Research and Analysis Division).

REFERENCES

Allen D, Duty S & Westerblad H (1993). Letter to the editors. *Journal of Muscle Research and Cell Motility* **14**, 543-544.

Allen DG, Lee JA & Westerblad H (1989). Intracellular calcium and tension during fatigue in isolated single muscle fibres from *Xenopus laevis. Journal of Physiology (London)* **415**, 433-458.

Altringham JD & Johnston IA (1985). Effects of phosphate on the contractile properties of fast and slow muscle fibres from an antarctic fish. *Journal of Physiology (London)* **368**, 491-500.

Bigland-Ritchie BR, Dawson NJ, Johansson RS & Lippold OCJ (1986). Reflex origin for the slowing of motoneurone firing rates in fatigue of human voluntary contractions. *Journal of Physiology (London)* **379**, 451-459.

Brenner B (1980). Effect of free sarcoplasmic Ca2+ concentration on maximum unloaded shortening velocity: measurements on single glycerinated rabbit psoas muscle fibres. *Journal of Muscle Research and Cell Motility* **1**, 409-428.

Caputo C, Edman KAP, Lou F & Sun Y-B (1994). Variation in myoplasmic Ca^{2+} concentration and relaxation studied by Fluo-3 in frog muscle fibres. *Journal of Physiology (London)* **478**, 137-148.

Cecchi G, Griffiths PJ & Taylor S (1986). Stiffness and force in activated frog skeletal muscle fibers. *Biophysical Journal* **49**, 437-451.

Chase PB & Kushmerick MJ (1988). Effects of pH on contraction of rabbit fast and slow skeletal muscle fibers. *Biophysical Journal* **53**, 935-946.

Cooke R, Franks K, Luciani GB & Pate E (1988). The inhibition of rabbit skeletal muscle contraction by hydrogen ions and phosphate. *Journal of Physiology (London)* **395**, 77-97.

Curtin NA & Edman KAP (1989). Effects of fatigue and reduced intracellular pH on segment dynamics in 'isometric' relaxation of frog muscle fibres. *Journal of Physiology (London)* **413**, 150-174.

Curtin NA & Edman KAP (1994). Force-velocity relation for frog muscle fibres: effects of moderate fatigue and of intracellular acidification. *Journal of Physiology (London)* **475**, 483-494.

Dantzig JA, Goldman YE, Millar NC, Lacktis J & Homsher E (1992). Reversal of the cross-bridge force-generating transition by photogeneration of phosphate in rabbit psoas muscle fibres. *Journal of Physiology (London)* **451**, 247-278.

Dawson MJ, Gadian DG & Wilkie DR (1978). Muscular fatigue investigated by phosphorus nuclear magnetic resonance. *Nature* **274**, 861-866.

Dawson MJ, Gadian DG & Wilkie DR (1980). Mechanical relaxation rate and metabolism studied in fatiguing muscle by phosphorus nuclear magnetic resonance. *Journal of Physiology (London)* **299**, 465-484.

Delay M, Ribalet B & Vergara J (1986). Caffeine potentiation of calcium release in frog skeletal muscle fibres. *Journal of Physiology (London)* **375**, 535-559.

Eberstein A & Sandow A (1963). Fatigue mechanisms in muscle fibres. In: Gutman E, Hnik P (eds.), *The Effect of Use and Disuse on Neuromuscular Functions*, pp. 515-526. Prague: Nakladatelstvi Ceskoslovenske akademie ved Praha.

Edman KAP (1979). The velocity of unloaded shortening and its relation to sarcomere length and isometric force in vertebrate muscle fibres. *Journal of Physiology (London)* **291**, 143-159.

Edman KAP (1988). Double-hyperbolic force-velocity relation in frog muscle fibres. *Journal of Physiology (London)* **404**, 301-321.

Edman KAP & Lou F (1990). Changes in force and stiffness induced by fatigue and intracellular acidification in frog muscle fibres. *Journal of Physiology (London)* **424**, 133-149.

Edman KAP & Lou F (1992). Myofibrillar fatigue *versus* failure of activation during repetitive stimulation of frog muscle fibres. *Journal of Physiology (London)* **457**, 655-673.

Edman KAP & Mattiazzi A (1981). Effects of fatigue and altered pH on isometric force and velocity of shortening at zero load in frog muscle fibres. *Journal of Muscle Research and Cell Motility* **2**, 321-334.

Edwards RHT, Hill DK & Jones DA (1975). Metabolic changes associated with the slowing of relaxation in fatigued mouse muscle. *Journal of Physiology (London)* **251**, 287-301.

Fabiato A & Fabiato F (1978). Effects of pH on the myofilaments and the sarcoplasmic reticulum of skinned cells from cardiac and skeletal muscles. *Journal of Physiology (London)* **276**, 233-255.

Ferenczi MA, Goldman YE & Simmons RM (1984). The dependence of force and shortening velocity on substrate concentration in skinned muscle fibres from *Rana temporaria*. *Journal of Physiology (London)* **350**, 519-543.

Fitts RH (1994). Cellular mechanisms of muscle fatigue. *Physiological Reviews* **74**, 49-94.

Fitts RH & Holloszy JO (1978). Effects of fatigue and recovery on contractile properties of frog muscle. *Journal of Applied Physiology* **45**, 899-902.

Garcia M, Gonzalez-Serratos H, Morgan JP, Perreault CL & Rozycka M (1991). Differential activation of myofibrils during fatigue in phasic skeletal muscle cells. *Journal of Muscle Research and Cell Motility* **12**, 412-424.

Garland SJ & McComas AJ (1990). Reflex inhibition of human soleus muscle during fatigue. *Journal of Physiology (London)* **429**, 17-27.

Godt RE & Nosek TM (1989). Changes of intracellular milieu with fatigue or hypoxia depress contraction of skinned rabbit skeletal and cardiac muscle. *Journal of Physiology (London)* **412**, 155-180.

Gonzalez-Serratos H, Garcia M, Somlyo A, Somlyo AP & McClellan G (1981). Differential shortening of myofibrils during development of fatigue. *Biophysical Journal* **33**, 224a.

Hnik K, Holas M, Krekule I, Kriz N, Mejsnar J, Smiesko V, Ujec E & Vyskocil F (1976). Work-induced potassium changes in skeletal muscle and effluent venous blood assessed by liquid ion-exchanger microelectrodes. *Pflügers Archiv* **362**, 84-95.

Huxley AF (1957). Muscle structure and theories of contraction. *Progress in Biophysics and Biophysical Chemistry* **7**, 255-318.

Huxley AF & Simmons RM (1971). Proposed mechanism of force generation in striated muscle. *Nature* **233**, 533-538.

Jones DA & Sacco P (1989). Failure of activation as the cause of fatigue in isolated mouse skeletal muscle. *Journal of Physiology (London)* **410**, 75P.

Juel C (1986). Potassium and sodium shifts during in vitro isometric muscle contraction, and the time course of the ion-gradient recovery. *Pflügers Archiv* **406**, 458-463.

Julian FJ & Morgan DL (1981). Variation of muscle stiffness with tension during tension gradients and constant velocity shortening in the frog. *Journal of Physiology (London)* **319**, 193-203.

Kanaya H, Takauyi M & Nagai T (1983). Properties of caffeine- and potassium-contractures in fatigued frog single twitch muscle fibres. *Japanese Journal of Physiology* **33**, 945-954.

Kawano T, Tanokura M & Yamada K (1988). Phosphorus nuclear magnetic resonance studies on the effect of duration of contraction in bull-frog skeletal muscle. *Journal of Physiology (London)* **407**, 243-261.

Kentish JC (1986). The effects of inorganic phosphate and creatine phosphate on force production in skinned muscles from rat ventricle. *Journal of Physiology (London)* **370**, 585-604.

Klein MG, Simon BJ & Schneider MF (1990). Effects of caffeine on calcium release from the sarcoplasmic reticulum in frog skeletal muscle fibres. *Journal of Physiology (London)* **425**, 599-626.

Lännergren J & Westerblad H (1989). Maximum tension and force-velocity properties of fatigued single *Xenopus* muscle fibres studied by caffeine and high K^+. *Journal of Physiology (London)* **409**, 473-490.

Lee JA, Westerblad H & Allen DG (1991). Changes in tetanic and resting $[Ca^{2+}]_i$ during fatigue and recovery of single muscle fibres from *Xenopus laevis*. *Journal of Physiology (London)* **433**, 307-326.

Lombardi V & Piazzesi G (1990). The contractile response during steady lengthening of stimulated frog muscle fibres. *Journal of Physiology (London)* **431**, 141-171.

Lynch GS & Williams DA (1994). The effect of lowered pH on the Ca^{2+}-activated contractile characteristics of skeletal muscle fibres from endurance-trained rats. *Experimental Physiology* **79**, 47-57.

Marsden CD, Meadows JC & Merton PA (1971). Isolated single motor units in human muscle and their rate of discharge during maximal voluntary effort. *Journal of Physiology (London)* **217**, 12-13P.

Metzger JM & Moss RL (1990). pH modulation of the kinetics of a Ca^{2+}-sensitive cross-bridge state transition in mammalian single skeletal muscle fibres. *Journal of Physiology (London)* **428**, 751-764.

Nassar-Gentina V, Passonneau JV, Vergara JL & Rapoport SJ (1978). Metabolic correlates of fatigue and of recovery from fatigue in single frog muscle fibers. *Journal of General Physiology* **72**, 593-606.

Podolsky RJ & Teichholz LE (1970). The relation between calcium and contraction kinetics in skinned muscle fibres. *Journal of Physiology (London)* **211**, 19-35.

Spande JJ & Schottelius BA (1970). Chemical basis of fatigue in isolated mouse soleus muscle. *American Journal of Physiology* **219**, 1490-1495.

Stienen GJM, Roosemalen MCM, Wilson MGA & Elzinga G (1990). Depression of force by phosphate in skinned skeletal muscle fibers of the frog. *American Journal of Physiology* **259**, C349-357.

Stienen GJM, Versteeg PGA, Papp Z & Elzinga G (1992). Mechanical properties of skinned rabbit psoas and soleus muscle fibres during lengthening: effects of phosphate and Ca^{2+}. *Journal of Physiology (London)* **451**, 503-523.

Westerblad H & Allen DG (1991). Changes of myoplasmic calcium concentration during fatigue in single mouse muscle fibers. *Journal of General Physiology* **98**, 615-635.

Westerblad H & Allen DG (1993). The contribution of $[Ca^{2+}]_i$ to the slowing of relaxation in fatigued single fibres from mouse skeletal muscle. *Journal of Physiology (London)* **468**, 729-740.

Westerblad H, Duty S & Allen DG (1993). Intracellular calcium concentration during low-frequency fatigue in isolated single fibers of mouse skeletal muscle. *Journal of Applied Physiology* **75**, 382-388.

Westerblad H, Lee JA, Lamb AG, Bolsover SR & Allen DG (1990). Spatial gradients of intracellular calcium in skeletal muscle during fatigue. *Pflügers Archiv* **415**, 734-740.

Woledge RC (1968). The energetics of tortoise muscle. *Journal of Physiology (London)* **197**, 685-707.

2

MECHANISMS OF EXCITATION-CONTRACTION COUPLING RELEVANT TO SKELETAL MUSCLE FATIGUE

D. G. Stephenson,[1] G. D. Lamb,[1] G. M. M. Stephenson,[2] and M. W. Fryer[3]

[1] School of Zoology, La Trobe University
Bundoora, Victoria 3083, Australia
[2] Department of Chemistry and Biology, Victoria University of Technology
Footscray, Victoria 3011, Australia
[3] School of Physiology and Pharmacology, University of New South Wales
Kensington, New South Wales 2033, Australia

ABSTRACT

This review examines recent progress in elucidation of the excitation-contraction coupling in skeletal muscle with particular reference to processes which may play an important role in muscle fatigue.

INTRODUCTION

For the purpose of this review excitation-contraction (E-C) coupling refers to the whole sequence of events which occur in the twitch skeletal muscle fiber between the generation of an action potential and the activation of the contractile apparatus (Sandow, 1965). This sequence of events comprises 1) the initiation and propagation of an action potential along the sarcolemma and into the transverse tubular system (t-system); 2) signal transmission from the t-system to the sarcoplasmic reticulum (SR) where Ca^{2+} is stored; 3) Ca^{2+}-release from the SR into the myoplasm and the rise in myoplasmic $[Ca^{2+}]$ ($[Ca^{2+}]_i$); and 4) Ca^{2+}-binding to the regulatory system and activation of the contractile apparatus.

The ionic composition of the t-system and of the myoplasmic environment, as well as the functional state of major participants in the E-C coupling may change markedly during and immediately after a period of intense stimulation of the muscle fiber. These cellular differences brought about by sustained activation of a muscle fiber can cause a decline in mechanical performance (power output, force, velocity of shortening) which is generally

Fatigue, Edited by Simon C. Gandevia et al.
Plenum Press, New York, 1995

known as muscle fatigue (Edwards, 1983; Bigland-Ritchie & Woods, 1984; Westerblad et al., 1991; Fitts, 1994).

There are a number of excellent recent reviews which explore aspects of the E-C coupling which are relevant to muscle fatigue (Ashley et al., 1991; Ebashi, 1991; Rios & Pizarro, 1991; Westerblad et al., 1991; Dulhunty, 1992; Lamb, 1992; Rios et al., 1992; Rüegg, 1992; Fitts, 1994; Franzini-Armstrong & Jorgensen, 1994; Meissner, 1994; Schneider, 1994; Melzer et al., 1995). The purpose of this overview is to provide an update on mechanisms of E-C coupling which can play an important role in muscle fatigue.

INITIATION AND PROPAGATION OF ACTION POTENTIALS ALONG THE SARCOLEMMA AND INTO THE T-SYSTEM

Under physiological conditions, skeletal muscle contracts in response to motoneuron stimulation. The motoneurons conduct action potentials to the neuromuscular junction where they cause release of acetylcholine from the nerve terminals. The transmitter activates nonselective cationic channels in the motor end-plate region of the fiber causing a local depolarization of the sarcolemma. This depolarization activates voltage-gated Na^+- (and K^+-) channels, triggering action potentials in the surface of twitch muscle fibers which spread out from the motor end plate along the sarcolemma and into the t-system.

The neuromuscular transmission does not appear to be blocked during muscle fatigue (Bigland-Ritchie et al., 1982), but there is clear evidence that repetitive stimulation of the muscle is associated with disturbances in sarcolemmal and t-system membrane excitability (for recent reviews see Sjøgaard, 1991; Fitts, 1994) which are caused by modifications in the functional properties of ionic channels and/or by changes in the ionic gradients for K^+, Ca^{2+}, Cl^- and Na^+. For example, Ca^{2+}-dependent K^+ and Cl^- channels become activated (Hille, 1992) when $[Ca^{2+}]_i$ rises above a certain level. Activation of these types of channels will always cause a repolarization or a hyperpolarization if the K^+ and Cl^- gradients across the membrane are not also modified. Since $[Ca^{2+}]_i$ between stimulations is more elevated in the fatigued fibers than in the resting fibers (Lee et al., 1991; Westerblad & Allen, 1991), it is possible that Ca^{2+}-dependent K^+ and Cl^- channels would be more activated during fatigue and thus contribute to the change in the shape of the action potential. Also, ATP-sensitive K^+-channels become activated in fatigued fibers due to a synergistic effect of decreased [ATP] and acidification (Fink & Lüttgau, 1976; Light et al., 1994), and mechanosensitive K^+- (and Ca^{2+}-) channels (Burton & Hutter, 1990) may be affected by swelling of the t-system and of the whole fiber (Gonzales-Serratos et al., 1978). Changes in ionic gradients across external membranes are brought about by increased ion flow through ionic channels in combination with altered activities of the Ca^{2+}-ATPase pump and the Na^+/K^+ pump, the latter of which is activated mainly by a rise in intracellular $[Na^+]$ and inhibited by a decrease in [ATP] below 1mM (see Fitts, 1994).

The extent to which the above complex modifications in surface and t-system membrane excitability contribute to fatigue is difficult to quantitate because it depends on a great number of parameters, including the precise geometry of the sarcolemmal and t-system membranes, diffusional pathways from the surface of the fiber, density of specific ion channels in the sarcolemma and t-system, density of Na^+/K^+ pumps in the t-system, pattern of stimulation, type of muscle fiber and type of contraction. Nevertheless, it is likely that altered membrane excitability within the fiber plays a major role in fatigue caused by high-frequency stimulation (>100Hz), where significant recovery of force takes place after a rest of only a few seconds, which is consistent with the time required for the diffusion of ions in and out of the t-system (evidence summarized by Westerblad et al., 1991). Several lines of evidence indicate that altered membrane excitability plays only a minor role in fatigue induced by prolonged intermittent stimulation. For example, the action potential was found to recover considerably faster than force after low-frequency fatiguing stimulation (e.g., Metzger & Fitts, 1986). Also, muscle fibers stimulated by action potentials at moderate frequency showed similar fatigability to fibers stimulated by voltage steps under voltage-clamp conditions using otherwise the same stimulation protocol (Györke, 1993). Finally, there is conclusive evidence that other factors, which more directly determine the level of force production, such as the sensitivity to Ca^{2+} of the contractile apparatus (Lee et al., 1991; Westerblad & Allen, 1991), and the average force produced per cross-bridge (Edman & Lou, 1990) are affected during fatigue (for further discussion see Fuglevand, Chapter 6).

SIGNAL TRANSMISSION FROM THE T-SYSTEM TO THE SR

The t-system and the SR are distinctly separate membrane compartments with the t-tubules forming, for most of their length, specialized junctions with two neighboring SR compartments and producing a structure which is known as a triad. Detection of t-system depolarization and signal transmission from the t-system to the SR takes place at these junctional regions between the t-system and the SR.

There is strong evidence (see recent reviews by Rios & Pizarro, 1991; Lamb, 1992; Franzini-Armstrong & Jorgensen, 1994; Schneider, 1994; Melzer et al., 1995) that the dihydropyridine (DHP) receptors found in the junctional domain of the t-tubules are the intramembrane molecules that serve as voltage sensors and are responsible for the asymmetric charge movement associated with depolarization to the contractile threshold. The DHP receptors appear as *tetrads* that are orderly arranged and there are suggestions that one tetrad consists of four DHP receptors (see review by Franzini-Armstrong & Jorgensen, 1994). At least some of the DHP receptors in skeletal muscle are slowly activating L-type Ca^{2+}-channels (there is dispute about the proportion of DHP receptors acting as Ca^{2+}-channels (see Lamb, 1992)), but Ca^{2+}-inflow through these channels is *not necessary* for contraction. Each DHP receptor is an oligomer consisting of five subunits α_1 (185kD), α_2 (143kD), β (54kD), γ (26kD), δ (30kD), where the α_1 subunit has the binding site for DHP and other Ca^{2+}-antagonists and also forms part of the L-type Ca^{2+}-channel when incorporated into lipid bilayers (see reviews by Catterall, 1991; Lamb, 1992). The structure of the α_1 subunit resembles that of the Na^+-channel with four *repeats* (I-IV), each consisting of 6 hydrophobic segments (s1-s6) thought to span the t-system membrane. The fourth segment (s4) in each of the four *repeats* is positively charged and is believed to be the voltage-sensitive element of the DHP receptor (Glossmann & Striessnig, 1990) responsible for the asymmetric charge movement in muscle (Rios & Pizarro, 1991; Schneider, 1994). The cytoplasmic loop between repeats II and III of the skeletal muscle DHP receptor appears to be essential for signal transmission to the SR (Tanabe et al., 1990). The β subunit of the DHP receptor plays an important regulatory role in the Ca^{2+}-channel function and it interacts with the α_1 subunit through the loop between *repeats* I and II (Pragnell et al., 1994). The functions of the other subunits α_2, γ and δ are not fully determined. Recent experiments using prolonged intermittent fatiguing stimulation have shown that charge movement is fatigue resistant (Györke, 1993), suggesting that the voltage sensor in the DHP receptor is not affected in fatigue.

There are multiple and distinct sites of phosphorylation on the α_1 and β subunits of the DHP receptor which are differentially phosphorylated by several protein kinases (Chang et al., 1991), and phosphorylation of certain sites alters the channel activity of the DHP receptors (Sculptoreanu et al., 1993). Since the state of phosphorylation of the DHP receptors may change following repetitive activation, it is conceivable that a change in the state of activation of DHP receptors, in their Ca^{2+} -channel function, may play some role in muscle fatigue (see also section on ([Ca2+] Rise in the Myoplasm)).

Electron microscope studies have revealed that the four DHP receptor particles comprising a tetrad are located in the immediate proximity of alternate triadic feet spanning the gap between the t-tubules and the SR membrane (Franzini-Armstrong & Jorgensen, 1994). The triadic feet are homotetrameric (~560kD/subunit) protein structures, with each subunit having several transmembrane segments at the carboxy-terminus and a large cytoplasmic domain. This homotetramer functions as a single Ca^{2+}-channel in the SR membrane and binds one molecule of the plant alkaloid ryanodine with very high affinity. Therefore, these tetrameric complexes are known either as the foot-spanning proteins, or as the ryanodine receptors, or as the SR Ca^{2+}-release channels. The SR Ca^{2+}-release channels are found mostly on the junctional region of the SR membrane but a small fraction have been identified in extrajunctional regions (see Dulhunty, 1992).

If it is true that a) the DHP receptors/voltage sensors activate the SR Ca^{2+}-release channels by some direct physical interaction (Chandler et al., 1976) (e.g., via the loop between *repeats* II and III - see above); and b) the DHP tetrads only oppose every alternate Ca^{2+}-release channel, it would imply that at least half of the Ca^{2+}-release channels are not under the direct control of the voltage sensors. Nevertheless, physiological experiments have indicated that, with the possible exception of a fast transient component, the SR Ca^{2+}-release in skeletal muscle *is* under the tight control of the DHP receptors (Jacquemond et al., 1991; Rios & Pizarro, 1991; Schneider, 1994). Therefore, it appears either that adjacent Ca^{2+}-release channels without direct voltage sensor control either rapidly inactivate or are non-functional.

Some reports suggest that the Ca^{2+}-release channels may be activated by a linking protein such as *triadin* (~95kD) (Kim et al., 1990; but see Franzini-Armstrong & Jorgensen, 1994 for a recent critical account) or via a diffusible second messenger such as inositol 1,4,5-trisphosphate (IP_3) or Ca^{2+}, which are known to be the major messengers for activation of Ca^{2+} release in smooth muscle (and non-muscle cells, Berridge & Irvine, 1989) and cardiac muscle (Fabiato, 1985; Stern & Lakatta, 1992), respectively. However, there is a growing amount of evidence indicating that IP_3 is unlikely to play a primary role in the physiological coupling between the DHP receptors and Ca^{2+}-release in skeletal muscle (see Schneider, 1994). G-protein activation, which is involved in the generation of IP_3, is not necessary for excitation-contraction coupling in skeletal muscle (Lamb & Stephenson, 1991b) and heparin, which blocks binding of IP_3 to its receptor, has little effect on the coupling in mammalian skeletal muscle and causes a unique activation-dependent block of coupling in anuran muscle fibers. This is inconsistent with a simple IP_3 second messenger hypothesis (Lamb et al., 1994c). Furthermore, the possibility that Ca^{2+} itself may be the primary factor in linking the voltage sensors to the Ca^{2+} release channels in the skeletal muscle is not likely (see Dulhunty, 1992; Melzer et al., 1995; Schneider, 1994). Experiments in our laboratory have shown that potent buffering of $[Ca^{2+}]_i$ to very low levels does not interrupt normal coupling (Lamb & Stephenson, 1990) emphasizing the view that Ca^{2+} is not necessary for signal transmission between the DHP receptor and the SR Ca^{2+}-release channel. However, it is likely that Ca^{2+} ions do have an important role in modulating the open time of the Ca^{2+}-release channel (see below).

An interesting possibility is that Mg^{2+} plays a key role in the coupling of the DHP receptor to the SR Ca^{2+}-release channel (Lamb & Stephenson, 1991a, 1992). Under resting physiological conditions, Mg^{2+} (~1mM) exerts a powerful inhibitory action on the SR Ca^{2+} release channels keeping them closed (Endo, 1985; Lamb & Stephenson, 1991a; Meissner, 1994) despite a strong stimulatory effect of ATP on channel opening. This inhibitory effect of Mg^{2+} is mainly due to Mg^{2+} binding to a low-affinity inhibitory site on the SR Ca^{2+}-release channel (Meissner, 1994; Lamb, 1993). It has been proposed that activation of the DHP receptors decreases the affinity of the inhibitory site for Mg^{2+} by a factor of 10-20 fold, causing dissociation of Mg^{2+} and removal of the Mg^{2+} block. Without Mg^{2+} inhibition, the Ca^{2+}-release channels open in the presence of ATP a substantial propoı tion of time, allowing Ca^{2+} release from the SR (Meissner et al., 1986). An increase in myoplasmic [Ca^{2+}] would cause Ca^{2+} binding to an activation site on the channel which in turn could further increase the fractional open time of the channel (see section on (Modulators of Ca^{2+} release)). In this way, Ca^{2+} can have a positive feedback effect on Ca^{2+} release. However, this type of Ca^{2+}-induced Ca^{2+}-release is totally under the control of the DHP receptor, by virtue of the Mg^{2+} inhibition which can be reimposed at any time by the deactivation of the DHP receptor. This proposal extends the earlier proposal made by Ashley and Moisescu (1973; see also Lüttgau & Stephenson, 1986l; Ashley et al., 1991) where the rate of Ca^{2+} release is a function of both $[Ca^{2+}]_i$ and the membrane depolarization and readily explains all findings to date. The model also reconciles and unifies the apparently contradictory hypotheses of *Ca^{2+}-induced Ca^{2+}-release* (Endo, 1985; Fabiato, 1985), which has received strong support by direct studies of the SR Ca^{2+}-release channel (Meissner, 1994; Györke et al., 1994), and of voltage control of the non-inactivating Ca^{2+}-release component which is well established (Rios & Pizarro, 1991; Schneider, 1994).

CA^{2+}-RELEASE INTO MYOPLASM AND THE RISE IN MYOPLASMIC [CA^{2+}]

After transmission of excitation from the t-system to the SR, Ca^{2+} is released from the SR into the myoplasm. The rate of Ca^{2+} release from the SR will directly depend on both the average open time of the SR Ca^{2+}-release channels and on the electrochemical gradient for Ca^{2+}. In the previous section, we have discussed how depolarization of the t-system may remove the Mg^{2+} block, causing activation of SR Ca^{2+}-release channels in the presence of ATP. Here we will discuss how factors which are known to change in fatigued fibers may further modulate the average probability of opening of the SR Ca^{2+} release channels and modify the electrochemical Ca^{2+} gradient across the SR membrane.

Modulators of SR Ca^{2+}-release channels relevant to fatigue

In vitro studies have indicated that the activity of the SR Ca^{2+}-release channels can be altered by many physiological factors including Mg^{2+}, adenine nucleotides, Ca^{2+}, pH, calmodulin, inorganic phosphate [P_i], phosphorylation, lipid metabolites, IP_3, polyamines and some muscle proteins (Meissner, 1994, as well as the membrane potential across the SR membrane (Stein & Palade, 1988). If the activation of the Ca^{2+}-release channels by the DHP receptors involves simply the removal of the Mg^{2+} block, then the channel opening time will be determined by the overall influence of all modulators present.

[Mg^{2+}]

As mentioned earlier, the binding of Mg^{2+} to a low-affinity inhibitory site on the SR Ca^{2+}-release channel powerfully inhibits channel opening. Mg^{2+} can also reduce the Ca^{2+} flow through the channel by competing with Ca^{2+} for a common binding site within the channel and by competing with Ca^{2+} for the high affinity Ca^{2+} activation site (Lamb, 1993; Meissner, 1994). We have shown that raising [Mg^{2+}] in myoplasm from 1mM to 3mM can have a marked inhibitory effect on the amount of Ca^{2+} released by depolarization and that at 10mM Mg^{2+} the transmission of excitation was completely blocked (Lamb & Stephenson,

1991a). This inhibitory effect of Mg^{2+} can play an important role later in fatigue when [Mg^{2+}] begins to rise well above resting levels (Westerblad & Allen, 1992).

[ATP]

The coupling between depolarization and Ca^{2+}-release was also blocked when [ATP] in the fiber was very low (<1 μM) but was not obviously affected when [ATP] was changed between 2 and 8mM (Lamb & Stephenson, 1991a). In extreme fatigue, the average [ATP] is unlikely to decrease below 2mM (for review see Fitts, 1994), but it is possible that the local [ATP] is much lower and quite different in the junctional area than in the rest of the myoplasm. This idea is supported by the experiments of Han and colleagues (1992), which have indicated that isolated skeletal muscle triads contain glycolytic enzymes that can synthesize ATP which is not readily exchangeable with the bulk myoplasmic ATP.

[Ca^{2+}]

Ca^{2+} ions may not be necessary for skeletal muscle coupling between the DHP receptors and the SR Ca^{2+}-release channels but they do play important roles in modulating channel opening. Thus, Ca^{2+} binding to a high-affinity activating site of the SR Ca^{2+}-release channels promotes opening, whereas Ca^{2+}-binding to the low-affinity Mg^{2+} inhibitory site prevents the channels from opening (see Lamb, 1993; Meissner, 1994). Recently, Györke and colleagues (1994) have shown that SR Ca^{2+}-release channels from skeletal muscle activate rapidly (in the absence of ATP and Mg^{2+}) following flash photolysis of caged Ca^{2+} and then deactivate slowly, a phenomenon which was interpreted to indicate channel adaptation. However, it is still not clear whether this *adaptation* is a consequence of a brief, large [Ca^{2+}] spike generated by the rapid photolysis of caged Ca^{2+}, rather than an intrinsic property of the Ca^{2+}-release channel (Lamb et al., 1994a).

The resting $[Ca^{2+}]_i$ in a fatigued fiber is elevated and this would cause an increase in the opening time of the SR Ca^{2+}-release channels particularly when activated by depolarization, assuming that the influence on the release channel of all other factors remains the same. However, the magnitude of the Ca^{2+} transient is normally reduced in a fatigued fiber, indicating that there is reduced Ca^{2+} flow through the channel which may be due, for example, to a reduced SR luminal [Ca^{2+}] (see below). A reduced SR luminal [Ca^{2+}] may also have a direct excitatory or inhibitory effect on the opening of the Ca^{2+}-release channels, but the results are not yet conclusive (see Meissner, 1994).

Raised $[Ca^{2+}]_i$ may also be important in a quite different way, as we have recently described a novel phenomenon in which raised myoplasmic [Ca^{2+}] uncouples the DHP receptor from the SR-Ca^{2+}-release channel, and coupling cannot be restored for the duration of the experiment (Lamb et al., 1994b). This phenomenon may underlie *low-frequency fatigue*, which occurs after prolonged exercise and persists for more than one day (Westerblad et al., 1993).

pH

Experiments with SR vesicles and isolated channels have also shown that pH can be a powerful modulator of the SR-Ca^{2+}-release channels (Rousseau & Pinkos, 1990; Meissner, 1994). The channels are open for a considerable amount of time at pH 8.0 but activation by Ca^{2+} is almost abolished by decreasing the pH below 6.5. Since muscle fatigue is often accompanied by a decrease in pH it has been suggested (see Fitts, 1994) that acidification of the myoplasm was directly responsible for the reduced Ca^{2+} release observed in fatigued fibers. However, in contrast to this potent inhibitory effect of acid pH on Ca^{2+}-activation of the SR Ca^{2+}-release channels, our experiments with functional skinned fibers showed that normal voltage-sensor control of Ca^{2+} release was virtually unaffected by pH in both mammalian and anuran muscle (pH 6.1 to 8.0; Lamb et al., 1992; Lamb & Stephenson, 1994). This result shows that the acidification observed in fatigued fibers is not directly responsible for the reduced Ca^{2+} release and highlights the need for examining channel release properties in functioning fibers as well as isolated systems.

Calmodulin

Calmodulin is known to exert a direct inhibitory effect on the SR Ca^{2+}-release channel at physiological concentrations (see Meissner, 1994) by reducing the channel opening time. The concentration of free calmodulin is likely to be lower in a fatigued fiber than in a rested fiber because the higher $[Ca^{2+}]_i$ in the fatigued fiber and the very high affinity constant of the Ca_4 -calmodulin complex for the regulatory units of the Ca^{2+}-calmodulin dependent protein kinases, will increase the amount of the bound form of calmodulin. Therefore, one would expect a net stimulatory effect on the opening time of the depolarization-activated Ca^{2+}-release channel in the fatigued fiber due to the reduced concentration of free calmodulin.

$[P_i]$

Another modulator of Ca^{2+}-release channel activity which plays multiple roles in muscle fatigue is P_i (see later). Recently, it has been shown that inorganic phosphate in the mM range concentrations has a stimulatory effect on the Ca^{2+}-release channels of skeletal muscle (Fruen et al., 1994). Since $[P_i]$ rises early during fatigue, this observation may explain why $[Ca^{2+}]_i$ between stimulations also increases early in fatigue (Lee et al., 1991; Westerblad & Allen, 1992).

Phosphorylation

Phosphorylation of the SR Ca^{2+}-release channel of skeletal muscle may play an important physiological role in muscle fatigue, where Ca^{2+}-calmodulin dependent protein kinases would be more active due to increased $[Ca^{2+}]_i$. In one patch-clamp study, phosphorylation caused inactivation of the SR Ca^{2+}-release channel (Wang & Best, 1992) while in other studies phosphorylation caused activation of the channel, seemingly by removal of the Mg^{2+} block (Hain et al., 1993) or by increasing the channel's sensitivity for Ca^{2+} and ATP (Herrmann-Frank & Varsányi, 1993). It is therefore difficult to predict at this stage whether phosphorylation of the skeletal muscle Ca^{2+}-release channel will contribute to an increase or decrease in depolarization-induced release of Ca^{2+} from the SR.

Fiber Volume

The volume of fatigued fibers can be as high as 180% of that at rest (Gonzales-Serratos et al., 1978), raising the possibility that a change in fiber morphology associated with this change in volume may affect Ca^{2+}-release. However, the coupling between the depolarization of the t-system and Ca^{2+}-release was not modified by changes in the myofilament lattice similar to those known to occur in fatigued fibers (Lamb et al., 1993).

Other Modulators

Of the other known modulators of SR Ca^{2+}-release channels (see Meissner, 1994), the fatty acids and their derivatives (El-Hayek et al, 1993) may play some role in muscle fatigue. This is because the skeletal muscle is known to switch from a fatty acid degradation pathway in the rested state to a glycogenolytic/glycolytic pathway during exercise (Stryer, 1988). Therefore, the concentration of the fatty acids and their derivatives which have a potent stimulatory effect on the Ca^{2+} release channel (El-Hayek et al., 1993) will be likely to be decreased in fatigue.

Modifications in the Electrochemical Ca^{2+} Gradient across the SR Membrane

An often overlooked, but most important factor determining the electrochemical Ca^{2+} gradient across the SR membrane and thus the rate of Ca^{2+}-release from the SR, is the $[Ca^{2+}]$ within the junctional SR. The SR luminal $[Ca^{2+}]$ is dependent on several parameters including the activity of the SR Ca^{2+} pump (Inesi & de Meis, 1989), rate of Ca^{2+} efflux from the SR as well as the binding properties of the sarcotubular Ca^{2+}-binding proteins (calsequestrin and calreticulin in the junctional SR and the 53 and 160kD glycoproteins in the longitudinal SR). The amount of calcium in the SR is not significantly altered by an increase in myoplasmic $[Mg^{2+}]$ from 1 to 3 mM, similar to that known to occur in fatigued fibers (Kabbara & Stephenson, 1994) but it may be affected by a decreased pH in fatigued fibers. An acidic pH inhibits the SR Ca^{2+}-pump, reduces the opening probability of the SR Ca^{2+}-release channels and lowers the affinity constant of calsequestrin (see Lamb et al., 1992 for references).

More importantly, the SR luminal $[Ca^{2+}]$ is dependent on the concentration of P_i. As $[P_i]$ builds up during muscle contraction, a threshold is reached above which a precipitate of $CaHPO_4$ is formed within the SR, causing a reduction in SR luminal $[Ca^{2+}]$. We have shown that 1) significant Ca precipitation occurred within the SR in the presence of P_i levels observed in fatigue; and 2) a stimulus-induced response was prolonged and of reduced amplitude when conditions were favorable for the formation of a $CaHPO_4$ precipitate (Fryer et al., 1995). The reduction in the amplitude of the response is due to the decrease of $[Ca^{2+}]$ in the SR lumen caused by precipitation, whereas the prolongation of the response is due to the $CaHPO_4$ precipitate acting as a slow Ca^{2+}-buffer.

$[Ca^{2+}]$ Rise in the Myoplasm

The magnitude and the time course of the $[Ca^{2+}]_i$ changes in the myoplasm following stimulation are the result of the interplay between Ca^{2+} release from the SR, Ca^{2+} uptake by the SR, Ca^{2+} entry through the DHP channels in the sarcolemma, Ca^{2+} removal by the Na^+/Ca^{2+} exchanger and the Ca^{2+} pump in the sarcolemma, and the Ca^{2+} movements associated with various Ca^{2+} binding sites in the myoplasm, including parvalbumin, troponin C, SR Ca^{2+} pump sites and regulatory myosin light chains (Lüttgau & Stephenson, 1986; Ashley et al, 1991; Schneider, 1994). The magnitude of the Ca^{2+} transient is only a very small fraction (~1%) of the total amount of Ca^{2+} released from the SR (Lüttgau & Stephenson, 1986; Ashley et al., 1991) and that the peak of the Ca^{2+} transient occurs before positive force generation (Claflin et al., 1994).

Late during fatigue, Ca^{2+} transients become more prolonged and of reduced magnitude (Westerblad & Allen, 1991). These changes in the time course and magnitude of the Ca^{2+} transient can be accounted for to a large extent by changes in the Ca^{2+} fluxes associated with the SR brought about by changes in the myoplasmic ionic composition with respect to $[P_i]$, pH, and $[Mg^{2+}]$ (see above).

It is of interest that Ca^{2+}-inflow through L-type Ca^{2+}-channels in the sarcolemma may significantly contribute to larger Ca^{2+}-transients and therefore cause a delay in the onset of fatigue when they are phosphorylated by cAMP dependent protein kinase, because then the channels show marked voltage-dependent potentiation (Sculptoreanu et al., 1993).

Ca^{2+}-BINDING TO THE REGULATORY SYSTEM AND ACTIVATION OF CONTRACTION

The contractile apparatus of vertebrate skeletal muscle becomes activated following the binding of Ca^{2+} ions to specific sites on the Ca^{2+} regulatory system comprising troponin C, troponin I, troponin T and tropomyosin (for reviews see Lüttgau & Stephenson, 1986; Ashley et al., 1991; Ebashi, 1991; Rüegg, 1992). Sensitivity to Ca^{2+} is conferred by troponin C but interactions between different myofibrillar components including myosin and actin (Lehrer, 1994) can have marked modulatory effects on the relationship between the isometric force response and $[Ca^{2+}]_i$. Of particular relevance to fatigued muscle fibers is the depression in sensitivity to Ca^{2+} of the contractile activation caused by an increase in $[P_i]$ (Cooke et al., 1988; Godt & Nosek, 1989; Rüegg, 1992; Fitts, 1994; Fryer et al., 1995), increase in $[Mg^{2+}]$ (Lamb & Stephenson 1991a, 1992; Rüegg, 1992) and decrease in pH (Metzger & Moss, 1987; Chase & Kushmerick, 1988; Lamb et al., 1992; Rüegg, 1992; Fitts; 1994). An increase in muscle temperature which occurs during intense exercise could further reduce the sensitivity to Ca^{2+} of the contractile response (Stephenson & Williams, 1981, 1985). However, it is generally ignored that other changes occurring in the fatigued fibers, such as the Ca^{2+}-calmodulin dependent phosphorylation of the myosin regulatory light chains (Rüegg, 1992; Stephenson & Stephenson, 1993) and the decrease in creatine phosphate concentration (Fryer et al., 1995) can actually enhance the Ca^{2+}-sensitivity of contractile activation. For example, we have shown that when the concentration of creatine phosphate and $[P_i]$ were varied reciprocally (as should occur *in vivo*), the inhibitory effects of P_i on the Ca^{2+} sensitivity of the contractile apparatus were greatly reduced.

CONCLUDING REMARKS

There are many events in the E-C coupling of a skeletal twitch muscle fiber which are altered by intense muscle fiber activity. Individually, these alterations could lead to either a decrease or a potentiation of the contractile response, but their net effect is a decrease in force and power output that is known as muscle fiber fatigue.

ACKNOWLEDGMENTS

Support from the National Health & Medical Research Council and the Research Council of Australia is gratefully acknowledged. Attendance of *G.M.M.S.* at the 1994 Bigland-Ritchie conference was supported, in part, by the American Physical Therapy Association (Section on Research), and the Muscular Dystrophy Association (USA).

REFERENCES

Ashley CC & Moisescu DG (1973). The mechanism of the free calcium change in single muscle fibres during contraction. *Journal of Physiology (London)* **231**, 23-25P.

Ashley CC, Mulligan IP & Lea TJ (1991). Ca^{2+} and activation mechanisms in skeletal muscle. *Quarterly Reviews of Biophysics* **24**, 1-73.

Berridge MJ & Irvine RF (1989). Inositol phosphates and cell signalling. *Nature* **341**, 197-205.

Bigland-Ritchie B, Kukulka OC, Lippold OCJ & Woods JJ (1982). The absence of neuromuscular transmission failure in sustained maximal voluntary contractions. *Journal of Physiology (London)* **330**, 265-278.

Bigland-Ritchie B & Woods JJ (1984). Changes in muscle contractile properties and neural control during human muscular fatigue. *Muscle & Nerve* **7**, 691-699.

Burton FL & Hutter OF (1990). Sensitivity to flow of intrinsic gating in inwardly rectifying potassium channel from mammalian skeletal muscle. *Journal of Physiology (London)* **424**, 253-261.

Catterall WA (1991). Excitation-contraction coupling in vertebrate skeletal muscle: a tale of two calcium channels. *Cell* **64**, 871-874.

Chandler WK, Rakowski RF & Schneider MF (1976). Effects of glycerol treatment and maintained depolarization on charge movement in skeletal muscle. *Journal of Physiology (London)* **254**, 285-316.

Chang CF, Gutiener LM, Meudina-Weilenmann C & Hosey MM (1991). Dihydropyridine-sensitive calcium channels from skeletal muscle. II, Functional effects of differential phosphorylation of channel subunits. *Journal of Biological Chemistry* **266**, 16395-16400.

Chase PB & Kushmerick MJ (1988). Effects of pH on contraction of rabbit fast and slow skeletal muscle fibres. *Biophysical Journal* **53**, 935-946.

Claflin DR, Morgan DL, Stephenson DG & Julian FJ (1994). The intracellular Ca^{2+}-transient and tension in frog skeletal muscle fibres measured with high temporal resolution. *Journal of Physiology (London)* **475**, 319-325.

Cooke R, Franks K, Luciani GB & Pate E (1988). The inhibition of rabbit skeletal muscle contraction by hydrogen ion and phosphate. *Journal of Physiology (London)* **395**, 77-97.

Dulhunty AF (1992). The voltage-activation of contraction in skeletal muscle. *Progress in Biophysics and Molecular Biology* **57**, 181-223.

Ebashi S (1991). Excitation-contraction coupling and the mechanism of muscle contraction. *Annual Review of Physiology* **53**, 1-16.

Edman KAP & Lou F (1990). Changes in force and stiffness induced by fatigue and intracellular acidification in frog muscle fibres. *Journal of Physiology (London)* **424**, 133-149.

Edwards RHT (1983). Biochemical bases of fatigue in exercise performance: catastrophe theory of muscular fatigue. In: Knuttgen HG (ed.), *Biochemistry of Exercise*, pp. 3-28. Champaign, IL: Human Kinetics.

El-Hayek R, Valdivia C, Valdivia HH, Hogan K & Coronado R (1993). Palmitoyl carnitine: Activation of the Ca^{2+} release channel of skeletal muscle sarcoplasmic reticulum by palmitoyl carnitine and related long chain fatty acids derivatives. *Biophysical Journal* **65**, 779-789.

Endo M (1985). Calcium release from sarcoplasmic reticulum. *Current Topics in Membranes and Transport* **25**, 181-230.

Fabiato A (1985). Time and calcium dependence on activation and inactivation of calcium induced release of calcium from the sarcoplasmic reticulum of a skinned canine cardiac Purkinje cell. *Journal of General Physiology* **85**, 247-289.

Fink R & Lüttgau HCh (1976). An evaluation of the membrane constants and the potassium conductance in metabolically exhausted muscle fibres. *Journal of Physiology (London)* **263**, 215-238.

Fitts RH (1994). Cellular mechanisms of muscle fatigue. *Physiological Reviews* **74**, 49-94.

Franzini-Armstrong C & Jorgensen AO (1994). Structure and development of E-C coupling units in skeletal muscle. *Annual Review of Physiology* **56**, 509-534.

Fruen BR, Mickelson JR, Shomer NH, Roghair TJ & Louis CF (1994). Regulation of the sarcoplasmic reticulum ryanodine receptor by inorganic phosphate. *Journal of Biological Chemistry* **269**, 192-198.

Fryer MW, Owen VJ, Lamb GD & Stephenson DG (1995). Effects of creatine phosphate and P_i on force development and Ca^{2+} movements in rat skinned skeletal muscle fibres. *Journal of Physiology (London)* **482**, 123-140

Glossmann H & Striessnig J (1990). Molecular properties of calcium channels. *Reviews in Physiology, Biochemistry and Pharmacology* **114**, 1-105.

Godt RE & Nosek TM (1989). Changes of intracellular milieu with fatigue or hypoxia depress contraction of skinned rabbit skeletal and cardiac muscle. *Journal of Physiology (London)* **412**, 155-180.

Gonzales-Serratos H, Somlyo AV, McClellan G, Shuman H, Borrero LM & Somlyo AP (1978). Composition of vacuoles and sarcoplasmic reticulum in fatigued muscle: electron probe analysis. *Proceedings of the National Academy of Sciences USA* **75**, 1329-1333.

Györke S (1993). Effects of repeated tetanic stimulation on excitation-contraction coupling in cut muscle fibres of the frog. *Journal of Physiology (London)* **464**, 699-710.

Györke S, Velez P, Suarez-Isla B & Fill M (1994). Activation of single cardiac and skeletal ryanodine receptor channels by flash photolysis of caged Ca^{2+}. *(Biophysical Journal)* **66**, 1879-1886.

Hain J, Schindler H, Nath S & Fleischer S (1993). Phosphorylation of the skeletal muscle calcium release channel removes block by magnesium ions. *Biophysical Journal* **64**, A151.

Han JW, Thieleczek R, Varsßnyi M & Heilmeyer LMG (1992). Compartmentalized ATP synthesis in skeletal muscle triads. *Biochemistry* **31**, 377-384.

Herrmann-Frank A & Varsßnyi M (1993). Enhancement of Ca^{2+} release channel activity by phosphorylation of the skeletal muscle ryanodine receptor. *FEBS Letters* **333**, 237-242.

Hille B (1992). *Ionic Channels of Excitable Membranes*, pp. 115-139. Sutherland, MA: Sinauer.

Inesi G & De Meis L (1989). Regulation of steady-state filling in sarcoplasmic reticulum. *Journal of Biological Chemistry* **264**, 5929-5936.

Jacquemond V, Csernock L, Klein MG & Schneider MF (1991). Voltage-gated and calcium-gated release during depolarization of skeletal muscle. *Biophysical Journal* **60**, 867-873.

Kabbara AA & Stephenson DG (1994). Effects of Mg^{2+} on Ca^{2+} handling by the sarcoplasmic reticulum in skinned skeletal and cardiac muscle fibres. *Pflügers Archiv* **428**, 331-339.

Kim KC, Caswell AH, Talvenheimo JA & Brandt NR (1990). Isolation of a terminal cisterna protein which may link the dihydropyridine receptor to the junctional foot protein in skeletal muscle. *Biochemistry* **29**, 9283-9289.

Lamb GD (1992). DHP receptors and excitation-contraction coupling. *Journal of Muscle Research and Cell Motility* **13**, 394-405.

Lamb GD (1993). Ca^{2+}-inactivation, Mg^{2+}-inhibition and malignant hyperthermia. *Journal of Muscle Research and Cell Motility* **14**, 554-556.

Lamb GD, Fryer MW & Stephenson DG (1994a). Technical comment: Ca^{2+}-induced Ca^{2+}-release in response to flash photolysis. *Science* **263**, 986-987.

Lamb GD, Junankar P & Stephenson DG (1994b). Abolition of excitation-contraction coupling in skeletal muscle by raised intracellular [Ca^{2+}]. *Proceedings of the Australian Physiological and Pharmacological Society* **25**, 76P.

Lamb GD, Posterino GS & Stephenson DG (1994c). Effects of heparin on excitation-contraction coupling in skeletal muscle fibres of toad and rat. *Journal of Physiology (London)* **474**, 319-329.

Lamb GD, Recupero E & Stephenson DG (1992). Effect of myoplasmic pH on excitation-contraction coupling in skeletal muscle fibres of the toad. *Journal of Physiology (London)* **448**, 211-224.

Lamb GD & Stephenson DG (1990). Control of calcium release and the effect of ryanodine in skinned muscle fibres of the toad. *Journal of Physiology (London)* **423**, 519-542.

Lamb GD & Stephenson DG (1991a). Effect of Mg^{2+} on the control of Ca^{2+} release in skeletal muscle fibres of the toad. *Journal of Physiology (London)* **434**, 507-528.

Lamb GD & Stephenson DG (1991b). Excitation-contraction coupling in skeletal muscle fibres of rat in the presence of GTPγS. *Journal of Physiology (London)* **444**, 65-84.

Lamb GD & Stephenson DG (1992). Importance of Mg^{2+} in excitation-contraction coupling in skeletal muscle. *News in Physiological Sciences* **7**, 270-274.

Lamb GD & Stephenson DG (1994). Effect of intracellular pH and [Mg^{2+}] on excitation-contraction coupling in skeletal muscle fibres of the rat. *Journal of Physiology (London)* **478**, 331-339.

Lamb GD, Stephenson DG & Stienen GJM (1993). Effects of osmolality and ionic strength on the mechanism of Ca^{2+} release in skinned skeletal muscle fibres of the toad. *Journal of Physiology (London)* **464**, 629-648.

Lee JA, Westerblad H & Allen DG (1991). Changes in tetanic and resting [Ca^{2+}]i during fatigue and recovery of single muscle fibres from *Xenopus laevis*. *Journal of Physiology (London)* **433**, 307-326.

Lehrer SS (1994). The regulatory switch of the muscle thin filament: Ca^{2+} or myosin heads? *Journal of Muscle Research and Cell Motility* **15**, 232-236.

Light PE, Comtois AS & Renaud JM (1994). The effect of glibenclamide on frog skeletal muscle: evidence for K^+_{ATP} channel activation during fatigue. *Journal of Physiology (London)* **475**, 495-507.

Lüttgau HCh & Stephenson DG (1986). Ion movements in skeletal muscle in relation to the activation of contraction. In: Andreoli TE, Hoffman JF, Fanestil DD Schultz SG (eds.), *Physiology of Membrane Disorders*, pp. 449-468. New York: Plenum Publishing Corporation.

Meissner G (1994). Ryanodine receptor/Ca^{2+} release channels and their regulation by endogenous effectors. *Annual Review of Physiology* **56**, 485-508.

Meissner G, Darling E & Eveleth J (1986). Kinetics of rapid Ca^{2+} release by sarcoplasmic reticulum. Effects of Ca^{2+}, Mg^{2+} and adenine nucleotides. *Biochemistry* **25**, 236-244.

Melzer W, Herrmann-Frank A & Lüttgau HCh (1995). The role of Ca^{2+} ions in excitation-contraction coupling in skeletal muscle fibres. *Biochimica et Biophysica Acta* In press.

Metzger JM & Fitts RH (1986). Fatigue from high- and low-frequency muscle stimulation: role of sarcolemma action potentials. *Experimental Neurology* **93**, 320-333.

Metzger JM & Moss RL (1987). Greater hydrogen ion induced depression of tension and velocity in skinned single fibres of rat fast than slow muscles. *Journal of Physiology (London)* **393**, 727-742.

Pragnell M, De Waard M, Mori Y, Tanabe T, Snutch TP & Campbell KP (1994). Calcium channel beta-subunit binds to a conserved motif in the I-II cytoplasmic linker of the alpha 1-subunit. *Nature* **368**, 67-70.

Rios E & Pizarro G (1991). Voltage sensor of excitation-contraction coupling in skeletal muscle. *Physiological Reviews* **71**, 849-908.

Rios E, Pizarro G & Stefani E (1992). Charge movement and the nature of signal transduction in skeletal muscle excitation-contraction coupling. *Annual Review of Physiology* **54**, 109-133.

Rousseau E & Pinkos J (1990). pH modulates conducting and gating behaviour of single calcium release channels. *Pflügers Archiv* **415**, 645-647.

Rüegg JC (1992). *Calcium in Muscle Contraction. Cellular and Molecular Physiology*, 2nd Edition, 354pp. Berlin, Heidelberg: Springer Verlag.

Sandow A (1965). Excitation-contraction coupling in skeletal muscle. *Pharmacological Reviews* **17**, 265-320.

Schneider MF (1994). Control of calcium release in functioning muscle fibres. *Annual Review of Physiology* **56**, 463-484.

Sculptoreanu A, Scheuer T & Catterall WA (1993). Voltage-dependent potentiation of L-type Ca^{2+} channels due to phosphorylation by cAMP-dependent protein kinase. *Nature* **364**, 240-243.

Sjøgaard G (1991). Role of exercise-induced potassium fluxes underlying muscle fatigue: a brief review. *Canadian Journal of Physiology and Pharmacology* **69**, 238-245.

Stein P & Palade P (1988). Sarcoballs: direct access to sarcoplasmic reticulum Ca^{2+}-channels in skinned frog muscle fibres. *Biophysical Journal* **54**, 357-363.

Stephenson DG & Williams DA (1981). Calcium-activated force responses in fast- and slow-twitch skinned muscle fibres from the rat. *Journal of Physiology (London)* **317**, 281-302.

Stephenson DG & Williams DA (1985). Temperature-dependent calcium sensitivity changes in skinned muscle fibres of rat and toad. *Journal of Physiology (London)* **360**, 1-12.

Stephenson GMM & Stephenson DG (1993). Endogenous MLC2 phosphorylation and Ca^{2+}-activated force in mechanically skinned skeletal muscle fibres of the rat. *Pflügers Archiv* **424**, 30-38.

Stern MD & Lakatta EG (1992). Excitation-contraction coupling in the heart: the state of the question. *FASEB Journal* **6**, 3092-3100.

Stryer L (1988). *Biochemistry*, 3rd Edition, pp.634-635. New York: Freeman and Co.

Tanabe T, Beam KG, Adams BA, Nicodome T & Numa S (1990). Regions of the skeletal muscle dihydropyridine receptor critical for excitation-contraction coupling. *Nature* **346**, 567-569.

Wang J & Best PM (1992). Inactivation of the sarcoplasmic reticulum by protein kinase. *Nature* **359**, 739-741.

Westerblad H & Allen DG (1991). Changes in myoplasmic calcium concentration during fatigue in single mouse muscle fibres. *Journal of General Physiology* **98**, 615-635.

Westerblad H & Allen DG (1992). Myoplasmic free Mg^{2+} concentration during repetitive stimulation of single fibres from mouse skeletal muscle. *Journal of Physiology (London)* **453**, 413-434.

Westerblad H, Duty S & Allen DG (1993). Intracellular calcium concentration during low-frequency fatigue in isolated single fibres of mouse skeletal muscle. *Journal of Applied Physiology* **75**, 382-388.

Westerblad H, Lee JA, Lännergren J & Allen DG (1991). Cellular mechanisms of fatigue in skeletal muscle. *American Journal of Physiology* **261**, C195-C209.

3

THE ROLE OF INTRACELLULAR ACIDOSIS IN MUSCLE FATIGUE

D. G. Allen,[1] H. Westerblad,[2] and J. Lännergren[2]

[1] Department of Physiology
University of Sydney
New South Wales 2006, Australia
[2] Department of Physiology and Pharmacology
Karolinska Institutet
S-171 77 Stockholm, Sweden

ABSTRACT

Muscle fatigue is often accompanied by an intracellular acidosis of variable size. The variability reflects the involvement of different metabolic pathways, the presence or absence of blood flow and the effectiveness of pH-regulating pathways. Intracellular acidosis affects many aspects of muscle cell function; for instance it reduces maximal Ca^{2+}-activated force and Ca^{2+} sensitivity, slows the maximal shortening velocity and prolongs relaxation. However, acidosis is not the only metabolic change in fatigue which causes each of the above, and there are important aspects of muscle fatigue (e.g., the failure of Ca^{2+} release) which do not appear to be caused by acidosis.

INTRODUCTION

During repetitive activity, muscle pH generally falls slowly acid due to accumulation of lactic acid. Simultaneously, muscle performance usually declines, and there is, therefore, often a correlation between the intracellular acidosis and fatigue. This much is generally accepted; the acidity of muscles in a fatigued deer was observed by Berzelius in 1807, and he showed that the acid involved was the same as that in sour milk (cited in Needham, 1971). A clear statement of the hypothesis that lactic acid accumulation is the cause of muscle fatigue was advanced by Hill and Kupalov (1929) though, of course, the possible mechanisms were unclear at that time. The questions we will address in the present review are: to what extent, and by what mechanism(s), does acidosis cause fatigue? The literature on this topic is large and often contradictory; there are many attempts with varying success to correlate fatigue with pH change, and many attempts to prove or disprove that a particular component of fatigue is caused by acidosis.

We believe that the true situation is conceptually simple though often complex in practice; acidosis does indeed cause a number of features of fatigue though in virtually all

Fatigue, Edited by Simon C. Gandevia et al.
Plenum Press, New York, 1995

cases there are additional causes for these features. Furthermore, the degree of acidosis is quite variable in different muscle types and in different muscle activities. If these points are accepted, much of the contradictory literature is easier to understand. Thus, there are now a number of well-documented cases in which many of the features of fatigue are observed e.g., reduced force, and slowed relaxation but there is no acidosis. Such observations do not, of course, disprove that acidosis is important in *other* types of fatigue, but raise interesting questions about the cause of the features of fatigue and about pH regulation in the muscle under study. Equally, the fact that an acidosis occurs in a particular muscle does not prove that it causes the reduced force that also occurs. We need a version of Koch's postulates to help us prove cause and effect. For instance, to prove that acidosis is the sole cause of the reduced force during fatigue, we need to show the following 1) acidosis and reduced force occur together during fatigue with similar time courses for both onset and recovery of fatigue; 2) a mechanism, usually established in a simpler preparation, linking acidosis and reduced force; 3) quantitative agreement between the magnitude of the pH change and the reduced force in the intact and simplified preparation; 4) when pH decreases in the absence of fatigue, it should cause a corresponding change in force; and 5) repetitive contractions produced in such a way that there is no pH change should not lead to any change in force. Once we accept that other mechanisms also reduce force, the approach becomes more difficult and relies more heavily on quantitative analysis.

Fortunately, recent experimental advances have minimized some of the difficulties in earlier approaches. It is now possible to measure the pH change and the mechanical activity in an isolated single fiber. This is advantageous since in multicellular preparations both the pH change and the force production may vary in different fiber types. It is also easy to change the pH of the fiber under study quickly and reversibly so that the consequences of acidosis independent of fatigue can be studied. Finally, in some preparations, it is possible to study fatigue with and without acidosis by inhibiting the key pH regulatory mechanism(s).

In the present account, we first describe the range of pH changes that can be observed in different muscles and activities. The differences in lactic acid production, in pH buffering and in proton extrusion mechanisms which lead to these differences are considered. In a separate section, the various muscle functions which are affected by acidosis are then considered. Many aspects of fatigue have been dealt with at length in earlier reviews and can be consulted particularly about issues other than pH (Westerblad et al., 1991; Fitts, 1994).

pH CHANGES DURING FATIGUE

In a recent review, Fitts (1994) lists a number of studies in which the resting pH_i was ~7.00 and the value recorded in fatigued muscle was around 6.3. These are representative values from the literature, and one way in which they can occur is if all the lactic acid produced by the muscle stays in the cell. If lactate accumulation is L (in mmol/l) and the buffer power over the relevant pH range is B (in mmol/l/pH unit), then the $\Delta pH_i = L/B$. Typical values of B are 40 mmol/l/pH unit (Curtin, 1987) so that an increase in intracellular lactic acid of 26 mmol/l would produce the observed acidosis of 0.7 units. Where we depart from Fitts (1994) is the implication that other values in the literature are either artifacts or minor departures from a general rule. In the next section, we will review the variations in the simple story given above and give examples from the literature which demonstrate these variations. As noted earlier, our conclusion is that the ΔpH_i can show large variations (between about 0 and 0.8 pH unit), and these variations are both functionally important and give valuable insights into fatigue mechanisms.

Source of the Protons

Many intracellular reactions involve the production or absorption of protons, but for the present purpose, we are concerned only with net reactions which occur to a large extent in working muscle. Muscles which are designed to use oxidative metabolism as their principle source of energy produce only CO_2 and H_20 as net products, but these can leave the cell rapidly and, therefore, do not usually accumulate. Muscles which are used in high intensity activity generally have inadequate capacity for oxidative metabolism and rely on anaerobic glycolysis, where the end product is lactic acid. Two further reactions which can contribute to ΔpH_i are the breakdown of PCr which absorbs a proton (Amorena et al., 1990) and the net breakdown of ATP which occurs when fatigue is severe (Nagesser et al., 1993) and which releases a proton (Smith et al., 1993).

One important source of variation between fiber types is the relative contribution of glycolytic *vs.* oxidative phosphorylation enzymes. This is the probable cause of the difference in ΔpH_i noted by Westerblad and Lännergren (1988) in the different fiber types of *Xenopus* muscle. They found that type II fibers had a much smaller ΔpH_i than type I fibers, and Van der Laarse and coworkers (1991) subsequently showed that the level of the oxidative enzyme, succinate dehydrogenase, was greater in type II fibers. A more dramatic example occurs in glycolysis when inhibited with iodoacetate, which eliminates the ΔpH_i (Sahlin et al., 1981), or in human myophosphorylase deficiency in which the ΔpH_i is small (Cady et al., 1989).

Buffer Power

If the buffer power of different fibers was shown to vary, this could be another possible source of ΔpH_i variations. The review by Roos and Boron (1981) gives a table of buffer power from different fiber types showing a wide variation; however, the only careful comparative study we know of showed no significant difference in deleted buffer power between type I and II fibers of *Xenopus* (Curtin, 1987). It is well known that the buffer power of any cell is greatly increased by the use of a CO_2/HCO^-_3 buffer (Roos & Boron, 1981), and use of different external solutions in *in vitro* experiments could contribute to the ΔpH_i by such a mechanism.

Removal of Protons

All cells have pH regulating mechanisms of varying degrees of complexity and efficacy. These generally lead to removal of protons at the same rate as they are produced. An intracellular acidosis develops only if the rate of proton production exceeds the rate of removal. This is the probable reason that low intensity, long term muscular activity in humans causes little or no acidosis (e.g., Vøllestad et al., 1988). However, in short bursts of maximal activity an acidosis develops because the proton removal rates cannot keep pace with the rate of production of lactic acid. This is particularly true of contractions sufficiently strong to collapse the arteries and cause ischemia or in human experiments in which ischemia is intentionally induced. It also applies to lesser extent in isolated muscle studies in which, even though superfusion continues, there is a long extracellular diffusion pathway from the cell membrane to the perfusate. In the above examples, even if the extrusion rate of protons across the surface membrane is adequate, the accumulation of extracellular protons inhibits the extrusion mechanism. This is, no doubt, partly why increasing the concentration of extracellular pH buffer can reduce the manifestations of fatigue (Mason et al., 1986).

Several studies have attempted to identify the most important pH regulatory pathways during fatigue. Juel (1988) studied mouse soleus muscle and found that lactate efflux (which is electroneutral and accompanied by a proton) carried part of the proton efflux and that the Na/H exchanger also contributed to proton efflux. A subsequent study on single, perfused mouse fibers found that there was no acidosis during repeated tetani, presumably because in the absence of extracellular barriers the extrusion mechanisms were sufficiently effective that there was no lactic acid accumulation (Westerblad & Allen, 1992a). This interpretation was confirmed when it was shown that blockage of the lactate transporter with cinnamate led to the development of a 0.4 pH unit acidosis in an identical stimulation protocol. This study is interesting because it is possible to compare the decline of force in a stimulation protocol which leads to fatigue both in the presence and absence of acidosis. Fig. 1 shows that in the absence of acidosis, force fell to 67 % over 50 tetani; whereas, after the addition of cinnamate, an acidosis of 0.4 pH unit developed and force fell to 21% over the same period. This experiment shows clearly that acidosis is only one of the mechanisms which lead to force decline. In these experiments, inhibiting the Na/H exchanger had little effect on the acidosis associated with activity; this does not prove that it is absent, only that it is unimportant under these conditions. A similar analysis of fatigue in the presence and absence of acidosis has been performed by Cady and coworkers (1989). Normal humans were compared with a subject with myophosphorylase deficiency; the latter showed no acidosis during ischemic muscle activity but fatigued faster than the normal subjects. This reinforces the argument that other metabolic changes also cause force decline and that these metabolic changes may be accelerated in the absence of glycolysis.

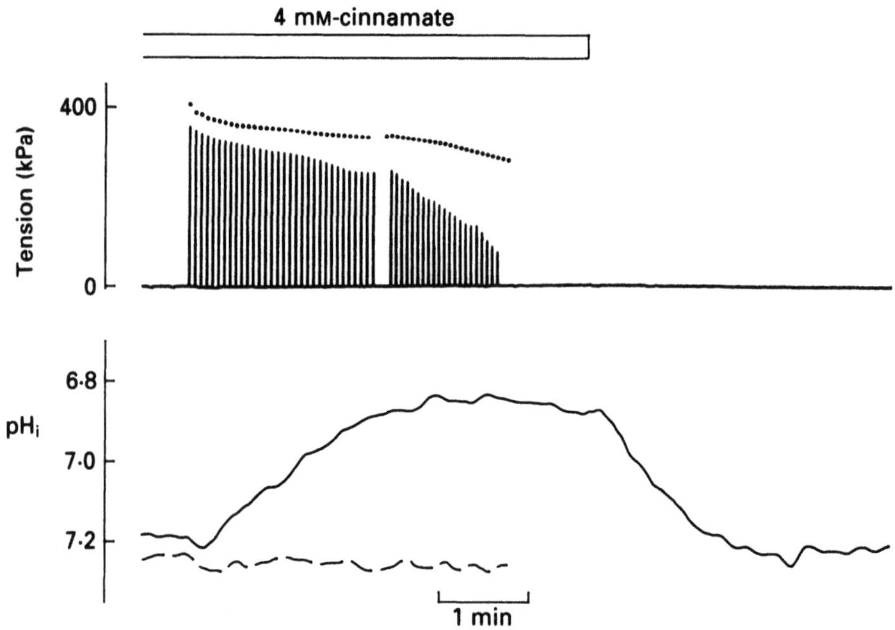

Figure 1. Force and intracellular pH in a single isolated mouse muscle fiber during repeated tetanic stimulation. The small circles show peak force and the dashed line shows intracellular pH in a control period of repeated tetani. Subsequently 4 mM α-cyano-4-hydroxycinnamate was applied to block the lactate transporter. The vertical lines show force and the full line shows the intracellular pH when repeated tetani were produced under these conditions. Note that in the absence of cinnamate there was a moderate fall of force but no intracellular acidosis. After inhibition of the lactate transporter, the decline of force was greater and faster and an intracellular acidosis of 0.4 pH-units developed (reproduced from Westerblad & Allen, 1992a).

CONSEQUENCES OF ACIDOSIS

The most obvious feature of fatigue is the accompanying decline in tetanic force. Eberstein and Sandow (1963) first established that failure of Ca^{2+} release was an important cause of the force decline by showing that caffeine and high K^+, which increase Ca^{2+} release, could overcome much of the force decline. There is also abundant evidence from skinned fiber studies that some of the metabolites that accumulate during fatigue can depress both the maximum Ca^{2+}-activated force and the myofibrillar Ca^{2+} sensitivity (reviewed in Westerblad et al., 1991; Fitts, 1994). In this section, we, therefore, first consider the contribution of ΔpH_i to each of these processes. A variety of other changes in muscle function can be observed in fatigue. Two changes that have important effects on the work output of fatigued muscles are a slowing of velocity of shortening and a reduction in the rate of relaxation. The effects of acidosis on these functions are considered next. Finally we consider the role of ΔpH_i on metabolic pathways. All enzymes are pH sensitive, so it is to be expected that pH might have an important influence here.

Failure of Ca^{2+} Release

Since the demonstration that fatigue involves a partial failure of excitation-contraction coupling, there has been speculation that acidosis interferes with the process (e.g., Nassar-Gentina et al., 1981). Support for this hypothesis was increased when Ma and coworkers (1988) showed that acidosis inhibited the opening of isolated sarcoplasmic reticulum (SR) Ca^{2+} channels in lipid bilayers. Although this is an attractive hypothesis, the arguments against it are strong. 1) Lamb and colleagues (1992), using skinned fibers with intact SR/t-tubular system, showed that depolarization-induced SR Ca^{2+} release is not affected by pH. On the other, hand the leakage of Ca^{2+} from the SR (in the absence of depolarization) was decreased by acidosis which is probably the physiological equivalent of the observations by Ma and colleagues (1988); 2) single mouse fibers show the normal decline in Ca^{2+} release in fatigue (Westerblad & Allen, 1991, 1993b), although fatigue in these fibers is not accompanied by any marked acidosis (Fig. 1). Thus, acidosis is probably not important for the declining tetanic $[Ca^{2+}]_i$ in fatigue; and 3) if intracellular acidosis is produced in single fibers by increasing extracellular CO_2 (Fig. 2), it causes a substantial *increase* in tetanic $[Ca^{2+}]_i$. The probable cause of this increase is a reduced troponin binding constant for Ca^{2+} so that the same Ca^{2+} release leads to a larger $[Ca^{2+}]_i$ but a smaller force (see subsequent section). At present, the cause of the failure of Ca^{2+} release during fatigue remains unknown (see Stephenson et al., Chapter 2). Two possibilities are that it is a consequence of the fall in intracellular ATP concentration, which is known to occur at about the same time (Westerblad & Allen, 1992b,c), or that it is a consequence of precipitation of calcium phosphate in the SR (Fryer et al., 1995).

Reduction of Maximal Isometric Force

One part of the force decline in fatigue cannot be overcome by increasing tetanic $[Ca^{2+}]_i$ with caffeine; and it has, therefore, been ascribed to reduced maximum force (Edman & Lou, 1990; Lännergren & Westerblad, 1991; Westerblad & Allen, 1991). This component dominates during the early phases of fatigue. Studies on skinned fibers have shown that, of the metabolic changes that can occur in fatigue, two make the major contribution to the decline of maximum force: acidosis (due to lactic acid accumulation) and accumulation of inorganic phosphate (P_i; due to PCr breakdown).

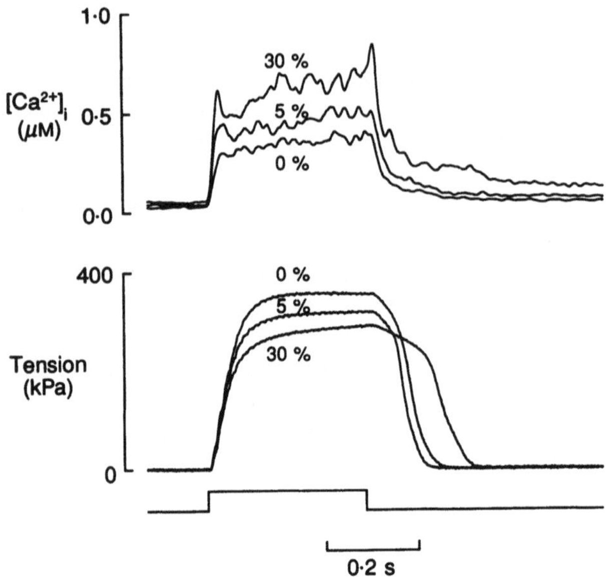

Figure 2. Intracellular calcium and force in 100 Hz tetani from isolated single fibers of mouse muscle. Intracellular pH was varied by changing the extracellular CO_2 concentration; 30 % CO_2 gives a $pH_i = 6.8$; 5 % CO_2 gives a $pH_i = 7.3$; 0 % CO_2 gives a $pH_i = 7.9$. Note that acidosis increases the $[Ca^{2+}]_i$ during a tetanus but reduces the tetanic force (reproduced from Westerblad & Allen, 1993a).

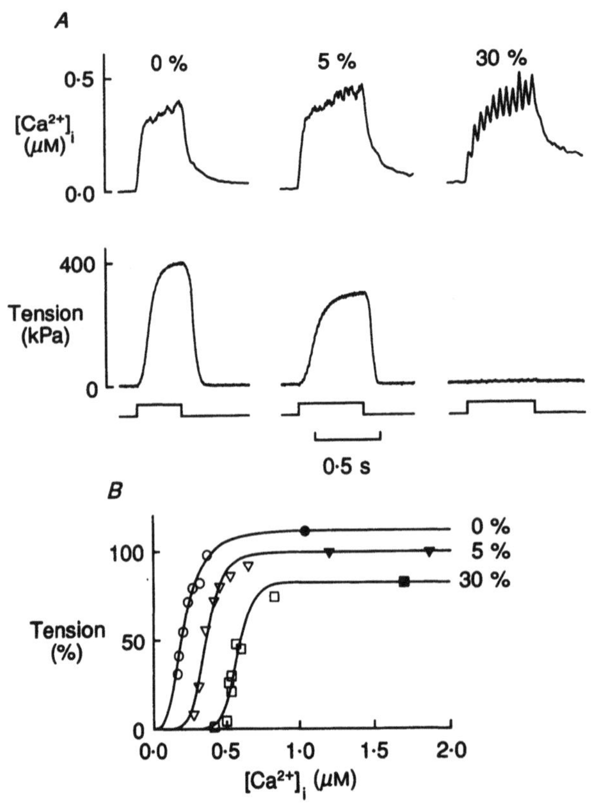

Figure 3. The relation between $[Ca^{2+}]_i$ and force in intact isolated mouse muscle fibers at various intracellular pH. *A* shows selected $[Ca^{2+}]_i$ and force records from tetani at various pH_i produced by changing extracellular CO_2 (0 % CO_2 gives a $pH_i = 7.9$; 5 % CO_2 gives a $pH_i = 7.3$; 30 % CO_2 gives a $pH_i = 6.8$). In this experiment, the stimulus frequency was varied at each pH_i so as to obtain a series of tetani with varying $[Ca^{2+}]_i$ during the tetanic contraction. To obtain very high $[Ca^{2+}]_i$, 10 mM caffeine was added (filled points in *B*). In *A*, records were selected with a roughly constant $[Ca^{2+}]_i$ during the tetanus to show the very different force obtained at different pH_i. *B* shows the relation between $[Ca^{2+}]_i$ and force at the various pH_i. Each point is the mean $[Ca^{2+}]_i$ and force taken from the plateau of one tetanus. Note the reduction in maximum Ca^{2+}-activated force and Ca^{2+} sensitivity in acid conditions (reproduced from Westerblad & Allen, 1993a).

It is well established that acidosis depresses maximal force in skinned fibers (e.g., Fabiato & Fabiato, 1978; Metzger & Moss, 1987; Godt & Nosek, 1989). To show that this also occurs in intact fibers, it is necessary to maximally activate the muscle. One approach to this problem is to construct $[Ca^{2+}]_i$ -force curves at different pH_i as illustrated in Fig. 3 (for details see legend to Fig. 3). Our results show a clear reduction in maximal force with acidosis in intact fibers, although the magnitude is somewhat smaller in intact than in skinned fibers. There are, however, studies (Adams et al., 1991) suggesting that acidosis has no effect on tetanic force in intact muscle. In this latter study, the maximal force was not assessed and the lack of an effect of acidosis can be partly explained by the fact that tetanic $[Ca^{2+}]_i$ increases with acidosis (Fig. 2). The latter would be expected to offset the reduction of force.

The exact mechanism by which acidosis reduces maximal force is not clear (Cooke et al., 1988; Kentish, 1991). Acidosis also reduces actomyosin ATPase rates (Cooke et al., 1988; Parkhouse, 1992) suggesting that crossbridge turnover rates are reduced. The fact that acidosis generally reduces ATPase less than force suggests that there is also a reduction in the force produced by each attached crossbridge (Orchard & Kentish, 1990). It seems probable that protons are interacting with multiple sites associated with the actomyosin ATPase.

Reduced Myofibrillar Ca^{2+} Sensitivity

In fatigue, there is often an acidosis and an accumulation of P_i. Skinned-fiber results have shown that both of these metabolic changes reduce the myofibrillar Ca^{2+} sensitivity (e.g., Fabiato & Fabiato, 1978; Metzger & Moss, 1987; Godt & Nosek, 1989). This reduction could occur by one of two mechanisms: either via competition for the Ca^{2+} binding site on troponin (Blanchard et al., 1984) or via a reduction in strongly bound crossbridges (Bremel & Weber, 1972; Güth & Potter, 1987). It was originally proposed that acidosis inhibited Ca^{2+} binding to troponin by competition between H^+ and Ca^{2+} but recent evidence suggests that a reduced number of strongly bound crossbridges may also contribute (Metzger & Moss, 1987). P_i, in contrast, does not change the Ca^{2+} binding to isolated troponin C (Kentish & Palmer, 1989); and, hence, its depression of Ca^{2+} sensitivity is thought to occur solely through reduced cooperativity due to a reduced number of strongly attached crossbridges (Millar & Homsher, 1990).

Changes of the myofibrillar Ca^{2+} sensitivity during fatigue have been assessed in single fibers from mouse and *Xenopus* (Allen et al., 1989; Lee et al., 1991; Westerblad & Allen, 1991, 1993b). In most of these studies, a clear-cut reduction of the sensitivity was observed. An exception was the earliest study (Allen et al., 1989) in which the photoprotein aequorin was used to measure $[Ca^{2+}]_i$. Aequorin light emission is reduced by Mg^{2+}, and it was subsequently shown that an increase in myoplasmic $[Mg^{2+}]$ ($[Mg^{2+}]_i$) occurs in fatigue (Westerblad & Allen, 1992b,c). When correction was made for the effect of $[Mg^{2+}]_i$ on the aequorin light emission, a reduction in myofibrillar Ca^{2+} sensitivity can be observed in these experiments as well. There are also metabolic changes in fatigue that may increase myofibrillar Ca^{2+} sensitivity (e.g., increased ADP, Cooke & Pate, 1985; and myosin light chain phosphorylation, Persechini et al., 1985).

Thus to summarize, several changes in fatigue may significantly affect the myofibrillar Ca^{2+} sensitivity: acidosis and increased P_i will reduce it, while increased ADP and myosin light chain phosphorylation may have the opposite effect. In fatigue, the net effect of these changes is generally a reduced Ca^{2+} sensitivity. This reduction is especially important for force production when tetanic $[Ca^{2+}]_i$ declines, i.e., towards the end of fatiguing stimulation.

The Rate of Relaxation

Relaxation is a complex process involving 1) Ca^{2+} uptake by the SR and buffers such as parvalbumin; 2) dissociation of Ca^{2+} from troponin; and 3) crossbridge detachment. It is well established that acidosis slows relaxation (Edman & Mattiazzi, 1981; their Figs. 2 and 4; Westerblad & Allen, 1993a; Westerblad & Lännergren, 1991) but it is also known that relaxation during fatigue can slow in the absence of acidosis (Cady et al., 1989; Westerblad & Allen, 1992a). There is evidence from both isolated SR and skinned fibers with functional SR that acidosis inhibits SR Ca^{2+} pumping. For instance a recent study by Lamb and colleagues (1992) estimated SR Ca^{2+} uptake rates by measuring mechanical responses in skinned fibers with intact t-tubule SR connections. They showed that SR pump rate fell roughly 2-fold between pH 7.1 and 6.6 and fell a further 2-fold between pH 6.6 and 6.1.

Such studies have led to the suggestion that the mechanism by which acidosis slows relaxation involves inhibition of the SR Ca^{2+} pump. Recent measurements of $[Ca^{2+}]_i$ during acidosis (Westerblad & Allen, 1993a) showed little change in the fast phase of $[Ca^{2+}]_i$ decline, perhaps because this process is dominated by the turnoff of SR permeability, and a pronounced increase in magnitude and slowing of the slow phase of $[Ca^{2+}]_i$ decline. SR pump analysis on the slow phase showed that the pump rate was reduced by about 3-fold between pH 7.3 and 6.8. Similar results were seen in studies on skinned fibers with intact SR (see above). However, to decide on the contribution of the reduced SR Ca^{2+} uptake to the slower relaxation, it is necessary to consider also the reduction of myofibrillar Ca^{2+}-sensitivity that is caused by acidosis (see earlier section). In our study, this was done by converting the $[Ca^{2+}]_i$ to the appropriate steady force by means of a previously determined relation between $[Ca^{2+}]_i$ and force (for details see legend to Fig 4). It can be seen from Fig. 4 that despite the elevated tetanic $[Ca^{2+}]_i$ and the enhanced slow decline, the Ca^{2+}-derived force in acidosis was little changed from control because of the reduced Ca^{2+} sensitivity and maximum force of the contractile proteins. Consequently it seems that the delay between the Ca^{2+}-derived force and the true force is increased in acidosis compared to control. This suggests that the

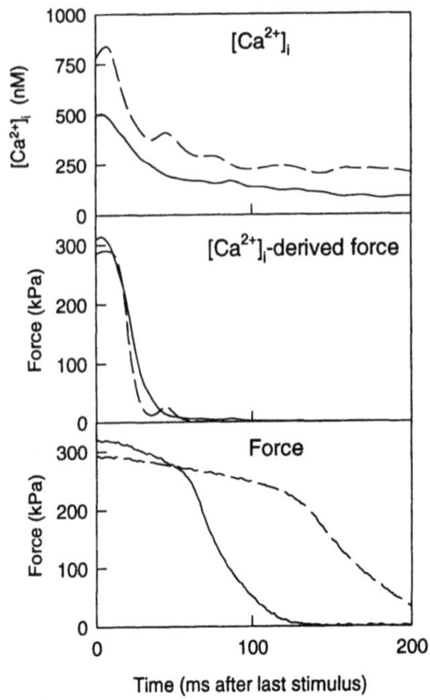

Figure 4. Relaxation in control and acid conditions following a tetanic contraction in an isolated mouse muscle fiber. In each panel the continuous line is control ($pH_i = 7.3$) while the dashed line shows results when an intracellular acidosis ($pH_i = 6.8$) was produced by increased extracellular CO_2. Upper panel shows the $[Ca^{2+}]_i$ with the last stimulus shown as zero time. Middle panel shows the Ca^{2+} -derived force in which $[Ca^{2+}]_i$ from the upper panel has been converted to force using the steady state $[Ca^{2+}]_i$ -force relation at the appropriate pH_i of the sort shown in Fig. 3*B*. Lower panel shows the measured force. Note that the $[Ca^{2+}]_i$ is larger and initially declines rapidly but later declines more slowly in acid conditions. However, because of the reduced maximum force and Ca^{2+} sensitivity in acid conditions, the $[Ca^{2+}]_i$-derived force is not greatly different under acid conditions. Thus the much slower decline of force in acid condition seems to be caused by slower detachment of crossbridges rather than the slower rate of $[Ca^{2+}]_i$ decline (modified from Westerblad & Allen, 1993a).

detachment rate of crossbridges is slowed in acidosis, which is consistent with a reduced myofibrillar ATPase in acidosis (Cooke et al., 1988; Parkhouse, 1992).

Shortening Velocity

Fatigue is generally accompanied by a reduction of shortening speed (DeHaan et al., 1989; Curtin & Edman, 1994; Westerblad & Lännergren, 1994). There is, however, a large variability in the magnitude of this slowing, which presumably depends on the type of preparation, stimulation regime etc. Also, there is no clear-cut relation between reductions in isometric tension and shortening velocity in fatigue. The latter can be explained by the fact that the metabolic changes in fatigue may have markedly different effects on isometric force compared to shortening velocity. A striking example of this is that increased ADP reduces the shortening velocity while isometric tension is increased (Cooke & Pate, 1985).

In an early study employing single, intact fibers of frog, Edman and Mattiazzi (1981) observed similar changes of the shortening velocity at zero load (V_o) in fatigue and in acidosis induced by exposure to CO_2. They concluded that reduced pH_i makes a large contribution to the reduction of shortening speed in fatigue. However, since pH_i was not measured in this study, it is not possible to say if the slowed shortening was exclusively due to acidosis or if other factors contributed.

Studies on skinned fibers support the idea that acidosis is important for the reduced shortening speed in fatigue. If we assume a typical pH_i change in fatigue of 0.7 pH-units (see above), this gives a reduction of V_o in skinned fibers of about 25% (Metzger & Moss, 1987; Cooke et al., 1988), which is similar to the slowing often observed in fatigue. This result indicates that acidosis may be responsible for all changes of the shortening velocity in fatigue. However, in a recent study we followed changes of V_o during fatigue produced by repeated short tetani in isolated *Xenopus* fibers (Westerblad & Lännergren, 1994). In these fibers, pH_i in fatigue is reduced by about 0.6 pH-units (Westerblad & Lännergren, 1988) which according to skinned fiber results would give a reduction of V_o of about 20%. However, the observed reduction in V_o was about 50%, thus suggesting that some other factor also depresses the shortening speed.

In conclusion, acidosis is of great importance for the reduced shortening speed in fatigue; but, at least in *Xenopus* fibers, there seems to be an additional factor which contributes to the slowing.

Metabolism

Enzymes are generally pH-sensitive, and even small changes of pH_i may have large impact on enzymatic activity and hence cellular metabolism. This is of particular importance in skeletal muscle fatigue which may be accompanied by a large decline of pH_i (see above). Acidosis reduces the glycogenolytic rate by inhibiting both glycogen phosphorylase and phosphofructokinase (e.g., Roos & Boron, 1981). This would lead to a reduced rate of ATP production, which might set an important limitation to muscle performance during fatigue. However, acidosis will also reduce the rate of ATP utilization by inhibiting myofibrillar ATPase (Cooke et al., 1988; Parkhouse, 1992) and SR Ca^{2+}-pumping (Fabiato & Fabiato, 1978; Lamb et al., 1992). Furthermore, a reduction of ATP derived from glycogenolysis will accelerate phosphocreatine (PCr) breakdown and eventually result in a net ATP breakdown, leading to an accumulation of P_i and AMP, both of which are known activators of glycogen phosphorylase and phosphofructokinase (Chasiotis et al., 1982). Thus, the overall consequences of acidosis on energy metabolism during fatigue are difficult to predict. In fatigue, the reduction of ATP is generally small (Fitts, 1994), showing that the inhibition of energy supply by acidosis is to a large extent counteracted by other mechanisms. Nevertheless, in frog muscle repeated contractions at long intervals are associated with a marked depletion of PCr under acidic conditions while PCr remained unchanged at normal pH_i (Nakamura & Yamada, 1992). Thus, at least under these experimental conditions the inhibition of ATP production by acidosis was not fully counteracted by a reduced rate of ATP utilization.

CONCLUDING REMARKS

Intracellular acidosis frequently, but not invariably, accompanies muscle fatigue. Its presence or absence can be understood in terms of the pH regulatory mechanisms of the muscle cell. The magnitude of the intracellular acidosis reflects the interplay between production of acid metabolites, the pH buffering by the cell and the activity of the pH_i regulating mechanisms. Intracellular acidosis has multiple actions on cell function, and the observed effects in whole muscles represent a complex summation of these effects. In a single cell, in which pH_i can be measured, it is often possible to separate the effects associated with acidosis from those associated with the other metabolic changes occurring in fatigue. In whole muscles, the activation pattern and the muscle fiber types are usually heterogenous, and it is likely that the changes in pH_i vary from cell to cell. These complications mean that detailed quantitative analysis of the role of acidosis is much more difficult.

ACKNOWLEDGMENTS

We thank Dr. Jennifer McDonagh for editorial comments on the manuscript. Work in the authors' laboratories is supported by grants from the National Health & Medical Research Council of Australia and the Medical Research Council of Sweden. Their attendance at the 1994 Bigland-Ritchie conference was supported, in part, by the Muscular Dystrophy Association (USA), and the American Physical Therapy Association (Section on Research; *D.G.A.*).

REFERENCES

Adams GR, Fisher MJ & Meyer RA (1991). Hypercapnic acidosis and increased $H_2PO_4^-$ concentration do not decrease force in cat skeletal muscle. *American Journal of Physiology* **260**, C805-C812.

Allen DG, Lee JA & Westerblad H (1989). Intracellular calcium and force in isolated single muscle fibres from *Xenopus*. *Journal of Physiology (London)* **415**, 433-458.

Amorena CE, Wilding TJ, Manchester JK & Roos A (1990). Changes in intracellular pH caused by high K^+ in normal and acidified frog muscle. *Journal of General Physiology* **96**, 959-972.

Blanchard EM, Pan BS & Solaro RJ (1984). The effect of acidic pH on the ATPase activity and troponin Ca^{2+} binding of rabbit skeletal myofilaments. *Journal of Biological Chemistry* **259**, 3181-3186.

Bremel RD & Weber A (1972). Cooperation within actin filaments in vertebrate skeletal muscle. *Nature* **238**, 97-101.

Cady EB, Elshove E, Jones DA & Moll A (1989). The metabolic causes of slow relaxation in fatigued human skeletal muscle. *Journal of Physiology (London)* **418**, 327-337.

Chasiotis D, Sahlin K & Hultman E (1982). Regulation of glycogenolysis in human muscles at rest and during exercise. *Journal of Applied Physiology* **53**,708-715.

Cooke R, Franks K, Luciani GB & Pate E (1988). The inhibition of rabbit skeletal muscle contraction by hydrogen ion and phosphate. *Journal of Physiology (London)* **395**, 77-97.

Cooke R & Pate E (1985). The effects of ADP and phosphate on the contraction of muscle fibers. *Biophysical Journal* **48**, 789-798.

Curtin NA (1987). Intracellular pH and buffer power of type 1 and 2 fibres from skeletal muscle of *Xenopus laevis*. *Pflügers Archiv* **408**, 386-389.

Curtin NA & Edman KAP (1994). Force-velocity relation for frog muscle fibres: effects of moderate fatigue and of intracellular acidification. *Journal of Physiology (London)* **475**, 483-494.

DeHaan A, Jones DA & Sargeant AJ (1989). Changes in velocity of shortening, power output and relaxation rate during fatigue of rat medial gastrocnemius muscle. *Pflügers Archiv* **413**, 422-428.

Eberstein A & Sandow A (1963). Fatigue mechanisms in muscle fibers. In: Gutman E, Hink P (eds.), *The Effect of Use and Disuse on Neuromuscular Functions*, pp. 515-526. Amsterdam: Elsevier.

Edman KAP & Lou F (1990). Changes in force and stiffness induced by fatigue and intracellular acidification in frog muscle fibres. *Journal of Physiology (London)* **424**, 133-149.

Edman KAP & Mattiazzi AR (1981). Effects of fatigue and altered pH on isometric force and velocity of shortening at zero load in frog muscle fibres. *Journal of Muscle Research and Cell Motility* **2**, 321-334.

Fabiato A & Fabiato F (1978). Effects of pH on the myofilaments and the sarcoplasmic reticulum of skinned cells from cardaic and skeletal muscles. *Journal of Physiology (London)* **276**, 233-255.

Fitts RH (1994). Cellular mechanisms of muscle fatigue. *Physiological Reviews* **74**, 49-94.

Fryer MW, Owen VJ, Lamb GD & Stephenson DG (1995). Effects of creatine phosphate and Pi on Ca^{2+} movements and tension development in rat skinned skeletal muscle fibres. *Journal of Physiology (London)* **482**, 123-140.

Godt RE & Nosek TM (1989). Changes of intracellular milieu with fatigue or hypoxia depress contraction of skinned rabbit skeletal and cardiac muscle. *Journal of Physiology (London)* **412**, 155-180.

Güth K & Potter JD (1987). Effect of rigor and cycling cross-bridges on the structure of troponin C and on the Ca^{2+} affinity of the Ca^{2+}-specific regulatory sites in skinned rabbit psoas fibers. *Journal of Biological Chemistry* **262**, 13627-13635.

Hill AV & Kupalov P (1929). Anaerobic and aerobic activity in isolated muscles. *Proceedings of the Royal Society (Series B)* **105**, 313-322.

Juel C (1988). Intracellular pH recovery and lactate efflux in mouse soleus muscles stimulated *in vitro*: the involvement of sodium/proton exchange and a lactate carrier. *Acta Physiologica Scandinavica* **132**, 363-371.

Kentish JC (1991). Combined inhibitory actions of acidosis and phosphate on maximum force production in rat skinned cardiac muscle. *Pflügers Archiv* **419**, 310-318.

Kentish JC & Palmer S (1989). Calcium binding to isolated bovine cardiac and rabbit skeletal troponin-C is affected by pH but not by caffeine or inorganic phosphate. *Journal of Physiology (London)* **417**, 160P.

Lamb GD, Recupero E & Stephenson DG (1992). Effect of myoplasmic pH on excitation-contraction coupling in skeletal muscle fibres of the toad. *Journal of Physiology (London)* **448**, 211-224.

Lännergren J & Westerblad H (1991). Force decline due to fatigue and intracellular acidification in isolated fibres from mouse skeletal muscle. *Journal of Physiology (London)* **434**, 307-322.

Lee JA, Westerblad H & Allen DG (1991). Changes in tetanic and resting $[Ca^{2+}]_i$ during fatigue and recovery of single muscle fibres from *Xenopus laevis*. *Journal of Physiology (London)* **433**, 307-326.

Ma J, Fill M, Knudson M, Campbell KP & Coronado R (1988). Ryanodine receptor of skeletal muscle is a gap junction-type channel. *Science* **242**, 99-102.

Mason MJ, Mainwood GW & Thoden JS (1986). The influence of extracellular buffer concentration and propionate on lactate efflux from frog muscle. *Pflügers Archiv* **406**, 472-479.

Metzger JM & Moss RL (1987). Greater hydrogen ion induced depression of tension and velocity in skinned single fibres of rat fast than slow muscles. *Journal of Physiology (London)* **393**, 727-742.

Millar NC & Homsher E (1990). The effect of phosphate and calcium on force generation in glycerinated rabbit skeletal muscle fibers. *Journal of Biological Chemistry* **265**, 20234-20240.

Nagesser AS, Van Der Laarse WJ & Elzinga G (1993). ATP formation and ATP hydrolysis during fatiguing, intermittent stimulation of different types of single muscle fibres from *Xenopus laevis*. *Journal of Muscle Research and Cell Motility* **14**, 608-618.

Nakamura T & Yamada K (1992). Effects of carbon dioxide on tetanic contraction of frog skeletal muscles studied by phosphorus nuclear magnetic resonance. *Journal of Physiology (London)* **453**, 247-259.

Nassar-Gentina V, Passonneau JV & Rapoport SI (1981). Fatigue and metabolism of frog muscle fibers during stimulation and in response to caffeine. *American Journal of Physiology* **241**, C160-C166.

Needham DM (1971). *Machina Carnis: The Biochemistry of Muscular Contraction and Its Historical Development.* Cambridge: Cambridge University Press.

Orchard CH & Kentish JC (1990). Effects of changes of pH on the contractile function of cardiac muscle. *American Journal of Physiology* **258**, C967-C981.

Parkhouse WS (1992). The effects of ATP, inorganic phosphate, protons, and lactate on isolated myofibrillar ATPase activity. *Canadian Journal of Physiology and Pharmacology* **70**, 1175-1181.

Persechini A, Stull JT & Cooke R (1985). The effect of myosin phosphorylation on the contractile properties of skinned rabbit skeletal muscle fibers. *Journal of Biological Chemistry* **260**, 7951-7954.

Roos A & Boron WF (1981). Intracellular pH. *Physiological Reviews* **61**, 297-434.

Sahlin K, Edström L, Sjöholm H & Hultman E (1981). Effects of lactic acid accumulation and ATP decrease on muscle tension and relaxation. *American Journal of Physiology* **240**, C121-C126.

Smith GL, Donoso P, Bauer CJ & Eisner DA (1993). Relationship between intracellular pH and metabolic concentrations during metabolic inhibition in isolated ferret heart. *Journal of Physiology (London)* **472**, 11-22.

Van Der Laarse WJ, Lännergren J & Diegenbach PC (1991). Resistance to fatigue of single muscle fibres from *Xenopus* related to succinate dehydrogenase and myofibrillar ATPase activities. *Experimental Physiology* **76**, 589-596.

Vøllestad NK, Sejersted OM, Bahr R, Woods JJ & Bigland-Ritchie B (1988). Motor drive and metabolic responses during repeated submaximal contractions in humans. *Journal of Applied Physiology* **64**, 1421-1427.

Westerblad H & Allen DG (1991). Changes in myoplasmic calcium concentration during fatigue in single mouse muscle fibres. *Journal of General Physiology* **98**, 615-635.

Westerblad H & Allen DG (1992a). Changes of intracellular pH during repeated tetani in single mouse skeletal muscle fibres. *Journal of Physiology (London)* **449**, 49-71.

Westerblad H & Allen DG (1992b). Myoplasmic free Mg^{2+} concentration during repetitive stimulation of single fibres from mouse skeletal muscle. *Journal of Physiology (London)* 453: 413-434.

Westerblad H & Allen DG (1992c). Myoplasmic $[Mg^{2+}]_i$ concentration in *Xenopus* muscle fibres at rest, during fatigue and during metabolic blockade. *Experimental Physiology* **77**, 733-740.

Westerblad H & Allen DG (1993a). The influence of pH on contraction, relaxation and $[Ca^{2+}]_i$ in intact single fibres from mouse muscle. *Journal of Physiology (London)* **466**, 611-628.

Westerblad H & Allen DG (1993b). The role of $[Ca^{2+}]_i$ in the slowing of relaxation in fatigued single fibres from mouse skeletal muscle. *Journal of Physiology (London)* **468**, 729-740.

Westerblad H & Lännergren J (1988). The relation between force and intracellular pH in fatigued, single *Xenopus* muscle fibres. *Acta Physiologica Scandinavica* **133**, 83-89.

Westerblad H & Lännergren J (1991). Slowing of relaxation during fatigue in single mouse muscle fibres. *Journal of Physiology (London)* **434**, 323-336.

Westerblad H & Lännergren J (1994). Changes of the force-velocity relation, isometric tension and relaxation rate during fatigue in intact, single fibres of *Xenopus* skeletal muscle. *Journal of Muscle Research and Cell Motility* **15**, 287-298.

Westerblad H, Lee JA, Lännergren J & Allen DG (1991). Cellular mechanisms of fatigue in skeletal muscle. *American Journal of Physiology* **261**, C195-C209.

ROLE OF INTERSTITIAL POTASSIUM

G. Sjøgaard[1] and A. J. McComas[2]

[1]Department of Physiology
National Institute of Occupational Health
Copenhagen, Denmark
[2]Department of Medicine, McMaster University
Hamilton, Ontario

ABSTRACT

Interstitial potassium concentration, $[K^+]$, is modulated during muscle activity due to a number of different mechanisms: diffusion and active transport of K^+ in combination with water fluxes. The relative significance of the various mechanisms for muscle function is quantified. The effect of interstitial $[K^+]$ locally on the single muscle fiber is discussed along with its effect on the cardiovascular and respiratory systems and its role in motor control. It is concluded that K^+ may play a significant role in the prevention as well as the development of fatigue.

INTRODUCTION

Normal function of muscle fibers is critically dependent on the electrolyte balance within the muscles, and especially on Ca^{2+} and Mg^{2+} in the different compartments within the cell, and on the concentration gradients of Na^+ and K^+ across the sarcolemma (plasmalemma). The ultimate force regulator is cytosolic Ca^{2+} which in turn is controlled by fluxes of Na^+ and K^+ across the membrane in relation to the action potential. Quantitatively the fluxes of Na^+ and K^+ per action potential are small but nevertheless, large changes in interstitial $[K^+]$ have been reported in many studies.

Muscle Activity Induces K^+ Loss from Muscle

Fenn first showed that muscle activity induces a K^+ loss (Fenn & Cobb, 1936; Fenn, 1937), and that this was proportional to the magnitude and frequency of muscle contraction (Fenn, 1938). De Lanne and colleagues were among the first to monitor plasma $[K^+]$ with time during and following muscle work (De Lanne et al., 1959). K^+ is highly permeable across the capillary membrane (Crone et al., 1978) and thus plasma $[K^+]$ in the venous effluent reflects interstitial $[K^+]$ in a well perfused muscle. The net loss from the muscle to the interstitial space can then be estimated as (plasma flow) x (arterial - venous)$[K^+]$. When

Fatigue, Edited by Simon C. Gandevia et al.
Plenum Press, New York, 1995

muscle blood flow is occluded (e.g., due to high tissue pressure during a contraction) no net K^+ loss from the muscle can occur. Nevertheless, intracellular K^+ will diffuse down its electrochemical gradient into the interstitial space, and using microelectrodes placed in the interstitial space dramatic increases in [K^+] have been reported (Hirche et al., 1980; Vyskocil et al., 1983).

Quantitative Data on K^+ Fluxes in Humans

Data have been obtained from whole body exercise in humans during which the subjects perceived fatigue or even became exhausted (Bergström et al., 1971; Bergström et al., 1973; Sejersted et al., 1984; Sejersted & Hallén, 1987; Medbø & Sejersted, 1985). The highest plasma [K^+] reported was 9 mmol l^{-1} during intense running (Medbø & Sejersted, 1990). An attempt to calculate the maximal rate of K^+ loss from the leg muscles per kg wet weight showed a value of 1.5 mmol kg ww^{-1} min^{-1} during running (Sejersted & Hallén, 1987) and 1.8 mmol kg ww^{-1} min^{-1} during bicycling (Vøllestad et al., 1994). However, for such calculations many assumptions have to be made and a better experimental design is needed to obtain more reliable data.

One such set-up is the knee-extensor model in which a single muscle group (mass of approx. 2.5 kg ww) can be studied during voluntary contractions. The K^+ balance was studied during various modes of exercise: dynamic and static from low to high intensities. The largest *total* K^+ loss occurred during submaximal dynamic exercise which could be maintained for 2 hr or more before exhaustion occurred, and amounted to around 20 mmol kg ww^{-1} (Sjøgaard et al., 1985). However, the highest *rate* of net K^+ loss occurred during high intensity dynamic contractions, which the subjects could only continue for around 5 min and was 1 mmol kg ww^{-1} min^{-1} (Juel et al., 1990; Sjøgaard et al., 1985). Interestingly, this value is similar to that estimated above for whole body exercise. The dynamic contractions were performed with a frequency of 1 Hz, and thus a K^+ loss of 15-17 μmol kg ww^{-1} per contraction was calculated. The venous effluent [K^+] was around 6.3-6.8 mmol l^{-1} and the arterial [K^+] was 5.5-5.8 mmol l^{-1}. These values are lower than those during whole body exercise and the difference is probably due to the smaller muscle mass in the knee-extensor model, allowing for a better clearance of K^+ into other tissues.

The largest increase in interstitial [K^+] probably occurs during static contractions where the tissue pressure impedes blood flow (Barcroft & Millen, 1939). The contraction-induced K^+ release is then trapped in the interstitial space where it accumulates throughout the contraction. This interpretation is in agreement with the highest interstitial [K^+] in man (15 mmol l^{-1}) being reported during intensive isometric contractions which could be sustained for only a few seconds (Vyskocil et al., 1983).

Quantitative Data on K^+ Fluxes in Animals

The most precise estimates of the magnitude of K^+ losses relative to muscle contractions are obtained by studying stimulation of animal muscles. The recalculation of data from a number of studies shows K^+ losses of 2.3-4.8 mmol kg ww^{-1} min^{-1} during continuous subtetanic stimulation (5 Hz), 0.9-1.5 mmol kg ww^{-1} min^{-1} during continuous tetanic stimulation, and 0.5-2.0 mmol kg ww^{-1} min^{-1} during intermittent tetanic stimulation (for references see Sjøgaard, 1991). If calculated per contraction during intermittent stimulation, the values ranged from 11-17 μmol kg ww^{-1} in 0.75-1.5 Hz trains, which comes close to the value for humans during 1 Hz voluntary contraction (see above). A more detailed analysis allows the K^+ loss per stimulation to be calculated in the electrical stimulation studies. At stimulation frequencies of 1-5 Hz the K^+ loss per stimulus was 3.9-7.5 μmol kg ww^{-1} , values which are in close agreement with those predicted from a model of K^+ efflux (Hazeyama & Sparks, 1979). During voluntary contractions in man the actual number of firings is not known but may be estimated: each contraction period during the above intense dynamic exercise lasted for less than 0.5 s and, with maximal firing rates of 30 Hz (Thomas et al., 1991), a reasonable assumption is approximately 10 discharges per contraction (Sjøgaard et al., 1985). On average the K^+ loss per stimuli is then 1.7 μmol kg ww^{-1}, a value which is relatively low compared to the animal data. The main explanation for this discrepancy may be that the K^+ efflux should be expressed relative to cell surface (cm^2) and not to cell mass (kg) in order to obtain data comparable between species.

Focus on Interstitial K^+ Concentration

The activity-induced K^+ flux from the intra- to the extracellular space of muscle affects the concentrations in both compartments and the transmembrane concentration gradients, which are of major functional significance. The reason for focusing on the interstitial [K^+] is that this compartment shows the largest changes. This characteristic is due to the interstitial space having a small volume and a low resting [K^+], relative to the surrounding muscle fibers. The fibers can therefore be envisaged as a large sink of K^+ which may readily leak into the interstitium.

We shall demonstrate that K^+ fluxes may be causally related to fatigue but at the same time point out that regulatory mechanism exist which are sufficiently potent to prevent a K^+-induced catastrophe from occurring (myotonia, paralysis). However, with extreme fatigue (i.e., exhaustion) the preventive mechanisms may fail as both muscle excitation and force decrease. This final stage may be considered as a safety mechanism which protects the muscle fiber against further energy depletion and thus metabolic exhaustion and cell destruction.

FACTORS AFFECTING INTERSTITIAL [K^+]

The interstitial compartment is surrounded by two sets of membranes, the sarcolemma (plasmalemma) and the capillary membrane. Estimates of changes in the amount of interstitial K^+ must be based on the net fluxes across both membranes (Fig. 1). Further, the interstitial [K^+] is affected by water fluxes across these membranes as well as by lymphatic drainage (Aukland & Reed, 1993). During exercise there is a net efflux of water from the intramuscular capillaries into the interstitial space, and subsequently into the muscle fibers as well. These water fluxes are induced primarily by changes in small-molecule osmotic pressure, although K^+ per se does not contribute significantly (Watson et al., 1993). It is this transfer of water which is responsible for the swelling and firmness of a muscle belly after intense or prolonged effort. In quadriceps muscles exercised to fatigue, the extracellular water may double per kg dry weight tissue, from 0.33 to 0.64 l kg dw^{-1} (Sjøgaard, 1986). Even if no net change took place in electrolyte flux, the increased interstitial volume would reduce [K^+] by half.

Efflux of K^+ Across Sarcolemma Into the Interstitial Space

The resting membrane potential (RMP) comes close to the K^+ equilibrium potential, E_K, for resting values of intracellular and extracellular [K^+] but does not completely balance the concentration dependent K^+ efflux. However, homeostasis is achieved by the Na^+/K^+-pump which, at rest, is activated to 2-6% of its maximal capacity (Clausen et al., 1987).

During muscle activity the increase in interstitial [K^+], reported in the above studies, is traditionally assumed to be caused by the action potential. More specifically, an outward flux of K^+ occurs through the delayed rectifier K^+ channels during the repolarization phase. Each action potential is associated with a net loss of K^+ which, in single frog muscle fibers,

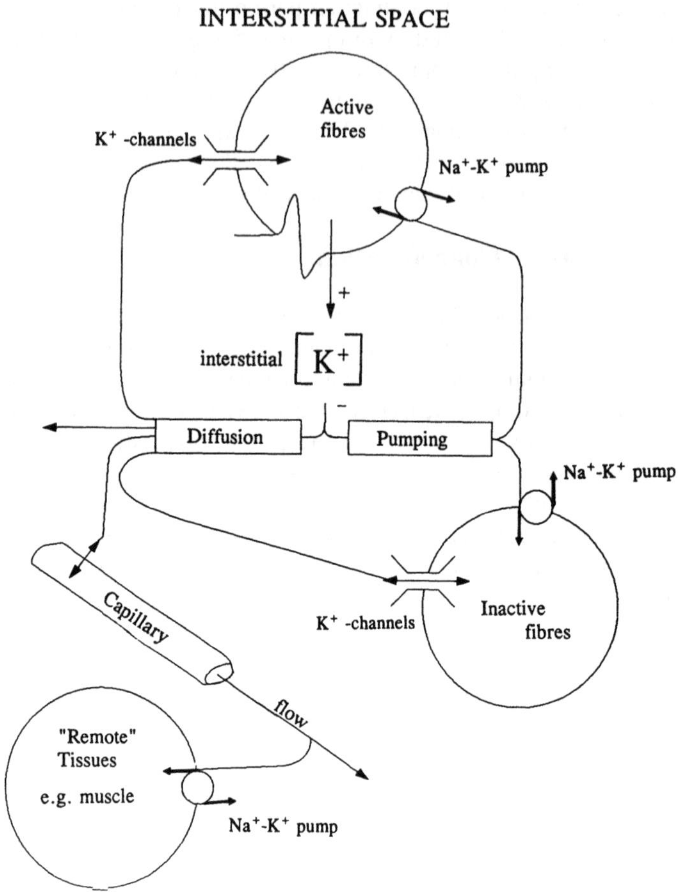

Figure 1. Possible pathways for K^+ fluxes.

was 9.6 pmol cm^{-3} at 21° C or about 5 μmol kg ww^{-1} (Hodgkin & Horowicz, 1959; see also above). Calculations of unidirectional fluxes from isotopic flux have been somewhat larger: 10.7 pmol cm^{-3} (or about 17 μmol kg ww^{-1}) in the rat diaphragm at 38° C (Creese et al., 1958) and 9.4 μmol kg ww^{-1} in the rat soleus at 30° C (Clausen & Everts, 1987). Recent evidence has been presented that, in addition to the delayed rectifier K^+ channels, the ATP-sensitive K^+ channels (K^+_{ATP}-channels) are involved in fatigue development and recovery, and that these last channels may have dual effects on K^+ fluxes (Renaud et al., 1994). During repetitive contractions the K^+_{ATP}-channels increase K^+ *efflux,* by contributing to the repolarization phase of the action potential. After the contractions cease, however, recovery of force is markedly delayed if the channels are blocked by glibenclamide, a finding which would be consistent with the channels normally allowing a net *influx* of K^+ to occur (see below). It is relevant to this interpretation that the K^+ conductance is increased 5-fold in single frog fibers which have been stimulated to exhaustion (Fink & Lüttgau, 1976) and that this increase is prevented by glibenclamide (Castle & Haylett, 1987). Also, although the K^+_{ATP}-channels have been reported not to open until ATP concentrations fall to 2 mmol l^{-1} (Spruce et al., 1985), a value well below those found in fatigue, the sensitivity of these channels would be greatly increased by a fatigue-induced rise in $[H^+]$. A further consideration is whether ATP is compartmentalized within the muscle fiber, such that the concentration close to the sarcolemma is lower than that in the remainder of the cytosol.

There are also Ca^{2+}-activated K^+ channels (K^+_{Ca}-channels) in the sarcolemma, and although some controversy still exists, recent studies have demonstrated that [Ca^{2+}] transients occur in the muscle large enough to activate K^+ efflux (Juel, 1988b). Juel has suggested that, in fatigue, these may contribute to K^+ efflux (Juel, 1988a; Juel, 1986). If so, this mechanism could account for the K^+ loss from the fibers being larger than the increase in Na^+ content, as found in a number of studies (Juel, 1986; Juel, 1988b; Hirche et al., 1980; Sjøgaard et al., 1985; Sjøgaard, 1983).

Another factor which may be important for the magnitude of changes in interstitial [K^+], and one which does not appear to have been considered previously, is the motor unit architecture of the muscle. Since the pioneering glycogen-depletion studies (Kugelberg & Edström, 1968; Brandstater & Lambert, 1973), it has been accepted that the muscle fibers of any one motor unit are normally intermingled with those of 20 or more others. One functional advantage of this arrangement is that, in submaximal contractions, it reduces interstitial [K^+] by allowing K^+ to diffuse from the vicinity of the active fibers into the spaces surrounding quiescent fibers. When the intermingling of motor units is disrupted, as in the fiber type grouping of chronic denervation, it is likely that [K^+] may rise to unusually high levels in the centers of fiber clusters.

Enhanced Pumping of K^+ Back into Fibers

Even in a quiescent muscle fiber, Na^+/K^+-pumping is required to compensate for the passive inward flux of Na^+ and associated loss of K^+ which are consequences of the permeability of the muscle plasmalemma. During exercise, however, the Na^+/K^+-pumps can increase their outputs significantly, possibly 20-fold (Clausen et al., 1987), particularly in the slow-twitch fibers (Everts et al., 1988). A potent stimulus for enhanced pumping is the rise in intracellular [Na^+] which accompanies impulse activity (Hodgkin & Horowicz, 1959; Sejersted & Hallén, 1987). A second factor is ß-adrenergic stimulation of the pump, since the enhanced activity is abolished by propranolol (Kuiack & McComas, 1992; cf. Everts et al., 1988). It is possible that the unmyelinated sympathetic axons which terminate directly on the muscle fibers (Barker & Saito, 1981) are especially important sources of norepinephrine. The varicosities of these fibers could release transmitter following both increased sympathetic nerve activity in exercise and passive depolarization by interstitial [K^+] and extracellular currents generated by the extrafusal muscle fibers. In addition, there is evidence to indicate that CGRP (calcitonin gene-related peptide), released from the motor nerve terminals, is another pump stimulant (Andersen & Clausen, 1993). Finally, the pump would be affected by epinephrine and norepinephrine released from the adrenal glands during exercise. Despite its enhanced activity, however, the Na^+/K^+-pump does not keep pace with the K^+ efflux. Some findings support the concept of a regulatory limitation of its activity (Medbø & Sejersted, 1990; Hallén et al., 1994) while others speak in favor of an insufficient capacity of the Na^+/K^+-pump when stimulation frequencies around 30 Hz are attained (Clausen & Everts, 1989).

A important consideration is that the enhanced pumping affects not only contracting fibers but quiescent ones also. Thus, in half-maximal contractions pump-induced hyperpolarizations were found which were as prominent in the relaxed fibers as in the active ones (Kuiack & McComas, 1992). There is evidence to indicate that fibers in quiescent muscles may also participate in the extra pumping since, in non-exercising limbs, the venous [K^+] may be lower than the arterial [K^+] (Sjøgaard, 1986; Lindinger et al., 1990).

Diffusion of K^+ across the Capillary Membrane

If the K^+ released from the muscle fibers was distributed only to the extracellular fluid, the plasma [K^+] would be expected to reach values close to 10 mmol l^{-1}, based on calculations of K^+ loss during several hours of intense exercise (Sjøgaard & Saltin, 1985). However, in such studies the venous effluent [K^+] never attained values higher than 5.5 mmol^{-1} indicating a limited increase in interstitial [K^+], and (v-a)[K^+] was maintained in the order of 0.2 mmol^{-1}. This finding is explained by an uptake of K^+ into inactive tissues e.g., resting muscle as mentioned above (Sjøgaard et al., 1988; Lindinger et al., 1990). Thus, a large muscle blood flow in combination with a relatively small exercising muscle mass (knee-extensors) allowed the transport to have sufficient capacity to clear K^+ from the interstitial space in the exercising muscle. During bicycle exercise, a redistribution of K^+ from exercising muscle to other compartments outside the vascular bed has also been reported (Vøllestad et al., 1994). One consequence of K^+ being accumulated remote from the exercising muscle is that it is no longer available for immediate reuptake into the fatiguing muscle fibers. Further, the recovery process may be delayed for this reason (Sjøgaard, 1990; Byström & Sjøgaard, 1991).

The significance of the K^+ wash-out by blood flow for muscle function is emphasized under conditions in which blood flow through the muscle had ceased as in steady contractions during which the intramuscular pressure exceeded the systolic blood pressure, or in the application of a tourniquet. One indication of this is the sudden recovery of the M-wave when a tourniquet is released from a previously exercised limb (McComas et al., 1993), attributable to the rapid increases in E_K and E_{RMP}. The removal of K^+ is facilitated by the hyperaemia which accompanies exercise and which may, in part, be mediated by K^+ (see below).

EFFECTS OF INCREASED INTERSTITIAL [K^+]

Changes in the interstitial [K^+] will have local effects, involving muscle fiber contractility and membrane excitability. However, reflex mechanisms may play an important role in development of fatigue (Fig. 2).

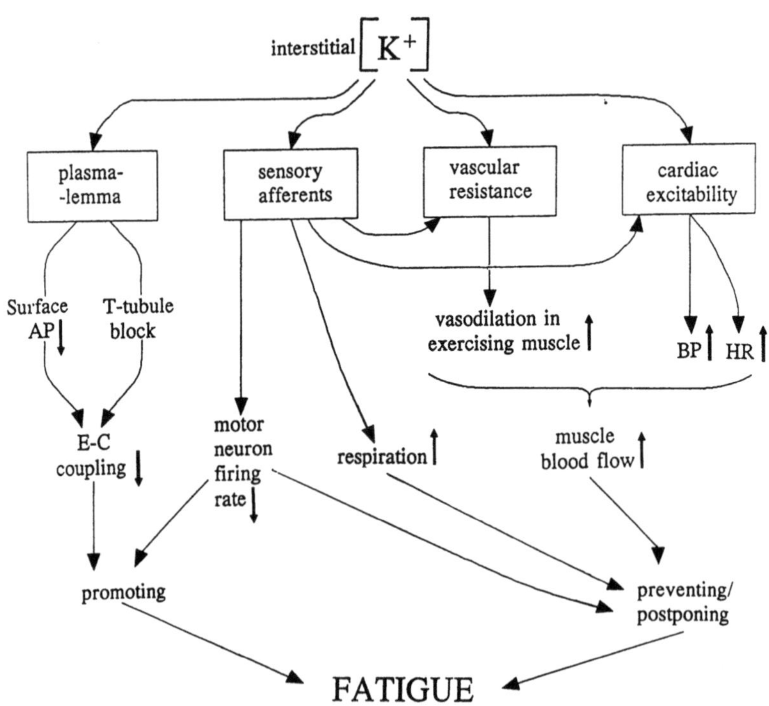

Figure 2. Mechanisms playing a role in interstitial [K^+] promoting or postponing fatigue.

Muscle Fiber Surface Plasmalemma

Increased concentrations of extracellular K^+ reduce contractile force (e.g., Clausen & Everts, 1991) and one mechanism for this effect is undoubtedly the depression of muscle fiber excitability. Thus in exercised human intercostal muscle fibers, a linear relationship has been demonstrated between the logarithm of the external $[K^+]$ and the RMP (resting membrane potential), except at low values of K^+ (Ludin, 1970). As the RMP falls, the fiber first enters a phase in which spontaneous action potentials may occur (myotonia); depolarizations greater than this cause impulse block, due to persisting inactivation of Na^+ channels. However, impulse block at the surface membrane may be prevented in muscle fibers during voluntary contractions due to the electrogenic component of the Na^+/K^+-pump.

Increased Electrogenic Component of the Na^+/K^+-Pump

As discussed above the Na^+/K^+-pump transports Na^+ and K^+ actively against their concentration gradients and is ultimately responsible for sustaining the entire RMP, by offsetting the ionic fluxes secondary to Na^+ leakage. However, it is also directly responsible for a small but significant fraction of the RMP. Thus, the observed RMP exceeds that calculated from the Goldman-Hodgkin-Katz equation by some -5 mV to -8 mV (Hicks & McComas, 1989). During maximal stimulated contractions the electrogenic contribution of the Na^+/K^+-pump can be as much as -30 mV and may even increase the membrane potential above its quiescent value. The hyperpolarization can be detected with intracellular microelectrodes in animal muscles (Hicks & McComas, 1989) and, in intact human muscles, is responsible for the enlargement of the M-wave (muscle compound action potential) evoked by maximal motor nerve stimulation, both during and following exercise (McComas et al., 1994). The extent of the hyperpolarization, as reflected in the size of the M-wave, is greater in some muscles than others; of the muscles examined, it is especially prominent in the biceps brachii, while animal studies suggest that fast-twitch motor units are more affected than slow-twitch (Enoka et al., 1992). Regardless of whether or not the M-wave is enlarged, the enhanced Na^+/K^+-pumping is capable of maintaining muscle fiber excitability for at least one minute of maximal contractions, even if the circulation is arrested (Fitch & McComas, 1985). The level of pumping, as reflected in the potentiation of the M-wave, appears to be proportional to the frequency of muscle excitation (Cupido, Galea & McComas, unpublished observations).

The electrogenic contribution of the Na/K^+-pump to the RMP may cause the RMP to become larger than E_K (the K^+ equilibrium potential), in which case a net K^+ flux back into the fiber will occur. Since the size of the flux will be proportional to the difference between RMP and E_K, the passive inward flux of K^+ will be especially large during exercise, because of the large electrogenic component of the Na/K^+-pump (see above). Early in exercise, when the membrane is hyperpolarized, the flux will be increased by opening of the inward rectifier channels (Adrian et al., 1969).

As fatigue sets in and the RMP begins to fall, those channels which are normally inhibited by ATP (K^+_{ATP}-channels) may start to open, again helping to keep the inward K^+ flux high in between action potentials and during recovery (see above and Renaud et al., 1994). Of particular importance is the sensitivity of these channels to H^+, such that a fall in intracellular pH, from 7.2 to 6.3, decreases the affinity of the channels for ATP 15-fold (Davies et al., 1992). Likewise, the K^+_{Ca}-channels would be expected to become increasingly important in fatigue, aiding in the net inward K^+ flux as long as the RMP exceeds E_K. When the hydrolysis of ATP is no longer able to drive the Na^+/K^+-pump sufficiently, or if the latter is inhibited by oxidative stress, there would be a net outward K^+ flux; this would occur because the RMP would fall below E_K, secondary to the small permeability of the plasmalemma for Na^+. It is questionable whether or not this stage is reached during voluntary contractions; the fact that M-wave is quite well preserved in some muscles during exercise (Bigland-Ritchie et al., 1982) would suggest that the Na^+/K^+-pump can maintain the RMP with only a modest reduction.

In contrast to the possible effects of K^+ on the surface plasmalemma, it is likely that the function in the T-tubules may be seriously jeopardized by a rising [K^+] (see below).

T-Tubule Plasmalemma

The situation regarding the T-tubules may be rather different to that of the inter-fiber space, since the very narrow lumens of the tubules restrict diffusion of K^+ (Hodgkin & Horowicz, 1959) and the K^+ permeability of the T-tubular membrane is approximately twice that of the surface membrane (Eisenberg & Gage, 1969). If the inter-fiber [K^+] can rise to 10-15 mM in an isometric contraction (Vyskocil et al., 1983), then the T-tubular [K^+] must be higher still. Further, the density of Na^+/K^+-pumps is lower in the T-tubules than at the fiber surface (Fambrough et al., 1987), so that the compensatory effects of pumping (see above) will be less evident. These factors could combine to block impulse propagation down the T-tubules from the surface plasmalemma, so that the myofibrils would become increasingly dependent on electrotonic spread of the surface action potential down the low-resistance T-tubular pathway. In unfatigued fibers of amphibians the size of the surface signal is just sufficient to activate the innermost myofibrils in the fiber by this mechanism (Adrian et al., 1969). However, were the surface potential to decline, as it does later in fatigue, the excitation of the central myofibrils would become increasingly precarious. Such a conclusion would be consistent with the results of Westerblad et al. in which progressively weaker Ca^{2+} transients occurred towards the centers of fatigued single fibers (Westerblad et al., 1990). In conclusion, then, the more important effects of a raised extracellular [K^+] are likely to involve excitation-contraction coupling via the T-tubules, rather than the excitability of the surface plasmalemma.

Cardiovascular and Respiratory System

When a subject begins to exercise, a number of physiological events take place in the cardiovascular system - the pulse quickens, blood pressure rises and, if the contractions are intermittent, blood flow increases through the muscles (Kiens et al., 1989; Saltin et al., 1987). Changes in heart rate and blood pressure can be mediated by central command as well as reflexes, the latter of which are elicited from afferent fibers originating in the active muscles (Goodwin et al., 1972). These fibers are either thinly myelinated (group III) or unmyelinated (group IV), and their free nerve endings must be responsive to some property of the contracting muscle - either deformation of the investing connective tissue or a change in the metabolic milieu of the muscle fibers (Kniffki et al., 1981). K^+ is one of several candidates for exciting the metabolically-sensitive fibers, and has been shown capable of activating group III and group IV muscle afferents, albeit transiently (Rybicki et al., 1985). Evidence in favor of such a role for K^+ in exercise includes the close parallelism between the relative magnitudes and time-courses of blood pressure and [K^+] respectively (Fallentin et al., 1992; Saltin et al., 1981).

Similar arguments to the above suggest that there is also a casual relationship between [K^+] and increased blood flow. For example, if Doppler ultrasonography is used to measure arterial blood velocity, and hence flow, then the changes in [K^+] and blood flow vary together during exercise (Kiens et al., 1989) although they may be dissociated in the recovery period (Jensen et al., 1993). In addition, the arterial perfusion of hyperkalaemic solutions produces relaxation of the arterial smooth muscle (Dawes, 1941). These observations do not, of course, deny the possibility that other factors may contribute to the hyperaemia of exercise (e.g., reduced pO_2, ATP, inorganic phosphate, and prostaglandins).

A third effect of K^+ on the cardiovascular system involves the heart. It is well known that intravascular injections of K^+ can produce cardiac arrest and use of this is made in cardiac surgery. In intensive exercise, such as treadmill running or cycle ergometry, arterial [K^+] concentrations as high as 8.2 mM have been found (Medbø & Sejersted, 1990) and would be expected to affect cardiac excitability, as indicated by the ECG. In patients with hyperkalaemia, plasma [K^+] values of 8 mmol l^{-1} or greater produce atrial depression and AV block, while 10 mmol l^{-1} [K^+] can cause ventricular arrhythmias, including fibrillation (Lipman et al., 1984). Taken together, these observations draw attention to the dangers of sudden intensive exercise in untrained subjects and may explain some cases of exercise-induced cardiac arrest, in the absence of occlusive arteriopathy.

Finally, it is known that K^+ stimulates the respiratory system also, increasing the ventilation rate (see, for example, Paterson et al., 1990). This effect of K^+ depends on a reflex in which the afferent fibers are excited by K^+-sensitive cells in the carotid body (Band & Linton, 1986).

Motor Control

There is now circumstantial evidence that the reduction in motoneuron firing rate, which occurs during fatiguing contractions, may have a reflex origin (see Bigland-Ritchie et al., Chapter 27; Bigland-Ritchie et al., 1986; Woods et al., 1987). Although the afferent limb of the reflex is thought to involve Group III and Group IV axons, the identities of the exciting stimuli remain uncertain; it is quite likely, however, that interstitial [K^+] is one factor, possibly activating both metaboreceptor and nociceptor fibers. This reduction in motoneuron activation is a process for blunting the rise in interstitial [K^+], such that the discharge rate declines and there are fewer releases of K^+ per unit time, which will prevent or postpone fatigue. In contrast to this an hypothesis has been suggested (Johansson & Sojka, 1991) that increased interstitial [K^+] may elicit a vicious circle via the gamma-loop. Evidence has been presented that activity in the group III but also group IV sensory afferents may have excitatory effects, especially on static but also on dynamic gamma-motoneurons. These effects are strong enough to increase firing in the muscle spindle afferents, which in turn will raise the activation level in the pool of alpha-motoneurons projecting to the extrafusal muscle fibers. This reflex mediated muscle activity leads to further K^+ fluxes and increased metabolism.

For as long as interstitial [K^+] is elevated an activity-induced increased cytosolic [Ca^{2+}] will persist according to the above hypothesis. This will be enhanced by the interstitial [K^+] per se inducing a Ca^{2+} uptake into the muscle fiber (Everts et al., 1993; Barnes, 1993). Elevated cytosolic [Ca^{2+}] will increase the respiration of the cell causing an even faster metabolic exhaustion (Barnes, 1993). Most critical to the muscle cell is that a maintained increased intracellular [Ca^{2+}] will induce an overload of the mitochondria causing further disruption of metabolism, and will also breakdown the cell membrane (Jackson et al., 1984). To prevent this catastrophe the inactivation of the cell membrane comprises a perfect mechanism. Thus muscle fatigue and safety mechanisms may be interrelated via interstitial [K^+].

ACKNOWLEDGMENTS

The authors acknowledge the research support of the Medical Research Council of Denmark (*G.S.*), and the Natural Sciences and Engineering Research Council (NSERC) of Canada (*A.J.McC.*). Attendance of the authors at the 1994 Bigland-Ritchie conference was supported, in part, by the National Institute of Occupational Health, Denmark (*G.S.*), NSERC (*A.J.McC.*), the United States Public Health Service (National Center for Medical Rehabilitation Research, National Institute of Child Health and Human Development; *G.S.*), and the Muscular Dystrophy Association (USA; *G.S.*).

REFERENCES

Adrian RH, Costantin LL & Peachey LD (1969). Radial spread of contraction in frog muscle fibres. *Journal of Physiology (London)* **204**, 231-257.

Andersen SLV & Clausen T (1993). Calcitonin gene-related peptide stimulates active Na^+-K^+ transport in rat soleus muscle. *American Journal of Physiology* **264**, C419-C429.

Aukland K & Reed RK (1993). Interstitial-lymphatic mechanisms in the control of extracellular fluid volume. *Physiological Reviews* **73**, 1-77.

Band DM & Linton RAF (1986). The effect of potassium on carotid body chemoreceptor discharge in the anaesthetized cat. *Journal of Physiology (London)* **381**, 39-47.

Barcroft H & Millen JLE (1939). The blood flow through muscle during sustained contraction. *Journal of Physiology (London)* **97**, 17-31.

Barker D & Saito M (1981). Autonomic innervation of receptors and muscle fibres in cat skeletal muscle. *Proceedings of the Royal Society of London, Series B-Biological Sciences* **B212**, 317-332.

Barnes WS (1993). Effects of Ca^{2+}-channel drugs on K^+-induced respiration in skeletal muscle. *Medicine and Science in Sports and Exercise* **25**, 473-478.

Bergström J, Guarnieri G & Hultman E (1971). Carbohydrate metabolism and electrolyte changes in human muscle tissue during heavy work. *Journal of Applied Physiology* **30**, 122-125.

Bergström J, Guarnieri G & Hultman E (1973). Changes in muscle water and electrolytes during exercise. In: Keul J. (ed.), *Limiting Factors of Physical Performance*, pp. 173-178. Stuttgart: Georg Thieme.

Bigland-Ritchie BR, Dawson NJ, Johansson RS & Lippold OCJ (1986). Reflex origin for the slowing of motoneurone firing rates in fatigue of human voluntary contractions. *Journal of Physiology (London)* **379**, 451-459.

Bigland-Ritchie B, Kukulka CG, Lippold OCJ & Woods JJ (1982). The absence of neuromuscular transmission failure in sustained maximal voluntary contractions. *Journal of Physiology (London)* **330**, 265-278.

Brandstater ME & Lambert EH (1973). Motor unit anatomy. In Desmedt JE (ed.), *New Developments in Electromyography and Clinical Neurophysiology*, pp. 14-22. Basel: Karger

Byström S & Sjøgaard G (1991). Potassium homeostasis during and following exhaustive submaximal static handgrip contractions. *Acta Physiologica Scandinavica* **142**, 59-66.

Castle NA & Haylett DG (1987). Effect of channel blockers on potassium efflux from metabolically exhausted frog skeletal muscle. *Journal of Physiology (London)* **383**, 31-43.

Clausen T & Everts ME (1987). Is the Na, K-pump capacity in skeletal muscle inadequate during sustained work? In: *Proceedings of the Vth International Conference on Na, K-ATPase. Århus, Denmark. June 14-19*, New York: Alan Liss Inc.

Clausen T & Everts ME (1989). Regulation of the Na, K-pump in skeletal muscle. *Kidney International* **35**, 1-13.

Clausen T & Everts ME (1991). K^+-induced inhibition of contractile force in rat skeletal muscle: role of active Na^+-K^+ transport. *American Journal of Physiology* **261**, C799-C807.

Clausen T, Everts ME & Kjeldsen K (1987). Quantification of the maximum capacity for active sodium-potassium transport in rat skeletal muscle. *Journal of Physiology (London)* **388**, 163-181.

Creese R, Hashish S & Scholes NW (1958). Potassium movements in contracting diaphragm muscle. *Journal of Physiology (London)* **143**, 307-324.

Crone C, Frøkjær-Jensen J, Friedman JJ & Christensen O (1978). The permeability of single capillaries to potassium ions. *Journal of General Physiology* **71**, 195-220.

Davies NW, Standen NB & Stanfield PR (1992). The effect of intracellular pH on ATP-dependent potassium channels of frog skeletal muscle. *Journal of Physiology (London)* **445**, 549-568.

Dawes GS (1941). The vaso-dilator action of potassium. *Journal of Physiology (London)* **99**, 224-238.

De Lanne R, Barnes JR & Brouha L. (1959). Changes in osmotic pressure and ionic concentrations of plasma during muscular work and recovery. *Journal of Applied Physiology* **14**, 804-808.

Eisenberg RS & Gage PW (1969). Ionic conductances of the surface and transverse tubular membranes of frog sartorius fibres. *Journal of General Physiology* **53**, 279-297.

Enoka RM, Trayanova N, Laouris Y, Bevan L, Reinking RM & Stuart DG (1992). Fatigue-related changes in motor unit action potentials of adult cats. *Muscle & Nerve* **14**, 138-150.

Everts ME, Lømo T & Clausen T (1993). Changes in K^+, Na^+ and calcium contents during in vivo stimulation of rat skeletal muscle. *Acta Physiologica Scandinavica* **147**, 357-368.

Everts ME, Retterstøl K & Clausen T (1988). Effects of adrenaline on excitation-induced stimulation of the sodium-potassium pump in rat skeletal muscle. *Acta Physiologica Scandinavica* **134**, 189-198.

Fallentin N, Jensen BR, Byström S & Sjøgaard G (1992). Role of potassium in the reflex regulation of blood pressure during static exercise in the human. *Journal of Physiology (London)* **451**, 643-651.

Fambrough DM, Wolitzky BA, Tamkim MM & Takeyasu K (1987). Regulation of the sodium pump in excitable cells. *Kidney International* **32** (suppl. 23), S97-S112.

Fenn WO (1937). Loss of potassium in voluntary contraction. *American Journal of Physiology* **120**, 675-680.

Fenn WO (1938). Factors affecting the loss of potassium from stimulated muscles. *American Journal of Physiology* **124**, 213-229.

Fenn WO & Cobb DM (1936). Electrolyte changes in muscle during activity. *American Journal of Physiology* **115**, 345-356.

Fink R & Lüttgau HC (1976). An evaluation of the membrane constants and the potassium conductance in metabolically exhausted muscle fibres. *Journal of Physiology (London)* **263**, 215-238.

Fitch S & McComas A (1985). Influence of human muscle length on fatigue. *Journal of Physiology (London)* **362**, 205-213.

Goodwin GM, McCloskey DI & Mitchell JH (1972). Cardiovascular and respiratory responses to changes in central command during isometric exercise at constant muscle tension. *Journal of Physiology (London)* **226**, 173-190.

Hallén J, Gullestad L & Sejersted OM (1994). K^+ shifts of skeletal muscle during stepwise bicycle exercise with and without ß-adrenoceptor blockade. *Journal of Physiology (London)* **477**, 149-159.

Hazeyama Y & Sparks HV (1979). A model of potassium ion efflux during exercise of skeletal muscle. *American Journal of Physiology* **236**, R83-R90.

Hicks A & McComas AJ (1989). Increased sodium pump activity following repetitive stimulation of rat soleus muscles. *Journal of Physiology (London)* **414**, 337-349.

Hirche H, Schumacher E & Hagemann H (1980). Extracellular K^+ concentration and K^+ balance of the gastrocnemius muscle of the dog during exercise. *Pflügers Archiv* **387**, 231-237.

Hodgkin AL & Horowicz P (1959). The influence of potassium and chloride ions on the membrane potential of single muscle fibres. *Journal of Physiology (London)* **148**, 127-160.

Jackson MJ, Jones DA & Edwards RHT (1984). Experimental skeletal muscle damage: The nature of the calcium-activated degenerative processes. *European Journal of Clinical Investigation* **14**, 369-374.

Jensen BR, Fallentin N, Sjøgaard G & Byström S (1993). Plasma potassium concentration and Doppler blood flow during and following submaximal handgrip contractions. *Acta Physiologica Scandinavica* **147**, 203-211.

Johansson H & Sojka P (1991). Pathophysiological mechanisms involved in genesis and spread of muscular tension in occupational muscle pain and in chronic musculoskeletal pain syndromes: A hypothesis. *Medical Hypotheses* **35**, 196-203.

Juel C (1986). Potassium and sodium shifts during in vitro isometric muscle contraction, and the time course of the ion-gradient recovery. *Pflügers Archiv* **406**, 458-463.

Juel C (1988a). Is a Ca^{2+}-dependent K^+ channel involved in the K^+ loss from active muscles? *Acta Physiologica Scandinavica* **132**, P26.

Juel C. (1988b). The effect of beta2-adrenoceptor activation on ion-shifts and fatigue in mouse soleus muscles stimmulated in vitro. *Acta Physiologica Scandinavica* **134**, 209-216.

Juel C, Bangsbo J, Graham T & Saltin B (1990). Lactate and potassium fluxes from human skeletal muscle during and after intense, dynamic, knee extensor exercise. *Acta Physiologica Scandinavica* **140**, 147-159.

Kiens B, Saltin B, Walløe L & Wesche J (1989). Temporal relationship between blood flow changes and release of ions and metabolites from muscle upon single weak contractions. *Acta Physiologica Scandinavica* **136**, 551-559.

Kniffki KD, Mense S & Schmidt RF (1981). Muscle receptors with fine afferent fibres which may evoke circulatory reflexes. *Circulation Research* **48** (suppl I), 25-31.

Kugelberg E & Edström L (1968). Differential histochemical effects of muscle contraction on phosphorylase and glycogen in various types of fibres: relation to fatigue. *Journal of Neurology, Neurosurgery and Psychiatry* **31**, 415-423.

Kuiack S & McComas AJ (1992). Transient hyperpolarization of non-contracting muscle fibres in anaesthetized rats. *Journal of Physiology (London)* **454**, 609-618.

Lindinger MI, Heigenhauser GJF, McKelvie RS & Jones NL (1990). Role of nonworking muscle on blood metabolites and ions with intense intermittent exercise. *American Journal of Physiology* **258**, R1486-R1494.

Lipman BS, Dunn M & Massie E (1984). *Clinical Electrocardiography*, pp. 268-271. Chicago: Year Book Medical Publishers Inc.

Ludin HP (1970). Microelectrode study of dystrophic human skeletal muscle. *European Neurology* **3**, 116-121.

McComas AJ, Galea V & Einhorn RW (1994). Pseudofacilitation: a misleading term. *Muscle & Nerve* **17**, 599-607.

McComas AJ, Galea V, Einhorn RW, Hicks AL & Kuiack S (1993). The role of the Na^+, K^+-pump in delaying muscle fatigue. In: Sargeant AJ, Kernell D (eds.), *Neuromuscular Fatigue*, pp. 35-43. Amsterdam: Royal Netherlands Academy of Arts and Sciences.

Medbø JI & Sejersted OM (1985). Acid-base and electrolyte balance after exhausting exercise in endurance-trained and sprint-trained subjects. *Acta Physiologica Scandinavica* **125**, 97-109.

Medbø JI & Sejersted OM (1990). Plasma potassium changes with high intensity exercise. *Journal of Physiology (London)* **421**, 105-122.

Paterson DJ, Friedland JS, Bascom DA, Clement ID, Cunningham DA, Painter R & Robbins PA (1990). Changes in arterial K^+ and ventilation during exercise in normal subjects and subjects with McArdle's syndrome. *Journal of Physiology (London)* **429**, 339-348.

Renaud JM, Light PE & Comtois AS (1994). The effect of glibenclamide on frog skeletal muscle: evidence for $K^+_{(ATP)}$ channel activation during fatigue. *Abstracts, Ontario Exercise Physiology Meeting*, (Toronto, Canada, February 1994).

Rybicki KJ, Waldrop TG & Kaufman MP (1985). Increasing gracilis muscle interstitial potassium concentrations stimulate group III and IV afferents. *Journal of Applied Physiology* **58**, 936-941.

Saltin B, Sjøgaard G, Gaffney FA & Rowell LB (1981). Potassium, lactate, and water fluxes in human quadriceps muscle during static contractions. *Circulation Research* **48** (suppl I), I-18-I-24.

Saltin B, Sjøgaard G, Strange S & Juel C (1987). Redistribution of K^+ in the human body during muscular exercise; its role to maintain whole body homeostasis. In: Shiraki K, Yousef MK (eds.), *Man in Stressful Environments. Thermal and Work Physiology*, pp. 247-267. Springfield, IL: CC Thomas.

Sejersted OM & Hallén J (1987). Na, K homeostasis of skeletal muscle during activation. In: Marconnet P, Komi P (eds.), *Muscle Function in Exercise and Training. Medicine and Science in Sports*, vol 26, pp. 1-11. Basel: Karger.

Sejersted OM, Medbø JI, Orheim A & Hermansen L (1984). Relationship between acid-base status and electrolyte balance after maximal work of short duration. In: Marconnet P, Poortmans JR, Hermansen L. (eds.), *Physiological Chemistry of Training and Detraining. Medicine and Science in Sports*, vol 17, pp. 40-55. Basel: Karger.

Sjøgaard G (1983). Electrolytes in slow and fast muscle fibers of humans at rest and with dynamic exercise. *American Journal of Physiology* **245**, R25-R31.

Sjøgaard G (1986). Water and electrolyte fluxes during exercise and their relation to muscle fatigue. *Acta Physiologica Scandinavica* **128** (suppl 556), 129-136.

Sjøgaard G (1990). Exercise-induced muscle fatigue: The significance of potassium. *Acta Physiologica Scandinavica* **140** (suppl. 593), 1-64.

Sjøgaard G (1991). Role of exercise-induced potassium fluxes underlying muscle fatigue: a brief review. *Canadian Journal of Physiology and Pharmacology* **69**, 238-245.

Sjøgaard G, Adams RP & Saltin B (1985). Water and ion shifts in skeletal muscle of humans with intense dynamic knee extension. *American Journal of Physiology* **248**, R190-R196.

Sjøgaard G & Saltin B (1985). Potassium redistribution within the body during exercise. *Clinical Physiology* **5** (suppl 4), 150.

Sjøgaard G, Savard G & Juel C (1988). Muscle blood flow during isometric activity and its relation to muscle fatigue. *European Journal of Applied Physiology and Occupational Physiology* **57**, 327-335.

Spruce AE, Standen NB & Stanfield PR (1985). Voltage-dependent ATP-sensitive potassium channels of skeletal muscle membrane. *Nature* **316**, 736-738.

Thomas CK, Bigland-Ritchie B & Johansson RS (1991). Force-frequency relationships of human thenar motor units. *Journal of Neurophysiology* **65**, 1509-1516.

Vyskocil F, Hnik P, Rehfeldt H, Vejsada R & Ujec E (1983). The measurement of Ke^+ concentration changes in human muscles during volitional contractions. *Pflügers Archiv* **399**, 235-237.

Vøllestad NK, Hallén J & Sejersted OM (1994). Effect of exercise intensity on potassium balance in muscle and blood of man. *Journal of Physiology (London)* **475**, 359-368.

Watson PD, Garner RP & Ward DS (1993). Water uptake in stimulated cat skeletal muscle. *American Journal of Physiology* **264**, R790-R796.

Westerblad H, Lee JA, Lamb AG, Bolsover SR & Allen DG (1990). Spatial gradients of intracellular calcium in skeletal muscle during fatigue. *Pflügers Archiv* **415**, 734-740.

Woods JJ, Furbush F & Bigland-Ritchie B (1987). Evidence for a fatigue-induced reflex inhibition of motoneuron firing rates. *Journal of Neurophysiology* **58**, 125-137.

SECTION II

Fatigue at the Neuromuscular Junction

In this section, axonal branch points, the neuromuscular junction, and the sarcolemma are considered as possible limiting sites in fatigue. Usually, the safety margins for propagation and transmission are sufficiently high at all three sites to ensure that adequate excitation reaches the t-tubules. Some controversy exists about the extent to which significant failure of neuromuscular transmission occurs during *normal* human muscle fatigue However, no doubt exists about the involvement of this *site* in pathological conditions such as myasthenia gravis. Three approaches to the study of neuromuscular transmission during fatiguing contractions arc prcscntcd.

In **Chapter 5,** Sieck and Prakash review both the pre- and postsynaptic mechanisms involved in normal neuromuscular transmission. In addition, they present data on the adaptations at these sites. Presynaptically, neuromuscular transmission failure may result from the inability of the axonal action potential to propagate into all its terminal branches, and from reductions in the frequency and size of quantal release. Failure of branch-point propagation is less well studied experimentally, but there is indirect evidence that it occurs, with a rise in interstitial potassium as a contributing cause. Postsynaptically, failure may occur because of a reduction in the number of acetylcholine receptors or because of an impairment of sarcolemmal excitability.

The likelihood that the sarcolemmal action potential fails to activate the muscle fiber fully is tackled in **Chapter 6** (Fuglevand). It is argued that for mammalian fast-twitch fibers in particular, the reduction in action potential amplitude during fatigue may mean that membrane depolarization is insufficient for full activation and that this mechanism would contribute to the decline in force. Given the known changes in action potential size (and area) produced by nerve stimulation in some studies of human fatigue, this remains a possibility. However, because it is difficult to maintain a supramaximal stimulus intensity for the requisite motor axons in human studies during vigorous exercise, a reduction in evoked action potential amplitude must be considered critically.

Chapter 7 (Trontelj & Stålberg) introduces the technique of single-fiber EMG which has been a powerful clinical tool in diagnosis of disorders at the neuromuscular junction. Rather than studying only motor units that can be recruited voluntarily at low forces, it is now possible to examine the responses of a wide range of single muscle fibers to axonal stimulation. This has increased the scope of the method as it permits the behavior of muscle fibers to be studied over a range of stimulus frequencies. Most end-plates in normal subjects have relatively stable jitter (i.e. the time difference between muscle action potentials in fibers from the *same* motor unit), although the jitter increases in some at stimulus frequencies greater than 15-20 Hz. This method allows the effective safety factor at the neuromuscular junction to be assessed and compared with that observed in pathological conditions. Under ischemic conditions, which occur during strong isometric contractions, jitter increases and, after 2000-4000 discharges some end-plates will eventually develop a conduction block and fatigue will increase.

These chapters highlight different aspects of neuromuscular transmission: force production can be jeopardized at one or more of the steps required for effective activation of muscle fibers. However, under *most* physiological conditions, these steps are designed with comfortable safely margins. Under some *in vivo* circumstances, perhaps for some motor units when motoneuron discharge rates are high and blood flow is impaired, fatigue might be enhanced through failure of neuromuscular transmission. Such circumstances are interesting because they provide insight into the design limits of the neuromuscular junction.

5

FATIGUE AT THE NEUROMUSCULAR JUNCTION

Branch Point *vs.* Presynaptic *vs.* Postsynaptic Mechanisms

G. C. Sieck and Y. S. Prakash

Departments of Anesthesiology and Physiology and Biophysics
Mayo Clinic and Foundation
Rochester, Minnesota 55905

ABSTRACT

There are several pre and postsynaptic sites where neuromuscular transmission failure (NTF) can occur, leading to peripheral muscle fatigue. Presynaptic sites of NTF include: axonal branch point conduction block; a failure of excitation-secretion coupling at the presynaptic terminal; reductions in quantal release of ACh; and reductions in quantal size. Postsynaptic sites of NTF include: cholinergic receptor desensitization; and reduced sarcolemmal excitability. Susceptibility to NTF increases with stimulation frequency and is most prevalent in fatigable fast-twitch motor units. In addition, susceptibility to NTF varies with age and with conditions of altered use.

INTRODUCTION

Muscle fatigue can be either central (failure of excitation of motoneurons) or peripheral (failure in the transmission of the neural signal, or a failure of the muscle to respond to neural excitation) in origin (Bigland-Ritchie et al., 1978, 1982). At the peripheral level, several pre- and postsynaptic mechanisms and sites are potentially implicated, including: a failure of action potential propagation along the axon; inadequate presynaptic release of acetylcholine (ACh); insufficient depolarization of the postsynaptic membrane; failure of action potential propagation along the sarcolemma; and, failure of excitation-contraction coupling. This review addresses possible mechanisms of peripheral fatigue that are proximal to a failure of excitation-contraction coupling and that can be generally categorized as neuromuscular transmission failure (NTF).

NEUROMUSCULAR TRANSMISSION FAILURE

The fact that NTF can occur under certain conditions is indisputable, but its contribution to muscle fatigue during normal motor behaviors remains controversial (e.g., Merton,

Fatigue, Edited by Simon C. Gandevia et al.
Plenum Press, New York, 1995

1954; Stephens & Taylor, 1972; Bigland-Ritchie et al., 1982). NTF failure has been detected using two general techniques: 1) comparison of the forces elicited by indirect nerve versus direct muscle stimulation; and 2) assessment of evoked endplate potentials (EPPS) or compound muscle action potentials (e.g., M waves). Both measurements have been used extensively to assess the incidence of NTF.

Comparison of Nerve versus Muscle Stimulation

The relative contribution of NTF to muscle fatigue can be estimated by comparing force loss during nerve stimulation to that elicited by direct muscle stimulation (thus bypassing the neural sites of failure) (Fig. 1). Using this procedure, the relative contributions of NTF and contractile failure to muscle fatigue can be partitioned, and estimated based on the following formula (Aldrich et al., 1986):

$$NTF = (F - MF)/(1 - MF)$$

where *NTF* is the relative contribution of NTF to fatigue, *F* is force loss during repetitive nerve stimulation and *MF* is force loss during direct muscle stimulation.

The relative contribution of NTF to muscle fatigue depends on the frequency of stimulation (Thesleff, 1959; Aldrich et al., 1986; Kuei et al., 1990; Johnson & Sieck, 1993), and this frequency dependence of NTF may explain the decrement in motoneuron discharge rate that occurs with sustained voluntary contractions (Bigland-Ritchie et al., 1979; Bigland-Ritchie & Lippold, 1979; Bigland-Ritchie, 1984; Woods et al., 1987). By decreasing motoneuron discharge rate, susceptibility to NTF can be reduced. Such a reduction in motoneuron discharge rate may be reflexive, emanating from small-diameter afferents within the muscle (Woods et al., 1987), and/or intrinsic to the motoneuron (Kernell, 1965; Kernell & Monster, 1982a; 1982b)(for further discussion of this issue, see Binder-Macleod, Chapters 16; Windhorst & Boorman, Chapter 17; Hagbarth & Macefield, Chapter 18; and Garland & Kaufman, Chapter 19).

Fiber type composition of a muscle is also an important determinant of the extent of NTF. Kugelberg and Lindegren (1979) suggested that muscles with a greater proportion of fast-twitch, glycolytic fibers may be more susceptible to NTF. In support, we used an *in vitro* nerve-diaphragm muscle preparation and the technique of glycogen depletion to demonstrate that type IIb/IIx fibers were the most susceptible to NTF, especially at higher rates of

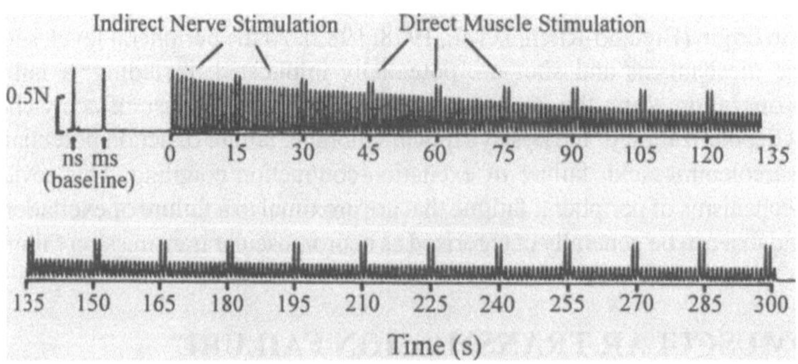

Figure 1. The relative contribution of NTF to muscle fatigue can be estimated by periodically superimposing direct muscle stimulation (ms) onto repetitive nerve stimulation (ns). In this case, the adult rat diaphragm was activated by stimulation of the phrenic nerve at 40 Hz in 330 ms duration trains repeated each s for 5 min. Every 15 s, direct muscle stimulation was superimposed. The difference between the forces generated by indirect nerve *vs.* direct muscle stimulation provides an index of NTF.

Table 1. Assessment of NTF in type-identified fibers of rat diaphragm by comparison of the extent of glycogen depletion

	10 Hz				75 Hz			
	Nerve		Muscle		Nerve		Muscle	
	% [Glycogen] ↓	# Depleted fibers	% [Glycogen] ↓	# Depleted fibers	% [Glycogen] ↓	# Depleted fibers	% [Glycogen] ↓	# Depleted fibers
I	38±5	86±5	53±6	95±2	38±6	80±9	67±5‡	98±1
IIa	46±5	84±5	59±6	89±7	36±4‡	67±10‡	67±5	96±2
IIb	51±6*	95±3*†	70±4*†	97±3	30±6‡	50±10*‡	80±4*†‡	99±1

%[Glycogen]↓: Decrement in glycogen (% change in optical density from control); #Depleted Fibers: Proportion (% of total) of "depleted" fibers with glycogen levels 2 SD lower than control values. All values are means ± SEM. * indicates significant difference ($p < 0.05$) from type I fibers; † indicates significant ($p<0.05$) difference from type IIa fibers; and ‡ indicates significant difference ($p<0.05$) from 10 Hz stimulation.

stimulation (Johnson & Sieck, 1993). The underlying premise was that during repetitive nerve stimulation, those fibers that are more susceptible to NTF will not be activated and therefore, will not display the depletion of glycogen stores resulting from exhaustive activation. During repetitive direct muscle stimulation, type IIb/IIx fibers displayed substantial and rapid depletion of their glycogen stores. In contrast, during repetitive nerve stimulation, type IIb/IIx fibers showed far less glycogen depletion. Differences in the depletion of glycogen stores between direct muscle and indirect nerve stimulation for type I and IIa fibers were much less pronounced (Table 1).

Susceptibility to NTF can also adapt under a variety of conditions of postnatal development and altered muscle use. For example, using an *in vitro* nerve-diaphragm muscle preparation in the rat, we demonstrated that adaptations in the extent of NTF occur during early postnatal development (Fournier et al., 1991). At all stimulation frequencies, NTF in the neonatal diaphragm muscle is much more pronounced than in the adult. As a result, in the neonate, diaphragm muscle fatigue induced by repetitive nerve stimulation (at 40 Hz) is more pronounced than that induced by direct muscle stimulation (Fig. 2). Using a similar procedure, Feldman and colleagues (1991) also found that the neonatal rat diaphragm is more susceptible to NTF than the adult.

Figure 2. Fatigue of the rat diaphragm was induced by either repetitive direct muscle simulation or indirect phrenic nerve stimulation at 40 Hz in 330 ms duration trains repeated each second for a 2-min period. A fatigue index was calculated as a ratio of the force generated after 2 min to the initial force. In neonates (0-7 days), fatigue induced by nerve stimulation was far more pronounced than that induced by direct muscle stimulation, indicating a greater contribution of NTF. This difference was less striking in the 21 day old and absent in the adult. (* indicates significant difference ($p<0.05$) between nerve and muscle stimulation).

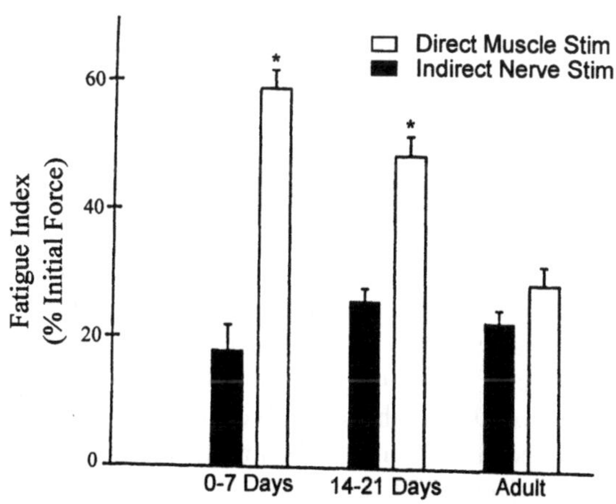

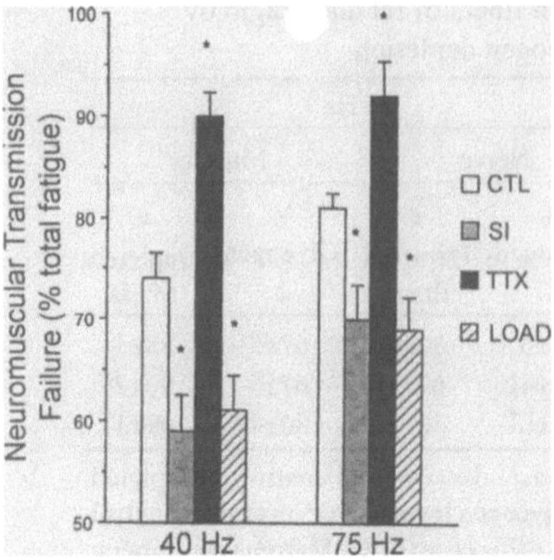

Figure 3. The relative contribution of NTF to total fatigue of the rat diaphragm muscle was estimated using the technique shown in Fig. 1. Following 2 weeks of inactivity induced by cervical (C_2) spinal cord hemisection, the relative contribution of NTF decreased. In contrast, after 2 weeks of inactivity induced by TTX blockade, NTF increased. In both cases, the intact contralateral hemidiaphragm increased its activity by 50% (compensatory loading), and after 2 weeks, the relative contribution of NTF to total fatigue decreased on the loaded side. (* indicates significant difference ($p<0.05$) from control values).

Recently, we have also observed that the extent of NTF in the rat diaphragm muscle is affected by prolonged periods of altered use (Miyata et al., 1994). For example, following two weeks of inactivity of the right hemidiaphragm, induced by cervical spinal cord transection to block descending inspiratory drive to the phrenic motoneuron pool (spinal isolation), the relative contribution of NTF to diaphragm fatigue induced by repetitive nerve stimulation at 40 and 75 Hz was diminished by 22% and 15%, respectively (Fig. 3).

In another model of diaphragm hemiparalysis, where axonal propagation of action potentials was blocked by continuous superfusion of the phrenic nerve with tetrodotoxin (TTX), the extent of NTF was found to be markedly increased (Fig. 3). The underlying basis for these discrepant effects of prolonged inactivity on the susceptibility of the diaphragm muscle to NTF remains unclear but may relate to the effects of mismatching motoneuronal and muscle activities. With TTX blockade of axonal propagation, the diaphragm muscle is paralyzed, but phrenic motoneuronal activity increases by approximately 50%. It is possible that, as a result, normal neurotrophic influences emanating from the motoneuron are altered with TTX blockade of axonal propagation. In contrast, with spinal isolation, diaphragm muscle paralysis is the result of inactivation of phrenic motoneurons. In this case, normal neurotrophic influences may persist because motoneuronal and muscle activities remain matched. This concept is supported by the fact that the muscle contractile and morphological adaptations that occur with TTX blockade are similar to those associated with phrenic denervation. In both cases, type IIb and IIx muscle fibers atrophy, specific force is reduced, and maximum unloaded shortening velocity slows. In contrast, only minimal muscle adaptations are found following two weeks of inactivity induced by spinal isolation. Since hemidiaphragm paralysis and its consequent mechanical effects are present in both the TTX and spinal isolation models, yet these two conditions are associated with divergent effects on susceptibility to NTF, it is doubtful that muscle inactivity *per se* causes the improvement in neuromuscular transmission following spinal isolation. This is supported by the fact that the compensatory overloading of the intact, contralateral hemidiaphragm was also associated with improved neuromuscular transmission (Fig. 3). It is more likely that an alteration in muscle use, whether an increase or a decrease in activity, is associated with an increase in neurotrophic influences that promote improvements in synaptic efficacy. Putative neurotrophic factors that might be involved in such adaptive responses would include calcitonin gene related peptide (CGRP) (Tsujimoto & Kuno, 1988) and/or ciliary neurotrophic factor (CNTF) (Helgren et al., 1994).

Assessment of Evoked Muscle Action Potentials - M Wave

The contribution of NTF to fatigue induced by voluntary contractions remains controversial. Merton (1954) utilized an assessment of evoked M waves as a means of estimating the extent of NTF during maximal voluntary contractions. He observed no reduction in M wave amplitude of the adductor pollicis muscle during more than 3 min of maximum voluntary contractions, in spite of a substantial loss in force. From these results, he concluded that NTF does not contribute to fatigue during maximal voluntary contractions. Subsequently, Bigland-Ritchie and colleagues (Bigland-Ritchie & Lippold, 1979; Bigland-Ritchie et al., 1979, 1982; Bellemare & Bigland-Ritchie, 1987; McKenzie et al., 1992) have conducted a number of similar studies assessing changes in M wave amplitude during voluntary contractions, and have concluded that NTF does not contribute to fatigue induced by sustained maximal voluntary contractions of human muscles. In contrast, other investigators have reported that M wave amplitude decreases during fatiguing maximal voluntary contractions of the human adductor pollicis (Naess & Storm-Mathisen, 1955) and first dorsal interosseus (Stephens & Taylor, 1972) muscles, suggesting to them that significant NTF does occur and does contribute to fatigue. Grob (1961) also found that repeated electrical stimulation of the ulnar nerve at frequencies greater than 25 Hz resulted in a reduction in M wave amplitude which paralleled the reduction in tetanic force of the first dorsal interosseus muscle. During inspiratory loaded breathing in unanesthetized sheep, Bazzy and Donnelly (1993) have also reported a reduction in M wave amplitude that paralleled force decline, suggesting a significant contribution of NTF to diaphragm fatigue. The reasons for these discrepant results are unclear but may relate to the experimental conditions and the type of voluntary or involuntary behaviors studied. However, it must be emphasized that changes in the M wave do not necessarily indicate a failure of neuromuscular transmission as they may be induced by changes in the sarcolemmal action potential (see Sjøgaard & McComas, Chapter 4).

Changes in evoked muscle action potentials have also been used to assess susceptibility of single muscle fibers and motor units to NTF. At the single fiber level, NTF can present itself as a failure to evoke an EPP in response to nerve stimulation and/or a reduction in EPP amplitude (Fournier et al., 1991). Using an *in vitro* nerve-diaphragm muscle preparation in rats, we recorded EPPS in response to repetitive stimulation at 10, 20, 40 and 75 Hz. In adult muscle, the incidence of a failure to evoke EPPS was found to be dependent on stimulation frequency, and was not appreciable until evoked at a stimulation rate of 75 Hz (Table 2). In contrast, in neonates the incidence of EPP failure was much more appreciable at all stimulation frequencies, and far more prevalent as compared to adults (Table 2). Presumably, the failure to evoke EPPS reflected a presynaptic axonal propagation failure,

Table 2. Postnatal changes in axonal propagation failure and EPP amplitude in rat diaphragm muscle fibers during repetitive stimulation at different rates

	Propagation failure rate (%)				% Decrement in EPP amplitude			
Age	10 Hz	20 Hz	40 Hz	75 Hz	10 Hz	20 Hz	40 Hz	75 Hz
0-7 days	4.4±1.7*	14.5±3.5*	40.5±3.8*	65.6±2.2*	52.2±10.8	70.0±8.2	87.2±7.7*	91.3±4.4*
14-21 days	0.0±0.0	2.0±1.0	16.7±4.2**	37.5±4.8**	63.7±9.7	61.2±5.2	78.5±3.1	77.4±2.8
Adult	0.0±0.0	2.5±0.5	4.0±2.4	26.0±5.1	61.3±7.8	62.8±6.2	73.2±2.7	75.1±1.3

Propagation failure rate was calculated as the proportion of absent evoked EPPS in the first 10 pulses of each stimulus train. The % decrement in EPP amplitude was that noted in the 10th stimulus pulse. Values are means ± SD. * indicates significant difference ($p<0.05$) from older animals. ** indicates significant difference ($p<0.05$) from adults.

most likely occurring at axonal branch points (see below). We also concluded that the higher incidence of axonal branch point failure in neonates reflected the increased axonal branching associated with polyneuronal innervation of muscle fibers. In both the adult and neonatal diaphragm, repetitive nerve stimulation was also associated with a decrement in EPP amplitude (Table 2). At both ages, this decrement was dependent on stimulation frequency. In contrast to the incidence of EPP failure, the decrement in EPP amplitude was greater in neonates only at higher rates of stimulation. The decrement in amplitude with repetitive stimulation could reflect either pre- or postsynaptic mechanisms (see below), but certainly synaptic efficacy would be reduced as a result.

As mentioned above, muscle fiber type is an important determinant of susceptibility to NTF. Fast-twitch glycolytic and/or type IIb/IIx fibers, which appear to be most susceptible to NTF (Kugelberg & Lindegren, 1979; Johnson & Sieck, 1993), comprise motor units that are more fatigable (Burke et al., 1973; Sieck et al., 1989). Accordingly, in motor unit studies, it has been demonstrated that fast-twitch fatigable (type FF) motor units are more susceptible to NTF than more fatigue-resistant fast-twitch (type FR) and slow-twitch (type S) units (Clamann & Robinson, 1985; Sandercock et al., 1985; Sieck & Fournier, 1990).

In the cat diaphragm, we estimated the extent of NTF in different motor unit types by assessing changes in the evoked motor unit action potential amplitude (M wave) during repetitive stimulation (Figs. 4 and 5).

Motor units were isolated by microdissection of cervical (C_4-C_6) ventral root filaments and classified as type F or S based on the presence or absence of "sag" in unfused tetani, respectively (Burke et al., 1973). Motor unit fatigue was assessed by repetitive stimulation at 40 Hz in 330 ms duration trains repeated each second. A fatigue index was

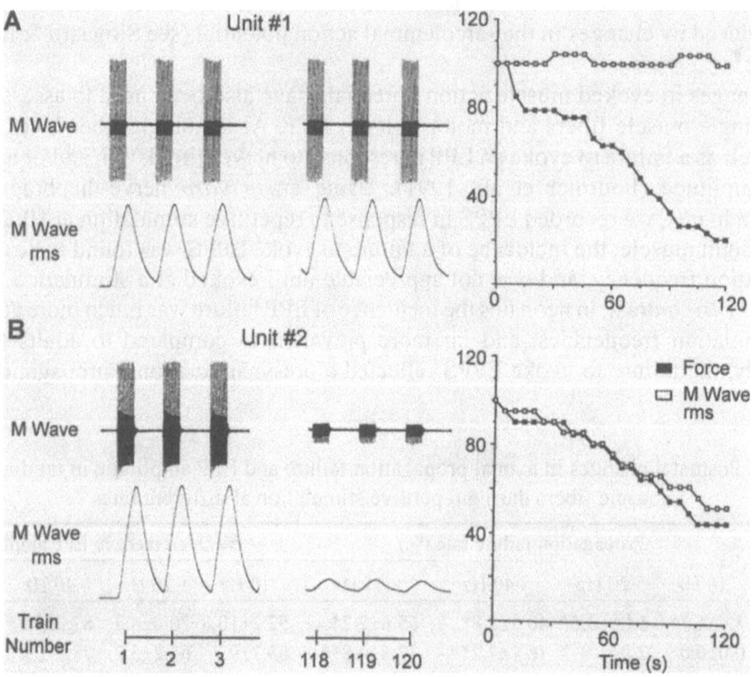

Figure 4. Evoked motor unit M waves were recorded in the adult cat diaphragm during repetitive stimulation at 40 Hz. The rms amplitude of the 13 evoked M waves in each train was calculated. Some motor units displayed very little change in evoked M waves during the 2-min stimulation period (*A*: unit #1) while other units displayed a marked reduction in M wave amplitude (*B*: unit #2).

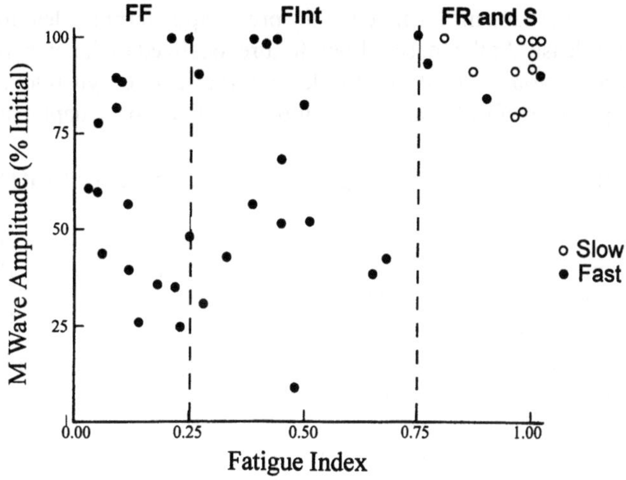

Figure 5. Changes in the ratio of motor unit M wave rms after 2 min stimulation to the initial M wave rms value (M wave index). A fatigue index was also calculated as the ratio of force generated after 2 min stimulation to the initial force. Fatigue resistant type S and FR units displayed very little NTF (M wave index > 0.75), while type FF and FInt units displayed variable extents of NTF.

then calculated as the ratio force generated after 2 min stimulation to the initial force (Burke et al., 1973). During repetitive stimulation, evoked unit M waves were recorded using fine-wire electrodes. The electromyographic (EMG) signals were amplified, band-pass filtered (20 Hz to 1 kHz), and then root mean square (rms) of the 13 M waves in each stimulus train was calculated based on the following transfer equation

$$V_{rms} - \sqrt{AVG(V^2_{EMG})}$$

where V_{rms} was the voltage output of the rms circuit, V_{EMG} was the voltage input of the EMG signal, and AVG was the averaging time constant (55 ms) of the rms circuit. Since the duration of the EMG signal (i.e., train duration of 330 ms) far exceeded the AVG duration (i.e., 55 ms), the rms value reached 99% of maximum output (i.e., rms plateau) during each stimulus train. Therefore, this rms calculation yielded an output signal that was directly related to the total power of the evoked EMG signal during each stimulus train. Changes in rms of motor unit M waves were used to assess the presence of NTF. In some motor units, a force decrement was dissociated from any NTF (Fig. 4*A*), while in other units the force decrement paralleled the decline in M wave amplitude (Fig. 4*B*).

In a population of motor units in the cat diaphragm, the extent of NTF was estimated by calculating the ratio of M wave rms after 2 min of stimulation to the initial M wave rms value. This M wave index was then compared to a fatigue index calculated as the ratio of motor unit force after 2 min stimulation to the initial force (Fig. 5). Type FR and S motor units (fatigue index>0.75) displayed little change in M wave index, indicating very little NTF. In contrast, type FF motor units (fatigue index < 0.25) and fatigue intermediate fast-twitch units (type FInt; fatigue index 0.25 to 0.75) displayed a range of decrement in M wave amplitudes.

PRESYNAPTIC SITES OF NEUROMUSCULAR TRANSMISSION FAILURE

Presynaptic sites where NTF might occur can be broadly categorized as being localized either to axons or to the presynaptic terminal. At the axon level, NTF can result from a failure of axonal propagation of action potentials. At the muscle fiber, axonal propagation failure would present itself as an absence of evoked EPPS. NTF can also result from mechanisms localized to the presynaptic terminal leading to inadequate neurotransmitter release. At the muscle fiber, failure localized to the presynaptic terminal would appear as a reduction in EPP amplitude and ineffective synaptic transmission, although such a depression of EPP amplitude could also reflect postsynaptic mechanisms of NTF (see below).

Failure of Axonal Propagation of Action Potentials

It has been recognized for a some time that action potentials may fail to propagate along each branch of a motor axon (Barron & Matthews, 1935). In experiments in the rat diaphragm, Krnjevic and Miledi (1958, 1959) observed that prolonged repetitive stimulation of the phrenic nerve resulted in a significant incidence of action potential propagation failure which they attributed to a failure of propagation at axonal branch points. They also noted that the incidence of presynaptic NTF was greater at higher rates of stimulation (Krnjevic & Miledi, 1959). Subsequently, a number of studies have also reported such axonal conduction blocks in the peripheral nervous system in a variety of species (Bittner, 1968; Parnas, 1972; Grossman et al., 1973, 1979; Hatt & Smith, 1976; Smith, 1980, 1983; Schiller & Rahamimoff, 1989). In a detailed study of the excitor axon innervating the abductor muscle of the crayfish, Smith (1980) reported that prolonged stimulation of the axon led to action potential propagation failure at axonal bifurcations, with blockage occurring first in the more peripheral terminal arborization of the axon and then progressively spreading centrally to regions where axonal caliber was larger (i.e., centripetal blocking pattern; Fig. 6). Therefore, points of axonal bifurcation are considered to be sites where failure of action potential generation and neurotransmission may occur (Raymond & Lettvin, 1978).

The potential mechanisms underlying the failure of axonal action potential propagation are summarized in Fig. 6. Accompanying action potential failure, there is also a prolonged depolarization of the axonal membrane and a decrease in the inward Na^+ current (Hatt & Smith, 1976). Axonal propagation failure can be reversed by hyperpolarization of the axon and/or by placing the preparation in a solution with low external K^+ levels. These observations suggested that a prolonged refractory period of the axon may be responsible for the failure of action potential generation and the consequent propagation failure at these axonal branch points. Alterations in perineural Na^+ and K^+ concentrations with repetitive axonal stimulation may therefore be responsible for axonal conduction block. In support, it has been shown that during repetitive stimulation, an accumulation of K^+ in the space surrounding the axon is associated with the development of conduction failure (Adelman et

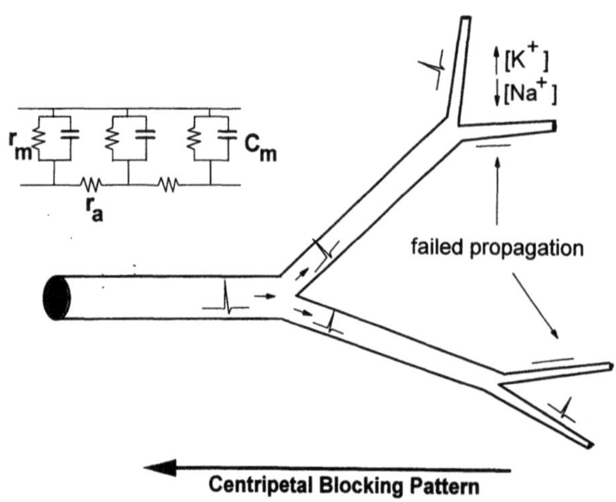

Figure 6. Potential mechanisms underlying axonal branch point failure include: changes in axonal geometry (higher axial resistivity, r_a and lower membrane capacitance C_m of daughter branches); and increased membrane refractoriness at smaller branches due to perineural accumulation of K^+ resulting from repetitive stimulation. As a result, axonal propagation failure displays a centripetal pattern.

al., 1973; Grossman et al, 1979; Smith, 1983). Shifts in ionic concentrations are more likely to occur in regions with smaller axonal sizes, since the surface-to-volume ratio is increased in these smaller axons, resulting in more pronounced changes in membrane potential due to ionic concentration changes (Smith, 1980).

Experimental observations of axonal conduction failure have led to a number of theoretical models that have focused on the contributions of axonal geometry on impulse propagation (Khodorov et al., 1969; Goldstein & Rall, 1974; Waxman, 1975; Swadlow et al., 1980; Stockbridge, 1988; Stockbridge & Stockbridge, 1988). Even if periaxonal accumulation of K^+ ions is ignored, changes in the excitability of the axonal membrane caused by differences in axonal geometry can lead to propagation failure in smaller axons, depending on impulse frequency in the parent axon (Stockbridge, 1988; Stockbridge & Stockbridge, 1988). For example, in a bifurcating axon, a short region of higher axial resistivity and lower capacitance in one of the daughter branches, compared to the rest of the adjacent axonal regions, can produce a frequency-dependent differential conduction along that daughter branch, in comparison to the other, *normal* branch. Even a very small change in resistivity and/or capacitance may result in axonal conduction being limited only to a band of stimulation frequencies, with varying probabilities of conduction block at different stimulation frequencies. Further, in shorter daughter branches, axonal conduction may be facilitated at higher frequencies of stimulation, compared to longer daughter branches. These complex interactions between the susceptibility to action potential propagation failure and axonal geometry may lead to NTF occurring in only a portion of muscle fibers within a motor unit.

The probability of blockade of action potentials at axonal branch points may be an important factor influencing the differences in neuromuscular transmission properties of different types of motor units. For example, type FF motor units, which are most susceptible to NTF (Fig. 5), tend to have more muscle fibers within the unit, i.e., a larger innervation ratio. Therefore, axons innervating type FF motor units with larger innervation ratios will have a greater number of axonal branches, and possibly, a greater probability of axonal conduction block as a result. Sandercock and colleagues (1985) recorded muscle fiber action potentials in fibers comprising motor units in the cat medial gastrocnemius muscle using microelectrodes, and reported that occasionally there was a failure to evoke action potentials in some muscle unit fibers. The incidence of such failures was more prevalent in type FF motor units as compared to other motor unit types. They also noted that the incidence of presynaptic NTF was dependent on the frequency of stimulation. We recorded evoked motor unit M waves in the cat diaphragm, and observed that in type FF and FInt motor units, abrupt transient changes in the waveform of M waves occurred (Sieck & Fournier, 1990). Presumably, the presynaptic failure of neuromuscular transmission occurred in only some muscle unit fibers. A presynaptic action potential conduction block at axonal branch points would explain these observations of abrupt changes in M wave waveform. Moreover, such a conduction block of only some muscle unit fibers during repetitive stimulation would reduce the effective innervation ratio of these motor units and lead to reductions in mechanical force (fatigue).

The higher incidence of NTF during early postnatal development in the rat diaphragm (Table 2) may also be influenced by axonal conduction block. During early postnatal development, muscle fibers are innervated by more than one motoneuron (polyneuronal innervation; Redfern, 1970; Bagust et al., 1974). Therefore, motor unit innervation ratios are greater during early postnatal development, and there are more axonal branch points where conduction block might occur. Subsequently, as a result of synapse elimination, motor unit innervation ratio decreases, and the number of axonal branch points is reduced accordingly. In the rat diaphragm, synapse elimination is complete by postnatal day 14 (Sieck & Fournier, 1991). A greater number of axonal branch points would increase the probability of axonal propagation failure in neonatal motor units (Table 2). Furthermore, the smaller axonal calibers in the neonate would increase axial resistivity and lower membrane capacitance. These intrinsic electrophysiological properties combined with possible heterogeneity of axonal calibers may be conducive to an increased incidence of axonal branch point conduction block. In addition, as mentioned above, shifts in ionic concentrations are more likely to occur around smaller axons.

Mechanisms of Failure at the Presynaptic Terminal

At the presynaptic terminal, NTF can result from a reduction in the release of ACh. Fig. 7 summarizes several potential mechanisms that might underlie a reduction in presynaptic ACh release.

By reducing extracellular Ca^{2+} at the frog neuromuscular junction, del Castillo and Katz (1954) demonstrated that Ca^{2+} influx is critical for neuromuscular transmission. Subsequently, a number of investigators have confirmed this dependence of neurotransmitter release on Ca^{2+} influx at the presynaptic terminal. Using a voltage clamp technique, Llinas and Heuser (1977) further demonstrated that Ca^{2+} influx at the presynaptic terminal is via voltage-dependent Ca^{2+} channels, and that there is very little delay between Ca^{2+} influx and transmitter release, suggesting that these Ca^{2+} channels were located in close proximity to the active zones of the nerve terminal. The entry of Ca^{2+} via voltage-dependent channels into the nerve terminal leads to a fusion of synaptic vesicles to the nerve terminal membrane, and the release of ACh into the synaptic cleft via a second messenger cascade. While the exact cascade of events is not entirely clear, studies using isolated, highly enriched preparations of synaptic vesicles and nerve terminal cytoplasm have suggested that calmodulin, a Ca^{2+} binding protein, plays a major role in regulating neurotransmitter release (DeLorenzo, 1981). Whether calmodulin or other second messengers involved in excitation-secretion coupling at the presynaptic terminal are affected during repetitive stimulation remains unclear. However, quantal (vesicular) release of ACh at the neuromuscular junction may be reduced by either a decrease in Ca^{2+} influx through voltage-gated Ca^{2+} channels or by a reduction in the Ca^{2+} sensitivity of processes within the nerve terminal that lead to vesicular release.

Since the study of del Castillo and Katz (1954), it has been recognized that during repetitive nerve stimulation of curarized nerve-muscle preparations, the change in EPP amplitude is dynamic. There is an initial facilitation of EPP amplitude, followed by a depression. Intracellular Ca^{2+} at presynaptic terminals can accumulate with repetitive stimu-

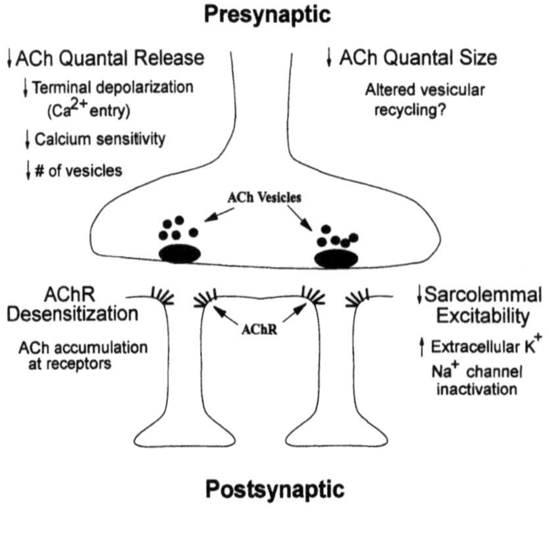

Figure 7. Possible pre- and postsynaptic sites where NTF might occur. At the presynaptic terminal, transmission failure might arise from a reduction in ACh quantal release and/or a decrease in quantal size. At the postsynaptic endplate, ACh receptors (AChR) may become desensitized and/or sarcolemmal excitability may be reduced.

lation, and this may form the basis for the initial facilitation of EPP amplitude, i.e., increased probability of quantal release and increased quantal content. The decrement in EPP amplitude with repetitive nerve stimulation has been attributed to either a reduction in the number of quanta of ACh being released by the presynaptic nerve terminal, or to a reduction in quantal size (i.e., the amount of ACh per vesicle as indicated by the average miniature endplate potential, MEPP, amplitude) (del Castillo & Katz, 1954; Jones & Kwanbunbumpen, 1970; Kurihara & Brooks, 1975; Smith, 1984) . A reduction in the number of vesicles released by the presynaptic terminal during repetitive nerve stimulation may be attributed to a reduction in Ca^{2+} influx, a decrease in Ca^{2+} sensitivity of excitation-secretion coupling or a depletion of the number of vesicles available for release. Repetitive stimulation of the nerve terminal may lead to an accumulation of extracellular K^+ due to limited diffusion possibilities (Frankenhaeuser & Hodgkin, 1956; Meech, 1974; Hatt & Smith, 1976; Smith, 1980). As a result, the terminal membrane would depolarize thereby initially increasing Ca^{2+} influx. However, the increase in intracellular Ca^{2+} may also increase K^+ conductance via Ca^{2+}-dependent K^+ channels, making the terminal membrane refractory to further depolarization (Meech, 1974). The net effect would be a reduction of Ca^{2+} influx and reduced neurotransmitter release.

Ca^{2+} sensitivity of excitation-secretion coupling may also be reduced by repetitive nerve stimulation. Volle and Branisteanu (1976) reported that during repetitive stimulation of the frog neuromuscular junction, there was a reduction in the number of quanta released, despite an apparent increase in intracellular Ca^{2+} as affected, experimentally, by increasing extracellular Ca^{2+}. This observation suggests that repetitive stimulation decreases the Ca^{2+} sensitivity of ACh release. In melanotrophs of the rat pituitary gland, Okano and colleagues (1993) demonstrated that vesicular release was reduced following repetitive exposure to extracellular Ca^{2+}, although intracellular Ca^{2+} was unaffected. They attributed the reduced Ca^{2+} sensitivity of vesicular release of histamine in mast cells to a depletion of guanosine triphosphatase (GTP). It is also possible that at the neuromuscular junction, depletion of GTP, or some other factor in a second messenger cascade, may also play a role in the reduction of Ca^{2+} sensitivity of ACh release following repetitive stimulation.

At the frog neuromuscular junction, Volle and Branisteanu (1976) also found that quantal release was reduced during repetitive nerve stimulation despite an increase in the probability of release (MEPP frequency). They concluded that the decrease in quantal release was due, at least in part, to a depletion of synaptic vesicles. There is morphological evidence of vesicular depletion at the presynaptic terminal following repetitive nerve stimulation. For example, based on electron microscopic (EM) evidence in the electric organ of the *narcine brasiliensis*, repetitive exhaustive stimulation of the nerve was reported to result in nearly a 50% reduction in the total number of synaptic vesicles (Boyne et al., 1975). Lentz and Chester (1982) used the uptake of horseradish peroxidase at the frog neuromuscular junction to follow presynaptic events during nerve stimulation, and found depletion of synaptic vesicles along with increased axolemmal infoldings and membranous cisternae. This suggests that vesicular recycling is impaired following repetitive stimulation.

In addition to a reduction in the number of quanta released, it is also possible that following repetitive stimulation there is a reduction in quantal size. At the neuromuscular junction of the rat diaphragm, Jones and Kwanbunbumpen (1970) reported that repetitive nerve stimulation caused both a reduction in the number of quanta released, and a reduction in quantal size. However, in this study, curare was used to abolish muscle contractions, which causes difficulties in the calculation of quantal size. In the rat diaphragm muscle, Kurihara and Brooks (1975) used a cut muscle preparation to abolish muscle contractions. They showed that the initial decrease in EPP amplitude during repetitive nerve stimulation (at 30 Hz) was entirely due to a reduction in the number of quanta released, without any decrement of quantal size. They concluded that the decrement in EPP amplitude during this initial period was not sufficient to cause a failure in neuromuscular transmission. However, following this initial period, these authors found that while quantal release continued to be reduced, there was also a linear decline in quantal size (i.e., average MEPP amplitude). They further concluded that during this latter stage of repetitive nerve stimulation, the reduction in EPP amplitude was sufficient to cause NTF and muscle fatigue. However, the underlying cause of a reduction in quantal size is unclear.

Quantal content and release may be affected in a variety of conditions where there are adaptations in the neuromuscular junction. For example, during early postnatal development there is a rapid growth of the neuromuscular junction in parallel with an increase in muscle fiber size. We recently observed that the postnatal growth of endplates on type I muscle fibers (comprising type S motor units), exceeded that dictated by muscle fiber growth, while the growth of type II fiber (comprising fast-twitch motor units) endplates matched the postnatal change in fiber size. During early postnatal development, there is a progressive decrease in MEPP amplitude (Diamond & Miledi, 1962, Kelly 1978), which may be attributed to a change in muscle fiber size and/or a decrease in quantal size. In an *in vitro* nerve diaphragm muscle preparation, Kelly (1978) reported that the reduction in MEPP amplitude during early postnatal development was accompanied by an increase in the quantal content of EPPS. Kelly (1978) also reported that the quantal content of EPPS decreased by over two-fold during prolonged repetitive stimulation at 10 Hz. The safety factor for neuromuscular transmission was also reported to be lower in the neonatal diaphragm as compared to adults, and at all ages, quantal content decreased with continued repetitive stimulation (Kelly, 1978). A reduction in MEPP amplitude occurs during aging (Gutmann et al., 1971; Kelly & Roberts, 1977; Kelly, 1978; Smith, 1984). In general, quantal content has been reported to be higher in older animals (Kelly & Roberts, 1977; Smith, 1984; Alshuaib & Fahim, 1990), suggesting an increase in quantal release, since MEPP amplitude is reduced.

During early postnatal development and in older animals, it has also been reported that MEPP frequency is lower (Kelly & Zacks, 1969; Gutmann et al., 1971; Kelly, 1978; Smith, 1984). It is likely that in neonates, the lower MEPP frequency reflects the smaller surface area of the neuromuscular junction and the associated reduction in active sites for ACh quantal release. Whether such a morphological basis underlies the lower MEPP frequency with aging remains unknown.

POSTSYNAPTIC SITES OF NEUROMUSCULAR TRANSMISSION FAILURE

There are two major postsynaptic sites where NTF can occur: at the motor endplate due to a desensitization of the cholinergic receptor (AChR); and at the sarcolemma due to a reduction in excitability. Krnjevic and Miledi (1959) provided evidence that NTF at the neuromuscular junction of the rat diaphragm muscle was due to a combination of a decrement in EPP amplitude and a reduction in sarcolemmal excitability, leading to a failure in the generation of action potentials in muscle fibers.

Desensitization of the Cholinergic Receptor

Katz and Thesleff (1957) found that prolonged exposures of the neuromuscular junction to ACh resulted in a reduction in endplate conductance due to a desensitization of the AChR. Desensitization involves a slow transition of AChR channels to non-conducting states due to the continued presence of ACh, with recovery only upon removal of stimulation (Katz & Thesleff, 1957; Sakmann et al., 1980; Feltz & Trautmann, 1982). As mentioned above, the decrement in EPP amplitude during repetitive stimulation could reflect a decrease in quantal release, but it could also reflect a desensitization of the AChR due to continued accumulation of ACh in the synaptic cleft, compounded by inadequate diffusion of the transmitter out of the synaptic cleft, and incomplete hydrolysis of the ACh by acetylcholinesterase (Katz & Thesleff, 1957), especially at higher frequencies of stimulation. At the frog neuromuscular junction in cut muscle fibers in the absence of acetylcholinesterase, Giniatullin and colleagues (1986) observed a 23% reduction in MEPP amplitude following repetitive stimulation at 10 Hz over a period of 60 s, indicating AChR desensitization. Albuquerque and colleagues (1986) have suggested that cyclic AMP may be involved in the phosphorylation of the AChR. During repetitive stimulation, changes in cAMP level may lead to AChR desensitization, and reduced EPP amplitudes.

Reduction in Sarcolemmal Excitability

Repetitive activation of the neuromuscular junction may reduce sarcolemmal excitability (Gruber, 1914; Rosenblueth, 1940; Krnjevic & Miledi, 1958; Metzger & Fitts, 1986; Roed, 1988; see also Fuglevand, Chapter 6). Edwards (1984) proposed that fatigue could be categorized as either high-frequency fatigue, characterized by a transient, rapidly recovering loss of force after high frequencies of stimulation, and low-frequency fatigue, characterized by prolonged loss in force. It was suggested that high-frequency fatigue was due, at least in part, to a failure of action potential propagation by the sarcolemma, possibly due to an accumulation of extracellular K^+, membrane depolarization and Na^+ channel inactivation leading to a reduction in Na^+ conductance (Bigland-Ritchie et al., 1979; Edwards, 1984). However, the influence of K^+ on sarcolemmal excitability during repetitive stimulation of muscle remains controversial. Juel (1988) showed that high-frequency fatigue of the mouse soleus and extensor digitorum longus muscles was associated with a decrease in action potential propagation velocity which was induced by increased extracellular K^+, but was nearly independent of moderate changes in the Na^+ gradient. Sjøgaard (1991) also suggested that an accumulation of K^+ occurred following repetitive stimulation. She attributed the increase in extracellular K^+ to the opening of Ca^{2+}-dependent K^+ channels. The likelihood that the increase in interstial K^+ contributes to fatigue is considered in Chapter 4, Sjøgaard and McComas.

While reduced sarcolemmal excitability may result from an accumulation of extracellular K^+, a number of studies have suggested that changes in extracellular K^+ concentration are not large enough to change sarcolemmal excitability. Using intracellular recordings in an isolated rat phrenic nerve-diaphragm muscle preparation, Metzger and Fitts (1986) found no change in resting membrane potential of muscle fibers following either high- or low-frequency stimulation, thus arguing against a significant accumulation of extracellular K^+. Instead, they suggested that the decline in amplitude of sarcolemmal action potentials during repetitive stimulation was most likely due to changes in extracellular Na^+. Renaud and Light (1992) also investigated the role of extracellular K^+ in muscle fatigue by testing whether increasing K^+ concentration in unfatigued frog sartorius muscle fibers caused a decrease in force similar to that observed during fatigue. However, accumulation of K^+ at the surface of the sarcolemma was not sufficiently high to reduce force during repetitive stimulation via blockade of muscle action potential propagation. However, they suggested that K^+ may play a role in causing failure at the level of the T-tubule; thus affecting excitation-contraction coupling.

In addition to changes in Na^+ and K^+ ionic concentrations at the sarcolemma, pH may also play a role in muscle fatigue. Renaud (1989) used ion-selective microelectrodes on frog sartorius muscle to measure pH during tetanic contractions elicited by field stimulation, and found that extracellular pH inhibits tetanic force recovery by acting on the outer surface of the sarcolemma. In the isolated hamster diaphragm, Brody and colleagues (1991) reported that changing extracellular pH initially results in decreased conduction velocity along the muscle membrane and a leftward shift in the median frequency of the EMG. However, with stimulation, the rate of decay of conduction velocity was less than the rate of decay of the median frequency. This suggested a complex interaction between pH and sarcolemmal excitability.

Sarcolemmal excitability is influenced by the intrinsic electrophysiological properties of muscle fibers. For example, larger muscle fibers will have a lower axial resistance and higher membrane capacitance, and this will affect the extent of depolarization for any given level of synaptic current. For this reason, the size of the motor endplate varies proportionately with fiber diameter (Nystrom, 1968; Wernig et al., 1986). Motor endplates on type I muscle fibers are larger than those on type II fibers (Ellisman et al., 1976; Nystrom, 1968; Fahim et al., 1984). Furthermore, the relationship between fiber diameter and endplate size is true only within a given fiber type since type I fibers are generally smaller than type II fibers. The matching of endplate size with fiber size is important in that it will ensure adequate synaptic current for action potential generation in muscle fibers (safety factor; ratio of EPP amplitude to threshold for action potential generation). The safety factor for type II muscle fibers is greater than that for type I fibers and, in both fiber types, sufficiently large to ensure reliable generation of action potentials. With repetitive stimulation, the safety factor for type I fibers shows little change while that for type II fibers displays a rapid decline (Gertler & Robbins, 1978).

The neuromuscular junction can adapt during postnatal development and under conditions of altered muscle use (Balice-Gordon & Lichtman, 1990, Balice-Gordon et al., 1990). For example, during early postnatal development in the rat diaphragm, motor endplates on both type I and II fibers become larger as muscle fiber size increases. However, the growth of motor endplates on type I fibers exceeds that predicted by the increase in fiber size alone while the growth of endplates on type II fibers matches the change in fiber diameter. Perhaps this disproportionate growth of motor endplates on type I and II fibers contributes to the increased susceptibility of type II fibers to NTF.

We have observed that following two weeks of diaphragm muscle inactivity induced by cervical spinal cord hemisection (spinal isolation) motor endplates on type II fibers become elongated and increase in area relative to fiber diameter (Fig. 8). This morphological change is associated with an improvement in neuromuscular transmission. Following two weeks of inactivity in the spinal isolated animals, type II fiber cross-sectional area was unaffected suggesting that sarcolemmal excitability was most likely unchanged. The improved neuromuscular transmission in the spinal isolated animals was therefore most likely attributable to an increase in synaptic current due to increased quantal release. In contrast, following two weeks of diaphragm muscle inactivity induced by TTX blockade of phrenic axonal propagation, type II fibers atrophied while motor endplate size remained unaffected. As a result, the relative size

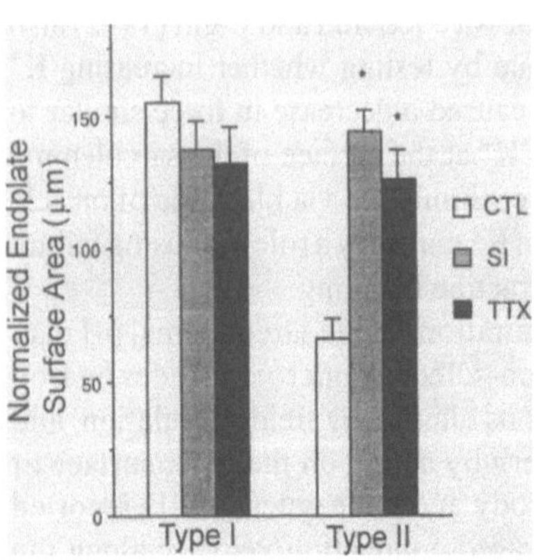

Figure 8. Surface area of motor endplates on type-identified rat diaphragm muscle fibers was normalized for fiber diameter. In controls, the normalized endplate surface area on type I fibers was larger than that on type II fibers. Following two weeks of inactivity induced by spinal isolation (SI) or TTX blockade, the normalized endplate surface area on type II fibers increased whereas that on type I fibers remained unchanged.

of motor endplates on type II fibers increased, as in the spinal isolated animals (Fig. 8). However, this increase in the relative size of motor endplates on type II fibers in the TTX animals was associated with an increased susceptibility to NTF. Since in the TTX animals, there was a reduction in cross-sectional area of type II muscle fibers, the sarcolemma should be more excitable if anything. Therefore, the increased susceptibility to NTF in the TTX animals most likely resulted from reduced quantal release of ACh.

SUMMARY AND CONCLUSIONS

A number of potential sites of transmission failure have been identified, although the relative contributions of these to muscle fatigue remain controversial. Blockade of axonal propagation of action potential, due to changes in axonal excitability and/or axonal geometry, may partly explain differences in the excitability of different motor unit types to NTF, and the greater incidence of transmission failure in neonatal muscles. Failure of ACh release at the presynaptic terminal may be due to reduced Ca^{2+} influx, decreased Ca^{2+} sensitivity of excitation-secretion coupling and/or a depletion of synaptic vesicles. At the postsynaptic membrane, AChR desensitization may be an important cause of NTF during repetitive stimulation. In addition, reduced sarcolemmal excitability may also play a role in NTF. Clearly, each of these potential mechanisms may adapt under a variety of conditions, and the relative contribution of each might vary depending on fiber type.

ACKNOWLEDGMENTS

The authors would like to acknowledge the invaluable contributions of Drs. Mario Fournier and Hirofumi Miyata. This work was supported by United States Public Health Service grants HL 34817 and HL 37680. Attendance of *G.C.S.* at the 1994 Bigland-Ritchie conference was supported, in part, by the University of Miami.

REFERENCES

Adelman WJ, Plati Y & Senft JP (1973). Potassium ion accumulation in a periaxonal space and its effect on the measurements of membrane potassium ion conductance. *Journal of Membrane Biology* **12**, 387-410.

Albuquerque EX, Deshpande SS, Aracava Y, Alkondon M & Daly JW (1986). Possible involvement of cyclic AMP in the expression of desensitization of the nicotinic acetylcholine receptor. A study with forskolin and its analogs. *FEBS Letters* **199**, 113-20.

Aldrich TK, Shander A, Chaudhry I & Nagashima H (1986). Fatigue of isolated rat diaphragm: role of impaired neuromuscular transmission. *Journal of Applied Physiology* **61**, 1077-1083.

Alshuaib WB & Fahim MA (1990). Aging increases calcium influx at motor nerve terminal. *International Journal of Developmental Neuroscience* **8**, 655-666.

Bagust J, Lewis DM & Westerman RA (1974). The properties of motor units in a fast and slow twitch muscle during postnatal development in the kitten. *Journal of Physiology (London)* **237**, 75-90.

Balice-Gordon RJ, Breedlove SM, Bernstein S & Lichtman JW (1990). Neuromuscular junctions shrink and expand as muscle fiber size is manipulated: in vivo observations in the androgen-sensitive bulbocavernosus muscle of mice. *Journal of Neuroscience* **10**, 2660-2671.

Balice-Gordon RJ & Lichtman JW (1990). In vivo visualization of the growth of pre- and postsynaptic elements of neuromuscular junctions in the mouse. *Journal of Neuroscience* **10**, 894-908.

Barron DH & Matthews BHC (1935). Intermittent conduction in the spinal cord. *Journal of Physiology (London)* **85**, 73-103.

Bazzy AR & Donnelly DF (1993). Diaphragmatic failure during loaded breathing: role of neuromuscular transmission. *Journal of Applied Physiology* **74**, 1679-1683.

Bellemare F & Bigland-Ritchie B (1987). Central components of diaphragm fatigue assessed by phrenic nerve stimulation. *Journal of Applied Physiology* **62**, 1307-1316.

Bigland-Ritchie B (1984). Muscle fatigue and the influence of changing neural drive. In: Loke J (ed.), *Symposium on Exercise: Physiology and Clinical Applications. Clinics in Chest Medicine* **5**, 21-34.

Bigland-Ritchie B, Jones DA, Hosking GP & Edwards RHT (1978). Central and peripheral fatigue in sustained maximum voluntary contractions of human quadriceps muscle. *Clinical Science and Molecular Medicine* **54**, 609-614.

Bigland-Ritchie B, Jones DA & Woods JA (1979). Excitation frequency and muscle fatigue: Electrical responses during human voluntary and stimulated contractions. *Experimental Neurology* **64**, 414-427.

Bigland-Ritchie B, Kukulka CG, Lippold OCJ & Woods JJ (1982). The absence of neuromuscular transmission failure in sustained maximal voluntary contractions. *Journal of Physiology (London)* **330**, 265-278.

Bigland-Ritchie B & Lippold OCJ (1979). Changes in muscle activation during prolonged maximal voluntary contractions. *Journal of Physiology (London)* **292**, 14-15P.

Bittner GD (1968). Differentiation of nerve terminals in crayfish opener muscle and its functional significance. *Journal of General Physiology* **51**, 731-758.

Boyne AF, Bohan TP & Williams TH (1975). Changes in cholinergic synaptic vesicle populations and the ultrastructure of the nerve terminal membranes of Narcine brasiliensis electric organ stimulated to fatigue in vivo. *Journal of Cell Biology* **67**, 814-25.

Brody LR, Pollock MT, Roy SH, DeLuca CJ & Celli B (1991). pH-induced effects on median frequency and conduction velocity of the myoelectric signal. *Journal of Applied Physiology* **71**, 1878-85.

Burke RE, Levine DN, Psairis P & Zajac FE (1973). Physiological types and histochemical profiles of motor units of cat gastrocnemius. *Journal of Physiology (London)* **234**, 723-748.

Clamann HP & Robinson AJ (1985). A comparison of electromyographic and mechanical fatigue properties in motor units of the cat hindlimb. *Brain Research* **327**, 203-219.

del Castillo J & Katz B (1954). The effect of magnesium on the activity of motor nerve endings. *Journal of Physiology (London)* **124**, 553-559.

DeLorenzo RJ (1981). The calmodulin hypothesis of neurotransmission. *Cell Calcium* **2**, 365-385.

Diamond J & Miledi R (1962). A study of foetal and new-born rat muscle fibers. *Journal of Physiology (London)* **162**, 393-408.

Edwards RHT (1984). New techniques for studying human muscle function, metabolism and fatigue. *Muscle & Nerve* **7**, 599-609.

Ellisman MH, Rash JE, Staehelin LA & Porter KR (1976). Studies of excitable membranes II. A comparison of specializations at neuromuscular junctions and nonjunctional sarcolemmas of mammalian fast and slow twitch muscle fibers. *Journal of Cell Biology* **68**, 752-774.

Fahim MA, Holley JA & Robbins N (1984). Topographic comparison of neuromuscular junctions in mouse slow and fast twitch muscles. *Neuroscience* **13**, 227-235.

Feldman JD, Bazzy AR, Cummins TR & Haddad GG (1991). Developmental changes in neuromuscular transmission in the rat diaphragm. *Journal of Applied Physiology* **71**, 280-286.

Feltz A & Trautmann A (1982). Desensitization at the frog neuromuscular junction: A biphasic process. *Journal of Physiology (London)* **332**, 257-272.

Fournier M, Alula M & Sieck GC (1991). Neuromuscular transmission failure during postnatal development. *Neuroscience Letters* **125**, 34-36.

Frankenhaeuser B & Hodgkin AL (1956). The after-effects of impulses in the giant nerve fibres of Loligo. *Journal of Physiology (London)* **131**, 341-376.

Gertler RA & Robbins N (1978). Differences in neuromuscular transmission in red and white muscles. *Brain Research* **142**, 160-164.

Goldstein SS & Rall W (1974). Changes of action potential shape and velocity for changing core conductor geometry. *Biophysical Journal* **14**, 731-757.

Grob D (1961). Muscular disease. *Bulletin of the New York Academy of Medicine* **37**, 809-834.

Grossman Y, Parnas I & Spira ME (1979). Differential conduction block in branches of a bifurcating axon. *Journal of Physiology (London)* **295**, 283-305.

Grossman Y, Spira ME & Parnas I (1973). Differential flow of information into branches of a single axon. *Brain Research* **64**, 379-386.

Gruber CM (1914). Studies in fatigue. IV. The relation of adrenalin to curare and fatigue in normal and denervated muscles. *American Journal of Physiology* **34**, 89-96.

Giniatullin RA, Baltser SK, Nikolskii EE & Magazanik LG (1986). Postsynaptic potentiation and desensitization of the myoneural synapse of the frog induced by rhythmic stimulation of a motor nerve. *Neirofiziologiia* **18**, 645-54.

Gutmann E, Hanlikova V & Vyskocil F (1971). Age changes in cross-striated muscle of the rat. *Journal of Physiology (London)* **216**, 331-343.

Hatt H & Smith DO (1976). Synaptic depression related to presynaptic axon conduction block. *Journal of Physiology (London)* **259**, 367-393.

Helgren ME, Squinto SP, Davis HL, Parry DJ, Boulton TG, Heck CS, Zhu Y, Yancopoulos GD, Lindsay RM & DiStefano PS (1994). Trophic effect of ciliary neurotrophic factor on denervated skeletal muscle. *Cell* **76**, 493-504.

Johnson BD & Sieck GC (1993). Differential susceptibility of diaphragm muscle fibers to neuromuscular transmission failure. *Journal of Applied Physiology* **75**, 341-348.

Jones SF & Kwanbunbumpen S (1970). Some effects of nerve stimulation and hemicholinium on quantal transmitter release at the mammalian neuromuscular junction. *Journal of Physiology (London)* **207**, 51-61.

Juel C (1988). Muscle action potential propagation velocity changes during activity. *Muscle Nerve* **11**, 714-719.

Kandel ER (1981). Calcium and the control of synaptic strength by learning. *Nature* **293**, 697-700.

Katz B & Thesleff S (1957). On the factors which determine the amplitude of the "miniature end-plate potential". *Journal of Physiology (London)* **137**, 267-278.

Kelly AM & Zacks SI (1969). The fine structure of motor endplate histogenesis. *Journal of Cell Biology* **42**, 154-169.

Kelly SS (1978). The effect of age on neuromuscular transmission. *Journal of Physiology (London)* **274**, 51-62.

Kelly SS & Roberts DV (1977). The effect of age on the safety factor in neuromuscular transmission in the isolated diaphragm of the rat. *British Journal of Anaesthesiology* **49**, 217-22.

Kernell D (1965). The adaptation and the relation between discharge frequencies and current strength of cat lumbosacral motoneurons stimulated by long-lasting injected current. *Acta Physiologica Scandinavica* **65**, 65-73.

Kernell D & Monster AW (1982a). Time course and properties of late adaptation in spinal motoneurons of the cat. *Experimental Brain Research* **46**, 191-196.

Kernell D & Monster AW (1982b). Motoneuron properties and motor fatigue. An intracellular study of gastrocnemius motoneurons of the cat. *Experimental Brain Research* **46**, 197-204.

Khodorov BI, Timin YN, Vilenkin Y & Gul'ko FB (1969). Theoretical analysis of the mechanisms of conduction of a nerve pulse over an inhomogeneous axon. I. Conduction through a portion with increased diameter. *Biofizika* **14**, 304-315.

Krnjevic K & Miledi R (1958). Failure of neuromuscular transmission in rats. *Journal of Physiology (London)* **140**, 440-461.

Krnjevic K & Miledi R (1959). Presynaptic failure of neuromuscular propagation in rats. *Journal of Physiology (London)* **149**, 1-22.

Kuei JH, Shadmehr R & Sieck GC (1990). Relative contribution of neuromuscular transmission failure to diaphragm fatigue. *Journal of Applied Physiology* **68**, 174-180.

Kugelberg E & Lindegren B (1979). Transmission and contraction fatigue of rat motor units in relation to succinate dehydrogenase activity of motor unit fibers. *Journal of Physiology (London)* **288**, 285-300.

Kurihara T & Brooks JE (1975). The mechanism of neuromuscular fatigue: A study of mammalian muscle using excitation-contraction coupling. *Archives of Neurology* **32**, 168-174.

Lentz TL & Chester J (1982). Synaptic vesicle recycling at the neuromuscular junction in the presence of a presynaptic membrane marker. *Neuroscience* **7**, 9-20.

Llinas RR & Heuser JE (1977). Depolarization-release coupling systems in neurons. *Neuroscience Research Program Bulletin* **15**, 555-687.

McKenzie DK, Bigland-Ritchie B, Gorman RB & Gandevia SC (1992). Central and peripheral fatigue of human diaphragm and limb muscles assessed by twitch interpolation. *Journal of Physiology (London)* **454**, 643-656.

Meech RW (1974). The sensitivity of *Helix aspersa* neurones to injected calcium ions. *Journal of Physiology (London)* **237**, 259-278.

Merton PA (1954). Voluntary strength and fatigue. *Journal of Physiology (London)* **123**, 553-564.

Metzger JM & Fitts RH (1986). Fatigue from high- and low-frequency muscle stimulation: role of sarcolemma action potentials. *Experimental Neurology* **93**, 320-333.

Miyata H, Zhan WZ, Prakash YS & Sieck GC (1994). Influence of inactivity on contribution of neuromuscular transmission failure to diaphragm fatigue. *Medicine and Science in Sports and Exercise* **26**, S167.

Naess K & Storm-Mathisen A (1955). Fatigue of sustained tetanic contractions. *Acta Physiologica Scandinavica* **34**, 351-366.

Nystrom B (1968). Postnatal development of motor nerve terminals in "slow-red" and "fast-white" cat muscles. *Acta Neurologica Scandinavica* **44**, 363-383.

Okano K, Monck JR & Fernandez JM (1993). GTP gamma S stimulates exocytosis in patch-clamped rat melanotrophs. *Neuron* **11**, 165-72.

Parnas I (1972). Differential block at high frequency of branches of a single axon innervating two muscles. *Journal of Neurophysiology* **35**, 903-914.

Raymond SA & Lettvin JA (1978). Aftereffects of activity in peripheral axons as a clue to nervous encoding. In: Waxman SG (ed.), *Physiology and Pathobiology of Axons*, pp. 203-225. New York: Raven Press.

Redfern PA (1970). Neuromuscular transmission in newborn rats. *Journal of Physiology (London)* **209**, 701-709.

Renaud JM (1989). The effect of lactate on intracellular pH and force recovery of fatigued sartorius muscles of the frog, *Rana pipiens*. *Journal of Physiology (London)* **416**, 21-47.

Renaud JM & Light P (1992). Effects of K+ on the twitch and tetanic contraction in the sartorius muscle of the frog, Rana pipiens. Implication for fatigue in vivo. *Canadian Journal of Physiology and Pharmacology* **70**, 1236-1246.

Roed A (1988). Fatigue during continuous 20 Hz stimulation of the rat phrenic nerve diaphragm preparation. *Acta Physiologica Scandinavica* **134**, 217-21.

Rosenblueth A (1940). The electrical excitability of mammalian striated muscle. *American Journal of Physiology* **129**, 22-38.

Sakmann B, Patlak J & Neher E (1980). Single acetylcholine-activated channels show burst-kinetics in presence of desensitizing concentrations of agonist. *Nature* **286**, 71-73.

Sandercock TG, Faulkner JA, Albers JW & Abbrecht PH (1985). Single motor unit and fiber action potentials during fatigue. *Journal of Applied Physiology* **58**, 1073-1079.

Schiller Y & Rahamimoff R (1989). Neuromuscular transmission in diabetes: response to high frequency activation. *Journal of Neuroscience* **9**, 3709-3719.

Sieck GC & Fournier M (1990). Changes in diaphragm motor unit EMG during fatigue. *Journal of Applied Physiology* **68**, 1917-1926.

Sieck GC & Fournier M (1991). Developmental aspects of diaphragm muscle cells. In: Haddad GG, Farber JP (eds.), *Developmental Neurobiology of Breathing*, pp. 375-478. New York: Dekker.

Sieck GC, Fournier M & Enad JG (1989). Fiber type composition of muscle units in the cat diaphragm. *Neuroscience Letters* **97**, 29-34.

Sjøgaard G (1991). Role of exercise-induced potassium fluxes underlying muscle fatigue: a brief review. *Canadian Journal of Physiology and Pharmacology* **69**, 238-45.

Smith DO (1980). Mechanisms of action potential propagation failure at sites of axon branching in the crayfish. *Journal of Physiology (London)* **301**, 243-259.

Smith DO (1983). Axon conduction failure under in vivo conditions in crayfish. *Journal of Physiology (London)* **344**, 327-333.

Smith DO (1984). Acetylcholine storage, release and leakage at the neuromuscular junction of mature adult and aged rats. *Journal of Physiology (London)* **347**, 161-76.

Stephens JA & Taylor A (1972). Fatigue of maintained voluntary muscle contractions in man. *Journal of Physiology (London)* **220**, 1-18.

Stockbridge N (1988). Differential conduction at axonal bifurcations. II. Theoretical basis. *Journal of Neurophysiology* **59**, 1286-1295.

Stockbridge N & Stockbridge LL (1988). Differential conduction at axonal bifurcations. I. Effect of electrotonic length. *Journal of Neurophysiology* **59**, 1277-1285.

Swadlow HA, Kocsis JD & Waxman SG (1980). Modulation of impulse conduction along the axonal tree. *Annual Review of Biophysics and Bioengineering* **9**, 143-179.

Thesleff S (1959). Motor end-plate "desensitization" by repetitive nerve stimuli. *Journal of Physiology (London)* **148**, 659-664.

Tsujimoto T & Kuno M (1988). Calcitonin gene-related peptide prevents disuse-induced sprouting of rat motor nerve terminals. *Journal of Neuroscience* **8**, 3951-3957.

Volle RL & Branisteanu DD (1976). Quantal parameters of transmitter release at the frog neuromuscular junction. *Naunyn Schmiedebergs Archives of Pharmacology* **295**, 103-8.

Waxman SG (1975). Integrative properties and design principles of axons. *International Review of Neurobiology* **18**, 1-40.

Wernig A, Jans H & Zucker H (1986). A parametric study of the neuromuscular junction during ontogenesis and under different external conditions. In: Katz B, Rahamimoff R (eds.), *Calcium, Neuronal Function and Transmitter Release*, pp. 413-430. Boston: Martinus Nijhoff Publishers.

Woods JJ, Furbush F & Bigland-Ritchie B (1987). Evidence for a fatigue-induced reflex inhibition of motoneuron firing rates. *Journal of Neurophysiology* **58**, 125-137.

6

THE ROLE OF THE SARCOLEMMA ACTION POTENTIAL IN FATIGUE

A. J. Fuglevand

The John B. Pierce Laboratory
New Haven, Connecticut 06519

ABSTRACT

A prevalent feature of neuromuscular fatigue is a decline in the extracellularly recorded myoelectric signal. One factor that could underlie this change is a decrease in the amplitude of the sarcolemmal action potential. Based on observed reductions in action potential amplitude without effect on force, it has been argued that changes in the action potential during sustained activity would be unlikely to contribute to fatigue. However, those observations were primarily from experiments in which 1) high frequency stimulation may have caused signal cancellation due to action potential overlap; or 2) sustained membrane depolarization may have directly activated excitation-contraction coupling. The relatively low and narrow range of membrane depolarization required for full activation of amphibian and slow-twitch mammalian fibers makes them resistant to incomplete activation if action potentials are depressed during fatigue. Mammalian fast-twitch fibers, on the other hand, require greater depolarization for full activation and also exhibit a greater decrease in action potential amplitude with fatigue. Therefore, it seems probable that fatigue-related decline in action potential amplitude in these fibers leads to incomplete activation and loss of force.

INTRODUCTION

The loss of force of skeletal muscle during sustained activity is often accompanied by a decline in the extracellular or surface-detected electromyographic signal. This occurs under a variety of experimental situations: during maximal voluntary contraction of human quadriceps (Bigland-Ritchie et al., 1983); in electrically evoked responses of human first dorsal interosseous following submaximal voluntary contractions (Fuglevand et al., 1993); during intermittent stimulation (brief 40 Hz trains) of cat extensor digitorum longus (Enoka et al., 1989); and during continuous 80 Hz stimulation of fast-twitch motor units of cat medial gastrocnemius (Clamann & Robinson, 1985). The fatigue-related fall in EMG could represent diminished excitation of muscle, or it might reflect ancillary adaptations that have little impact on force development. The interpretation of the EMG decline, therefore, seems crucial for understanding the mechanisms of neuromuscular fatigue.

Fatigue, Edited by Simon C. Gandevia et al.
Plenum Press, New York, 1995

The decline in EMG responses to nerve stimulation could be due to impairment of transmission at axonal branch points or at the neuromuscular junction (see Sieck, Chapter 5). In addition, the decrease in EMG during sustained voluntary contraction could indicate a diminished output from the motorneuron pool. These possibilities clearly represent a decrease in muscle excitation. Whether fatigue-related diminution of muscle excitation directly promotes force loss, however, is an unresolved question which depends on how the input-output properties of motor units adapt during different fatigue tasks (Enoka & Stuart, 1992).

Another factor that may underlie the decline in EMG is a reduction in the amplitude of the sarcolemmal action potential. Intra- and extracellular recordings of single muscle fibers have shown the sarcolemmal action potential to decrease during prolonged activity (Sandercock et al., 1985; Lännergren & Westerblad, 1986; Metzger & Fitts, 1986; Radicheva et al., 1986; Balog et al., 1994). Activity-associated alteration in the transmembrane distribution of electrolytes (Hirche et al., 1980; Vyskocil et al., 1983; Sjøgaard, 1990, 1991; Fujimoto & Nishizono, 1993) can cause progressive depolarization of the resting membrane potential (Kwiecinski et al., 1984; Juel; 1986; Lindinger & Heigenhauser, 1991) which slows (Juel, 1988) and decreases the amplitude of the action potential (Jones, 1981; Jones & Bigland-Ritchie, 1986; Lännergren & Westerblad, 1986). If membrane depolarization is large enough, action potentials may fail to propagate altogether (Sandercock et al., 1985; Juel, 1988; Balog et al., 1994). Whether this occurs in voluntary contraction, however, has been extremely difficult to establish. Also, it is unclear if a diminished sarcolemmal action potential *per se* could fail to fully actuate the voltage-sensor calcium-release system of the t-tubule. This review explores the possibility that fatigue-related changes in the sarcolemmal action potential may contribute to diminished muscle excitation.

SIGNAL CANCELLATION OF EXTRACELLULAR POTENTIALS WITH HIGH-FREQUENCY STIMULATION

Some evidence indicates that reduction in action potential amplitude is not associated with a loss in muscle force. Action potentials recorded extracellularly in frog muscle fibers decline significantly with stimulation frequencies ≥ 50 Hz without a concomitant decrease in force (Lüttgau, 1965). Similarly, in slow-twitch motor units of the cat hindlimb, there is a substantial depression of EMG responses with continuous 80 Hz stimulation prior to any reduction in force (Clamann & Robinson, 1985). These observations suggest that action potential size can vary over a large range with little effect on excitation-contraction coupling and force. Other studies, however, have shown that during intermittent stimulation with 40 Hz trains, reduction of action potential amplitude is accompanied with decline in force in mammalian fast-twitch fatigable motor units and muscle with a high proportion of fast-twitch fibers (Reinking et al., 1975; Gardiner & Olha, 1987; Enoka et al., 1989). This discrepancy may be partly related to species and fiber-type differences in the relation between membrane potential and activation (described in a later section) and may also be an effect of extracellular recording of action potentials when stimulating at high rates.

Action potential amplitude decays steeply with increased distance between active fiber and extracellular electrode (Gath & Stålberg, 1978; Albers et al., 1989; Fuglevand et al., 1992). In addition, the duration of the biphasic action potential, as detected by extracellular electrodes, increases with electrode-fiber distance (Fuglevand et al., 1992). Therefore, when stimulating at high frequencies and recording EMG responses with extracellular or surface electrodes, it is possible for the positive phase of one action potential to overlap in

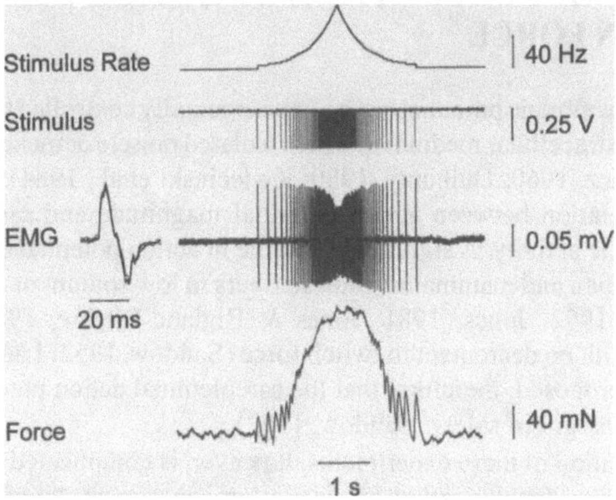

Figure 1. Example of electromyographic signal cancellation due to action potential overlap. Inset: five superimposed surface-detected EMG responses to intraneural stimulation (~1 Hz) of a single motor unit in human flexor digitorum superficialis muscle. The same unit was activated by a train of stimuli in which frequency increased from 5 to 80 Hz and then returned to 5 Hz. For frequencies above about 45 Hz, there was a progressive decay in EMG amplitude which was restored when frequency decreased. This pattern of change in the EMG was likely caused by progressive signal cancellation due to increasing overlap in biphasic action potentials with a duration of about 22 ms (from unpublished data of Macefield, Fuglevand and Bigland-Ritchie).

time with the negative phase of an adjacent potential. This causes signal cancellation and reduces the magnitude of the detected EMG response.

An example of this effect is presented in Fig. 1. A single motor unit of human flexor digitorum superficialis was stimulated via an intraneural microelectrode with a train in which frequency continuously increased from 5 to 80 Hz and then decreased back to 5 Hz. As would be anticipated for a biphasic motor unit potential with a duration of 22 ms (inset, Fig. 1) there was a progressive decline in the action potential amplitude due to overlap as frequency increased above 45 Hz. The immediate and essentially symmetric recovery of action potential amplitude as frequency declined indicates that the depressed action potential was an artifact of signal cancellation associated with high-frequency stimulation.

A consistent feature of fatigue is a broadening of the action potential associated with slowing of propagation velocity (Juel, 1988; Lindström et al., 1977). An increase in the duration of the action potential usually precedes reduction in amplitude or can occur in the absence of change in amplitude (Bigland-Ritchie et al., 1982, Duchateau & Hainaut, 1987, Kranz et al., 1983). Therefore, the progressive decline in action potential amplitude seen when stimulating at relatively high frequencies (Clamann & Robinson, 1985; Lüttgau, 1965) could be due partly to a gradual widening of the action potential leading to greater temporal overlap and signal cancellation of the EMG. A step decrease in amplitude between the first and second action potentials in high-frequency trains (Lüttgau, 1965; his Figs. 2, 3, 7, 8, 9, 12) and a nearly instantaneous recovery of the action potential upon cessation of high frequency stimulation (Lüttgau, 1965, his Fig. 5; Clamann & Robinson, 1985, their Fig. 6) are indicative of some degree of overlap cancellation. Therefore, to identify the physiological association between EMG and force in fatigue, it is important to ensure that interstimulus intervals are longer than the duration of the extracellularly detected action potential.

NON-FATIGUING MANIPULATION OF ACTION POTENTIAL AND ITS EFFECT ON FORCE

The resting membrane potential can be experimentally controlled by varying the ionic composition of the extracellular medium in which isolated muscle or muscle fibers are bathed (Hodgkin & Horowicz, 1960; Dulhunty, 1980; Kwiecinski et al., 1984). It is then possible to investigate the relation between action potential magnitude and mechanical response independent of muscle activity. A significant decline in action potential amplitude has been described for amphibian and mammalian muscle fibers in low sodium or elevated potassium solutions (Sandow, 1952; Jones, 1981; Jones & Bigland-Ritchie, 1986; Lännergren & Westerblad, 1986) with no decrement in twitch force (Sandow, 1952; Lännergren & Westerblad, 1986). It was proposed, therefore, that the sarcolemmal action potential operates as a trigger with a wide margin of safety (Sandow, 1952).

The interpretation of these experiments, however, is complicated by the direct effect of sustained membrane depolarization on excitation-contraction coupling. Muscle fibers exposed to high potassium develop force in the absence of propagated action potentials (Hodgkin & Horowicz, 1960; Lüttgau, 1963; Caputo, 1972) and the magnitude of the force contracture can be as large or larger than that obtained with tetanic stimulation (Hodgkin & Horowicz, 1960; Dulhunty & Gage, 1985). High extracellular potassium depolarizes the t-tubular membrane which directly activates the voltage sensor and induces sustained calcium release from the sarcoplasmic reticulum (Dulhunty, 1992). Thus, although experimental depolarization of the membrane diminishes the magnitude of the action potential transient, it can at the same time activate excitation-contraction coupling directly. Maintenance, indeed potentiation (Sandow, 1952; Chapman, 1969; Lännergren & Westerblad, 1986) of the force response under these conditions, therefore, may be less a demonstration of an action potential safety factor and more a reflection of direct activation of the t-tubular membrane.

MUSCLE ACTIVATION BY STEADY STATE DEPOLARIZATION

The difficulty associated with experimental control of the transient action potential has prompted investigations in which the membrane potential is held constant by adjusting the concentration of extracellular potassium or by voltage clamp (Caputo & de Bolaños, 1979; Dulhunty, 1992). A significant property of muscle activation revealed in these experiments is the dependence of developed force on membrane potential. Membrane depolarization by 30 - 40 mV from the resting potential is required to initiate a just-detectable mechanical response in amphibian and slow twitch mammalian fibers (Dulhunty, 1982; Hodgkin & Horowicz, 1960). To elicit a mechanical response in mammalian fast twitch fibers requires an additional 10-15 mV depolarization (Dulhunty, 1980; Dulhunty & Gage, 1983). Also, the threshold for mechanical response increases as the duration of the depolarization decreases (Dulhunty, 1980, 1982). Thus, further depolarization of about 10 mV is needed to attain mechanical threshold for brief pulses of duration similar to that of an action potential.

Beyond threshold depolarization, force increases as a sigmoid function of membrane potential (Hodgkin & Horowicz, 1960; Lüttgau & Oetliker, 1968; Dulhunty, 1980; Garcia et al., 1991). The membrane potential required to achieve half-maximum force under steady-state depolarization is about -48 mV for amphibian muscle (frog semitendinosus, Hodgkin & Horowicz, 1960), -25 mV for mammalian slow twitch (rat soleus, Dulhunty & Gage, 1983, 1985), and -10 mV for mammalian fast-twitch muscle fibers (rat extensor digitorum longus, Dulhunty & Gage, 1983; 1985). Maximal force is attained at steady-state

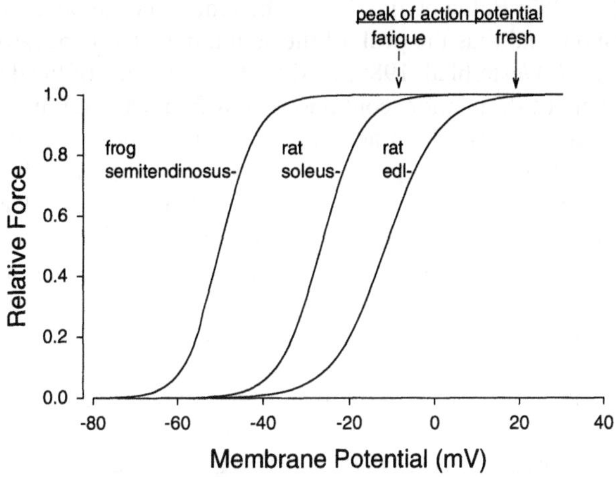

Figure 2. Relation between steady-state membrane potential and force for different muscles. The solid arrow indicates the approximate overshoot amplitude of action potentials in unfatigued muscle fibers. With fatigue, the action potential amplitude may decline (dashed arrow) which would likely have minimal effect on frog or mammalian slow-twitch fibers but could cause a significant decrement in force in mammalian fast-twitch fibers (frog semitendinosus curve estimated from data of Hodgkin and Horowicz, 1960; rat soleus and extensor digitorum longus (edl) curves from Dulhunty and Gage, 1985).

membrane potentials of about -25 mV for amphibian, -10 mV for mammalian slow twitch, and +15 mV for mammalian fast twitch fibers (Fig. 2). For brief, action potential-like depolarization it is probable that these curves would be shifted toward slightly more positive potentials (Heistracher & Hunt, 1969).

Although not yet fully identified, the mechanisms that underlie the marked dependence of force on membrane potential seem related to the activation of the voltage-sensors in the t-tubule. The movement of charged elements within the t-tubular membrane appears to provoke calcium release from the sarcoplasmic reticulum (Schneider & Chandler, 1973; Kovács et al., 1979). The magnitude of calcium release depends on the number of charged elements transposed to activating positions in the membrane (Chandler et al., 1975). The movement of this intramembranous charge has been quantified and, like force, is sigmoidally related to membrane potential (Schneider & Chandler, 1973; Kovács et al., 1979; Dulhunty & Gage, 1983).

For present purposes, two features of the data represented in Fig. 2 warrant particular attention. First, activation of muscle depends critically on the magnitude (and duration) of the membrane depolarization; and, therefore, does not operate as a simple trigger or *all-or-nothing* phenomenon. Second, the relation between developed force and membrane potential varies with fiber type, muscle, and species. In fresh muscle fibers, the overshoot of the intracellular action potential is to about + 20 mV (solid arrow, Fig. 2) and does not differ substantially across fiber types or muscles (19.5 mV in frog semitendinosus, Balog et al., 1994; 25.8 mV in frog sartorius, Light et al., 1994; 20.1 mV in rat soleus, 25.8 mV in rat extensor carpi radialis, Hanson, 1974; 23.0 mV in rat diaphragm, Metzger & Fitts, 1986; 16.8 mV in human intercostal, Kwiecinski et al., 1984). Action potentials of this magnitude should be more than adequate to fully activate amphibian and slow-twitch mammalian fibers while just sufficient to completely activate mammalian fast-twitch fibers (Fig. 2).

Many studies have shown the intracellular action potential to decline as a consequence of sustained activity (Hanson & Persson, 1971; Benzanilla et al., 1972; Hanson, 1974; Metzger & Fitts, 1986; Radicheva et al., 1986; Westerblad & Lännegren, 1986; Lännergren

& Westerblad, 1986, 1987; Balog et al., 1994). The reduction can be large enough so as to eliminate the overshoot and thus the peak of the action potential is negative (Benzanilla et al., 1972; Lännergren & Westerblad, 1986, 1987). The magnitude of the depression appears to depend on the pattern of stimulation; continuous high-frequency stimulation causes a rapid and marked decline in the intracellular action potential (e.g., peak amplitude to -28 mV in 14 s with 70 Hz stimulation, Lännergren & Westerblad, 1986) while less of a change in amplitude occurs with intermittent high-frequency stimulation (e.g., peak amplitude to +7.3 mV with 100 ms trains at 150 Hz every 1 s for 5 min, Balog et al., 1994).

Although less well documented, there is a greater decline in the intracellular action potential in fast-twitch as compared to slow-twitch fibers (Hanson, 1974). Extracellular recordings of muscle and motor unit action potentials provide additional indirect support for fiber-type differences in susceptibility for action potential decline. Fast-twitch fatigable motor units and muscles comprised of a high proportion of fast-twitch fibers exhibit a substantial decline in action potential amplitude during prolonged activity whereas little change is seen in fatigue-resistant motor units and muscles (Reinking et al., 1975; Kugelberg & Lindegren, 1979; Sandercock et al., 1985; Gardiner & Olha, 1987; Enoka et al., 1989; Hamm et al., 1989; Larsson et al., 1991).

The crucial issue is whether a reduction in the action potential would impair fiber activation. If the peak of the intracellular action potential declined to -10 mV during sustained activity (dashed arrow, Fig. 2) this would have little or no effect on force production in amphibian or mammalian slow-twitch fibers. It would, however, significantly reduce force for mammalian fast-twitch fibers. Fast-twitch fibers, therefore, appear susceptible to incomplete activation, not only because of the more positive position of their activation curves, but also because of their greater propensity for depressed action potentials (as described above). Furthermore, fatigue shifts the steady-state activation curves toward more positive potentials (by about 10 mV, Lüttgau & Oetliker, 1968; Garcia et al., 1991) thereby enhancing the probability of diminished activation.

Finally, Pagala and colleagues (1994) compared changes in electrically evoked tetanic force to potassium contracture force in mouse extensor digitorum longus following 3 min of intermittent stimulation with 30 Hz trains. Tetanic force declined by 50% whereas there was no reduction in potassium contracture force. These findings suggest that the reduction in force with electrical stimulation was due to an inadequate activation of the t-tubular membrane because direct depolarization of that membrane with high potassium revealed no deficit in force. Those observations are consistent with the proposal that fatigue-related depression of the sarcolemmal action potential in fast twitch muscle fibers may contribute to incomplete activation and force loss.

ACKNOWLEDGMENTS

The author wishes to thank Dr. Michael Walsh for helpful comments on the manuscript. Some of the work was supported by United States Public Health Service grants NS 14576 and HL 30062 (to Brenda Bigland-Ritchie). Attendance of the author at the 1994 Bigland-Ritchie conference was supported, in part, by the University of Miami.

REFERENCES

Albers BA, Put JHM, Wallinga W & Wirtz P (1989). Quantitative analysis of single muscle fibre action potentials recorded at known distances. *Electroencephalography and Clinical Neurophysiology* **73**, 245-253.

Balog EM, Thompson LV & Fitts RH (1994). Role of sarcolemma action potentials and excitability in muscle fatigue. *Journal of Applied Physiology* **76**, 2157-2162.

Benzanilla F, Caputo C, Gonzalez-Serratos H & Venosa RA (1972). Sodium dependence of the inward spread of activation in isolated twitch muscle fibres of the frog. *Journal of Physiology (London)* **223**, 507-523.

Bigland-Ritchie B, Johansson R, Lippold OCJ & Woods JJ (1983). Contractile speed and EMG changes during fatigue of sustained maximal voluntary contractions. *Journal of Neurophysiology* **50**, 313-324.

Bigland-Ritchie B, Kukulka CG, Lippold OCJ & Woods JJ (1982). The absence of neuromuscular transmission failure in sustained maximal voluntary contractions. *Journal of Physiology (London)* **330**, 265-278.

Caputo C (1972). The time course of potassium contractures of single muscle fibres. *Journal of Physiology (London)* **223**, 483-505.

Caputo C & de Bolaños PF (1979). Membrane potential, contractile activation and relaxation rates in voltage clamped short muscle fibres of the frog. *Journal of Physiology (London)* **289**, 175-189.

Chandler WK, Schneider MF, Rakowski RF & Adrian RH (1975). Charge movements in skeletal muscle. *Philosophical Transactions of the Royal Society of London, Series B* **270**, 501-505.

Chapman JB (1969). Potentiating effect of potassium on skeletal muscle twitch. *American Journal of Physiology* **217**, 898-902.

Clamann HP & Robinson AJ (1985). A comparison of electromyographic and mechanical fatigue properties in motor units of the cat hindlimb. *Brain Research* **327**, 203-219.

Duchateau J & Hainaut K (1987). Electrical and mechanical failure during sustained and intermittent contractions in humans. *Journal of Applied Physiology* **58**, 942-947.

Dulhunty AF (1980). Potassium contractures and mechanical activation in mammalian skeletal muscles. *Journal of Membrane Biology* **57**, 223-233.

Dulhunty AF (1982). Effects of membrane potential on mechanical activation in skeletal muscle. *Journal of General Physiology* **79**, 233-251.

Dulhunty AF (1992). The voltage-activation of contraction in skeletal muscle. *Progress in Biophysics and Molecular Biology* **57**, 181-223.

Dulhunty AF & Gage PW (1983). Asymmetrical charge movement in slow- and fast-twitch mammalian muscle fibres in normal and paraplegic rats. *Journal of Physiology (London)* **341**, 213-231.

Dulhunty AF & Gage PW (1985). Excitation-contraction coupling and charge movement in denervated rat extensor digitorum longus and soleus muscles. *Journal of Physiology (London)* **358**, 75-89.

Enoka RM, Rankin LL, Stuart DG & Volz KA (1989). Fatigability of rat hindlimb muscle: associations between electromyogram and force during a fatigue test. *Journal of Physiology (London)* **408**, 251-270.

Enoka RM & Stuart DG (1992). Neurobiology of muscle fatigue. *Journal of Applied Physiology* **72**, 1631-1648.

Fuglevand AJ, Winter DA, Patla AE & Stashuk D (1992). Detection of motor unit action potentials with surface electrodes: influence of electrode size and spacing. *Biological Cybernetics* **67**, 143-153.

Fuglevand AJ, Zackowski KM, Huey KA & Enoka RM (1993). Impairment of neuromuscular propagation during human fatiguing contractions at submaximal forces. *Journal of Physiology (London)* **460**, 549-572.

Fujimoto T & Nishizono H (1993). Involvement of membrane excitation failure in fatigue induced by intermittent submaximal voluntary contraction of the first dorsal interosseous muscle. *Journal of Sports Medicine and Physical Fitness* **33**, 107-117.

Garcia MDC, Gonzalez-Serratos H, Morgan JP, Perreault CL & Rozycka M (1991). Differential activation of myofibrils during fatigue in phasic skeletal muscle cells. *Journal of Muscle Research and Cell Motility* **12**, 412-424.

Gardiner PF & Olha AE (1987). Contractile and electromyographic characteristics of rat plantaris motor unit types during fatigue in situ. *Journal of Physiology (London)* **385**, 13-34.

Gath I & Stålberg E (1978). The calculated radial decline of the extracellular action potential compared with in situ measurements in the human brachial biceps. *Electroencephalography and Clinical Neurophysiology* **44**, 547-552.

Hamm TM, Reinking RM & Stuart DG (1989). Electromyographic responses of mammalian motor units to a fatigue test. *Electromyography and Clinical Neurophysiology* **29**, 485-494.

Hanson J (1974). The effects of repetitive stimulation on the action potential and the twitch of rat muscle. *Acta Physiologica Scandinavica* **90**, 387-400.

Hanson J & Persson A (1971). Changes in the action potential and contraction of isolated frog muscle after receptive stimulation. *Acta Physiologica Scandinavica* **81**, 340-348.

Heistracher P & Hunt CC (1969). The relation of membrane changes to contraction in twitch muscle fibres. *Journal of Physiology (London)* **201**, 589-611.

Hirche H, Schumacher E & Hagemann H (1980). Extracellular K^+ concentration and K^+ balance of the gastrocnemius of the dog during exercise. *Pflügers Archiv* **387**, 231-237.

Hodgkin AL & Horowicz P (1960). Potassium contractures in single muscle fibres. *Journal of Physiology (London)* **153**, 386-403.

Jones DA (1981). Muscle fatigue due to changes beyond the neuromuscular junction. In: Edwards RHT (ed.), *Human Muscle Fatigue: Physiological Mechanisms. Ciba Foundation Symposium*, pp. 178-196. London: Pitman Medical.

Jones DA & Bigland-Ritchie B (1986). Electrical and contractile changes in muscle fatigue. In: Saltin B (ed.), *International Series on Sport Sciences. Biochemistry of Exercise VI*, pp. 377-392. Champaign IL: Human Kinetics.

Juel C (1986). Potassium and sodium shifts during in vitro isometric muscle contraction, and the time course of the ion-gradient recovery. *Pflügers Archives* **406**, 458-463.

Juel C (1988). Muscle action potential propagation velocity changes during activity. *Muscle & Nerve* **11**, 714-719.

Kovács L, Ríos E & Schneider MF (1979). Calcium transients and intramembrane charge movement in skeletal muscle fibres. *Nature* **279**, 391-396.

Kranz H, Williams AW, Cassel J, Caddy DJ & Silberstein RB (1983). Factors determining the frequency content of the electromyogram. *Journal of Applied Physiology* **55**, 392-399.

Kugelberg E & Lindegren B (1979). Transmission and contraction fatigue of rat motor units in relation to succinate dehydrogenase activity of motor unit fibers. *Journal of Physiology (London)* **288**, 285-300.

Kwiecinski H, Lehmann-Horn F & Rudel R (1984). The resting membrane parameters of human intercostal muscle at low, normal, and high extracellular potassium. *Muscle & Nerve* **7**, 60-65.

Lännergren J & Westerblad H (1986). Force and membrane potential during and after fatiguing, continuous high-frequency stimulation of single *Xenopus* muscle fibers. *Acta Physiologica Scandinavica* **128**, 359-368.

Lännergren J & Westerblad H (1987). Action potential fatigue in single skeletal muscle fibres of *Xenopus*. *Acta Physiologica Scandinavica* **129**, 311-318.

Larsson L, Edström L, Lindegren B, Gorza L & Schiaffino, S (1991). MHC composition and enzyme-histochemical and physiological properties of a novel fast-twitch motor unit type. *American Journal of Physiology* **261**, C93-C101.

Light PE, Comtois AS & Renaud JM (1994). The effect of glibenclamide on frog skeletal muscle: evidence for K^+_{ATP} channel activation during fatigue. *Journal of Physiology (London)* **475**, 495-507.

Lindinger MI & Heigenhauser GJF (1991). The roles of ion fluxes in skeletal muscle fatigue. *Canadian Journal of Physiology and Pharmacology* **69**, 246-253.

Lindström L, Kadefors R & Petersén I (1977). An electromyographic index for localized muscle fatigue. *Journal of Applied Physiology* **43**, 750-754.

Lüttgau HC (1963). The action of calcium ions on potassium contractures of single muscle fibres. *Journal of Physiology (London)* **168**, 679-697.

Lüttgau HC (1965). The effect of metabolic inhibitors on the fatigue of the action potential in single muscle fibres. *Journal of Physiology (London)* **178**, 45-67.

Lüttgau HC & Oetliker H (1968). The action of caffeine on the activation of the contractile mechanism in striated muscle fibres. *Journal of Physiology (London)* **194**, 51-74.

Metzger JM & Fitts RH (1986). Fatigue from high- and low-frequency muscle stimulation: role of sarcolemma action potentials. *Experimental Neurology* **93**, 320-333.

Pagala M, Ravindran K, Amaladevi B, Namba T & Grob D (1994). Potassium and caffeine contractures of mouse muscles before and after fatiguing stimulation. *Muscle & Nerve* **17**, 852-859.

Radicheva N, Gerilovsky L & Gydikov A (1986). Changes in the muscle fibre extracellular action potentials in long-lasting (fatiguing) activity. *European Journal of Applied Physiology* **55**, 545-552.

Reinking RM, Stephens JA & Stuart DG (1975). The motor units of cat medial gastrocnemius: problem of their categorisation on the basis of mechanical properties. *Experimental Brain Research* **23**, 301-313.

Sandercock TG, Faulkner JA, Albers JW & Abbrecht PH (1985). Single motor unit and fiber action potentials during fatigue. *Journal of Applied Physiology* **58**, 1073-1079.

Sandow A (1952). Excitation-contraction coupling in muscular response. *Yale Journal of Biology and Medicine* **25**, 176-201.

Schneider MF & Chandler WK (1973). Voltage dependent charge movement in skeletal muscle: a possible step in excitation-contraction coupling. *Nature* **242**, 244-246.

Sjøgaard G (1990). Exercise-induced muscle fatigue: the significance of potassium. *Acta Physiologica Scandinavica Supplement* **593**, 1-63.

Sjøgaard G (1991). Role of exercise-induced potassium fluxes underlying muscle fatigue: a brief review. *Canadian Journal of Physiology and Pharmacology* **69**, 238-245.

Vyskocil F, Hník P, Rehfeldt H, Vejsada R & Ujec E (1983). The measurement of K_e^+ concentration changes in human muscles during volitional contractions. *Pflügers Archiv* **399**, 235-237.

Westerblad H & Lännergren J (1986). Force and membrane potential during and after fatiguing, intermittent tetanic stimulation of single *Xenopus* muscle fibers. *Acta Physiologica Scandinavica* **128**, 369-378.

SINGLE FIBER ELECTROMYOGRAPHY IN STUDIES OF NEUROMUSCULAR FUNCTION

J. V. Trontelj[1] and E. Stålberg[2]

[1] University Institute of Clinical Neurophysiology
University Medical Center of Ljubljana, Ljubljana, Slovenia
[2] Department of Clinical Neurophysiology
University Hospital
Uppsala, Sweden

ABSTRACT

Single-fiber electromyography (SFEMG) allows precise study of the microphysiology of the human motor unit under normal conditions. The physiological parameters that can be quantified include impulse transmission along the intramuscular axon collaterals, pre- and post synaptic events at the neuromuscular junction, and muscle fiber membrane properties. This chapter illustrates some of the advantages of SFEMG in studies of neuromuscular fatigue in normal muscle, as well as in disorders of neuromuscular transmission, and conditions associated with disturbed muscle fiber depolarization-repolarization.

INTRODUCTION

Single-fiber electromyography (SFEMG) is based on extracellular recordings of single muscle fiber action potentials with a small electrode, 25 μm in diameter, contained in a side port of a steel cannula which is inserted into muscle. The average uptake radius of the electrode is about 300 μm. Owing to this high spatial resolution, single and multiple muscle fiber action potentials of individual motor units can be reliably recognized and differentiated from others, allowing recordings over prolonged periods of time and in different experimental circumstances. During repetitive discharges, the single fiber action potentials have a well-defined and reproducible shape that justifies time measurements with an accuracy as high as 0.1 μs.

These recording advantages have been exploited to study morphological and functional details of the motor unit, ranging from action potential parameters of single muscle fibers to discharge characteristics of ventral horn cells and cortico-spinal tract axons. Microstimulation of individual muscle fibers and motor axons, either within the muscle or nerve trunk, has conferred additional advantages and enlarged the scope of research problems that can be studied with this method. A detailed description of the SFEMG technique, as well

Fatigue, Edited by Simon C. Gandevia et al.
Plenum Press, New York, 1995

as a comprehensive review of its uses and findings in health and diseased muscle, has been published (Stålberg & Trontelj, 1994)

In this chapter, we discuss some SFEMG findings pertinent to studies of muscle fatigue. These include analysis of the safety of transmission in the axonal tree and across the neuromuscular junction (NMJ) in ischemia and certain neuromuscular disorders, changes in muscle fiber propagation velocity, and muscle fiber de- and repolarization disturbances.

TRANSMISSION SAFETY IN THE INTRAMUSCULAR NERVE TREE

When recordings are made from three or more muscle fibers during voluntary activation, SFEMG offers the opportunity to study transmission in the intramuscular axon-collateral tree. In studies of axon reflexes in normal subjects, no failure of transmission is detectable at the nodes of Ranvier. In conditions associated with peripheral reinnervation, two or more components in a multi-unit recording may show simultaneous intermittent blocking. These potentials also show a large concomitant jitter relative to other parts of the action potential complex. This is usually due to an intermittent block in the common axonal branch supplying those muscle fibers from which the blocked action potentials are recorded. The concomitant jitter in relation to the non-blocking spikes in the multi-unit recording thus results from unreliable propagation of the axonal impulse in this common branch (Stålberg & Thiele, 1972).

As with the neuromuscular junction (NMJ) blocking encountered in myasthenia gravis, such axonal blocking may increase during repetitive discharge, and with an increase in stimulation rate. Both may produce a decrement in the surface-recorded EMG responses to repetitive stimulation. Axonal blocking may also show a response to edrophonium; that is, decreased jitter and less frequent blocking (Stålberg & Thiele, 1972) from which it follows that the presence of decrement and a positive edrophonium effect are not absolute proof of a synaptic transmission defect. Also, the blocking of even a single axonal impulse may result in a prolonged reduction of tension produced by the axon's motor unit, due to an inverse *catch-like effect* (Burke et al., 1970).

TRANSMISSION SAFETY AT THE NORMAL NEUROMUSCULAR JUNCTION

When a nerve fiber is stimulated repetitively at suprathreshold strength, and responses are recorded from a single muscle fiber, there is a latency variability in the order of tens of microseconds. This is called *jitter* (Ekstedt, 1964). Its main source in normal muscle is at the NMJ. As discussed elsewhere (Stålberg & Trontelj, 1994), it is due largely to small fluctuations in the firing threshold of single muscle fibers, resulting in a variable neuromuscular transmission time. Minor variations in the amplitude and slope of the end-plate potential are also contributing factors.

In pathology, additional mechanisms may contribute to the jitter. For example, there may be an abnormally low amplitude of the EPP due to decreased postsynaptic sensitivity, exaggerated variability of EPP amplitude due to impaired mechanism of transmitter release and/or disintegrated EPP shape due to asynchronous transmitter release. Thus, a combination of factors determines the amount of both normal and abnormal jitter.

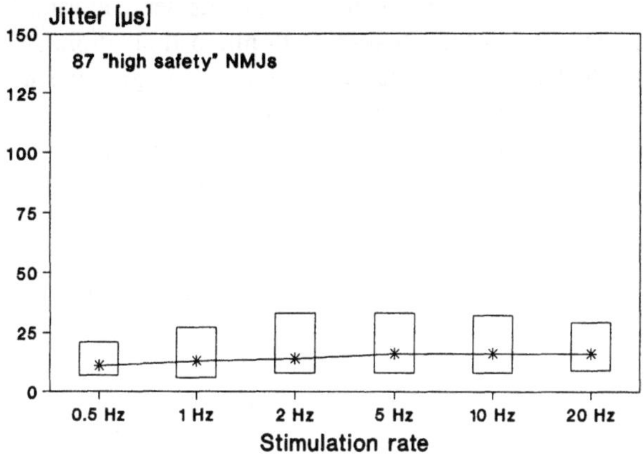

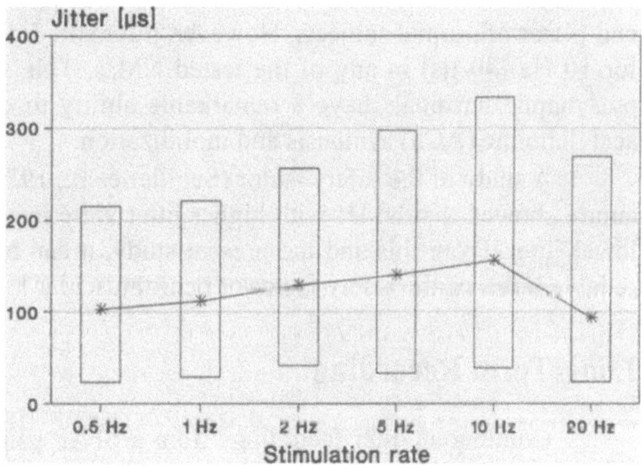

Figure 1. Jitter measurements at different stimulation rates in normal muscle. Median (*asterisk*) and range (*bar*) of values provided. *Top: low jitter, high safety* NMJs; n = 87; jitter ≤ 20µs). *Bottom: high jitter, low safety* NMJs; n = 33; jitter > 20µs). Note increase in jitter as the stimulation rate is raised from 0.5 to 5 Hz, and a decrease between 10 to 20 Hz (reprinted with permission from Stålberg & Trontelj, 1994; their Fig. 7.14).

Jitter can be studied in voluntarily activated muscle. In this case, the action potential of another muscle fiber of the same motor unit serves as the time reference for its measurement.

Changes in Jitter Related to Stimulation Rate

The effect of the degree of activity on jitter is of interest in normal muscle, and even more so in cases of disturbed neuromuscular transmission. Most normal end-plates display a rather uniform jitter (≤ 20 µs) at different stimulation rates within the physiological range. We denote these as *high safety factor endplates*. In a recent study of 120 NMJs in the extensor digitorum communis muscle (EDC) of 10 normal adults, the mean jitter values showed relatively little change when tested at different stimulation rates (Fig. 1). Jitter was slightly but significantly smaller at the lowest stimulation rate (0.5 Hz), as compared with a range of rates from 2 to 50 Hz. The majority of endplates exhibited this behavior. However, some maintained remarkably uniform, rather low jitter throughout the whole range of stimulation frequencies (2 to 50 Hz). Not all muscle fibers could follow stimulus frequencies >15-20 Hz. In such instances, the amplitude of the single fiber action potential (SFAP) rapidly diminished, and its rise time and latency increased. These changes progressed until the SFAP could no longer be recognized. When the stimulation rate was subsequently reduced, the SFAP reverted to its previous state. This reversible state is attributed to a progressive failure and subsequent recovery of the electrogenic Na^+-K^+ pump (McComas & Einhorn, 1990). Failure of the this pump is probably responsible for much of the decrement of the electrical response of a muscle to supramaximal stimulation of its nerve at rates ≥ 50 Hz (Desmedt, 1973). However, in contrast to the situation with microstimulation, supramaximal tetanic activation of a muscle tends to produce ischemia and progressive hypoxia. These may eventually compromise neuromuscular transmission.

Approximately 25% of randomly sampled end-plates in the Fig. 1 study showed changes in jitter related to stimulation (i.e., activation) rate. Jitter at most stimulation rates exceeded 20 μs (*low safety factor end-plates*). The typical pattern was a progressive increase in jitter to values > 30 μs as the stimulation rate was increased from 0.5 to 5 Hz. The range of jitter values decreased significantly at stimulation rates of 10, and particularly 20 Hz. Conversely, end-plates with moderate jitter values at low rates only rarely showed a significant increase (depression) at 20 Hz. Both patterns were more commonly seen in the *low safety* normal (and moderately abnormal) NMJs, as well as in the blocked NMJs of patients with myasthenia. The pattern exhibited by endplates with moderate jitter (no change) was much more common than that which decreased at 20 Hz. At 50 Hz (not shown in Fig. 1), a substantial increase in jitter occurred in over two thirds of both *low* and *high safety* end-plates of normal subjects. However, jitter still did not exceed the upper limit of normal for 10 Hz (40 μs) in any of the tested NMJs. This suggests that at these high rates, the presynaptic terminals have a remarkable ability to cope with the increased demand for acetylcholine (ACh) synthesis and mobilization.

A study of the safety factor (Schiller et al., 1975) of individual NMJs using regional curare showed that NMJs with higher jitter were more sensitive to curare than those with lower jitter. From this and the present study, it can be concluded that the jitter value is a reliable index of the safety factor of neuromuscular transmission.

Long-Term Recording

Continuous jitter recordings from a SFAP pair is possible for up to three hours at stimulation rates between 10 and 15 Hz. In normal muscle there is little change in jitter throughout this period. For stimulation rates > 30 Hz, jitter does not change appreciably for at least 10 min, by which time the test muscle becomes fatigued.

NEUROMUSCULAR TRANSMISSION DURING ISCHEMIA

To study the effects of ischemia, Dahlbäck and colleagues (1970) applied a sphygmomanometer to the subject's arm above the elbow, and the cuff was inflated to 200 mm Hg. The EDC muscle contracted weakly and SFEMG recordings were performed before, during and after (release of cuff) ischemia . The cuff was deflated when total blocking of one or the other action potential was evident. Following a few minutes of continuous activity during ischemia, the jitter increased rapidly, with one or the other action potential blocked intermittently; first rarely, and then more frequently until total block had occurred. The time to blocking was shorter with high activation rates. Approximately 2000-4000 discharges were required before the onset or blocking. After ischemia, the blocked action potentials recovered quickly, and jitter became close to normal within a few minutes. If a second period of ischemia was applied within 1 min, the time to impulse blocking was shorter than on the first occasion. Thus, in ischemia, block of neuromuscular transmission precedes the failure of depolarization/repolarization. This is in contrast to the rather normal jitter associated with progressive SFAP decrement during high frequency microstimulation in the normally perfused muscle.

NEUROMUSCULAR TRANSMISSION IN DISEASE

Myasthenia Gravis

The electrophysiological mechanisms underlying disturbed neuromuscular transmission in this disease have been well elucidated in *in vitro* studies (Elmqvist et al., 1964; Lambert et al., 1976, Engel et al., 1976). While the number of ACh quanta released from the presynaptic terminal per nerve impulse (quantum content of the end-plate potential (EPP)) is normal, the EPP amplitude is reduced due to a deficiency of postsynaptic ACh receptors, as well as to blocking of the receptors by IgG antibodies.

During activity, the deficiency and blocking of ACh receptors are partially counteracted by an increased mobilization rate of ACh vesicles in the presynaptic nerve terminal. This results in a short lived potentiation (Magleby & Zengel, 1976), followed within seconds to minutes by an even more pronounced postactivation depression of the EPP. In a repetitive nerve stimulation test, the size of the EPP response represents the net result of these two opposing processes.

The typical SFEMG findings in a patient with myasthenia (Stålberg et al., 1971; Stålberg et al., 1974; Stålberg et al., 1976; Sanders, 1987) include: 1) normal jitter values; 2) jitter values above the normal range but without impulse blocking; and 3) as dependent on the severity of the disease, recordings with increased jitter and intermittent impulse blocking. The latter usually first appears in association with a jitter near 100 μs (Fig. 2). Of 445 patients with myasthenia, the EDC muscle was abnormal in 99% of those with moderate or severe generalized disease and in 75% of the patients in clinical remission (Stålberg & Trontelj, 1994).

Jitter that is initially abnormal may increase during continuous activity particularly at increasing stimulation rates. However, when the stimulation rate subsequently decreases, the jitter and degree of blocking also decrease. In contrast, jitter that is initially normal changes little during activity. Sometimes a decrease in jitter and blocking is seen at increasing activation rates, and, occasionally, some muscle fibers belonging to the motor unit under study may become activated whereas previously, they had been in a state of persistent block.

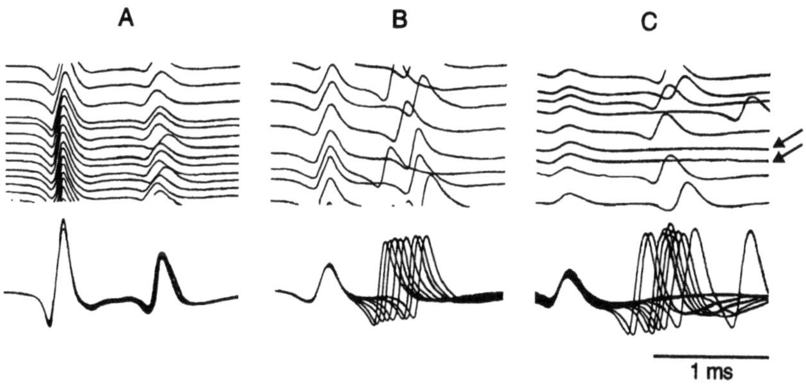

Figure 2. SFEMG jitter recordings from the EDC muscle of a patient with myasthenia gravis. The oscilloscope sweep is triggered by the first action potential and the jitter is the interval variability between the two single fiber action potentials. Upper traces show individual sweeps and the lowest trace is the superimposed sweeps. *A*, normal jitter; *B*, increased jitter but no impulse blocking; *C*, increased jitter and occasional blocking. In the lower part, the action potential discharges are superimposed (reprinted from Trontelj & Stålberg, 1991 by permission of John Wiley & Sons, Inc.).

Changes in Neuromuscular Transmission Efficiency Related to Changes in Activation Rate

These changes can be conveniently studied at the individual end-plates by use of SFEMG measurements. Valuable additional information, not on the affected postsynaptic apparatus, but rather on the function of the nerve terminal, may be provided. Indeed, myasthenia may be regarded as a valuable model to study the normal function of the motor nerve terminal.

A recent study (Trontelj & Stålberg, 1991; Fig. 3A) in 10 myasthenic patients observed the response of 58 NMJs to different stimulation rates. The lowest stimulation rates, 0.5 and 1.0 Hz, produced the lowest jitter and incidence of blocking which increased at stimulation rates of 2, 5, and 10 Hz. However, increasing the stimulation rate to 20 Hz reduced the amount of jitter and blocking to the level observed at the low stimulation rates. This improvement is considered to be due to an increase in ACh released per stimulus; i.e., intratetanic potentiation (Fig. 3).

Functionally, the most important part of the stimulation frequency spectrum investigated lies between 10 and 20 Hz. Most of the activity of motor units in the limb muscles occurs in this range during sustained contractions involving submaximal effort (Stålberg et al., 1976). The most significant intratetanic potentiation is expected to occur at these frequencies and it follows that the prominent potentiation observed at the NMJs with large jitter might be an adaptive presynaptic mechanism to safeguard function of an affected or a normal *low safety* NMJ.

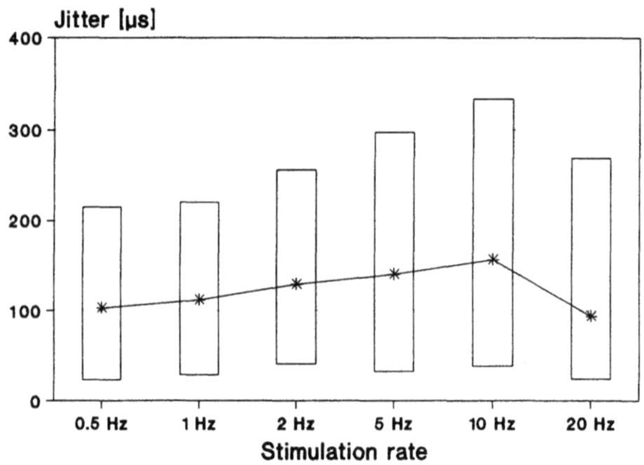

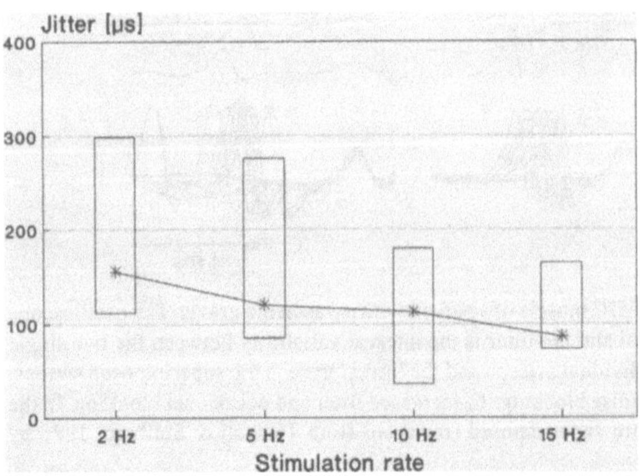

Figure 3. The effect of intramuscular electrical stimulation at different frequencies. Median and range values as in Fig. 1. *Top:* Jitter changes at 40 NMJs in myasthenia gravis. *Bottom:* Jitter changes at 18 NMJs in Lambert-Eaton myasthenic syndrome (reprinted with permission from Stålberg & Trontelj, 1994; their Figs. 15.30 and 15.33).

However, in a severely affected muscle, a proportion of end-plates may be unable to follow a stimulation rate of 2 Hz or more since they develop a persistent block within the first few stimuli (Trontelj & Stålberg, 1993). Also, a few NMJs in myasthenic muscle demonstrated moderate jitter at low and intermediate rates with an increased jitter at 20 Hz. These suggest insufficient transmitter synthesis and/or mobilization at the higher rate.

Lambert-Eaton Myasthenic Syndrome (LEMS)

This rare neuromuscular transmission disorder is often associated with bronchogenic carcinoma (Eaton & Lambert, 1957). It is caused by an impaired neurotransmitter release, resulting from an autoimmune attack on calcium channels in the presynaptic nerve terminal (Lambert & Elmqvist, 1971). Even in clinically strong muscles, the jitter is often grossly abnormal. The most striking effect is a dramatic reduction of jitter and, in particular, blocking as the stimulation rate is increased from 1 - 2 Hz to 10 - 20 Hz (Trontelj & Stålberg, 1991). Many NMJs (often up to one-third or more in a given muscle) are in a state of complete conduction block. This is only overcome when the stimulation rate is increased above 10-20 Hz. This seems to be the SFEMG hallmark of LEMS (Trontelj & Stålberg, 1993).

PROPAGATION VELOCITY OF SINGLE MUSCLE FIBERS

Measurement of the propagation velocity of the single muscle fiber action potential along the muscle fiber (Stålberg, 1966) requires use of a multi-electrode recording with two arrays of recording surfaces. To calculate the propagation velocity, the time interval is measured between potentials recorded from the muscle fiber at the two recording sites. The range of normal muscle fiber propagation velocity values lies between 1.5 and 6.5 m/s. It varies from muscle to muscle and even within a given muscle, the major determinant being the diameter of the muscle fiber (Håkansson, 1956).

Propagation Velocity during Repetitive Activity

Propagation velocity usually decreases during the first minute of continuous single fiber activation (Stålberg, 1966; Fig. 4). This decrease can amount to 50% of the initial value over a period of 3 min. When the stimulation period is interrupted, the amount the propagation velocity recovers depends on the length of the interruption.

When using concentric needle EMG recording (Lindström et al., 1977), such a decrease in propagation velocity explains the change in power spectrum observed when the higher stimulation frequencies decrease, and the lower frequencies increase during continuous activity. It also explains, in part, changes in the number of turns in analysis of the EMG's interference pattern (Fuglsang-Frederiksen & Rönager, 1988). Using arrays of surface electrodes, it is possible to study the mean conduction velocity of fibers in *individual motor units* and to determine the direction of action potential propagation (Masuda & Sadoyama, 1986). A specialized method for analyzing the surface EMG (Lindström et al., 1977) provides an indirect measure of the mean propagation velocity in the *entire active muscle*. This value decreases during continuous activity. A similar decrease in velocity is seen during high frequency stimulation of individual muscle fibers, in which an increase in latency is also observed.

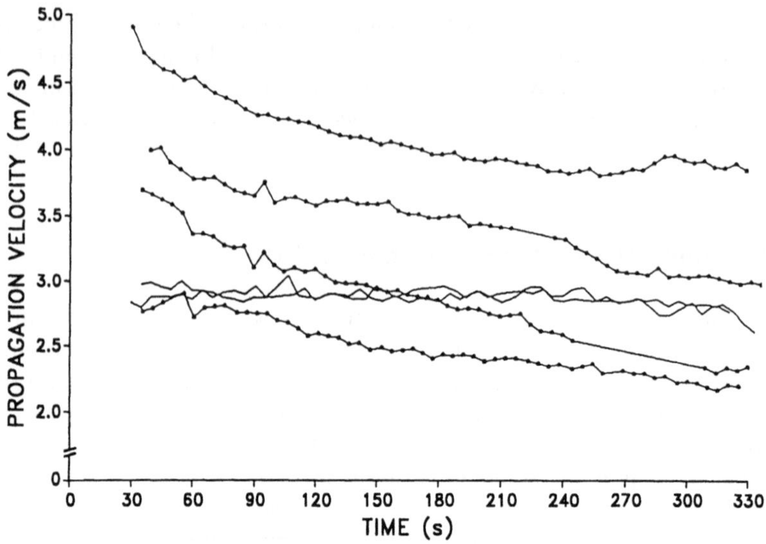

Figure 4. Propagation velocity of five muscle fibers during repetitive stimulation. Lines connected by asterisks show decrease in velocity of four fibers discharging at 11, 16, 12 and 9 discharges/s (top to bottom, respectively). A fifth fiber (the two continuous lines) was tested at stimulation rates of 6/s and 16/s. It exhibited no decrease in propagation velocity (reprinted with permission from Stålberg & Trontelj, 1994; their Fig. 10.2).

MUSCLE FIBER DEPOLARIZATION AND CONDUCTION IN MYOTONIA

In myotonic disorders, muscle force may decline during ongoing contraction, particularly when started after a period of rest. This is most pronounced in the recessive (Becker's) form of myotonia congenita, in which weakness may be more disturbing for the patient than stiffness. Weakness is also seen in the dominant (Thomsen's) form.

Repetitive electrical stimulation (5 Hz) of the ulnar nerve in a patient with Becker's type of myotonia congenita produced a decline of twitch force to zero within 5 s. Force gradually recovered after 20 s of continuous stimulation (Stålberg & Trontelj, 1994). The initial decline in force was associated with a decrement of the amplitude of the compound muscle action potential to a very low value, but not to zero. This indicates, in part, undisturbed neuromuscular transmission and preserved muscle fiber depolarization with impulse conduction, but failing excitation-contraction coupling (Ricker & Meinck, 1972; Buchthal & Rosenfalck, 1963). At a stimulation frequency of 15 Hz, paralysis occurred within 3 s and lasted about 80 s. However, this happened only when the muscle was allowed to rest for about 15 min before the stimulation period. If nerve stimulation was performed after a voluntary contraction of 5 min, there was barely any decrement in the amplitude of the muscle action potentials.

A stimulation SFEMG study was performed in 5 patients, 3 with the Thomsen type and 2 with sporadic myotonia, with characteristics of the Becker type (Trontelj, 1986; Trontelj et al., 1987). Forty-five muscle fibers were studied at various stimulation rates. In any individual, there was a large range in degree and slope of action potential decrement. For most of the fibers in patients with Thomsen's type of myotonia congenita, stimulation at 20 Hz produced a rather rapid decrement in amplitude to between 40 and 70% within the first half second. In the two patients with probable Becker's type of myotonia congenita, profound action potential decrements could be observed at relatively low stimulation

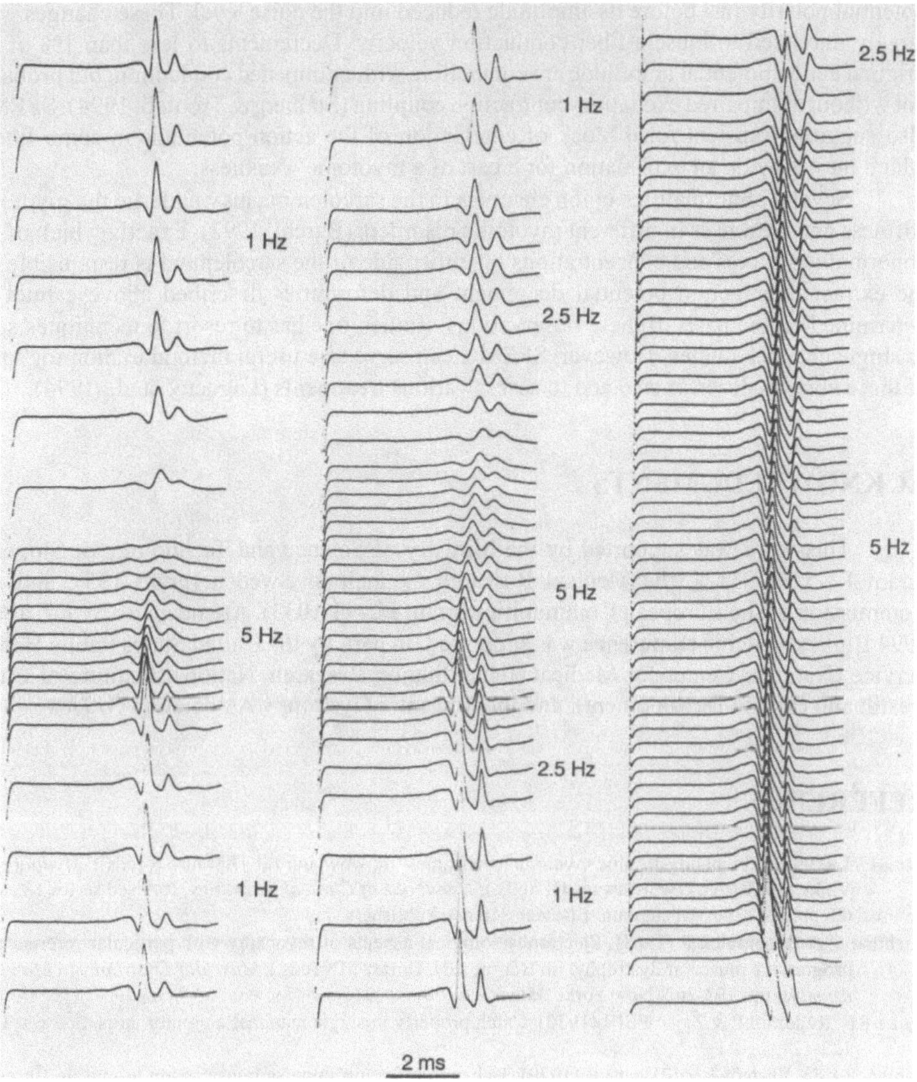

Figure 5. Profound single fiber action potential decrement in a patient with probable Becker's type of myotonia congenita. Repeated trains of stimuli separated by short pauses of 2 s show successively smaller decrement (related to the *warm-up* phenomenon) (reprinted with permission from Stålberg & Trontelj, 1994; their Fig. 15.41).

frequencies. In several fibers, action potential amplitudes were reduced to zero after just 10-30 s of stimulation at 2 Hz. This degree of decrement was not commonly observed in the three patients with the Thomsen type. When a series of stimulations were repeated at intervals as short as 2-10 s, the decrement usually became progressively less prominent, thereby indicating a warm-up effect (Fig. 5). However, upon occasion, the opposite behavior was seen.

In addition to amplitude decrement, the SFEMG recordings in patients with Becker's type of myotonia congenita showed prominent changes in action potential shape. A progressively longer negative after-potential was followed, in some cases, by a reversal of action potential polarity just before its amplitude reduced into the noise level. These changes were largely unrelated to muscle fiber conduction velocity. Decrements to less than 1% of the original action potential amplitude are compatible with unimpeded conduction; but probably not without unimpaired excitation-contraction coupling (Stålberg & Trontelj, 1994). SFEMG also suggests transient focal block of conduction of the action potentials in some fibers, which may provide an explanation for a part of a myotonic weakness.

Several abnormalities in ion channels in the sarcolemma may underlie the myotonic stiffness and weakness in different myotonic disorders (Barchi, 1992). Exactly which of the abnormal ion fluxes and concentrations on either side of the sarcolemma is responsible for the extracellular action potential decrement and deformities described above cannot be determined on the basis of these observations. Rather, one has to resort to techniques such as single channel studies. However, SFEMG can serve as a useful method to monitor some of these abnormalities *in vivo* and to assess various treatments (Lagueny et al., 1994).

ACKNOWLEDGMENTS

This work was supported by the Ministry of Science and Technology of Slovenia (grant P3-5285-0312), the Medical Research Council of Sweden (grant 135), and the Commission of the European Communities (grant PL-93-1033). Attendance of *J.V.T.* at the 1994 Bigland-Ritchie conference was supported, in part, by the United States Public Health Service (National Center for Medical Rehabilitation Research, National Institute of Child Health and Human Development), and the Muscular Dystrophy Association (USA).

REFERENCES

Barchi RL (1992). The nondystrophic myotonic syndromes. In: Rowland LP, DiMauro S (eds.), *Myopathies*, Vinken PJ, Bruyn GW, Klawans HL (eds.), *Handbook of Clinical Neurology*, Revised series 18,. vol. 62, pp. 261-286. Amsterdam: Elsevier Science Publishers.

Buchthal F & Rosenfalck P (1963). Electrophysiological aspects of myopathy with particular reference to progressive muscular dystrophy. In: Bourne GH, Golarz MN (eds.), *Muscular Dystrophy in Man and Animals*, pp. 193-262. New York: Hafner.

Burke RE, Rudomin P & Zajac FEIII (1970). Catch property in single mammalian motor units. *Science* **168**, 122-124.

Dahlbäck L-O, Ekstedt J & Stålberg E (1970). Ischemic effect on impulse transmission to muscle fibres in man. *Electroencephalography and Clinical Neurophysiology* **29**, 579-591.

Desmedt JE (1973). The neuromuscular disorder in myasthenia gravis. I. Electrical and mechanical responses to nerve stimulation in hand muscles. In: Desmedt JE (ed.), *New Developments in Electromyography and Clinical Neurophysiology*, vol. 1, pp. 241-304. Basel: Karger.

Eaton LM & Lambert EH (1957). Electromyography and electric stimulation of nerves in diseases of motor unit: observations on the myasthenic syndrome associated with malignant tumors. *Journal of American Medical Association* **163**, 1117-1124.

Ekstedt J (1964). Human single muscle fiber action potentials. *Acta Physiologica Scandinavica* **61** (suppl. 226), 1-96.

Elmqvist D, Hofmann WW, Kugelberg J & Quastel DM (1964). An electrophysiological investigation of neuromuscular transmission in myasthenia gravis. *Journal of Physiology (London)* **174**, 417-434.

Engel AG, Tsujihata M, Lambert JM & Lennon VA (1976). Experimental autoimmune myasthenia gravis; a sequential and quantitative study of the neuromuscular junction ultrastructure and electrophysiologic correlations. *Journal of Neurophathology and Experimental Neurology* **35**, 569-587.

Fuglsang-Frederiksen A & Rönager J (1988). The motor unit firing rate and the power spectrum of EMG in humans. *Electroencephalography and Clinical Neurophysiology* **70**, 68-72.

Håkansson CH (1956). Conduction velocity and amplitude of the action potential as related to the circumference in the isolated fibre of frog muscle. *Acta Physiologica Scandinavica* **37**, 14-34.

Lagueny A, Marthan R, Sheuermans P, Le Collen P, Ferrer X, Julien J (1994). Single fiber EMG and spectral analysis of surface EMG in myotonia congenita with and without transient weakness. *Muscle & Nerve* **17**, 248-250.

Lambert EH & Elmqvist D. (1971). Quantal components of the end-plate potential in the myasthenic syndrome. *Annals of the New York Academy of Sciences* **183**, 183-199.

Lambert EH, Lindstrom JM & Lennon VA (1976). End-plate potentials in experimental autoimmune myasthenia gravis in rats. *Annals of the New York Academy of Sciences* **274**, 300-318.

Lindström L, Kadefors R & Petersen I (1977). An electromyographic index for localised muscle fatigue. *Journal of Applied Physiology* **43**, 750-754.

Magleby KI & Zengel JE (1976). Long term changes in augmentation, potentiation and depression of transmitter release as a function of repeated synaptic activity at the frog neuromuscular junction. *Journal of Physiology (London)* **257**, 471-494.

Masuda T & Sadoyama T (1986). The propagation of single motor unit potentials detected by a surface electrode array. *Electroencephalography and Clinical Neurophysiology* **63**, 590-598.

McComas AJ & Einhorn RW (1990). Electric responses of human muscles to prolonged repetitive stimulation. *Muscle & Nerve* **13**, 879.

Ricker K & Meinck HM (1972). Muscular paralysis in myotonia congenita. *European Neurology* **7**, 221-227.

Sanders DB (1987). The electrodiagnosis of myasthenia gravis. *Annals of New York Academy of Sciences* **505**, 539-556.

Schiller HH, Stålberg E & Schwartz MS (1975). Regional curare for the reduction of the safety factor in human mot end-plates studied with single fibre electromyography. *Journal of Neurology, Neurosurgery and Psychiatry* **38**, 805-809.

Stålberg E (1966). Propagation velocity in single human muscle fibres in situ. *Acta Physiologica Scandinavica* **70** (suppl. 287), 1-112.

Stålberg E, Ekstedt J & Broman A (1971). The electromyographic jitter in normal human muscles. *Electroencephalography and Clinical Neurophysiology* **31**, 429-438.

Stålberg E, Ekstedt J & Broman A (1974). Neuromuscular transmission in myasthenia gravis studied with single fibre electromyography. *Journal of Neurology, Neurosurgery and Psychiatry* **37**, 540-547.

Stålberg E & Thiele B (1972). Transmission block in terminal nerve twigs: a single fibre electromyographic finding in man. *Journal of Neurology, Neurosurgery and Psychiatry* **35**, 52-59.

Stålberg E & Trontelj JV (1994). *Single Fiber Electromyography. Studies in Healthy and Diseased Muscle.* New York: Raven Press.

Stålberg E, Trontelj JV & Schwartz MS (1976). Single-muscle-fiber recording of the jitter phenomenon in patients with myasthenia gravis and in members of their families. *Annals of the New York Academy of Sciences* **274**, 189-202.

Trontelj JV (1986). Intramuscular stimulation: Technique and findings in the normal and abnormal motor unit. *Muscle & Nerve* **9**, S83.

Trontelj JV, Mihelin M & Stålberg E (1987). Extracellularly recorded single muscle fibre responses to electrical stimulation in myotonia congenita. *Electroencephalography and Clinical Neurophysiology* **66**, S106.

Trontelj JV & Stålberg E (1991). Single motor end-plates in myasthenia gravis and LEMS at different firing rates. *Muscle & Nerve* **14**, 226-232.

Trontelj JV & Stålberg E (1993). Stimulation SFEMG: A stimulation rate of 2 Hz might be inadequate to find the recording position (a reply). *Muscle & Nerve* **16**, 564-565.

SECTION III

Fatigue of Single Motor Units

A motor unit is no more than a collection of relatively homogeneous muscle fibers innervated by a single motoneuron. The force exerted by a single motor unit depends not only on the state of neuromuscular transmission and the intracellular environment but also on the discharge properties of the motoneuron. This section evaluates the behavior of the interface between the CNS and the peripheral motor. This perspective is important because it defines the constraints imposed on the CNS "upstream" of the motoneuron pool.

A functional match is presumed to exist between the electrical properties of motoneurons and the mechanical behavior of the muscle fibers they innervate. The intrinsic properties of motoneurons that affect their recruitment and repetitive firing characteristics are reviewed in **Chapter 8** (Sawczuk and colleagues). Particular attention is given to factors affecting the threshold current required to activate motoneurons of different types and those that influence the duration of the motoneuron's after-hyperpolarization. This duration appears to parallel the duration of the twitch force produced by the motor unit. Thus, both durations are long in the least fatigable, slow twitch, type S motor units and their motoneurons but brief in fatigable, fast-twitch, type F motor units and motoneurons. Clearly, factors that control the excitability of a motoneuron and its after-hyperpolarization are important for understanding how it will behave under the influence of the powerful net excitatory drive during strong contractions. The biophysical properties that underlie the early and late adaptation of motoneuron-discharge rate to sustained input are discussed. Because the adaptation is greater in type F than type S units, these properties may be tuned to optimize the tension-firing frequency relation in individual motoneurons. Furthermore, the extent to which a motoneuron may take up one or more stable *states* under different tasks is currently the subject of much investigation.

In **Chapter 9**, Kernell continues to explore how the motoneuron's discharge to a constant input current matches the response of its muscle fibers to different discharge rates. He reviews evidence that activity-dependent mechanisms at the level of both the motoneuron (transformation of input current to discharge rate) and the muscle fiber (transformation of discharge rate to force) may be designed to minimize fatigue and to minimize the central command (or *effort*) required for force production by a muscle or motor unit. The sigmoidal relation between isometric force and discharge rate may be distorted by the exact pattern of activation of the motor unit leading to a mixture of both potentiation (i.e. an increase in force) and fatigue (i.e. a reduction in force). Another recurrent theme surfaces here, namely the simultaneous mix of exercise-induced, intra- and extracellular factors, some of which increase force and others that decrease it.

In the next two chapters (**Chapter 10**, Thomas; **Chapter 11**, Elek and Dengler), the concepts developed earlier in the section are translated into studies of the properties of human motor units. Several methods are available to study single motor units in humans and, not surprisingly, none is perfect. Thomas reviews the data obtained with spike-triggered averaging and the different forms of motor axonal stimulation. The analysis highlights the differences between the techniques, the mutability of the force-frequency relationship for single motor units, some apparent differences in fatigability between human motor units and those in other species, and how little is known about the contractile properties of human motor units during normal use. Elek and Dengler compare spike-triggered averaging and intramuscular stimulation for the first dorsal interosseous muscle of human subjects. Data are reported for patients with hereditary motor and sensory neuropathy (type 1) and amyotrophic lateral sclerosis. Predictably, values for twitch contraction times and relaxation times are shorter when assessed with spike-triggered averaging. The suitability of the two methods for use in patients is discussed, with the yield being higher with spike-triggered averaging. However, in patients in whom voluntary contractions for long periods are precluded or who may have increased motor unit synchronization, intramuscular stimulation may be preferable.

INTRINSIC PROPERTIES OF MOTONEURONS

Implications for Muscle Fatigue

A. Sawczuk,[1] R. K. Powers,[2] and M. D. Binder[2]

[1] Department of Oral Biology
[2] Department of Physiology and Biophysics
University of Washington
Seattle, Washington 98195

ABSTRACT

The following is a brief review of the intrinsic properties of motoneurons that contribute to their recruitment and rate modulation. Our emphasis is on properties that may either accelerate or delay the onset of muscular fatigue. In general, intrinsic motoneuron properties are regulated in a way that minimizes energy expenditure. The correlation of recruitment threshold with motoneuron type ensures that the most fatigable motor units are reserved for the most forceful contractions. The variation in minimum firing rates arising from variations in AHP characteristics ensures that motoneurons begin to fire at rates that are matched to the force producing characteristics of their muscle units. Further, it is possible that spike-frequency adaptation contributes to optimization of the tension (force)-firing frequency (T-f) transform of individual motor units.

INTRODUCTION

The notion of a precise, functional match between the intrinsic electrical properties of motoneurons and the mechanical properties of the muscle fibers they innervate was formalized by Henneman and colleagues as the *size principle* of motor unit recruitment (Henneman et al., 1965; Henneman & Olson, 1965). One advantage of size-ordered recruitment arises from the inverse relation between motor unit size and fatigue-resistance, so that larger, fatigable motor units are normally reserved for brief, forceful contractions (Henneman & Olson, 1965). Subsequent research has confirmed the generality of this size-based recruitment order and has provided insights into the underlying physiological mechanisms (Burke, 1981; Henneman & Mendell, 1981; Heckman & Binder, 1990; Binder et al., 1995). Once recruited, the firing rate of motoneurons increases as a function of increasing depolarizing synaptic input. The intrinsic mechanisms controlling motoneuron spike-frequency are also matched to motor unit mechanical properties, so that rate modulation normally occurs over a range that is most efficacious in changing motor unit force (Kernell, 1983; Binder et al., 1995). Finally, both the spike-frequency of motoneurons and the mechanical output of

Fatigue, Edited by Simon C. Gandevia et al.
Plenum Press, New York, 1995

motor units depend upon their activation history (Binder et al., 1995). These history-dependent properties of motoneurons may also be matched to those of their motor units. The following review will briefly cover the intrinsic properties of motoneurons that contribute to their recruitment and rate modulation, with particular emphasis on those properties that may be implicated in muscle fatigue.

INTRINSIC MOTONEURON PROPERTIES UNDERLYING RECRUITMENT

Motoneurons are recruited when the somatic membrane potential is displaced from its resting value (V_r) to the threshold value for initiating an action potential (V_{thr}). The amount of injected or synaptic current (rheobase; I_{rh}) needed to recruit a motoneuron will depend upon the difference between the threshold and resting voltages, and upon the effective input resistance of the cell (R_N): $I_{rh} = (V_{thr} - V_r)/R_N$

The intrinsic properties important for recruitment are those determining V_r, V_{thr} and R_N. Input resistances measured in cat lumbar motoneurons *in vivo* exhibit about a 10-fold range from about 0.4 - 4.0 MΩ (e.g., Gustafsson & Pinter, 1984b; Zengel et al., 1985). Average R_N values are largest in type S, smaller in type FR, and smallest in type FF motoneurons (Fleshman et al., 1981; Zengel et al., 1985). The finding that the range of variation of I_{rh} exceeds that of R_N may result from a tendency for V_{thr} to be lowest in low rheobase, high input resistance (presumably type S and FR) motoneurons (Gustafsson & Pinter, 1984a; Carp, 1992).

Once the voltage threshold for spike initiation is exceeded, action potentials arise from the influx of sodium at the initial segment of the motoneuron. Initial segment sodium currents are first observed at voltages 10 mV above the resting potential, whereas somatic sodium currents are not observed until the membrane is depolarized to 20 mV above rest and are maximal at 30 mV of depolarization (Schwindt & Crill, 1982). The somatic voltage at which the initial segment spike is elicited (V_{thr}) is not fixed, but can exhibit both rapid (Calvin, 1974) and slow fluctuations (Schwindt & Crill, 1982) during repetitive discharge. This behavior is known as accommodation, and is thought to be due, in part, to subthreshold changes in the amount of sodium channel inactivation (Frankenhaeuser & Vallbo, 1964; Vallbo, 1964; Schlue et al., 1974). Experiments on cat motor axons suggest inactivation time constants in the range of 2 - 4 ms (Richter et al., 1974), but additional slower inactivation processes may underlie the slow increase in firing level observed during repetitive discharge (Schwindt & Crill, 1982). Variations in V_{thr} among different motoneurons may reflect variations in the relative density of initial segment sodium channels or perhaps a difference in their accommodative properties (Burke & Nelson, 1971).

The motoneuron action potential in vertebrates is followed by a prolonged afterhyperpolarization (AHP), lasting 50-200 ms. This process has been called the medium-duration AHP (mAHP; Nishimura et al, 1989; Viana et al., 1993) to distinguish it from a distinct rapid hyperpolarization often observed in the same motoneurons (Hounsgaard et al., 1988; Nishimura et al., 1989; Viana et al., 1993), and a much slower process observed in other types of neurons (cf. Nishimura et al., 1989). There is now abundant supportive evidence that this medium-duration AHP is mediated by an apamin-sensitive, calcium-activated potassium conductance (G_{KCa}; Krnjevic et al., 1979; Zhang & Krnjevic, 1987; Hounsgaard et al., 1988; Mosfeldt-Laursen & Rekling, 1989; Nishimura et al., 1989; Viana et al., 1993; Chandler et al., 1994). The magnitude and duration of the AHP vary according to motoneuron size (Zwaagstra & Kernell, 1980) and motor unit type: larger and longer AHP*s* are recorded in type S than in type F motoneurons (Kernell, 1983; Zengel et al., 1985). It is not known whether these differences in AHP characteristics reflect differences in the properties or density of G_{KCa} channels or differences in the various factors controlling the time course of the calcium concentration in the vicinity of the channels.

REPETITIVE FIRING

If the magnitude of injected or synaptic current is increased above that needed to elicit a single action potential, repetitive discharge ensues. Subsequent firing rate modulation depends upon the intrinsic relation between injected or synaptic current (I) and the spike frequency (f), which is linear over a wide range of currents: $f = f_0 + (f/I) * (I - I_0)$, where I_0 is the minimum current needed to elicit steady repetitive discharge (generally about 1.5 times the current needed to produce a single action potential, I_{rh}; Kernell, 1965c); f_0 is the minimum firing rate; and f/I is the slope of the frequency-current relation. In some motoneurons, the steady-state relation between spike frequency and injected current (f-I) can be characterized by a single linear equation over the entire range of applied currents, whereas in others, the steady-state f-I relation consists of two or three linear segments of different slope (primary, secondary and tertiary ranges of firing; cf. Schwindt & Crill, 1984).

The conductance underlying the AHP is one of the primary determinants of the steady-state repetitive discharge properties of motoneurons. The minimum steady spike-frequency is approximately equal to the reciprocal of AHP duration (Kernell, 1965c; Jodkowski et al., 1988) and the steady-state frequencies attained at the end of the primary and secondary ranges of firing are also correlated with the reciprocal of AHP duration (Kernell, 1965c; see also Kernell, Chapter 9, Fig. 1). Finally, comparisons of steady-state f-I relations before and after pharmacological manipulations of the AHP conductance (Hounsgaard et al., 1988; Nishimura et al., 1989; Viana et al., 1993; Chandler et al., 1994), indicate that the steady-state f-I slope is inversely proportional to the magnitude of the AHP conductance.

A variety of evidence suggests that two other intrinsic properties make important contributions to motoneuron spike-frequency behavior 1) a low threshold, persistent inward current; and 2) variations in V_{thr} resulting from the accommodative properties of the initial segment. The persistent inward current was first identified in cat lumbar motoneurons based on the clamp current records during slow depolarizing voltage-clamp ramp commands (Schwindt & Crill, 1977). The steady-state current-voltage (I-V) curve exhibits an N-shape starting with a region of negative slope conductance at about 10-20 mV above the resting potential, which subsequently reverses to a positive slope due to increasing activation of outward conductances. Voltage-clamp commands encompassing the region of negative slope conductance reveal a slowly-activating inward current that is enhanced by external Ba^{++} (Schwindt & Crill, 1980a), and is not blocked by intracellular injection of lidocaine derivatives that block the fast sodium currents (Schwindt & Crill, 1980a,b). Based on this evidence, the persistent inward current in cat motoneurons is probably calcium mediated, although in other motoneurons, a negative slope region in the I-V curve is probably due to a persistent sodium current (Mosfeldt-Laursen & Rekling, 1989; Nishimura et al., 1989; Rekling, 1992; Chandler et al., 1994).

The presence of a persistent inward current is likely to have important effects on steady-state discharge behavior provided that the somatic membrane potential can traverse voltage ranges over which this current is significantly activated. The increase in V_{thr} at increasing rates of discharge allows this condition to occur (Schwindt & Crill, 1982; Powers, 1993). Although relatively little inward current activation may take place at the membrane voltages traversed at the lowest spike-frequencies, the persistent inward current is likely to be continuously activated throughout the interspike interval at higher firing rates. The upward inflection in the steady-state f-I curve that occurs in some cells (i.e., secondary range firing; Kernell, 1965b; Schwindt, 1973; Schwindt & Crill, 1982) may result from the predominance of the inward current. Once activated, the inward current component may continue to predominate even when the activating stimulus is withdrawn, leading to repetitive discharge that outlasts the depolarizing synaptic or injected current. This switching between a quiescent state and sustained repetitive firing in response to transient inputs has been termed bistable firing behavior (cf. Hounsgaard and Kiehn, 1989). It is thought to depend upon a reduction in the potassium conductance underlying the AHP, leading to a change in the relative balance of inward and outward currents (Schwindt & Crill, 1980c; Hounsgaard et al., 1984; Hounsgaard & Kiehn, 1989).

A number of aspects of the current-to-frequency transduction vary systematically across the motoneuron pool. The 10-fold variation in rheobase (I_{rh}) is associated with a similar range in the minimum current needed to elicit steady repetitive discharge (I_0 Kernell & Monster, 1981). Variations in the *f-I* relations among different motoneuron types also arise from type-related differences in AHP characteristics (*vide supra*). Due to their longer AHP durations, type S motoneurons begin to discharge at significantly lower rates than type F motoneurons (Kernell, 1979), and the steady-state frequencies attained at the end of the primary and secondary ranges of firing are also lower in motoneurons with long AHP durations (presumably type S motoneurons; Kernell, 1965c). In contrast, the primary and secondary range *f-I* slopes do not vary systematically across the motoneuron pool (Kernell, 1979). Nonetheless, the systematic differences between minimum and maximum primary range spike-frequency are matched to differences in motor unit tension-frequency (*T-f*) relations, so that the rate modulation within the primary range modulates force along the steepest portion of the muscle unit's tension-frequency curve (Kernell, 1965c; Kernell, 1979).

SPIKE-FREQUENCY ADAPTATION

Motoneuron firing rate decreases as a function of time after the onset of a current step in a process called spike-frequency adaptation. In motoneurons, adaptation has a rapid, *initial* phase (Kernell, 1965a,b; Sawczuk et al., 1995), followed by a slow, gradual decline that continues throughout the duration of firing which may be several minutes (*late adaptation*; Granit et al., 1963; Kernell, 1965a; Kernell & Monster, 1982a; Spielmann et al., 1993; Sawczuk et al., 1995). We have recently conducted a detailed, quantitative analysis of the complete time course of spike-frequency adaptation in rat hypoglossal motoneurons (Sawczuk et al., 1995). As illustrated in Fig. 1, the *initial*, rapid phase of adaptation is linear and is complete within the first few interspike intervals. In 50% of the trials, the slow decline in spike-frequency over a 60 s period could be characterized by an exponential function with a single time constant. However, in the other 50% of trials, the slow decline in firing had two components: an *early* phase of adaptation that was typically complete within the first 2 s of discharge; and a *late* phase of adaptation that continued for the duration of firing (Fig. 1). The pattern of adaptation displayed in Fig. 1 was quite typical, with more than 90% of the total decrease in spike-frequency occurring in the first 2 s, about 6% in the 2 s to 26 s period, and the remaining 4% occurring after 26 s. Although most of the spike-frequency adaptation occurs within the first two seconds of discharge, there remains a further 40% reduction in firing rate during the *late* phase of adaptation (i.e., the firing rate at 26 s is about 60% of that at 2 s). These results are similar to those reported in cat lumbar motoneurons (Kernell & Monster, 1982a; Lindsay et al., 1986). The magnitude of *initial* adaptation was correlated with the initial spike-frequency (i.e., the reciprocal of the first interspike interval). The magnitudes of the *early* and *late* phases of adaptation were correlated with the firing frequency reached at the end of *initial* adaptation. In rat hypoglossal motoneurons, neither

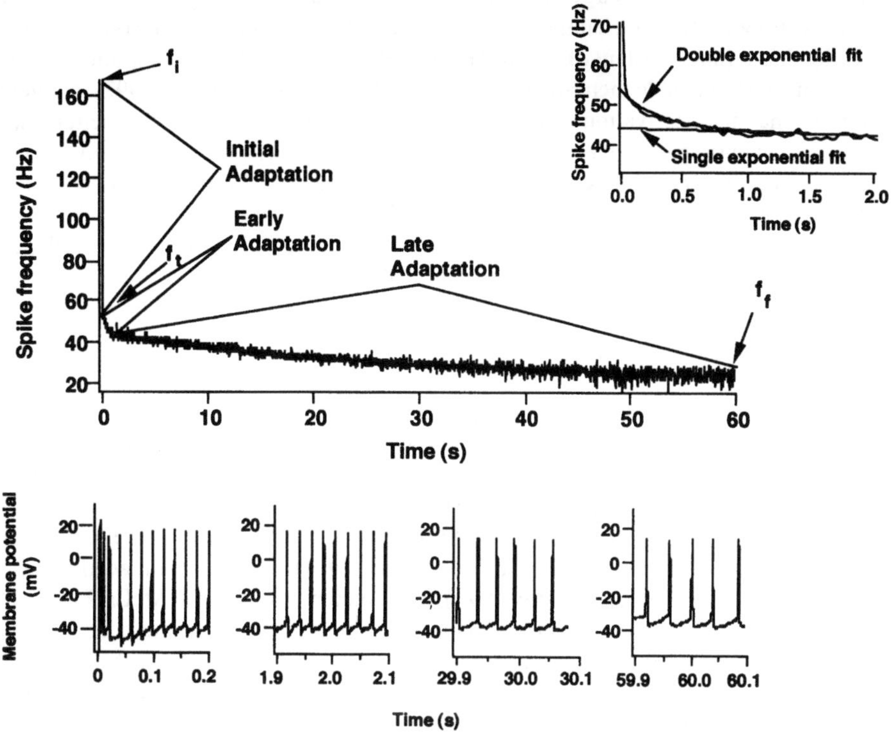

Figure 1. Three phases of spike-frequency adaptation during repetitive discharge of rat hypoglossal motoneurons. An *initial,* rapid decline in firing frequency (f_i - f_t) was complete within three interspike intervals. The subsequent decline in spike-frequency (f_t - f_f) can be fit with the sum of two exponential functions. (As illustrated in the upper right inset, a single exponential curve-fit failed to match the data from f_t to 750 ms. An additional exponential function was required to match the first 2 s of discharge.) The *initial* adaptation (f_i - f_t) was linear with slope of -2.2 x 10^3 Hz/s. The *early* adaptation had a time constant (t_E) of 0.3 s, and the *late* adaptation had a time constant (t_L) of 27.0 s. Beneath the spike-frequency-time curve are 200 ms samples of repetitive discharge taken at four time periods during the trial. f_i = initial firing frequency based on the first interspike interval; f_t = firing rate at transition between the *initial,* linear phase of adaptation and the subsequent exponential decline in firing rate; f_f = firing rate at conclusion of the trial (modified from Sawczuk et al., 1995).

the magnitude nor the time course of the three phases were correlated with other membrane properties such as input resistance, rheobase or repetitive firing threshold (Sawczuk et al., 1995; see, however, Kernell and Monster, 1982a; Spielmann et al., 1993).

MECHANISMS UNDERLYING SPIKE-FREQUENCY ADAPTATION

Initial Adaptation

It has been previously shown that increases in AHP amplitude are well correlated with initial adaptation in motoneurons (Kernell & Sjoholm, 1973; Jack et al., 1975; Baldissera et al., 1978; Barrett et al., 1980; Schwindt & Crill, 1982). Summation of AHP currents could occur if the calcium concentration in the vicinity of the G_{KCa} channels does not return to resting levels at the end of the interspike intervals. *Initial* adaptation is also associated with an increase in V_{thr} (Schwindt & Crill, 1982). As mentioned earlier, changes in V_{thr} during repetitive discharge may result from subthreshold changes in the inactivation of initial segment sodium channels. Although the fast component of sodium channel inactivation has a time constant of < 5 ms (Hille, 1992), a much slower component has been described in some excitable membranes (Brismar, 1977; Howe & Ritchie, 1992), suggesting that sodium channel inactivation could be involved in all phases of spike-frequency adaptation. Finally, depending upon its activation range and kinetics, a hyperpolarization-activated, mixed cation current (I_h) could contribute to *initial* adaptation: If it were partly activated at rest, it would provide an inward current that would increase spike-frequency at the onset of a current step and then deactivate at the more depolarized voltages encountered during maintained repetitive discharge (cf. Spain et al., 1991).

Early and Late Adaptation

The slow decline in spike-frequency seen during the later phases of adaptation could reflect a progressive increase in outward currents, a progressive decrease in inward currents, or some combination of these two mechanisms. Slow increases in the calcium concentration near the G_{KCa} channels would prolong the increase in G_{KCa} beyond the period of *initial* adaptation. Such increases in calcium concentration might reflect the saturation of intracellular calcium sequestering systems (Barrett et al., 1980) or release of calcium from intracellular stores triggered by a second messenger (Zhang, 1990) or by calcium itself (Sah & McLachlan, 1991). Recent computer simulations (Powers, unpublished) suggest that the time course of *early* adaptation is consistent with the saturation of an intracellular calcium buffer with slow kinetics (cf. Sah, 1992). Alternative mechanisms underlying a progressive increase in outward current include 1) the activation of an electrogenic Na^+-K^+ pump (Sokolove & Cooke, 1971; Kernell & Monster, 1982a; French, 1989); 2) activation of an M-current (Adams et al., 1986; Marrion, 1993); and 3) activation of a Na^+-activated potassium current (Schwindt et al., 1989). The first mechanism is not likely to be responsible for *late* adaptation, since *late* adaptation has been shown to be unaffected by blockade of the Na^+-K^+ pump with ouabain in rat hypoglossal motoneurons (Sawczuk & Binder, 1992; Sawczuk, 1993). It is not yet known if significant M-currents or Na^+ - activated potassium currents are present in motoneurons.

A slow decrease in spike-frequency could also reflect a slow inactivation of inward currents. Slow inactivation processes in initial segment and somatic sodium channels would be expected to lead to a progressive increase in V_{thr} along with changes in spike shape (cf. Schwindt & Crill, 1982; Lindsay et al., 1986; Sawczuk, 1993). A progressive increase in V_{thr} would act to lengthen interspike intervals even in the absence of changes in G_{KCa} or other potassium conductances, since a given rate of rise of membrane depolarization would take longer to cross the spike threshold.

Finally, more complex interactions between inward and outward currents might underlie *late* adaptation. For example, reducing or eliminating G_{KCa} by replacing external Ca^{++} with Mn^{++} leads to an increase in spike-frequency for a given current level, and yet the magnitude of *late* adaptation is increased (Sawczuk, 1993; Sawczuk & Binder, 1993). It is possible that under these conditions, an increase in the mean level of membrane depolarization during the interspike interval accelerates the development of sodium channel inactivation.

FUNCTIONAL SIGNIFICANCE OF SPIKE-FREQUENCY ADAPTATION

Although spike-frequency adaptation is observed in many types of neurons, its function remains largely speculative. The most obvious functional consequence of spike-frequency

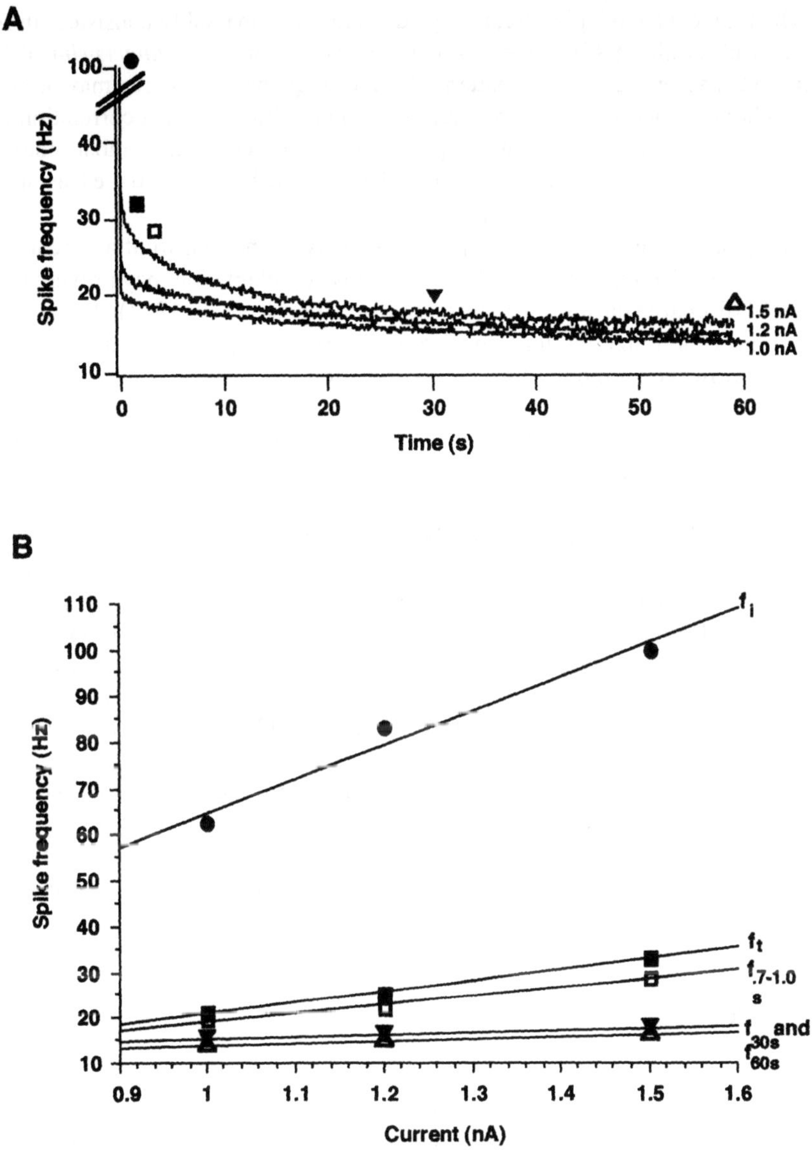

Figure 2. The slope of the spike-frequency-current (*f-I*) relation of motoneurons changes during adaptation. *A*, firing frequency *vs.* time curves obtained from a single rat hypoglossal motoneuron in response to three different levels of injected current. The relation between firing rate and the level of injected current was measured at five different times in the trials (circles: f_i = initial firing rate; filled squares: f_t = firing rate at the transition from *initial* to the later phases of adaptation; open squares: $f_{.7\text{-}1.0s}$ = average rate from 0.7 to 1.0 s; filled triangles: f_{30s} = firing rate at 30 s; open triangles: f_{60s} = firing rate at 60 s). *B*, *f-I* relations obtained from trials illustrated in *A*. Firing rate was linearly related to current intensity in all cases, but the *f-I* slope varied with time from 73.5 Hz/nA measured at f_i to 4.4 Hz/nA measured at 60 s (modified from Sawczuk et al., 1995).

adaptation in motoneurons is its profound influence on their frequency-current relations (Granit et al., 1963; Kernell, 1965a; Sawczuk et al., 1995). The slope of the *f-I* relation decreases dramatically during *initial* adaptation (Granit et al., 1963). The decrease in *f-I* slope continues throughout *early* adaptation (Kernell, 1965a), and a true steady-state value is not reached until after the *early* phase of adaptation (Sawczuk et al., 1995; their Fig. 2).

The time course of spike-frequency adaptation is remarkably consistent in repeated trials (Sawczuk et al., 1995) suggesting that adaptation may be a *fingerprint* of intrinsic motoneuron behavior. As such, the pattern of spike-frequency adaptation may be correlated with a number of other motor unit properties including fatigability. A correlation between the extent of adaptation and motor unit type has been reported for cat lumbar motoneurons, with the greatest amount of adaptation observed in the type FF units and the least in the type S units (Kernell & Monster, 1982a; Spielmann et al., 1993).

Since only a few, high frequency spikes may be needed to attain near maximal force in some motor units in the cat, the *initial* phase of adaptation may prevent excessive and wasteful discharge during short bursts of discharge (Kernell, 1983). Thus, adaptation may function to optimize the tension (force)-frequency (*T-f*) transduction of individual motor units. However, although this *T-f* optimization can be clearly demonstrated under controlled, isometric conditions when single units are stimulated electrically (e.g., Burke et al., 1970; Zajac & Young, 1980), *initial* adaptation has not been a consistent finding from recordings of cat motor unit activity during locomotion (Hoffer et al., 1981; Brownstone et al., 1992).

The *late* phase of adaptation may be matched to the progressive increase in twitch contraction time that occurs during prolonged contractions and allows motor units to maintain a given force level with a progressively decreasing activation rate (Bigland-Ritchie et al., 1983a,b; Bigland-Ritchie & Woods, 1984; Botterman & Cope 1988). The match between the progressive decrease in motoneuron spike-frequency and the increase in motor unit contraction time has been suggested as a strategy to optimize force production in the presence of fatiguing conditions (Kernell & Monster, 1982b; Marsden et al., 1983; Enoka & Stuart, 1992; Spielmann et al., 1993; see also Stuart & Callister, 1993). Although there is evidence to suggest that the decline in motoneuron firing frequency is produced in part by reflex inhibition (Bigland-Ritchie et al., 1986; Woods et al., 1987; Garland et al., 1988; Stuart & Callister, 1993) and /or a decline in excitatory synaptic input from muscle spindle afferents (Macefield et al., 1991), it is likely that *late* adaptation also makes a contribution (see, however, Stuart & Callister, 1993).

SUMMARY

In general, intrinsic motoneuron properties are regulated in a way that minimizes energy expenditure and muscle fatigue. The variation of I_0 according to motoneuron type ensures that the most fatigable motor units will be reserved for the most forceful contractions. The variation in minimum spike-frequency (f_0) arising from variations in AHP characteristics ensures that motoneurons begin to fire at rates that are matched to the force-producing characteristics of their muscle units. Further, variations in the three phases of spike-frequency adaptation may contribute to optimization of the tension-firing frequency (*T-f*) transductions of individual motor units during both phasic and sustained contractions.

ACKNOWLEDGMENTS

The experimental work in our laboratory is supported by United States Public Health Service grants, DE 00161, NS 31925, NS 26840, and NS 01650. Attendance of *A.S.* and *M.D.B.* at the 1994 Bigland-Ritchie conference was supported, in part, by the University of Miami.

REFERENCES

Adams PR, Jones SW, Pennefather P, Brown DA, Koch C & Lancaster B (1986). Slow synaptic transmission in frog sympathetic ganglia. *Journal of Experimental Biology* **124**, 259-285.

Baldissera F, Gustafsson B & Parmiggiani F (1978). Saturating summation of the afterhyperpolarization conductance in spinal motoneurones: a mechanism for 'secondary range' repetitive firing. *Brain Research* **146**, 69-82.

Barrett EF, Barrett JN & Crill WE (1980). Voltage sensitive outward currents in cat motoneurons. *Journal of Physiology (London)* **304**, 251-276.

Bigland-Ritchie B, Dawson NJ, Johansson RS & Lippold OC (1986). Reflex origin for the slowing of motoneurone firing rates in fatigue of human voluntary contractions. *Journal of Physiology (London)* **379**, 451-459.

Bigland-Ritchie B, Johansson R, Lippold OC, Smith S & Woods JJ (1983a). Changes in motoneurone firing rates during sustained maximal voluntary contractions. *Journal of Physiology (London)* **340**, 335-346.

Bigland-Ritchie B, Johansson R, Lippold OC & Woods JJ (1983b). Contractile speed and EMG changes during fatigue of sustained maximal voluntary contractions. *Journal of Neurophysiology* **50**, 313-24.

Bigland-Ritchie B & Woods JJ (1984). Changes in muscle contractile properties and neural control during human muscular fatigue. *Muscle & Nerve* **7**, 691-699.

Binder MD, Heckman CJ & Powers RK (1995). The physiological control of motoneuron activity. In: Rowell LF, Shepherd JT (eds.), Smith JL (sec. ed.), *Handbook of Physiology: Integration of Motor, Respiratory and Metabolic Control During Exercise, sec. A, Neural Control of Movement,* pp. 00-00. Bethesda: American Physiological Society. In press

Botterman BR & Cope TC (1988). Motor-unit stimulation patterns during fatiguing contractions of constant tension. *Journal of Neurophysiology* **60**, 1198-1214.

Brismar T (1977). Slow mechanism for sodium permeability inactivation in myelinated nerve fibre of *Xenopus laevis*. *Journal of Physiology (London)* **270**, 283-297.

Brownstone RM, Jordan LM, Kriellaars DJ, Noga BR & Shefchyk SJ (1992). On the regulation of repetitive firing in lumbar motoneurones during fictive locomotion in the cat. *Experimental Brain Research* **90**, 441-55.

Burke RE (1981). Motor units: anatomy, physiology, and functional organization. In: Brookhart JM, Mountcastle VB (eds.), Brooks VB (vol. ed.), *Handbook of Physiology, sec. 1, vol. II, pt 1. The Nervous System: Motor Control.*, pp. 345-422. Bethesda, MD: American Physiological Society.

Burke RE, Dum RP, Fleshman JW, Glenn LL, Lev-Tov A, O'Donovon MJ & Pinter MJ (1982). An HRP study of the relation between cell size and motor unit type in cat ankle extensor motoneurons. *Journal of Comparative Neurology* **209**, 17-28.

Burke RE & Nelson PG (1971). Accommodation to current ramps in motoneurons of fast and slow twitch motor units. *International Journal of Neuroscience* **1**, 347-356.

Burke RE, Rudomin P & Zajac FE (1970). Catch property in single mammalian motor units. *Science* **168**, 122-124.

Calvin WH (1974). Three modes of repetitive firing and the role of threshold time course between spikes. *Brain Research* **59**, 341-346.

Carp JS (1992). Physiological properties of primate lumbar motoneurons. *Journal of Neurophysiology* **68**, 1121-1132.

Chandler SH, Hsaio C, Inoue T & Goldberg LJ (1994). Electrophysiological properties of guinea pig trigeminal motoneurons recorded in vitro. *Journal of Neurophysiology* **71**, 129-145.

Enoka RM & Stuart DG (1992). Neurobiology of muscle fatigue. *Journal of Applied Physiology* **72**, 1631-1648.

Fleshman JW, Munson JB, Sypert GW & Friedman WA (1981). Rheobase, input resistance, and motor-unit type in medial gastrocnemius motoneurons in the cat. *Journal of Neurophysiology* **46**, 1326-1338.

Frankenhaeuser B & Vallbo AB (1964). Accomodation in myelinated nerve fibres of Xenopus laevis as computed on the basis of voltage clamp data. *Acta Physiologica Scandinavica* **63**, 1-20.

French AS (1989). Ouabain selectively affects the slow component of sensory adaptation in an insect mechanoreceptor. *Brain Research* **504**, 112-114.

Garland SJ, Garner SH & McComas AJ (1988). Reduced voluntary electromyographic activity after fatiguing stimulation of human muscle. *Journal of Physiology (London)* **401**, 547-556.

Granit R, Kernell D & Shortess GK (1963). Quantitative aspects of repetitive firing of mammalian motoneurones, caused by injected currents. *Journal of Physiology (London)* **168**, 911-931.

Gustafsson B & Pinter MJ (1984a). An investigation of threshold properties among cat spinal alpha-motoneurones. *Journal of Physiology (London)* **357**, 453-83.

Gustafsson B & Pinter MJ (1984b). Relations among passive electrical properties of lumbar alpha-motoneurones of the cat. *Journal of Physiology (London)* **356**, 401-31.

Heckman CJ & Binder MD (1990). Neural mechanisms underlying the orderly recruitment of motoneurons. In: Binder MD, Mendell LM (eds.), *The Segmental Motor System*, pp. 182-204. New York: Oxford University Press.

Henneman E & Mendell LM (1981). Functional organization of motoneuron pool and its inputs. In: Brookhart JM, Mountcastle VB (sec. eds.), Brooks VB (vol. ed.), *Handbook of Physiology, sec. 1, vol. II, pt 1, The Nervous System: Motor Control.*, pp. 423-507. Bethesda, MD: American Physiological Society.

Henneman E & Olson CB (1965). Relations between structure and function in the design of skeletal muscle. *Journal of Neurophysiology* **28**, 581-598.

Henneman E, Somjen G & Carpenter DO (1965). Excitability and inhibitability of motoneurons of different sizes. *Journal of Neurophysiology* **28**, 599-620.

Hille B (1992). *Ionic Channels of Excitable Membranes* (2nd Ed.). Sunderland, MA: Sinauer Assoc. Inc.

Hoffer JA, O'Donovan MJ, Pratt CA & Loeb GE (1981). Discharge patterns of hindlimb motoneurons during normal cat locomotion. *Science* **213**, 466-7.

Hounsgaard J, Hultborn H, Jespersen B & Kiehn O (1984). Intrinsic membrane properties causing a bistable behaviour of alpha-motoneurones. *Experimental Brain Research* **55**, 391-394.

Hounsgaard J & Kiehn O (1989). Serotonin-induced bistability of turtle motoneurones caused by a nifedipine-sensitive calcium plateau potential. *Journal of Physiology (London)* **414**, 265-282.

Hounsgaard J, Kiehn O & Mintz I (1988). Response properties of motoneurones in a slice preparation of the turtle spinal cord. *Journal of Physiology (London)* **398**, 575-89.

Howe JR & Ritchie JM (1992). Multiple kinetic components of sodium channel inactivation in rabbit Schwann cells. *Journal of Physiology (London)* **455**, 529-566.

Jack JJB, Noble D & Tsien RW (1975). *Electric Current Flow in Excitable Cells*. Oxford: Clarendon Press.

Jodkowski JS, Viana F, Dick TE & Berger AJ (1988). Repetitive firing properties of phrenic motoneurons in the cat. *Journal of Neurophysiology* **60**, 687-702.

Kernell D (1965a). The adaptation and the relation between discharge frequency and current strength of cat lumbosacral motoneurones stimulated by long-lasting injected currents. *Acta Physiologica Scandinavica* **65**, 65-73.

Kernell D (1965b). High frequency repetitive firing of cat lumbosacral motoneurones stimulated by long-lasting injected currents. *Acta Physiologica Scandinavica* **65**, 74-86.

Kernell D (1965c). The limits of firing frequency in cat lumbosacral motoneurones possessing different time course of afterhyperpolarization. *Acta Physiologica Scandinavica* **65**, 87-100.

Kernell D (1979). Rhythmic properties of motoneurones innervating muscle fibres of different speed in m. gastrocnemius medialis of the cat. *Brain Research* **160**, 159-62.

Kernell D (1983). Functional properties of spinal motoneurons and gradation of muscle force. In: Desmedt JE (ed.), *Motor Control Mechanisms in Health and Disease*, pp. 213-226. New York: Raven Press.

Kernell D & Monster AW (1981). Threshold current for repetitive impulse firing in motoneurones innervating muscle fibres of different fatigue sensitivity in the cat. *Brain Research* **229**, 193-196.

Kernell D & Monster AW (1982a). Time course and properties of late adaptation in spinal motoneurones of the cat. *Experimental Brain Research* **46**, 191-196.

Kernell D & Monster AW (1982b). Motoneurone properties and motor fatigue. An intracellular study of gastrocnemius motoneurones of the cat. *Experimental Brain Research* **46**, 197-204.

Kernell D & Sjoholm H (1973). Repetitive impulse firing: comparisons between neurone models based on 'voltage clamp equations' and spinal motoneurones. *Acta Physiologica Scandinavica* **87**, 40-56.

Krnjevic K, Lamour Y, MacDonald JF & Nistri A (1979). Effects of some divalent cations on motoneurones in cats. *Canadian Journal of Physiology and Pharmacology* **57**, 944-56.

Lindsay A, Heckman CJ & Binder MD (1986). Analysis of late adaptation in cat motoneurons. *Society for Neuroscience Abstracts* **12**, 247.

Macefield G, Hagbarth KE, Gorman R, Gandevia SC & Burke D (1991). Decline in Spindle Support to alpha-Motoneurones During Sustained Voluntary Contractions. *Journal of Physiology (London)* **440**, 497-512.

Marrion NV (1993). Selective reduction of one mode M-channel gating by muscarine in sympathetic neurons. *Neuron* **11**, 77-84.

Marsden CD, Meadows JC & Merton PA (1983). 'Muscular Wisdom' that minimizes fatigue during prolonged effort in man: Peak rates of motoneuron discharge and slowing of discharge during fatigue. In: Desmedt JE (eds.), *Motor Control Mechanisms in Health and Disease*, pp. 169-211. New York: Raven Press.

Mosfeldt-Laursen A & Rekling JC (1989). Electrophysiological properties of hypoglossal motoneurons of guinea-pigs studied in vitro. *Neuroscience* **30**, 619-637.

Nishimura Y, Schwindt PC & Crill WE (1989). Electrical properties of facial motoneurons in brainstem slices from guinea pig. *Brain Research* **502**, 127-142.

Powers RK (1993). A variable-threshold motoneuron model that incorporates time-dependent and voltage-dependent potassium and calcium conductances. *Journal of Neurophysiology* **70**, 246-262.

Powers RK & Binder MD (1985). Distribution of oligosynaptic group I input to the cat medial gastrocnemius motoneuron pool. *Journal of Neurophysiology* **53**, 497-517.

Rekling JC (1992). Interaction between thyrotropin-releasing hormone (TRH) and NMDA-receptor-mediated responses in hypoglossal motoneurones. *Brain Research* **578**, 289-96.

Richter DW, Schlue WR, Mauritz KH & Nacimiento AC (1974). Comparison of membrane properties of the cell body and the initial part of the axon of phasic motoneurones in the spinal cord of the cat. *Experimental Brain Research* **21**, 193-206.

Sah P (1992). Role of calcium influx and buffering in the kinetics of Ca(2+)-activated K+ current in rat vagal motoneurons. *Journal of Neurophysiology* **68**, 2237-2247.

Sah P & McLachlan EM (1991). Ca^{2+}-activated K^{+} currents underlying the afterhyperpolarization in guinea pig vagal neurons: a role for Ca^{2+}-activated Ca^{2+} release. *Neuron* **7**, 257-264.

Sawczuk A (1993) *Adaptation in Sustained Motoneuron Discharge.* PhD Dissertation, Seattle: University of Washington.

Sawczuk A & Binder MD (1992). Reduction of Na^{+}-K^{+} activity does not reduce the late adaptation of motoneuron discharge. *Society for Neuroscience Abstracts* **18**, 512.

Sawczuk A & Binder MD (1993). Reduction of the AHP with Mn^{++} decreases the initial adaptation, but increases the late adaptation in rat hypoglossal motoneuron discharge. *The Physiologist* **36**, A23.

Sawczuk A, Powers RK & Binder MD (1995). Spike-frequency adaptation studied in hypoglossal motoneurons of the rat. *Journal of Neurophysiology* **73**, In press

Schlue WR, Richter DW, Mauritz KH & Nacimiento AC (1974). Mechanisms of accommodation to linearly rising currents in cat spinal motoneurones. *Journal of Neurophysiology* **37**, 310-315.

Schwindt P & Crill WE (1977). A persistent negative resistance in cat lumbar motoneurons. *Brain Research* **120**, 173-178.

Schwindt PC (1973). Membrane-potential trajectories underlying motoneuron rhythmic firing at high rates. *Journal of Neurophysiology* **36**, 434-439.

Schwindt PC & Crill WE (1980a). Effects of barium on cat spinal motoneurons studied by voltage clamp. *Journal of Neurophysiology* **44**, 827-846.

Schwindt PC & Crill WE (1980b). Properties of a persistent inward current in normal and TEA-injected motoneurons. *Journal of Neurophysiology* **43**, 1700-1724.

Schwindt PC & Crill WE (1980c). Role of a persistent inward current in motoneuron bursting during spinal seizures. *Journal of Neurophysiology* **43**, 1296-1318.

Schwindt PC & Crill WE (1982). Factors influencing motoneuron rhythmic firing: results from a voltage-clamp study. *Journal of Neurophysiology* **48**, 875-90.

Schwindt PC & Crill WE (1984). Membrane Properties of Cat Spinal Motoneurons. In: Davidoff RA (ed.), *Handbook of the Spinal Cord,* pp. 199-242. New York: Marcel Dekker.

Schwindt PC, Spain WJ & Crill WE (1989). Long-lasting reduction of excitability by a sodium-dependent potassium current in cat neocortical neurons. *Journal of Neurophysiology* **61**, 233-44.

Sokolove PG & Cooke IM (1971). Inhibition of impulse activity in a sensory neuron by an electrogenic pump. *Journal of General Physiology* **57**, 125-163.

Spain WJ, Schwindt PC & Crill WE (1991). Post-inhibitory excitation and inhibition in layer V pyramidal neurones from cat sensorimotor cortex. *Journal of Physiology (London)* **434**, 609-626.

Spielmann JM, Laouris Y, Nordstrom MA, Robinson GA, Reinking RM & Stuart DG (1993). Adaptation of cat motoneurons to sustained and intermittent extracellular activation. *Journal of Physiology (London)* **464**, 75-120.

Stuart DG & Callister RJ (1993). Afferent and spinal reflex aspects of muscle fatigue: Issues and speculations. In: Sargeant AJ, D. Kernell (eds.), *Neuromuscular Fatigue*, pp. 169-180. Amsterdam: Royal Netherlands Academy of Sciences.

Vallbo AB (1964). Accommodation related to the inactivation of the sodium permeability in single myelinated nerve fibres from *Xenopus laevis. Acta Physiologyica Scandinavica* **61**, 429-444.

Viana F, Bayliss DA & Berger AJ (1993). Calcium conductances and their role in the firing behavior of neonatal rat hypoglossal motoneurons. *Journal of Neurophysiology* **69**, 2137-2149.

Woods JJ, Furbush F & Bigland-Ritchie B (1987). Evidence for a fatigue-induced reflex inhibition of motoneuron firing rates. *Journal of Neurophysiology* **58**, 125-137.

Zajac FE & Young JL (1980). Properties of stimulus trains producing maximum tension-time area per pulse from single motor units in medial gastrocnemius muscle of the cat. *Journal of Neurophysiology* **43**, 1206-1220.

Zengel JE, Reid SA, Sypert GW & Munson JB (1985). Membrane electrical properties and prediction of motor-unit type of medial gastrocnemius motoneurons in the cat. *Journal of Neurophysiology* **53**, 1323-1344.

Zhang L (1990). Effects of inositol, 1,4,5-trisphosphate injections into cat spinal motoneurons. *Canadian Journal of Physiology and Pharmacology* **68**, 1062-1068.

Zhang L & Krnjevic K (1987). Effects of intracellular injections of phorbol ester and protein kinase C on cat spinal motoneurons in vivo. *Neuroscience Letters* **77**, 287-292.

Zwaagstra B & Kernell D (1980). The duration of after-hyperpolarization in hindlimb alpha motoneurones of different sizes in the cat. *Neuroscience Lettters* **19**, 303-307.

9

NEUROMUSCULAR FREQUENCY-CODING AND FATIGUE

D. Kernell

Department of Medical Physiology
University of Groningen
Bloemsingel 10, 9712 KZ Groningen, The Netherlands

ABSTRACT

In daily life, muscle fatigue often becomes noticeable as an apparent decline in the efficiency of force production by central commands, making it necessary to increase drive (or "effort") to produce a constant motor output. Such aspects of fatigue may be caused by changes in the way in which synaptic messages arriving at the motoneurons are translated into forces by the muscle fibers. Therefore, an understanding of these neuromuscular gradation mechanisms is essential for any analysis of motor fatigue. A brief general review is given of 1) how muscle fibers transduce motoneuronal discharge rates into force; 2) how synaptic currents are transduced into motoneuronal discharge rates; 3) how activity-dependent changes in the neuromuscular transduction mechanisms contribute to neuromuscular fatigue; and 4) how the matching between the transduction mechanisms of motoneurons and those of their muscle fibers may help to optimize neuromuscular gradation efficiency and decrease the severity of fatigue.

INTRODUCTION

In voluntary motor behavior, the word *fatigue* has two common-sense connotations: either that a certain action requires an increasing amount of effort to be continued, or that any such continuation is rendered impossible because the necessary force cannot any longer be produced. The sense of effort is likely somehow to reflect (probably via central *corollary discharges*; McCloskey, 1981) the *amount* of central motor drive descending down to the spinal motoneurons from the brain (for further comments, see Enoka & Stuart, 1992). The force produced by a given amount of descending motor drive will depend on: 1) how the descending synaptic input is translated into motoneuronal discharge (recruitment procedures and stimulus current-to-spike-frequency transduction); 2) how, for the recruited cells, the motoneuronal discharge is translated into relative muscle-fiber force (spike-frequency-to-tension (force) transduction); and 3) the maximum force-generating capacities of the muscle fibers, determining what the absolute force of contraction will be for a given relative degree of activation. Hence, it is obvious that even if the potential maximal muscle force remains

Fatigue, Edited by Simon C. Gandevia et al.
Plenum Press, New York, 1995

normal, fatigue may still be present because of, for instance, changes in the transduction properties of motoneurons and muscle fibers (factors 1-2). The physiological significance of the frequency-transduction properties of muscle cannot be well understood without also considering those of their motoneurons. Therefore, although the present brief general review is primarily focused on muscle properties, some aspects of motoneuron physiology will be dealt with as well.

Firstly, an account will be given of the unfatigued tension-frequency relation of muscle. Secondly, the frequency-encoding properties of motoneurons will be briefly described. Thirdly, the potentiating and/or fatiguing effects of preceding use on these neuromuscular transduction properties will be analyzed and, lastly, the possible fatigue-alleviating effects of the matching of transduction properties between motoneurons and muscle fibers will be discussed.

TENSION-FREQUENCY RELATION IN MUSCLE

Effects of Steady-Rate Activation on Force

It has been known since long that, when tested with constant-rate bursts, there is a markedly sigmoid relation between relative isometric force and the rate of activation (*T-f relation*; Fig. 1*B*) in whole muscles (e.g., Cooper & Eccles, 1930; Edwards et al., 1977; Cooper et al., 1988) as well as in single muscle units (e.g., Kernell et al., 1975, 1983; Botterman et al., 1986; Thomas et al., 1991; Powers & Binder, 1991). The rate-position of this relationship is in itself an important expression of isometric contractile speed: the faster the muscle (unit) twitch, the further is the steep portion of the *T-f* curve shifted toward higher rates. One practical measure for this aspect of contractile speed is, for instance, the stimulation rate (or interval) required for producing 50% of maximum force. Within a given muscle, this *midforce interval* tends to be

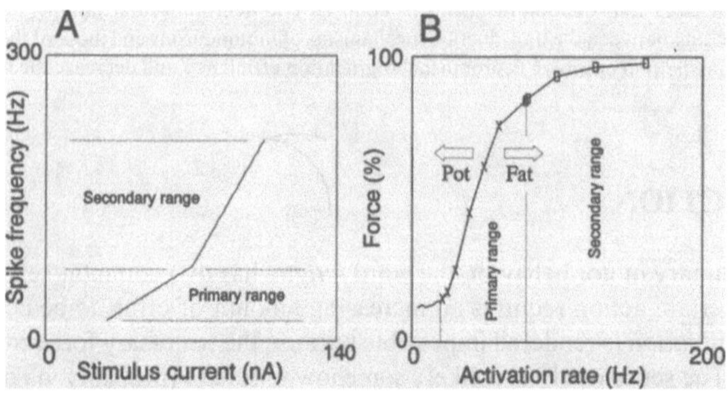

Figure 1. Relationship between motoneuronal and muscle fiber transduction properties (schematic). *A*, motoneuronal relation between discharge rate and the intensity of a steady activating current (*f-I curve*), e.g., as applied via an intracellular microelectrode. *B*, muscle unit relation between isometric force and activation rate (*T-f curve*). Numerical dimensions on x- and y-axes of *A* and *B* are those characteristic for relatively fast units of cat hindlimb muscles, for *B* as studied at the muscle length giving a maximal twitch force. A rightward shift of the *T-f* curve will cause less force for a given intermediate spike frequency (*B*, arrow labeled *Fat*; *fatigue*) and a leftward shift gives the opposite effect (arrow labeled *Pot*; *potentiation*). The two linear portions of the *f-I* curve are referred to as *primary* and *secondary* ranges (*A*, Kernell, 1965a). For stimulus currents exceeding those of the linear secondary range, the *f-I* relationship flattens off or declines (see dotted lines) and, finally, firing typically stops (*spike inactivation*). Lower and upper limits of spike-frequency for primary and secondary range indicated by lines in *A* (horizontal) and *B* (vertical). Firing rates are obtained after initial phase of adaptation.

highly correlated with twitch time course (e.g., time-to-peak, half-relaxation time; Kernell et al., 1983; Botterman et al., 1986; Thomas et al., 1991). Units of different muscles may, however, differ in their precise relationship between *T-f* curve and twitch time-course (Kernell et al., 1983; Botterman et al., 1986). The steep portion of the *T-f* curve has a higher slope for slow than for faster muscles and units (Cooper & Eccles, 1930; Kernell et al., 1983; Botterman et al., 1986). It should be noted that most of the published measurements of *T-f* relations concern isometric contractions rather than shortening or lengthening ones (for recent data on non-isometric contractions, see Heckman et al., 1992).

To understand the genesis of the sigmoid *T-f* relation, the apparent force *summation* has repeatedly been analyzed for two or more single-pulse stimuli at various intervals (e.g., Cooper & Eccles, 1930; Burke et al., 1976; Parmiggiani & Stein, 1981). One of the main results of such analysis is that the *summation* is markedly non-linear with, at the most effective intervals, typically more total force-time area being produced than the algebraic sum of the individual twitches. The apparent summation is likely to depend on the kinetics of calcium-associated mechanisms (e.g., release, removal, etc.), and also on changes of muscle fiber stiffness (cf. Parmiggiani & Stein, 1981), that take place after consecutive muscle action potentials in a stimulus train. The apparent fusion frequency (no oscillations visible) may be considerably lower than the stimulation rate needed for maximal force; one should not use the term *fused contraction* as being equivalent to *maximal tetanic contraction*. The presence of apparently fused but non-maximal tetanic contractions (note importance of measuring at sufficiently high gain, see e.g., Buller & Lewis, 1965) might partly depend on filtering properties of tissue in series with the contractile elements, rendering small mechanical sarcomere oscillations invisible at the tendon.

Length Effects

Muscle length is known to influence isometric muscle speed such that, over most of its possible working lengths in the body, the twitches of a muscle become progressively faster at shorter lengths (e.g., Rack & Westbury, 1969; Stephens et al., 1975). Correspondingly, also the *T-f* curve shifts toward higher rates as a muscle is shortened within its physiological working range (Rack & Westbury, 1969).

Temperature Effects

Significant effects on twitch-speed and *T-f* curve take place as changes occur in muscle temperature. Such effects might well be of importance under fatiguing circumstances when muscle temperature becomes elevated during work. As compared to cold muscles, warmer muscles have faster twitches and a *T-f* curve shifted toward higher rates (Ranatunga, 1982), thus increasing the amount of motor drive needed for a given *relative* force. The effect on absolute force is, however, complex because a warmer muscle may also have a higher maximal tetanic force (Ranatunga, 1982). Furthermore, in concentric contractions, the increased shortening speed of a warmer muscle apparently has beneficial results in that it results in a greater maximal power output (Sargeant, 1987).

Variable-Rate, Tension-Frequency Relations and "Catch-Like" Effects

One or a few brief initial intervals may have a prolonged enhancing effect on the forces evoked by subsequent firing at lower rates. Such a *catch-like* facilitation of force production is more long-lasting in slow than in fast units (Burke et al., 1976), but effects of this kind may be relevant in units of any contractile speed (e.g., Bevan et al., 1992) and such phenomena are also known to exist in man (e.g., Binder-Macleod & Barker, 1991).

STIMULUS CURRENT-TO-SPIKE-FREQUENCY TRANSDUCTION IN MOTONEURONS

General Considerations

A first requirement for using the rate gradation of force is that motoneurons can indeed vary their discharge rate within a range adequate for the *T-f* relations of their muscle fibers (see also Sawczuk et al., Chapter 8). Motoneurons receive thousands of synapses on the receptive membrane regions, i.e., on their soma and dendrites. When activated by a presynaptic action potential, a single synapse produces only a small and rather short-lasting pulse of current in the motoneuron. Normally occurring repetitive motoneuron discharges are driven by the relatively steady sum of numerous such unitary pulses, arriving asynchronously at the motoneuron. Due to their intrinsic membrane properties, the motoneurons (and most other kinds of nerve cell) are capable of transducing such steady currents into maintained repetitive discharges (reviews: Kernell, 1992; Binder et al., 1993). This current-to-frequency transduction can be investigated with currents injected through an intracellular microelectrode. In spinal motoneurons of the cat, the *frequency-current curve (f-I curve)* generally has a shape such as that shown in Fig. 1*A*, whereby it should be noted that firing within the lower *primary* range seems to be more stable than that within the upper *secondary* one. The shape and slope of the *f-I* curve depend on the combined effects of several membrane properties, including an important contribution of the calcium-dependent, potassium-conductance changes that lead to the occurrence of a long-lasting hyperpolarizing afterpotential following an action potential (afterhyperpolarization, AHP; for further references, see Kernell, 1992). The duration of the AHP is a major determinant of the minimum rate of maintained discharge (Kernell, 1965b; for other contributing factors, see Carp et al., 1991), and its time-course and size are important for the steepness of the *f-I* relation, i.e., for the motoneuronal *gain*.

Modifications of *f-I* Gain

In various species, the conductance changes underlying the AHP of motoneurons can be affected by transmitter substances, such as serotonin (review: Binder et al., 1993). This provides the synaptic means for changing the *f-I* gain. During *fictive locomotion*, maintained in decerebrate cats with brain stem stimulation, motoneurons have been seen to discharge at a relatively high rate in a state of apparently very low *f-I* gain (i.e., *f* not varied by extra

Figure 2. *A*, late phase of spike-frequency adaptation in motoneurons. Measurements of discharge rate from motoneurons of cat medial gastrocnemius, as obtained during intracellular stimulation with a weak steady current of 5 nA above the threshold for rhythmic firing. Mean data for 6 cells that were all discharging steadily during 180 s or more (based on data from Kernell & Monster, 1982a). *B*, simultaneous presence of fatigue and potentiation. Normalized measurements of isometric force from 1st deep lumbrical muscle of cat while the muscle nerve was electrically stimulated with pulse rates of alternating 360 ms bursts set to 121 Hz (fused contractions, filled symbols) and 30 Hz (non-fused contractions, open symbols), respectively. Horizontal dotted line drawn through initial force value obtained with 30 Hz stimulation. Note continuous fall in upper curve (fatigue) while there is an initial phase of growth in the lower curve (potentiation) (based on data from Kernell et al., 1975).

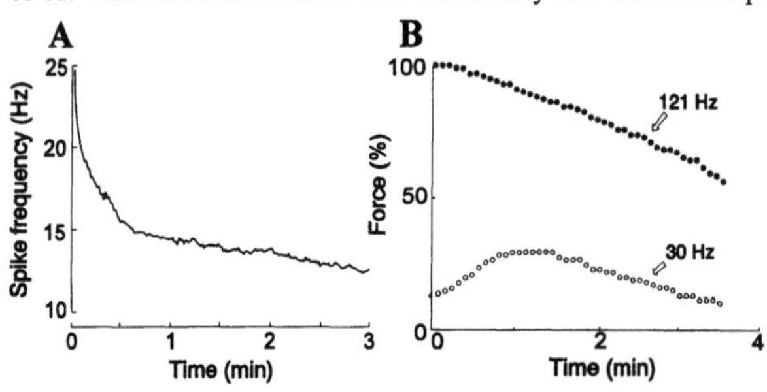

injected current) and with AHPs of substantially decreased amplitude (Brownstone et al., 1992). It is, however, still unknown whether this interesting behavior represents a common functional state of these cells.

Spike-Frequency Adaptation

As a motoneuron is activated with a step of steady injected current, it shows two phases of spike-frequency adaptation: 1) the *initial phase of adaptation*, during which there is a rapid decline of discharge rate as well as of *f-I* slope; and 2) the *late phase of adaptation* during which a slowly progressive frequency decline takes place without practically any change of *f-I* slope (Kernell & Monster, 1982a, their Fig. 2*A*; see also Sawczuk et al., Chapter 8). The initial phase mainly occupies the first few intervals of a discharge, and most of the late phase takes place during the subsequent 30-60 s. AHP properties (*AHP summation*) are important for the initial phase, but probably not for the late phase (see Kernell & Monster, 1982a). Both phases of adaptation are markedly more pronounced at high than at low levels of activity and, as will be further commented upon below, they are both of direct interest in relation to time-dependent muscle properties such as fatigue. When activated by the same (small) amount of steady injected current, the late phase of adaptation is much more marked for fast than for slow motoneurons (probably largely a reflection of their differences in absolute discharge rate, Kernell & Monster, 1982b; see also Spielmann et al., 1993).

POTENTIATING AND/OR FATIGUING EFFECTS OF ACTIVITY ON NEUROMUSCULAR ENCODING PROPERTIES

Steady-Rate Muscle Activation

There are two types of effect on the *T-f* relation during continued contractions, as produced by a constant pattern of submaximal activation (e.g., in a fatigue test): 1) *potentiation*, a force-increasing effect; and 2) *fatigue*, a force-decreasing effect. In submaximal contractions, either one of these effects may manifest itself as a shift of the *T-f* curve: leftward along the rate-axis for potentiation, rightward for fatigue (see arrows labeled *Pot* and *Fat* in Fig. 1*B*). In submaximal or maximal contractions, fatigue may also be caused by a change in force-generating capabilities of the muscle, e.g., by a drop of maximal tetanic force. In most instances, potentiation- and fatigue-associated effects are probably more or less simultaneously present, and the net result will depend on their sum. The simultaneous presence of both of these aspects of muscle reaction is easily demonstrated by activating the muscle alternatingly with low and high rates (Kernell et al., 1975, their Fig. 2*B*; Cooper et al., 1988; Bevan et al., 1992; for further references, see Enoka & Stuart, 1992).

During moderate to strong on-going motor activity, muscle contractions may become markedly slower (e.g., Bigland-Ritchie et al., 1983b), in itself an effect that might impede normal muscle use in rapidly alternating movements. The slowing is also, however, potentially beneficial in that it acts to shift the *T-f* curve toward lower rates; i.e., it produces a potentiating effect (arrow *Pot* in Fig. 1*B*). Thus, from the gradually increasing to the gradually decreasing portion of a prolonged, variable-force muscle contraction, the *T-f* curve is typically shifted toward lower rates (Binder-Macleod & Clamann, 1989). For a similarly arranged series of intermittent contractions caused by steady-rate bursts of different frequencies, net potentiation effects on the *T-f* relation are typically mainly seen for units that have a relatively high degree of fatigue-resistance (Kernell et al., 1975; Thomas et al., 1991; Powers & Binder, 1991). In the presence of a fatigue-induced decline in maximal tetanic force, the leftward potentiation-shift of the *T-f* curve will help to maintain submaximal forces without having to increase motoneuronal activity, i.e., potentiation may be seen as a (temporary) fatigue-alleviating mechanism (see Fig. 2*B*, note delay of about 3 min before net fatigue is seen for the non-fused contractions).

At the end of (a series of) intense fatiguing contractions there might not only be a rate-shift of the *T-f* curve but also a relatively greater loss of force-generating effects at high than at low rates (cf. Fig. 2*B*), i.e., the slope of the *T-f* curve will also decline (e.g., Kernell et al., 1975). In extreme cases, the *T-f* curve may even temporarily flatten out to nearly zero slope (Nagesser et al., 1992).

In addition to the acute slowing and potentiating *T-f* effects associated with ongoing activity, there are also often long-lasting (delayed) post-activation effects going in the opposite direction. These tend to cause a marked rightward shift of the *T-f* curve such that low-rate activation now produces relatively *less* contractile force (*low-frequency fatigue*; Edwards et al., 1977; Jami et al., 1983; Cooper et al., 1988; Powers & Binder 1991; see arrow *Fat* in Fig. 1*B*). Interestingly, these effects may continue to increase for some time after the end of the fatiguing exercise, and typically, they markedly outlast the fatigue-associated decline of maximum tetanic force.

Variable-Rate Activation of Muscle

Imposed stimulation experiments in animals and humans indicate that the relative and absolute force decline seen during a fatigue test might differ between constant-rate activation bursts and those having brief initial interval(s), evoking *catch-like* muscle effects (Binder-Macleod & Barker, 1991; Bevan et al., 1992). Variable-rate patterns of activation may be of practical value for minimizing muscle fatigue in the prosthetic use of *functional electrical stimulation* (FES).

Motoneuronal "Fatigue-Reactions"

The late phase of spike-frequency adaptation represents, in itself, a fatigue-like reaction of the motoneuron; firing rate declines continuously in the presence of a constant drive (Fig. 2*A*). Thus, this reaction might contribute to central aspects of motor fatigue. However, it may also play a role in processes *alleviating* fatigue (see further below). During late adaptation, the *f-I* relation is essentially shifted in parallel along the *I*-axis (rightward shift).

NEUROMUSCULAR MATCHING OF TRANSDUCTION PROPERTIES: RELEVANCE FOR ALLEVIATING FATIGUE

Rate-Matching Between *f-I* Relations of Motoneurons and *T-f* Relations of Muscle Units

The middle, steep region of the *T-f* curve represents the range of rates for which a variation of activation frequency gives a marked change of muscle output. Within this portion of the *T-f* curve, the conditions are optimal for the rate-gradation of force. In rate gradation of force, endurance (i.e., resistance to fatigue) will presumably be promoted by keeping motoneuronal discharge rates as low as possible. Hence, it would be advantageous for endurance to recruit motoneurons at firing rates corresponding to the low end of the steep portion of the *T-f* curve (Fig. 1). This is also how the *f-I* and *T-f* relationships are typically matched (cf. *A* and *B* of Fig. 1, motoneuron-muscle unit speed-match; Kernell, 1965b, 1992). At least in anaesthetized animals, the minimal rate of maintained motoneuronal firing takes place at intervals equal to the duration of the AHP (Kernell, 1965b). The lower end of the *T-f* curve is situated at the activation rate at which consecutive twitches start to sum. This occurs at a stimulus interval about equal to the total twitch duration. Thus, the motoneuron-muscle unit speed-match is a consequence of the fact that the total duration of AHP in motoneurons tends to be very similar to the total duration of the twitch in their muscle fibers (Bakels & Kernell, 1993a,b). In muscles with a wide variation of unit contractile speed, a systematic co-variation between AHP and twitch speed has been observed (e.g., medial gastrocnemius; cat, Zengel et al., 1985; rat, Gardiner & Kernell, 1990 and Bakels & Kernell, 1993a) in muscles with a more restricted speed range, similar *average* AHP and twitch durations may occur in the absence of a systematic co-variation (rat tibialis anterior, Bakels & Kernell, 1993b). Muscles with units showing a systematic co-variation between *f-I* and *T-f* properties apparently exist also in humans; for example, in the short toe extensor, extensor digitorum brevis, voluntarily activated motor units were reported to show lower minimal firing rates for units with a slow twitch than for those with a faster one (Grimby et al., 1979).

The matching between motoneuronal and muscle fiber properties does not only concern the lower limits; also, the maximal discharge rates of motoneurons seem well adapted to those needed for producing forces up to the (near-)maximal ones (Kernell, 1965b, 1992). Furthermore, at high activation rates, the steep slope of the *f-I* curve tends to compensate for the relatively flat slope at the upper end of the *T-f* curve (cf. Fig. 1).

Dynamic Matching of Motoneuronal Frequency to Muscle Length and Temperature

As was described above, the *T-f* relation has the complicating property that it is markedly dependent on muscle length. Attempts to find out whether, in voluntarily activated muscles, motoneuronal discharge behaviors alter correspondingly has met with mixed results. Positive findings were obtained in two studies on *minimal firing rates* (recruitment rates), which were found to be higher during shortening than lengthening contractions (Tax et al., 1989), and during short- than long-length isometric contractions (Vander Linden et al., 1991). The mechanisms underlying these adaptations are still unknown (possibly synaptic effects of monoamines on membrane properties). When studying *mean* EMG spike-frequencies of multiple units at medium to high levels of isometric force, no relationship was found between this parameter and muscle length (Bigland-Ritchie et al., 1992a). Similarly, in a study concerning the effects of muscle temperature, the decreased speed of a cold *vs.* a warmer muscle was not associated with a systematic change of mean motoneuronal discharge rates (Bigland-Ritchie et al., 1992b).

Initial Spike-Frequency Adaptation

The initial phase of adaptation means that motoneurons may easily be made to fire at very high frequencies for brief initial durations, reaching rates much faster than those possible in steady firing (in cats, up to several hundred Hz; Kernell, 1965a, 1992). This fits well to the fact that such very high rates are also needed for a maximal rate of rise of force in skeletal muscle (Buller & Lewis, 1965); by virtue of the transduction properties of motoneurons, such high, initial spike rates can be produced without having to use excessive amounts of central drive (and effort) (Baldissera et al., 1987). The ease with which a few brief initial discharge intervals may be produced by direct intracellular activation of motoneurons fits well to the findings that such brief initial intervals are often found in quick voluntary contractions (Desmedt & Godaux, 1977). However, this facet of motoneuronal behavior is not only of interest for the *rapidity* of force increase; one or a few brief initial spike intervals may also have a prolonged enhancing effect on the forces evoked by subsequent firing at lower rates (*catch-like* effects). As was pointed out above, a series of contractions produced by such variable-rate patterns might show less force decline than those evoked by steady-rate bursts.

Possible Protective Value of Late Spike-Frequency Adaptation

During a maximal voluntary contraction (MVC), the gradual slowing of contractile speed leads to a gradual decrease in the motoneuronal firing rates needed for maximum muscle activation (Bigland-Ritchie et al., 1983b); and, indeed, their spike-frequencies decline correspondingly (Bigland-Ritchie et al., 1983a; Marsden et al., 1983). This has a protective value in helping to maintain an adequate neuromuscular transmission and excitation-contraction coupling throughout continuous high-force contractions (Jones et al., 1979; *muscular wisdom*, Marsden et al., 1983). In human MVCs, part of the drop in discharge rate is caused by reflex mechanisms (e.g., Woods et al., 1987; Macefield et al., 1993; further references: Enoka & Stuart, 1992). However, if present also in humans, late adaptation of the motoneurons (Fig. 2*A*) would contribute as well, perhaps mainly in initial phases of MVCs.

Motoneuronal Recruitment Gradation

Although not a main issue of the present chapter, the spike-frequency-gradation of motoneuronal activity represents only one of the two best-known mechanisms for the voluntary control of muscle force, the other one being a gradation in the number of recruited motoneurons and muscle units. With regard to the *recruitment-hierarchy* used in force-gradation in most contractions, recruitment tends to start with the most fatigue-resistant units and proceed toward those with gradually less endurance (e.g., Henneman & Mendell, 1981; Kernell, 1992; see also Sawczuk et al., Chapter 8). As has been repeatedly pointed out, this fractionated muscle use helps to optimize muscle endurance in normal motor behavior, e.g., by letting weak postural contractions be executed by units with an adequately high degree of fatigue-resistance. In addition to such hierarchy considerations, *recruitment-gain* is a variable of potential importance for other facets of motor control. This gain (i.e., the *ease* with which synaptic inputs may alter the number of recruited motoneurons) may be influenced by steady background activity of excitatory and/or inhibitory synaptic systems (Kernell & Hultborn, 1990). Hypothetically, fatigue-associated alterations of such background synaptic *biasing* might lead to changes in the effort needed to drive a motoneuron pool, i.e., to changes of conceivable relevance for neuromuscular fatigue. Furthermore, the recruitment gain would be important for determining the balance between recruitment and rate modulation when both these mechanisms are used in parallel in contractions of weak to moderate force. For instance, in long-lasting postural contractions, a high degree of endurance would be expected to be promoted by emphasizing recruitment gradation (keeping motoneuronal discharge rates low) for the motoneurons concerned, i.e., by having particularly closely spaced recruitment thresholds for the most easily recruited units (corresponding distribution of intrinsic electrical thresholds seen for gastrocnemius motoneurons; Bakels & Kernell, 1994).

CONCLUDING REMARKS

The considerations of this brief review emphasize that: 1) the frequency-transduction properties of motoneurons and muscles have to be considered when analyzing neuromuscular fatigue; 2) these transduction properties have to be viewed *together* for motoneurons and their muscle fibers in order to understand their physiological significance; and 3) the manner in which motoneuronal and muscle fiber properties are combined (*matched*) helps, in a significant manner, to decrease the severity of neuromuscular fatigue.

ACKNOWLEDGMENTS

This work was supported, in part, by the Netherlands Organization for Scientific Research (NWO).

REFERENCES

Bakels R & Kernell D (1993a). Matching between motoneurone and muscle unit properties in rat medial gastrocnemius. *Journal of Physiology (London)* **463**, 307-324.

Bakels R & Kernell D (1993b). "Average" but not "continuous" speed-match between motoneurons and muscle units of rat tibialis anterior. *Journal of Neurophysiology*, **70**, 1300-1306.

Bakels R & Kernell D (1994). Threshold-spacing in motoneurone pools of rat and cat: possible relevance for manner of force gradation. *Experimental Brain Research* **102**, 69-74.

Baldissera F, Campadelli P & Piccinelli L (1987). The dynamic response of cat gastrocnemius motor units investigated by ramp current injection into their motoneurones. *Journal of Physiology (London)* **387**, 317-330.

Bevan L, Laouris Y, Reinking RM & Stuart DG (1992). The effect of the stimulation pattern on the fatigue of single motor units in adult cats. *Journal of Physiology (London)* **449**, 85 108.

Bigland-Ritchie BR, Furbush FH, Gandevia SC & Thomas CK (1992a). Voluntary discharge frequencies of human motoneurons at different muscle lengths. *Muscle & Nerve* **15**, 130-137.

Bigland-Ritchie B, Johansson R, Lippold OCJ, Smith S & Woods JJ (1983a). Changes in motoneurone firing rates during sustained maximal voluntary contractions. *Journal of Physiology (London)* **340**, 335-346.

Bigland-Ritchie B, Johansson R, Lippold OCJ & Woods JJ (1983b). Contractile speed and EMG changes during fatigue of sustained maximal voluntary contractions. *Journal of Neurophysiology* **50**, 313-324.

Bigland-Ritchie B, Thomas CK, Rice CL, Howarth JV & Woods JJ (1992b). Muscle temperature, contractile speed, and motoneuron firing rates during human voluntary contractions. *Journal of Applied Physiology* **73**, 2457-2461.

Binder MD, Heckman CJ & Powers RK (1993). How different afferent inputs control motoneuron discharge and the output of the motoneuron pool. *Current Opinion in Neurobiology* **3**, 1028-1034.

Binder-Macleod SA & Barker CB (1991). Use of a catchike property of human skeletal muscle to reduce fatigue. *Muscle & Nerve* **14**, 850-857.

Binder-Macleod SA & Clamann HP (1989). Force output of cat motor units stimulated with trains of linearly varying frequency. *Journal of Neurophysiology* **61**, 208-217.

Botterman BR, Iwamoto GA & Gonyea WJ (1986). Gradation of isometric tension by different activation rates in motor units of cat flexor carpi radialis muscle. *Journal of Neurophysiology* **56**, 494-506.

Brownstone RM, Jordan LM, Kriellaars DJ, Noga BR & Shefchyk SJ (1992). On the regulation of repetitive firing in lumbar motoneurones during fictive locomotion in the cat. *Experimental Brain Research* **90**, 441-455.

Buller AJ & Lewis DM (1965). The rate of tension development in isometric tetanic contractions of mammalian fast and slow skeletal muscle. *Journal of Physiology (London)* **176**, 337-354.

Burke RE, Rudomin P & Zajac FE (1976). The effect of activation history on tension production by individual muscle units. *Brain Research* **109**, 515-529.

Carp JS, Powers RK & Rymer WZ (1991). Alterations in motoneuron properties induced by acute dorsal spinal hemisection in the decerebrate cat. *Experimental Brain Research* **83**, 539-548.

Cooper RG, Edwards RHT, Gibson H & Stokes MJ (1988). Human muscle fatigue: Frequency dependence of excitation and force generation. *Journal of Physiology (London)* **397**, 585-599.

Cooper S & Eccles JC (1930). The isometric responses of mammalian muscles. *Journal of Physiology (London)* **69**, 377-385.

Desmedt JE & Godaux E (1977). Ballistic contractions in man: characteristic recruitment pattern of single motor units of the tibialis anterior muscle. *Journal of Physiology (London)* **264**, 673-693.

Edwards RHT, Hill DK, Jones DA & Merton PA (1977). Fatigue of long duration in human skeletal muscle after exercise. *Journal of Physiology (London)* **272**, 769-778.

Enoka RM & Stuart DG (1992). Neurobiology of muscle fatigue. *Journal of Applied Physiology* **72**, 1631-1648.

Gardiner PF & Kernell D (1990). The "fastness" of rat motoneurones: time-course of afterhyperpolarization in relation to axonal conduction velocity and muscle unit contractile speed. *Pflügers Archiv* **415**, 762-766.

Grimby L, Hannerz J & Hedman B (1979). Contraction time and voluntary discharge properties of individual short toe extensor motor units in man. *Journal of Physiology (London)* **289**, 191-201.

Heckman CJ, Weytjens JLF & Loeb GE (1992). Effect of velocity and mechanical history on the forces of motor units in the cat medial gastrocnemius muscle. *Journal of Neurophysiology* **68**, 1503-1515.

Henneman E & Mendell LM (1981). Functional organization of motoneuron pool and its inputs. In: Brookhart JM, Mountcastle VB (sec. eds.), Brooks VB (vol. ed.), *Handbook of Physiology, sec. 1, vol. II, pt 1, The Nervous System: Motor Control.*, pp. 423-507. Bethesda, MD: American Physiological Society.

Jami L, Murthy KSK, Petit J & Zytnicki D (1983). After-effects of repetitive stimulation at low frequency on fast-contracting motor units of cat muscle. *Journal of Physiology (London)* **340**, 129-143.

Jones DA, Bigland-Ritchie B & Edwards RHT (1979). Excitation frequency and muscle fatigue: Mechanical responses during voluntary and stimulated contractions. *Experimental Neurology* **64**, 401-413.

Kernell D (1965a). High-frequency repetitive firing of cat lumbosacral motoneurones stimulated by long-lasting injected currents. *Acta Physiologica Scandinavica* **65**, 74-86.

Kernell D (1965b). The limits of firing frequency in cat lumbosacral motoneurones possessing different time course of afterhyperpolarization. *Acta Physiologica Scandinavica* **65**, 87-100.

Kernell D (1992). Organized variability in the neuromuscular system: A survey of task-related adaptations. *Archives Italiennes de Biologie* **130**, 19-66.

Kernell D, Ducati A & Sjöholm H (1975). Properties of motor units in the first deep lumbrical muscle of the cat's foot. *Brain Research* **98**, 37-55.

Kernell D, Eerbeek O & Verhey BA (1983). Relation between isometric force and stimulus rate in cat's hindlimb motor units of different twitch contraction time. *Experimental Brain Research* **50**, 220-227.

Kernell D & Hultborn H (1990). Synaptic effects on recruitment gain: a mechanism of importance for the input-output relations of motoneurone pools? *Brain Research* **507**, 176-179.

Kernell D & Monster AW (1982a). Time course and properties of late adaptation in spinal motoneurones in the cat. *Experimental Brain Research* **46**, 191-196.

Kernell D & Monster AW (1982b). Motoneurone properties and motor fatigue. An intracellular study of gastrocnemius motoneurones of the cat. *Experimental Brain Research* **46**, 197-204.

Macefield VG, Gandevia SC, Bigland-Ritchie B, Gorman RB & Burke D (1993). The firing rates of human motoneurones voluntarily activated in the absence of muscle afferent feedback. *Journal of Physiology (London)* **471**, 429-443.

Marsden CD, Meadows JC & Merton PA (1983). "Muscular wisdom" that minimizes fatigue during prolonged effort in man: peak rates of motoneurone discharge and slowing of discharge during fatigue. In: Desmedt J.E. (ed.), *Motor Control Mechanisms in Health and Disease*, pp.169-211. New York: Raven Press.

McCloskey DI (1981). Corollary discharges: motor commands and perception. In: Brookhart JM, Mountcastle VB (sec. eds.), Brooks VB (vol. ed.), *Handbook of Physiology, sec. 1, vol. II, pt 2, The Nervous System: Motor Control*, pp. 1415-1447. Bethesda, MD: American Physiological Society.

Nagesser AS, Van der Laarse WJ & Elzinga G (1992). Metabolic changes with fatigue in different types of single muscle fibres of *Xenopus Laevis*. *Journal of Physiology (London)* **448**, 511-523.

Parmiggiani F & Stein RB (1981). Nonlinear summation of contractions in cat muscles. II. Later facilitation and stiffness changes. *Journal of General Physiology* **78**, 295-311.

Powers RK & Binder MD (1991). Effects of low-frequency stimulation on the tension-frequency relations of fast-twitch motor units in the cat. *Journal of Neurophysiology* **66**, 905-918.

Rack PMH & Westbury DR (1969). The effects of length and stimulus rate on tension in the isometric cat soleus muscle. *Journal of Physiology (London)* **204**, 443-460.

Ranatunga KW (1982). Temperature-dependence of shortening velocity and rate of isometric tension development in rat skeletal muscle. *Journal of Physiology (London)* **329**, 465-483.

Sargeant AJ (1987). Effect of muscle temperature on leg extension force and short-term power output in humans. *European Journal of Applied Physiology* **56**, 693-698.

Spielmann JM, Laouris Y, Nordstrom MA, Robinson GA, Reinking RM & Stuart DG (1993). Adaptation of cat motoneurons to sustained and intermittent extracellular activation. *Journal of Physiology (London)* **464**, 75-120.

Stephens JA, Reinking RM & Stuart DG (1975). The motor units of cat medial gastrocnemius: Electrical and mechanical properties as a function of muscle length. *Journal of Morphology* **146**, 495-512.

Tax AAM, Van der Gon JJD, Gielen CCAM & Van den Tempel CMM (1989). Differences in the activation of m. biceps brachii in the control of slow isotonic movements and isometric contractions. *Experimental Brain Research* **76**, 55-63.

Thomas CK, Bigland-Ritchie B & Johansson RS (1991). Force-frequency relationships of human thenar motor units. *Journal of Neurophysiology* **65**, 1509-1516.

Vander Linden DW, Kukulka CG & Soderberg GL (1991). The effect of muscle length on motor unit discharge characteristics in human tibialis anterior muscle. *Experimental Brain Research* **84**, 210-218.

Woods JJ, Furbush F & Bigland-Ritchie B (1987). Evidence for a fatigue-induced reflex inhibition of motoneuron firing rates. *Journal of Neurophysiology* **58**, 125-137.

Zengel JE, Reid SA, Sypert GW & Munson JB (1985). Membrane electrical properties and prediction of motor-unit type of medial gastrocnemius motoneurons in the cat. *Journal of Neurophysiology* **53**, 1323-1344.

10

HUMAN MOTOR UNITS STUDIED BY SPIKE-TRIGGERED AVERAGING AND INTRANEURAL MOTOR AXON STIMULATION

C. K. Thomas

The Miami Project to Cure Paralysis
University of Miami School of Medicine
Miami, Florida 33136

ABSTRACT

When low-threshold motor units are activated at low rates during sustained, weak voluntary contractions, most unit force profiles exhibit fatigue but some show force potentiation. These data, obtained by spike-triggered averaging, are compared to the fatigue resistance of human motor units activated at twitch and tetanic rates by intraneural motor axon stimulation. With the latter technique, representative sampling of the motor units from one muscle group shows that unit force fatigue or potentiation at submaximal frequencies, and contractile rate changes, dictate the shifts in unit force-frequency relationships. More diverse fatigue protocols, and when possible, careful comparisons of data obtained by both these techniques, are needed to further our understanding of the force and frequency changes of single motor units during voluntary and stimulated exercise.

INTRODUCTION

Many studies involving mammalian (cat, rat and rabbit) limb muscles have explored the loss of force generating capacity or fatigability by single motor units. In contrast, human studies have regularly examined fatigue at the whole muscle level during either voluntary or stimulated contractions. These differences between studies relate in part to the physiological and technical difficulties of selectively activating single human motor units and the recording of the associated contractile properties. In human studies, four methods have been used: 1) spike triggered averaging (Buchthal & Schmalbruch, 1970; Stein et al., 1972; Milner-Brown et al., 1973a); 2) intramuscular microstimulation (Taylor & Stephens, 1976; see Elek & Dengler, Chapter 11); 3) percutaneous nerve stimulation (Sica & McComas, 1971); and 4) intraneural motor axon stimulation (Westling et al., 1990). This chapter reviews the contributions the techniques of spike-triggered averaging and intraneural motor axon

Fatigue, Edited by Simon C. Gandevia et al.
Plenum Press, New York, 1995

stimulation have made to understanding of muscle fatigue. Some of the reasons for the paucity of data are highlighted. They are followed by a brief discussion of how comparisons of data from voluntary and stimulation protocols conducted at the whole muscle and single motor unit level are crucial to further our understanding of the mechanisms underlying muscle fatigue.

SPIKE-TRIGGERED AVERAGING

General Considerations

The concept of triggering off a steadily firing nerve or muscle potential and averaging the signals linked to that activity to extract even the weakest of associations (Mendell & Henneman, 1968) was extended to human experiments by Buchthal and Schmalbruch (1970). These investigators described the contraction times and fiber types of groups of muscle fibers for numerous human muscles such as biceps and triceps brachii as well as gastrocnemius, soleus, tibialis anterior, and platysma. With further refinement (Stein et al., 1972), evaluations were made of the contractile properties of human muscle at the single motor unit level and during voluntary contractions. In these studies, the subjects were given feedback of the activity of a single motor unit and asked to fire that unit at a slow steady rate (<12 Hz) for about two min. Provided other active motor units responded asynchronously relative to the test unit, the electrical and mechanical responses of the test unit could be averaged from the whole muscle responses. This spike-triggered averaging technique was used to show that the weakest units were activated during low force voluntary contractions and as the voluntary contraction grew in strength, stronger units were recruited. Unit strength was inversely related to twitch contraction time (time to peak force; Milner-Brown et al., 1973a,b).

These experiments were important in many ways. First, orderly recruitment of motor units by force output (or size) during voluntary contractions supported the predictions made from studies examining the reflex recruitment of motor units in cat hindlimb muscles (Henneman et al., 1965). Subsequent studies have shown similar patterns of recruitment in other limbs and jaw muscles (Desmedt & Godaux, 1977b; Goldberg & Derfler, 1977; Yemm, 1977), during ballistic contractions (Desmedt & Godaux 1977a; cf. Grimby & Hannerz, 1974), when muscles act in different directions (Thomas et al., 1986 cf. Person, 1974; Thomas et al., 1978; Desmedt & Godaux, 1981; ter Haar Romeny et al., 1982), during reflex contractions (Calancie & Bawa, 1985), in relation to axon conduction velocity (Freund et al., 1975; Dengler et al., 1988) as well as during dynamic contractions (Thomas et al., 1987a). Recruitment by increasing order of force output also occurs in reinnervated muscles after nerve section and resuture (Milner-Brown et al., 1974) and particularly when the reinnervated muscles are synergistic (Thomas et al., 1987b), in muscles of individuals with amyotrophic lateral sclerosis (Dengler et al., 1990), after chronic spinal cord injury (Stein et al., 1990; Thomas, unpublished data), and following immobilization (Duchateau & Hainaut, 1990). Alternative recruitment schemes have been described during speech (Smith et al., 1981), eccentric contractions (Nardone & Schieppati, 1988; Nardone et al., 1989; Howell et al., 1994), when proprioceptive afferents from the active muscles are blocked (Hannerz & Grimby, 1979), and when cutaneous stimulation is superimposed on a voluntary contraction (Stephens et al., 1978; Datta & Stephens, 1981). Thus, the spike-triggered averaging technique has spawned a number of studies that examine recruitment phenomena. Furthermore, there have been detailed examinations of various factors which can affect the data. These include: the influence of twitch fusion at different unit firing rates (Monster & Chan 1977; Calancie & Bawa, 1986; Thomas et al., 1990a), the changes in twitch parameters in relation to the short-term unit history (Nordstrom et al., 1989), whether the actual twitch can be predicted from the averaged, partially fused force profile (Lim et al., 1995), as well as the implications of unit synchrony on unit force (Milner-Brown et al., 1973a; Yue et al., 1995). At first glance, the characterization of the contractile properties of human motor units during voluntary contractions also provided the framework to explore how motor units respond during fatiguing protocols. However, surprisingly few studies have examined these responses.

Contributions to Understanding Muscle Fatigue

In two studies individuals fired single motor units in the first dorsal interosseus or masseter muscle at low rates (10 Hz) for five or 15 min, respectively (Stephens & Usherwood, 1977; Nordstrom & Miles, 1990). The ratio of the final to initial twitch force was used as a fatigue index. In the first dorsal interosseous, units were recruited up to 60% of maximum voluntary contraction (MVC) force. Low threshold units (recruited below 15% MVC) were fatigue resistant (< 25% force reduction) whereas units activated above 25% MVC force were highly fatigable (> 75% force reduction). In the masseter study, units were recruited up to 26% MVC force. Units showed one of three fatigue responses: 1) a slow progressive decline in twitch force over 15 min; 2) an initial increase in twitch amplitude which was maintained for 15 min; or 3) a dramatic force reduction during the first four min, then force stabilization. Overall, the units were usually more fatigable after 15 compared to 6 min of activation. However, the force loss was not well correlated with the initial twitch force or contraction time.

In general, these studies have sampled units with low recruitment thresholds and have described changes in the magnitude of the partially fused force without regard to alterations in contraction rate. Changes in the relaxation phase have been particularly difficult to assess using spike-triggered averaging because of force fusion. Moreover, the nature of the orderly recruitment of motor units during voluntary contractions has made it largely impossible to either control the activation history of each unit. Even so, some of these spike-triggered averaging results, and those from intramuscular microstimulation studies (Gydikov et al., 1976; Stephens & Usherwood, 1977; Garnett et al., 1979; Young & Mayer, 1981) are discussed as if the units have been fatigued by classical protocols (Burke et al., 1973). Such comparisons imply, dangerously, that the actual protocol is unimportant in determining the fatigue response.

Why Have So Few Studies Used Spike-Triggered Averaging to Explore How Motor Unit Properties Change with Use?

Various factors underlie the paucity of fatigue-related studies which have used the spike-triggered averaging technique. Four variables will be reviewed here briefly. They include: 1) the averaging process; 2) the problems associated with deriving unit responses from an active muscle; 3) the limits imposed by force fusion; and, 4) the consequences of the contractile speed changes which accompany use.

When signals are averaged over long time periods (e.g., minutes) to extract unit properties, the temporal resolution of the data becomes diluted. It can also mask unit potentiation (see below) and dramatically change the extent of fatigue. Many human masseter units which were activated at 10 Hz for 15 min showed an initial period of potentiation followed by a decline in force (Nordstrom & Miles, 1990). When the ratio of the final to initial force (fatigue index) was computed over one min rather than three min intervals, the population of motor units seemed much more fatigue resistant. While this associated force potentiation with use has been reported previously in both whole muscle and motor unit studies, it has largely been ignored by studies that have used pre-potentiation of the muscle by tetanic stimulation. But it is clear from these human and other studies (see below) that its effect can be substantial when motor units are activated submaximally.

A second problem relates to extracting single motor unit activity from an active muscle. It is unclear whether the forces of different motor units summate linearly or non-linearly. Even when firing asynchronously, there may be interaction between units. Similarly, short term synchrony can occur between motor units (Datta & Stephens, 1980), but the magnitude of synchrony, and alterations in unit synchrony during fatigue are unknown. Moreover, as the contraction intensifies, the electrical interference pattern makes it much more difficult to selectively identify the same single motor unit potential. As a consequence, most spike-triggered averaging studies have favored the sampling of low threshold motor units. In those muscles where recruitment is prominent at low forces, for example, small hand muscles, large fractions of the population may actually be sampled. However, for large muscles like biceps brachii where recruitment can continue up to near maximal forces (Kukulka & Clamann, 1981), unit responses to fatigue have been left unexamined. Furthermore, as more units are sampled during an experiment, earlier contractions may have already involved some motor units, altering their apparent responses to fatigue.

Another variable is whether activating different motor units at the same rate (e.g., 10 Hz) is the most appropriate comparison in terms of understanding actual muscle function. Since recruitment thresholds and force-frequency relationships differ dramatically between units, it is unlikely that a unit which is recruited at 10% MVC force and another which is activated at 60% MVC will be active simultaneously at 10 Hz. The lower threshold unit may have been activated for some time before the higher threshold unit is even recruited. Thus, clearer information on the interactions between motor unit recruitment and frequency modulation during fatiguing voluntary contractions may be attained by examining unit behavior at fixed levels of force instead of at constant firing rates. When measuring motor unit contractile properties by spike-triggered averaging, however, the feasibility of implementing this suggestion is limited. The ability to average the unit forces depends on limited twitch fusion. The increases in force fusion at higher firing rates are well documented. Similarly, fusion varies with unit contractile rate (Monster & Chan 1977; Calancie & Bawa, 1986; Thomas et al., 1990a).

Another important feature which confounds the use of spike-triggered averaging in studies of fatigue is the changes in contractile speed which accompany use. Although slowing of muscle relaxation is well documented in response to whole muscle fatigue, changes in twitch contraction time are not. However, at the motor unit level, studies in cat hindlimb and human thenar muscles not only show that both occur, but that units can have quite divergent responses (Dubose et al., 1987; Thomas et al., 1991b; see below). Because of these changes in twitch speed, the relative fusion of force at a given frequency (e.g., 10 Hz) will be altered. If a unit is activated at the same absolute rate both before and after fatigue, its force may simply rise or fall depending on the alterations in twitch contraction and/or half-relaxation time.

Fig. 1 illustrates this point by showing data recorded from a single motor unit before and after a 60 s MVC of the triceps brachii muscle. The protocol was performed by a 22 year old man who suffered a spinal cord injury at C5-6 3 years before. Since the triceps muscle is now partially paralyzed, and few motor units remain under voluntary control, the sparse interference pattern made it possible to follow the same motor unit during the entire protocol. Two tests were performed immediately before and after the sustained MVC. The unit was first fired at a slow, steady rate and the associated force averaged. Second, the unit firing

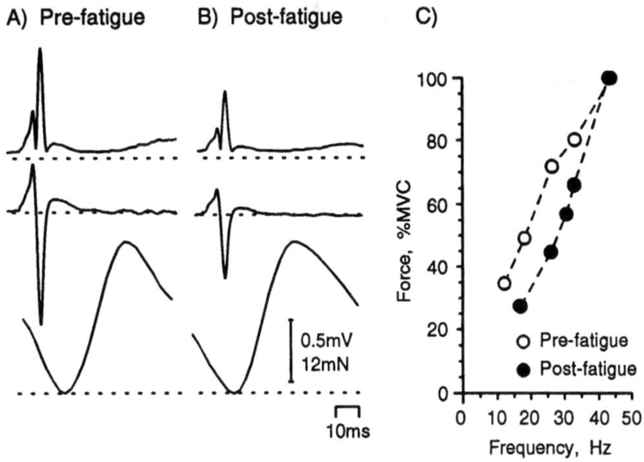

Figure 1. Averaged, rectified surface EMG and force from one triceps brachii motor unit before (*A*) and after (*B*) a 60 s MVC. *C*, whole muscle force *vs*. unit frequency measured during brief voluntary contractions performed prior to and after a 60 s MVC (open and closed symbols, respectively). Note the reduction in surface EMG and the apparent absence of change in the force profile with fatigue (*A*,*B*). However, any submaximal contraction level required a higher unit firing rate after fatigue.

rate was measured during brief (3-5s), submaximal and maximal contractions (Thomas et al., 1994).

During the fatiguing contraction, the whole muscle force and surface EMG declined and was accompanied by a 40% reduction in unit firing rate (43 Hz to 27 Hz). Within two min, maximal firing frequency had almost recovered. However, the force-frequency relationship recorded by voluntary effort moved horizontally to the right such that higher firing rates were required to produce the same relative force. For example, to exert 50% of maximal whole muscle force, the unit was activated at 20 Hz before and 28 Hz after the fatigue test. Since the relative positions of these force-frequency relationships are closely linked to unit force and speed in various cat muscles and human thenar muscles (Kernell et al., 1983; Thomas et al., 1991a), these data suggest twitch force and contraction time have been reduced. However, as shown in Fig. 1*A*, the force and contraction time of the motor unit were comparable before and after fatigue when assessed using spike triggered averaging (30.5 mN, 26.3 ms and 31.3 mN, 27.0 ms respectively). Thus, any reductions in unit force and contraction time due to fatigue may simply have been offset by the gains in each parameter resulting from reductions in twitch fusion (Thomas et al., 1994). At present there is no easy way of determining initial twitch force and speed from the partially fused profile obtained by spike triggered averaging (Lim et al., 1995). Changes in muscle compliance and the pattern of muscle co-activation may also occur with fatigue. Thus, the complexity of interpreting data obtained by spike-triggered averaging after fatigue is apparent.

Conclusion

Despite all the potential pitfalls apparent in using the spike-triggered averaging technique to examine how motor unit mechanical properties change during fatigue, two important features much be acknowledged. The technique has focused attention on unit fatigue during low-force voluntary contractions thereby emphasizing alterations in *unfused* force profiles. Twitch parameters, often considered to be unreliable because of their susceptibility to potentiation have important implications for understanding force-frequency relationships. Moreover, spike-triggered averaging remains the only technique currently available to examine motor unit properties during voluntary contractions. As such it remains a valuable method which may be better understood if its use could be carefully coupled to alternative techniques that permit the stimulation of single motor units at both twitch and tetanic rates.

INTRANEURAL MOTOR AXON STIMULATION

General Considerations

The recent development of measuring human motor unit contractile properties in response to intraneural motor axon stimulation (Johansson et al., 1988a, 1988b; Westling et al., 1990) is analogous to various methods used in animals in that single motor axons are stimulated some distance from the muscle, and each allows measurements of axon conduction velocities and the full time course of the twitch and tetanic forces. In humans, stimulation of motor axons within the main nerve trunk depends on the use of conventional microneurographic techniques. The appropriate nerve must first be located by stimulating through a tungsten microelectrode (typically 0.2mm diameter, insulated except for the tip, electrode impedance: 200-400 kΩ measured with low current at 1 kHz). With further minute manipulation it is then possible to stimulate and record force and EMG selectively from a single motor unit. However, baseline fluctuations due to respiration and pulse pressure waves feature prominently when the signals are amplified sufficiently to measure unit twitch forces. Thus, three adaptations are necessary for successful recordings (Fig. 2). First, stimulation has to be selective. Second, stimulation of the motor axon must be time locked to the pulse pressure cycle so that it is delivered during a period of minimum baseline fluctuation. Third, the evoked force signals of individual units should be recorded only after the baseline is reset electronically. With simultaneous use of all these methods, individuals force responses can be recorded without signal averaging (cf. McKeon & Burke, 1983).

One of the important findings from these experiments involving intraneural stimulation of motor axons to thenar muscle has been that the distribution of measured axonal conduction velocities suggests that it is possible to sample the entire range of large diameter myelinated axons in the nerve. Thus, for one human muscle group, it was possible to define for the first time, the entire range of twitch and tetanic forces, their contraction and relaxation rates as well as their axonal conduction velocities. When F-responses were encountered (two unitary EMG responses to a single stimulus), proximal conduction velocities could also be evaluated. Moreover, the measurement of the force responses in two directions demonstrated that individual units within the same muscle group generate force in widely different directions. Each of these unit parameters could be measured readily from single sweeps (Westling et al., 1990), a feature which makes this method ideal for examining fatigue of single human motor units in response to various stimulation patterns. Similarly, microneurographic data has been recorded in experiments involving strong voluntary contrac-

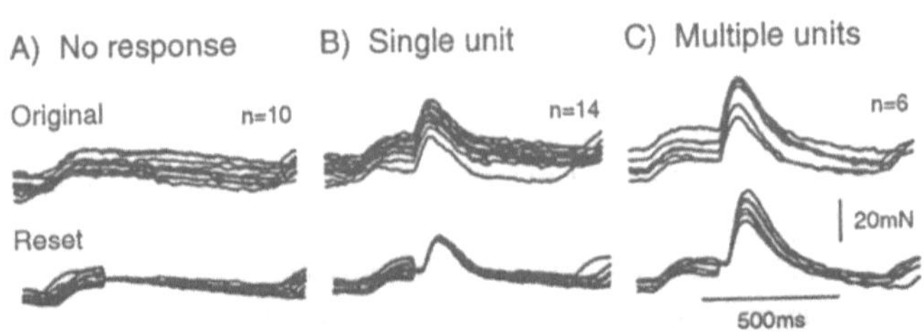

Figure 2. Sequential superimposed force records with and without baseline reset when no (*A*), a single (*B*) and several motor axons are stimulated (*C*). Note that even without stimulation of an axon (*A*), the pulse pressure waves produce a characteristic rapid rise in the baseline followed by a slower relatively linear decline. Provided stimuli are delivered during periods of relatively low baseline fluctuation, and the baseline is reset electronically just before the stimulation, unitary responses or graded responses from multiple units become obvious (modified from Westling et al., 1990).

tions (Vallbo, 1970), so it should also be possible to measure motor unit contractile responses before and after voluntary contractions. Such experiments still need to be performed.

To date, the contractile properties of human motor units have been examined by intraneural stimulation of motor axons to thenar (median nerve; Westling et al., 1990; Thomas et al., 1990a,b, 1991a,b) and toe extensor muscles (common peroneal nerve; Bigland-Ritchie et al., 1993; Fuglevand et al., 1993). In each experimental series, the twitch forces spanned a narrow range compared to that reported using spike-triggered averaging, and intramuscular microstimulation. Twitch contraction and half relaxation times were generally faster for thenar units than for toe extensor units or for units measured using alternative human methods. This was even after the significant slowing resulting from post-tetanic potentiation and fatigue (Thomas et al., 1990b, 1991b).

The force-frequency relationships measured for single human thenar motor units were similar to the corresponding force-frequency functions reported for various whole human muscles (Marsh et al., 1981; Gandevia & McKenzie, 1988) in that twitch fusion began at low stimulation frequencies (already between 5 and 8 Hz), half maximal tetanic force was reached at 12 ± 4 Hz (mean ± SD), and maximal force output was usually obtained between 30 and 50 Hz. A few units reached maximal tetanic force between 50 and 100 Hz (Thomas et al., 1991a). Similar results have been found for human toe extensor motor units although half maximal force was attained at 10 ± 1 Hz and full fusion sometimes occurred at 20 Hz (Bigland-Ritchie et al., 1993). Thus, force modulation for these human units occurred over a narrow frequency range and one which is typical of the motor unit firing rates recorded during submaximal and maximal voluntary contractions of various human muscles (Bellemare et al., 1983; Bigland-Ritchie et al., 1986).

As expected (Cooper & Eccles, 1930), gradation of force with stimulus frequency was highly dependent on unit contractile rate. Half initial maximal tetanic force was evoked at lower stimulation frequencies for units with slow twitch contraction and half relaxation times, as found for cat limb motor units (Kernell et al., 1975, 1983; Botterman et al., 1986). However, for both human thenar and toe extensor motor units, there was limited separation between the force-frequency curves of even the fastest and slowest units which is consistent with the narrow range of measured twitch contraction times (37 to 70 ms and 50 to 107 ms respectively). In cat limb muscles, twitch contraction times may vary by more than 2 fold, force is graded over much wider frequency ranges and force-frequency curves of fast and slow units fail to overlap (Kernell et al., 1975, 1983; Botterman et al., 1986). Similarly, axon conduction velocities for human thenar units are slower and cover a narrower range compared to values for various hindlimb muscles in animals (Westling et al., 1990).

Clearly there is a limited range of human twitch properties, axonal conduction velocities and frequencies over which force is modulated. These differences may reflect variations between distal *vs.* limb muscles rather than discrepancies between species. Similarly, the stronger and slower force profiles recorded by spike-triggered averaging compared to intraneural motor axon stimulation are not explained readily by sample bias or twitch fusion. At present, no stimulation studies have actually imitated the data which is typically recorded by spike-triggered averaging. That is, synchronous stimulation at fixed rates is dramatically different from the conditions where contractile responses are averaged during low force voluntary contractions. With spike-triggered averaging many other units are active simultaneously and asynchronously whereas with intraneural stimulation all but one motor unit are relaxed. Recording from an actively contracting muscle may increase twitch force because of the higher resting tension. Thus, some of the differences in results may depend on series compliance. Studies that measure unit force responses by spike-triggered averaging and use the same pattern of unit firing to stimulate the motor axon would be most helpful in explaining some of these differences. Such studies may be feasible using surface stimulation (Doherty & Brown, 1994).

Contributions to Understanding Muscle Fatigue

Intraneural stimulation of thenar motor axons has revealed features of the underlying muscle anatomy. The most fatigable units exerted more force in abduction. Fatigue resistant units produced more force in flexion. This gradient of fatigability with the angle of force exertion seems appropriate for the function of human thenar muscles. Abductor pollicis brevis typically acts during the spacing of the fingers prior to grasping objects (Johansson & Westling, 1988). This is an intermittent action and one that generally does not involve external loads. In contrast, the thumb flexor or opponens muscles serve to maintain the grip force for long periods of time, a capacity which would seem well matched to the high fatigue resistance of these units.

Thenar motor units (Thomas et al., 1991b) were only subjected to one fatigue protocol: intermittent 40 Hz stimulation for two min (Burke et al., 1973). Most units (72%) lost less than 25% force, or potentiated indicating how fatigue resistant these units were. The remaining units lost between 25-75% force showing intermediate fatigability. There were no units with fatigue indices less than 0.25, which would classify them as *fatigable* using traditional criteria. Overall, there was no significant change in the mean twitch or maximal tetanic force with fatigue. Nine units were subjected to a second fatigue run after several other additional tetanic protocols had been applied. This additional activation induced only small changes in force output. Thus, this method of assessing fatigability demonstrated that human thenar motor units were remarkably fatigue resistant (Thomas et al., 1991b).

In contrast to many motor units in muscles of other mammalian species, most thenar motor units produced maximal tetanic forces with 40 Hz stimulation. For many of these units there was an obvious change in the shape of the force profile with fatigue because of the marked slowing in the contraction and particularly the relaxation phases. This slowing meant that consecutive force responses began to fuse together (Thomas et al., 1991b), phenomena that have been observed but rarely emphasized in some other mammalian motor units (Burke et al., 1973). In terms of *tetanic* force responses, three trends were apparent. Some thenar units lost force and showed contractile slowing (28%). Others lost minimal force in the presence of contractile slowing (36%). The remaining units displayed little change in either force or contractile speed (36%). Those units with the strongest maximal tetanic forces fatigued and slowed the most. Units that lost *twitch* force with fatigue (55%) tended to slow in the twitch contraction phase but speed up during relaxation. The 45% of units that showed twitch potentiation with fatigue showed the opposite contractile rate trends (Thomas et al., 1991b).

Some other studies report twitch potentiation and fatigue (Garnett et al., 1979; Young & Mayer, 1981; Nordstrom & Miles, 1990). Contractile slowing however was only reported for opponens pollicis units as measured by intramuscular microstimulation (Gydikov et al., 1976). In these units, slowing was greatest for units with fast initial contraction and half-relaxation times, as observed for human thenar motor units (Thomas et al., 1990b, 1991b). Similar results were found for cat medial gastrocnemius units (Dubose et al., 1987).

After the fatigue test, the frequencies needed to evoke 50% of the initial maximal force were not significantly different for human thenar units (Thomas et al., 1991b). However, there were diverse responses for different units in that five units showed a potentiated force output at all submaximal stimulation frequencies (potentiating units) while seven units showed a force loss at all stimulation frequencies (fatiguing units). Fig. 3 shows the force-frequency curves before and after fatigue for two typical units. These units produced initial maximal tetanic forces of 95 and 75 mN and stimulation rates of 14 and 10 Hz evoked half these maximal tetanic forces (*A*,*B* respectively). In response to fatigue, one unit lost force at all frequencies; therefore, needing higher rates of stimulation to produce any given absolute force. A stimulation frequency of 24 Hz was required to evoke half the

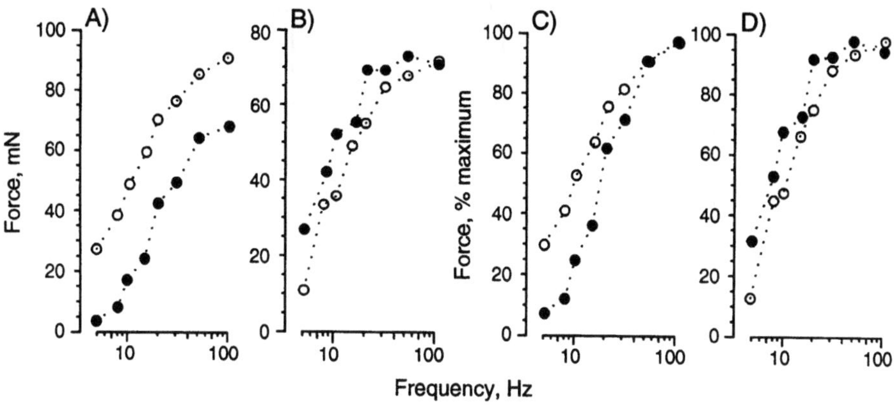

Figure 3. Initial (open symbols) and post-fatigue (closed symbols) force-frequency relationships for two thenar motor units (*A* and *B*, respectively). *C* and *D* show the corresponding relationships expressed relative to maximal tetanic force. Fatigue was induced by intermittent 40 Hz stimulation for two min (modified from Thomas et al., 1991a).

initial maximal tetanic force (*C*). In contrast, the maximal tetanic force for the unit shown in *B* was similar in both force-frequency tests. However, there were increases in the force evoked by submaximal frequencies. Half of this unit's maximal tetanic force could now be evoked at 7 Hz compared to 10 Hz before the test for fatigability (*D*).

These divergent unit responses could not be explained by variable activation histories. Rather, the changes in stimulation rate needed to evoke submaximal forces with activation history depended on changes in force output *and* the time course of these changes. The maximal potentiated twitch response was reached after one or two tetanic tests for all the fatigable units. During the fatigue test, both the twitch and tetanic force of these units declined. In contrast, the other units produced similar maximal tetanic forces before and after fatigue and continued to show potentiation of force at all submaximal frequencies (Thomas et al., 1991b). Thus the effects of potentiation seem to carry over to unfused force responses as observed by others (Olson & Swett, 1971; Kernell et al., 1975; Burke et al., 1976).

These data reaffirm the influence that twitch and subtetanic force responses have on the unit force-frequency curve. At low activation rates, a combination of contraction *and* relaxation phase speed changes are probably important as well. However, contractile slowing occurs with both potentiation and fatigue. Thus, the slowing which accompanies twitch potentiation, rather than fatigue, may dictate the direction in which these force-frequency curves move.

What Is the Relationship Between Twitch and Tetanic Responses to Fatigue

Most human studies have examined how twitch forces change with fatigue but most animal studies have first pre-potentiated the muscle with tetanic stimulation, then examined the fatigue response to tetanic stimulation. Direct comparisons between the twitch and tetanic forces in response to the same fatigue test are necessary to understand these different approaches. Fig. 4 compares for human thenar motor units, the twitch and tetanic fatigue indices (ratio of final to initial value) for force, maximal contraction rate (MCR) and half-relaxation time (HRT; *A-C* respectively). In each plot, the dashed line represents where the data would fall if both the twitch and tetanic responses changed by comparable amounts.

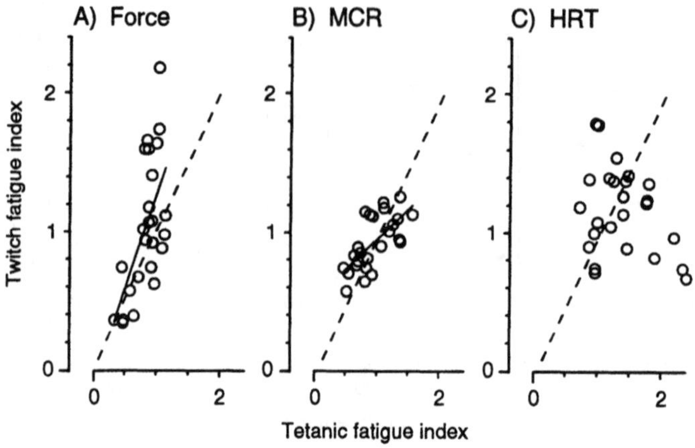

Figure 4. Twitch *vs*. tetanic fatigue index (final to initial ratio) for force (*A*), maximal contraction rate (*B*) and half relaxation time (*C*). Data falling on the dashed lines indicate comparable change in each parameter. The solid lines indicate a regression between twitch and tetanic fatigue indices.

The changes in the twitch and tetanic responses were correlated (Fig. 4*A*, $r=0.61$, $p<0.01$). However, the tetanic forces of some of the fatigue resistant units showed mild potentiation and most of these units showed even greater potentiation of twitch forces. Thus, tetanic force responses may be unchanged after fatigue while the twitch force is potentiated. Since the fatigable units also underwent a period of twitch potentiation that was over prior to the fatigue test, the twitch to tetanic force relationship depends heavily on the unit activation history.

Changes in twitch and tetanic maximal contraction rates with fatigue were also correlated (Fig. 4*B*, $r=0.67$, $p<0.01$). Thus, for the fatigable units, both these rates slow, along with the twitch contraction time. The converse is the case for fatigue resistant units. Relaxation measures from tetanic responses were not correlated to the corresponding measures taken from twitches (Fig. 4*C*), so twitch relaxation parameters have little or no value for predicting the corresponding tetanic parameters. Thus care must be taken when force and relaxation properties measured from twitches are used to draw inferences about tetanic contractile properties and vice versa (Thomas et al., 1991b).

Why Have So Few Studies Used Intraneural Motor Axon Stimulation to Explore How Motor Unit Properties Change with Use

Intraneural stimulation of motor axons and measurement of the associated unit contractile properties is a new technique. Numerous other reasons could underlie the small data pool. First, it is technically difficult to locate a peripheral nerve with a fine tungsten electrode. When the nerve lies deep within a limb, even the smallest of adjustments in the angle of the needle can dramatically alter the location of the electrode tip. Patience, as well as care are not only needed for nerve location, but also to ensure an absence of nerve damage. Second, long experiments (up to six hours) require the relaxed and cooperative participation of the subject. Any movements by the subject can dislodge the position of the stimulating electrode and thus disrupt recordings. Descriptions of what the subject feels, particularly during the initial search for the nerve, are also helpful in guiding the experimenter towards the nerve. When perception to touch has been lost, for example after certain spinal cord injuries, visible criteria must be used (e.g., muscle contraction). Third, special equipment is needed to electronically reset the force baseline to minimize fluctuations from respiration and pulse pressure waves (see Fig. 2). Force transducers must be sensitive enough to register the weak twitch forces of single motor units without averaging. It is also preferable to use a constant current stimulator. Finally, anatomical constraints can limit the choice of test nerves and muscles. Experimental conditions are more stable if the stimulation site is some distance from where the muscle recordings are made (e.g., thenar motor axons in the median nerve can be accessed above the elbow). The particular motor axons of interest must be sufficiently spaced within the nerve to permit selective stimulation. For example, thenar motor axons are dispersed across the entire median nerve trunk above the elbow but tend to focus in only one of two nerve fascicles near the wrist. Isolation of thenar motor axons by the surrounding axons and the ratio of sensory to motor axons in the median nerve are two reasons proposed for the success of these earlier recordings. In summary, all these factors translate into low data yields. Despite this, the technique has much promise for examining single human motor unit fatigue. Future studies would benefit by varying activation rates and protocols and coupling these to voluntary contraction protocols.

CONCLUSION

With spike-triggered averaging, only a limited number of twitch parameters are measured. Force fusion during weak voluntary contractions, which results in reductions in twitch amplitudes and contraction times, means that the spike-triggered averaging method is only appropriate for measuring force responses at low firing rates. During sustained, low force voluntary contractions of masseter or first dorsal interosseous muscles, there is either decline or potentiation of unit twitch force. However, recording force profiles with spike-triggered averaging at the same rate before and after fatigue produces data which is difficult to interpret. Any changes in contractile speed with fatigue will alter how the twitches sum together and the apparent force profile. Other factors to consider include the inability to control unit activation history, alterations in muscle compliance, and muscle co-contraction.

Studies with intraneural stimulation of motor axons have allowed the full time course of twitch and tetanic forces to be recorded for a representative sample of units from one muscle group. When activated intermittently and maximally at rates which could be defined, human thenar units either showed twitch potentiation and tetanic fatigue resistance or a decline in both twitch and tetanic force. After fatigue, these units needed lower and higher rates respectively, to produce any given submaximal force. These data not only emphasize how subtetanic force responses influence force-frequency unit behavior but also illustrate that twitch and tetanic force relationships depend on activation history. Caution is therefore required when making inferences about tetanic forces from twitch responses and vice versa. Similarly, it would seem equally dangerous to impose unit classifications traditionally used for motor units in mammalian muscles on human units if the protocols used to activate the units vary, and if the evaluation criteria are based primarily on twitch parameters.

Other factors also make it difficult to compare data obtained during evoked *vs.* voluntary contractions. For example, most animal studies and human motor axon stimulation studies have evoked contractions at fixed stimulation rates (cf. Botterman & Cope, 1988). Motor unit firing rates are modulated continuously during voluntary contractions. Second, the same fatigue protocol (Burke et al., 1973) has been used monotonously. It does not reflect usual muscle activation because it is impossible, by voluntary effort, to maximally activate a human muscle for a fraction of a second, each second for two min. Third, stimulation drives motor units synchronously. Motor unit activity during voluntary contractions is generally asynchronous. Moreover, one unit is typically stimulated whereas numerous units are often active during voluntary effort. This may change muscle compliance. All these poorly understood phenomena argue for an infusion of new experiments that aim to define the diversity of unitary responses and the differences between voluntary and stimulated contractions. Such information will enhance our appreciation of whole muscle responses, the balance and importance of recruitment and rate modulation in force production, the interactions between muscle potentiation and fatigue, and whether motor unit behavior is organized around groups of motor units within or across muscles.

ACKNOWLEDGMENTS

The author wishes to thank Drs. Mary Pocock and James Broton for constructive comments on the manuscript. The author's laboratory is supported by United States Public Health Service grant NS 30226, and The Miami Project to Cure Paralysis. Attendance of the author at the 1994 Bigland-Ritchie conference was supported, in part, by NS 30226.

REFERENCES

Bellemare F, Woods JJ, Johansson R & Bigland-Ritchie B (1983). Motor-unit discharge rates in maximal voluntary contractions of three human muscles. *Journal of Neurophysiology* **50**, 1380-1392.

Bigland-Ritchie B, Cafarelli E & and Völlestad NK (1986). Fatigue of submaximal static contractions. *Acta Physiologica Scandinavica Supplement* **556**, 137-148.

Bigland-Ritchie B, Fuglevand AJ & Macefield VG (1993). Force modulation by rate-coding in single motor units of human toe extensor muscles. *Society for Neuroscience Abstract* **19**, 154.

Botterman BR & Cope TC (1988). Motor-unit stimulation patterns during fatiguing contractions of constant tension. *Journal of Neurophysiology* **60**, 1198-1214.

Botterman BR, Iwamoto GA & Gonyea WJ (1986). Gradation of isometric tension by different activation rates on motor units of cat flexor carpi radialis muscle. *Journal of Neurophysiology* **56**, 494-506.

Buchthal F & Schmalbruch H (1970). Contraction times and fibre types in intact human muscle. *Acta Physiologica Scandinavica* **79**, 435-452.

Burke RE, Levine DN, Tsairis P & Zajac FE (1973). Physiological types and histochemical profiles in motor units of the cat gastrocnemius. *Journal of Physiology (London)* **234**, 723-748.

Burke RE, Rudomin P & Zajac FE (1976). The effect of activation history on tension production by individual motor units. *Brain Research* **109**, 515-529.

Calancie B & Bawa P (1985). Voluntary and reflexive recruitment of flexor carpi radialis motor units in humans. *Journal of Neurophysiology* **53**, 1194-1200.

Calancie B & Bawa P (1986). Limitations of the spike-triggered averaging technique. *Muscle & Nerve* **9**, 78-83.

Cooper S & Eccles JC (1930). The isometric response of mammalian muscles. *Journal of Physiology (London)* **69**, 377-385.

Datta AK & Stephens JA (1980). Short-term synchronization of motor unit firing in human first dorsal interosseous. *Journal of Physiology (London)* **308**, 19-20P.

Datta AK & Stephens JA (1981). The effects of digital nerve stimulation on the firing of motor units in human first dorsal interosseous muscle. *Journal of Physiology (London)* **318**, 501-510.

Dengler R, Konstanzer A, Küther G, Hesse S, Wolf W, & Struppler A (1990). Amyotrophic lateral sclerosis: macro-EMG and twitch forces of single motor units. *Muscle & Nerve* **13**, 545-550.

Dengler R, Stein RB, & Thomas CK (1988). Axonal conduction velocity and force of single human motor units. *Muscle & Nerve* **11**, 136-145.

Desmedt JE & Godaux E (1977a). Fast motor units are not preferentially activated in rapid voluntary contractions in man. *Nature* **267**, 717-719.

Desmedt JE & Godaux E (1977b). Ballistic contractions in man: characteristic recruitment patterns of single motor units of the tibialis anterior muscle. *Journal of Physiology (London)* **264**, 673-693.

Desmedt JE & Godaux E (1981). Spinal motoneuron recruitment in man: rank deordering with direction but not with speed of voluntary movement. *Science* **214**, 933-936.

Doherty TJ & Brown WF (1994). A method for the longitudinal study of human thenar motor units. *Muscle and Nerve* **17**, 1029-1036.

Dubose L, Schelhorn TB & Clamann HP (1987). Changes in contractile speed of cat motor units during activity. *Muscle & Nerve* **10**, 744-752.

Duchateau J & Hainaut K (1990). Effects of immobilization on contractile properties, recruitment and firing rates of human motor units. *Journal of Physiology (London)* **422**, 55-65.

Freund H-J, Büdingen H-J & Dietz V (1975). Activity of single motor units from human forearm muscles during voluntary isometric contractions. *Journal of Neurophysiology* **38**, 933-946.

Fuglevand AJ, Macefield VG & Bigland-Ritchie B (1993). Twitch properties of human toe extensor motor units. *Society for Neuroscience Abstract* **19**, 154.

Gandevia SC & McKenzie DK (1988). Activation of human muscles at short muscle lengths during maximal static efforts. *Journal of Physiology (London)* **407**, 599-613.

Garnett RAF, O'Donovan MJ, Stephens JA & Taylor A (1979). Motor unit organisation of human medial gastrocnemius. *Journal of Physiology (London)* **287**, 33-43.

Goldberg LJ & Derfler B (1977). Relationship among recruitment order, spike amplitude, and twitch tension of single motor units in human masseter muscle. *Journal of Neurophysiology* **40**, 879-890.

Grimby L & Hannerz J (1974). Differences in recruitment order and discharge pattern of motor units in the early and late flexion reflex components in man. *Acta Physiologica Scandinavica* **90**, 555-564.

Gydikov A, Dimitrov G, Kosarov D & Dimitrova N (1976). Functional differentiation of motor units in human opponens pollicis muscle. *Experimental Neurology* **50**, 36-47.

Hannerz J & Grimby L (1979). The afferent influence on the voluntary firing range of individual motor units in man. *Muscle & Nerve* **2**, 414-422.

Henneman E, Somjen G & Carpenter DO (1965). Functional significance of cell size in spinal motoneurons. *Journal of Neurophysiology* **28**, 560-580.

Howell JN, Fuglevand AJ, Walsh ML & Bigland-Ritchie B (1994). Motor unit firing during shortening and lengthening contractions. *Society for Neuroscience Abstract* **20**, 1759.

Johansson RS, Thomas CK, Westling G & Bigland-Ritchie B (1988a). A new method for examining contractile properties of human single motor units. *European Journal of Physiology* **411**, S1 R196.

Johansson RS & Westling G (1988). Coordinated isometric muscle commands adequately and erroneously programmed for the weight during lifting task with precision grip. *Experimental Brain Research* **71**, 59-71.

Johansson RS, Westling G, Thomas CK & Bigland-Ritchie B (1988b). Recording human single motor unit properties: a new method. *Society for Neuroscience Abstract* **14**, 1232.

Kernell D, Ducati A & Sjöholm H (1975). Properties of motor units in the first deep lumbrical muscle of the cat's foot. *Brain Research* **98**, 37-55.

Kernell D, Eerbeek O & Verhey BA (1983). Relation between isometric force and stimulation rate in cat's hindlimb motor units of different twitch contraction time. *Experimental Brain Research* **50**, 220-227.

Kukulka CG & Clamann HP (1981). Comparison of the recruitment and discharge properties of motor units in human brachial biceps and adductor pollicis during isometric contractions. *Brain Research* **219**, 45-55.

Lim KY, Thomas CK & Rymer WZ (1995). Computational methods for improving estimates of motor unit twitch contraction properties. *Muscle & Nerve* **18**, 165-174.

Marsh E, Sale D, McComas AJ & Quinlan J (1981). Influence of joint position on ankle dorsiflexion in humans. *Journal of Applied Physiology* **51**, 160-167.

McKeon B & Burke D (1983). Muscle spindle discharge in response to contraction of single motor units. *Journal of Neurophysiology* **49**, 291-302.

Mendell LM & Henneman E (1968). Terminals of single Ia fibers: Distribution within a pool of 300 homonymous motor neurons. *Science* **160**, 96-98.

Milner-Brown HS, Stein RB & Lee RG (1974). Pattern of recruiting human motor units in neuropathies and motor neurone disease. *Journal of Neurology, Neurosurgery & Psychiatry* **37**, 665-669.

Milner-Brown HS, Stein RB & Yemm R (1973a). The contractile properties of human motor units during voluntary isometric contractions. *Journal of Physiology (London)* **228**, 285-306.

Milner-Brown HS, Stein RB & Yemm R (1973b). The orderly recruitment of human motor units during voluntary isometric contractions. *Journal of Physiology (London)* **230**, 359-370.

Monster AW & Chan H (1977). Isometric force production by motor units in the extensor digitorum communis muscle in man. *Journal of Neurophysiology* **40**, 1432-1443.

Nardone A, Romano C & Schieppati M (1989). Selective recruitment of high-threshold human motor units during voluntary isotonic lengthening of active muscles. *Journal of Physiology (London)* **409**, 451-471.

Nardone A & Schieppati M (1988). Shift of activity from slow to fast muscle during voluntary lengthening contractions of the triceps surae muscle in humans. *Journal of Physiology (London)* **395**, 363-381.

Nordstrom MA & Miles TS (1990). Fatigue of single motor units in human masseter. *Journal of Applied Physiology* **68**, 26-34.

Nordstrom MA, Miles TS & Veale JL (1989). Effect of motor unit firing pattern on twitches obtained by spike-triggered averaging. *Muscle & Nerve* **12**, 556-567.

Olson CB & Swett CP (1971). Effect of prior activity on properties of different types of motor units. *Journal of Neurophysiology* **34**, 1-16.

Person RS (1974). Rhythmic activity of a group of human motoneurons during voluntary contraction of a muscle. *Electroencephalography* **36**, 585-595.

Sica REP & McComas AJ (1971). Fast and slow twitch units in a human muscle. *Journal of Neurology, Neurosurgery & Psychiatry* **34**, 113-120.

Smith A, Zimmerman GN & Abbas PJ (1981). Recruitment patterns of motor units in speech production. *Journal of Speech and Hearing Research* **24**, 567-576.

Stein RB, Brucker BS & Ayyar DR (1990). Motor units in incomplete spinal cord injury: electrical activity, contractile properties and the effects of biofeedback. *Journal of Neurology, Neurosurgery & Psychiatry* **53**, 880-885.

Stein RB, French AS, Mannard D & Yemm R (1972). New methods for analyzing motor function in man and animals. *Brain Research* **40**, 187-192.

Stephens JA Garnett R & Buller NP (1978). Reversal of recruitment order of single motor units produced by cutaneous stimulation during voluntary muscle contraction in man. *Nature* **272**, 362-364.

Stephens JA & Usherwood TP (1977). The mechanical properties of human motor units with special reference to their fatiguability and recruitment threshold. *Brain Research* **125**, 91-97.

Taylor A & Stephens JA (1976). Study of human motor unit contractions by controlled intramuscular microstimulation. *Brain Research* **117**, 331-335.

Ter Haar Romeny BM, Denier van der Gon JJ & Gielen CAM (1982). Changes in recruitment order of motor units in the human biceps muscle. *Experimental Neurology* **78**, 360-368.

Thomas CK, Bigland-Ritchie B & Johansson RS (1991a). Force-frequency relationships of human thenar motor units. *Journal of Neurophysiology* **65**, 1509-1516.

Thomas CK, Bigland-Ritchie B, Westling G & Johansson RS (1990a). A comparison of human thenar motor unit properties studied by intraneural motor axon stimulation and spike-triggered averaging. *Journal of Neurophysiology* **64**, 1347-1351.

Thomas CK, Johansson RS & Bigland-Ritchie B (1991b). Attempts to physiologically classify human thenar motor units. *Journal of Neurophysiology* **65**, 1501-1508.

Thomas CK, Johansson RS, Westling G & Bigland-Ritchie B (1990b). Twitch properties of human thenar motor units measured in response to intraneural motor axon stimulation. *Journal of Neurophysiology* **64**, 1339-1346.

Thomas CK, Pocock ME & Evans JJ (1994). Fatigue and post-fatigue responses to sustained MVCs after chronic cervical spinal cord injury. *Neural and Neuromuscular Aspects of Muscle Fatigue Abstracts* **B20**, 39.

Thomas CK, Ross BH & Calancie B (1987a). Human motor-unit recruitment during isometric contractions and repeated dynamic movements. *Journal of Neurophysiology* **57**, 311-324.

Thomas CK, Ross BH & Stein RB (1986). Motor-unit recruitment in human first dorsal interosseous muscle for static contractions in three different directions. *Journal of Neurophysiology* **55**, 1017-1029.

Thomas CK, Stein RB, Gordon T, Lee RG & Elleker MG (1987b). Patterns of reinnervation and motor unit recruitment in human hand muscles after complete ulnar and median nerve section and resuture. *Journal of Neurology, Neurosurgery & Psychiatry* **50**, 259-268.

Thomas JS, Schmidt EM & Hambrecht FT (1978). Facility of motor unit control during tasks defined directly in terms of unit behaviors. *Experimental Neurology* **59**, 384-395.

Vallbo, AB (1970). Discharge patterns in human muscle spindle afferents during isometric voluntary contractions. *Acta Physiologica Scandinavica* **78**, 315-333.

Westling G, Johansson RS, Thomas CK & Bigland-Ritchie B. (1990). Measurement of contractile and electrical properties of single human motor units in response to intraneural motor axon stimulation. *Journal of Neurophysiology* **64**, 1331-1338.

Yemm R (1977). The orderly recruitment of motor units of the masseter and temporal muscles during voluntary isometric contraction in man. *Journal of Physiology (London)* **265**, 163-174.

Young JL & Mayer RF (1981). Physiological properties and classification of single motor units activated by intramuscular microstimulation in the first dorsal interosseous muscle in man. In: Desmedt JE (ed.), *Motor Unit Types, Recruitment and Plasticity in Health and Disease. Progress in Clinical Neurophysiology* **9**, pp. 17-25. Basel: Karger.

Yue G, Fuglevand AJ, Nordstrom MA & Enoka RM (1995). Limitations of the surface-EMG technique for estimating motor unit synchronization. *Biological Cybernetics* In press.

HUMAN MOTOR UNITS STUDIED BY INTRAMUSCULAR MICROSTIMULATION

J. M. Elek and R. Dengler

Department of Neurology and Clinical Neurophysiology
Medical School of Hannover
D-30623 Hannover, Germany

ABSTRACT

We recorded twitch parameters of motor units of the human first dorsal interosseus muscle applying low-rate intramuscular microstimulation of motor axons (IMS) or spike-triggered averaging (STA). The values of contraction time, half-relaxation time and maximal rate of rise of force were significantly smaller in the STA studies. The reduction corresponded to the effect expected from partial twitch fusion. Twitch amplitudes, however, were not smaller than those in the IMS studies, indicating that other factors, such as motor unit synchronization, compensate for the effects of partial fusion.

INTRODUCTION

Single motor units in human muscles can be easily studied electromyographically by assessing their action potentials. The associated motor unit contractions, however, are more difficult to measure, although they represent the more relevant parameter from a functional point of view and could contribute to the pathophysiological understanding of diseases of the motor system (Young & Mayer, 1982, Dengler et al., 1990, Yang et al., 1990). Several methods have been developed to assess motor unit twitch parameters in human muscles. The two techniques most widely used are spike-triggered averaging (STA) (Stein et al., 1972) and intramuscular microstimulation of individual axons of motor nerves (IMS; Taylor & Stephens, 1976). Intraneural microstimulation of motor axons (Thomas et al., 1990a, 1990b) is another elegant technique to assess motor unit contractions and has been reviewed by Thomas, Chapter 10.

STA is carried out during voluntary motor unit activity and does not involve stimulation. It is easy to apply and provides both contraction parameters and the voluntary recruitment threshold of a motor unit (Stein et al., 1972; Milner-Brown et al., 1973a, 1973b; Stephens & Usherwood, 1977).

However, it has some methodological limitations: 1) motor unit twitches may be distorted by partial fusion of twitch responses (Monster & Chan, 1977; Andreassen &

Fatigue, Edited by Simon C. Gandevia et al.
Plenum Press, New York, 1995

Bar-On, 1983; Calancie & Bawa, 1986, Nordstrom et al., 1989); 2) synchronization of simultaneously firing motor units may lead to an overestimation of twitch forces (Milner-Brown et al., 1973a); and 3) sampling of motor units may be biased towards slow, fatigue-resistant, low-threshold motor units (Calancie & Bawa, 1986). The technique of low-rate IMS circumvents the problems associated with STA (Elek et al., 1992). We, therefore, compared twitch parameters of a representative number of normal motor units obtained by STA and IMS in a human hand muscle. In addition, data from patients with diseases with various forms of chronic partial denervation are presented.

MATERIAL AND METHODS

Thirty-seven normal subjects were investigated, 12 by STA (6 men and 6 women, mean age 36.5 yrs., range 23 to 63 yrs.) and 25 by IMS (15 men and 10 women, mean age 36 yrs., range 23 to 87 yrs.). Studies were approved by the local ethics committee and all subjects gave informed consent.

Force Measurements

The first dorsal interosseus muscle of the hand was investigated. The same set-up was used for STA- and IMS-experiments. An isometric strain gauge was positioned at the radial side of the index finger with a lever arm of 6.5 cm from the metacarpo-phalangeal joint, and measuring the contraction force of the first dorsal interosseus muscle in the direction of abduction only.

STA-Experiments

A total of 236 motor units was investigated using STA. Subjects produced stable isometric contractions at force levels sufficient to recruit a motor unit and keep it firing regularly (tonic recruitment threshold, up to 60% of maximal voluntary force) at low rate (8 to 12 Hz). Motor unit action potentials were picked up by the single fiber lead of a macro-EMG electrode (Stalberg & Fawcett, 1982) and by a pair of surface electrodes. The single-fiber signal triggered a computer to average the surface EMG potentials (sampling rate 5 kHz) and the twitch contractions (sampling rate 3 kHz) of the motor unit (200 to 500 trials). Twitches were recorded using a high-gain, AC-coupled version of the force signal (bandwith 0.3 to 100 Hz). The surface EMG potentials were used to identify different motor units.

IMS-Experiments

251 motor units were studied. A bipolar needle electrode was inserted into the first dorsal interosseus muscle at the border between the proximal and middle third to stimulate single motor axons. Surface electrodes were placed over the muscle belly and the knuckle of the index finger to record the electromyographic responses. The position of the needle electrode and the stimulus intensity (up to 3 mA, duration 0.05 to 0.1 ms) were adjusted to evoke a reproducible all-or-none EMG potential representing the response of a single motor unit. Stimuli were applied at a low rate (0.9 Hz) to obtain individual twitch contractions. Force signals were low-pass filtered (0 to 100 Hz) and amplified at high gain. The evoked EMG responses were monitored continuously and all trials with irregularities, e.g. missing and occasionally occurring F-responses, were rejected on-line. Twenty to 50 trials were averaged.

In 10 subjects with particularly stable experimental conditions, stimuli could also be applied at higher rates (up to 10 Hz in 20 motor units, up to 12 Hz in 15 motor units and up

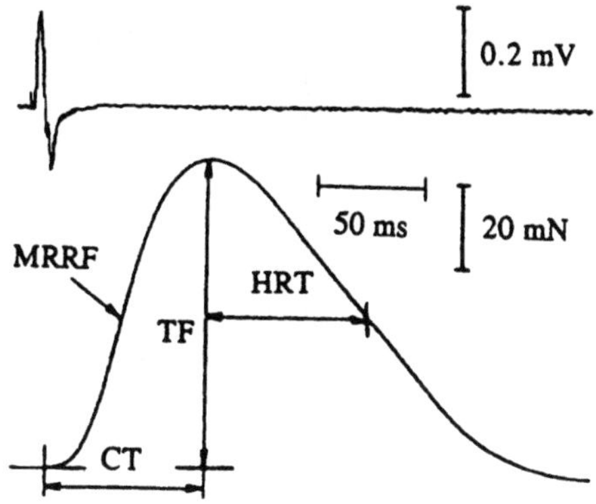

Figure 1. Averaged EMG response (*upper trace*) and twitch contraction (*lower trace*) of a single motor unit obtained by IMS at 0.9 Hz: twitch force (TF); maximum rate of rise of force (MRRF); contraction time (CT); half relaxation time (HRT).

to 14 Hz in 10 motor units) to induce partial fusion of twitches mimicking the situation in STA. Averaging was started when the contraction had stabilized after 3 to 5 stimuli. Parameters were measured from the unfused force fluctuations (trough to peak).

Twitch parameters

The following parameters were assessed off-line for twitches in both STA and IMS studies (Fig. 1): twitch force (TF), contraction time (CT), half relaxation time (HRT) and maximal rate of rise of force (MRRF).

Statistical Procedures

For statistical comparison of twitch parameters, medians were compared using the Mann-Whitney test in case of skewed distributions (twitch force and maximal rate of rise of force, see below); for normal distributions (contraction and half-relaxation times, see below) means were compared with student's t-test.

RESULTS

Comparison of Motor Unit Twitch Parameters Obtained by STA and IMS

Twitch force: the distributions of twitch forces obtained by STA and IMS are illustrated in Fig. 2, *left panel*. They appeared fairly similar and were both skewed to the right. The mean and median values of the *STA twitch force* were slightly larger than those of the *IMS twitch force* (Table 1). The difference, however, was not significant (Mann-Whitney test). Another slight difference was seen in the low force spectrum. With STA, 13 motor units (6.5%) revealed twitch forces smaller than 1 mN, while the smallest twitch force value obtained by IMS was 1 mN.

Maximal rate of rise of force: the distributions of maximal rate of rise of forces resembled those of the twitch forces (Fig. 2, *right panel*). The median values obtained by STA, however, were significantly smaller than the IMS-values ($p < 0.05$, Mann-Whitney test, Table 1).

Contraction and half-relaxation times: Both parameters showed an approximately Gaussian distribution for both STA and IMS (Fig. 3). The mean values obtained by STA,

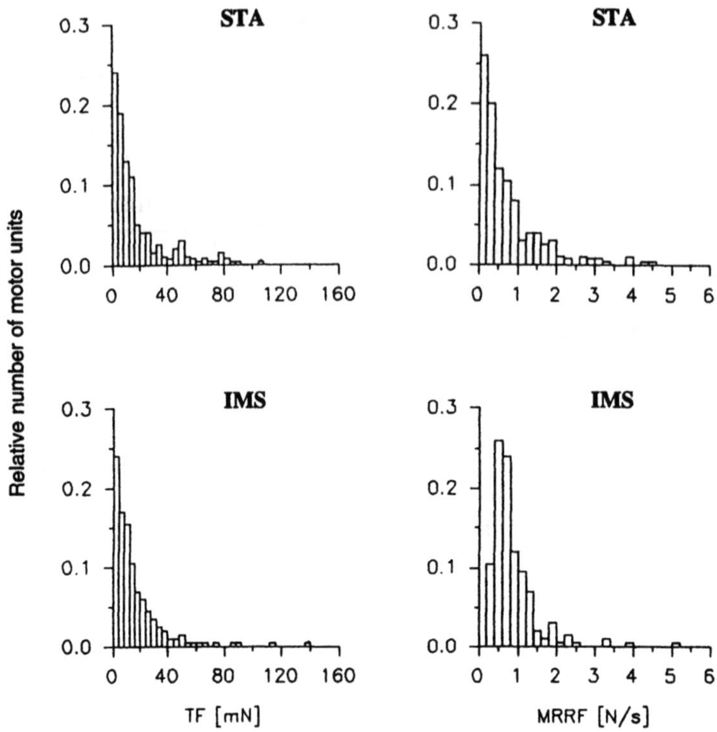

Figure 2. Distributions of twitch forces (TF, left) and maximal rate of rise of force values (MRRF, *right*) obtained by STA (*top*) and by IMS (*bottom*).

however, were significantly smaller for both contraction time and half-relaxation time than the corresponding IMS values ($p < 0.05$, t-test, Table 1).

Effect of Stimulus Rate on Twitch Parameters in IMS

Fig. 4 illustrates the effect of stimulus rate (0.9 to 14 Hz) on twitch parameters obtained with IMS. Values were normalized to the corresponding values at 0.9 Hz stimulation. A reduction of all twitch parameters with increasing stimulus rate was found beyond 4 Hz . Most prominent was the decrease of twitch force reaching a relative value of 0.45 ± 0.20 at 10 Hz while maximal rate of rise of force proved to be fairly robust with 0.88 ± 0.18.

Table 1. Motor unit twitch parameters obtained by IMS and STA

	IMS (n = 251)		STA (n = 236)		Expected Values for 10 Hz	
	Mean ± SD	Median	Mean ± SD	Median	Mean	Median
TF (mN)	(14±15)	9	(17.7±19.8)	10.3	(6.3)	4.1
MRRF (N/s)	(0.79±0.53)	0.61	(0.76±0.84)	0.42*	(0.70)	0.54
CT (ms)	64±14	(63)	47.3±12.8*	(44.8)	42	(42)
HRT (ms)	61±16	(59)	33.9±10.3*	(33.6)	36	(35)

*significant ($p < 0.05$) difference between values obtained by IMS and STA.

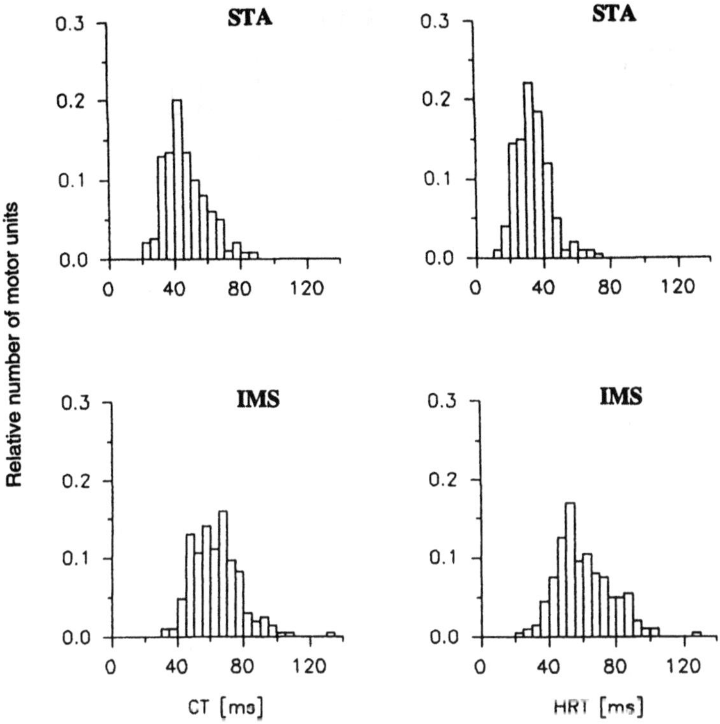

Figure 3. Distributions of contraction times (CTs, *left*) and half-relaxation time (HRTs, *right*) obtained by STA (*top*) and by IMS (*bottom*).

Contraction time and half-relaxation time decreased moderately to 0.66 ± 0.10 and 0.59 ± 0.15, respectively (at 10 Hz).

The above relative values were multiplied by the mean and median values to give *expected* twitch parameters at 10 Hz for the whole sample of motor units studied with IMS. The resulting *expected* mean and median values (Table 1, right columns) were then compared with the STA data (Table 1, middle columns) which were collected at very similar, voluntary firing rates of 8 to 12 Hz. The *expected* values for contraction time and half-relaxation time, to a lesser extent maximal rate of rise of force corresponded nicely to those obtained by STA. The twitch forces obtained by STA, however, were much larger than the *expected* values and resembled those recorded by IMS at low stimulus rate.

MOTOR UNIT TWITCH PARAMETERS IN DISEASES WITH CHRONIC PARTIAL DENERVATION

We investigated 31 patients with chronic partial denervation using low-rate IMS: 8 patients with hereditary motor and sensory neuropathy type I, age 13 to 52 yrs., mean 40 yrs.), 6 patients with spinal muscular atrophy (age 21 to 51 yrs., mean 36 yrs.) and 17 patients with amyotrophic lateral sclerosis (age 28 to 70 yrs., mean 50 yrs.).

Maximal voluntary force of abduction of the first dorsal interosseus muscle was reduced in all patient groups as compared to controls (14 - 55 N, mean 26.4 ± 10.2 N); hereditary motor and sensory neuropathy type I 0.4 - 33 N, mean 10.9 ± 10.6 N; spinal

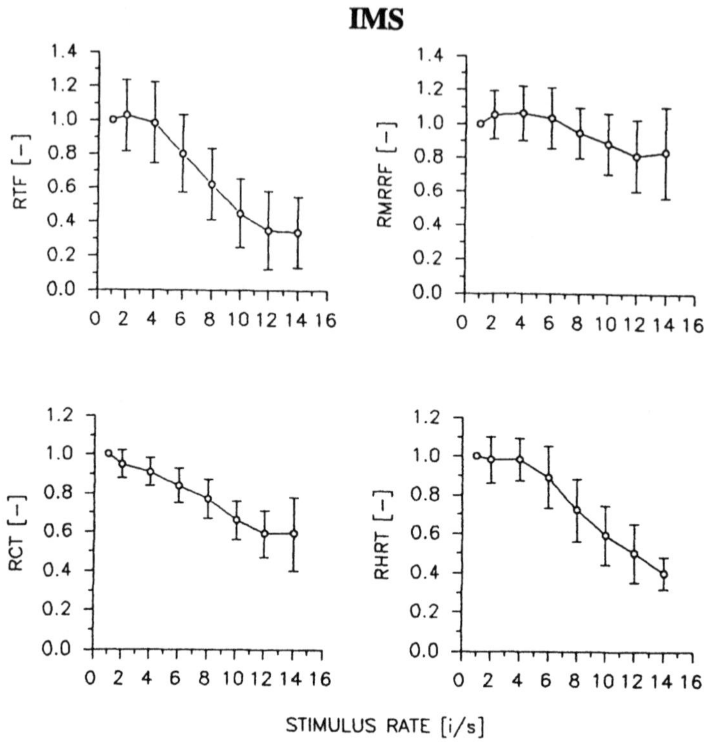

Figure 4. Mean values and standard deviations (20 motor units) of relative motor unit twitch parameters at different stimulus rates: RTF, RMRRF, RCT and RHRT - *relative* twitch parameters normalized to individual unfused twitch values of TF, MRRF, CT and HRT at 0.9 Hz, respectively.

muscular atrophy 0.5 - 17 N, mean 8.4 ± 6.8 N; amyotrophic lateral sclerosis 0.4 - 28 N, mean 9.2 ± 7.9 N.

Motor Unit Twitch Parameters in Hereditary Motor and Sensory Neuropathy Type I

The mean and median values of twitch parameters are summarized in Table 2 for 64 units. Overall, twitch forces did not differ significantly from controls, maximal-rate-of-rise-

Table 2. Motor unit twitch parameters in hereditary motor and sensory neuropathy type I (HMSN I)

	controls (n = 251)	HMSN I (n = 64)	MVC > 5 N (n = 46)	MVC ≤ 5 N (n = 18)
TF (mN)	9	9	12*	4*
MRRF (N/s)	0.61	0.46*	0.46*	0.30*
CT (ms)	64±14	74±20*	71±16*	83±27*
HRT (ms)	61±16	70±23*	71±24*	71 ± 24*

Median values for TF and MRRF; mean ± SD for CT and HRT
*significant difference ($p < 0.05$) compared to controls.

Table 3. Motor unit twitch parameters in spinal muscular atrophy (SMA)

	Controls (n = 251)	SMA (n = 32)	SMA (MVC > 5 N) (n = 28)	SMA(MVC ≤5 N) (n = 4)
TF (mN)	9	15*	20*	9
MRRF (N/s)	0.61	0.83*	0.88	0.82
CT (ms)	64±14	59±10	59±9	58±18
HRT (ms)	61±16	57±16	58±17	42±6

Median values for TF and MRRF, means ± SD for CT and HRT.
*Significant difference ($p < 0.05$) compared to controls.

of-force medians were significantly reduced, and mean contraction and half-relaxation times were significantly prolonged.

Patients were then subdivided in those with a severely affected first dorsal interosseus muscle with a maximal voluntary contraction (MVC) force of 5 N or less and those with a slightly or moderately affected first dorsal interosseus muscle (MVC > 5 N). In slightly/moderately affected muscles twitch force medians were significantly larger than in controls. The severely affected muscles, however, showed significantly smaller twitch force medians with a more pronounced decrease in maximal-rate-of-rise-of-force medians and prolongation of contraction time means (Table 2).

Motor Unit Twitch Parameters in Spinal Muscular Atrophy

In the 32 motor units investigated, twitch force and maximal-rate-of-rise-of-force medians were significantly increased compared to controls. Interestingly, twitch force medians were also increased in the subgroup with MVC > 5 N, but not in the subgroup with MVC > 5 N. No difference was found for contraction and half-relaxation times (Table 3).

Motor Unit Twitch Parameters in Amyotrophic Lateral Sclerosis

In the amyotrophic lateral sclerosis group (112 motor units), twitch force medians were significantly increased, contraction times were slightly but significantly prolonged. In both subgroups (MVC > 5 N and MVC ≤ 5 N) twitch force medians were about equal and significantly increased, contraction time means were only prolonged in the subgroup with MVC force ≤ 5 N (Table 4).

Table 4. Motor unit twitch parameters in amyotrophic lateral sclerosis (ALS)

	controls (n = 251)	ALS (n = 112)	(MVC >5 N) (n = 67)	(MVC ≤5 N) (n = 45)
TF (mN)	9	17*	16*	17*
MRRF (N/s)	0.61	0.69	0.71	0.54
CT (ms)	64±14	69±21*	65±23	75±16*
HRT (ms)	61±16	62±15	61±16	64±14

Median values for TF and MRRF, means ± SD for CT and HRT.
*Significant difference ($p < 0.05$) compared to controls.

DISCUSSION

STA is easier to apply and has a somewhat higher yield (about 8 motor units per hour of investigation). It has, however, methodological limitations as pointed out in the introduction. Furthermore, it requires that the subject drive a motor unit at a low and constant rate which may be difficult for patients with motor disorders. It should also be kept in mind that STA only allows investigation of motor units that can be activated voluntarily. This may be of importance when investigating patients with peripheral or central weakness. IMS is more difficult to apply and has a somewhat lower yield (about 4-5 motor units per hour of investigation), particularly in patients with partially denervated muscles. It is carried out with the muscle completely relaxed; and, therefore, needs no assistance from the subject and precludes motor unit synchronization. Motor units are investigated independently of whether they can be recruited voluntarily or not. Employing low stimulus rates, IMS also circumvents twitch distortion associated with partial twitch fusion.

It is generally assumed that the all-or-nothing response evoked by IMS is produced by a single entire motor unit. On the other hand it appears possible that IMS may not evoke a response of all muscle fibers of a given motor unit due to stimulation of terminal axon branches. Although this cannot be ruled out entirely, it is unlikely to play an essential role (see Elek et al., 1992). An argument against partial activation of motor units with IMS is the finding that frequently occurring F-responses (in about 30% of motor units) show the same wave forms as the preceding M-responses (Dengler et al., 1992). Therefore, we assume that motor axons proximal to terminal branching are the site of stimulation. From this argument, a distortion of the twitches, particularly of the rising phase, due to unphysiological antidromic action potential propagation from a stimulated terminal axon branch into others also appears very improbable.

Direct muscle fiber stimulation occurs occasionally and produces small short-latency EMG responses that characteristically increase continuously, not stepwise, with increasing stimulus intensity (EMG amplitude not above 25μV). The associated forces are extremely small (below 1 mN) and often unmeasurable. It cannot be excluded, however, that stimulation of individual motor axons also activates muscle fibers in the close vicinity of the stimulation electrode that belong to other motor units that could result in a slight overestimate of twitch forces.

We compared motor unit twitch parameters in the human first dorsal interosseus muscle obtained by STA and IMS. Our results show relevant differences only for the time dependent parameters: The values of maximal rate of rise of force, contraction time and half-relaxation time are significantly lower with STA than with IMS, whereas the values of twitch force reveal no significant difference. This finding is unexpected with regard to the results of the IMS experiments employing higher stimulus rates where partial fusion decreased the twitch force. The values of the time dependent twitch parameters in the STA studies correspond well with the high-rate IMS results (10 Hz) which stresses the role of partial fusion of twitches in STA. In the case of twitch force, however, other factors must counteract or compensate the effects of partial fusion in STA.

Milner-Brown and colleagues (1973a) compared twitch parameters of a small number of motor units in the human first dorsal interosseus muscle studied by STA and IMS. They did not find relevant differences for twitch forces and contraction times whereas half-relaxation times were underestimated by STA. In a larger sample of about 200 motor units in human thenar muscles, Stein and Yang (1990) obtained about the same twitch forces with STA and IMS. The predicted effect of partial twitch fusion was also not evident in the data of Thomas and coworkers (1990a). To their surprise, they found *STA twitch forces* significantly stronger and *STA contraction times* significantly longer than the corresponding values obtained by intraneural microstimulation.

Which factors other than partial fusion influence twitch parameters in STA? An important candidate is synchrony between firing times of different motor units in a voluntarily active muscle (Nordstrom et al., 1989). In fact, even in very smooth muscle contractions, simultaneously active motor units do not only show chance synchronization but frequently also show a more than chance synchronization of the so called short-term type (Sears & Stagg, 1976, Dengler et al., 1984). In favor of the role of motor unit synchronization is our finding that the weakest motor units have smaller twitch force values when STA is used. These motor units are recruited early (Milner-Brown et al., 1973b) when only few motor units are active and when motor unit synchronization would be negligible. Therefore, partial fusion dominates in this situation and causes lower twitch forces than obtained with methods using low rate motor axon stimulation.

Differences in the activation history of motor units during STA and IMS may also influence twitch contractions obtained by both techniques. For STA, twitch contractions are averaged following up to 500 motor unit discharges, i.e., over a period of at least 1 min given a mean firing rate of about 10 Hz. In addition, this motor unit may have already been active during the averaging of previously investigated motor units. This may lead to a potentiation of twitch contractions (Thomas et al., 1990b) and could also partly explain the unexpectedly high twitch forces obtained with STA as compared to low-rate IMS where twitches are averaged following 25 to 50 stimuli at 1/s. On the other hand fatigue develops during STA even in normal subjects after about 5 min of constant motor unit activation at 10 Hz (Stephens & Usherwood, 1977), whereas fatigue is less likely to occur during low-rate IMS. This should be taken into account when STA is used to investigate patients who might fatigue more rapidly.

Another factor that may mask the effects of partial fusion on twitch forces in STA is a more-than-linear summation of contraction forces of simultaneously active motor units as has been described by Clamann and Schelhorn (1988). In cat hindlimb muscles, they found that the combined forces of two motor units exceeded the algebraic sum of the separate forces by an average of 5 to 12%, probably due to mechanical interaction of contracting muscle fibers.

It should further be considered that mechanical conditions (viscous and elastic properties of muscle, tissue and tendon) are different in STA and IMS. While motor unit twitches are superimposed on a more or less stable background tension during STA, IMS is carried out in a relaxed muscle. The lower stiffness of muscle and tendon in the relaxed state may lead to longer contraction times and reduced twitch forces with IMS. In animal experiments, however, McPhedran and colleagues (1965) stated that motor unit twitch parameters were not distorted when recorded from a relaxed muscle, so the effect of these mechanical factors is probably weak.

The distribution of twitch forces is skewed to the right with both STA and IMS; i.e., the motor units with the smallest forces are the most frequent. Whereas such a distribution can result from a sampling bias in favor of small, i.e., low-threshold motor units in STA; this can hardly be the case with IMS. One would rather expect a bias towards large, i.e., strong motor units with IMS because they have thicker axons and should be more readily excited. Other factors, however, such as the distance of the axon from the stimulating electrode may be more important. Similar to our study, Stein and Yang (1990) did not find significant differences in twitch forces and in amplitudes of motor unit potentials obtained by STA and IMS, so the same group of motor units appears to be sampled by either technique.

To illustrate the use of IMS for clinical studies we obtained in patients with chronic partial denervation. In brief, we found an increase of twitch forces which corresponds to an enlargement of motor units due to collateral axonal sprouting. The prolonged contraction times probably reflect slowed conduction velocities of sprouted axons, leading to a less synchronized contraction of muscle fibers which in the case of hereditary motor and sensory neuropathy type I also resulted in prolonged half-relaxation times. In patients with severely denervated muscles, however, twitch forces were reduced. This confirms findings of a previous study using STA (Dengler et al., 1990) which also showed a decrease of twitch forces in severely denervated muscles although macro EMG potentials were not decreased. An impaired function of the contractile apparatus was hypothesized, although an increased fatigability during the STA procedure could not be ruled out. The present data obtained with low-rate IMS clearly exclude the latter possibility.

Investigations of fatigue in human muscles with either STA or IMS have only rarely been carried out. The use of both techniques for fatigue studies is limited as the recording (STA) or stimulating electrode (IMS) inserted into the muscle is very easily shifted during fatiguing contractions. Using STA, Stephens and Usherwood (1977) studied motor unit twitch contractions in the human first dorsal interosseus muscle before and after a fatiguing paradigm (voluntary motor unit activation at about 10 Hz over 5 min). According to their results, motor units recruited at higher strengths had higher twitch forces, faster contraction times and were highly fatigable which is in line with results from animal experiments. In motor units of the human masseter muscle Nordstrom and Miles (1990) did not find a correlation of twitch forces, contraction times and fatigability with STA (fatiguing paradigm: voluntary activation of a motor unit at 10 Hz over 15 min) which they attributed to the unusual histochemical features of the masseter muscle. Fatigue tests with IMS have been carried out by Young and Mayer (1982) in the first dorsal interosseus muscle of normal subjects and patients with hemiplegia. They classified motor units as S, FF and FR based on twitch parameters, sag (8-15 Hz stimulation) and fatigability (30 Hz stimulation for 300 ms repeated once per second over 2 min). In hemiplegic patients, they also found slow-twitch fatigable motor units which they called SF. Comparing data obtained in controls and patients they inferred an increased fatigability of motor units in long-term spastic hemiparesis (see also McComas et al., Chapter 36).

ACKNOWLEDGMENTS

The authors' laboratory has been supported by the Federal Ministry for Research and Technology, Germany (BMFT), and the German Research Society (DFG). Attendance of the authors at the 1994 Bigland-Ritchie conference was supported, in part, by the Medical School of Hannover, the United States Public Health Service (National Center for Medical Rehabilitation Research, National Institute for Child Health and Human Development; *R.D.*), and the Muscular Dystrophy Association (USA).

REFERENCES

Andreassen S & Bar-On E (1983). Estimation of motor unit twitches. *IEEE Transactions of Biomedical Engineering* **30**, 742-748.

Calancie B & Bawa P (1986). Limitations of the spike-triggered averaging technique. *Muscle & Nerve* **9**, 78-83.

Clamann HP & Schelhorn TB (1988). Nonlinear force addition of newly recruited motor units in the cat hindlimb. *Muscle & Nerve* **11**, 1079-1089.

Dengler R, Konstanzer A, Küther G, Hesse S, Wolf W & Struppler A (1990). Amyotrophic lateral sclerosis: macro- EMG and twitch forces of single motor units. *Muscle & Nerve* **13**, 545-550.

Dengler R, Kossev A, Wohlfarth K, Schubert M, Elek J & Wolf W (1992). F-waves and motor unit size. *Muscle & Nerve* **15**, 1138-1142.

Dengler R, Wolf W, Birk P & Struppler A (1984). Synchronous discharges in pairs of steadily firing motor units tend to form clusters. *Neuroscience Letters* **47**, 167-172.

Elek JM, Kossev A, Dengler R, Schubert M, Wohlfarth K & Wolf W (1992). Parameters of human motor unit twitches obtained by intramuscular microstimulation. *Neuromuscular Disorders* **2**, 261-267.

McPhedran AM, Wuerker RB & Hennemann E (1965). Properties of motor units in a homogeneous red muscle (m. soleus) of the cat. *Journal of Neurophysiology* **28**, 71-84.

Milner-Brown HS, Stein RB & Yemm R (1973a). The contractile properties of human motor units during voluntary isometric contractions. *Journal of Physiology (London)* **228**, 285-306.

Milner-Brown HS, Stein RB & Yemm R (1973b). The orderly recruitment of human motor units during voluntary isometric contractions. *Journal of Physiology (London)* **230**, 359-370.

Monster AW & Chan H (1977). Isometric force production by motor units of extensor digitorum communis muscle in man. *Journal of Neurophysiology* **40**, 1432-1443.

Nordstrom MA & Miles TS (1990). Fatigue of single motor units in human masseter muscle. *Journal of Applied Physiology* **68**, 26-34.

Nordstrom MA, Miles T & Veale J (1989). Effect of motor unit firing pattern on twitches obtained by spike-triggered averaging. Muscle & Nerve **12**, 556-567.

Sears TA & Stagg D (1976). Short-term synchronization of intercostal motoneurone activity. *Journal of Physiology (London)* **263**, 357-381.

Stalberg E & Fawcett PRW (1982). Macro-EMG in healthy subjects of different ages. *Journal of Neurology, Neurosurgery & Psychiatry* **45**, 870-878.

Stein RB, French AS, Mannard A & Yemm R (1972). New methods for analysing motor function in man and animals. *Brain Research* **49**, 187-192.

Stein RB & Yang JF (1990). Methods for estimating the number of motor units in human muscles. *Annals of Neurology* **28**, 487-495.

Stephens JA & Usherwood TP (1977). The mechanical properties of human motor units with special reference to their fatiguability and recruitment threshold. *Brain Research* **125**, 91-97.

Taylor A & Stephens JA (1976). Study of human motor unit contractions by controlled intramuscular microstimulation. *Brain Research* **117**, 331-335.

Thomas CK, Bigland-Ritchie B, Westling G & Johansson RS (1990a). A comparison of human thenar motor-unit properties studied by intraneural motor-axon stimulation and spike-triggered averaging. *Journal of Neurophysiology* **64**, 1347 -1351.

Thomas CK, Johansson RS, Westling G & Bigland-Ritchie B (1990b). Twitch properties of human thenar motor units measured in response to intraneural motor-axon stimulation. *Journal of Neurophysiology* **64**, 1339-1346.

Yang JF, Stein RB, Jhamandas J & Gordon T (1990). Motor unit numbers and contractile properties after spinal cord injury. *Annals of Neurology* **28**, 496-502.

Young YL & Mayer RF (1982). Physiological alterations of motor units in hemiplegia. *Journal of Neurological Sciences* **54**, 401-412.

[illegible] A, Wolf W, Dirk P & Struppler A (1984) Synchronous discharges in pairs of steadily firing motor units tend to form clusters. *Neuroscience Letters* 47, 167-172.

[illegible] BH, Losch A, Dengler R, Schulz M, Wohlfarth K & Wolf W (1994) Parameters of human motor unit twitches obtained by intramuscular microstimulation. *Neuromuscular Disorders* 2, 261-267.

[illegible] AM, Westera RF & [illegible] (1985) [illegible]

[illegible] Stein RB & [illegible] (1979) The contractile properties of human motor units [illegible]. *Journal of Physiology* 296, [illegible].

[illegible] HS, [illegible] RB & [illegible] R (1973) The orderly recruitment of human motor units during voluntary isometric contractions. *Journal of Physiology* 230, 359-370.

[illegible] (1977) [illegible] by motor units of [illegible] 40, [illegible].

[illegible]

[illegible] (1980) [illegible] motor unit firing patterns [illegible]. [illegible] 12, 359-361.

[illegible]

[illegible]

[illegible]

SECTION IV

Fatigue Studied with NMR Techniques

This section contains four chapters that rely on the benefits afforded by non-invasive measurements with nuclear magnetic resonance spectroscopy (NMR). It begins with a global assessment of the metabolic homogeneity of muscle fiber types and ends with a glimpse of the future measurements which will be possible with NMR.

In **Chapter 12**, Kushmerick uses data from ^{31}P NMR and muscle histochemistry to examine fatigue in muscles containing a range of different fiber types. Muscle-specific differences in the levels of high-energy phosphates are less in humans than in commonly used experimental animals. Furthermore, the constancy of the *ratio* of phosphocreatine to ATP acts to normalize the bioenergetic function of different muscle cells. Hence, large-scale measurements of cellular bioenergetics provided by NMR are a valid way to examine comparatively homogeneous human muscles.

The metabolic responses of human muscles to different types of exercise are considered in **Chapter 13** (Vøllestad). There are limits in the extent to which data from *in vitro* studies can be applied to *in vivo* studies. These limits arise because the type and duration of the fatiguing exercise and the pattern of neural drive to the muscle influence the extent to which excitation-contraction coupling or ATP supply cause muscle fatigue. Of particular interest are observations on repeated submaximal contractions in which force can be impaired with little change in phosphocreatine, inorganic phosphate or pH. Phosphocreatine levels are low at exhaustion, but these levels are already reached minutes before exercise was voluntarily stopped. Evidence is presented that the energy cost of contractions (assessed by oxygen consumption or heat output) increases progressively during isometric exercise but not bicycling or running.

An integrated approach to studying human muscle fatigue is presented in **Chapter 14** (Miller and colleagues). They review the studies that have applied the Nank technique to human muscles especially during fatiguing exercise. Force, EMG and metabolites have been measured during fatigue and recovery in a range of exercise protocols in both control subjects and patients with neuromuscular disorders. As in the preceding chapter, the extent to which excitation-contraction coupling can be reliably inferred from *in vivo* studies of human muscles is of paramount importance. Initial glycogen stores and arterial pH contribute little to the variability between subjects in high-intensity exercise performance and this confirms the view that the fiber-type composition of each subject powerfully determines the exercise performance. By measurement of the force produced in both voluntary and stimulated contractions, the added contribution of central fatigue to reductions in force was evaluated. The goal of these studies is to apportion the fatigue in a particular task to the various intracellular components, as well as those that may develop at the neuromuscular junction or within the CNS.

In **Chapter 15,** Bertocci outlines the future directions for NMR methods that may be applied to study fatigue. In addition to its established role in assessment of the high-energy phosphates, it is possible to follow the precise incorporation of substrates and the regulation of the citric acid cycle with ^{13}C NMR spectroscopy. An advantage is that ^{13}C is a stable, non-radioactive carbon isotope. By its use, it is now possible to monitor the changing use of substrates during exercise and their availability. $^{23}Na^+$ and $^{39}K^+$ magnetic resonance spectroscopy can reveal the changing distribution of sodium and potassium ions across the muscle cell membrane. This may require infused reagents which shift the resonance frequency of extracellular $^{23}Na^+$ so that intra- and extracellular distributions can be estimated. Finally, magnetic resonance imaging can be used to assess whether muscle is active or not, and it can be adapted to depict accurately movement of fatiguing muscles in real time.

12

BIOENERGETICS AND MUSCLE CELL TYPES

M. J. Kushmerick

Departments of Radiology, Bioengineering, and Physiology
University of Washington
Seattle, Washington 98195

ABSTRACT

Potential complexities in biochemical and bioenergetic interpretation due to fiber type heterogeneity are not significant for human muscle. Paradigms for understanding muscle bioenergetics then can be understood from a set of basic premises of biochemical energy balance 1) ATP provides the energy for all forms of muscle work; 2) chemical energy is stored in cells as phosphocreatine, a biochemical capacitor; 3) the sum of the coupled ATPases sets the demand side of the balance and defines energetic states; and 4) this demand is supplied by aerobic metabolism and the products of the coupled ATPases provide control signals for regulation of energy balance. We speculate that cytoplasmic signals at work in energy balance may also control muscle plasticity.

INTRODUCTION

A variety of muscle cell types have been identified in vertebrate animals, including humans. This chapter assesses the bioenergetic distinctions among those stereotyped muscle fiber types and whether those distinctions make a significant difference for analyzing muscle bioenergetics; it then outlines the concept of energy balance for muscle bioenergetics, and lastly shows how energy balance offers a useful paradigm for understanding muscle function.

FIBER TYPES

Properties of individual muscles cells are classified on the basis of a number of anatomical, physiological and biochemical criteria into various categories of fibers called fiber types (Saltin & Gollnick, 1983) 1) fatigue-sensitivity; 2) recruitment order of individual motor units; 3) metabolic repertoire, namely oxidative and glycolytic; 4) mechanical speed, namely fast- and slow-twitch types; 5) myofibrillar motor, namely myosin types 1 and 2 (and further subtypes based on light chain components and variants of the heavy chains); 6) other protein isoforms, which include a large number troponin subunits, calcium pump in the sarcoplasmic reticulum, creatine kinase, and enzymes of intermediary metabolism, etc.

Fatigue, Edited by Simon C. Gandevia et al.
Plenum Press, New York, 1995

One notes immediately that any one or a group of these characteristics defines only some aspects of muscle structure and function. Thus they do not necessarily generate the same categories of *fiber types*, and so different classification schemes result depending on the criteria chosen. Moreover the several types classically described are stereotypes; but in fact there is a continuum of fibers and their properties (Pette & Staron, 1990). Additional distinctions can be made on the basis of ion channel subunits, membrane receptors, and many other cellular and molecular characteristics still being discovered. Thus the number of known isoforms of proteins in the thick and thin filaments alone is sufficient to produce a very large number of possible cell types (Pette & Staron, 1990). Nonetheless individual fibers and whole muscles with predominantly one characteristic can be obtained and their energy metabolism and other functions can be studied (Crow & Kushmerick, 1982; Metzger & Moss, 1987; Sweeney et al., 1988; Bottinelli et al., 1991; Hofmann et al., 1991; Kushmerick et al., 1992; Larsson & Moss, 1993).

In addition to the categories of characteristics listed above, muscle cells also differ in the concentrations of the metabolites involved in energy metabolism, for which the phrase *high energy phosphates* was introduced (Lipmann, 1941). Measurements of metabolite composition by ^{31}P NMR demonstrated a higher phosphocreatine (PCr) and lower inorganic phosphate (Pi) contents in the fast-twitch muscle compared with a slow-twitch muscle (Meyer et al., 1985) confirming the results of biochemical assays of muscle extracts of fast-twitch and slow-twitch muscles (Meyer et al., 1980; Crow & Kushmerick, 1982). Chemical analyses of single fiber segments dissected from resting and stimulated rat *plantaris* and *soleus* muscles (Hintz et al., 1982) showed differences in metabolite contents at rest which correlated with the fiber type as well as with the extent of PCr and ATP splitting during stimulation. A clear difference was found in the composition of high-energy phosphates and myosin heavy chain composition in rodent muscles (Kushmerick et al., 1993). Slow type 1 myosin heavy chain correlated with lower PCr and total creatine content and higher Pi content than was observed in fast type 2 myosin-containing muscle. In anatomically different muscles of the human, leg small but definite differences are measured in high-energy phosphate metabolite content. ^{31}P NMR data indicates a higher Pi and lower PCr content in the *soleus* than in the *gastrocnemius* by the use of localized spectroscopic techniques to sample regions totally within each muscle (Vandenborne et al., 1993a). But the magnitude of the muscle-specific differences in humans is definitely lower than in the animal experiments. This is due in part to optimization of biological materials to demonstrate big differences in experimental animal preparations. The ^{31}P NMR data confirm analyses of single fibers from human muscle (sorted histochemically) which showed significantly small differences in PCr contents between types 1 and 2 fibers (Edstrom et al., 1982; Söderlund & Hultman, 1991; Greenhaff et al., 1993), and no differences in the ATP content. The conclusion must be that, in a given human muscle, the range of differences in metabolite composition is smaller than predicted from the extreme examples from laboratory animals, and that this narrower range is due to a narrower distribution of cell types.

With respect to muscle properties as a chemo-mechanical machine and to quantities related to bioenergetics, it is known that there is approximately a 4-fold range in the maximal mechanical power and ATPase rates and a similar range in metabolic capacities (including oxidative phosphorylation) over the range of skeletal muscle cells. Moreover, these rates are graded on the intensity of muscle activation, from rest to maximal, for a more than 10-fold range. Finally due to the compositional differences in the *high-energy phosphate* content, all of these chemical changes with muscle performance begin from different starting values.

Thus we see that for whatever property, there is significant possibility for a distribution of cell types (and therefore heterogeneity) with respect to functional performance. This suggests the pattern and kinetics of bioenergetics of muscle activity could be highly complex. If so, it would be difficult to interpret macroscopic measurements in terms of integrated biochemical and molecular mechanisms as has been attempted in a number of studies in human and animal limb muscles at rest and during exercise (Gollnick et al., 1973; Meyer & Terjung, 1979; Taylor et. al., 1983; Arnold et al., 1984; Kushmerick & Meyer, 1985; Chance et al., 1986; Ren & Hultman, 1989). Non-invasive measurements by ^{31}P NMR spectroscopy are perhaps the most powerful because they can be repeated with high time resolution in human limb muscle exercise and yield information on the concentration of metabolically active, as opposed to bound, bioenergetic compounds of interest. The ratio Pi/PCr or Pi/(PCr + Pi) is commonly used to infer the bioenergetic status of the muscle in human exercise. In addition to studies of *high-energy phosphate*, studies of glucose and glycogen metabolism using ^{13}C NMR spectroscopy (Bloch et al., 1994; Price et al., 1994) similarly make cellular and molecular conclusions from macroscopic measures. Is it possible that these interpretations and conclusions are fallacious in principle because of macroscopic sampling of heterogeneous cell types?

No! There are good biological reasons why the overall bioenergetic functions of a given muscle are interpretable in biochemical and molecular mechanisms. First, the bioenergetic quantities are self-normalizing as indicated by several authors (Connett, 1988; Meyer, 1989; Conley, 1994). Although the total creatine, PCr and ATP concentrations can differ many fold in different muscle cells, the ratio of PCr/ATP is relatively constant (Connett, 1988; Conley, 1994). The implication of this constancy is that cells have a similar ADP concentration and chemical potential of ATP, at least much more alike than expected from the 2-fold absolute concentration differences. Second, the coupling between mechanical power output and chemical power input means that the mechanical output dominates the rate of energy utilization. Because cells with higher ATPase rates have higher PCr and ATP content, different cells each working, for example at 50% of maximal power, will have similar relative intracellular changes in ADP and chemical potential. Third, the steady rate of ATP synthesis (and thus of oxygen consumption), depends on the rate of ATP utilization. The implication of the two preceding points is that there is a scaling of signals controlling oxidative metabolism, the predominant source of *high-energy phosphates* in the cells. In fact, heterogeneity has been difficult to demonstrate in the bioenergetics of single human muscles. It has been shown that in strong voluntary exercise of calf muscles a heterogeneity of pH (as measured by a splitting of the Pi peak in NMR spectra) can be detected (Vandenborne et al., 1991), but this was due to sampling of more than one muscle (anatomical heterogeneity) when localization techniques were not done. With proper localization, individual muscles were sampled and muscle-specific differences in pH could be shown (Vandenborne et al., 1993b) and each muscle was relatively homogeneous. The rate of PCr re-synthesis following three identical activations of the forearm musculature in the same individual was highly repeatable (approximately ± 10%) and mono-exponential, but the individuals differed among themselves by almost 200% (Blei et al., 1993). The implication is that individuals differ because their component fibers differ. Moreover, the PCr breakdown rate, also highly repeatable, differed in individuals and the magnitude of the breakdown rate was inversely correlated with the content of Pi (Blei et al., 1993). Also individuals can be grouped on the basis of the kinetics of recovery of PCr following graded voluntary exercise (Mizuno et al., 1994) and show recovery kinetics that were explainable on the basis of the histochemical fiber types of biopsies from the same region sampled by the NMR experiment. These results (Blei et al., 1993; Mizuno et al., 1994) show that the inter-individual NMR variation in energy metabolism is related closely to fiber type composition of the specific muscle. The conclusion from this evidence is that any given human muscle can be best represented as a unimodal continuum of properties of a relatively small range of cell types. Therefore macroscopic measurements of cellular bioenergetics can be reliably interpreted in terms of intracellular and molecular mechanisms with the full realization of differences across muscles of the same individual, and in the same muscle of different individuals. We now consider some rules that apply to all muscles.

A PARADIGM FOR ENERGY BALANCE IN MUSCLE FUNCTION

The classical approach to describe and understand energy balance was defined by A.V. Hill by the application of equilibrium thermodynamics (energy conservation) to contracting muscle. The conceptual basis of myothermal energy balance states that the total energy change between two states of a muscle is exactly equal to the sum of the heat output and the work done in that change. The many successes of this approach (Wilkie, 1960; Kushmerick, 1983; Woledge et al., 1986) have been displaced by elaboration of so many individual reactions, including highly exothermic and endothermic ones, that the task of quantifying their extents of reaction and enthalpies led to unmanageable experimental difficulties and uncertain correlations of the details of the chemical processes with myothermal events. An alternative, a biochemical energy balance, was elaborated that organizes the functional integration of the major metabolic pathways in maintaining a chemically defined energy balance (Kushmerick, 1977).

Lipmann's work suggested how this task might be done. He used the phrase *high-energy phosphate* bond to point out that, with all the intricacy of metabolic pathways, there is only a small set of common biochemicals involved in energy transducing mechanisms, e.g. ATP (Lipmann, 1941). Energy transductions in muscle (molecular electro-chemical and chemo-osmotic machines and motors) synthesize or dissipate ATP, the cell's source of chemical potential energy, in a way that couples the energy to the metabolic, electrical, osmotic and mechanical work. The extent of energy dissipation uncoupled to ATP is negligible. The magnitude of ATP hydrolysis uncoupled to the molecular transducers (in analogy to the proton leak dissipating chemical energy in the mitochondria without coupling to ATP synthesis (Brown, 1992)) is unknown. It is estimated to be very small on the basis of the low rate of metabolism in muscles at rest. Although a small amount of ATP is stored in the cell (on the order of 5 mM), ATP must be produced by metabolism concurrent with work performance. The reason is that typical rates of ATP utilization in active muscle (on the order of 1 mM per s) would otherwise exhaust the muscle's source of chemical potential energy in several seconds. We know that steady states of metabolic activity readily occur over a wide range of muscle functions from a basal minimum up to some maximum. Beyond this range of energy balance, called the buffering range (Connett, 1988), the integrated operation disintegrates into dysfunction for which we use terms like fatigue, pathology, disease, etc. Energy balance describes those properties of energy using and generating mechanisms and their control that sets the supply of chemical energy to match the demands of chemical transducing machines.

The following statements and working hypotheses provide a relatively simple scheme by which one can integrate the diverse component mechanisms of bioenergetic systems. These are tentative rules and hypotheses for all types of muscles and suggest novel experimental approaches. This set of biochemical rules are not *a priori* predictable from first principles of physics because the mechanisms involved are the products of evolution.

ATP Provides the Energy for All Forms of Muscle Work

There is no other form of energy coupling known; muscle is not a heat engine or fuel cell. The essential feature of muscle as a chemo-mechanical system is that free energy of chemical reactions is coupled to mechanical performance by the actomyosin motor. Chemical input power matches the mechanical output power. Actomyosin interactions generating force and doing work (moving a load) and ion pumps doing electrical and osmotic work all rely on energy available from ATP, thereby coupling the exergonic process of ATP splitting to the endergonic processes of cellular work.

Chemical Energy Is Stored in Cells - Concept of a Biochemical Capacitor

Certain forms of chemical potential energy are biochemically inter-convertible by means of near equilibrium reactions. Perhaps the best known example is creatine kinase (Meyer et al., 1984; Wallimann et al., 1992) which catalyses the inter-conversion between ATP and PCr:

$$ATP + Creatine \leftrightarrow PCr + ADP + H^+.$$

This reaction defines a chemical energy capacitance because the content of PCr is effectively a capacitor of chemical energy by the nature of its near equilibration with ATP and ADP. Note that the rate of *charging* and *discharging* this capacitor may not be the simple exponential form of a typical electrical RC circuit because the flux is defined by the rate equations for the enzyme catalyzed reaction. Muscle cells have concentrations of PCr higher than ATP. Yet none of the chemical energy transducing molecular machines is known to use PCr directly for coupling chemical energy to work production; PCr is a substrate for only one enzyme, creatine kinase. Thus, as a capacitor, PCr represents chemical potential energy previously synthesized by metabolism. Thus endergonic processes can be temporarily coupled to free energy dissipation without the requirement for simultaneous oxidative metabolism provided the limits of the chemical energy capacitance, the supply of PCr, is not exceeded. The range of ATP chemical potential over which PCr and creatine kinase is an effective buffer was distinguished from a *depleting range* (Connett, 1988) in which there is depletion of ATP, loss of total adenine nucleotides via AMP deaminase, fatigue and cellular dysfunction.

The Sum of the Coupled ATPases Sets the Demand Side of the Balance and Defines Energetic States

The overall metabolic rate of a cell changes because its ATPase rates change; the converse is also true, viz. that if a measure of the overall metabolic rate increases then the steady-state sum of the ATPases must have increased. In this sense, the rate of the coupled free energy dissipation mechanisms is the primary and causal mechanism in bioenergetics. ATP dissipation (coupled to performing work) is in turn activated by signals external to this molecular motor; that is, the motor is not controlled by the chemical potential of ATP. In this view neither the availability of extra oxygen or substrate *per se* nor an increase in the magnitude of the chemical potential stored in ATP and PCr concentrations will increase the metabolic rate. Only a change in the rate of the ATPases does that.

The Coupled ATPases Provide Control Signals for Regulation of Energy Balance

How is the remarkable constancy of the concentration of cellular metabolites achieved in the face of varied levels of demand? One possible way is that the signals which activate the ATP utilizing mechanisms also provide information to the coupled processes synthesizing ATP. A common external signal could result in energy balance by acting on the ATP utilizing and synthesizing mechanisms in parallel. This kind of mechanism is essentially the role envisioned for calcium activation of mitochondrial dehydrogenases (McCormack & Denton, 1990; McCormack & Denton, 1993) and possibly the ATPsynthetase itself (Das & Harris, 1990). Although such feedforward mechanisms exist, it is still mechanistically necessary that a signal (or signals) derived from the coupled chemo-mechanical machine acts in feedback regulation of ATP synthesis. Because energy balance is a cellular process, it is a better teleological strategy to have the regulation self contained. The primary signal molecules must then be one or more of the chemical products of ATP-coupled molecular machines: Pi, ADP, H^+ or creatine. These internal signals in turn regulate intracellular ATP synthesis by feedback signals (the concept is independent of the mechanism for whether control of oxidative phosphorylation is kinetic or thermodynamic - consideration beyond the scope of this essay). The presence of feedback control does not negate the existence of feedforward ones. The logical necessity of feedback mechanisms is thus derived from the fact that coupled ATP dissipating mechanisms are essential to achieve energy balance if the result of the feedforward signals were in the slightest way imbalanced.

APPLICATION OF THESE PRINCIPLES TO QUESTIONS OF HUMAN MUSCLE BIOENERGETICS

Homogeneity versus Heterogeneity Revisited

Reasons given above remove major concerns about our ability to study biochemical and molecular mechanisms in whole muscle by NMR and other sampling methods. We now consider how the principles in the preceding section apply to bioenergetically different muscles, and that insights are thereby provided into the function of muscles with a range of properties. A mechanically faster muscle has a greater rate of coupled ATPase than a slower muscle. Similarly muscles differ in their mitochondrial content such that the maximal rate of substrate oxidation and oxidative phosphorylation of slow muscle is greater than fast muscle. Energy balance is satisfied even when a muscle transiently depends on its *high-energy phosphate* capacitor PCr, instead of new ATP synthesis (Meyer et al., 1984; Connett, 1988). Mechanical power output is only possible if there is sufficient chemical potential energy available. When the magnitude of the ATPases exceeds the maximum of the ATP synthesis and the PCr capacitor is discharged - its function fails - we commonly refers to this state as fatigue wherein some or all of the same feedback signals for ATP synthesis are inhibitory for ATP utilizing mechanisms. That extreme state is out of the continuous range of regulation we have been considering, and was called the depleting phase (Connett, 1988). Muscles which have a smaller demand by their ATPases and a larger capacity for oxidative re synthesis may never achieve that state under physiological conditions (examples include *soleus* and cardiac muscle). Oxidative phosphorylation will increase as if feedback error signals were absent for three reasons 1) the size of the error signal will be small and perhaps within the noise of our experimental measures; 2) the effective *gain* in the system is large so that small signals produce big effects; and 3) the effectiveness of a feedforward mechanism may be greater in some muscles than in others. For all these reasons the control mechanisms may appear qualitatively different in cardiac *vs.* skeletal muscle (Heineman & Balaban, 1990) whereas the same mechanisms and principles can operate with only quantitative differences. Depending on which view is taken, one's experimental design may differ. One insight of the view proposed here is that a common regulatory scheme can be considered in general, with differences between muscle types being quantitative, not qualitative.

The Chemical Potential of ATP Provides the Energy for Muscle Power

Further important questions follow from this statement. Does the power of the actomyosin motor depend on the magnitude of the chemical potential? It appears not. In isometric conditions, the force produced per unit ATP utilized was independent of the concentrations of ATP, PCr, Cr, Pi and pH. That is, there was a constant ATP utilization proportional to the integral of isometric force independent of the changing content of high energy phosphate compounds over a range in which most of the PCr was depleted in amphibian (Kushmerick & Paul, 1976) and mouse *soleus* muscle (Crow & Kushmerick, 1982). Thus the cost per unit mechanical output was constant although the chemical potential decreased. Experiments answering similar questions about the quantities of work done per unit ATP utilization (Woledge et al., 1986; their Fig. 4.37) and the comparison of chemical power input per unit of mechanical power output for working muscle (Kushmerick & Davies, 1969) indicate a matching of ATP utilization to mechanical work output, also in constant proportion although the chemical potential decreased substantially over the range of measurements. There is evidence that the *soleus* muscle may differ from others (Crow & Kushmerick, 1982; Meyer et al., 1985), so these issues merit reinvestigation.

Mechanisms of Energy Balance May Control Muscle Plasticity

The phenotype of adult muscle is highly adaptable and is a function of the historical pattern and intensity of usage (Pette & Vrbova, 1992). An argument from parsimony extends the mechanisms for control of ATP synthesis to mechanisms for control of gene expression. The observation that the slow-twitch glycolytic muscle stereotype has not been observed in Nature suggests there may be an interaction between quantitative energy balance in bioenergetics and muscle phenotype. Feedforward control of ATP synthesis relegates the control of muscle phenotype to mechanisms outside of the muscle; many of these external influences are known (e.g., thyroid hormone) and are certainly important. However, normal muscle activity has strong effects on isoform expression. This effect of muscle use on phenotype makes control of gene expression by feedback signals originating in energy balance an attractive possibility. In some systems, hypoxic states are strong stimuli to altered gene expression (Goldberg et al., 1988; Firth et al., 1994). Alterations of the cellular *high-energy phosphate* content by creatine analogs is associated with altered expression of myosin heavy chains (Moerland et al., 1989), metabolic enzyme profiles (Shoubridge et al., 1985) and mechanical function (Meyer et al., 1986). These several observations suggest a speculative scheme in which altered bioenergetic states play a role in the processes of maintaining and altering protein expression to achieve the various phenotypes of adult muscle. This chapter ends with the speculation that if this kind of control occurs where the bioenergetic state effects gene expression and muscle phenotype, there would be a simple explanation (despite what likely are complex molecular mechanisms) for the relatively uniform distribution of cell types in a given limb muscle. This distribution is highly plastic (within genetic constraints) depending on muscle use, that is the integrated history of its ATPases.

ACKNOWLEDGMENTS

This work is largely due to the experimental and conceptual efforts of Kevin Conley and Michael Blei who continue to collaborate on this project. The authors' work is supported by United States Public Health Service grants AR 41928, AG 10853, and AR 01914. Attendance of *M.J.K.* at the 1994 Bigland-Ritchie conference was supported, in part, by the University of Miami.

REFERENCES

Arnold DL, Matthews PM & Radda GK (1984). Metabolic recovery after exercise and the assessment of mitochondrial function in vivo in human skeletal muscle by means of 31P NMR. *Magnetic Resonance in Medicine* **1**, 307-315.

Blei ML, Conley KE, Odderson IR, Esselman PC & Kushmerick MJ (1993). Individual variation in contractile cost and recovery in human skeletal muscle. *Proceeding of the National Academy of Science, USA* **90**, 7396-7400.

Bloch G, Chase JR, Meyer DB, Avison MJ, Shulman GI & Shulman RG (1994). In vivo regulation of rat muscle glycogen resynthesis after intense exercise. *American Journal of Physiology (Endocrine Metabolism)* **266**, E85-E91.

Bottinelli R, Schiaffino S & Reggiani C (1991). Force-velocity relations and myosin heavy chain isoform compositions of skinned fibres from rat skeletal muscle. *Journal of Physiology (London)* **437**, 655-672.

Brown GC (1992). Control of Respiration and ATP synthesis in mammalian mitochondria and cells. *Biochemical Journal* **284**, 1-13.

Chance B, Leigh J, Kent J, McCully K, Nioka S, Clark BJ & Maris JM (1986). Multiple controls of oxidative metabolism in living tissues as studied by phosphorus magnetic resonance. *Proceedings of the National Academy of Science, USA* **83**, 9458-9462.

Conley KE (1994). Cellular energetics during exercise. *Advances in Veterinary Science and Comparative Medicine* **38A**, 1-39.

Connett RJ (1988). Analysis of metabolic control: new insights using scaled creatine kinase model. *American Journal of Physiology* **254**, R949-R959.

Crow MT & Kushmerick MJ (1982). Chemical energetics of slow- and fast-twitch muscles of the mouse. *Journal of General Physiology* **79**, 147-166.

Das AM & Harris DA (1990). Regulation of the mitochondrial ATP synthase in intact rat cardiomyocytes. *Biochemical Journal* **266**, 355-361.

Edstrom L, Hultman E, Sahlin K & Sjoholm H (1982). The contents of high-energy phosphates in different fibre types in skeletal muscles from rat, guinea-pig, and man. *Journal of Physiology (London)* **332**, 47-58.

Firth JD, Ebert BL, Pugh CW & Ratcliffe PJ (1994). Oxygen-regulated control elements in the phosphoglycerate kinase 1 and lactate dehydrogenase A genes: Similarities with the erythropoietin 3' enhancer. *Proceedings of the National Academy of Science, USA* **91**, 6496-6500.

Goldberg MA, Dunning SP & Bunn HF (1988). Regulation of the erythropoietin gene: evidence that the oxygen sensor is a heme protein. *Science* **242**, 1412-1414.

Gollnick PD, Armstrong RB, Sembrowich WL, Shepherd RE & Saltin B (1973). Glycogen depletion pattern in human skeletal muscle fibers after heavy exercise. *Journal of Applied Physiology* **34**, 615-618.

Greenhaff PL, Söderlund K, Ren JM & Hultman E (1993). Energy metabolism in single human muscle fibres during intermittent contraction with occluded circulation. *Journal of Physiology (London)* **460**, 443-453.

Heineman FW & Balaban RS (1990). Control of mitochondrial respiration in the heart in vivo. *Annual Review of Physiology* **52**, 523-542.

Hintz, CS, Chi M M-Y, Fell RD, Ivy JL, Kaiser KK, Lowry CV & Lowry OH (1982). Metabolite changes in individual rat muscle fibers during stimulation. *American Journal of Physiology* **242**, C218-C228.

Hofmann PA, Hartzell HC & Moss RL (1991). Alterations in Ca^{2+} sensitive tension due to partial extraction of C-protein from rat skinned cardiac myocytes and rabbit skeletal muscle fibers. *Journal of General Physiology* **97**, 1141-1163.

Kushmerick MJ (1977). Energy balance in muscle contraction: A biochemical approach. In: Sanadi R (ed.), *Current Topics in Bioenergetics*, vol. 6, pp. 1-15. New York: Academic Press.

Kushmerick MJ (1983). Energetics of muscle contraction. In: Peachey L, Adrian R, Geiger SR (eds.), *Handbook of Physiology, Skeletal Muscle*, pp. 189-236. Bethesda: American Physiological Society.

Kushmerick MJ & Davies RE (1969). The chemistry, efficiency and power of maximally working sartorius muscles. *Proceedings of the Royal Society, London B*, 315-353.

Kushmerick MJ & Meyer RA (1985). Chemical changes in rat leg muscle by phosphorus nuclear magnetic resonance. *American Journal of Physiology* **248**, C542-C549.

Kushmerick MJ, Moerland TS & Wiseman RW (1992). Mammalian skeletal muscle fibers distinguished by contents of phosphocreatine, ATP, and Pi. *Proceedings of the National Academy of Science, USA* **89**, 7521-7525.

Kushmerick MJ, Moerland TS & Wiseman RW (1993). Two classes of mammalian skeletal muscle fibers distinguished by metabolite content. *Advances in Experimental Medicine and Biology* **332**, 749-761.

Kushmerick MJ & Paul RJ (1976). Aerobic recovery metabolism following a single isometric tetanus in frog sartorius muscle at 0°C. *Journal of Physiology (London)* **254**, 693-709.

Larsson L & Moss RL (1993). Maximum velocity of shortening in relation to myosin isoform composition in single fibres from human skeletal muscles. *Journal of Physiology (London)* **472**, 595-614.

Lipmann F (1941). Metabolic generation and utilization of phosphate bond energy. In: Nord FF, Werkman CH (eds.), *Advances in Enzymology*, vol 1, pp. 99-162. New York: Intersciences Publishers, Inc.

McCormack JG & Denton RM (1990). Ca^{2+} as a second messenger within mitochondria in the heart and other tissues. *Annual Review of Physiology* **52**, 451-466.

McCormack JG & Denton RM (1993). The role of intramitochondrial Ca^{2+} in the regulation of oxidative phosphorylation in mammalian tissues. *Biochemical Society Transactions* **21**, 793-799.

Metzger JM & Moss RL (1987). Shortening velocity in skinned single muscle fibers. *Biophysical Journal* **52**, 127-8131.

Meyer RA (1989). Linear dependence of muscle phosphocreatine kinetics on total creatine content. *American Journal of Physiology* **257**, C1149-C1157.

Meyer RA, Brown TR, Krilowicz BL & Kushmerick MJ (1986). Phosphagen and intracellular pH changes during contraction of creatine-depleted rat muscle. *American Journal of Physiology* **250**,C264-C274.

Meyer RA, Brown TR, Kushmerick MJ (1985). Phosphorus nuclear magnetic resonance of fast- and slow-twitch muscle. *American Journal of Physiology* **248**, C279-C287.

Meyer RA, Dudley GA & Terjung RL (1980). Ammonia and IMP in different skeletal muscle fibers after exercise in rats. *Journal of Applied Physiology* **49**, 1037-1041.

Meyer RA, Sweeney HL & Kushmerick MJ (1984). A simple analysis of the "phosphocreatine shuttle." *American Journal of Physiology* **246**, C365-C377.

Meyer RA & Terjung RL (1979). Differences in ammonia and adenylate metabolism in contracting fast and slow muscle. *American Journal of Physiology* **237**, C111-C118.

Mizuno M, Secher NH & Quistorff B (1994). 31P-NMR spectroscopy, rsEMG, and histochemical fiber types of human wrist flexor muscles. *Journal of Applied Physiology* **76**, 531-538.

Moerland TS, Wolf NG & Kushmerick MJ (1989). Administration of a creatine analogue induces isomyosin transitions in muscle. *American Journal of Physiology* **257**, C810-C816.

Pette D & Staron RS (1990). Cellular and molecular diversity of mammalian skeletal muscle fibers. *Reviews of Physiology, Biochemistry & Pharmacology* **116**, 1-76.

Pette D & Vrbova G (1992). Adaptation of mammalian skeletal muscle fibers to chronic electrical stimulation. *Reviews of Physiology, Biochemistry & Pharmacology* **120**, 115-202.

Price TB, Rothman DL, Taylor R, Avison MJ, Shulman GI & Shulman RG (1994). Human muscle glycogen resynthesis after exercise: Insulin-dependent and -independent phases. *Journal of Applied Physiology* **76**, 104-111.

Ren J-M & Hultman E (1989). Regulation of glycogenolysis in human skeletal muscle. *Journal of Applied Physiology* **67**, 2243-2248.

Saltin B & Gollnick PD (1983). Skeletal muscle adaptability: significance for metabolism and performance. In: Geiger SR, Adrian RH (eds.), Peachey LD (sec. ed.), *Handbook of Physiology, sec. 10, Skeletal Muscle*, pp. 555-631. Bethesda, MD: American Physiological Society.

Shoubridge EA, Challiss RAJ, Hayes DJ & Radda GK (1985). Biochemical adaptation in the skeletal muscle of rats depleted of creatine with the substrate analogue B-guanidinopropionic acid. *Biochemical Journal* **232**, 125-131.

Söderlund K & Hultman E (1991). ATP and phosphocreatine changes in single human muscle fibers after intense electrical stimulation. *American Journal of Physiology (Endocrine Metabolism)* **261**, E737-E741.

Sweeney HL, Kushmerick MJ, Mabuchi K, Sreter FA & Gergely J (1988). Myosin alkali light chain and heavy chain variations correlate with altered shortening velocity of isolated skeletal muscle. *Journal of Biological Chemistry* **263**, 9034-9039.

Taylor DJ, Bore PJ, Styles P, Gadian DG & Radda GK (1983). Bioenergetics of intact human muscle. A 31P nuclear magnetic resonance study. *Molecular Biology & Medicine* **1**, 77-94.

Vandenborne K, McCully K, Kakihira H, Prammer M, Bolinger L, Detre JA, De Meirleir K, Walter G, Chance B & Leigh JS (1991). Metabolic heterogeneity in human calf muscle during maximal exercise. *Proceedings of the National Academy of Science, USA* **88**, 5714-5718.

Vandenborne K, Walter G, Goelman G, Ploutz L, Dudley G & Leigh JS (1993a). Phosphate content of fast and slow twitch muscles. *Proceedings of the Society of Magnetic Resonance in Medicine* **3**, 1140.

Vandenborne K, Walter G, Leigh JS & Goelman G (1993b). pH heterogeneity during exercise in localized spectra from single human muscles. *American Journal of Physiology (Cell Physiology)* **265**, C1332-C1339.

Wallimann T, Wyss M, Brdiczka D, Nicolay K & Eppenberger HM (1992). Intracellular compartmentation, structure and function of creatine kinase isoenzymes in tissues with high and fluctuating energy demands: the 'phosphocreatine circuit' for cellular energy homeostasis. *Biochemical Journal* **281**, 21-40.

Wilkie DR (1960). Thermodynamics and the interpretation of biological heat measurements. *Progress in Biophysics and Biophysical Chemistry* **10**, 260-298.

Woledge RC, Curtin NA & Homsher E (eds.) (1986). *Energetic Aspects of Muscle Contraction*. New York: Academic Press.

METABOLIC CORRELATES OF FATIGUE FROM DIFFERENT TYPES OF EXERCISE IN MAN

N. K. Vøllestad

Department of Physiology
National Institute of Occupational Health
Box 8149 Dep, N-0033 Oslo, Norway

ABSTRACT

It is well established that muscle fatigue, defined as a decline in maximal force generating capacity, is a common response to muscular activity. To what extent metabolic factors contribute to the reduced muscle function is still debated. Metabolic effects can affect muscle through different processes, either through a reduced ATP supply or by effects on EC-coupling or crossbridge dynamics. Observations from in vitro experiments are often extrapolated to interpret fatigue mechanisms from measurements performed in vivo, without recognizing that the biochemical reactions involved can be quite different depending upon such factors as activation pattern, mode and duration of exercise. During repeated submaximal contractions, there is a negligible accumulation of H^+ and inorganic phosphate, and hence fatigue must be ascribed to other factors. Substrate depletion might contribute to exhaustion, but cannot explain the gradual loss of maximal force. Curiously, the energetic cost of contraction increases progressively during repeated isometric but not during concentric contractions. With contractions involving high-force or high power output, fatigue is better related to $H_2PO_4^-$ than to pH, but still other factors seem to play a role.

INTRODUCTION

This chapter describes the metabolic responses during different types of exercise and relates them to fatigue. Metabolic and biochemical aspects of fatigue involve energy release and utilization, and the consequences of substrate degradation and electrolyte shifts. While all of these factors may be important in fatigue, only the first two factors will be examined in this chapter. The effects of pH and electrolyte balance are covered elsewhere (Allen et al., Chapter 3; Sjøgaard & McComas, Chapter 4).

To generate force and power, ATP is hydrolyzed by myosin ATPase. In addition, ATP is utilized in the re-establishment of electrolyte homeostasis after ion fluxes associated with propagation of the action potential and Ca^{2+} release from sarcoplasmic reticulum. Adequate resynthesis of ATP, therefore, is important in the maintenance of cellular function. During

Fatigue, Edited by Simon C. Gandevia et al.
Plenum Press, New York, 1995

prolonged exercise, oxidative phosphorylation of carbohydrates and free fatty acids are the main sources for energy. Although the products of this process are not known to affect force, the availability of substrates or the deterioration of mitochondrial function may contribute to muscle fatigue. With an inadequate oxygen supply, such as during sustained high-force contractions, glycogen and phosphocreatine (PCr) are rapidly broken down, and lactic acid and inorganic phosphate (Pi) accumulate. The possible role of these metabolites and substrate supply in the development of fatigue is still debated, and I will review some of the available data with emphasis upon how they relate to fatigue from voluntary activation in humans.

FATIGUE AND EXHAUSTION

The term *muscle fatigue* is used to denote a decline in the maximal contractile force of the muscle. Defined this way, fatigue may be assessed from brief maximal voluntary contractions or tetanically stimulated contractions. Most importantly, fatigue may occur during exercise at submaximal levels without being reflected in the performance. Fatigue is different to exhaustion, where the latter is defined as the moment in time when the expected force level cannot be maintained.

Muscle fatigue can occur with all types of muscular activity (i.e., isometric, shortening and lengthening contractions), and with both short-lasting intense and prolonged exercise. The various modes and intensities of exercise involve different metabolic processes, and the importance of biochemical changes to the decline in force may vary accordingly. To illustrate these possibilities, this chapter focuses on metabolic responses to repeated submaximal contractions and to exercise involving high-intensity force or power. These examples indicate that it is difficult to extrapolate the biochemical changes seen under one condition to other exercise protocols.

ANALYTICAL TECHNIQUES

Several different analytical methods have been used to study the effects of biochemical factors on performance. While some early data were obtained by manipulating substrate levels or metabolic pathways (Lundsgaard, 1930), these approaches were limited and initial biochemical analysis could only be carried out on stimulated animal muscles. With the introduction of needle biopsies in the sixties (Bergström, 1962), a more detailed picture of the metabolic changes in human muscle during various types of exercise was obtained. Substrate degradation and accumulation of metabolic products could readily be determined. The biopsy technique, however, has several limitations. First, only a limited number of samples can be obtained from each muscle. Second, the time resolution is poor. And thirdly, there is evidence that biochemical processes occur in the muscle when the sample is cut (Söderlund & Hultman, 1986). Hence, interpretation of small metabolic changes measured with the biopsy technique requires caution.

These disadvantages are not present when metabolites are assessed using ^{31}P nuclear magnetic resonance spectroscopy (^{31}P-MRS). Relative changes in PCr, Pi and ATP can be determined from the spectra, and from these ADP, $H_2PO_4^-$ and pH can be calculated (see Bertocci, Chapter 15). This nonivasive technique is based on phosphorous spectra recorded from muscles by surface coils and usable spectra may be obtained in a few seconds (Degroot et al., 1993). The greatest disadvantage of this method is that it has been applied almost exclusively to the study of phosphorous compounds in relation to fatigue. In the future, further developments in MRS may expand to analyze other aspects of metabolism. Another disadvantage of MRS is that it restricts the experimental design to the performance of a task within a strong magnetic field. Hence, the exercise-induced changes in metabolites have mainly been studied during isometric contractions, or dynamic contractions of smaller muscle groups. A third disadvantage of the ^{31}P-MRS is that it is not possible to calibrate the quantification of metabolites determined from the recorded spectra. To compensate for this, high-energy phosphates are most often expressed in relation to total phosphate. A fourth disadvantage is that the observed changes is concentrations will always be an average of the muscle fibers within the field of view. Details about cellular changes cannot be obtained when only a small portion of the muscle fibers are active.

In a recent study, Bangsbo and coworkers (1993) compared changes in metabolites during fatigue in parallel studies using ^{31}P-MRS and biochemical analyses of muscle biopsies. Similar changes in PCr were observed with the two methods, but a somewhat larger decline in ATP was determined biochemically after high-force contractions. The increase in ADP was much more pronounced when determined by ^{31}P-MRS than biochemically. This discrepancy was caused by the fact that the biochemical analysis measures total ADP whereas free ADP is determined from ^{31}P-MRS spectra. Hence, changes in PCr and for most purposes ATP can be reliably assessed with either method. However, only ^{31}P-MRS is sensitive for changes in free ADP.

The knowledge gained over the last three decades using different experimental models and analytical techniques has shown that the biochemical changes associated with fatigue vary over a wide range. Most investigators concur that several different mechanisms may play a role depending on the type and intensity of exercise and the fiber-type composition of the involved muscles.

PROLONGED EXERCISE - REPETITIVE SHORTENING OR ISOMETRIC CONTRACTIONS

During prolonged repeated submaximal contractions, the maximal force generating capacity declines continually (Bigland-Ritchie et al., 1986b; Vøllestad et al., 1988). At the onset of submaximal exercise only a fraction of the muscle fibers in the working muscles are activated. This has been shown by biopsy studies from several laboratories (Gollnick et al., 1973; Gollnick et al., 1974; Vøllestad et al., 1984; Vøllestad & Blom, 1985; Bigland-Ritchie et al., 1986a). When performing histochemical staining for the glycogen concentration in the fibers, it has been shown that only type I fibers are recruited from onset of bicycle exercise at about 40% of the maximal oxygen uptake (VO_2max) (Vøllestad & Blom, 1985). With increasing intensity, a larger fraction of the muscle fibers are recruited at the beginning. Interestingly, we observed a linear rise in the number of active muscle fibers as the exercise intensity increased (Vøllestad & Blom, 1985). If submaximal exercise is prolonged for an hour or more, exhaustion of the glycogen stores may occur in some fibers (Gollnick et al., 1973; Vøllestad et al., 1984). Due to the lack of glycogen, these fibers will therefore have a decreased capacity to release ATP. As exercise approaches exhaustion, an increasing number of fibers show total glycogen depletion.

One appealing hypothesis is thus that exhaustion is caused by an insufficient rate of ATP release in the muscle fibers. Several lines of evidence, however, argue against such a causative relationship. Maximal force generating capacity is determined by brief MVCs or tetanically stimulated contractions. The ATP hydrolysis in these contractions may amount to 1-2 mmol/kg ww (Edwards, 1976; Katz et al., 1986), indicating that only a minor fraction of the PCr and ATP stores are utilized under the test contractions. In keeping with this, several reports based on ^{31}P-MRS show that only small, if any, changes occur in ATP during short lasting high-force contractions (Miller et al., 1988; Quistorff et al., 1992). It might be argued that the fall in ATP concentration is much larger close to the crossbridges than in other parts of the cell. Hence, the ATP concentration determined as an average for the cell or muscle does not reflect the true conditions at the force-generating sites. However, as recently reviewed by Fitts (1994), it is unlikely that even in localized parts of the cell the ATP concentration should drop to levels below 50 μM which is shown to be sufficient for maximal isometric tension in rabbit psoas fibers (Cooke & Bialek, 1979). It may thus be concluded that ATP availability is not the main cause of fatigue under these conditions.

It is well established that lowered pH or elevated concentrations of inorganic phosphate (Pi) reduces the force generated by the crossbridges (Cooke & Pate, 1985; Donaldson & Hermansen, 1978). The effects of these factors during prolonged submaximal exercise are probably marginal because only a small fraction of the energy is released by anaerobic pathways. Hence, the accumulation of lactic acid or Pi is less than that required to affect force generation significantly (Vøllestad et al., 1988).

More recently, different research groups have studied fatigue using repeated submaximal isometric contractions. In several series of experiments, we have examined changes during quadriceps contractions held at 30% MVC for 6 s with 4 s rest between. The MVC force fell gradually during the entire exercise period reaching 40-60% of control at exhaustion, whereas serial muscle biopsies revealed only marginal changes in muscle glycogen, PCr, ATP and lactic acid for the first 30 min (Vøllestad et al., 1988). At exhaustion, however, almost total depletion of the PCr store was observed, while glycogen was reduced by about 30%. These observations indicate that also with repetitive isometric contractions the gradual decline in maximal force generating capacity during repeated shortening or isometric contractions must be explained by other factors than those connected to substrate depletion or accumulation of Pi and H^+.

Further evidence for the lack of a uniform relationship between fatigue and metabolite changes have been provided by experiments using ^{31}P-MRS. Fig. 1 shows results obtained from two subjects during repeated isometric contractions at 40% MVC with the quadriceps muscle. In one subject (dotted line), PCr fell by almost 90% within 10 min from the start of exercise, which was continued for more than 20 min with depleted PCr stores. Even though he was totally depleted of PCr, only marginal changes were seen in pH. In the other subject (solid line), PCr fell continually until exhaustion, and no changes were seen in pH. Fatigue developed gradually and by an equal amount for the two subjects. In addition to emphasizing that fatigue can develop without large changes in PCr, Pi and pH, these results show that constant force contractions can be carried out for a long time almost in the absence of PCr. This is at variance with our earlier data from repeated isometric contractions, in which exhaustion seemed to be clearly related to PCr depletion (Vøllestad et al., 1988). An example of the metabolic changes in one of these subjects is given in Fig. 2, illustrating that PCr was maintained at almost resting levels for 101 min, but totally depleted at exhaustion which was reached after 104 min. Also ATP fell rapidly by 2.3 mmol/kg ww in the last 3 min of exercise. The discrepancy between the data obtained in the two series of experiments is not understood, but could be related to differences in the angle of the knee (and thus muscle length) or to circulatory effects due to differences in body position (supine vs. sitting).

Interestingly, there is one distinct difference in the metabolic response to cycling or running compared with repeated isometric contractions. When this type of dynamic exercise is kept at a constant intensity, the oxygen consumption (VO_2) rises initially, and reaches a new steady level within a minute or two. Hence, the energy cost of contraction appears to remain constant with time. In contrast to this, repeated submaximal isometric contractions at 30% MVC with the quadriceps muscles results in an initial rise in VO_2 followed by a continual further rise until exhaustion (Vøllestad et al., 1990). Similar results have been reported by Sahlin and coworkers (1992), who also showed that the oxygen cost of isometric contractions remained elevated for an hour after end of exercise.

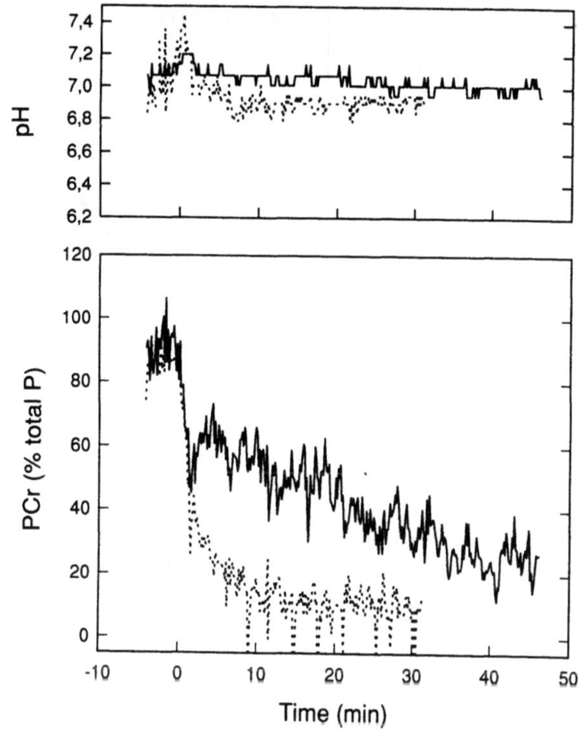

Figure 1. Changes in pH and PCr determined in the quadriceps muscle every 9 s by ^{31}P-MRS during intermittent isometric contractions at 40% MVC. The exercise was carried out with the subjects in a supine position and with each contraction held for 6 s with 4 s rest between. Data for two subjects shown by dotted and solid lines. Some of the deflections are due to maximal test contractions carried out regularly.

The causes for the gradual rise in VO_2 during repeated isometric contractions are not clarified. There are two alternative explanations. Either the ATP demand (ATP turnover:force ratio) increases, or the mitochondrial ATP production becomes less efficient (decreased P:O ratio). The latter mechanism would predict that oxygen consumption during any type of exercise involving the fatigued muscles would be elevated. Sahlin and coworkers (1992) tested this by comparing the VO_2 during bicycle exercise before and after fatigue from repeated isometric contractions. They observed no change in VO_2 during bicycling, indicating that the mitochondrial respiration was not altered. The most probable explanation for the increased VO_2 thus appears to be that the ATP turnover:force ratio increases. Two lines of

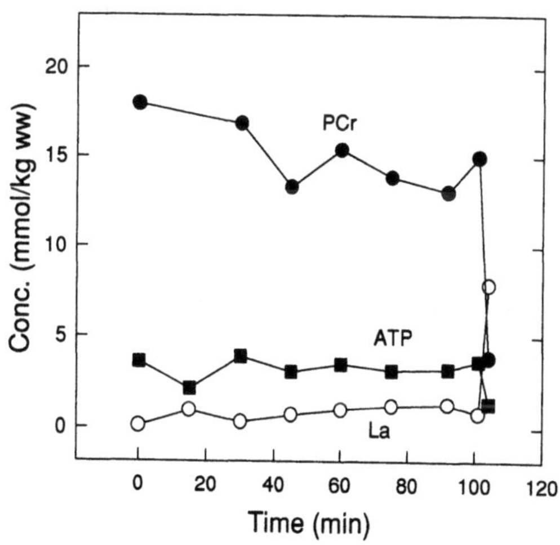

Figure 2. Changes in muscle concentrations of PCr, ATP and lactate (La) determined from biopsies taken from the quadriceps muscle of one subject. Intermittent isometric knee extension contractions at 30% MVC were carried out until exhaustion (104 min). The subject was seated in a chair and each contraction lasted 6 s with 4 s rest between.

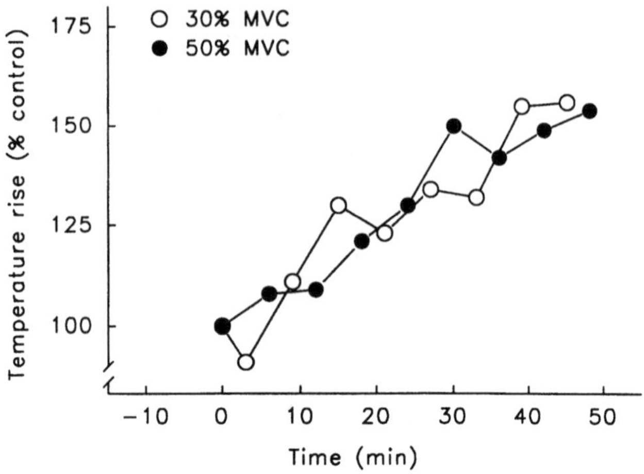

Figure 3. Temperature rise in the vastus lateralis muscle during isometric test contractions performed every 4th min of an intermittent isometric knee extension protocol (6 s contraction and 4 s rest) held at 30% and 50% MVC for 6 s with 4 s rest between. The parallel increase in the rate of temperature rise during the test contractions at the two force levels (with more type II fibers active at 50% MVC) indicate that the recruitment pattern plays a minor role for the increased energy utilization (reproduced with permission from Sejersted & Vøllestad, 1993).

evidence from recent experiments in our laboratory support this conclusion. In a separate set of experiments, we examined the rate of heat accumulation during constant force test contractions for 10-15 s carried out at regular intervals during 60 min of repeated 6 s isometric contractions. As shown in Fig. 3, the rate of temperature rise increased gradually as fatigue developed. Because the O_2 stored in the muscle is insufficient to support 10-15 s contractions at 30-50% MVC, heat accumulation after the first few seconds probably reflects the rate of anaerobic energy release and thus should be unrelated to mitochondrial respiration. In another series of experiments, we examined the changes in relaxation rate, because this parameter seems to be closely related to the ATP turnover (Edwards et al., 1975). Despite profound fatigue (MVC fell by 30-40%), the rate of relaxation rose gradually (Vøllestad et al., 1995). During repeated 30% MVC contractions carried out till exhaustion, the half-relaxation time of twitches and tetanic contractions decreased by 27 %. The exercise-induced increase in rates of heat accumulation and relaxation are in keeping with an increased ATP utilization. It must be emphasized, however, that there is no direct evidence for this conclusion. More research is needed to identify the mechanisms behind the increased metabolic rate during repeated isometric contractions. In addition, it should be sought out why the metabolic response to repeated concentric and isometric contractions differs.

HIGH-INTENSITY DYNAMIC EXERCISE AND HIGH-FORCE ISOMETRIC CONTRACTIONS

In contrast to prolonged submaximal exercise, exhaustive short lasting exercise requires extensive activation of most fibers in the working muscles, and fatigue develops rapidly. During sustained maximal voluntary contractions, the MVC force will decline by 50% within 1-3 min (Miller et al., 1988; Wilson et al., 1988; Degroot et al., 1993). Using an isokinetic ergometer or a nonmotorized treadmill, it has been shown that maximal power falls by about 50% during 30-45 s of concentric contractions (Thorstensson & Karlsson, 1976; McCartney et al., 1983; Cheetham et al., 1986).

Under these conditions, the ATP released from aerobic processes is insufficient to match the energy demand. Hence, force and power output must rely heavily on ATP from glycolysis and PCr degradation. Accordingly, a number of studies have shown that lactic acid increases by 20-30 mmol/kg ww and PCr is reduced by more than 80% (Hermansen & Vaage, 1977; Katz et al., 1986; Cheetham et al., 1986; Miller et al., 1988; Wilson et al., 1988). Inorganic phosphate increases as a mirror image of PCr, whereas only small and insignificant changes are reported for ATP (Miller et al., 1988; Quistorff et al., 1992). Because MRS has a relatively good time resolution, it has been possible to compare the temporal changes in maximal force generating capacity and metabolite concentration during sustained contractions. Several recent studies have shown that the fall in PCr and rise in Pi parallels the reduction in maximal force over the first 1-2 min. Thereafter, the levels of PCr and Pi almost stabilizes while force continues to decline (Wilson et al., 1988; Degroot et al., 1993). In contrast, pH increases initially, before a steady decline begins (Miller et al., 1988; Wilson et al., 1988; Degroot et al., 1993). An illustration of these changes are shown in Fig. 4.

Studies of skinned fibers have shown that both low pH and increased Pi may reduce maximal force by 50% (Donaldson & Hermansen, 1978; Cooke & Pate, 1985). Accordingly, both of these factors have been assumed to be responsible for fatigue. Several lines of evidence, however, indicate that a low pH is not necessarily an important mechanism of muscle fatigue. By comparing the relationship between muscle force and metabolite levels during maximal concentric contractions of different speeds, Wilson and coworkers (1988) found that there were no uniform relation between pH and force, whereas the relationship between $H_2PO_4^-$ and force remained unaffected by the duty cycle. Hence, they concluded that fatigue was attributed more to high levels of $H_2PO_4^-$ and less to a low pH. Other studies with different contraction protocols have arrived at the same conclusion (Miller et al., 1988; Le Rumeur et al., 1990; Degroot et al., 1993). Furthermore, the recovery of muscle force is

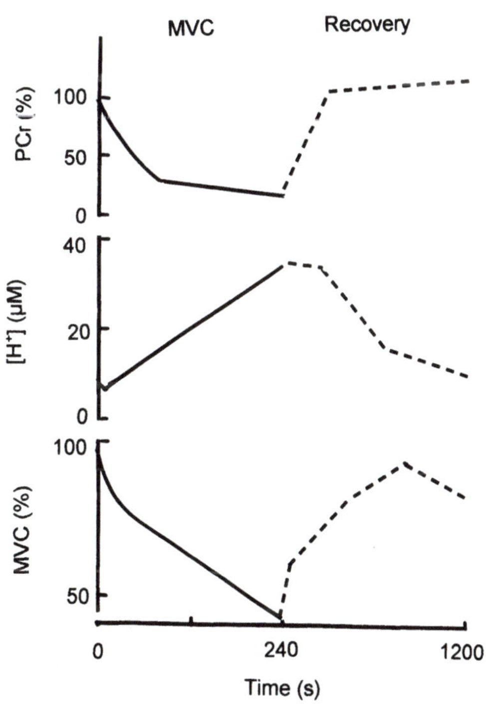

Figure 4. Changes in concentrations of PCr and H^+ of calf muscle during and after 4 min of sustained MVC (adapted from Degroot et al., 1993).

more closely related to $H_2PO_4^-$ than to pH because MVC force increases immediately after the end of exercise, while pH remains depressed or continues to decline for a minute or two before slowly returning to pre-exercise levels (Miller et al., 1988; Degroot et al., 1993).

Based on the differences in glycolytic capacity between type I (slow twitch) and II (fast twitch) fibers, it is often assumed that anaerobic energy release occurs predominantly in type II fibers. Accordingly, one would expect pH and Pi to influence the force generating capacity in type I less than in type II. From muscle biopsies obtained before and after short lasting bicycle exercise of various intensities we calculated the glycogenolytic rate of muscle fibers of various types (Vøllestad et al., 1992). At three different intensities (120-200% VO_2max), the glycogen breakdown in type I fibers was about 30% lower than in type II fibers. During exercise at the highest intensity (200% VO_2max), the glycogenolytic rate in type I fibers was twice the rate observed in type II fibers during exercise at 120% VO_2max. This observation suggests that in human muscle marked reductions in pH may occur in both fiber types, in keeping with single fiber analysis of lactate content (Essén & Häggmark, 1975) Because PCr is reported to decline by 80% or more in the quadriceps muscle during intense exercise, PCr depletion must be substantial in type I fibers. This is shown for electrically stimulated contractions (Söderlund et al., 1992). Hence, it appears that fatigue in both fiber types can potentially be due to low pH or PCr depletion. In contrast to the fast glycogenolytic rate that may be observed in type I fibers during voluntary activation, electrically stimulated contractions of the quadriceps muscle fails to stimulate glycogenolysis markedly in type I fibers (Söderlund et al., 1992). The causes for the different metabolic response to voluntary and electrically stimulated contractions are unresolved. However, these data emphasize that metabolic regulation may be task dependent.

CONCLUSION

The available data shows that fatigue related changes vary markedly between different tasks. Furthermore, correlations between fatigue and known metabolic changes all indicate that the metabolic factors are insufficient to fully explain the changes in maximal force. There is now substantial evidence that other factors also contribute to fatigue. In particular, during high-force contractions or exercise at high power output, electrolyte shifts over the sarcolemma and between sarcoplasmic reticulum and cytosol can be dramatically altered (Sejersted, 1992). Future studies should examine the interaction between electrolyte balance and metabolic changes.

ACKNOWLEDGMENTS

The author wishes to thank Eirik Saugen, Dr. Henry Gibson and Prof. Richard H. T. Edwards for stimulating discussions. Her work is supported by the Research Council of Norway. Her attendance at the 1994 Bigland-Ritchie conference was supported, in part, by the American Physical Therapy Association (Research and Analysis Division), and the Muscular Dystrophy Association (USA).

REFERENCES

Bangsbo J, Johansen L, Quistorff B & Saltin B (1993). NMR and analytic biochemical evaluation of CrP and nucleotides in the human calf during muscle contraction. *Journal of Applied Physiology* **74**, 2034-2039.

Bergström J (1962). Muscle electrolytes in man. *Scandinavian Journal of Clinical and Laboratory Investigation Supplement (Oslo)* **14**, 9-88.

Bigland-Ritchie B, Cafarelli E & Vøllestad NK (1986a). Fatigue of submaximal static contractions. *Acta Physiologica Scandinavica* **128**, 137-148.

Bigland-Ritchie B, Furbush F & Woods JJ (1986b). Fatigue of intermittent submaximal voluntary contractions: central and peripheral factors. *Journal of Applied Physiology* **61**, 421-429.

Cheetham ME, Boobis LH, Brooks S & Williams C (1986). Human muscle metabolism during sprint running. *Journal of Applied Physiology* **61**, 54-60.

Cooke R & Bialek W (1979). Contraction of glycerinated muscle fibers as a function of the ATP concentration. *Biophysical Journal* **28**, 241-258.

Cooke R & Pate E (1985). The effects of ADP and phosphate on the contraction of muscle fibers. *Biophysical Journal* **48**, 789-798.

Degroot M, Massie B, Boska M, Gober J, Miller RG & Weiner MW (1993). Dissociation of $[H^+]$ from fatigue in human muscle detected by high time resolution ^{31}P-NMR. *Muscle &Nerve* **16**, 91-98.

Donaldson SKB & Hermansen L (1978). Differential, direct effects of H^+ on Ca^{2+}-activated force of skinned fibers from the soleus, cardiac and adductor magnus muscles of rabbits. *Pflügers Archiv* **376**, 55-65.

Edwards RHT (1976). Metabolic changes during isometric contractions of the quadriceps muscle. In: Jokl E (ed.), *Medicine and Sport*, vol. 9, pp. 114-131. Basel: Karger.

Edwards RHT, Hill DH & Jones DA (1975). Metabolic changes associated with the slowing of relaxation in fatigued mouse muscle. *Journal of Physiology (London)* **251**, 287-301.

Essén B & Häggmark T (1975). Lactate concentration in type I and II muscle fibres during muscle contraction in man. *Acta Physiologica Scandinavica* **95**, 344-346.

Fitts RH (1994). Cellular mechanisms of muscle fatigue. *Physiological Reviews* **74**, 49-94.

Gollnick PD, Armstrong RB, Saubert CW, IV, Sembrowich WL, Sherpherd RE & Saltin B (1973). Glycogen depletion patterns in human skeletal muscle fibers during prolonged work. *Pflügers Archiv* **344**, 1-12.

Gollnick PD, Piehl K & Saltin B (1974). Selective glycogen depletion pattern in human muscle fibers after exercise of varying intensity and at varying pedalling rates. *Journal of Physiology (London)* **241**, 45-57.

Hermansen L & Vaage O (1977). Lactate disappearance and glycogen synthesis in human muscle after maximal exercise. *American Journal of Physiology* **233**, E422-E429.

Katz A, Sahlin K & Henriksson J (1986). Muscle ATP turnover rate during isometric contraction in humans. *Journal of Applied Physiology* **60**, 1839-1842.

Le Rumeur E, Le Moyec L, Toulouse P, Le Bars R & de Certaines JD (1990). Muscle fatigue unrelated to phosphocreatine and pH: an "in vivo" 31-P NMR spectroscopy study. *Muscle & Nerve* **13**, 438-444.

Lundsgaard E (1930). Untersuchungen über muskelkontraktionen ohne Milchsäurebildung. *Biochemische Zeitshrift* **217**, 162-175.

McCartney N, Heigenhauser GJF, Sargeant AJ & Jones NL (1983). A constant-velocity cycle ergometer for the study of dynamic muscle function. *Journal of Applied Physiology* **55**, 212-217.

Miller RG, Boska MD, Moussavi RS, Carson PJ & Weiner MW (1988). 31P nuclear magnetic resonance studies of high energy phosphates and pH in human muscle fatigue. *Journal of Clinincal Investigation* **81**, 1190-1196.

Quistorff B, Johansen L & Sahlin K (1992). Absence of phosphocreatine resynthesis in human calf muscle during ischaemic recovery. *Biochemical Journal* **291**, 681-686.

Sahlin K, Cizinsky S, Warholm M & Höberg J (1992). Repetitive static muscle contractions in humans - a trigger of metabolic and oxidative stress? *European Journal of Applied Physiology and Occupational Physiology* **64**, 228-236.

Sejersted OM (1992). Electrolyte imbalance in body fluids as a mechanism of fatigue. In: Lamb DR, Gisolfi CV (eds.), *Energy Metabolism in Exercise and Sport (Perspectives in Exercise Science and Sports Medicine)*, pp 149-207. Carmel, IN: Brown & Benchmark.

Sejersted OM & Vøllestad NK (1993). Physiology of muscle fatigue and associated pain. In: Vaerøy H, Merskey H (eds.), *Progress in fibromyalgia and myofascial pain*, pp. 41-51. Amsterdam: Elsevier Science Publications.

Söderlund K, Greenhaff PL & Hultman E (1992). Energy metabolism in type I and type II human muscle fibres during short term electrical stimulation at different frequencies. *Acta Physiologica Scandinavica* **144**, 15-22.

Söderlund K & Hultman E (1986). Effects of delayed freezing on content of phosphagens in human skeletal muscle biopsy samples. *Journal of Applied Physiology* **61**, 832-835.

Thorstensson A & Karlsson J (1976). Fatiguability and fibre composition of human skeletal muscle. *Acta Physiologica Scandinavica* **98**, 318-322.

Vøllestad NK & Blom PCS (1985). Effect of varying exercise intensity on glycogen depletion in human muscle fibres. *Acta Physiologica Scandinavica* **125**, 395-405.

Vøllestad NK, Sejersted I, Saugen E (1995) Increased relaxation rates during and following intermittent submaximal isometric contractions. *Clinical Physiology* In press.

Vøllestad NK, Sejersted OM, Bahr R, Woods JJ & Bigland-Ritchie B (1988). Motor drive and metabolic responses during repeated submaximal contractions in man. *Journal of Applied Physiology* **64**, 1421-1427.

Vøllestad NK, Tabata I & Medbø JI (1992). Glycogen breakdown in different human muscle fibre types during exhaustive exercise of short duration. *Acta Physiologica Scandinavica* **144**, 135-141.

Vøllestad NK, Vaage O & Hermansen L (1984). Muscle glycogen depletion patterns in type I and subgroups of type II fibres during prolonged severe exercise in man. *Acta Physiologica Scandinavica* **122**, 433-441.

Vøllestad NK, Wesche J & Sejersted OM (1990). Gradual increase in leg oxygen uptake during repeated submaximal contractions in humans. *Journal of Applied Physiology* **68**, 1150-1156.

Wilson JR, McCully KK, Mancini DM, Boden B & Chance B (1988). Relationship of muscular fatigue to pH and diprotonated P_i in humans: a ^{31}P-NMR study. *Journal of Applied Physiology* **63**, 2333-2339.

MECHANISMS OF HUMAN MUSCLE FATIGUE

Quantitating the Contribution of Metabolic Factors and Activation Impairment

R. G. Miller,[1] J. A. Kent-Braun,[2] K. R. Sharma,[3] and M. W. Weiner[4]

[1] Departments of Neurology, California Pacific Medical Center and University of California
San Francisco, California 94118
[2] Magnetic Resonance Unit, VA Medical Center, Department of Radiology, University of California
San Francisco, California 94121
[3] Department of Neurology, University of Miami
Miami Florida 33136
[4] Magnetic Resonance Unit, VA Medical Center, Departments of Medicine, Radiology and Psychiatry, University of California
San Francisco, California 94121

ABSTRACT

Both metabolic factors and impairment of activation appear to play a role in human muscle fatigue. By measuring force, EMG and metabolites during fatiguing exercise and recovery, we are attempting to estimate the contribution of the different factors which produce fatigue.

INTRODUCTION

Thirteen years have elapsed since the first studies utilizing 31 phosphorous (^{31}P) nuclear magnetic resonance spectroscopy (NMR) were applied to healthy human skeletal muscle to study metabolism and fatigue (Chance et al., 1981). In the intervening years, there has been increasing access to NMR techniques and as an enlarging literature attests, the application of NMR has proven extremely useful in broadening our understanding of mechanisms of fatigue and muscle bioenergetics. In this paper, we will attempt to review the lessons learned from NMR studies in terms of the metabolic basis of muscle fatigue. We will highlight how these studies, along with force and EMG measures, have helped to clarify the role of central fatigue and excitation-contraction-coupling impairment as important mecha-

Fatigue, Edited by Simon C. Gandevia et al.
Plenum Press, New York, 1995

nisms of fatigue. An expanded review of the NMR technique may be found in Kent-Braun and colleagues (1995). In the 1981 CIBA symposium on fatigue, there emerged two camps of investigators: those who subscribed primarily to the theory that metabolic changes are the principal cause of muscular fatigue, and the other group who postulated that electrophysiological changes (activation impairment) are the major mechanism of muscle fatigue (Edwards 1981; Wilkie 1981). We have attempted to analyze simultaneously electrophysiological and metabolic measures of fatigue, and these studies have indicated that both mechanisms play important roles. The complex interaction between altered muscle metabolism and impaired muscular activation will be the major focus of the present review.

ADVANTAGES AND LIMITATIONS OF NMR

A super-conducting magnet is required for NMR studies of human skeletal muscle during fatiguing exercise. These examinations require both a spectrometer and non-magnetic exercise equipment, which allows for isolation of the exercise primarily to the specific muscle group studied by NMR. Data is usually obtained with the use of a surface coil of a size and shape that determines the volume of tissue under investigation. The major advantage of NMR is that continuous biochemical data can be obtained from exercising human muscle without causing discomfort or altering the exercise conditions in any way. In the past, muscle biopsies were required to obtain such information but these involved considerable discomfort and some interruption of the exercise protocol.

In NMR, radio frequency waves are transmitted through a non-magnetic coil in the presence of a static magnetic field to sample the relative concentration of metabolites that contain various nuclei (e.g., phosphorous, proton, or carbon). The frequency is plotted on the X axis, and the peak height and area of the resonance provide information about the identity and concentration of each compound (Fig. 1). The area of the peak is proportional to the concentration of the molecule. Sampling rates for individual spectra vary from 15-60 sec. Very fast spectra have been obtained in as short as 2 sec (DeGroot et al., 1993).

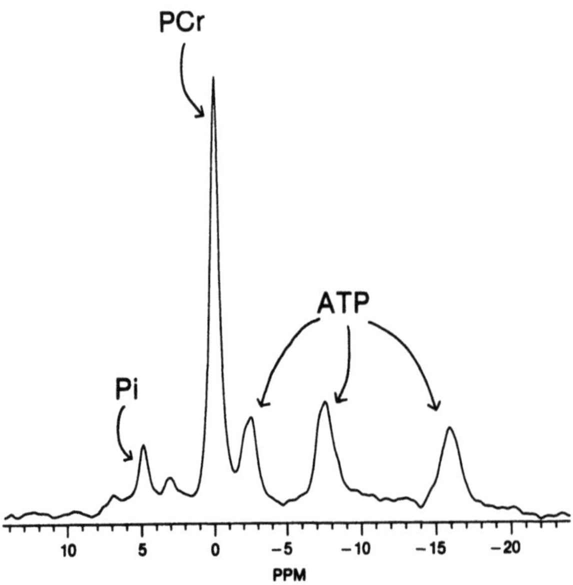

Figure 1. Phosphorus-31 magnetic resonance spectrum from human tibialis anterior muscle. The spectrum, obtained at rest, clearly shows the resonances (peaks) from inorganic phosphate (Pi), phosphocreatine (PCr), and the three phosphorus peaks from adenosine triphosphate (ATP). The phosphomonoester region is to the left of Pi and the phosphodiester region is to the right of Pi. This spectrum was acquired over 4 min with a 3 cm x 5 cm elliptical surface coil and has 15 Hz of line broadening. A broad component to the baseline was removed with convolution difference (line broadening = 350 Hz) (reproducted with permission from Kent-Braun et al., 1994).

There are important limitations to this technique. As the magnetic bore size is increased, the isocenter of the magnet is enlarged but the muscle under study must be located in the isocenter of the magnet; this is critical to reduce noise. Although continuous and non-invasive measures of muscle metabolism can be obtained, the absolute quantitation of metabolite concentration is difficult. Metabolite concentrations are usually expressed in relative terms and with some imprecision. Moreover, because it is impossible to restrict the sample volume, all determinations are approximate and sometimes involve more than a single muscle and often less than the entire muscle in question. In addition, because all human muscles represent a mixture of slow and fast fibers, there is only an average expression of metabolic changes; and it is not possible to distinguish between the changes within different fiber types in human muscle. In our experience, these studies are extremely difficult to schedule when there is competition for magnet time to carry out clinical imaging studies. In the ideal situation, the magnet is completely committed for experimental investigation with NMR, as is the case in our laboratory. Even under these conditions, there is usually intense competition for experimental time in the magnet.

MUSCLE ENERGY METABOLISM

A representative spectrum from resting human muscle using ^{31}P NMR is shown in Fig. 1. The peak for phosphocreatine (PCr) is the most prominent peak in the center of the spectrum, and the inorganic phosphate peak (Pi) is readily identifiable to the left. There are three peaks of adenosine triphosphate (ATP), and these can be easily identified as well. Intracellular pH can be calculated from the chemical shift of the Pi peak to the right towards PCr (a distance expressed in parts per million). The adenosine diphosphate (ADP) concentration is small but can be calculated from the spectra as well. Muscle contraction requires the release of energy from ATP; and in the process of energy release, ADP and inorganic phosphate are formed. With decreasing intracellular pH, the concentration of an important product of ATP hydrolysis, the monovalent form of inorganic phosphate ($H_2PO_4^-$), is increased. This compound and other products of ATP hydrolysis have the potential to inhibit muscular contraction as will be described below. The creatine kinase reaction is important in both energy production and utilization, because ADP, PCr and hydrogen ions (H^+) are combined to produce ATP and creatine. The creatine kinase equilibrium reaction may be used to estimate the intracellular ADP. Chance and colleagues (1985) demonstrated that a linear relationship existed between Pi/PCr and work rate during relatively low exercise intensities. This linear relationship had previously been observed in preparations of isolated mitochondria and suggested that ADP regulation of oxidative phosphorylation is important under these experimental conditions. Chance and colleagues (1986) also described a hyperbolic relationship between work level and *energy cost* (Pi/PCr) wherein unlimited amounts of oxygen and substrate are available. Because Pi/PCr can be utilized to estimate intracellular [ADP], the dynamics of ADP control of oxidative phosphorylation can be calculated, and the capability of the muscle for oxidative phosphorylation and metabolism can be estimated. Endurance training increases the capacity for oxidative phosphorylation as measured by this method (Kent-Braun et al., 1990).

Important validation of the overall accuracy of NMR techniques was provided by Wilkie and coworkers (1984) who found the total Pi and PCr in muscle was similar using muscle biopsy and NMR. The level of Pi was slightly higher while that of PCr was somewhat lower in biopsy tissue, a finding that was attributed to some continuing hydrolysis of PCr prior to freezing biopsy tissue. These studies also demonstrated that ischemia produced a modest reduction in PCr and a slight increase in ADP and pH. Muscle contraction, however, was required to produce the onset of glycolysis that was not activated by increased ADP alone. These studies provided substantial support for the value of NMR measurements in evaluating metabolism in skeletal muscle.

The role of limiting oxygen supply to muscle has been studied with a blood pressure cuff to reduce muscle blood flow. With ischemia, both a higher Pi/PCr and lower pH were observed at each work load in healthy individuals, suggesting the important role of oxygen during muscular exercise (Wiener et al., 1986b). Muscle deoxymyoglobin has also been measured during exercise using proton spectroscopy (Wang et al., 1990). Deoxymyoglobin is not present in resting muscle; but with occlusion of the circulation, deoxymyoglobin can be detected. Exercise under conditions of ischemia produced a prominent desaturation of myoglobin, which peaked after approximately 6 min of circulatory occlusion during muscular exercise. The measurement of myoglobin desaturation during exercise may provide additional useful information about oxygen limitation and blood flow during fatiguing exercise. In another series of experiments, less tissue deoxygenation during exercise was found when warm-up preceded the fatiguing muscular exercise (Wiener et al., 1986a); lower steady state levels of Pi/PCr and higher steady state levels of pH were attained when subjects warmed up prior to exercise compared with no warm up.

These observations suggested to us that the experimental conditions may be extremely important in determining not only the experimental results but also the variability between subjects during fatiguing exercise carried out with NMR. We therefore studied the variability of fatigue in patients with a standardized exercise protocol of the tibialis anterior muscle and also examined the factors that might contribute to this variability (Miller et al., 1995). Between control subjects, there was marked variability in the degree of muscular fatigue in the tibialis anterior (Fig. 2.) The speed of recovery was also variable, but the degree of recovery was rather uniform. To evaluate the effects of blood glucose and also glycogen storage upon these results, we examined the effect of 48 hours of fasting and also 72 hours of carbohydrate loading to separately determine the effects of a major dietary manipulation upon the variability of muscle fatigue. There were no significant differences between the fasting state, the carbohydrate-loaded state, and the control condition in these subjects. These results suggest that, at least for high-intensity, short-duration exercise, the variability between subjects is independent of dietary factors. We also examined the effect of serum pH upon the variability of muscular fatigue by administering 22 grams of ammonium chloride orally which lowered the serum pH in arterialized venous blood to levels as low as 7.25. There was no significant difference in muscular fatigability induced by systemic acidification compared to controls. These findings suggest that the muscle buffering capability is sufficient to neutralize the impact of any changes in serum pH. We also examined the variability of muscular fatigue in the same subject on different days, and found excellent

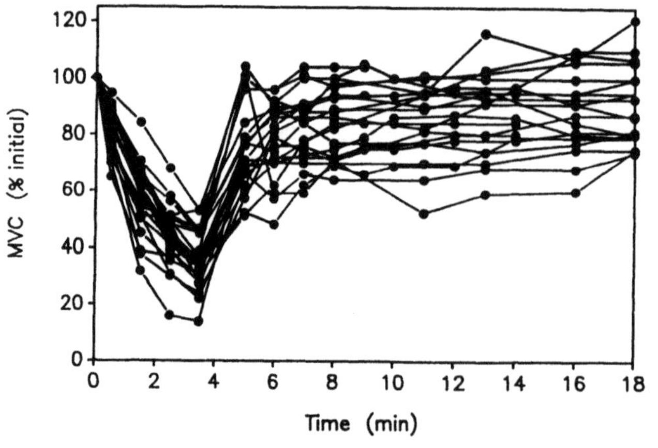

Figure 2. Maximal voluntary contraction (MVC) during and following a 4-min maximal sustained contraction of the tibialis anterior (reproduced from Miller et al., 1995, by permission of John Wiley & Sons, Inc.).

reproducibility for each subject. The major factors that were not manipulated in our studies, and which probably contribute to the variability of muscular fatigue between subjects, are fiber-type composition of the muscle which is controlled by genetic factors and the level of physical conditioning. It is likely that these are the major determinants of variability in muscular fatigability between human subjects, and both of these factors deserve more systematic examination in the future.

METABOLIC FACTORS IN MUSCLE FATIGUE

In our prior work, we define fatigue as the reduction in maximal force generating capability of the muscle during exercise. NMR is an excellent technique for continuously and non-invasively quantitating the changes in muscle metabolism during fatiguing exercise. In 1986, Taylor and colleagues demonstrated that some depletion of ATP can occur with high intensity exercise in normal human subjects when PCr depletion exceeds 80% (Taylor et al., 1986). Intracellular pH had fallen to 6.2, indicating the intense degree of muscular exercise. The calculated free energy of ATP hydrolysis was similar in moderate exercise and during high intensity exercise. Following intense exercise, the rate of recovery of Pi, PCr, and pH was slower than after moderate exercise. The recovery of ATP was slower than that of other metabolites. Thus, intense fatiguing exercise could lead to a pronounced depletion of high-energy phosphates, and also produce slowing of metabolic recovery.

Over the past decade, we have carried out experiments to examine the relationship between muscle fatigue and changes in various metabolites during different types of exercise (Miller et al., 1987; Miller et al., 1988; Boska et al., 1990; Weiner et al., 1990; DeGroot et al., 1993). We monitored the time course of the changes in force, intracellular pH, PCr, compound muscle action potential amplitude, and the rectified integrated electromyogram (RIEMG) in the exercising adductor pollicis muscle of healthy volunteer subjects (Miller et al., 1987). During fatiguing exercise, maximal voluntary force fell to only 10% of initial; and intracellular pH fell to 6.4, while PCr was almost completely depleted. Both the neuromuscular efficiency (force/RIEMG) and the compound muscle action potential amplitude decreased during fatigue. However, the duration of the compound muscle action potential amplitude increased, and the overall area of the potential was preserved so that changes in muscle membrane excitability and neuromuscular transmission were probably not responsible for fatigue in this experimental paradigm. The pattern of recovery (Fig. 3) suggested three components of fatigue: first there was a rapidly recovering alteration of muscle membrane excitation and impulse propagation as reflected by changes in the compound muscle action potential; second, there was a more slowly recovering alteration in the metabolic state of the muscle; and third, there was a long duration failure of excitation-contraction coupling as reflected by impaired neuromuscular efficiency and reduced twitch tension.

In a subsequent study, we compared the metabolic and physiological changes during fatigue resulting from intermittent and sustained exercise where the source of energy was largely from aerobic and anaerobic metabolism, respectively (Miller et al., 1988). The rapid force and metabolite changes in the adductor pollicis during sustained exercise are shown in Figs. 4*A* and 4*C*. During intermittent exercise, there were more gradual changes in both force and metabolites (Figs. 4*B* and 4*D*). However, the relationships between force and both H^+ and $H_2PO_4^-$ were closer than those between force and either PCr or Pi. This was a finding that was consistent in both exercise protocols, and led us to conclude that H^+ and $H_2PO_4^-$ may both play important roles in the development of muscle fatigue.

The close relationships between force and both H^+ and $H_2PO_4^-$ were also observed in the less fatigable tibialis anterior muscle, and demonstrated the constant nature of this

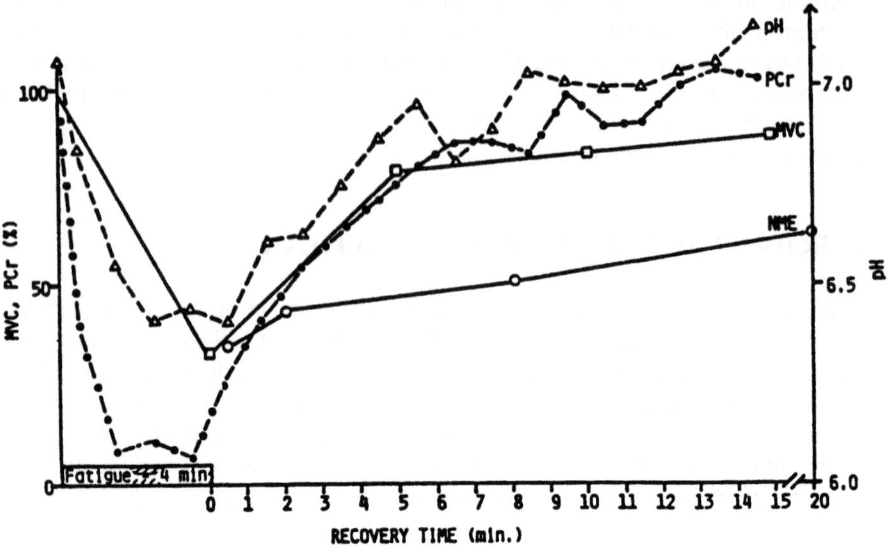

Figure 3. Recovery of neuromuscular efficiency, phosphocreatine, pH, and maximal voluntary contraction after a 4 min fatigue test (reproduced from Miller et al., 1987, by permission of John Wiley & Sons, Inc.).

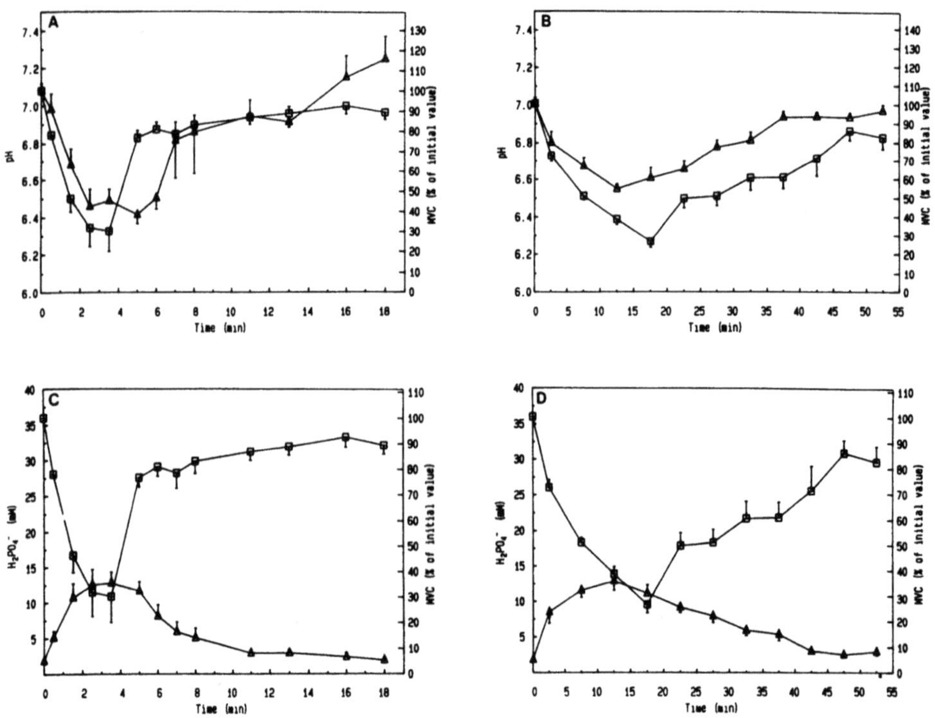

Figure 4. pH (Δ) and MVC (❐) plotted as a function of time during 4 min of sustained (*A*) and intermittent (*B*) exercise. $H_2PO_4^-$ (Δ) and MVC (❐) plotted as a function of time during 4 min of sustained (*C*) and intermittent (*D*) exercise (reproduced from Miller et al., 1988, by copyright permission of The Society for Clinical Investigation).

relationship (Weiner et al., 1990). A subsequent study of the relationship between the recovery of force and metabolites following fatiguing exercise also suggested that $H_2PO_4^-$ was the metabolic measure most closely related to maximal force generating capacity(Boska et al., 1990). Thus, we provided evidence for an important role of diprotonated Pi ($H_2PO_4^-$) in the development of muscle fatigue. This is not to say that intracellular pH and total Pi are not important determinants of muscle fatigue, but the single metabolite that correlated most closely with muscular fatigue was the build-up of $H_2PO_4^-$. Other studies have documented a strong linear relationship between fatigue and $H_2PO_4^-$ in the wrist flexor muscles of healthy human subjects (Cady et al., 1989a). Thus in 3 muscles and with 3 different exercise protocols, there is evidence that the relationship with $H_2PO_4^-$ and fatigue was stronger and more consistent than the relationship between pH and fatigue.

Cady and co-workers (Cady et al., 1989b; see also Hultman et al., 1981) examined the relationship between changes in muscle metabolism and muscular fatigue. They found a linear relationship between $H_2PO_4^-$ and fatigue in the first dorsal interosseous muscle of healthy subjects. No linear relationship was found, however, in a subject with McArdle's disease who, because of defective glycogenolysis, could not utilize muscle glycogen (Cady et al., 1989b). These investigators also noted a dissociation between fatigue and pH that was particularly prominent during the recovery after fatiguing exercise. They also provided evidence that there is a pH-independent component of muscular fatigue that may be related to the slowed rate of force relaxation observed during muscular fatigue. This was demonstrated by a 50% slowing of relaxation in fatigued muscle at a time when pH remained unchanged.

The examination of muscle oxidative potential was also the focus of a recent study of the ankle dorsiflexor muscle group (Kent-Braun et al., 1993a). A sequence of three different metabolic phases was found during the transition from rest to a fatiguing level of muscular exercise (Fig. 5).

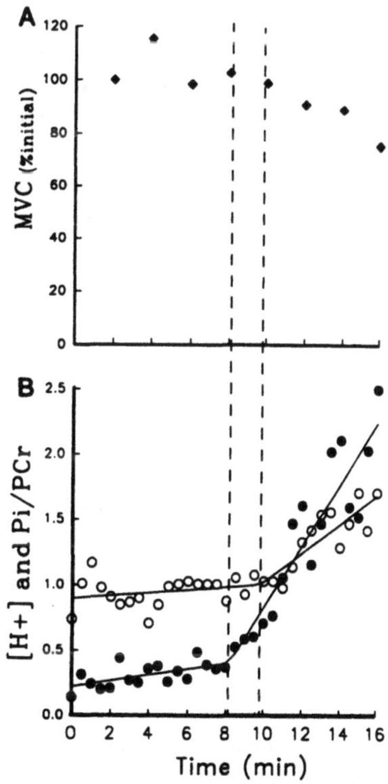

Figure 5. Changes in MVC (*A*) and Pi/PCr and [H^+] (*B*) *vs.* time in 1 subject. Bilinear fit is included to illustrate the 3 phases of metabolism delineated by inflection points for Pi/PCr (●) and [H^+] (X10^{-7},○) *vs.* time. Dashed vertical lines indicate location of inflection points (reproduced with permission from Kent-Braun et al., 1993a).

The length of the first phase, during which energy for the muscle contraction was produced primarily by oxidative phosphorylation, was closely related to the oxidative potential of the muscle (i.e., the initial slope of force *vs.* Pi/PCr). This first phase was followed by an intermediate phase, during which Pi/PCr was increasing at a more rapid rate but H^+ remained relatively unchanged. In a third phase, during which glycolytic sources of ATP began to contribute more significantly to energy production, there was an inflection in H^+ with continued steady increases in Pi/PCr. It was the third, non-steady state phase that was associated with the production of muscular fatigue. These studies demonstrated the utility of ^{31}P NMR for the study of relationships between oxidative and glycolytic metabolism during progressive muscular exercise. By comparing the time course of metabolic changes involved primarily in either oxidative or glycolytic metabolism, insight can be gained into the interplay between these pathways and the development of fatigue.

Thus, numerous studies of muscle fatigue in healthy human subjects have demonstrated the importance of altered metabolites (e.g., pH, Pi, $H_2PO_4^-$) in the development of muscular fatigue during various types of exercise. However, fatigue may also develop due to activation impairment, especially during intermittent long duration exercise. This is discussed in more detail below.

CENTRAL FATIGUE

Impaired firing rate modulation or reduced motor unit recruitment during fatiguing exercise is often referred to as central fatigue. Any breakdown in muscular activation that is proximal to the point of stimulation of the motor nerve is generally considered central. Impaired central activation of muscle during exercise is measured in two different ways. During a maximal voluntary contraction, superimposed electrical stimulation of the motor nerve, which may produce added force from incompletely activated muscle fibers, provides a quantitative measure of central activation failure. When the muscle is fully activated, the superimposed electrical stimulus produces no added force (Bigland-Ritchie et al., 1986a; Lloyd et al., 1991). Previous workers have variously utilized the superimposed twitch, paired stimuli, and brief trains of tetanic stimuli (e.g., Gandevia & McKenzie, 1985; Bigland-Ritchie et al., 1986ab; Lloyd et al., 1991; Kent-Braun et al., 1993b; Sharma et al., 1994b). In a comparative study, we found that a train of 50 Hz tetanic stimulation is a more sensitive method of detecting added force (unpublished data). Therefore, we have utilized this measure as a primary means of evaluating central fatigue.

An additional technique utilizes a comparison of the decline in maximal voluntary contraction strength and the decline in tetanic tension during fatiguing exercise. When fatigue is entirely peripheral to the point of electrical stimulation of the motor nerve, the decline in MVC and tetanic tension should be identical. When fatigue is central, at least in part, there should be a greater decline in MVC compared with the decline in tetanic tension. We have used both of these techniques to evaluate central fatigue in a number of studies that will now be summarized.

In the pre-fatigued tibialis anterior muscle of sedentary healthy subjects, we found during a maximal voluntary contraction that there was no added force (Kent-Braun et al., 1993b). During a sustained 4-min maximal isometric voluntary contraction, there was a small degree of added force with increasing fatigability. At the end of the 4th min, when fatigue was significant, the added force averaged 15% of the voluntary force generated by the muscle. Larger degrees of added force were found in a study of healthy children (age 6-10) using the same isometric sustained contraction protocol (Sharma et al., 1995c). The children were quite healthy but had some difficulty with concentration and sustaining motivation. They demonstrated substantial added force at the end of 4 min, indicating significant central impairment of activation of the muscle. In both children and adults, there was a tendency for the added force to steadily increase during the 4 min of sustained contraction, suggesting that some central fatigue is commonly found in healthy subjects during arduous, sustained muscular fatiguing exercise. Surprisingly, and by contrast, central fatigue in an age-matched group of boys with Duchenne muscular dystrophy was much less at the end of 4 min of fatiguing exercise compared with the healthy children (Sharma et al., 1995c) suggesting that boys with Duchenne muscular dystrophy have improved activation of muscle either as a compensation for muscular weakness or because they are more used to the testing procedures.

In contrast, in a recent study of patients with chronic fatigue syndrome, we found much greater central fatigue in patients compared with sedentary adults who served as age-matched control subjects (Kent-Braun et al., 1993b; cf. Lloyd et al., 1991). No other neuromuscular abnormality was found in evaluating muscular fatigue in chronic fatigue syndrome patients. Specifically, during fatiguing exercise, neither the decline in voluntary muscle force generation nor the change in tetanic tension were significantly different from controls. The compound muscle action potential amplitude was preserved during fatiguing exercise in both patients and controls. Metabolic changes were, if anything, less in patients than in controls, consistent with reduced activation of muscle by patients during exercise. These studies document the presence of a central component of fatigue in some patients with chronic fatigue syndrome, although the precise cause of the central fatigue remains unclear. Possibilities include reduced attention, muscular pain, deconditioning, and reduced motivation.

We also found that a unique type of central fatigue was produced by rapid repetitive muscular contractions (Miller et al., 1993). A series of repetitive isometric contractions was carried out with the ankle dorsiflexor muscles in normal human subjects beginning with a rate of 24/min and gradually increasing every 3 min to a maximum frequency of 72 contractions per min. The subject was instructed to execute each contraction *as fast as possible* to reach the target force of 40% MVC. Throughout the 15-min exercise period, there was a steady decline in maximal voluntary contraction strength down to approximately 50% of initial force, and a decrease in intracellular pH to 6.5. The changes in spectra, both during exercise and recovery, are shown in Fig. 6.

During the first 1 or 2 min, the changes in metabolism and maximal voluntary contraction strength were slight. As exercise increased, the metabolic changes steadily developed. By contrast, there was a decrease in the speed of voluntary force generation within the first min of fatiguing exercise (Figs. 7 and 8).

At this time, there was only a minimal alteration in metabolites and MVC. Moreover, the speed of tension generated in the tetanus was unchanged at this stage, and the speed of tension generated in the twitch was actually increased due to twitch potentiation. But, the duration of the EMG burst associated with the rapid voluntary movements was prolonged by approximately 20%-25% (Fig. 8). Thus, during rapid repetitive movements, there are two components of fatigue: a steadily progressive change in metabolism and a decline in maximal force generation, but an earlier slowing of force generation during rapid voluntary muscle contractions that is not present during electrically stimulated muscle contraction in either the twitch or the tetanus. Because the slowed speed of voluntary force generation is correlated with a prolonged EMG burst duration, there must be a slowing of recruitment and/or firing rate modulation of motor unit potentials during a rapid voluntary contraction. This type of central fatigue, which develops so early in this exercise protocol, may be relevant to many activities of daily living including piano playing, keypunching, typing, etc. Further studies of rapid repetitive movements are indicated both in healthy individuals and in various disease states where the ability to execute rapid movements is often selectively impaired.

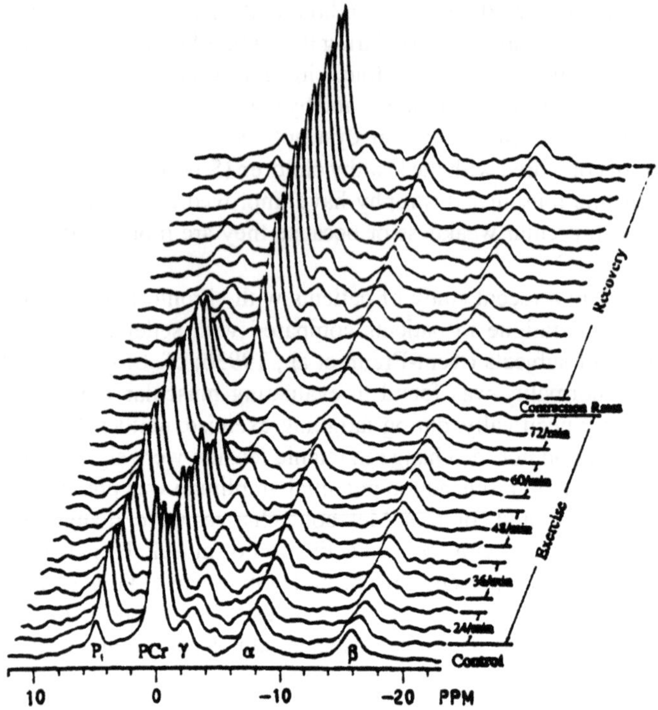

Figure 6. A stacked plot of the 1-min ^{31}P NMR spectra acquired during a typical rapid exercise protocol (reproduced with permission from Miller et al., 1993).

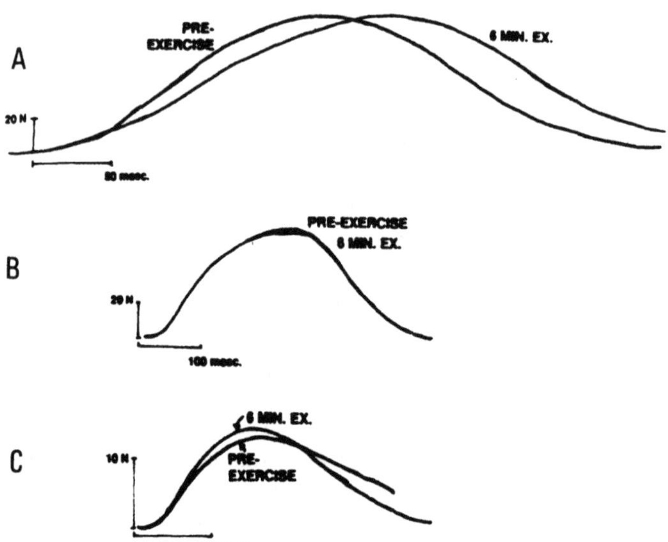

Figure 7. Superimposed force recordings from the adductor pollicis of a representative subject during rapid voluntary contractions to a 40% MVC target (*A*), tetanic stimulation (*B*), and a single twitch (*C*). Each recording was made at the beginning and again after 6 min of rapid repetitive contractions (reproduced with permission from Miller et al., 1993).

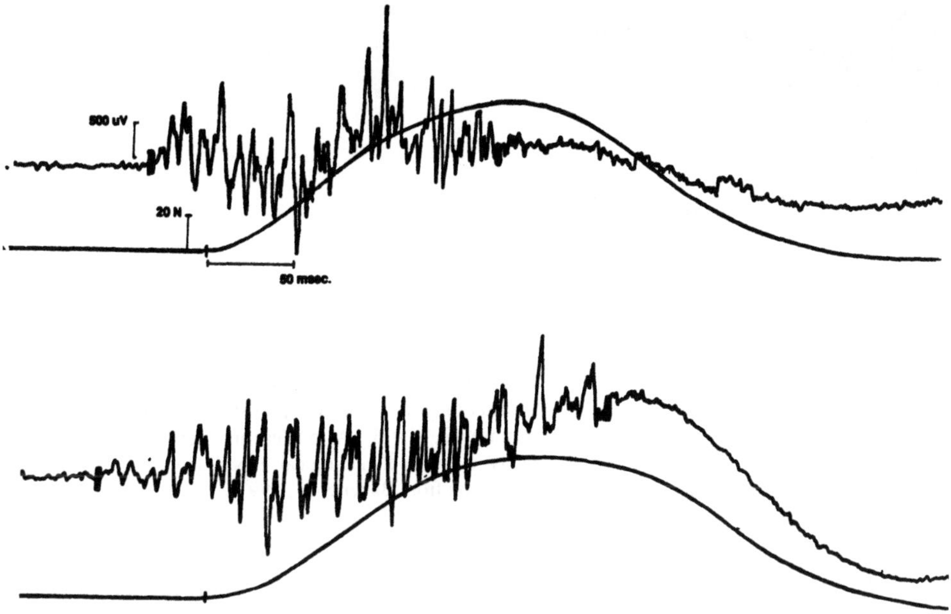

Figure 8. Force and EMG recordings at the beginning (top) and after 1 min of rapid repetitive contractions (bottom) with the adductor pollicis muscle from one subject. EMG signals were recorded with surface electrodes (bandpass filter settings 1.6 Hz to 16 kHz). Records are aligned at the onset of the increase in force (reproduced with permission from Miller et al., 1993).

EXCITATION-CONTRACTION-COUPLING IMPAIRMENT

The other likely source of activation impairment, besides central fatigue, is excitation-contraction coupling. However, this is the most difficult aspect of neuromuscular function to quantitate during fatiguing exercise. Although it is impossible in healthy human subjects to measure the impulse propagation in the transverse tubule or the amount and rate of calcium released by the sarcoplasmic reticulum, indirect measures of excitation contraction coupling are available. Probably the most reliable measure of impaired excitation contraction coupling is the existence of a long duration fatigue following the completion of muscular exercise. Long duration fatigue has also been called low-frequency fatigue, and the phenomenon was first attributed to excitation-contraction-coupling impairment by Edwards and colleagues (1977). These investigators found that after exhaustive exercise there was a long-lasting impairment of the force produced by low-frequency stimulation, with an earlier recovery of the force produced by high-frequency stimulation. In a subsequent study of intermittent low intensity exercise, Bigland-Ritchie and colleagues observed a form of fatigue that was associated with very little alteration in metabolites as analyzed by muscle biopsy (Bigland-Ritchie, et al., 1986b). A particularly pronounced depression of the twitch and a more modest depression of maximal voluntary force generation was noted and attributed to excitation-contraction-coupling impairment. In a subsequent study, we observed that low intensity intermittent exercise could produce not only fatigue associated with very little metabolic alteration as examined by NMR spectroscopy, but also with some long lasting impairment of muscular force generation (Moussavi et al., 1989). Again, this pattern was attributed to excitation-contraction-coupling impairment.

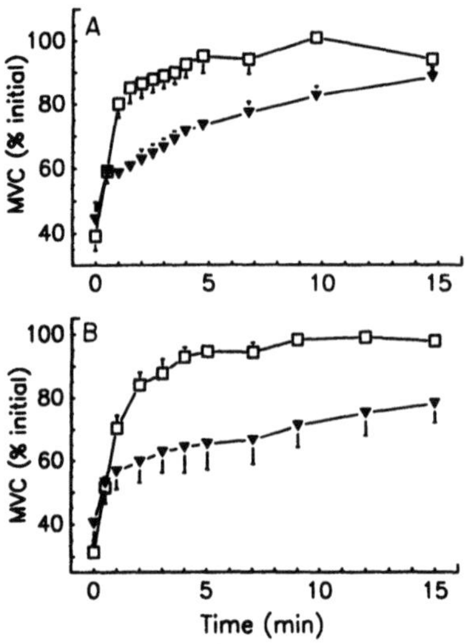

Figure 9. Maximal voluntary contraction (MVC) force recovery after long-duration exercise (▼; n = 5) and short-duration exercise (❑; n = 6). A, nuclear magnetic resonance experiments. *B*, neurophysiological experiments (reproduced with permission from Baker et al., 1993).

More recently, our group has compared the recovery from short duration exercise during two min of maximal sustained contraction under ischemic conditions with a longer exercise protocol (20-25 min) under aerobic conditions (Baker et al., 1993). The speed of recovery of both tetanic tension and maximal voluntary contraction was markedly slowed in the long duration exercise compared to short duration exercise (Fig. 9). The recovery of twitch tension was particularly slow in the subjects who carried out long duration exercise. By contrast, the recovery of metabolites (PCr, Pi, and $H_2PO_4^-$) was prompt and complete in both groups. The PCr recovery was, if anything, slower in the subjects who performed short-duration exercise. In both groups, metabolic recovery was complete after 10 min at a time when the recovery of both MVC and tetanic tension was incomplete.

The evoked compound muscle action potential amplitude was not significantly reduced following either type of exercise. The amplitude of the surface EMG signal was somewhat depressed after both types of exercise, but was more markedly reduced after short duration exercise. Within the first min, there was a rapid recovery of the excessive EMG impairment associated with the short duration exercise.

These observations document the presence of a long-lasting fatigue that is particularly striking for low-frequency force (twitch tension). The absence of any alteration of the compound muscle action potential suggests that neither neuromuscular transmission impairment nor altered muscle membrane excitability can explain the long-lasting fatigue. Moreover, the prompt recovery of metabolites following both types of exercise suggests that this is not a defect in energy metabolism induced by long-duration exercise. Similarly, the comparable decline in tetanic tension and maximal voluntary strength strongly suggests that fatigue was peripheral rather than central in origin.

The contribution of metabolic changes to long-duration fatigue was estimated from the time course of recovery of inorganic phosphate and maximal voluntary contraction strength (Fig. 10).

Following short-duration exercise, the time course of recovery of inorganic phosphate and muscular force are almost identical suggesting that the recovery of force depended primarily upon the recovery of metabolites after short-duration exercise. By contrast, the recovery of force was delayed following long-duration exercise compared with the recovery of inorganic phosphate and other metabolites. As already indicated, the absence of change in the compound muscle action potential suggests that the neuromuscular junction and

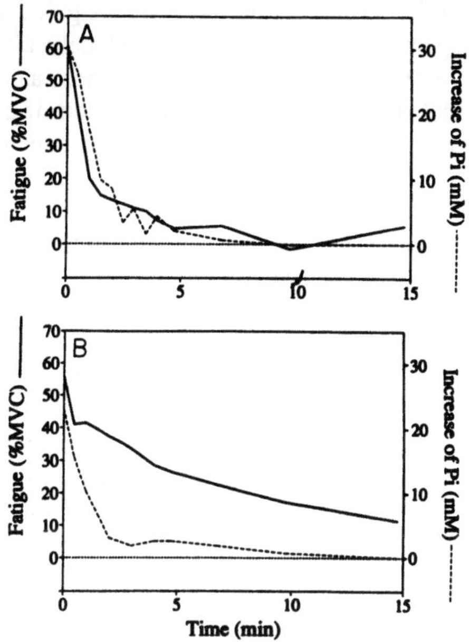

Figure 10. Mean level of fatigue (expressed as %MVC reduction) and mean P_i elevation after short-duration exercise (*A*) and long-duration exercise (*B*) (reproduced with permission from Baker et al., 1993).

muscle membrane did not contribute significantly to the muscle fatigue. The central component of fatigue was judged to be insignificant because the parallel and similar changes were observed in tetanic tension and MVC during fatigue, and also because the recovery of force was independent of any change in EMG. Thus, the contribution of excitation-contraction-coupling impairment was estimated as the remaining mechanism of fatigue unaccounted for by the other mechanisms elaborated above (Baker et al., 1993).

We attempted to estimate the quantitative contribution of metabolic changes, excitation-contraction-coupling impairment, altered neuromuscular transmission, and central fatigue to the overall muscular fatigue induced by both short-duration exercise and long-duration exercise (Fig. 11).

At the conclusion of fatiguing short-duration exercise, most of the fatigue could be explained by metabolic changes. Some alteration of excitation-contraction coupling could also be present, but was not identifiable using our techniques. Because of the reduced surface EMG signal, a slight degree of central fatigue was also suggested in both types of exercise. Similarly, because of the very slight reduction in compound muscle action potential amplitude, a mild contribution of neuromuscular-transmission impairment was possible. However,

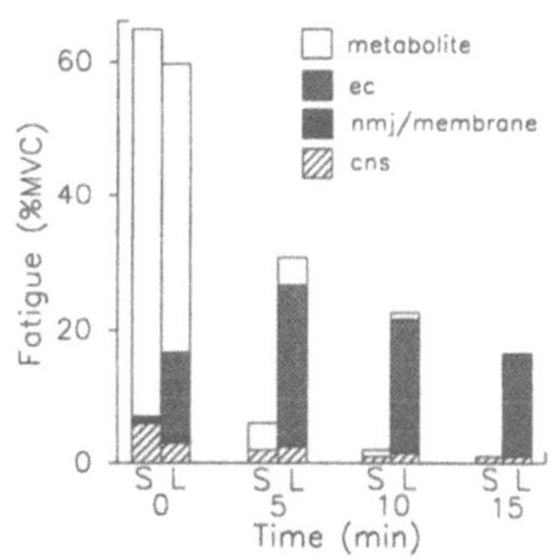

Figure 11. Histogram of recovery from fatigue after short-duration (S) and long-duration exercise (L) at 0, 5, 10, and 15 min of recovery. Relative contributions of 4 different fatigue mechanisms are indicated by shading in each column. EC, excitation-contraction coupling; NMJ, neuromuscular junction; CNS, central nervous system. The mean level of fatigue (expressed as % MVC reduction) is indicated on the ordinate (reproduced with permission from Baker et al., 1993).

in the long-duration exercise, a significant impairment of excitation-contraction coupling appeared to be present immediately after exercise; and this appeared to become more prominent during the recovery phase as indicated in Fig. 11. We have probably underestimated, if anything, the contribution of excitation-contraction-coupling impairment in the present study, particularly at the conclusion of exercise. Precise quantitation of these parameters is clearly not possible using the methods of this study, but in a general way it helps us to begin to focus on the role of individual mechanisms of fatigue at different stages of fatigue and recovery and in different types of exercise, both in health and in disease. The role of excitation-contraction-coupling impairment appears to be even more prominent in some neuromuscular diseases. In patients with post-polio syndrome and amyotrophic lateral sclerosis, we have found even more marked impairment of excitation-contraction coupling (Sharma et al., 1994b; Sharma et al., 1995a).

In summary, NMR spectroscopy in conjunction with force and EMG measurements has been useful in allowing us to broaden our understanding of the mechanisms of fatigue in health and disease. As emphasized by Bertocci, Chapter 15, when NMR allows measurement of calcium, sodium and potassium levels as well as glycogen, high-energy phosphates, and intracellular pH, a much more comprehensive understanding of fatigue in human muscle will be possible.

ACKNOWLEDGMENTS

The authors' research has been supported, in part, by grants from the Multiple Sclerosis Society (USA), and the Muscular Dystrophy Association (USA).

REFERENCES

Baker AJ, Kostov KG, Miller RG & Weiner MW (1993). Slow force recovery after long duration exercise: Metabolic and activation factors in muscle fatigue. *Journal of Applied Physiology* **74**, 2294-2300.

Bigland-Ritchie B, Cafarelli E & Vollestad NK (1986a). Fatigue of submaximal static contractions. *Acta Physiologica Scandinavica* **128** (suppl. 556), 137-148.

Bigland-Ritchie B, Furbush F & Woods JJ (1986b). Fatigue of intermittent submaximal voluntary contractions: central and peripheral factors. *Journal of Applied Physiology* **61**, 421-429.

Boska MD, Moussavi RS, Carson PJ, Weiner MW & Miller RG (1990). The metabolic basis of recovery after fatiguing exercise in human muscle. *Neurology* **40**, 240-244.

Cady EB, Elshove H, Jones DA & Moll A (1989a). The metabolic causes of slow relaxation in fatigued human skeletal muscle. *Journal of Physiology (London)* **418**, 327-337.

Cady EB, Jones DA, Lynn J & Newham DJ (1989b). Changes in force and intracellular metabolites during fatigue of human skeletal muscle. *Journal of Physiology (London)* **418**, 311-325.

Chance B, Eleff S, Leigh JS, Sokolow D & Sapega A (1981). Mitochondrial regulation of phosphocreatine-phosphate ratios in exercising human muscle: A gated ^{31}P NMR study. *Proceedings of the National Academy of Sciences of the United States of America* **78**, 6714-6718.

Chance B, Leigh JS, Clark BJ, Maris J, Kent J, Nioka S & Smith D (1985). Control of oxidative metabolism and oxygen delivery in human skeletal muscle: A steady-state analysis of the work/energy cost transfer function. *Proceedings of the National Academy of Sciences of the United States of America* **82**, 8384-8388.

Chance B, Leigh JS, Kent J, McCully K, Nioka S, Clark BJ, Maris JM & Graham T (1986). Multiple controls of oxidative metabolism in living tissues as studied by phosphorus magnetic resonance. *Proceedings of the National Academy of Sciences of the United States of America* **83**, 9458-9462.

DeGroot M, Massie BM, Boska M, Gober J, Miller RG & Weiner MW (1993). Dissociation of [H^+] from fatigue in human muscle detected by high time resolution ^{31}P-NMR. *Muscle & Nerve* **16**, 91-98.

Edwards RHT (1981). Human muscle function and fatigue. In: Porter R, Whelan J (eds.), *Human Muscle Fatigue: Physiological Mechanisms, Ciba Foundation Symposium No. 82*, pp. 1-18. London:Pitman Medical.

Edwards RHT, Hill DK, Jones A & Merton PA (1977). Fatigue of long duration in human skeletal muscle after exercise. *Journal of Physiology (London)* **272**, 769-778.

Gandevia SC, McKenzie DK (1985). Activation of the human diaphragm during maximal static efforts. *Journal of Physiology (London)* **367**, 45-56.

Hultman E, Sjöholm H, Sahlin K & Edström L (1981). Glycolytic and oxidative energy metabolism and contraction characteristics of intact human muscle. In: Porter R, Whelan J (eds.), *Human Muscle Fatigue: Physiological Mechanisms, Ciba Foundation Symposia No. 82*, pp. 19-40. London: Pitman Medical.

Kent-Braun JA, McCully KK & Chance B (1990). Metabolic effects of training in humans: a ^{31}P-MRS study. *Journal of Applied Physiology* **69**, 1165-1170.

Kent-Braun JA, Miller RG, Weiner MW (1994). Magnetic resonance spectroscopy studies of human muscle. *Radiologic Clinics of North America* **32**, 313-335

Kent-Braun JA, Miller RG & Weiner MW (1993a). Phases of metabolism during progressive exercise to fatigue in human skeletal muscle. *Journal of Applied Physiology* **75**, 573-580.

Kent-Braun JA, Miller RG & Weiner MW (1995). Magnetic resonance spectroscopy: studies of human skeletal muscle metabolism in health and disease. In: Holloszy JO (ed.), *Exercise and Sport Science Reviews* **23**, pp. 305-347. Baltimore: Williams and Wilkins.

Kent-Braun JA, Sharma KR, Weiner MW, Massie B & Miller RG (1993b). Central basis of muscle fatigue in chronic fatigue syndrome. *Neurology* **43**, 125-131.

Lloyd AR, Gandevia SC & Hales JP (1991). Muscle performance, voluntary activation, twitch properties and perceived effort in normal subjects and patients with chronic fatigue syndrome. *Brain* **114**, 85-98.

Miller RG, Boska MD, Moussavi R, Carson PJ & Weiner MW (1988). ^{31}P Nuclear magnetic resonance studies of high energy phosphates and pH in human muscle fatigue. Comparison of aerobic and anaerobic exercise. *Journal of Clinical Investigation* **81**, 1190-1196.

Miller RG, Carson PJ, Moussavi RS, Green A, Baker A, Boska MD & Weiner MW (1995). Factors which influence alterations of phosphates and pH in exercising human skeletal muscle: measurement error, reproducibility, and effects of fasting, carbohydrate loading, and metabolic acidosis. *Muscle & Nerve* **18**, 60-67.

Miller RG, Giannini D, Milner-Brown HS, Layzer RB, Koretsky AP, Hooper D & Weiner MW (1987). Effects of fatiguing exercise on high-energy phosphates, force and EMG: Evidence for three phases of recovery. *Muscle & Nerve* **10**, 810-821.

Miller RG, Moussavi RS, Green AT, Carson PJ & Weiner MW (1993). The fatigue of rapid repetitive movements. *Neurology* **43**, 755-761.

Moussavi RS, Carson PJ, Boska MD, Weiner MW & Miller RG (1989). Nonmetabolic fatigue in exercising human muscle. *Neurology* **39**, 1222-1226.

Sharma KR, Weiner MW, Miller RG, Kent-Braun J, Majumdar S, Huang Y & Mynhier M (1995a). Physiology of fatigue in amyotrophic lateral sclerosis. *Neurology* **45**, 733-740.

Sharma KR, Kent-Braun J, Mynhier MA, Weiner MW & Miller RG (1994b). Excessive muscular fatigue in the postpoliomyelitis syndrome. *Neurology* **44**, 642-646.

Sharma KR, Mynhier M & Miller RG (1995c). Muscular fatigue in Duchenne muscular dystrophy. *Neurology* **45**, 306-310.

Taylor DJ, Styles P, Matthews PM, Arnold DA, Gadian DG, Bore P & Radda GK (1986). Energetics of human muscles: Exercise- induced ATP depletion. *Magnetic Resonance in Medicine* **3**, 44-54.

Wang Z, Noyszewski EA & Leigh JS (1990). In vivo MRS measurement of deoxymyoglobin in human forearms. *Magnetic Resonance in Medicine* **14**, 562-567.

Wiener DH, Fink LI, Maris J, Jones RA, Chance B & Wilson JR (1986a). Abnormal skeletal muscle bioenergetics during exercise in patients with heart failure: Role of reduced muscle blood flow. *Circulation* **73**, 1127-1136.

Wiener DH, Maris J, Chance B & Wilson JR (1986b). Detection of skeletal muscle hypoperfusion during exercise using phosphorus-31 nuclear magnetic resonance spectroscopy. *Journal of the American College of Cardiology* **7**, 793-799.

Weiner MW, Moussavi RS, Baker AJ, Boska MD & Miller RG (1990). Constant relationships between force, phosphate concentration, and pH in muscles with differential fatigability. *Neurology* **40**, 1888-1893.

Wilkie DR (1981). Shortage of chemical fuel as a cause of fatigue: studies by nuclear magnetic resonance and bicycle ergometry. In: Porter R, Whelan J (eds.), *Human Muscle Fatigue: Physiological Mechanisms, Ciba Foundation Symposium No. 82*, pp. 102-119. London: Pitman Medical.

Wilkie DR, Dawson MJ, Edwards RHT, Gordon RE & Shaw D (1984). ^{31}P NMR studies of resting muscle in normal human subjects. In: Pollack, GH & Sugi, H (eds.) *Contractile Mechanisms in Muscle* , pp.333-347. New York: Plenum Publishing Corporation.

EMERGING OPPORTUNITIES WITH NMR

L. A. Bertocci

Institute for Exercise and Environmental Medicine, Presbyterian Hospital of Dallas
Dallas Texas 75231 and
Rogers Magnetic Resonance Center, Department of Radiology
UT Southwestern Medical Center
Dallas Texas 75235

ABSTRACT

Several nuclear magnetic resonance (NMR) methods with potential as tools for the study of neuromuscular fatigue are described briefly. ^{13}C MR spectroscopy (MRS) is presented as a means of studying the regulation of substrate (fuel) flux into the citric acid cycle. ^{23}Na and ^{39}K MRS can be used to study the distribution of sodium and potassium ions across the cell membrane. ^{1}H MR imaging (MRI) takes advantage of the relationship between the environment of water ^{1}H and the mechanisms of nuclear spin relaxation to make static and cine images which present spatial (anatomical) information that can be tied to essential physiological, biochemical, or biophysical phenomena of fatiguing muscle.

INTRODUCTION

Since the discovery of the physical property of nuclear magnetic resonance (NMR; Bloch et al., 1946, Purcell et al., 1946) new applications continue to be found for NMR. Although most of the early NMR work was done using ^{1}H magnetic resonance spectroscopy (MRS) in NMR systems with narrow bore resistive magnets up to 1.5T in strength, it was the development of larger, superconducting magnets that has made it possible for most of these new applications and, in particular, has made it possible to apply NMR techniques to the study of intact organs *in vivo*. The application of NMR to living systems, ranging from chemical analyses of organic and bio-organic compounds to the more recent studies of whole micro-organisms, intact tissues, through to entire animals and human clinical patients, represents an extraordinary advance in modern biological and clinical research.

Several NMR techniques have begun to be applied to the study of neuromuscular fatigue. Foremost is the application of ^{31}P MRS to measure the changes in relative concentration of high energy phosphates (ATP, ADP, and PCr) or pH that occur during steady-state and fatiguing exercise. However, there are other NMR techniques in use in other experimental settings which, if properly adapted to the fatiguing neuromuscular unit, may provide even

Fatigue, Edited by Simon C. Gandevia et al.
Plenum Press, New York, 1995

more fundamental biological information about neuromuscular fatigue. Several such emerging opportunities for the study of neuromuscular fatigue will be described in this chapter. Although by no means an exhaustive list, the more promising methods will be described theoretically, with examples of their current utility and suggestions for their future applications.

SPECTROSCOPY

^{13}C MRS

Carbon atoms are the basic building blocks of nearly all organic compounds, from the CO_2 produced during basic respiration to the amino acid sequences of large molecular weight proteins. To make the energy required for muscle to perform useful work, carbon-containing compounds (fuels, or more properly called substrates) must flow through the pathways of intermediary metabolism where in a series of biochemical reactions, they are ultimately converted to CO_2, H_2O, and energy. Insofar as muscle fatigue can be considered to be an excess of energy demand *vs.* an inadequate amount of energy supply, the study of these reactions is critically important to the understanding of the causes of muscle fatigue.

These reactions have been studied primarily using labeled compounds, either substrates or substrate analogs (for example [1-^{14}C]glucose, [1 or 2-^{14}C]acetate or [1, 2, or 3-^{14}C]pyruvate) (Katz, 1985) and or 2-deoxy-d-[2,6-^{3}H]glucose) (Fushiki et al., 1991) labeled radioactively or with non-radioactive stable isotopes (for example [U-^{13}C]glucose (Katz & Lee, 1991; Lee et al., 1991). The progress of the chemical reactions of interest can be monitored by several methods, most often by some combination of one-time or serial measurements of the $^{14}CO_2$, $^{3}H_2O$ or other ^{3}H- or ^{14}C- containing compounds that are produced. Alternatively, ^{13}C labeled compounds can be introduced and incorporation into the products can be determined by mass spectrometry and mass isotopomer analysis (Katz & Lee 1991; Lee et al., 1991).

Unfortunately, there are disadvantages to all of these methods, such as 1) the need to handle radioactively labeled compounds; 2) the difficulty in discriminating between different metabolic fates or pathways; and 3) the small number of discrete results can be determined from a single experiment. As an alternative to such techniques, ^{13}C MRS has been applied more recently to the study of intermediary metabolism *in vivo*. The first demonstrations of this technique were in bacteria (Walker et al., 1982) and the isolated, perfused rat liver (Cohen, 1983), although more rigorous mathematical treatments (Chance et al., 1983) of the resultant spectra have been developed which allow a great deal of information to be extracted from a single spectrum (Malloy et al., 1990; Jeffrey et al., 1991).

Isotopomer analysis of glutamate (O_2C-$^{2}CHNH_2$-$^{3}CH_2$$^{4}CH_2$-$CO_2$), either the non-steady state analysis using the C4 alone or the steady-state analysis using C2, C3, and C4 in combination, provides a straightforward measurement of the relative contribution of various substrates to the citric acid cycle. This information can be used to compare the relative flux of acetyl-CoA to the TCA cycle from 1) glycogenolysis *vs.* fatty acid β-oxidation; 2) labeled *vs.* unlabeled substrates; and 3) energy production *vs.* anaplerosis. The principal advantages of ^{13}C MRS compared with more traditional approaches to studying substrate regulation (e.g., measuring plasma concentrations of substrates, calculation of respiratory quotient, ^{14}C radioisotope studies) include 1) safety from use of radioactive isotopes (^{13}C is a naturally occurring, stable isotope of carbon); 2) convenience (a large amount of information is available from a single spectrum; 3) simultaneous measurement of the contribution of multiple substrates to the citric acid cycle; 4) measurement of energy producing and anaplerotic activities; and 5) both steady-state and non-steady state analyses.

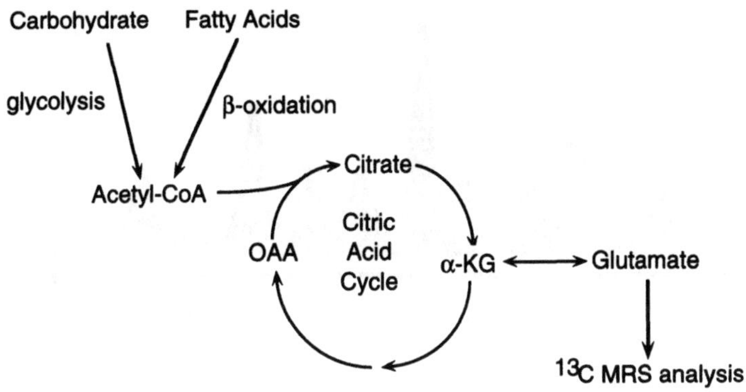

Figure 1. A simplified schematic of the biochemical rationale for the design of the proposed ^{13}C MRS experiments.

The relevant properties of the magnetic resonance signal from a detectable atom include 1) the effect of the chemical environment of the detectable atom on the frequency at which it resonates; and 2) the proportionality between the intensity of the spectra peaks and the number of detectable atoms in an individual chemical environment. In the case of an organic molecule such as glutamate, each carbon experiences a unique chemical environment and thus a separate spectral peak. However, only the ^{13}C atoms, but not the naturally abundant ^{12}C atoms, are detectable by MRS.

As exogenously supplied ^{13}C atoms are incorporated into the TCA cycle (Fig. 1), individual carbon atoms of each intermediate molecule may become labeled. The position of the label depends on the labeling pattern of the compound entering the TCA cycle and the isotopic mixing that occurs with multiple turns of the cycle. Although labeling occurs in all TCA cycle intermediates, the incorporation of label into α-ketoglutarate is most relevant because α-ketoglutarate is in chemical equilibrium with glutamate and the equilibrium conditions are such that the normal concentration of glutamate is much greater than α-ketoglutarate, or any other TCA cycle intermediate, making it easily visible in a ^{13}C spectrum. Thus, the labeling pattern of glutamate (which is the same as α-ketoglutarate) provides all the necessary information about entry of ^{13}C labeled substrate into the TCA cycle. When the C4 of glutamate has been labeled with ^{13}C, the presence or absence of an adjacent ^{13}C atom (in C3 and/or C5) alters its chemical environment and gives rise to a unique signal (Fig. 2). The different combinations of label in C3, C4, and C5 combine to produce a spectrum of 9 different peaks, which can be analyzed to deduce the fractional contributions of the different label combinations (Fig. 3).

Figure 2. The isotopomer combinations resulting from the incorporation of [3-^{13}C]lactate and [1,2-^{13}C]acetate into glutamate. Solid circles indicate label which is detectable in the spectral region of the glutamate C4. Label incorporation results in the nine line C4 glutamate spectrum from the four different isotopomers.

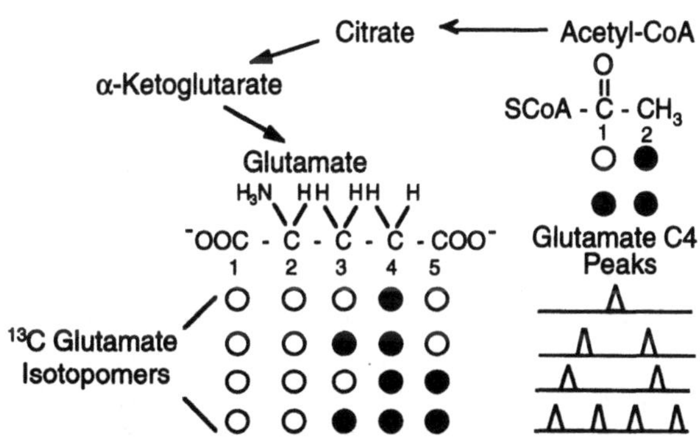

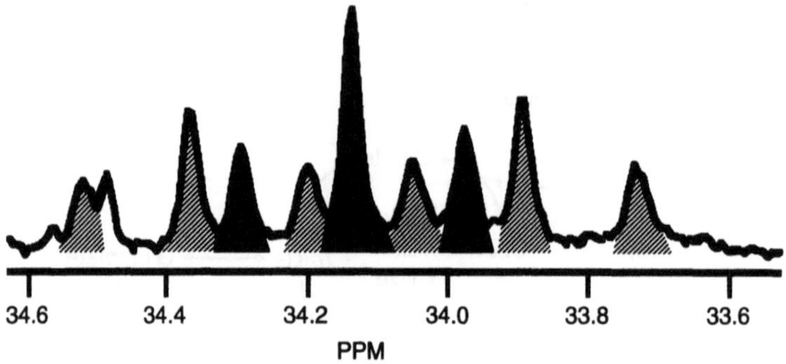

Figure 3. A representative spectrum from the C4 of glutamate. The abscissa, in units of parts per million (ppm) hertz, describes the chemical shift differences in resonance frequencies between the individual peaks. The peaks arising from incorporation of [3-^{13}C]lactate are shaded; those from [1,2-^{13}C]acetate are hatched.

By way of example, the spectrum in Fig. 3 is the 9 line resonance arising from the C4 of glutamate after incorporation of [1,2-^{13}C]acetate and [3-^{13}C]lactate. As a consequence of the label mixing, the 3 peaks arising from the incorporation of [3-^{13}C]lactate can be distinguished from the 6 peaks arising from incorporation of [1,2-^{13}C]acetate. Analysis of one such spectrum provides information about 1) the relative contribution to the TCA cycle of [3-^{13}C]lactate compared with [1,2-^{13}C]acetate and unlabeled sources; 2) the relative contribution of [3-^{13}C]lactate to anaplerosis; and 3) the relative rate of anaplerosis to TCA cycle flux. Comparison of the total area under the peaks arising from the C2, C3, and C4 of glutamate provides information about how close the system was to metabolic steady state (Malloy et al., 1987; Malloy et al., 1990).

Although this technique has been used primarily for the study of metabolic regulation in the heart and liver, the technique appears to have potential for the study of skeletal muscle fatigue. For example, the technique could be used to examine the pattern of substrate utilization as the muscle makes a transition between rest and exercise, particularly as the exercise leads to fatigue. It has been known for many years that skeletal muscle alters its relative utilization of carbohydrates *vs.* fats as it goes from rest to exercise, that this alteration is extreme during fatiguing exercise, and that the capacity to perform maximal exercise is attenuated when certain fuels are unavailable (Hedman, 1957; Gollnick et al., 1981; Brooks & Mercier, 1994). Therefore, one could use a ^{13}C MRS protocol to study the relationship between skeletal muscle fatigue and the flux through one or more contributing metabolic pathways.

It is well known that skeletal muscle uses fat, mostly in the form of fatty acids in the blood, as its primary fuel during rest. The energy it makes from fat comes from the process of β-oxidation in which a long chain fatty acid is broken down into 2-carbon units, and then completely oxidized to CO_2, H_2O, and chemical energy. During exercise, the fraction of energy derived from carbohydrate oxidation increases. Although there are several different carbohydrates that produce pyruvate, most pyruvate is produced from glycogenolysis. Oxidation of the pyruvate so derived occurs following decarboxylation and activation to acetyl-CoA where is then has the same fate as acetyl-CoA derived from fat.

With the preceding as the biochemical basis for an experiment, how could one use ^{13}C MRS to measure the differences in the pattern of substrate utilization between rest and exercise or between non-fatiguing and fatiguing exercise? This could be done, for example by studying the depletion of endogenous glycogen and the subsequent repletion with [1,6-^{13}C]glucose. Glycogenolysis of the labeled glucosyl units will produce [3-^{13}C]pyruvate

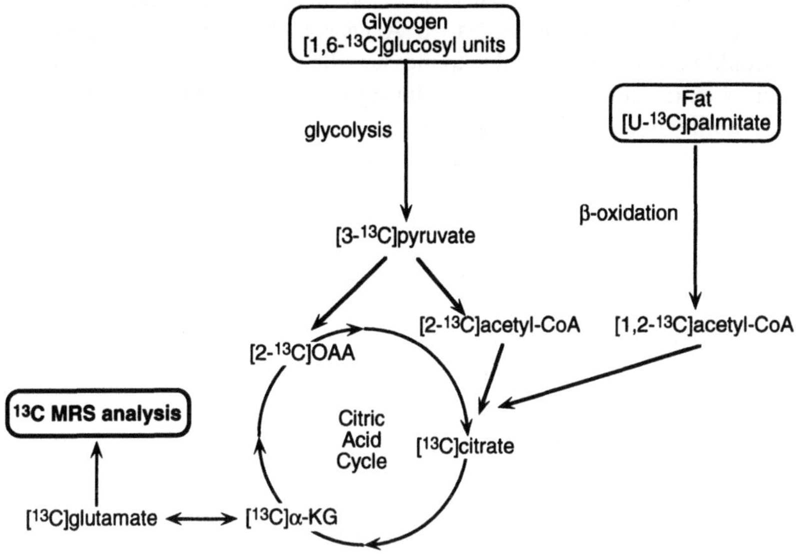

Figure 4. A specific substrate competition experiment where ^{13}C label is introduced as glycogen preloaded with [1,6-^{13}C]glucosyl units and as exogenous [U-^{13}C]palmitate. This results in label entering the citric acid cycle non-oxidatively as [2-^{13}C]oxaloacetate, oxidatively via pyruvate as [2-^{13}C]acetyl-CoA, or oxidatively via β-oxidation as [1,2-^{13}C]acetyl-CoA.

and, following oxidative decarboxylation by the pyruvate dehydrogenase complex, [2-^{13}C]acetyl-CoA. Incorporation of [2-^{13}C]acetyl-CoA into the citric acid cycle via oxidative pathways would label the C3 and C4 of glutamate. Similarly, two carbon degradation of the labeled palmitate and subsequent activation will produce [1,2-^{13}C]acetyl-CoA. Incorporation of [1,2-^{13}C]acetyl-CoA into the citric acid cycle via oxidative pathways would label the C3, C4, and C5 of glutamate. By examination of the labeling pattern in the glutamate C4 alone, one can compare the relative rates of incorporation of labeled glycogen, palmitate, and unlabeled sources. By examining the patterns and the relative amount of labeling in the C2, C3, and C4 of glutamate, one can also determine the relative contributions of oxidative *vs.* non-oxidative (anaplerotic) entry into the citric acid cycle if the system is in isotopic steady state. Thus, one can determine if there is a change in the pattern of substrate utilization during fatiguing exercise; or conversely, by altering, and monitoring, the relative availability of substrates, one can determine the effects on exercise capacity or the tendency for fatigue.

^{23}Na and ^{39}K MRS

The ability of any cell to function properly depends on the maintenance, of its intracellular ionic composition. For nerve and muscle cells, the maintenance of basal trans-membrane sodium (Na^+) and potassium (K^+) gradients is the basis for the initiation and propagation of action potentials. Each muscle cell contraction is initiated by an action potential, which results in a net influx of Na^+, and efflux of K^+ across the cell membrane. Although this ion flux is countered by the Na^+/K^+ ATPase, an electrogenic ATP-requiring transmembrane protein which actively transports Na^+ out of, and K^+ into, the cell against a bulk concentration gradient, extended periods of exercise result in net increases in intracellular Na^+ (Sréter, 1963; Blum et al., 1988; Lindinger & Heigenhauser, 1988) and extracellular K^+ (Sréter 1963; Ahlborg et al., 1967; Saltin et al., 1981; Clausen, 1990; Lindinger &

Heigenhauser, 1991). To the extent that a continuous maintenance of the normal distribution of $[Na^+]_{out}$ and $[K^+]_{in}$ depends on the Na^+/K^+ ATPase, and thus on the energetic state of the muscle, it would be valuable to be able to measure the relative amounts of these ions in the intracellular *vs* extracellular space *in vivo* during fatiguing exercise. Of particular interest would be the ability to relate the changes in the steady state distribution of Na^+ or K^+ in the presence of the measurable changes in cellular energetics as seen using ^{31}P MRS.

Let us first consider the measurement of sodium by MRS. The sodium in muscle tissue has several properties that are attractive to the MR scientist attempting to measure tissue sodium. First, ^{23}Na has a very high natural abundance, such that nearly 100% of all naturally occurring sodium exists as ^{23}Na isotope (Friebolin, 1993). Second, the sensitivity of the ^{23}Na nucleus is approximately 9.25×10^{-2} relative to ^1H for equal numbers of nuclei and at constant field (Friebolin, 1993). This makes it the second most NMR-sensitive nucleus found in biological tissues. Third, sodium is present in muscle tissue in high concentration. In typical muscle tissue, $[Na^+]_{out}$ is $\geq$ 100mM, and $[Na^+]_{in}$ is about 10 mM. Thus, the combination of high natural abundance, high relative sensitivity, and high concentration make the observation *in vivo* of biological sodium by magnetic resonance an attractive prospect.

Unfortunately, it is only in relatively rare cases that it is of value to measure the total amount of sodium in a tissue; often the values of interest are the individual values of intracellular and extracellular sodium. The normal concentration of sodium in the extracellular space is almost an order of magnitude greater than that in the intracellular space. In contrast, the intracellular volume is usually a much larger fraction of the total space than is the extracellular space. For example, skeletal muscle is about 95% intracellular space. Thus, if one performs a simple ^{23}Na MRS experiment on skeletal muscle, one will get a single spectral peak arising from a combination of the >100mM concentration of $[Na^+]_{out}$ in the 5% extracellular volume co-resonating with the $\leq$ 10mM concentration of $[Na^+]_{in}$ in the 95% intracellular volume. To separate the signals, more sophisticated techniques must be applied that take advantage of some of the properties of the $^{23}Na^+$ nucleus *in vivo*. There are two fundamental strategies used to cope with this problem of co-resonance: use of shift reagents to move one resonance away from the other or use of a quantum filter to attenuate one of the signals.

The general mechanism of action of a shift reagent is two fold 1) to cause a shift in the resonance frequency, due in largest part to the paramagnetic component of the reagent; and 2) to shorten the T_1 and/or T_2 of the nucleus by providing another pathway for relaxation of the perturbed nuclear spins. Depending on the nature of the shift reagent, either the paramagnetic or relaxation effect will predominate. Although there are cases where a shortening of relaxation times is of paramount importance, such as gadolinium-based shift reagents in MRI, the paramagnetic effect is most important in detecting the transmembrane distribution of the ^{23}Na nucleus *in vivo*.

Shift reagents bind reversibly to the extracellular Na^+ and cause their magnetic resonance frequency to shift. Shift reagents are usually large inorganic molecules that are combined with one or more strongly paramagnetic atoms that can interact with the ion of interest, in this case $^{23}Na^+$. The interaction of the $^{23}Na^+$ and the paramagnetic atom in the complex causes an alteration in the normal interaction between the $^{23}Na^+$ and the external magnetic field. If the paramagnetic atom complex is large and polar, it will not cross the muscle cell membrane. Thus, only the extracellular $^{23}Na^+$ moieties will interact with the shift reagent, only resonances arising from the extracellular $^{23}Na^+$ will be shifted, and one will see two different spectral peaks in a ^{23}Na MR spectrum.

Fig. 5 is an example of the use of the shift reagent Thulium 1,4,7,10-tetraazacyclo-dodecane-1,4,7,10-tetrakis(methylene phosphonate) (TmDOTP), a complex organic ligand containing the lanthanide thulium as the paramagnetic agent. In this experiment, TmDOTP

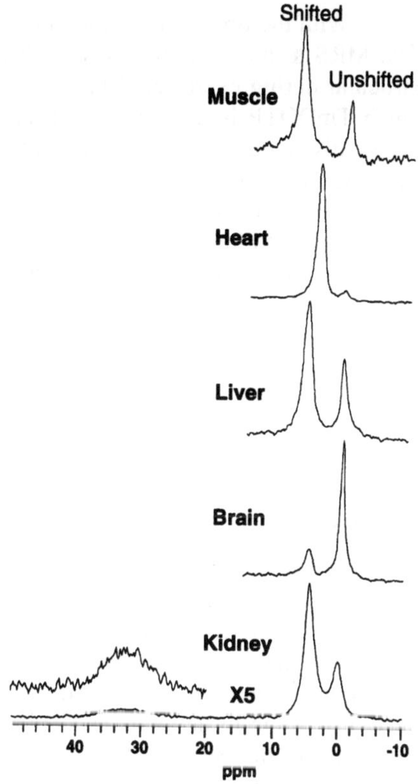

Figure 5. Examples of the effect of the shift reagent $TmDOTP^{-5}$ on ^{23}Na spectra collected from rat muscle, heart, liver, brain, and kidney. In each case, the effect of the interaction between the $TmDOTP^{-5}$ and the $^{23}Na^{+}$ ions is to move the resonance downfield by about 5 ppm (reproduced with permission from Bansal, 1994).

was infused into a rat via a jugular catheter. After allowing time for the reagent to attain a steady-state concentration in the body, ^{23}Na MR spectra were collected from several different tissues. Because the TmDOTP is too large and hydrophilic to cross a cell membrane, the shifted peak arises from the ion in the vascular space and the unshifted peak from the ion in the intracellular space.

The other method to differentiate the intracellular and extracellular Na^+ is to take advantage of the way in which sodium undergoes transitions between its nuclear spin states. Sodium is a quadripolar nucleus, and can undergo nuclear transitions between its four possible spin states of 3/2, 1/2, -1/2, and -3/2. These transitions can be made singly or in double or triple quantum steps. By altering the RF pulse sequence imposed on the tissue, it is possible to select only the double or triple step changes in spin state. In principle, such an experiment, designed to detect only double or triple quantum transitions, selects only for an individual pool of sodium atoms. Such a selection is useful only if an individual pool of sodium represents a single biological, physiological, or anatomical pool. For example, it is assumed that the intracellular pool of sodium is composed of Na^+ which is bound to large proteins or membranes or is contained in a viscous environment where multiple quantum transitions would predominate. Conversely, it is assumed that the vascular pool of sodium is composed of Na^+ which is unbound, where Na^+ is allowed to tumble relatively freely, and where single quantum transitions would predominate. Although the results of such experiments are rarely unambiguous, and require further validation, comparisons between the spectral distributions of sodium detected with a standard one-pulse experiment *vs.* those detected using a multiple quantum filter might provide information about the distribution of the ion across the membrane.

With the preceding as the biochemical basis for an experiment, how could one use ^{23}Na MRS to measure the transmembrane distribution of Na^+ in skeletal muscle, and in particular during exercise? By means of illustration, let us consider a specific example in which TmDOTP is infused to a steady state in the rat and the hindquarter muscles are electrically stimulated using several different patterns of varying or increasing intensity. The difference in the relative areas of the spectral peaks at these different stimulation rates would then represent the steady-state distribution of sodium and would reflect the ability of the Na^+/K^+ pump to keep up with the influx of Na^+ and efflux of K^+ mediated by the action potentials associated with the stimulation protocol. Similarly, the same variables could be measured using a multiple quantum filter experiment. Thus, one can determine if exercise changes the pattern of ion movement as a function of any number of variables associated with fatigue, including exercise intensity, exercise training, muscle fiber type, local blood flow, or any other such perturbation, and all these measurements can be made in the intact tissue.

IMAGING

One of the most revolutionary developments in the technology of modern medical science is magnetic resonance imaging (MRI). It provides more detail and tissue-type information than any other imaging modality available. In the text that follows, two aspects of MRI will be described. First, the use of MRI to discriminate active from inactive muscle and to use this information to make biophysical and physiological inferences about ion and water movements. Second, the possibility of using cine MRI, currently used primarily in the study of the beating myocardium, to study actively contracting skeletal muscle.

^1H MRI

The hydrogen nucleus (also referred to as the hydrogen ion, H^+, or proton) has the largest relative sensitivity of any MR detectable nucleus. Its combination of large concentration in living tissues, large gyromagnetic ratio, and large natural isotopic abundance, make it the nucleus of choice for making MR images. Because the general principles of magnetic resonance imaging can be learned from any number of excellent textbooks, it is not the intent to provide a treatise on the details of MRI. However, it is necessary to be reminded that the physical basis of the imaging process depends on the interaction between magnetic field strength, the intrinsic resonance properties of the proton, and the interaction(s) between the proton and its biophysical environment. Based on the potential for all these types of interactions, one can produce MRIs as remarkably informative as they are by making the proper adjustments to the collection techniques that take advantage of one or more of these physical properties so that the anatomical feature of interest is appropriately emphasized.

One of the biophysical properties involved in MRI that can be exploited is relaxation, and specifically the speed and mechanism of the relaxation of the nuclear spins that have been perturbed during the process of acquiring an MRI. By altering the details of the MRI acquisition process, one can create an image that selectively emphasizes protons undergoing relaxation at specific rates or via specific mechanisms and to use these images to differentiate different tissue types (such as fat *vs.* muscle or tissue *vs.* blood). Nuclear spins decay according to two different relaxation mechanisms 1) spin lattice (also called longitudinal) relaxation, described by the relaxation parameter T_1, and 2) spin-spin (also called transverse) relaxation, described by the relaxation parameter T_2. The difference in T_2 relaxation of water protons in different tissue compartments is the basis for using MRI to examine muscle during rest and exercise.

The use of such T_2 weighted images is based on the likelihood that the relaxation properties of the nuclear spins of water protons should depend on the interaction of the protons and their environment. Although this idea generates little argument, the exact nature of the mechanism at work is controversial: it has been argued that the increase in image signal intensity results from water movement into the interstitial space because the transverse relaxation (T_2) time of extracellular, and in particular interstitial, protons should be longer than that of intracellular protons (Archer et al., 1992). The converse argument has also been proposed: that the increase in image signal intensity results from water movement into the intracellular space because the transverse relaxation (T_2) time of intracellular protons should be longer than that of extracellular protons (Fisher et al., 1990). In either case, if there is a decrease in the concentration of water in one space (with its own characteristic T_2), and an accompanying increase in concentration of water in another space (with its own, and different, T_2), it will be reflected as a change in the signal intensity of the image pertaining to one space or another. And more specifically, if changes in image signal intensity can be associated with a known movement of water during exercise, the T_2 characteristics of the protons in the two different spaces can be used to explain the phenomenon in biophysical terms.

One promising application of this general technique is the use of MRI to selectively detect exercised *vs.* non-exercised skeletal muscle (Fleckenstein et al., 1988). Initially, this was used to identify the spatial distribution of skeletal muscles recruited during forearm exercise (Fleckenstein et al., 1989). This identification has been particularly valuable in that it has provided a means of performing routine ^{31}P MRS examinations of skeletal muscle with confidence that the ^{31}P MR spectra are being collected from actively exercising muscle (Fleckenstein et al., 1991; Bertocci et al., 1993). But more importantly, it has provided a means for determining the extent to which skeletal muscles are recruited during exercise as well as some of the biophysical and physiological consequences of this contractile activity.

Another application is the use of MRI to examine the movement of water between compartments in skeletal muscle during exercise. Cell membranes are very permeable to water molecules and water moves across muscle cell membranes along an ionic concentration gradient so that osmotic balance is maintained. If there is a change in the distribution of osmotically active moieties in one of the tissue compartments, water will move in parallel so that osmotic balance is maintained. For example, if exercise causes glycogenolysis, and glycolytic flux increases, the intracellular concentration of hexose and triose phosphates will increase. Water will move into the cell to maintain osmotic balance, the fraction of intracellular tissue water will increase, and the MRI signal intensity associated with intracellular water will also increase. Conversely, when repeated action potentials cause a net loss of intracellular osmolality due to extrusion of K^+ greater than can be handled by the Na^+/K^+ ATPase, or when heavy exercise causes a net extrusion of lactate ions, the extracellular concentration of ions will increase. Water will move out of the cell to maintain osmotic balance, the fraction of extracellular tissue water will increase, and the MRI signal intensity associated with extracellular water will also increase. Thus, the nature of the change in the MRI signal intensity as a result of skeletal muscle exercise can be used to make biophysical and physiological conclusions about the movement of ions into and out of skeletal muscle cells.

This type of approach has been used extensively in recent years by several different research groups, and a proportionality between exercise intensity and an increase in MRI intensity has been demonstrated. However, some questions remain concerning the meaning of this proportionality. In general, the question of the exact definition of exercise intensity, and the real biophysical and physiological meaning of the increase in image intensity remain to be resolved. The exercise protocols used in these studies varies from group to group, but they all follow the same general pattern of graded rhythmic exercise. The results have been similarly general: as the intensity (determined by various means) of the applied force is increased, so does the image intensity. The increase in image intensity has been assumed to be due to an increase in the concentration of interstitial (Archer et al., 1992) or intracellular (Fisher et al., 1990) water.

Following the initial demonstration of this technique to discriminate regions of active *vs.* inactive muscle (Fleckenstein, 1988), it was demonstrated that the increase in signal intensity was proportional to exercise intensity (Fisher et al., 1990). In this study, 1.5 min bouts of dorsiflexion exercise were used in combination with volume changes caused by venous occlusion. Most of the increase in image signal intensity was associated with increasing force generation at a given rate of limb flexion. There was a similar result when the force output was held constant but the limb flexion rate was increased from 1 to 2 to 3 Hz (Jenner et al., 1994). Thus, the increase in signal intensity was proportional to power output in the range observed. When the exercise was designed so that the power output was relatively constant, but the duration was varied, it was observed that the increase in signal intensity reached a plateau (Fleckenstein et al., 1993).

Several hypotheses have been proposed to explain these results. One explanation is the contributions from bulk water movement. This has been studied using 1) partial vascular occlusion (external blood pressure cuff at diastolic pressure) to increase total limb water (Fisher et al., 1990); 2) total vascular occlusion (external blood pressure cuff at suprasystolic pressure) to allow water movement between compartments but to prevent any change in total limb water (Archer et al., 1992); and 3) hyperemic reflow (Archer et al., 1992). These results are generally consistent with a small, but detectable, contribution of bulk water to the increase in image signal intensity.

Another explanation is the movement of water out of the vascular space and into the interstitial or intracellular space as a result of changes in osmotic gradients. This has been studied using 1) exercise routines which result in differing amounts of glycolytic flux, resulting in differing amounts of accumulation of hexose and triose phosphates (Fleckenstein et al., 1993, Jenner, 1994); and 2) exercise in patients with metabolic disorders which preclude the accumulation of lactate (Fleckenstein et al., 1991). These results are consistent with osmotic drive being the primary contribution to exercise-induced increases in signal intensity.

Where does this lead in the future? Apart from the obvious use of ^{1}H MRI to examine the biophysical and physiological basis of water movement in muscle tissue, one can imagine several other possible uses for this technique. One is the use of MRI to determine skeletal muscle fiber type non-invasively. Although skeletal muscle fiber type has traditionally been determined histologically from a tissue biopsy, the ability to use MR to estimate this non-invasively is very attractive. Such efforts have already been employed using MRS (Kushmerick et al., 1992, Blei et al., 1993, Mizuno et al., 1994). It would be good to be able to take advantage of the superior spatial and temporal resolution of the proton by using ^{1}H MRI to make such measurements. Additionally, it may be possible to use MRI to study motor recruitment patterns and the regulation of motor control. MRI has already been used to demonstrate the difference between voluntary exercise and cutaneous evoked contractions (Adams et al., 1993). Such studies can supplement other measures of muscle fiber recruitment, such as EMG. Given the superior MR sensitivity of the hydrogen nucleus, it is likely that other uses for this technology will be found.

Cine

Cine MRI is a relatively recent technical development which owes its existence in part to the continuing efforts of MR physicists, and in part to the development of computers with sufficient computational ability to maintain continuous updates of the image information in near-real time. So far, this has been applied to dynamic internal events such as the beating heart. Relevant to the study of neuromuscular fatigue, the ability to use MRI to monitor anatomical events in real time may someday provide the ability to study fatiguing muscle in situ. Although no such work has been done in skeletal muscle, cine MRI has provided an important new method for viewing the anatomy and mechanics of the beating myocardium.

As examples of its power, cine MRI has been used to 1) make a precise geometrical analysis of the ventricular wall (Sacks et al., 1993); 2) examine left ventricular dysfunction (Dell 'Italia et al., 1993); and 3) perform wall motion analysis (Matheihssen et al., 1993). These examples indicate that cine MRI can act as a live real-time imaging modality that can be used to accurately assess the mechanics and the anatomy of moving tissue. To the extent that skeletal muscle is a moving tissue, the same principles could be applied to the study of muscle. It is relatively easy to imagine skeletal muscle movements being observed in real time and correlated to their related neurological or biochemical regulation, particularly as muscle undergoes fatigue. All that remains is the actual design and conduct of the experiments.

SUMMARY

In this brief text, two types of MRS and two types of MRI have been discussed in the context of their possible future applications to the study of neuromuscular fatigue. Although the metabolic aspects of ^{13}C and ^{23}Na MRS make them more likely to produce important information in the near future, it is the high relative sensitivity of the proton which makes the possible applications of ^{1}H MRI more likely to produce important information in the more distant future. In particular, when the biophysics of water movement and relaxation patterns is resolved, T_1 and T_2 weighted imaging experiments may result in fundamental advances in our understanding of the dynamics of water and ion movements. Also, use of cine MR to obtain real-time views of actively contracting muscle may provide similarly fundamental advances in our understanding of the processes which regulate the contractile process. The future is as bright as our imaginations.

ACKNOWLEDGMENTS

The author's work and laboratory is supported by the National Aeronautics and Space Administration (NASA; grant NAGW 3582), United States Public Health Service grants RR 02584, HL 06296 and HL 07360, the Department of Radiology, University of Texas (UT) Southwestern Medical Center, and the Institute for Exercise and Environmental Medicine, Presbyterian Hospital of Dallas. Attendance of the author at the 1994 Bigland-Ritchie conference was supported, in part, by the Department of Radiology, UT Southwestern Medical Center, the Institute for Exercise and Environmental Medicine, Presbyterian Hospital of Dallas, and the University of Miami.

REFERENCES

Adams GR, Harris RT, Woodward D & Dudley GA (1993). Mapping of electrical muscle stimulation using MRI. *Journal of Applied Physiology* **74**, 532-537.

Ahlborg B, Bergström J, Ekelund LG & Hultman E (1967). Muscle glycogen and muscle electrolytes during prolonged physical exercise. *Acta Physiologica Scandinavica* **70**, 129-142.

Archer BT, Fleckenstein JL, Bertocci LA, Haller RG, Barker B, Parkey RW & Peshock RM (1992). Effect of perfusion on exercised muscle: MR imaging evaluation. *Journal of Magnetic Resonance Imaging* **2**, 407-413.

Bansal N (1994). In vivo ^{23}Na MRS with a shift reagent. In: Weatherall PT, Deslauriers R (eds.), *MR Pulse, Newsletter of the Society of Magnetic Resonance vol. 1*, pp. 18-19. Berkeley, CA: Society of Magnetic Resonance.

Bertocci LA, Lewis SF & Haller RG (1993). Lactate infusion in human muscle PFK deficiency. *Journal of Applied Physiology* **74**, 1342-1347.

Blei ML, Conley KE & Kushmerick MJ (1993). Separate measures of ATP utilization and recovery in human skeletal muscle. *Journal of Physiology (London)* **465**, 203-222.

Bloch F, Hansen W & Packard M (1946). Nuclear induction. *Physics Reviews* **69**, 127.

Blum H, Schnall MD, Chance B & Buzby GP (1988). Intracellular sodium flux and high-energy phosphorus metabolites in ischemic skeletal muscle. *American Journal of Physiology* **255**, C377-C384.

Brooks GA & Mercier J (1994). Balance of carbohydrate and lipid utilization during exercise: the "crossover" concept. *Journal of Applied Physiology* **76**, 2253-2261.

Chance EM, Seeholzer SH, Kobayashi K & Williamson JR (1983). Mathematical analysis of isotope labeling in the citric acid cycle with applications to ^{13}C NMR studies in perfused rat hearts. *Journal of Biological Chemistry* **258**, 13785-13794.

Clausen T (1990). Significance of Na^+-K^+ pump regulation in skeletal muscle. *News in Physiological Sciences* **5**, 148-151.

Cohen SM (1983). Simultaneous ^{13}C and ^{31}P NMR studies of perfused rat liver. *Journal of Biological Chemistry* **258**, 14294-14308.

Dell 'Italia LJ, Blackwell GG, Urthaler F & Pearce DJ (1993). A stable model of left ventricular dysfunction in an intact animal assessed with high fidelity pressure and cinemagnetic resonance imaging. *Cardiovascular Research* **27**, 974-979.

Fisher MJ, Meyer RA, Adams GR, Foley JM & Potchen EJ (1990). Direct relationship between proton T_2 and exercise intensity in skeletal muscle MR images. *Investigative Radiology* **25**, 480-485.

Fleckenstein JL, Bertocci LA, Nunnally RL & Peshock RM (1989). Exercise-enhanced MR imaging of variations in forearm muscle anatomy and use: importance in MR spectroscopy. *American Journal of Roentgenology* **153**, 693-698.

Fleckenstein JL, Canby RC, Parkey RW & Peshock RM (1988). Acute effects of exercise on MR imaging of skeletal muscle in normal volunteers. *American Journal of Radiology* **151**, 231-237.

Fleckenstein JL, Haller RG, Lewis SF, Archer BT, Barker BR, Payne J, Parkey RW & Peshock RM (1991). Absence of exercise-induced MRI enhancement of skeletal muscle in McArdle's disease. *Journal of Applied Physiology* **71**, 961-969.

Fleckenstein JL, Watamull D, McIntire DD, Bertocci LA, Chason DP & Peshock RM (1993). Muscle proton T_2 relaxation times and work during repetitive maximal voluntary exercise. *Journal of Applied Physiology* **74**, 2855-2859.

Fleckenstein JL, Weatherall PT, Bertocci LA, Ezaki M, Haller RG, Greenlee R, Bryan WW & Peshock RM (1991). Locomotor system assessment by muscle magnetic resonance imaging. *Magnetic Resonance Quarterly* **7**, 79-103.

Friebolin H (1993). *Basic One- and Two-Dimensional NMR Spectroscopy.* New York, NY: VCH Publishers.

Fushiki T, Kano T, Inoue K & Sugimoto E (1991). Decrease in muscle glucose transporter number in chronic physical inactivity in rats. *American Journal of Physiology* **260**, E403-E410.

Gollnick PD, Pernow B, Essén B, Jansson E & Saltin B (1981). Availability of glycogen and plasma FFA for substrate utilization in leg muscle of man during exercise. *Clinical Physiology* **1**, 27-42.

Hedman R (1957). The available glycogen in man and the connection between rate of oxygen intake and carbohydrate usage. *Acta Physiologica Scandinavica* **40**, 305-321.

Jeffrey FMH, Rajagopal A, Malloy CR & Sherry AD (1991). ^{13}C-NMR: a simple yet comprehensive method for analysis of intermediary metabolism. *Trends in Biological Science* **16**, 5-10.

Jenner G, Foley JM, Cooper TG, Potchen EJ & Meyer RA (1994). Changes in magnetic resonance images of muscle depend on exercise intensity and duration, not work. *Journal of Applied Physiology* **76**, 2119-2124.

Katz J (1985). Determination of gluconeogenesis in vivo with ^{14}C labeled substrates. *American Journal of Physiology* **248**, R391-R399.

Katz J & Lee PWN (1991). Application of mass isotopomer analysis for determination of pathways of glycogen synthesis. *American Journal of Physiology* **261**, E332-E336.

Kushmerick MJ, Moerland TS & Wiseman RW (1992). Mammalian skeletal muscle fibers distinguished by contents of phosphocreatine, ATP, and Pi. *Proceedings of the National Academy of Sciences* **89**, 7521-7525.

Lee PWN, Soruo S & Bergner E (1991). Glucose isotope, carbon recycling, and gluconeogenesis using [U-^{13}C]glucose and mass isotopomer analysis. *Biochemical Medicine and Metabolic Biology* **45**, 298-309.

Lindinger MI & Heigenhauser GJF (1988). Ion fluxes during tetanic stimulation in isolated perfused rat hindlimb. *American Journal of Physiology* **254**, R117-R126.

Lindinger MI & Heigenhauser GJF (1991). The role of ion fluxes in skeletal muscle fatigue. *Canadian Journal of Physiology and Pharmacology* **69**, 246-253.

Malloy CR, Sherry AD & Jeffrey FMH (1987). Carbon flux through citric acid cycle pathways in perfused heart by ^{13}C NMR spectroscopy. *FEBS Letters* **212**, 58-62.

Malloy CR, Sherry AD & Jeffrey FMH (1990). Analysis of tricarboxylic acid cycle of the heart using ^{13}C isotope isomers. *American Journal of Physiology* **259**, H987-H995.

Matheihssen NA, de Roos A, Doornbos J, Reiber JH, Waldman GJ & van der Wall EE (1993). Left ventricular wall motion analysis in patients with acute myocardial infarction using magnetic resonance imaging. *Magnetic Resonance Imaging* **11**, 485-492.

Mizuno M, Secher NH & Quistorff B (1994). ^{31}P-NMR spectroscopy, rsEMG, and histochemical fiber types of human wrist flexor muscles. *Journal of Applied Physiology* **76**, 531-538.

Purcell E, Torrey H & Powel R (1946). Resonance absorption by nuclear magnetic moments in solids. *Physics Reviews* **69**, 37-38.

Sacks MS, Chuong CJ, Templeton GH & Peshock RM (1993). In vivo 3-D reconstruction and geometric characterization of the right ventricular free wall. *Annals of Biomedical Engineering* **21**, 263-275.

Saltin B, Sjøgaard G, Gaffney FA & Rowell LB (1981). Potassium, lactate, and water fluxes in human quadriceps muscle during static contractions. *Circulation Research (Supplement I)* **48**, 18-24.

Sréter FA (1963). Cell water, sodium, and potassium in stimulated red and white mammalian muscles. *American Journal of Physiology* **205**, 1295-1298.

Walker TE, Han CH, Kollman VH, London RE & Matwiyoff NA (1982). ^{13}C nuclear magnetic resonance studies of the biosynthesis by microbacterium ammoniaphilum of L-glutamate selectively enriched with carbon-13. *Journal of Biological Chemistry* **257**, 1189-1195.

SECTION V

The Case for Segmental Motor Mechanisms

Neural mechanisms at the spinal level are explored in this section. The dominant question is to determine if the control system for motoneuronal discharge is intrinsically *wise*. If so, what are the segmental constraints on its actions, and what peripheral inputs contribute to its operation?

First, in **Chapter 16,** Binder-Macleod addresses the potential role in fatigue of muscle wisdom and the catch-like property. For muscle wisdom, evidence is presented that during voluntary, fatiguing contractions subtle slowing in the discharge frequency of motoneurons optimizes force and thus minimizes fatigue. An important distinction is made between the force optimization observed during voluntary *vs.* imposedcontractions (by electrical stimulation). However, because some patterns of motor unit discharge in exercise do not lead to the expected slowing of muscle relaxation times, alternative strategies may also be considered *wise*. The catch-like property is an example of how a subtle change in the pattern of imposed electrical stimulation can produce more force in both control and fatiguing contractions. Establishment of the boundary conditions for such stimulation patterns and the extent to which catch-like effects are observable in voluntary contractions are issues of current interest. However, imposed stimulation is an end in itself when designing patterns of functional electrical stimulation to reduce the fatigue of patients with motor impairments.

Windhorst and Boorman (**Chapter 17**) review the segmental machinery for reflexly changing motor unit behavior. They point out that little is known about the strategies used by the CNS to mitigate against the effects of fatigue, although there is evidence that fatigue may disturb the normal patterns of motor unit recruitment. Potentiation of motor unit force and feedforward and feedback signals acting on premotoneuronal networks can optimize force. As well as the descending inputs to motoneurons, there are many interneurons carrying information from low- and high-threshold mechanoreceptors and chemoreceptors in the muscles. Similarly, inputs from Golgi tendon organs and recurrent inhibition seem likely to be important in the moment-to-moment regulation of motoneuron discharge rate, although their individual effects are not simple to differentiate *in vivo*. Furthermore, the effects of presynaptic inhibition are strong and likely to be distributed differently to modulate the reflex effectiveness of inputs from spindle group Ia and II afferents and tendon organ (group Ib) afferents. Ultimately, it will help to have more estimates of the net behavior and gain of the various reflex paths during muscle fatigue and information on the interactions between segmental mechanisms and pattern generators for locomotion.

The fusimotor system activates muscle spindle afferents during voluntary contractions under relatively isometric conditions. Hence, it could modulate the homonymous excitation provided to motoneurons by muscle spindle afferents. Removal of the reflex facilitation (by partial or complete nerve blocks) reduces the maximal firing rates of human motoneurons. In **Chapter 18**, Hagbarth and Macefield explore the evidence that this disfacilitation of motoneurons abolishes the progressive decline in motoneuronal firing rate during maximal voluntary isometric contractions. A surprising conclusion from any analysis of the role of the fusimotor system in human muscle fatigue is that it fails to act as a sufficiently powerful servo to offset the effects of fatigue in the extrafusal muscle fibers. Such a role had previously been suggested as one of the prime functions of this system. In the initial seconds of a strong voluntary contraction, the declining muscle spindle input and increasing presynaptic inhibition will contribute to the early decline in firing rate of motoneurons.

In **Chapter 19** (Garland & Kaufman), the properties of small-diameter afferents innervating muscle are reviewed and various lines of evidence are presented for a reflex inhibition of motoneurons during fatiguing isometric exercise. There is no consensus on the extent to which a reduction in discharge rate during maximal voluntary contractions is due to reflex inhibition, or the exact spinal and supraspinal sites involved in the inhibition. Indirect evidence favors a role for recurrent inhibition. It would be helpful to know how motoneurons discharge would be affected if descending supraspinal drives were stimulated maximally; or, alternatively, how motoneurons would behave in the absence of feedback from of small-diameter afferents. Unfortunately, the exact effects of muscle fatigue on the discharge of unmyelinated (group IV) muscle afferents are not yet available, but it is likely that those sensitive to metabolic factors are activated later in a strong sustained voluntary contraction.

Importantly, this section reveals areas in which data from human and animal experiments are either discrepant or inconclusive. Further work could well focus on these discrepancies along with the changes in afferent properties with fatigue, the central (particularly segmental) effects of these changes, and the extent to which supraspinal and other descending drives can modify the segmental behavior.

VARIABLE-FREQUENCY STIMULATION PATTERNS FOR THE OPTIMIZATION OF FORCE DURING MUSCLE FATIGUE

Muscle Wisdom and the Catch-Like Property

S. A. Binder-Macleod

Department of Physical Therapy
University of Delaware
Newark, Delaware 19716

ABSTRACT

Muscle wisdom is the process whereby the activation rates of motor units are modulated by the central nervous system to optimize the force during sustained voluntary contractions. During maximal voluntary contractions the activation rates decline as the muscle fatigues. No similar decline has been observed during submaximal contractions. Subsequent chapters explore the potential mechanisms for muscle wisdom. In this chapter, a historical background on the development of ideas on muscle wisdom is first presented. Next, artificial wisdom, the procedure used to optimize force during an electrically imposed tetanus by progressively reducing the stimulation frequency as the muscle fatigues, is discussed. Finally, recent studies are described in which fatigue was delayed and reduced by the use of variable-frequency stimulus trains that elicit the catch-like property of muscle.

INTRODUCTION

It has long been known that the activation frequency of skeletal muscle affects the generation, maintenance, and decline (fatigue) of force. In the control (unfatigued) state, there is a sigmoidal relationship between the stimulation frequency and the peak force produced by both whole muscles (Cooper & Eccles, 1930) and single motor units (Kernell et al., 1983). Low frequencies produce unfused twitches; intermediate frequencies produce a steep, near-linear relationship between force and frequency, and relatively high frequencies are needed to produce maximal forces (see also Kernell, Chapter 9). In contrast to high-frequency stimulation, low-frequency stimulation produces relatively less fatigue during sustained or intermittent contractions (Bigland-Ritchie et al., 1979; Jones et al., 1979; Binder-Macleod et al., 1995). Interestingly, the discharge rate of a motor unit during voluntary contractions exhibits short-term (i.e., pulse to pulse) and long-term variations (i.e.,

Fatigue, Edited by Simon C. Gandevia et al.
Plenum Press, New York, 1995

changes over seconds to minutes) (Marsden et al., 1971; Bigland-Ritchie et al., 1983a; Maton & Gamet, 1989). Variations in the motor unit activation rate have been shown to maximize force (Burke et al., 1976; Binder-Macleod & Clamann, 1989; Bevan et al., 1992) and minimize fatigue (Marsden et al., 1976, Jones et al., 1979; Binder-Macleod & Barker, 1983; 1991) during imposed electrical activation of muscle. The remainder of this chapter explores the ways in which the modulation in activation rate of motor units can be used to maximize force during both voluntary and imposed contractions.

HISTORICAL BACKGROUND ON MUSCLE WISDOM

At the 1968 meeting of the Physiological Society, Marsden, Merton and Meadows coined the term *muscle wisdom*. This term describes the process whereby the activation rates of the motor units are modulated by the central nervous system (CNS) to optimize the force during sustained contractions (Marsden et al., 1976, 1983; for review, see: Enoka & Stuart, 1992; Stuart & Callister, 1993). The experimental evidence for their conclusions can be directly linked to a series of experiments conducted by these investigators, and their contemporaries, studying both imposed electrical activation of the human skeletal muscle and the activation pattern of motor units during fatiguing voluntary contractions. In 1954 Merton showed that, for the human adductor pollicis muscle, the force decline during a sustained maximal voluntary contraction was due to failure within the muscle (cf., Bigland-Ritchie et al., 1978). Merton (1954) also showed for the first time that voluntary contractions were able to produce comparable forces to those produced by maximal tetani. These were important findings because they suggested that decreased central drive was not the cause of fatigue, but rather, that the CNS was able to maximally activate muscle. The following year Næss and Storm-Mathisen (1955), also studying the human adductor pollicis muscle, showed that a stimulation train with a pulse frequency that elicited a near-maximal tetanic contraction (50 pps) produced a greater rate of decline in force than a maximal voluntary contraction. Næss and Storm-Mathisen (1955) also showed that higher rates of stimulation (200 pps), while producing slightly greater initial forces, were less effective at maintaining force. In 1971, Marsden and colleagues showed that the compound action potential discharge rate (i.e., activation rate) of motor units during prolonged, maximal, voluntary contractions declined over time, beginning at ~60-100 pps initially and slowing to ~20 pps by 30 s.

In 1976 Marsden, Merton and Meadows reported that an *artificial wisdom* procedure, where the frequency used to elicit a tetanus is progressively reduced as the muscle fatigues, produced a force comparable to that obtained during sustained, voluntary, maximal contractions. They also noted that the fatigue experienced under ischemic conditions was related to the number of motor impulses, rather than the duration or force of contraction. Marsden and colleagues (1976) concluded that, based on their observations and those of Næss and Storm-Mathisen, to obtain the "most prolonged and powerful voluntary contractions the activation rate of the motor units should slow down consistent with maintaining a fully fused tetanus as the contraction proceeds." Dietz (1978), also an early proponent of the notion that the decline in motor unit activation rate during a maximal voluntary contraction actually helped to optimize force output (i.e., muscle wisdom) suggested that the decline of discharge frequency "took an optimal course because higher as well as lower discharge rates would reduce the force development."

Finally, in an often cited 1983 publication, Marsden and colleagues developed more fully their theory of muscle wisdom. They argued that the force developed during a sustained contraction depends on "a compromise between activation of tension generation and a concomitant aggravation of activation failure." During the onset of a contraction high activation rates are needed to produce maximal force; however, maintenance of these high rates results in rapid fatigue. In addition, they noted that during a prolonged contraction the "time scale of the contractile mechanism slows down, so that the rate of innervation needed to activate it optimally falls progressively as the contraction proceeds." The authors argued that activation failure produces the need for a decreased activation rate while contractile slowing allows the activation rate to decline while still eliciting *the best out of the muscle*.

While most of the results reported by Marsden and colleagues have been supported by more recent studies, caution must be advised in generalizing their findings (cf., Stuart & Callister, 1993). It should be noted that most of their results were based on the study of just one to three, highly trained subjects (i.e., the authors). Also, many of their observations were made on motor units in adductor pollicis that received an aberrant innervation from the median nerve (Marsden et al., 1971). Furthermore, to allow the recording of the EMG of single motor units during maximal efforts, the authors used chemical or ischemic blocking of the ulnar nerve at the elbow to reduce the number of active motor units. Although these conditions may have biased their results, the observations and conclusions drawn by Marsden and colleagues have served as an important impetus for the study of muscle fatigue.

REDUCTION IN THE ACTIVATION RATE OF MOTOR UNITS DURING VOLUNTARY CONTRACTIONS

Since the original work by Marsden and colleagues (1971), several investigators have found similar declines in the activation rate of motor units during maximal voluntary contractions. Bigland-Ritchie and colleagues (1983a) have reported that during a ~60 s isometric MVC of the human adductor pollicis brevis, the mean activation rate of motor units fell by 50%, from ~27 to ~15 Hz. During this time, there was no decline in the muscle's compound action potential (M-wave) elicited by a single-shock stimulus to the muscle nerve. This suggested that the decline in motor unit activation rate was attributable to a corresponding decline in motoneuron discharge rate rather than a progressive block of neuromuscular conduction (also see Thomas et al., 1989; cf. Bellemare & Garzaniti, 1988). Grimby and colleagues (1981), who reported comparable declines in human short big toe extensor and anterior tibialis muscles, have suggested that this progressive decline in motor unit activation rate may protect peripheral neuromuscular propagation and hence elicit more force over time than if continuous high frequency activation was maintained. Furthermore, Bigland-Ritchie and colleagues (1981; 1983a; 1983b), and Woods and colleagues (1987) have argued that the decline in motor unit activation rate may not only serve to reduce fatigue but may also allow more effective modulation of the force of voluntary contractions by rate coding during fatigue; that is, as the contractile (including force-relaxation) speed declines, the activation frequency required for full tetanic force declines. If there was not a parallel decline in the motoneuronal firing rate, the activation rate of motor units would become supra-tetanic, and rate coding as a means of force modulation would be ineffective (Bigland-Ritchie, 1981; her Fig. 6). Interestingly, during submaximal, fatiguing contractions the motor units activation rate has been shown to increase (Bigland-Ritchie et al., 1986; Maton & Gamet, 1989; Dorfman et al., 1990) or remain stable (Maton & Gamet, 1989).

A CAVEAT ON MUSCLE WISDOM: EFFECTS OF FATIGUE ON THE NORMALIZED FORCE-STIMULUS FREQUENCY RELATION

Muscle wisdom proposes that force is optimized during fatiguing contractions. If correct, such optimization should be reflected in the profile of the normalized relation between force and activation rate (*T-f* relation). As reviewed in Sawczuk et al., Chapter 8 and Kernell, Chapter 9, the *T-f* relation is subject to alteration by fatigue. For example, the *T-f* relation is influenced by fatigue-induced contractile slowing, effects brought on by low- *vs.* high-frequency stimulation, and motor unit type.

Contractile Slowing

Such slowing with fatigue is well documented (Fitts, 1994). It appears to be primarily the result of an increase in force-relaxation time of the muscle (Bigland-Ritchie et al., 1992). The longer contraction time allows a greater summation of forces at subfusion stimulus frequencies and produces a shift in the *T-f* relation to the left (for further explanation see Sawczuk et al., Chapter 8). That is, the steep portion of the *T-f* relation is shifted toward lower rates. However, the assumption that fatigue must result in a concurrent shift in the *T-f* relation toward lower frequencies has been challenged (for review see Enoka & Stuart, 1992; Fitts, 1994); the confounding variables include the relative presence of low-frequency fatigue and the motor-unit type distribution of the muscle.

Low-Frequency Fatigue

This term refers to a long-lasting (min to hr) selective force reduction in response to low-frequency stimulation (e.g., ≤ 20 Hz) (Edwards et al., 1977; 1981). It is seen following the fatigue brought on by voluntary and imposed contractions and is thought to be due to an impairment in excitation-contraction coupling (Edwards et al., 1977; Jones, 1981). Low-frequency fatigue results in an attenuation of the twitch force, a shift in the steep portion of *T-f* relation toward higher frequencies, and little change in the frequencies needed to produce a fused contraction or a maximal tetanus.

Most studies on human muscle have observed significant low-frequency fatigue and a shift in the *T-f* relation to higher stimulus rates (Edwards et al., 1977; Cooper et al., 1988; Jones et al., 1989; Stokes et al., 1989; Binder-Macleod & McDermond, 1992; cf., Thomas et al., 1991). In contrast, *in vitro* work on the frog semitendinosus (Thompson et al., 1992) and rat diaphragm muscles (Metzger & Fitts, 1987) have shown much greater attenuation of forces at the higher rates of stimulation (i.e., > 60 pps) than at the lower rates of stimulation and a shift in the *T-f* relation to the left. These contrasting results can probably be explained by differences in the experimental protocols used to produce the fatigue (e.g., the stimulation frequency used to fatigue the muscle), the conditions of the preparation during the testing (e.g., muscle temperature and blood supply), and the motor-unit- and fiber-type distribution of the muscles.

Motor-Unit Type

Studies on fast-twitch single motor units in the cat (Powers & Binder, 1991) have shown short-term potentiation in response to imposed low-frequency (20 Hz) stimulation of type FR units immediately after fatiguing stimulation. This effect was not present in FI and FF motor units. In addition, a longer-lasting depression in maximal tension and a delayed onset (after ≥ 30 min) depression of low-frequency responses (i.e., low-frequency fatigue) was seen in all types of fast motor units. These changes resulted in a shift to the right in their *T-f* relation, beginning ≥ 30 min after the fatiguing protocol. In contrast, FR units first showed a shift to the left in the *T-f* relation immediately after fatiguing protocol, and ≥ 30 min later, a shift to the right after fatiguing stimulation. Furthermore both Dubose and colleagues (1987) and Gordon and colleagues (1990) have shown that type S motor units in cat hindlimb muscles are far less susceptible than FF units to a fatigue-induced slowing in their force-relaxation times. These results, together with those of Powers and Binder (1991), infer that muscle wisdom effects are more likely to be a feature of high- rather than low-force contractions. This possibility is consistent with the finding that during fatiguing, submaximal voluntary contractions, activation rates of human motor units have been shown to increase (Bigland-Ritchie et al., 1986; Maton & Gamet, 1989; Dorfman et al., 1990) or remain stable (Maton & Gamet, 1989). This increase in discharge rate, along with additional recruitment (Fallentin et al., 1993; Garland et al., 1994) is needed to maintain the targeted force (generally 20%-30% of the maximal voluntary contraction) as the force generating ability of the fatiguing muscle declines (Bigland-Ritchie et al., 1986). On a functional level, any shifts in the *T-f* relation that occurred with fatigue were insufficient to alleviate the need for this increase in motor unit activation rate.

BOUNDARY CONDITIONS FOR THE USE OF MUSCLE WISDOM

Though a number of studies have recently explored the mechanisms underlying muscle wisdom (for review see Windhorst & Boorman, Chapter 17; Hagbarth & Macefield, Chapter 18; and Garland & Kaufman, Chapter 19), as recently noted by Stuart and Callister (1993), the boundary conditions of muscle wisdom remain relatively unexplored. To date, all reports that have shown a decline in the motor unit activation rates have used sustained, maximal, isometric contractions (Marsden et al., 1971; Grimby et al., 1981; Bigland-Ritchie et al., 1983a). Generally, no such decline in activation rate has been seen with submaximal contractions (Bigland-Ritchie et al., 1986; Maton & Gamet, 1989, Dorfman et al., 1990; Garland et al., 1994). The relative amount of slowing in activation rate for the different types of motor units has remained relatively unexplored (cf, Powers & Binder, 1991). Perhaps the lack of decline in activation rate during submaximal contractions may be related to differences in the types of motor units recruited during maximal versus submaximal contractions. As noted by Enoka and Stuart (1992), the muscle wisdom theme needs to be extended to dynamic conditions to determine if the results observed during isometric conditions hold true for a wide variety of natural movements. Similarly, the study of the changes in activation rate during fatiguing intermittent and rhythmic contractions are also worthy of further study (see Enoka & Stuart, 1992). Despite the need for further experiments to explore the boundary conditions for muscle wisdom, the case is compelling for force optimization by variable-frequency stimulation during evoked fatiguing contractions, particularly when the effects of such stimulation bring on the catch-like property of muscle.

REDUCTION IN THE ACTIVATION RATE OF MOTOR UNITS DURING EVOKED CONTRACTIONS

Artificial Wisdom

In 1976, Marsden and colleagues coined this term to describe the procedure used to optimize force during an electrically elicited tetanus by progressively reducing the stimulation frequency as the muscle fatigues. Using this approach they, and others, have shown that the forces produced by the imposed stimulation can approach those achieved during sustained, voluntary, maximal contractions (see Fig. 1) (Marsden et al., 1976, 1983; Jones et al., 1979). More recently, Binder-Macleod and Guerin (1990) showed that reducing the stimulation frequency from 60 to 30 pps as the muscle fatigued evoked more force from the human quadriceps femoris muscle during intermittent, submaximal contractions than that

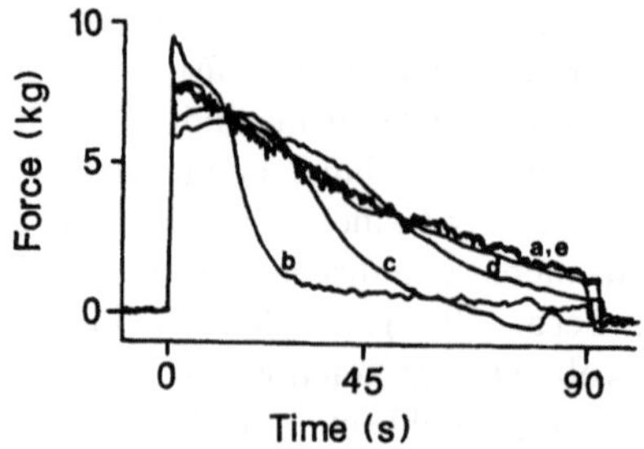

Figure 1. Effects of *artificial wisdom* on the force output of human adductor pollicis muscle. Traces represent the forces produced by a sustained voluntary contraction (a, bold, irregular line); evoked responses to 100 pps (b), 50 pps (c), and 35 pps (d) constant-frequency trains; and the response evoked by a progressive reduction in stimulation frequency (e). In e (artificial wisdom), stimuli were delivered at 60 pps for 8 s, 45 pps for 17 s, 30 pps for 15 s, and 20 pps for 55 s. At the onset of stimulation, the 100 pps train evoked the most force and the 35 pps train the least. Throughout the contraction, the *artificial wisdom* procedure produced forces that closely matched the voluntary contraction (modified from data presented by Marsden et al., 1983; their Fig. 4).

achieved by maintaining a constant frequency of electrical stimulation (60 pps). To explain the greater maintenance of force, Marsden and colleagues showed that the fatigue decline was related to the number of motor impulses, rather than the duration or force of contraction. More recently reports, have shown that the pattern of stimulation (e.g., the train frequency) and the force of contraction both affect the rate of fatigue (Garland et al., 1988; Binder-Macleod et al., 1995). Nevertheless, it appears clear that progressively reducing the stimulation frequency as the muscle fatigues is more effective at maintaining the force than a constant, high-rate of stimulation.

The Catch-Like Property of Skeletal Muscle

This term describes the force augmentation that occurs when an initial, brief, high-frequency burst (containing ~2-4 pulses) of stimuli is included at the beginning of an imposed subtetanic train of pulses (Burke et al., 1970, 1976; Binder-Macleod & Clamann, 1989). Burke and colleagues were the first to demonstrate a catch-like property in mammalian motor units (Burke et al., 1970, 1976; cf. Wilson & Larimer, 1968). The term catch-like was used to distinguish this phenomenon from the true catch property of molluscan muscle in which force production does not require continued stimulation of the muscle (Burke et al., 1976). A catch-like property has been observed in single motor units (Burke et al., 1970, 1976; Zajac & Young, 1980a; Bevan et al., 1992) as well as whole human (Binder-Macleod & Barker, 1991) and animal muscles (Binder-Macleod & Barrish, 1992).

The catch-like property is a fundamental property of the muscle; the tension enhancement is not due to the activation of additional muscle fibers (see Burke et al., 1970; Bevan et al., 1992). Increased release of Ca^{2+} from the sarcoplasmic reticulum has been proposed as a potential mechanism for the greater force (i.e., the sum of the forces produced by the burst are greater than the sum of forces produced by an equivalent number of twitches) (Duchateau & Hainaut, 1986b). This greater summation could explain the force augmentation seen with the catch-like property. However, as shown by Duchateau and Hainaut (1986a, 1986b), the presence of positive-staircase potentiation (i.e., a progressive increase in force elicited in response to identical stimuli) can eliminate this greater summation. Although positive-staircase potentiation reduces the augmentation associated with an imposed variable-frequency stimulus train that exploits the catch-like property of muscle (i.e., an *optimized* train) (Burke et al., 1976; Bevan et al., 1992), highly potentiated motor units

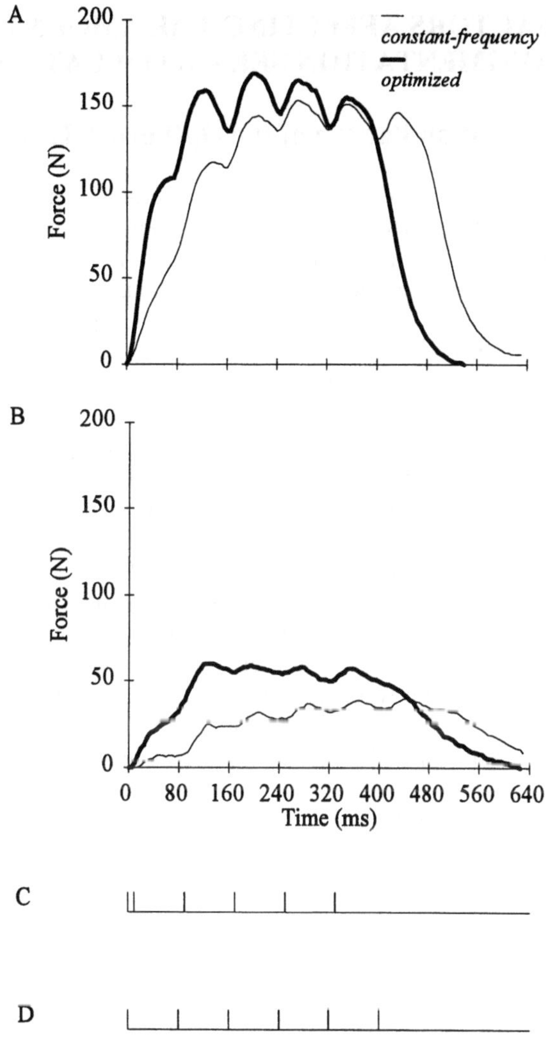

Figure 2. Characteristic evoked forces to *optimized* and *constant-frequency* trains from human quadriceps femoris muscle in the control (nonfatigued) (*A*) and fatigued (*B*) states. Both the *optimized* (*C*) and *constant-frequency* (*D*) trains contained six pulses. The *optimized* train contained an initial interpulse interval of 5 ms and four subsequent intervals = 80 ms; the *constant-frequency* train had all interpulse intervals = 80 ms. Due to marked slowing in the rate of tension development with the *constant-frequency* train as the muscle fatigued and the *catch-like property* of skeletal muscle, the *optimized* train produced greater force augmentation when the muscle was fatigued than in the control state.

(Burke et al., 1976; Bevan et al., 1992) and whole muscle (Binder-Macleod & Barker; 1991; Binder-Macleod & Barrish, 1992) can still show significant augmentation. Thus, although increased release of Ca^{2+} from the sarcoplasmic reticulum may contribute to the catch-like property under certain circumstances (see Burke et al., 1976), significant force augmentation can be observed in the apparent absence of augmented Ca^{2+} release.

Increased muscle stiffness may also explain the force augmentation seen with the catch-like property. Parmiggiani and Stein (1981) have likened the series elastic component of the muscle to an elastic band at slack length. The slack in the series elastic component must be taken up before force is produced. The stiffer the muscle the more substantial the development of tension in response to each stimulus pulse. The initial, high-frequency burst of the *optimized* train rapidly increases the force and muscle stiffness, thereby reducing the slack in the series elasticity that will be encountered by subsequent stimuli (see Fig. 2). Thus, the force augmentation due to the catch-like property may be explained largely by the ability of the initial burst to increase muscle stiffness and reduce the time-dependent rate of tension development (Cooper & Eccles, 1930; Binder-Macleod & Barrish, 1992).

FACTORS AFFECTING THE AMOUNT OF FORCE AUGMENTATION SEEN WITH CATCH-INDUCING TRAINS

Activation Pattern of the Optimized Train

To identify the pattern of interpulse intervals that evokes the greatest force-time integral from single motor units or whole muscle most investigators have *built* pulse trains one pulse at a time (Stein & Parmiggiani, 1979; Zajac & Young, 1980a; Parmiggiani & Stein, 1981; Duchateau & Hainaut, 1986a; Hennig & Lømo, 1987). Muscles were first stimulated with two-pulse trains to determine the interpulse interval that produced the maximal force. Next, three-pulse trains were tested. The first interpulse interval was set at the duration that was optimal for the two-pulse train and the duration of the third interpulse interval was varied. This process was continued until a train with the desired number of optimized interpulse intervals was completed. Surprisingly, although both fast and slow motor units and muscles have been studied, most studies have shown that the optimal train contained one or two short-duration (5-10 ms) intervals initially. In contrast, the optimal sequence after these initial, short-durations intervals has been shown to vary as a function of the twitch contraction time of the muscle (Burke et al., 1976; Zajac & Young, 1980a). For cat medial gastrocnemius motor units, these optimal interpulse intervals corresponded to a repetitive train (i.e., all interval durations were equal) with their durations equal to ~1.8 x twitch contraction time of the muscle (Burke et al., 1976; Zajac & Young, 1980a). Recently, we have found for whole muscle that the constant-frequency portion of the *optimized* train had durations of ~1.0 x twitch contraction time for rat gastrocnemius muscle and 1.2 x twitch contraction time for human quadriceps femoris muscle (see Figs. 3 and 4).

Motor Unit Type

Though not studied systematically, Burke and colleagues (1976) demonstrated that slow-twitch motor units showed greater augmentation than fast-twitch units (Burke et al., 1976; their Figs. 6-7). Recently, we compared the forces produced with *optimized* and *constant-frequency* trains in fast (gastrocnemius) and slow (soleus) rat muscles (Binder-Macleod & Barrish, 1992; Binder-Macleod & Landis, 1994) and found results consistent with those of Burke and colleagues. In the control condition (non-fatigued, highly potentiated state) the soleus muscle produced a ~17% greater force-time integral when stimulated with the *optimized* trains than when stimulated with the *constant-frequency* trains (Binder-Macleod & Barrish, 1992). In contrast, the gastrocnemius muscle showed no augmentation when stimulated under similar conditions (Binder-Macleod & Landis, 1994) (see Fig. 3*A*). However, as noted below, the gastrocnemius muscle did show augmentation with the *optimized* train when the muscle was fatigued.

Activation History of the Motor Unit

Burke and colleagues (1976) showed that during isometric contractions of single motor units in cat hindlimb muscles, the catch-like force augmentation was greater for unpotentiated than potentiated units (potentiation is here defined as the increase in the force due to prior activation; see also Kernell, Chapter 9). Bevan and colleagues (1992) recently studied the effects of *optimized* stimulus trains on fast-twitch fatigable motor units from cat tibialis posterior muscles during fatiguing contractions. The interpulse intervals of the *constant-frequency* trains were 1.8 x the twitch contraction time of the units, which resulted in a stimulus frequency range of 19-37 Hz. The *optimized* trains

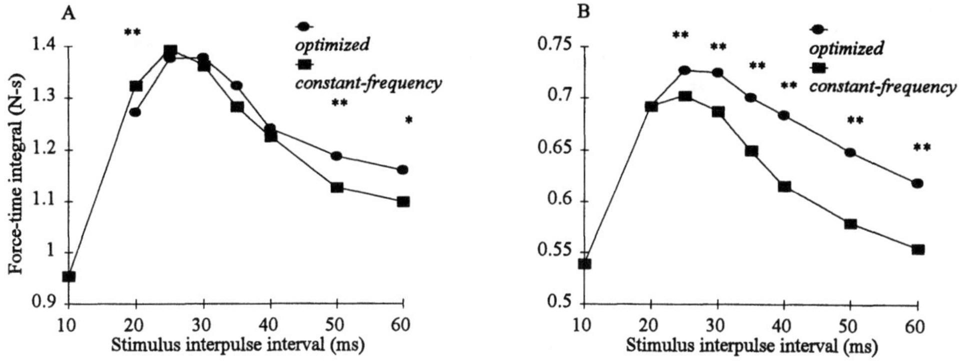

Figure 3. Effects of *optimized* stimulus trains on the force output (force-time integral) of potentiated and non-fatigued (*A*) and fatigued (*B*) rat gastrocnemius muscles. In this and Fig. 4, stimulus interpulse interval (X-axis) refers to the constant-frequency stimuli of the full *constant-frequency* trains and the last four intervals of the *optimized* trains. In the control, non-fatigued state, force was augmented when the interpulse interval of the constant-frequency component of the *optimized* train was ≥ 2 x the twitch contraction time of the muscle (≥ 50 ms), which was not changed appreciably by fatigue. When the muscle was fatigued, force was augmented for the *optimized* train when its later interpulse intervals were ≥ 25 ms. Comparisons were made at each stimulus interpulse interval, using a paired t-test (* = $p \leq 0.05$; ** = $p \leq 0.001$).

began with a high-frequency burst consisting of three pulses with an interpulse interval of 10 ms (100 Hz) followed by a constant interpulse-interval train that was adjusted so the total train had the same number of pulses, duration, and average frequency as the *constant-frequency* train for each muscle. They found that at the onset of stimulation, (i.e., prior to fatigue and before the muscle was fully potentiated) the *optimized* trains markedly increased the force-time integral produced by the muscle. After ~30 s of stimulation, when the muscle was maximally potentiated, the *optimized* trains produced slightly less force than the *constant-frequency* trains. However, as the muscle began to fatigue, the *optimized* trains again augmented the force compared with the *constant-frequency* trains. Thus, it appears that fast-twitch fatigable motor units primarily show augmentation with *optimized* trains when the muscle is either unpotentiated or fatigued. We have observed that the effects of imposed *optimized* stimulus trains in human quadriceps femoris (contains ~60% fast-twitch motor units) and whole rat gastrocnemius (~100% fast-twitch motor units) muscles are similar to those reported by Bevan and colleagues. In the non-fatigued, but highly potentiated state we observed little augmentation for either muscle. However, as the muscles fatigued, the optimized trains were able to significantly augment forces (Binder-Macleod & Barker, 1991; Binder-Macleod & Baadte, 1992; Binder-Macleod & Landis, 1994).

Recently, we completed a series of studies that attempted to determine the effects on the force of using various *optimized* trains, covering a wide range of train frequencies. These studies on rat gastrocnemius and human quadriceps femoris muscles compared the force produced by a variety of 6-pulse *constant-frequency* trains *vs. optimized* trains with the first two pulses of the latter separated by a brief interpulse interval (10 ms for rat muscle, and 5 ms for human muscle). The remaining four pulses within the *optimized* trains were separated by interpulse intervals identical to those used in the *constant-frequency* trains. (see Figs. 3 and 4). Data were collected prior to fatigue, and during repetitive, fatiguing contractions (at 1 stimulus train/s). For rat gastrocnemius in the control (nonfatigued) state, the *optimized* trains only augmented forces when the stimulus interpulse intervals of the last four intervals in the train were ≥ ~ 2 x the twitch contraction times of the muscles (i.e., ≥ 50 ms; see Fig.

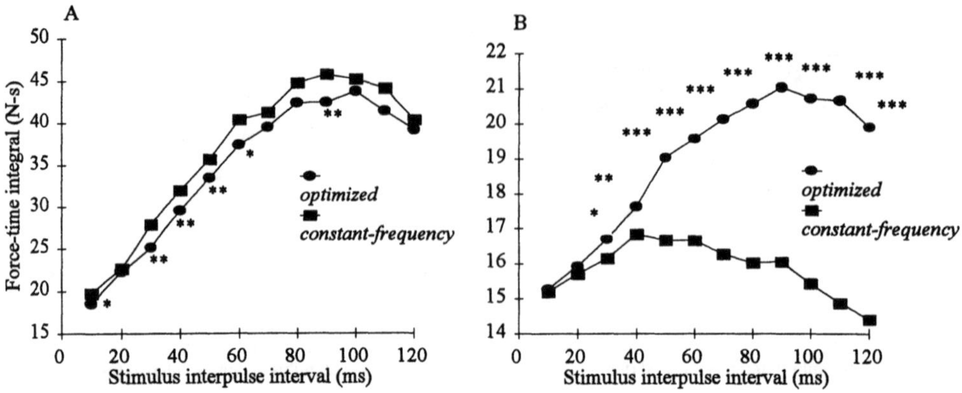

Figure 4. Effects of *optimized* stimulus trains on the force output (force-time integral) of potentiated and non-fatigued (*A*) and fatigued (*B*) human quadriceps femoris muscle. In the control, non-fatigued state, no force augmentation was produced by the *optimized* trains at any of the frequencies tested (i.e., with their later, constant-frequency intervals ≤ 1.5 x 80 ms, the twitch contraction time). When the muscle was fatigued, twitch contraction times were reduced from ~ 80 ms to ~ 40 ms. Force augmentation was produced by the *optimized* trains at all constant-frequency intervals ≥ 20 ms. The comparison is as in Fig. 3. (* = $p \leq 0.05$; ** = $p \leq 0.01$; ***= $p \leq 0.001$).

3). The one *constant-frequency* train that had an interpulse interval < 25 ms produced greater force than the corresponding optimizing train. Similarly, for the human quadriceps femoris, in the control state, *constant-frequency* trains produced greater forces than *optimized* trains for almost all interpulse intervals < ~ 80 ms (the muscles' twitch contraction time; note that no measurement was made of stimulus interpulse intervals > 1.5 x this contraction time; see Fig. 4).

In the fatigued state, very different results were observed. For the rat gastrocnemius, the twitch contraction time (~ 25 ms) showed little change when the muscle was fatigued, though the *optimized* trains produced force augmentation for all trains with interpulse intervals ≥ 25 ms. The stimulus trains with 20-ms intervals revealed no significant difference in the forces produced by the two train types. For the human quadriceps femoris, the twitch contraction time decreased from ~80 ms to ~40 ms with fatigue, and for all trains with interpulse intervals ≥ 30 ms, the *optimized* trains produced significantly more force than the *constant-frequency* trains. Interestingly, although previous studies have shown that catch-inducing trains can augment forces from low-frequency subtetanic trains, these data show that when muscles are fatigued catch-inducing trains can produce greater forces than even the *best constant-frequency* train.

Anisometric Contractions

As noted by Enoka and Stuart (1992), the muscle wisdom theme needs to be extended to dynamic conditions to determine if the results observed during isometric conditions hold true for a wide variety of natural movements. Similarly, studies should examine the use of variable-frequency trains that exploit the catch-like property of muscle. Unfortunately, there is a dearth of information on both of these issues. Callister and colleagues (1992) reported for a turtle hindlimb muscle that during shortening and lengthening contractions, *optimized* stimulus trains produced slightly less force augmentation than during isometric contractions. We have recently reported (Binder-Macleod & Lee, 1993) that in the potentiated, non-fatigued state, *optimized* stimulus trains produced modest increases in the force-time integral of the human quadriceps muscle during shortening contractions, and this augmentation increased as the muscle became fatigued. In contrast, no augmentation was seen during lengthening contractions in the control (non-fatigued) or fatigued state. Clearly, substantial additional work is needed on anisometric contractions.

RELEVANCE OF THE CATCH-LIKE PROPERTY TO VOLITIONAL CONTRACTIONS

High-frequency bursts (i.e., doublets or triplets) of motor unit EMG activity at the onset of a train of activity have been reported to occur during volitional contractions in human (Maton & Gamet, 1989) and animals (Zajac & Young, 1980b; Hennig & Lømo, 1985, 1987). In addition, motor unit activation rates during volitional contractions in human appear to be sufficiently low to allow these high-frequency bursts to elicit a catch-like effect (Bellemare et al., 1983; Bigland Ritchie et al., 1983a). Based on the appearance of a doublet during walking in high-decerebrate cat Zajac and Young (1980b) reasoned that, if during normal locomotion there was a constraint on the number of pulses within a train to reduce fatigue, then a motor unit should fire with the activation pattern that optimizes the force. However, several studies have shown that doublets occur infrequently during walking in normal cats (Hoffer et al., 1987) and that the incidence of doublets decreased as the treadmill speed increased in both high-decrebrate (Zajac & Young, 1980b) and normal cats (Hoffer et al., 1987). In addition, doublets are seen in some subpopulations of motor units and not others in the rat hindlimb (Hennig & Lømo, 1985, 1987). These data suggest that there must be some functional advantages to utilizing the catch-like property for some motor unit types and under some circumstances. Unfortunately, there are not enough data available to determine which parameters determine when the catch-like property is used during volitional activation of motor units (Bevan et al., 1992; Binder-Macleod & Barrish, 1992).

CONCLUSION

In conclusion, it appears that selected patterns of variable-frequency stimulation can optimize force output during sustained and prolonged intermittent contractions. These stimulation patterns have been shown to be effective during both voluntary and imposed high-force contractions. However, the boundary conditions for such patterns remain an open issue. The stimulus patterns that exploit the catch-like property have been shown to reduce and delay the fatigue produced by imposed contractions, with their role in voluntary movement also an open issue. However, imposed stimulus patterns that exploit the catch-like property of muscle have significant implications for the use of functional electrical stimulation in rehabilitation and to our overall understanding of segmental motor control.

ACKNOWLEDGMENTS

The author would like to thank Dr. Lynn Snyder-Mackler for her criticism of a draft of this chapter. The work was supported by United States Public Health Service (USPHS) grant AR 441264. The author's attendance at the 1994 Bigland-Ritchie conference was supported, in part, by AR 441264, and the University of Miami.

REFERENCES

Bellemare F & Garzaniti N (1988). Failure of neuromuscular propagation during human maximum voluntary contraction. *Journal of Applied Physiology* **64**, 1084-1093.

Bellemare F, Woods JJ, Johansson R & Bigland-Ritchie B (1983). Motor unit discharge rates in maximal voluntary contractions of three human muscles. *Journal of Neurophysiology* **50**, 1380-1392.

Bevan L, Laouris Y & Reinking RM (1992). The effect of the stimulation pattern on the fatigue of single motor units in adult cats. *Journal of Physiology (London)* **449**, 85-108.

Bigland-Ritchie B (1981). EMG/force relations and fatigue of human voluntary contractions. *Exercise Sport Sciences Reviews* **9**, 75-117.

Bigland-Ritchie B, Cafarelli E & Vøllestad NK (1986). Fatigue of submaximal static contractions. *Acta Physiologica Scandinavica* **128**, 137-148.

Bigland-Ritchie B, Johansson R, Lippold OCJ, Smith S & Woods JJ (1983a). Changes in motoneuron firing rates during sustained maximal voluntary contractions. *Journal of Physiology (London)* **340**, 335-346.

Bigland-Ritchie B, Johansson R, Lippold OCJ & Woods JJ (1983b). Contractile speed and EMG changes during fatigue of sustained maximal voluntary contractions. *Journal of Neurophysiology* **50**, 313-324.

Bigland-Ritchie B, Jones DA, Hosking GP & Edwards RHT (1978). Central and peripheral fatigue in sustained maximum voluntary contractions of human quadriceps muscle. *Clinical Science and Molecular Medicine* **54**, 609-614.

Bigland-Ritchie B, Jones DA & Woods JJ (1979). Excitation frequency and muscle fatigue: electrical responses during human voluntary and stimulated contractions. *Experimental Neurology* **64**, 414-427.

Bigland-Ritchie B, Thomas CK, Rice CL, Howarth JV & Woods JJ (1992). Muscle temperature, contractile speed, and motorneuron firing rates during human voluntary contractions. *Journal of Applied Physiology* **73**, 2457-2461.

Binder-Macleod SA & Baadte SA (1992). Identification of optimal interpulse interval (IPI) patterns for activation of fatigued human quadriceps femoris muscle. *Society for Neuroscience Abstracts* **18**, 1557.

Binder-Macleod SA & Barker CB (1991). Use of the catchlike property of human skeletal muscle to reduce fatigue. *Muscle & Nerve* **14**, 850-857.

Binder-Macleod SA & Barrish WJ (1992). Force response of rat soleus muscle to variable-frequency train stimulation. *Journal of Neurophysiology* **68**, 1068-1078.

Binder-Macleod SA & Clamann HP (1989). Force-frequency relations of cat motor units during linearly varying dynamic stimulation. *Journal of Neurophysiology* **61**, 208-217.

Binder-Macleod SA & Guerin T (1990). Preservation of force output through progressive reduction of stimulation frequency in the human quadriceps femoris muscle. *Physical Therapy* **70**, 619-625.

Binder-Macleod SA, Halden EE & Jungles KA (1995). Effects of stimulation intensity on the physiological responses of human motor units. *Medicine and Science in Sports and Exercise* **27**, 556-565.

Binder-Macleod SA & Landis LJ (1994). Effects of train frequency and fatigue state on the catchlike property in the rat gastrocnemius muscle. *Society for Neuroscience Abstracts* **20**, 1204.

Binder-Macleod SA & Lee SCK (1993). Catchlike property of human muscle during shortening and lengthening contractions. *Society for Neuroscience Abstracts* **19**, 155.

Binder-Macleod SA & McDermond LR (1992). Changes in the force-frequency relationship of the human quadriceps femoris muscle following electrically and voluntarily induced fatigue. *Physical Therapy* **72**, 95-104.

Burke RE, Rudomin P & Zajac FE (1970). Catch properties in single mammalian motor units. *Science* **168**, 212-214.

Burke RE, Rudomin P & Zajac FE (1976). The effect of activation history on tension production by individual muscle units. *Brain Research* **109**, 515-529.

Callister RJ, Laidlaw DH, Reinking RM & Stuart DG (1992). The catch-like property of turtle muscle: effect of shortening, lengthening, and fatiguing isometric contractions. *Society for Neuroscience Abstracts* **18**, 1559.

Cooper RG, Edward RHT, Gibson H & Stokes MJ (1988). Human muscle fatigue: frequency dependence of excitation and force generation. *Journal of Physiology (London)* **397**, 585-599.

Cooper S & Eccles JC (1930). The isometric response of mammalian muscles. *Journal of Physiology (London)* **69**, 377-385.

Dietz V (1978) Analysis of the electrical muscle activity during maximal contraction and the influence of ischaemia. *Journal of the Neurological Sciences* **37**, 187-197.

Dorfman LJ, Howard JE & McGill KC (1990). Triphasic behavioral response of motor units to submaximal fatiguing exercise. *Muscle & Nerve* **13**, 621-628.

Dubose L, Schelhorn TB & Clamann HP (1987). Changes in contractile speed of cat motor units during activity. *Muscle & Nerve* **10**, 744-752.

Duchateau J & Hainaut K (1986a). Nonlinear summation of contractions in striated muscle. I. Twitch potentiation in human muscle. *Journal of Muscle Research and Cell Motility* **7**, 11-17.

Duchateau J & Hainaut K (1986b). Nonlinear summation of contractions in striated muscle. II. Potentiation of intracellular Ca^{2+} movements in single barnacle muscle fibers. *Journal of Muscle Research and Cell Motility* **7**, 18-24.

Edwards RHT (1981). Human muscle function and fatigue. In: Porter R, Whelan J (eds.), *Human Muscle Fatigue, Physiological Mechanisms,* pp. 1-18. London (Ciba Foundation symposium 82): Pitman Medical.

Edwards RHT, Hill DK, Jones DA & Merton PA (1977). Fatigue of long duration in human skeletal muscle after exercise. *Journal of Physiology (London)* **272**, 769-778.

Enoka RM & Stuart DG (1992). Neurobiology of muscle fatigue. *Journal of Applied Physiology* **72**, 1631-1648.

Fallentin N, Jorgensen K & Simonsen EB (1993). Motor unit recruitment during prolonged isometric contractions. *European Journal of Applied Physiology & Occupational Physiology* **67**, 335-341.

Fitts RH (1994). Cellular mechanisms of muscle fatigue. *Physiological Reviews* **74**, 49-94.

Garland SJ, Enoka RM, Serrano LP & Robinson GA (1994). Behavior of motor units in human biceps brachii during a submaximal fatiguing contraction. *Journal of Applied Physiology* **76**, 2411-2419

Garland SJ, Garner SH & McComas AJ (1988). Relationship between numbers and frequencies of stimuli in human muscle fatigue. *Journal of Applied Physiology* **65**, 89-93.

Gordon DA, Enoka RM, Karst GM, & Stuart DG (1990). Force development and relaxation in single motor units of adult cats during a standard fatigue test. *Journal of Physiology (London)* **421**, 583-594.

Grimby L, Hannerz J & Hedman B (1981). The fatigue and voluntary discharge properties of single motor units in man. *Journal of Physiology (London)* **316**, 545-554.

Hennig R & Lømo T (1985). Firing patterns of motor units in normal rats. *Nature* **314**, 164-166.

Hennig R & Lømo T (1987). Gradation of force output in normal fast and slow muscles of the rat. *Acta Physiologica Scandinavica* **130**, 133-142.

Hoffer JA, Sugano N, Loeb G, Marks WB, O'Donovan MJ, & Pratt CA (1987). Cat hindlimb motoneurons during locomotion. II. Normal activity patterns. *Journal of Neurophysiology* **57**, 530-553.

Jones DA (1981). Muscle fatigue due to changes beyond the neuromuscular junction. In: Porter R, Whelan J (eds.), *Human Muscle Fatigue, Physiological Mechanisms,* pp. 178-196. London (Ciba Foundation symposium 82): Pitman Medical

Jones DA, Bigland-Ritchie B & Edwards RHT (1979). Excitation frequency and muscle fatigue: Mechanical responses during voluntary and stimulated contractions. *Experimental Neurology* **64**, 401-413.

Jones DA, Newham DJ & Torgan C (1989). Mechanical influences on long-lasting human muscle fatigue and delayed-onset pain. *Journal of Physiology (London)* **412**, 415-427.

Kernell D, Eerbeek O & Verhey BA (1983). Relation between isometric force and stimulation rate in cat's hindlimb motor units of different twitch contraction times. *Experimental Brain Research* **50**, 220-227.

Marsden CD, Meadows JC & Merton PA (1971). Isolated single motor units in human muscle and their rate of discharge during maximal voluntary effort (abstract). *Journal of Physiology (London)* **217**, 12P-13P.

Marsden CD, Meadows JC & Merton PA (1976). Fatigue of human muscle in relation to the number and frequency of motor impulses (abstract). *Journal of Physiology (London)* **258**, 94P-95P.

Marsden CD, Meadows JC & Merton PA (1983). "Muscular wisdom" that minimizes fatigue during prolonged effort in man: peak rates of motoneuron discharge and slowing of discharge during fatigue. *Advances in Neurology* **39**, 169-211.

Maton B & Gamet D (1989). The fatigability of two agonistic muscles in human isometric voluntary submaximal contraction: an EMG study. *European Journal of Applied Physiology* **58**, 369-374.

Merton PA (1954). Voluntary strength and fatigue. *Journal of Physiology (London)* **128**, 553-564.

Metzger JM & Fitts RH (1987). Fatigue from and high- and low-frequency muscle stimulation: contractile and biochemical alterations. *Journal of Applied Physiology* **62**, 2075-2082.

Næss K & Storm-Mathisen A (1955). Fatigue of sustained tetanic contradictions. *Acta Physiologica Scandinavica* **34**, 351-366.

Parmiggiani F & Stein RB (1981). Nonlinear summation of contractions in cat muscle. II. Later facilitation and stiffness changes. *Journal of General Physiology* **78**, 295-311.

Powers RK & Binder MD (1991). Summation of motor unit tensions in the tibialis posterior muscle of the cat under isometric and nonisometric conditions. *Journal of Neurophysiology* **66**, 1838-1846.

Stein RB & Parmiggiani F (1979). Optimal motor patterns for activating mammalian muscle. *Brain Research* **175**, 372-376.

Stokes MJ, Edwards RHT & Cooper RG (1989). Effect of low frequency fatigue on human muscle strength and fatigability during subsequent stimulated activity. *European Journal of Applied Physiology* **59**, 278-283.

Stuart DG & Callister RJ (1993). Afferent and spinal reflex aspects of muscle fatigue: Issues and speculations. In: Sargent AJ, Kernell D (eds.), *Neuromuscular Fatigue*, pp. 169-180. Amsterdam: Royal Netherlands Academy of Arts and Sciences.

Thomas CK Bigland-Ritchie B & Johansson RS (1991). Force-frequency relationship of human thenar motor units. *Journal of Neurophysiology* **65**, 1509-1516.

Thomas CK, Woods JJ & Bigland-Ritchie B (1989). Impulse propagation and muscle activation in long maximal voluntary contractions. *Journal of Applied Physiology* **67**, 1835-1842.

Thompson LV, Balog EM, Riley DA & Fitts RH (1992). Muscle fatigue in frog semitendinosus: Alterations in contractile function. *American Journal of Physiology (Cell Physiology)* **262**, C1500-C1506.

Wilson DM & Larimer JL (1968). The catch properties of ordinary muscle. *Proceedings of the National Academy of Science, USA* **61**, 909-916.

Woods JJ, Furbush F & Bigland-Ritchie B (1987). Evidence for a fatigue-induced reflex inhibition of motoneuron firing rates. *Journal of Neurophysiology* **58**, 125-137.

Zajac FE & Young JL(1980a). Properties of stimulus trains producing maximum tensiontime area per pulse from single motor units in medial gastrocnemius muscle of the cat. *Journal of Neurophysiology* **43**, 1206-1220.

Zajac FE & Young JL (1980b). Discharge properties of hindlimb motoneurons in decerebrate cats during locomotion induced by mesencephalic stimulation. *Journal of Neurophysiology* **43**, 1221-1235.

OVERVIEW: POTENTIAL ROLE OF SEGMENTAL MOTOR CIRCUITRY IN MUSCLE FATIGUE

U. Windhorst and G. Boorman

Departments of Clinical Neurosciences and Medical Physiology
The University of Calgary
Faculty of Medicine
3330 Hospital Drive N.W., Calgary, Alberta T2N 4N1, Canada

ABSTRACT

This chapter reviews several mechanisms that the CNS may use to mitigate muscle fatigue, including intrinsic motoneuron properties and feedback systems. The emphasis is on the effects of sensory inputs on spinal cord interneurons including: Renshaw cells; Ib inhibitory interneurons; interneurons mediating presynaptic inhibition; Ia inhibitory interneurons; and interneuronal networks constituting central pattern generators for locomotion. This exercise brings out how little is known about the operation of these circuits in dealing with muscle fatigue.

INTRODUCTION

Muscle fatigue is a phenomenon of immense practical importance in normal individuals, particularly in physically demanding occupations and in athletic competition, and in patients with various types of disorders affecting the CNS and/or peripheral neuromuscular apparatus. Whereas the first part of this statement is everyday experience, the latter is not as evident because other phenomena and symptoms often supersede fatigue. For example, in patients with upper motoneuron disease after stroke, multiple sclerosis and spinal cord injury, leg muscle fatigability is enhanced, possibly resulting from conversion of many fatigue-resistant motor units into fatigable ones after prolonged disuse (Lenman et al., 1989; Miller et al., 1990), although there are reports to the contrary (for review: Enoka & Stuart, 1992). In the rare cases of mitochondrial myopathies leading to lactacidosis (Edwards et al., 1982), the predominant features are symptoms of severe exhaustion, weakness and muscle fatigue during exercise. The lactacidosis should be a strong stimulus to high-threshold muscle afferents which may reflexly alter motoneuron discharge. Muscle fatigue is also an important limiting factor in situations in which functional electrical stimulation (FES) is used to help restore function after neurological damage (Binder-Macleod & Barker 1991).

Fatigue, Edited by Simon C. Gandevia et al.
Plenum Press, New York, 1995

Despite its ubiquity, little is yet known about the mechanisms contributing to muscle fatigue and the means available to the CNS to mitigate its effects (Enoka & Stuart, 1992; Stuart & Callister, 1993). This is due in only minor part to the lack of a unanimously accepted definition of muscle fatigue (Clamann, 1990; Enoka & Stuart, 1992; Bigland-Ritchie, 1993). Rather, the key problem is that multiple processes and mechanisms are involved at different levels, from higher-order CNS structures to electrical and biochemical alterations within muscle fibers (Clamann, 1990; Enoka & Stuart, 1992). Whatever muscle fatigue's underlying causes, it becomes manifest as ". . . any reduction in the force-generating capacity (measured by the maximum voluntary contraction), regardless of the task performed" (Bigland-Ritchie & Woods 1984). However, nature appears to have evolved a number of mechanisms for maintaining force output at the optimum achievable at any time in the course of fatigue and thus mitigating its effects. Some such mechanisms are intrinsic to the fatiguing muscle fibers themselves, whereas others (like effects of sensory feedback) are located in the CNS. Thus, study of these mechanisms not only reveals specifics about muscle fatigue, but provides a window into the way the CNS optimizes the performance of its neuromuscular periphery. What is now known about such mechanisms is orders of magnitude less than what has yet to be discovered. For this reason, the present chapter deals with problems rather than solutions. It focuses on the potential role of sensory feedback in adjusting static and dynamic properties of spinal circuits which may be involved in optimizing force output during muscle fatigue.

Like the mechanisms giving rise to muscle fatigue, those counteracting it at the segmental level are complex and multi-faceted. They include motor unit potentiation, the diversity and recruitment order of the motor unit population, and the operation of premotoneuronal networks.

MOTOR UNIT POTENTIATION

The first protective mechanism resides in the muscle fibers. It involves *potentiation*, i.e., an increase in motor unit force output due to preceding activation (see Kernell, Chapter 9). This is illustrated in Fig. 1, showing that injection of long-lasting depolarizing current into cat lumbar α-motoneurons caused them to discharge, first at a high rate and then they adapted to lower rates.

Motoneurons labeled *S unit* (see below) did not adapt nearly as much as did *FF unit* motoneurons. Similar differences in adaptation have been seen with extracellular current application, even with repeated intermittent current pulses (Spielmann et al., 1993). Furthermore, Fig. 1*B* shows that the average force declined much more in FF motor units than in S units. Thirdly, the force of FF units first declined steeply and then exhibited an intermediate hump which probably resulted from force potentiation. The high initial firing rate of FF motor units is well suited to potentiate their twitches after some seconds. Hence, motor units may show both fatigue and potentiation in parallel (Gordon et al., 1990; Powers & Binder, 1991; Bevan et al., 1992; see also Thomas, Chapter 10), the latter transiently opposing the former. In addition, selected firing patterns with initial short interspike intervals may optimize force output in the face of fatigue (Binder-Macleod & Barker 1991; Bevan et al., 1992), a form of *pattern optimization* (see Binder-Macleod, Chapter 16)

MOTOR UNIT POPULATION: DIVERSITY OF MOTOR UNIT TYPES AND RECRUITMENT ORDER

Implicit in Fig. 1 is the finding that there are different types of motor units, and that they are differentially fatigable (type S *vs.* FR *vs.* FF; Burke et al., 1973; Burke, 1981;

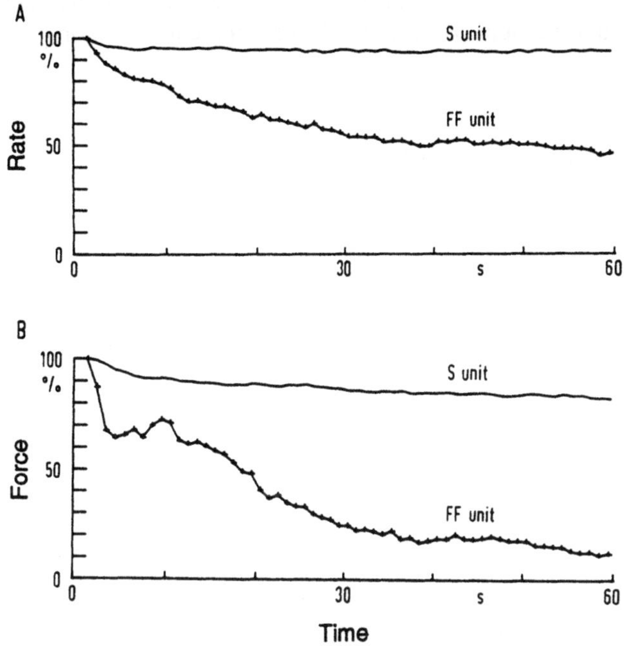

Figure 1. *A*, plot of the firing rate (%) versus time (s) for first minute of motoneuron discharges produced by depolarizing current of 5 nA above threshold for rhythmic firing. For consecutive seconds of discharge, mean firing rates have been connected by straight lines. Firing rates given as percentage of rate during 2nd second of discharge (first value plotted = 16.1 pulses per second (pps) for the S unit and 28.9 pps for the FF unit). Twitch contraction time was 53 ms for the S unit and 25 ms for the FF unit. The standard fatigue index according to Burke and colleagues (1973) was 0.97 for the S unit and 0.07 for the FF unit. *B*, contractile force produced by the discharges in *A*. Force given as percentage of value for 2nd second of discharge. In nonfused contractions, mean force was calculated over time periods of 1 s (reproduced from Kernell & Monster, 1982; their Fig. 5, by permission of Springer-Verlag).

Kernell, 1993; for details see Kernell, Chapter 9; Thomas, Chapter 10). The type S motor units have the longest duty periods throughout the day and are active in most muscle activities of moderate strength and speed. This activity pattern is in keeping with the finding that their discharge lasts longer during sustained depolarization (Kernell and Monster, 1982; Spielmann et al., 1993). Faster contracting, more fatigable motor units come into play during stronger and faster contractions. Thus, the functionally ordered recruitment from type S to FR to FF motor units of different size, usually referred to as Henneman's *size principle* (Henneman, 1957; see also Binder & Mendell, 1990; Kernell, 1992; Sawczuk et al., Chapter 8; Kernell, Chapter 9; Sargeant & Jones, Chapter 24; and Bigland-Ritchie, Chapter 26), is another protective mechanism and may be called *recruitment optimization* (Kernell, 1993). However, when strong contractions have to be sustained, the problem of high fatigability of strong motor units has to be dealt with. Any fatigue-reducing mechanism should, therefore, focus on fast-twitch motor units (FF in particular). In fact, at the single unit level, F (i.e., FR + FF) motor units exhibit faster and more extensive adaptation of discharge rate to maintained currents. At the population level, the recruitment patterns found normally during non-fatiguing contractions could conceivably be altered. Indeed, evidence is accumulating that recruitment of motor units, which during non-fatigued increasing muscle contractions usually occurs in the above orderly fashion, may change during fatigue, such that low-threshold motor units may not be recruited (Enoka & Stuart, 1992), with the underlying mechanisms yet to be determined. Presumably, both intrinsic motoneuron properties and synaptic inputs have to be taken into account (Burke, 1981), requiring consideration of premotoneuronal networks.

PREMOTONEURONAL NETWORKS

The third line of defense against fatigue resides in the action of premotoneuronal networks, including five components: 1) spinal tracts descending from higher motor centers; 2) neuromodulatory systems; 3) reflex circuits; 4) intrinsic spinal feedback circuits; and 5) locomotor pattern generators

The *first category* includes a number of tracts mediating motor commands to motoneurons.

The *second category* refers to descending systems from brainstem structures such as the locus coeruleus and raphé nuclei, which modulate the excitability of many spinal neurons including those of the central pattern generator (CPG).

The *third category* is the largest and most heterogeneous, including pathways from a variety of muscle receptor afferents which exert effects on α-motoneurons, directly or most often indirectly, via interneurons (see below). These include mono- and oligosynaptic connections from muscle spindle group Ia and II afferents, the classical autogenetic di- and trisynaptic pathways, as well as more complex connections from Golgi tendon organ (group Ib) afferents, and various pathways from higher-threshold afferents which may sense some aspects of muscle fatigue via mechanical and/or chemical stimuli (see Garland & Kaufman, Chapter 19).

The *fourth category* primarily encompasses recurrent inhibition, involving Renshaw cells which receive inputs from many species of muscle afferent.

The *fifth category* encompasses interneuronal networks which constitute the CPG for locomotion and allied movements. The CPG cyclically modulates reflex pathways and, in turn, is influenced by muscle afferents and may thus play a role in fatigue-compensating mechanisms during cyclic motor activities.

There is no clear delineation between the above circuits, requiring that we first provide a global approach and then dissect out single mechanisms.

A GLOBAL APPROACH: NEGATIVE FORCE FEEDBACK SYSTEMS

Negative feedback circuits have been designed both by nature and man to compensate for various kinds of disturbances interfering with the execution of selected tasks. For example, activation of a skeletal muscle by its motoneurons is co-modulated by feedback from receptors monitoring muscle fiber length and force; that is, muscle spindles and Golgi tendon organs, respectively. With qualifications to be discussed below, the afferents from these receptors can have actions on synergistic and homonymous motoneurons that provide negative feedback (Fig. 2). Under selected conditions, such actions allow compensation for disturbances, which may arise either externally, internally, or both (Houk & Rymer, 1981). An example of an internal disturbance is muscle fatigue, during which event properties of the controlled *plant* (i.e., muscle) change in response to continuing activation (Houk et al., 1970).

When the force-producing capability of muscle declines, a force feedback would be well suited (Houk et al., 1970). There are at least two parallel loops that could contribute to force feedback. First and foremost, Golgi tendon organs are very sensitive force sensors, and

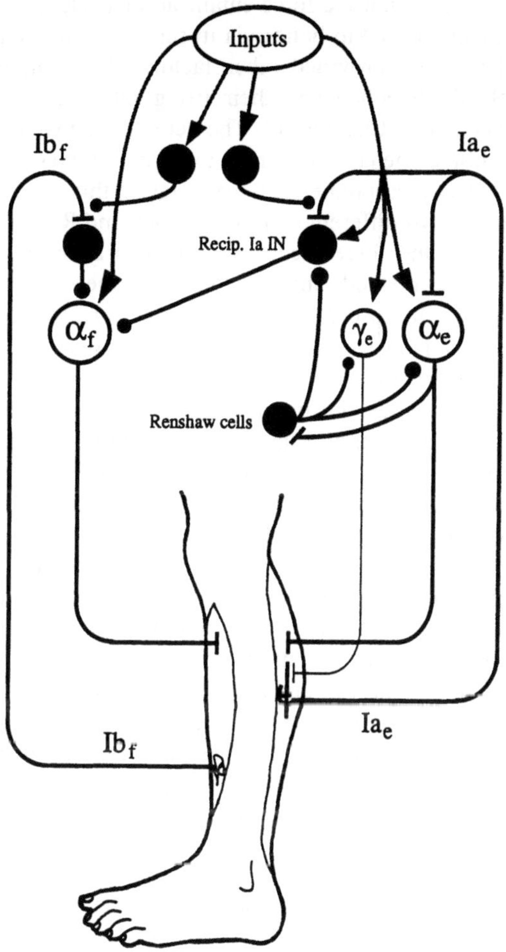

Figure 2. Schematic diagram of selected spinal reflex circuits involving group Ia and Ib afferents. A pair of antagonist muscle groups are shown for each circuit. Excitatory neurons are represented by open circles and their synaptic terminals by T-shaped lines. Inhibitory neurons are represented by filled circles. Muscle spindles are symbolized as a straight line with a coil around its middle portion. They consist of specialized muscle fibers with a separate motor innervation, as indicated by the thin line originating from the extensor γ-motoneuron (labeled γ_e) in the spinal cord. Thick sensory nerve fibers, denoted Ia_e, originate from the spiral sensory ending on extensor muscle spindles and carry information to the spinal cord where they contact several types of neuron, with monosynaptic excitatory connections to α-motoneurons (α_e) of their own (homonymous) muscle and muscles of similar (synergistic) function. When ankle extensor muscles are stretched, their Ia_e axons are excited and, in turn, excite homonymous/synergistic motoneurons which induce contraction in the stretched muscles, thereby opposing stretch (negative feedback). Extensor Ia_e axons also inhibit antagonist flexor motoneurons via reciprocal Ia inhibitory interneurons (Recip. Ia IN), and vice versa (latter not illustrated). Renshaw cells receive their main excitatory input from a particular motoneuron pool (here extensor) and inhibit homonymous and synergistic α-and less frequently and strongly, γ-motoneurons. Recip. Ia INs are excited by homonymous and synergistic Ia fibers, and other Renshaw cells (in particular, those predominantly related to antagonist motoneurons); mutual inhibition, not shown. Ib afferents (here of flexor origin: Ib_f) from Golgi tendon organs are located at musculo-tendinous junctions. There axons di- or trisynaptically inhibit homonymous and synergistic motoneurons (*autogenetic inhibition.*). The pattern of Ib afferent connections is actually more complicated, involving alternative excitatory pathways to synergistic motoneurons (particularly those of extensor muscles). These may be switched on during the stance phase of locomotion, along with other complex reflex effects (see text). α- and γ-motoneurons, Recip. Ia INs, Renshaw cells and other interneurons receive a variety of sensory and descending, excitatory and inhibitory inputs (upper oval labeled *Inputs*), as symbolized by the dashed arrows. These inputs include afferents from high-threshold muscle afferents.

under certain circumstances (see below), their Ib afferents exert inhibitory effects on homonymous and synergistic motoneurons according to the classical concept of *autogenetic inhibition* (Fig. 2, left side). Thus, when force declines due to fatigue, Ib afferent excitation should be reduced, and thereby also autogenetic inhibition. Descending motor commands should then be disinhibited and muscle excitation enhanced. In addition, mono-and oligosynaptic excitation of homonymous and synergistic motor units from muscle spindle afferents (Fig. 2, right side) might assist, because this loop also constitutes a negative feedback system.

Muscle spindles monitor the length of their in-parallel skeletomotor muscle fibers. Even when the whole muscle is in isometric contraction (i.e., at constant overall length), a decline in muscle force entails an increase in internal muscle length because muscle fibers are connected to bones via in-series elastic elements which shorten with decreasing force. Therefore, provided other factors like fusimotor innervation remain constant, spindles should lengthen and their firing rate increase during muscle fatigue (see Hagbarth & Macefield, Chapter 18). The net outcome should be increased motoneuron excitation. The problem then is whether the loop gain of the system is large enough to reduce fatigue effects. Recent measurements in man suggest this to be the case. Based on several assumptions and using two different methods, Kirsch and Rymer (1992) estimated the open-loop gain in the compound force feedback system to average between 1.3 and 4.6 (first method) or 8.3 (second method) and concluded that these unexpectedly high loop gains may significantly reduce the susceptibility of the overall neuromuscular system to fatigue by 56-89%.

However, this simple model poses a number of problems which, in part, are related to the precise nature of the fatiguing task, and, in part, to the interference of additional reflex actions. Thus, during maximal isometric muscle contractions in man when no recruitment of additional motor units is assumed to occur, motor unit discharge rates decline, contrary to expectations based on negative feedback operation. Only in sustained or repeated sub-maximal contractions can such recruitment be observed (Vøllestad et al., 1988; Fuglevand et al., 1993). Furthermore, muscle spindle firing in man appears not to increase, but to decline during fatiguing muscle contraction (see Hagbarth & Macefield, Chapter 18). This may have detrimental effects on the gain in the autogenetic Ib afferent pathway because in the cat about 40% of the Ib inhibitory interneurons with Ib input from extensor muscles are also excited by spindle Ia afferents (Jankowska et al., 1981). Therefore, since Ia input is declining, these interneurons' sensitivity to Ib input is likely to decrease as well. The patterns of changes in Ib afferent discharge during fatiguing muscle contractions in humans are not yet known. Furthermore, the gain of force feedback can also be altered, at a number of sites, by modulating influences arising, in particular, from high-threshold afferents in the fatiguing muscle(s) (for details see Garland & Kaufman, Chapter 19). Finally, central effects of Ia and Ib afferents are subject to presynaptic inhibition (see below), which might also change during fatigue. In summary, the neuromuscular processes during fatigue are more complex than captured in global feedback models, necessitating another approach.

MUSCLE WISDOM: RATE OPTIMIZATION AND ITS MECHANISMS

Of increasing visibility and appeal, is the formulation of a hypothesis based on system optimization, and termed *muscle wisdom* (for details, see Binder-Macleod, Chapter 16). The idea is currently based on results using imposed (electrical) stimulation and voluntary contractions of human subjects. The issue of its generalization is still open (Enoka and Stuart, 1992)

Motor Unit Firing Rate Adaptation

During fatiguing maximal voluntary isometric muscle contractions, human motor units reduce their firing rate over several tens of seconds from an initially high value (Bigland-Ritchie et al., 1983a,b; Marsden et al., 1983; Bigland-Ritchie & Woods, 1984; Gandevia et al., 1990); for exceptions see Enoka and Stuart (1992). From simple control considerations, this decrease is counter-intuitive because, in view of the monotonic (sigmoidal) steady-state force-frequency relation (e.g., Clamann, 1990; Kernell, 1992; for further details, see Kernell, Chapter 9), compensation for force reduction would seem to call for a firing rate increase. Decrease in force-producing capability during continuing muscle exercise being inevitable, the only possibility is to minimize it. Why then would a decrease in firing rate optimize force output?

During progressing fatigue, twitch contraction relaxation slows in whole muscle, motor units and muscle fibers (Bigland-Ritchie et al., 1983b; Dubose et al., 1987; reviewed in Bigland-Ritchie, 1993). This has several consequences. Contractile slowing reduces fusion frequency and, hence, the need for high activation rates which are particularly fatiguing (Marsden et al., 1983; Binder-Macleod & Guerin, 1990; Clamann, 1990; Binder-Macleod & Barker, 1991; reviewed in Gandevia, 1993). Thus, to maintain the optimal force output achievable, activation rates should decrease over time (see also Botterman & Cope, 1988). Also, the reduced firing rate may help prevent failure of spike propagation at motor axon branching points (see Fuglevand et al., 1993; Fuglevand, Chapter 6).

Therefore, it has been hypothesized that the rate reduction is an adaptive optimizing mechanism (i.e., muscle wisdom; Bigland-Ritchie et al., 1983a; Marsden et al., 1983; Enoka & Stuart, 1992; see also Binder-Macleod, Chapter 16). An implicit corollary is that each motor unit should display its particular rate reduction profile tailored to its individual contractile and fatigue properties. Another corollary is that, since fatigability of muscle depends on muscle length (Fitch & McComas, 1985) and other peripheral factors, the optimal adjustment of motoneuron firing patterns would best be accomplished by taking account of peripheral circumstances through muscle afferent feedback (see below).

What requires optimization for muscle wisdom to occur is a fatiguing motor unit's discharge pattern, in terms of both its mean rate (*frequency optimization*; see Kernell 1993) and, potentially, its variability (*pattern optimization,* for further details, see Binder-Macleod, Chapter 16). Several possible mechanisms need to be considered, including: descending commands, motoneuron properties, and reflex effects onto motoneurons from fatiguing muscle(s); that is, as mediated by mono- and oligosynaptic excitatory connections, Renshaw cells, Ib inhibitory interneurons, other inhibitory interneuronal systems, and interneurons involved in CPGs.

Descending Motor Commands

Voluntary motoneuron activation declines with fatigue (McKenzie et al., 1992; Gandevia 1993; for further details, see Gandevia et al., Chapter 20). This may affect motoneuron discharge not only directly, but also indirectly via interneurons intercalated in reflex and other pathways.

Motoneuron Properties

Motoneuron properties may contribute to late firing-rate adaptation (Fig. 1*A*), through a variety of potential mechanisms that are discussed in Sawczuk et al., Chapter 8. Appropriately, late adaptation is more prevalent in F-type than in S-type motoneurons (Fig. 1A; Kernell & Monster, 1982; Spielmann et al., 1993). If these mechanisms were intrinsic to the motoneuron and unalterable by external (synaptic-input) influences, late adaptation could not be matched to mechanical and metabolic circumstances encountered by the motor unit's muscle fibers. For instance, the strength and time course of fatigue depends on muscle length, a peripheral variable (Fitch & McComas, 1985). However, at least after-hyperpolarization is mutable: for example, by descending monoaminergic systems favoring plateau potentials and perhaps other descending commands (see above), and during fictive locomotion in decerebrate cats, where it is reduced and does not significantly affect motoneuron discharge patterns (Brownstone et al., 1992). Likewise it may not be an important feature of motoneuronal discharge during natural movements, including maximal voluntary muscle contractions (Stuart & Callister, 1993). But the possibility remains that sensory feedback from the fatiguing muscle alters after-hyperpolarization to contribute to appropriate motoneuronal adaptation (Windhorst & Kokkoroyiannis, 1991).

Afferent Feedback from the Fatiguing Muscle

To match the discharge patterns of motor units to their changing contractile properties, afferent feedback appears to be required (for details, see Garland & Kaufman, Chapter 19). Bigland-Ritchie and colleagues (1986), Woods and colleagues (1987) and Garland and colleagues (1988) working on humans, and Hayward and colleagues (1988) working on cats provided indirect evidence that feedback from the fatiguing muscle may inhibit homonymous and synergistic motoneurons and thus reducing their firing rate. There are two possible sources of such feedback: large-diameter mechano-afferents, and small-diameter polymodal afferents.

The first group includes afferent feedback to excite motoneurons mono- or oligosynaptically, in particular, Ia afferents from primary muscle spindle endings (Fig. 2) and group II afferents from secondary muscle afferents. Furthermore, group Ib afferents from Golgi tendon organs whose effects are mediated via Ib inhibitory and other interneurons (Fig. 2). The second group requires the mediation of interneurons, including: 1) Renshaw cells; 2) Ib inhibitory interneurons receiving input from Ib afferents; 3) interneurons mediating presynaptic inhibition; 4) Ia interneurons mediating reciprocal inhibition; 5) interneurons mediating polysynaptic effects from muscle spindle afferents; and 6) interneurons establishing *private* pathways from small-diameter muscle afferents to motoneurons.

Large-Diameter Muscle Afferents

Large-diameter afferents from fatiguing muscles may act in two different ways to reduce motoneuron firing rate during fatigue. Human muscle spindle afferents reduce their mean discharge rate in the course of fatiguing muscle contractions, possibly due to a reduction in fusimotor drive to the spindles (Macefield et al., 1991; for details, see Hagbarth & Macefield, Chapter 18). The ensuing disfacilitation of homonymous and synergistic motoneurons might contribute to the decline of their firing rate. Group II afferents from muscle spindles may add to these effects. The effect of muscle fatigue on Golgi tendon organ responsiveness is not clear, nor is their reflex effect (reviewed in Enoka & Stuart, 1992; Stuart & Callister, 1993). The changes in mean discharge rate, particularly of Ia and Ib afferents, have been hypothesized to be supplemented by much subtler means of signaling reflected in their precise temporal patterns of discharge.

Due to their high sensitivity to small muscle fiber length changes, spindle Ia afferents exhibit prominent transient responses, usually including rate reductions, to twitch contractions even of single motor units (Windhorst, 1988; Windhorst & Kokkoroyiannis, 1991). Golgi tendon organs respond with firing rate increases to single motor unit contractions (for review, see Windhorst, 1988). Large-diameter muscle afferents would thus appear optimally suited to monitor the change in motor unit contractile properties during fatigue (decay in amplitude and slowing of contraction), as hypothesized by Windhorst & Kokkoryiannis (1991). Surprisingly, however, even if large-diameter afferents changed their discharge pattern in response to changing contractile properties, this does not appear to significantly affect motoneuron firing rates. Alterations in contractile muscle properties brought about by changes in muscle temperature and length do not affect motoneuron firing rates (Bigland-Ritchie, 1993).

Small-Diameter Muscle Afferents

Group III-IV (and some group II non-spindle) polymodal afferents may monitor the mechanical state of the muscle and be sensitized by metabolic substances released by fatiguing muscle fibers (Mense, 1986; for further details, see Garland & Kaufman, Chapter 19). One hypothesis proposes that these afferents are primarily responsible for the inhibition exerted onto homonymous and synergistic motoneurons from fatiguing muscles (see Bigland-Ritchie, 1993). Indeed, acute animal experiments have demonstrated that selected afferents in these groups were excited by products of muscle metabolism, such as K^+, H^+, lactic acid and arachidonic acid (Rotto & Kaufmann, 1988; Sinoway et al., 1993; see also Garland & Kaufman, Chapter 19). Other acute animal experiments have shown that afferents in these groups were excited after forceful, long-sustained fatigue-inducing contractions (Hayward et al., 1991). It was suggested (Hayward et al., 1988; 1991) that these afferents could inhibit homonymous and synergistic motoneurons via Ib inhibitory interneurons or Renshaw cells which receive excitatory input from high-threshold muscle afferents (see Windhorst, 1988; Jankowska, 1992).

The above reflex mechanisms are not mutually exclusive, but may predominantly act at different times during an ongoing fatiguing contraction. In fact, the high-threshold muscle afferents studied by Hayward and colleagues (1991) were excited at relatively long latency after the onset of sustained contraction. Initially, the sensitivity of these afferents to mechanical stimuli was actually reduced. This led Gandevia (1993) to suggest that the decline in motoneuronal firing rate over the first few seconds of sustained contractions may primarily reflect a reduction in muscle spindle facilitation, whereas the inhibition from small-diameter muscle afferents may occur subsequently.

SPINAL INTERNEURONS: INTEGRATIVE CENTERS

In an anatomical but only limited functional sense, motoneurons are the *final common path* to muscle. Rather, interneurons and propriospinal neurons comprise premotoneuronal networks which perform most of the integrative computation. Convergence of descending motor commands and sensory afferent information on to these networks enables them to modulate, up-date and adapt motor commands to motoneurons according to the prevailing state of the peripheral mechanical apparatus, and vice versa. This appears to be a general principle, applying to all interneurons identified to date, including Renshaw cells, Ib inhibitory interneurons, reciprocal Ia inhibitory interneurons, interneurons mediating presynaptic inhibition, etc. (Jankowska, 1992). As an example, Fig. 4 (below) depicts the convergence of inputs onto the *traditional* Ib interneuron.

Hayward and colleagues (1991) emphasized two types of interneurons for potentially mediating inhibitory reflex effects onto motoneurons from small-diameter muscle afferents, Renshaw cells and Ib inhibitory interneurons. These authors gave priority to the Renshaw cell pathway because they felt that this pathway might be involved in specifically regulating motoneuron discharge rate (Hayward et al., 1991; see also Windhorst & Kokkoroyiannis, 1991).

RECURRENT INHIBITION

Hayward and colleagues (1991) presumed that in the course of ongoing fatiguing contractions, the Renshaw cell burst discharge after each motoneuron spike is strengthened and prolonged by high-threshold muscle afferent feedback, so as to prolong motoneuron interspike intervals. This presumption requires that a motoneuron's action potential can influence the timing of its own next spike via recurrent inhibition; that is, via a high-gain recurrent inhibitory feedback loop from each motoneuron back onto itself. This assumption is by no means self-evident. Each motoneuron excites many Renshaw cells, each of which, in turn, receives input from many motoneurons and distributes its output to many motoneurons. As such, the response of Renshaw cells to asynchronous input from many motoneurons should be a somewhat erratic discharge without producing a prolonged burst to a single motoneuron's individual action potentials, as proposed by Hayward and colleagues (1991). This problem of convergence-divergence is potentially obviated by *focusing* and/or *synchronization*. For focusing, recurrent inhibition should be strongest between adjacent motoneurons (Windhorst, 1989; Windhorst & Kokkoroyiannis, 1991), and there is experimental evidence that neighboring motoneurons are coupled via stronger recurrent inhibition than more remote motoneurons (Hamm et al., 1987). Moreover, individual motoneurons are subject to significant self-inhibition (van Keulen, 1981). For synchronization to play a role, small subsets of adjacent motoneurons should discharge synchronously; and therefore, increase the strength and duration of the burst discharges of local Renshaw cells. These possibilities are not mutually exclusive. Provided the above prerequisites are met, the mechanism suggested by Hayward and colleagues (1991) would also be appealing insofar as fast fatigable (FF) motoneurons provide the strongest input to Renshaw cells (Hultborn et al., 1988; Windhorst, 1988). Their motoneuron discharges could thus produce the Renshaw cell burst discharges required to delay each of the subsequent spikes, with previously recruited smaller motoneurons continuing to provide background excitation to the fatiguing muscle.

If recurrent inhibition contributes to the decline of motoneuron rate during sustained maximal contractions, it could do so by virtue of its own intrinsic dynamics and/or by a change in its gain brought on by descending commands and/or sensory feedback from fatiguing muscle. To address the first possibility, Windhorst and Kokkoroyiannis (1991) recorded the discharges of cat Renshaw cells to motor axon stimulation at time-varying rates simulating motoneuron adaptation during fatigue (Fig. 3*A*). Typically, during such a test, the relation between motoneuron and Renshaw cell discharge rates (Fig. 3*B*) was nonlinear. Recurrent inhibition was strong over the initial few hundred milliseconds and could conceivably have contributed to the early adaptation of motoneuron firing rate. Subsequently, recurrent inhibition was relatively depressed for several seconds, but, thereafter, became stronger again. This latter intrinsic increase in gain could have contributed to the prolonged decline in motoneuron discharge rate. More generally, this peculiar input-output relationship shows the need to obtain quantitative data, if precise regulation of motoneuron discharge patterns by interneurons is to be understood.

Renshaw cells receive additional input from various sources (Windhorst 1988). For example, activation of a motoneuron pool by descending tracts is often associated with an inhibition of Renshaw cells. Thus, if during maintained fatiguing muscle contractions, descending motor activation declines (McKenzie et al., 1992; Gandevia et al., 1993; Gandevia, 1993; for details see Gandevia et al., Chapter 20), the concomitant reduction in descending inhibition of Renshaw cells may increase recurrent inhibition, which, in turn, may contribute to a decline in motoneuron firing rate. Also, with excitation of small-diameter muscle afferents, the dynamically changing strength of recurrent inhibition could be appropriately augmented during fatigue so as to optimally adapt motoneuron discharge, in line with a suggestion of Hayward and colleagues (1991). However, there is little relevant data on these possibilities. Kukulka and colleagues (1986), using an indirect H-reflex method in humans, provided some evidence that recurrent inhibition and/or summation of motoneuron after-hyperpolarizations increased over the first 30 s of maximal sustained triceps surae contractions. How this change comes about, which afferents are involved and how they affect the Renshaw cell dynamics shown in Fig. 3 are not yet known. In principle, afferent effects onto Renshaw cells could be exerted via two mechanisms. Firstly, the mean discharge rate (bias) of Renshaw cells could be elevated; and secondly, their dynamic response to changing motoneuronal input (see Fig. 3) could be altered. Alternatively, or in addition, motoneuron after-hyperpolarization could be modulated (see above).

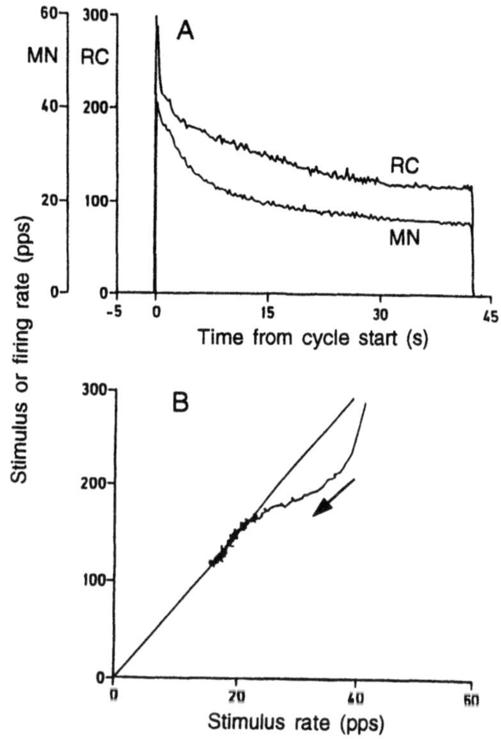

Figure 3. Response of a Renshaw cell to time-varying motor axon stimulation simulating adapted motoneuron firing. In an anesthetized cat with dorsal roots L6-S1 cut, the biceps posterior muscle nerve was stimulated with a pulse train of 42-43 s duration. It started at a high initial rate (ca. 45 pps), and decayed exponentially with a fast (10 ms) and a slow (20 s) time constant towards a final rate of 16 pps. This train was repeated 10 times every 75 s, and the cycle histogram (bin width 200 ms) displaying the average stimulus rate over time is shown as the lower curve labeled MN in part *A*. The cycle-averaged firing rate elicited in the Renshaw cell (RC) is shown by the upper curve labeled RC (same bin width). The wriggled thin line in part *B* is a plot derived from the cycle histograms in part *A* of instantaneous RC discharge rate *vs.* stimulus rate with time progressing as indicated by the arrow. The curved part of this dynamic input-output relationship covers the first 8-10 s. The straight line has been drawn through the origin, and the later part of this relationship corresponding to times beyond 8-10 s (time direction indicated by arrow). During this later phase within the cycle, the RC's discharge rate was close to proportional to the stimulus rate; that is, the simulated motor axon activation rate. The slope of this line represents the input-output gain of the system transforming motor axon activation rate into RC discharge rate. As compared to the later period, this gain undergoes dynamic changes during the first few seconds (up to 10 s), starting from an initially high to a declining gain which then increases again to settle to a *static* value represented by the slope of the straight line.

EFFECTS FROM Ib AFFERENTS

Golgi tendon organ Ib afferents from one muscle exert reflex effects that spread widely to many motor nuclei of a limb and may be inhibitory or excitatory, this effect in part depending on movement context (Jami, 1992; Jankowska, 1992; see also below). Thus, there are excitatory and inhibitory Ib interneurons.

The responses of Ib interneurons to muscle contraction before and during muscle fatigue are neither well known, nor easily predicted. About half of them receive excitatory input from muscle spindle Ia afferents (see above). They also receive input from high-threshold muscle afferents (Jankowska, 1992). As such, firing rates of Ib interneurons might decrease due to decreasing inputs from Ia and Ib afferents during ongoing fatiguing muscle contraction (Fig. 4), but increase due to augmented inputs from high-threshold muscle afferents. The balance of these effects in determining mean discharge rates is hard to predict, as is modulation of the gain of Ib interneurons in response to decrementing and slowing contractions of individual motor units (see above). Therefore, the responses of Ib interneurons need to be determined experimentally. Another important issue is that the different convergent inputs (cf. Fig. 4) by no means distribute homogeneously to each single neuron, entailing a fractionation of the interneuron pool (Harrison & Jankowska, 1985). Such a functional diffentiation could principally be used to re-distribute reflex effects to different motor unit combinations depending on peripheral circumstances; for example, inhomogeneous distribution of fatigue in muscles with different motor unit compositions. But again, this

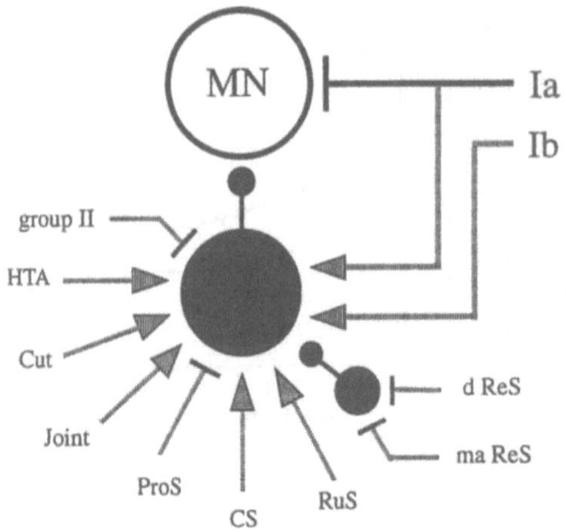

Figure 4. Convergence of segmental afferent and spinally descending inputs onto lumbar group Ib inhibitory interneurons. Note that, as mentioned before, only about 40% of cat Ib inhibitory interneurons receive excitatory Ia afferent input (Jankowska et al., 1981). There are, therefore, subpopulations of Ib interneurons with different input patterns (Jankowska 1992). Dashed arrowheads, mixed excitatory and inhibitory effects. Dashed arrowshafts, oligo- or polysynaptic pathways. Abbreviations: CS, corticospinal; Cut, cutaneous; HTA, high-threshold muscle and joint afferents; ProS, propriospinal; d ReS, dorsal reticulospinal; ma ReS, monaminergic reticulospinal; RuS, rubrospinal; VS, vestibulospinal (modified from Baldissera et al., 1981; Harrison & Jankowska, 1985).

has yet to be studied. Similar functional fractionation may apply to other interneuron pools, such as reciprocal Ia inhibitory interneurons (less pronounced: E. Jankowska, personal communication) and interneurons mediating presynaptic inhibition.

PRESYNAPTIC INHIBITION

An additional complication making predictions difficult results from the fact that the synaptic efficacy of spindle group Ia and II, and tendon organ Ib afferents can be altered via presynaptic inhibition elicited by activity in the same groups of afferents, by high-threshold muscle afferents, by signals in descending tracts, and by the CPGs (see below) (Jankowska, 1992). Since descending signals and muscle afferent feedback change during fatiguing muscle contractions, so too should signal transmission from some afferent classes to spinal networks. Presynaptic inhibition generally has a long duration of the order of a few hundreds of milliseconds. One of the major advantages of pre-*vs.* postsynaptic inhibition is its potential for separate control of subsets of synaptic inputs to a neuron, but this possibility is not to be always exploited. For example, GABAergic interneurons can simultaneously exert presynaptic inhibition of primary afferents and postsynaptic inhibition of motoneurons (Rudomin, 1990; Jankowska, 1992); therefore, irrespective of their effect on synaptic transmission, last-order presynaptic inhibitory interneurons provide for yet another postsynaptic inhibitory pathway to motoneurons from muscle afferents affected by muscle fatigue.

The pattern of presynaptic inhibition is complex. Group Ia and II muscle spindle afferents and group Ib Golgi tendon organ afferents appear to be presynaptically inhibited by separate populations of interneurons with slightly different input patterns (Rudomin, 1990; Jankowska, 1992). As with Ib inhibitory interneurons, this arrangement may generate functional fractionation that could be used to advantage in re-shaping spatio-temporal motor activity during fatigue.

RECIPROCAL INHIBITION

Thus far, emphasis has been placed on afferent effects from fatiguing muscle onto synergistic motoneurons. However, reciprocal inhibition of antagonists (Fig. 2) will be affected as well, because reciprocal Ia inhibitory interneurons receive virtually the same convergent inputs as do Ib inhibitory interneurons (Fig. 4), particularly from Ia afferent input and high-threshold muscle afferent input. This may not matter when only agonist muscle groups are undergoing fatiguing contractions as in most of the experimental paradigms discussed above. However, when fatigue of alternating rhythmic movements such as locomotion is considered, there is no information about reciprocal inhibition, or other networks, including those generating the basic rhythmic pattern.

LOCOMOTION

It is everyday knowledge that muscle fatigue and its attendant problems arise during rhythmic movements, such as locomotion and mastication. Their basic patterns are generated actively by the CNS by way of CPGs. Even the isolated lumbosacral spinal cord is capable, upon proper pharmacological pretreatment, to produce quite elaborate motor output patterns which are quite akin to normal ones. This applies even if movement and hence sensory afferent feedback is abolished by paralysis (*fictive locomotion*; see below). Despite intense efforts, the network constituting each CPG has not yet been identified in mammals and, therefore, no data are available on the effects of sensory feedback from fatiguing muscle.

Motoneurons, Renshaw cells and reciprocal Ia inhibitory interneurons (Fig. 2) are presumably included in the CPG, their interactions including mutual inhibition and many other control components found in lower animals' CPGs (Getting, 1989); cf., Pratt & Jordan, 1987; Gelfand et al., 1988). The discharge patterns of Renshaw cells and Ia inhibitory interneurons are rhythmically modulated during locomotion, and they are also subject to effects from high-threshold muscle afferents, which are responsive to muscle fatigue (for details, see Garland & Kaufman, Chapter 19). In 1912, Graham-Brown postulated that two tonically excited *half-centers* of interneurons, each driving flexors or extensors at the same joint, would produce the alternating locomotor pattern by reciprocally inhibiting each other. Candidates for Graham-Brown's half-centers might be the interneurons which mediate the flexion reflex response to noxious and other low-threshold stimuli (Jankowska et al., 1967). Like the half-centers, these interneurons mutually inhibit each other Jankowska et al., 1967a,b and, undoubtedly, receive input from muscle afferents excited during muscle fatigue. If these interneurons can be demonstrated to be part of CPGs, then the CPGs, themselves, should be directly affected by muscle fatigue.

Changes in environmental conditions, errors of posture and movement, and alterations in internal system properties such as fatigue require that centrally generated patterns be adapted on a short-term basis. CPGs and sensory feedback must interact rapidly, with sensory information modulating CPG activity and, vice versa, with descending controls superimposed, as well.

Sensory feedback plays important roles in timing locomotor activity and the transition between different phases of rhythmic movement, including reinforcement of ongoing motor output and adjustment to environmental conditions (Pearson, 1993). For example, in the presence of sensory feedback, cat locomotion is more normal and stable, less fatigable, and more easily adjustable to a wider range of speeds (Grillner, 1985). But the mechanisms for these effects are not known. Nonetheless, these effects must change during fatigue, either by a primary change in discharge patterns, or by modulatory effects on intercalated interneurons.

In the phase switch from extension to flexion, it is likely that cat mid-lumbar interneurons are implicated. These receive monosynaptic excitatory input from group II muscle spindle (and other) afferents from hip muscles (Jankowska, 1992). Another set of interneurons which may be involved in switching from extension to flexion are those mediating reciprocal inhibition. During human walking, the ankle dorsiflexors are stretched in the late stance phase; (i.e., between heel off and toe off), when their Ia muscle spindle afferents are excited and thereby help activate their own motoneurons as well as reciprocally inhibit the antagonist soleus motoneurons (Capaday et al., 1990). A third interneuronal system of potential importance involves those mediating reflex effects from Golgi tendon organ afferents. Whereas in quiescent preparations, extensor Ib afferents exert widespread inhibition on leg extensor motoneurons (Baldissera et al., 1981), this effect is reduced or abolished during locomotion. Instead, extensor Ib afferents activate extensor motoneurons and inhibit ipsilateral flexor motoneurons, effects that vary in size according to the phase of the locomotor step cycle (Pearson & Collins, 1993; Gossard et al., 1994). Such declining extensor force and Ib discharge in the late stance phase may release flexors from inhibition and facilitate initiation of the swing phase (Pearson, 1993). Since muscle fatigue affects all the above afferent systems, it should disturb locomotor timing, as, for example, when fatigue makes walking less stable and secure.

The locomotor evidence suggests that its CPGs influence the sign and strength of various reflexes as a means of selecting sensory information that is appropriate for the phase of the particular step (Dietz, 1992; Pearson, 1993). Substantial evidence is on hand that *reflexes can be modulated in size, or even reversed in effect, as dependent on the phase of the step*. (Forssberg et al., 1977; Yang & Stein, 1990). Such phase-dependent reflex control and reversal is illustrated in Fig. 5. It has three features at the segmental level (i.e., suprasegmental effects are not shown).

1. The first effect (*presynaptic inhibition*, 1 in Fig. 5) occurs at sites immediately after the entrance of sensory afferents into the spinal cord. Its strength varies as the locomotor phase (Dubuc et al., 1988; Dueñas et al., 1990; Gossard & Rossignol, 1990). Such phase control is evident in even the presynaptic inhibitory modulation of the monosynaptic Ia afferent-motoneuronal connection.
2. The CPG may modulate any premotor interneuron intercalated between primary afferents and motoneurons (e.g., first and last order, 2 and 3, respectively in Fig. 5), including Renshaw cells and reciprocal Ia inhibitory interneurons. Since the latter receive convergent input from many sensory afferents, transmission of this information to motoneurons is phasically gated. Furthermore, autogenetic Ib inhibition from extensor muscles may be depressed and replaced with excitation during locomotion. It is also appropriate to consider γ-motoneurons as premotoneurons, insofar as they integrate many segmental afferent and descending influences, and transmit such influences via muscle spindles and their afferents to α-motoneurons and back onto themselves.
3. Motoneurons and interneurons receive locomotor drive potentials from the CPG. During the hyperpolarizing phases of these potentials, which are caused in part by inhibitory inputs from the CPG, excitatory sensory inputs are effectively shunted. Thus, the motoneurons themselves contribute to locomotor reflex modulation (4 in Fig. 5).

In summary, reflex modulation plays an important role in normal locomotor activity and the mechanisms involved are strongly affected by muscle fatigue.

SUMMARY ON THE SPINAL SEGMENTAL REGULATION OF MOTOR UNIT FIRING RATE

The CNS appears to have mechanisms available for mitigating the effects of muscle fatigue, at lease in the short term. Many of them shape the firing patterns of motor units in

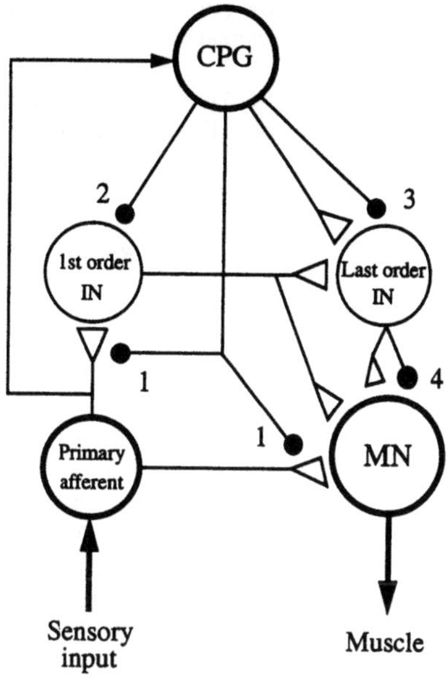

Figure 5. Scheme of the possible sites of modulation of reflex transmission by the locomotor CPG. The sites include: *1*, terminals of primary sensory afferents (for presynaptic inhibition); *2*, inhibition of the first-order interneuron; *3*, postsynaptic excitation and inhibition of the last-order interneurons which connect monosynaptically (site 4) with motoneurons. Effects of muscle afferents on the CPG are also indicated (modified from Sillar, 1991; his Fig. 1).

order to optimize their force output. It would make physiological sense that these mechanisms primarily target fast-fatigable motor units (Stuart & Callister, 1993). However, these mechanisms are multifarious as to their location and principles of operation, and they understandably cooperate in a complex way. To unravel these mechanisms is an enormous task for experimenters. Of special immediate concern is to measure signal transmission in spinal interneurons and motoneurons and the modulating effects exerted by those of the muscle afferents that are affected by muscle fatigue. The elements and networks requiring further detailed study include: 1) α-motoneurons; 2) γ-motoneurons; 3) recurrent inhibition via Renshaw cells; 4) Ib inhibitory interneurons; 5) reciprocal Ia inhibitory interneurons; 6) interneurons mediating presynaptic inhibition; and 7) other interneurons whose effects on motoneurons, such as those possibly involved in locomotor pattern generation. This list is by no means exhaustive. As such work is undertaken, theoretical concepts and computer models will need to be developed in parallel, in order to integrate experimental data.

ACKNOWLEDGMENTS

This work was supported by the Medical Research Council of Canada (*U.W.*), and the Alberta Heritage Foundation for Medical Research, Canada. *G.B.* held a postdoctoral fellowship from the Alberta Paraplegic Foundation, Canada. Attendance of *U.W.* at the 1994 Bigland-Ritchie conference was supported, in part, by the University of Miami.

REFERENCES

Baldissera F, Hultborn H & Illert M (1981). Integration in spinal neuronal systems. In: Brooks VB (ed.), *Handbook of Physiology, Vol. II, Part 1, The Nervous System*, pp. 509-595. Bethesda: American Physiological Society.

Bevan L, Laouris Y, Reinking RM & Stuart DG (1992). The effect of the stimulation pattern on the fatigue of single motor units in adult cats. *Journal of Physiology (London)* **449**, 85-108.

Bigland-Ritchie BR (1993). Regulation of motorneuron firing rates in fatigue. In: Sargeant AJ, Kernell D (eds.), *Neuromuscular Fatigue*, pp 147-155. Amsterdam: Royal Netherlands Academy of Arts and Sciences.

Bigland-Ritchie B, Dawson NJ, Johansson RS & Lippold OCJ (1986). Reflex origin for the slowing of motoneurone firing rates in fatigue of human voluntary contractions. *Journal of Physiology (London)* **379**, 451-459.

Bigland-Ritchie B, Johansson RS, Lippold OCJ & Woods JJ (1983a). Changes in motoneurone firing rates during sustained maximal voluntary contractions. *Journal of Physiology (London)* **340**, 335-346.

Bigland-Ritchie B, Johansson RS, Lippold OCJ & Woods JJ (1983b). Contractile speed and EMG changes during fatigue of sustained maximal voluntary contractions. *Journal of Neurophysiology* **50**, 313-324.

Bigland-Ritchie B & Woods JJ (1984). Changes in muscle contractile properties and neural control during human muscular fatigue. *Muscle & Nerve* **7**, 691-699.

Binder MD & Mendell LM (eds.) (1990). *The Segmental Motor System.* New York: Oxford University Press

Binder-Macleod SA & Barker CB (1991). Use of a catch-like property of human skeletal muscle to reduce fatigue. *Muscle & Nerve* **14**, 850-857.

Binder-Macleod SA & Guerin T (1990). Preservation of force output through progressive reduction of stimulation frequency in human quadriceps femoris muscle. *Physical Therapy* **70**, 619-625.

Botterman BR & Cope TC (1988). Motor-unit stimulation patterns during fatiguing contractions of constant tension. *Journal of Neurophysiology* **60**, 1198-1214.

Brownstone RM, Jordan LM, Kriellaars DJ, Noga BR & Shefchyk SJ (1992). On the regulation of repetitive firing in lumbar motoneurones during fictive locomotion in the cat. *Experimental Brain Research* **90**, 441-455.

Burke RE (1981). Motor units: anatomy, physiology, and functional organization. In: Brooks VB (ed.), *Handbook of Physiology, Vol. II, Part 1, The Nervous System*, pp 354-422. Bethesda: American Physiological Society.

Burke RE, Levine DN, Tsairis P & Zajac FE (1973). Physiological types and histochemical profiles in motor units of the cat gastrocnemius. *Journal of Physiology (London)* **234**, 723-748.

Capaday C, Cody FWJ & Stein RB (1990). Reciprocal inhibition of soleus motor output in humans during walking and voluntary tonic activity. *Journal of Neurophysiology* **64**, 607- 616.

Clamann HP (1990). Changes that occur in motor units during activity. In: Binder MD, Mendell LM (eds.), *The Segmental Motor System*, pp. 239-257. New York: Oxford University Press.

Dietz V (1992). Human neuronal control of automatic functional movements: interaction between central programs and afferent input. *Physiological Reviews* **72**, 33-69.

Dubose L, Schelhorn TB & Clamann HP (1987). Changes in contractile speed of cat motor units during activity. *Muscle & Nerve* **10**, 744-752.

Dubuc R, Cabelguen J-M & Rossignol S (1988). Rhythmic fluctuations of dorsal root potentials and antidromic discharges of primary afferents during fictive locomotion in the cat. *Journal of Neurophysiology* **60**, 2014-2036.

Dueñas SH, Loeb GE & Marks WB (1990). Monosynaptic and dorsal root reflexes during locomotion in normal and thalamic cats. *Journal of Neurophysiology* **63**, 1467-1476.

Edwards RHT, Wiles CM, Gohil K, Krywawych S & Jones DA (1982) Energy metabolism in human myopathy. In: Schotland DL (ed.), *Disorders of the Motor Unit*, pp. 715-726. New York: Wiley.

Enoka RM & Stuart DG (1992). Neurobiology of muscle fatigue. *Journal of Applied Physiology* **72**, 1631-1648.

Fitch S & McComas A (1985). Influence of human muscle length on fatigue. *Journal of Physiology (London)* **362**, 205-213.

Forssberg H, Grillner S & Rossignol S (1977). Phasic gain control of reflexes from the dorsum of the paw during spinal locomotion. *Brain Research* **132**, 121-139.

Fuglevand AJ, Zackowski KM, Huey KA & Enoka RM (1993). Impairment of neuromuscular propagation during human fatiguing contractions at submaximal forces. *Journal of Physiology (London)* **260**, 549-572.

Gandevia SC (1993). Central and peripheral components to human isometric muscle fatigue. In: Sargeant AJ, Kernell D (eds.), *Neuromuscular Fatigue*, pp. 156-164. Amsterdam: Royal Netherlands Academy of Arts and Sciences.

Gandevia SC, Macefield G, Burke D & McKenzie DK (1990). Voluntary activation of human motor axons in the absence of muscle afferent feedback. The control of the deafferented hand. *Brain* **113**, 1563-1581.

Gandevia SC, Macefield VG, Bigland-Ritchie B, Gorman R & Burke D (1993). Motoneuronal output and gradation of effort in attempts to contract acutely paralyzed leg muscles in man. *Journal of Physiology (London)* **474**, 411-427.

Garland SJ, Garner SH & McComas AJ (1988). Reduced voluntary electromyographic activity after fatiguing stimulation of human muscle. *Journal of Physiology (London)* **401**, 547-556.

Gelfand IM, Orlovsky GN & Shik ML (1988). Locomotion and scratching in tetrapods. In: Cohen AH, Rossignol S, Grillner S (eds.), *Neural Control of Rhythmic Movements in Vertebrates*, pp 167-199. New York: Wiley.

Getting PA (1989). Emerging principles governing the operation of neural networks. *Annual Review of Neuroscience* **12**, 185-204.

Gordon DA, Enoka RM & Stuart DG (1990). Motor-unit force potentiation in adult cats during a standard fatigue test. *Journal of Physiology (London)* **421**, 569-582.

Gossard J-P, Brownstone RM, Barajon I & Hultborn H (1994). Transmission in a locomotor-related group Ib pathway from hindlimb extensor muscles in the cat. *Experimental Brain Research* **98**, 213-228.

Gossard J-P & Rossignol S (1990). Phase-dependent modulation of dorsal root potentials evoked by peripheral nerve stimulation during fictive locomotion in the cat. *Brain Research* **537**, 1-13.

Graham-Brown T (1912). The factors in rhythmic activity of the nervous system. *Proceedings Royal Society London B* **85**, 278-289.

Grillner S (1985). Neural control of vertebrate locomotion - central mechanisms and reflex interaction with special reference to the cat. In: Barnes WJP, Gladden MH (eds.), *Feedback and Motor Control in Invertebrates and Vertebrates*, pp. 35-56. London: Croom Helm.

Hamm TM, Sasaki SI, Stuart DG, Windhorst U & Yuan C-U (1987). Distribution of single-axon recurrent inhibitory post-synaptic potentials in a single spinal motor nucleus in the cat. *Journal of Physiology (London)* **388**, 653-664.

Harrison PJ & Jankowska E (1985). Sources of input to interneurones mediating group I non- reciprocal inhibition of motoneurones in the cat. *Journal of Physiology (London)* **361**, 379-401.

Hayward L, Breitbach D & Rymer WZ (1988). Increased inhibitory effects on close synergists during muscle fatigue in the decerebrate cat. *Brain Research* **440**, 199-203.

Hayward L, Wesselmann U & Rymer WZ (1991). Effects of muscle fatigue on mechanically sensitive afferents of slow conduction velocity in the cat triceps surae. *Journal of Neurophysiology* **65**, 360-370.

Henneman E (1957). Relation between size of neurons and their susceptibility to discharge. *Science* **126**, 1345-1346.

Houk JC & Rymer WZ (1981). Neural control of muscle length and tension. In: Brooks VB (ed.), *Handbook of Physiology, Vol II, Part 1, The Nervous System*, pp. 257-323. Bethesda: American Physiological Society.

Houk JC, Singer JJ & Goldman MR (1970). An evaluation of length and force feedback to soleus muscles of decerebrate cats. *Journal of Neurophysiology* **33**, 784-811.

Jami L (1992). Golgi tendon organs in mammalian skeletal muscle: functional properties and central actions. *Physiological Reviews* **72**, 623-666.

Jankowska E (1992). Interneuronal relay in spinal pathways from proprioceptors. *Progress in Neurobiology* **38**, 335-378.

Jankowska E, Johanisson T & Lipski J (1981). Common interneurons in reflex pathways from group Ia and Ib afferents of ankle extensors in the cat. *Journal of Physiology (London)* **310**, 381-402.

Jankowska E, Jukes MGM, Lund S & Lundberg A (1967a). The effect of DOPA on the spinal cord. 5. Reciprocal organization of pathways transmitting excitatory action to alpha motoneurones of flexors and extensors. *Acta Physiologica Scandinavica* **70**, 369-388.

Jankowska E, Jukes MGM, Lund S & Lundberg A (1967b). The effect of DOPA on the spinal cord. 6. Half-centre organization of interneurones transmitting effects from the flexor reflex afferents. *Acta Physiologica Scandinavica* **70**, 389-402.

Kernell D (1992). Organized variability in the neuromuscular system: a survey of task- related adaptations. *Archives Italiennes de Biologie* **130**, 19-66.

Kernell D (1993) Neuromuscular fatigue and the differentiation of motoneurone and muscle unit properties. In: Sargeant AJ, Kernell D (eds.), *Neuromuscular Fatigue*, pp 139-146. Amsterdam: Royal Netherlands Academy of Arts and Sciences.

Kernell D & Monster AW (1982). Motoneurone properties and motor fatigue. An intracellular study of gastrocnemius motoneurones of the cat. *Experimental Brain Research* **46**, 197-204.

Kirsch RF & Rymer WZ (1992). Neural compensation for fatigue-induced changes in muscle stiffness during perturbations of elbow angle in human. *Journal of Neurophysiology* **68**, 449-470.

Kukulka CG, Moore MA & Russell AG (1986). Changes in human α-motoneuron excitability during sustained maximum isometric contractions. *Neuroscience Letters* **68**, 327-333.

Lenman AJR, Tulley FM, Vrbova G, Dimitrijevic MR & Towle JA (1989). Muscle fatigue in some neurological disorders. *Muscle & Nerve* **12**, 938-942.

Macefield G, Hagbarth K-E, Gorman R, Gandevia SC & Burke D (1991). Decline in spindle support to α-motoneurones during sustained voluntary contractions. *Journal of Physiology (London)* **440**, 497-512.

Marsden CD, Meadows JC & Merton PA (1983). "Muscular wisdom" that minimizes fatigue during prolonged effort in man: peak rates of motoneuron discharge and slowing of discharge during fatigue. In: Desmedt JE (ed.), *Motor Control Mechanisms in Health and Disease*, pp. 169-211. New York: Raven Press.

McKenzie DK, Bigland-Ritchie B, Gorman RB & Gandevia SC (1992). Central and peripheral fatigue of human diaphragm and limb muscles assessed by twitch interpolation. *Journal of Physiology (London)* **454**, 643-656.

Mense S (1986). Slowly conducting afferent fibers from deep tissues - neurobiological properties and central nervous actions. In: Ottoson D (ed.), *Progress in Sensory Physiology*, Vol. 6, pp. 139-219. Berlin: Springer-Verlag.

Miller RG, Green AT, Moussavi RS, Carson PJ & Weiner MW (1990) Excessive muscular fatigue in patients with spastic paraparesis. *Neurology* **40**, 1271-1274

Pearson KG (1993) Common principles of motor control in vertebrates and invertebrates. *Annual Review of Neuroscience* **16**, 265-297.

Pearson KG & Collins DF (1993). Reversal of the influence of group Ib afferents from plantaris on activity in medial gastrocnemius muscle during locomotor activity. *Journal of Neurophysiology* **70**, 1009-1017.

Powers RK & Binder MD (1991). Effects of low-frequency stimulation on the tension- frequency relations of fast-twitch motor units in the cat. *Journal of Neurophysiology* **66**, 905-918.

Pratt CA & Jordan LM (1987). Ia inhibitory interneurons and Renshaw cells as contributors to the spinal mechanisms of fictive locomotion. *Journal of Neurophysiology* **57**, 56-71.

Rotto DM & Kaufmann MP (1988). Effect of metabolic products of muscular contraction on discharge of group III and IV afferents. *Journal of Applied Physiology* **64**, 2306-2313.

Rudomin P (1990). Presynaptic inhibition of muscle spindle and tendon organ afferents in the mammalian spinal cord. *Trends in Neurosciences* **13**, 499-505.

Sillar KT (1991) Spinal pattern generation and sensory gating mechanisms. *Current Opinion in Neurobiology* **1**, 583-589.

Sinoway LI, Hill, JM, Pickar, JG & Kaufman, MP (1993). Effects of contraction and lactic acid on the discharge of group III muscle afferents in cats. *Journal of Neurophysiology* **69**, 1053-1059.

Spielmann JM, Laouris Y, Nordstrom MA, Robinson GA, Reinking RM & Stuart DG (1993). Adaptation of cat motoneurons to sustained and intermittent extracellular activation. *Journal of Physiology (London)* **464**, 75-120.

Stuart DG & Callister RJ (1993). Afferent and spinal reflex aspects of muscle fatigue: issues and speculations. In: Sargeant AJ, Kernell D (eds), *Neuromuscular Fatigue*, pp. 169- 180. Amsterdam: Royal Netherlands Academy of Arts and Sciences.

Vøllestad NK, Sejersted OM, Bahr R, Woods JJ & Bigland-Ritchie B (1988). Motor drive and metabolic responses during repeated submaximal voluntary contractions in man. *Journal of Applied Physiology* **64**, 1421-1427.

Windhorst U (1988) *How Brain-like is the Spinal Cord? Interacting Cell Assemblies in the Spinal Cord.* Berlin: Springer-Verlag.

Windhorst U & Kokkoroyiannis T (1991). Interaction of recurrent inhibitory and muscle spindle afferent feedback during muscle fatigue. *Neuroscience* **43**, 249-259.

Woods JJ, Furbush F & Bigland-Ritchie B (1987). Evidence for a fatigue-induced reflex inhibition of motoneuron firing rates. *Journal of Neurophysiology* **58**, 125-137.

Yang JF & Stein RB (1990). Phase-dependent reflex reversal in human leg muscles during walking. *Journal of Neurophysiology* **63**, 1109-1117.

THE FUSIMOTOR SYSTEM

Its Role in Fatigue

K-E. Hagbarth[1] and V. G. Macefield[2]

[1] Department of Clinical Neurophysiology
University Hospital
S-751 85 Uppsala
Sweden
[2] Prince of Wales Medical Research Institute
Randwick, Sydney 2031
Australia

ABSTRACT

Several lines of evidence point to an important role of the fusimotor system in the "muscle-wisdom" phenomenon during peripheral fatigue of some human voluntary contractions: 1) muscle afferents provide a net amplification of skeletomotor output, with the only known afferent species capable of this being the muscle spindle; 2) muscle spindle firing rates decline during constant-force voluntary contractions, so fusimotor support to skeletomotor output decreases; 3) this waning support can be offset by application of high-frequency vibration to the fatiguing muscle, which excites spindle endings; and 4) the progressive decline in motor unit firing rates during maximal voluntary contractions is abolished by blocking muscle afferent inputs, and it is argued that, at least in the initial stages of a contraction, this must be due to a progressive withdrawal of spindle support.

INTRODUCTION

The functional role of the fusimotor system in muscle fatigue has been the subject of debate for several decades. This is understandable given the potent excitatory inputs that spindle afferents provide to alpha motoneurons and hence to force production. Yet the roles of muscle spindles in motor control have varied with the fashions of the time, and so fatigue theories have had to adapt to encompass the prevailing views on fusimotor function. Therefore, before dealing with the issue of muscular fatigue it is worth reviewing the shifts in these basic hypotheses and the controversies that still exist in this field of research.

Fatigue, Edited by Simon C. Gandevia et al.
Plenum Press, New York, 1995

BASIC HYPOTHESES ON FUSIMOTOR INVOLVEMENT IN MOTOR CONTROL

The *length follow-up servo theory* from the 1950's was largely based on experiments performed on decerebrate cats. In this theory, an important role was attributed to the gamma motoneurons, regarded at the time as the initial elements in a servo-controlled system which, via the muscle spindle endings and their stretch reflex afferents, drive the skeletomotor output to achieve desired muscle length in voluntary contractions (Eldred et al., 1953; Merton, 1953). Although this was an attractive theory, subsequent human-based studies and experiments in awake and alert animals led to its dismissal (see Matthews, 1981). Partly as a result of microneurographic studies of muscle spindle activity in man, it has become clear that the skeletomotor output during voluntary contractions is not servo-driven, but rather servo-assisted by fusimotor-induced activity in muscle spindle afferents. This widely accepted *servo-assistance theory* implies that during voluntary contractions, the fusimotor-biased spindle endings provide not only a tonic autogenetic reflex support to the skeletomotor output (combined with a tonic inhibitory influence on antagonistic motoneurons) but by virtue of the high dynamic sensitivity of primary endings to stretch, the fusimotor-biased spindle endings can also exert a reflex influence on the precise timing of motor unit discharges and, thereby, provide reflex compensation for unexpected perturbations.

Much interest has been directed to factors which may influence the efficacy of this servo-assistance and the gain of the stretch reflex. Such factors may include 1) the relative strength of the fusimotor output; 2) the extent to which spindle unloading during concentric contractions counteracts the fusimotor-induced *internal* activation of spindle endings; 3) the extent to which the history of movements and contractions affects the inherent slackness of intrafusal muscle fibers; 4) variations in pre-motoneuronal excitability of segmental and supraspinal stretch reflex arcs; and 5) the level of excitability of alpha motoneurons.

Studies in alert cats indicate that the participation of the fusimotor system in motor performance is task-dependent; i.e., fusimotor outputs may or may not be accompanied by concurrent skeletomotor outputs, depending on the *motor-set* as determined by the task to be performed (Loeb & Hoffer, 1981; Prochazka & Wand, 1981). By contrast, human microneurography studies have so far not shown any consistent signs of such task-dependent dissociations between skeletomotor and fusimotor outputs (Vallbo et al., 1979; Burke, 1981; Hagbarth, 1993; but see Aniss et al., 1990). As judged by these studies, factors 4) and 5) must be awarded the primary roles in task-dependent variations in fusimotor-driven servo assistance of skeletomotor output.

HYPOTHESES ON FUSIMOTOR INVOLVEMENT IN MUSCLE FATIGUE

As exemplified by the above discussion, theories based on animal experiments do not always agree with those derived from studies in humans. Such discrepancies are particularly evident when considering theories dealing with involvement of the fusimotor system in muscle fatigue. Conflicting interpretations may partly be explained by the diversity of definitions of muscle fatigue, partly by the variety of experimental paradigms. If muscle fatigue is defined simply as "... any reduction in the force- or velocity-generating capacity of a muscle that is alleviated by rest" (Gandevia et al., 1992), then, strictly speaking, fatigue sets in at the start of a contraction, not just at the end. In the following, we will, therefore, deal not only with that later stages of fatigue, evidenced by a reduced force output, but also with the earlier stages when a gradually increasing effort is required to maintain a constant force.

While the *length follow-up servo theory* was still in vogue, the fusimotor system was envisaged as the initiator of a powerful servo which provided stretch reflex compensation for a decline in force during fatigue of extrafusal muscle fibers (Merton, 1954). That was before it was realized that, rather than an increase, a decrease in motor unit firing rates is actually required to optimize the force output of fatiguing muscles (Bigland-Ritchie et al., 1979; Marsden et al., 1983; Bigland-Ritchie et al., 1983; Binder-Macleod & Guerin, 1990). There is now convincing evidence that as extrafusal contractile fatigue develops during a sustained maximal voluntary contraction (MVC), there is a concurrent decline in motor unit firing rates (Grimby et al., 1981; Marsden et al., 1983; Bigland-Ritchie et al., 1983; Bigland-Ritchie & Woods, 1984; Bigland-Ritchie et al., 1986; Woods et al., 1987; Gandevia et al., 1990). This decline has been regarded as an expression of *muscle wisdom*, since it apparently serves to ensure appropriate economical activation of fatiguing muscle (Marsden et al., 1983): as twitch relaxation times increase during fatigue, the fusion frequency decreases such that lower firing rates are required for individual motor units to generate their maximal force (Bigland-Ritchie & Woods, 1984). In sustained submaximal voluntary contractions when increasing effort is required to maintain a desired force, the whole muscle EMG activity increases even though some motor units may show declines in firing rate (Bigland-Ritchie et al., 1986). The increase in EMG, which reflects progressive recruitment of new motor units, typically continues until the desired force can no longer be maintained, at which point motor unit firing rates decline largely in parallel with the force even though signs of *motor unit rotation* can be seen (Person, 1974). While recognizing that a reduced voluntary drive will necessarily contribute to a reduction in motor unit firing rates during a sustained voluntary contraction, the subsequent discussion will consider only segmental contributions to muscle fatigue; the phenomenon of *central fatigue* will be addressed in later Chapters 20-23.

The mechanisms underlying *muscle wisdom* are controversial, and have been discussed by Windhorst and Boorman, Chapter 17). Here, it suffices to say that the two leading explanations for the phenomenon are based on a changing reflex input to the motoneuron pool:

1. As a fatiguing contraction proceeds, there is a gradual disfacilitation of alpha motoneurons due to a progressive withdrawal of fusimotor-driven spindle support to the skeletomotor output. This theory derives its main support from experiments on conscious humans trying to maintain a desired voluntary force. The most direct evidence comes from microneurographic recordings of impulse activity in identified muscle afferents and motor axons in human subjects. There are distinct advantages to studying muscle fatigue in humans: activation of muscle volitionally will access segmental (and suprasegmental) reflex circuitry in a physiological manner (e.g., Hultborn et al., 1987; Baldissera & Pierrot-Deseilligny, 1989; Meunier & Pierrot-Deseilligny, 1989; Burke et al., 1992), and motoneurons will fire with an intrinsic variability that may in itself optimize the production of force (see Bevan et al., 1992).
2. The alpha motoneurons are inhibited by group III and IV muscle afferents, activated by the accumulation of metabolites in the fatiguing muscle. The most direct evidence for this theory comes from experiments on reduced animals, in which muscle fatigue is induced not volitionally but by electrical nerve stimulation (Kniffki et al., 1981; Cleland et al., 1982). This idea has received indirect support from experiments in human subjects (Bigland-Ritchie et al., 1986; Kukulka et al., 1986; Woods et al., 1987; Garland et al., 1988; Garland &

McComas, 1990; Garland, 1991; see Garland & Kaufman, Chapter 19). However, the studies supporting the alternative *disfacilitation theory* stand on their own, and the two theories need not be considered mutually exclusive. Both sources of peripheral input to the motoneuron pool, large-diameter as well as small-diameter intramuscular afferents, could contribute to the decline in motoneuronal discharge during muscular fatigue, but their relative contributions may well be expected to differ during the development of fatigue.

MUSCLE AFFERENT FEEDBACK PROVIDES A NET AMPLIFICATION OF SKELETOMOTOR OUTPUT

Irrefutable evidence indicates that muscle afferent feedback causes a net facilitation of volitionally generated skeletomotor output in human subjects:

1. Peak motor unit firing rates during brief MVC's of nonfatigued muscles are reduced by partial nerve blocks following perineural injection of local anesthetic, which preferentially blocks small-diameter axons before effects on large-diameter fibers can be observed (Fig. 1). This was taken as evidence that normally activity in such fibers (efferent and/or afferent) has a net facilitatory effect on skeletomotor output (Hagbarth et al., 1986; Bongiovanni et al., 1990; Bongiovanni & Hagbarth, 1990). It was also found that the decline in motor unit firing rates normally seen during sustained MVCs was absent during the partial nerve blocks, indicating that with intact innervation the net facilitatory effect of small-fiber activity gradually fades away as muscle fatigue develops. Muscle vibration, known to be a potent stimulus for primary spindle endings, was used to test the hypothesis that the normal fatigue-induced decline in skeletomotor output primarily depends on a decline in fusimotor-driven spindle activity. This hypothesis was supported by the findings that short periods (<10 s) of muscle vibration counteracted both the fatigue-induced declines in motor unit firing rates during sustained MVCs and the reductions in peak motor unit firing rates induced by partial nerve blocks.

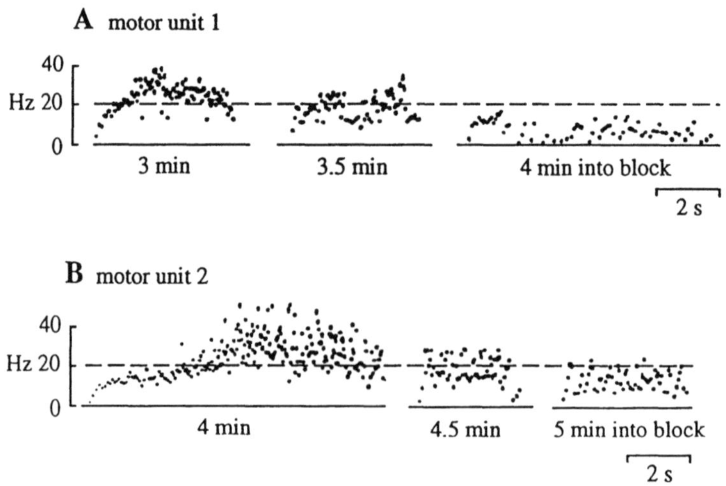

Figure 1. Instantaneous frequency plots of two single motor units recorded from tibialis anterior during maximal voluntary contractions performed during the development of paralysis following injection of local anesthetic around the peroneal nerve (modified from Hagbarth et al., 1986).

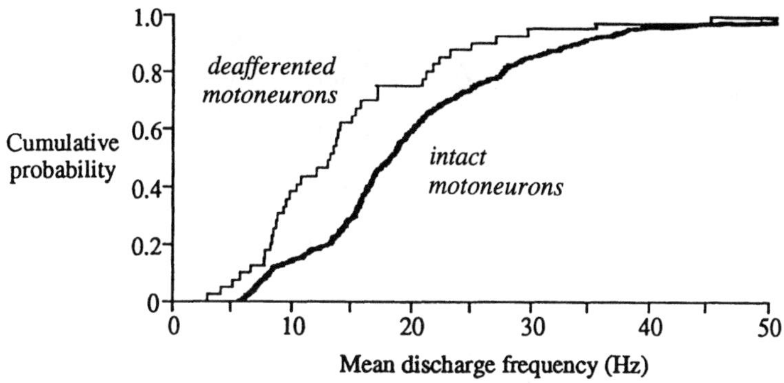

Figure 2. Firing rate distributions for tibialis anterior motoneurons from minimal to maximal levels of voluntary drive. *Deafferented motoneurons*, data from 39 single motor axons recorded proximal to a complete block of the peroneal nerve. *Intact motoneurons*, data from 595 single motor units. At all levels of voluntary drive, firing rates were lower in the absence of muscle afferent feedback (modified from Macefield et al., 1993).

However, longer periods of high-frequency muscle vibration had opposite effects, accentuating the fatigue-related declines in skeletomotor output. This effect was tentatively ascribed to Ia presynaptic inhibition or exhaustion of Ia excitatory synapses (Bongiovanni & Hagbarth, 1990).

2. Efferent activity in human motor axons has been recorded from the ulnar and peroneal nerves during acute deafferentation of the target muscles by complete anesthetic block of nerve conduction distal to the recording site (Gandevia et al., 1990; Gandevia et al., 1992; Gandevia et al., 1993; Macefield et al., 1993). In the absence of muscle afferent feedback, subjects were able to recruit single motoneurons and to grade their discharge frequency, but in attempted maximal efforts the firing rates of single motor axons were lower than those of normally innervated motor units recorded in separate experiments. Firing rates were reduced by approximately one-third at all levels of volitional drive, from minimal to maximal efforts (Macefield et al., 1993). The overall range of firing rates expressed by intact and deafferented motoneurons was the same, indicating that muscle afferent feedback provides a net amplification of skeletomotor output from the spinal cord without affecting the dynamic range over which the alpha motoneurons discharge volitionally. This is illustrated in Fig. 2. These experiments also showed that in the complete absence of muscle afferent feedback, individual motoneurons did not show the progressive decline in firing rates exhibited by normally-innervated motor units during sustained maximal efforts, supporting the reflexogenic nature of the decline in motoneuron discharge frequency during muscle fatigue. These results, and those outlined in the preceding paragraph, are consistent with the view that the reflex facilitatory influence provided by muscle afferents declines as fatigue develops (Gandevia et al., 1990; Gandevia et al., 1992; Gandevia et al., 1993; Macefield et al., 1993).

MUSCLE SPINDLE AFFERENT FEEDBACK DOES DECLINE DURING FATIGUE

Direct proof that muscle spindle feedback does decline during a constant-force voluntary contraction came from microneurographic recordings of the discharge of single

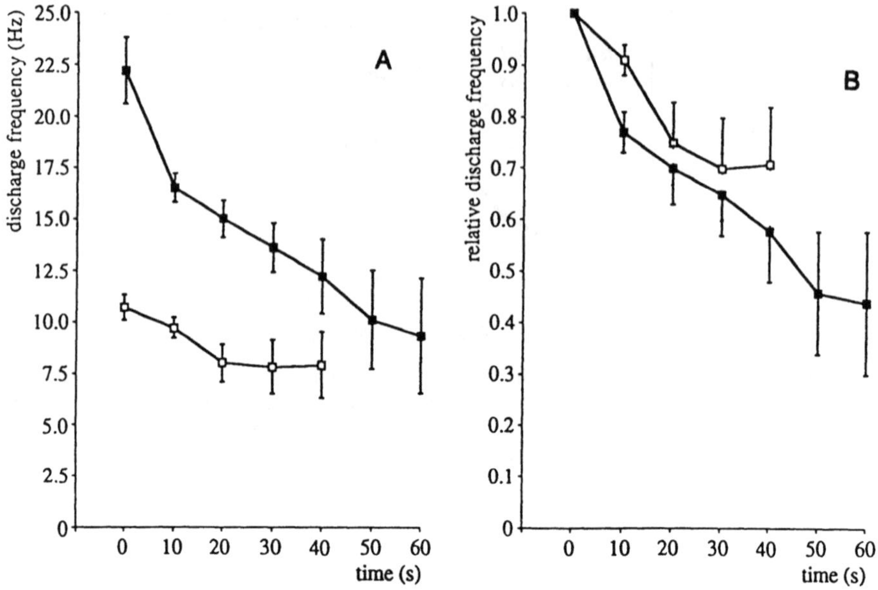

Figure 3. Decline in discharge frequency of 11 muscle spindles in tibialis anterior as a function of initial firing rate at the onset of an isometric voluntary contraction. The data were divided into units with initial frequencies above 15 Hz (closed symbols) and below 15 Hz (open symbols). The data are expressed as absolute changes in *A* and relative changes in *B* (from Macefield et al., 1991).

muscle spindle afferents from tibialis anterior and extensor digitorum longus (Macefield et al., 1991). The discharge of most muscle spindle afferents decreased by some 50% during contractions of the parent muscles sustained for at least 60 s, when increasing effort was required to maintain the isometric force at the target level. This reduction began within the first few seconds of the contraction, even before the target force had been reached, and continued progressively while the surface EMG activity of the muscle gradually increased. The decline in spindle discharge frequency was more prominent with contractions at higher force levels, when firing rates were higher, and for contractions performed later in the experiment after many sustained contractions. Significantly, the decline in firing rates of single motor units during maximal voluntary contractions is also greater for units discharging at higher initial frequencies (Bigland-Ritchie et al., 1983). The decline in spindle discharge during a contraction is shown in Fig. 3. For technical reasons, it has so far not been possible to record muscle spindle activity in man during maximal voluntary contractions, but the experiments described in the preceding paragraph indicate that fusimotor support is equally important in maximal as well as submaximal voluntary contractions.

These studies indicate that *muscle wisdom* can at least partly be explained by a progressive withdrawal of tonic support via the fusimotor loop, but the mechanisms responsible for this waning input remain conjectural. Possibly, muscle ischemia with accompanying accumulation of metabolites produces failure in excitation-contraction coupling or failure in the contractile capacity of intrafusal muscle fibers. The observed decline in afferent spindle discharge might also be explained by a declining fusimotor output or by adaptation processes in muscle spindle receptors. In addition, central processes such as presynaptic inhibition of Ia terminals may contribute to a reduction in fusimotor assistance as muscle fatigue develops during sustained voluntary contractions.

EFFECTS OF MUSCLE FATIGUE ON SPINDLE-MEDIATED REFLEX ASSISTANCE

Seemingly conflicting reports have been presented concerning fatigue-induced changes in the efficacy of stretch and unloading reflexes (Darling & Hayes, 1983; Hunter & Kearney, 1983; Häkkinen & Komi, 1983; Windhorst et al., 1986; Kirsch & Rymer, 1987, 1992; Balestra et al., 1992; Duchateau & Hainaut, 1993). One possible explanation for these discrepancies is that some investigators have been mainly concerned with reflex changes occurring during submaximal constant-force contractions, when there is a gradual overall increase in EMG activity, whereas others have paid more attention to the later stages of fatigue when skeletomotor output declines. Considering the principle of *automatic gain compensation* (Matthews, 1986), by which the relative amplitude of reflex modulation matches the ongoing level of EMG, it is to be expected that with the increase in EMG activity during sustained, submaximal constant-force contractions, there will be a corresponding increase in EMG recorded in phasic stretch and unloading reflexes. Whereas, reductions of reflex EMG responses are likely to occur later when the overall decline in motor unit firing rates can no longer be prevented. In agreement with this principle, some investigators have observed early enhancements and late reductions of reflex responses to muscle perturbations during sustained isometric contractions (Marsden et al., 1983; Darling & Hayes, 1983).

In order to avoid variations in reflex efficacy which are due to variations in the pre-existing level of skeletomotor activity, it is essential that this level is similar in control (non-fatigued) conditions as when the reflexes are tested during the different stages of fatigue. In recent experiments where this criterion was fulfilled, both stretch and unloading reflexes (especially their short-latency components) were found to be reduced at a stage of fatigue just before the desired force could no longer be maintained. At this stage, there was also a marked increase in the EMG activity of antagonistic muscles, a finding interpreted as a sign of reduced Ia reciprocal inhibition (Hagbarth, in preparation).

DIRECT RECORDINGS FROM FUSIMOTOR NEURONS DURING MUSCLE FATIGUE

Due to technical difficulties, there have been no recordings from human fusimotor neurons during the development of muscle fatigue, the most direct recordings being from the resultant changes in muscle spindle discharge (Macefield et al., 1991). However, there have been a few recent studies in which the fatigue-related changes in discharge behavior of single fusimotor neurons have been recorded from experimental animals. Interestingly, though, whereas the human studies referred to above indicate that the support via the fusimotor loop declines during fatiguing contractions, data derived from experiments on reduced animals point in an opposite direction. Based on their findings when studying fatigue-induced changes in fusimotor output and afferent spindle activity in decerebrate or spinal cats, Ljubisavljevic and Anastasijevic (1995) arrived at the following hypothesis: ". . . the same muscle afferent discharge, elicited in groups III and IV afferents by muscle fatigue itself would at the same time inhibit skeletomotor neurons by direct reflex action and lend them support through the gamma loop". In the following, we review the evidence leading up to this hypothesis.

Animal experiments have provided convincing evidence that known products of muscle metabolism can increase the discharge rate in group III and IV intramuscular afferents (Kumazawa & Mizumura, 1977; Kniffki et al., 1978; Mense & Stahnke, 1983; Kaufman et al., 1983; Rotto & Kaufman, 1988; Lagier-Tessonnier et al., 1993; Sinoway et al., 1993).

However, not all group III and IV afferents are stimulated during fatiguing contractions: small-diameter mechanoreceptive afferents have been shown to exhibit a decline in their discharge rate during the initial phase of fatigue, although chemosensitive afferents become spontaneously active late in a maintained contraction (Hayward et al., 1991; Sinoway et al., 1993). This would strongly suggest that any contribution small diameter intramuscular afferents may make to reflexogenic muscle fatigue would be towards the end of a contraction, rather than during the initial part. Their contribution might also depend on the ischemic condition of a muscle, such that metabolites are not washed out from the intramuscular environment. A logical consequence of this is that these afferents may not be involved in the generation of fatigue associated with submaximal contractions, in which the vascular supply of the contracting muscle is not compromised.

Several investigators have reported that in reduced animals, activity evoked in group III and IV muscle afferents has an autogenetic excitatory influence on fusimotor neurons (Appelberg et al., 1983; Jovanovic et al., 1990; Ljubisavljevic et al., 1992a,b). In their studies on decerebrate or spinal cats, Ljubisavljevic and coworkers (1992a,b) found that an increase in discharge rate of fusimotor neurons develops in parallel with muscle fatigue during long-lasting muscle contractions elicited by tetanic stimulation of muscle nerves or ventral roots. The bearing of these results on human muscle fatigue is difficult to evaluate since segmental reflex machineries are not likely to operate in a similar way in decerebrate or spinal cats as in healthy human beings. Reported differences in fatigue-induced fusimotor responses when comparing spinal with decerebrate preparations or preparations with or without wide limb denervations (Anastasijevic et al., 1995) provide good examples of how reflex responses can vary with the state of excitability in segmental reflex arcs and in internuncial networks. Moreover, in cats anaesthetized with chloralose, the excitability in central neural networks may well be different from that in conscious human subjects (cf. Johansson et al, 1993). Other matters of concern in experiments of this type are the difficulties associated with distinguishing both between fatigue-related and non fatigue-related reflex responses and between nociceptive and non-nociceptive reactions.

CONFOUNDING EFFECTS OF THIXOTROPIC CHANGES IN MUSCLE SPINDLE BEHAVIOR

Further difficulties arise when interpreting changes in spindle afferent activity resulting from long-lasting tetanic stimulation of muscle nerves or ventral roots. In particular, it may be hard to distinguish between fatigue-induced changes and changes due to thixotropic (actomyosin cross-bridge dependent) after-effects in extrafusal and intrafusal muscle fibers. In a study performed by Nelson and Hutton (1985) on the triceps surae in anaesthetized cats, sustained contractions of these muscles (elicited by 7x threshold electrical stimulation of ventral roots) were followed by a significant increase in resting discharge and dynamic sensitivity of both group Ia and group II spindle afferents. In addition, resting muscle force and passive peak muscle stiffness were consistently higher following contraction. These post-tetanic effects were believed to depend on increased inherent stiffness of intrafusal and extrafusal muscle fibers, due to persisting actomyosin bonds. It was also proposed that accompanying adjustments in resting discharge and dynamic sensitivity of spindle endings can provide reflex compensation for fatigue-induced force declines (cf. Ribot-Ciscar et al., 1991). However, as claimed by Proske and colleagues (Gregory et al., 1987; Gregory et al., 1990; Polus et al., 1991), the thixotropic after-effects of contractions may be either an enhancement or a reduction of inherent muscle stiffness, dependent on whether a muscle has previously been held at a short length or been exposed to stretch. Recent studies in man confirm this claim (Hagbarth et al., 1995). The history-dependent changes in muscle spindle behavior may well account for some of the apparently conflicting reports on involvement of the fusimotor system in muscle fatigue.

CONCLUDING REMARKS

To summarize, it is questionable whether studies performed on decerebrate or spinal cats can ever shed light on the behavior of fusimotor neurons during fatiguing contractions in man. Further microneurographic studies on conscious human subjects are needed to elucidate this issue, particularly with respect to the possible contributions of group III and IV afferents to fatigue. Results obtained with this technique so far indicate that the amplification of skeletomotor output via the fusimotor loop declines as muscle fatigue develops. It remains to be determined whether this declining support is the main cause of the *muscle wisdom* phenomenon, but the parallel time courses of the changes in spindle discharge and motor unit firing lend strong support to this idea. As pointed out by Gandevia and colleagues (1990), it is likely that waning excitation via the fusimotor loop (due to a reduction in muscle spindle input and increase in pre-synaptic inhibition) is most prominent in the early phases of fatigue, whereas progressive reflex inhibition mediated by small diameter afferents contributes more later in prolonged contractions. Certainly all available experimental evidence supports this view. Evidence of *active inhibition* of alpha motoneurons during fatigue could come from knowing the firing rates of single motor units in tibialis anterior towards the end of a sustained maximal voluntary contraction; only MVCs of 10 s duration have been studied in this muscle so far (Bigland-Ritchie et al., 1992). If these firing rates turn out to be lower than the known peak firing rates of deafferented motoneurons supplying this muscle (Macefield et al., 1993) then active inhibition must be responsible. Conversely, if peak firing rates are not lower then the decline in firing rates could be explained fully by a progressive disfacilitation of alpha motoneurons resulting from withdrawal of fusimotor support. Observed enhancements in the efficacy of stretch and unloading reflexes during muscle fatigue can be explained in terms of *automatic gain compensation* (Matthews, 1986) but do not exclude the possibility that there actually is a reduction in the spindle-mediated neural messages impinging on the motoneuron pool, such that motor output becomes more *alpha-dominated* as fatigue processes develop.

ACKNOWLEDGMENTS

The authors' laboratories are supported by the Medical Research Council of Sweden and the National Health & Medical Research Council (NH & MRC) of Australia. Attendance of V.G.M. at the 1994 Bigland-Ritchie conference was supported, in part, by the NH & MRC, the American Physical Therapy Association (Section on Research), and the Muscular Dystrophy Association (USA).

REFERENCES

Anastasijevic R, Jovanovic K & Ljubisavljevic M (1995). Fusimotor responses to fatiguing contractions of a muscle in non-denervated cat hindlimb. In: Stuart DG, Gantchev GN, Gurfinkel VS, Wiesendanger M (eds.), *Motor Control VII*, pp. 00-00. Tucson: Motor Control Press. In press

Aniss AM, Diener HC, Hore J, Burke D & Gandevia SC (1990). Reflex activation of muscle spindles in human pretibial muscles during standing. *Journal of Neurophysiology* **64**, 671-679.

Appelberg B, Hulliger M, Johansson H & Sojka P (1983). Actions on gamma-motoneurones elicited by electrical stimulation of group III muscle afferents in the hind limb of the cat. *Journal of Physiology (London)* **335**, 275-292.

Baldissera F & Pierrot-Deseilligny E (1989). Facilitation of transmission in the pathway of non-monosynaptic Ia excitation to wrist flexor motoneurones at the onset of voluntary movement in man. *Experimental Brain Research* **74**, 437-439.

Balestra C, Duchateau J & Hainaut K (1992). Effects of fatigue on the stretch reflex in a human muscle. *Electroencephalography and Clinical Neurophysiology* **85**, 46-52.

Bevan L, Laouris Y, Reinking RM & Stuart DG (1992). The effect of the stimulation pattern on the fatigue of single motor units in adult cats. *Journal of Physiology (London)* **449**, 85-108.

Bigland-Ritchie B, Cafarelli E & Vøllestad NK (1986). Fatigue of submaximal static contractions. *Acta Physiologica Scandinavia Supplement* **556**, 137-148.

Bigland-Ritchie B, Johansson R, Lippold OCJ, Smith S & Woods JJ (1983). Changes in motoneurone firing rates during sustained maximal voluntary contractions. *Journal of Physiology (London)* **340**, 335-346.

Bigland-Ritchie B, Jones DA & Woods JJ (1979). Excitation frequency and muscle fatigue: electrical responses during voluntary and stimulated contractions. *Experimental Neurology* **64**, 414-427.

Bigland-Ritchie B & Woods JJ (1984). Changes in muscle contractile properties and neural control during human muscular fatigue. *Muscle & Nerve* **7**, 691-699.

Bigland-Ritchie BR, Dawson NJ, Johansson RS & Lippold OC (1986). Reflex origin for the slowing of motoneurone firing rates in fatigue of human voluntary contractions. *Journal of Physiology (London)* **379**, 451-459.

Bigland-Ritchie BR, Furbush FH, Gandevia SC & Thomas CK (1992). Voluntary discharge frequencies of human motoneurons at different muscle lengths. *Muscle & Nerve* **15**, 130-137.

Binder-Macleod SA & Guerin T (1990). Preservation of force output through progressive reduction of stimulation frequency in human quadriceps femoris muscle. *Physical Therapeutics* **70**, 619-625.

Bongiovanni LG & Hagbarth KE (1990). Tonic vibration reflexes elicited during fatigue from maximal voluntary contractions in man. *Journal of Physiology (London)* **423**, 1-14.

Bongiovanni LG, Hagbarth KE & Stjernberg L (1990). Prolonged muscle vibration reducing motor output in maximal voluntary contractions in man. *Journal of Physiology (London)* **423**, 15-26.

Burke D (1981). The activity of human muscle spindle endings during normal motor behavior. In: Porter R (ed.), *Neurophysiology IV. International Review of Physiology*, pp. 91-126. Baltimore, MD: University Park Press.

Burke D, Gracies JM, Meunier S & Pierrot-Deseilligny E (1992). Changes in presynaptic inhibition of afferents to propriospinal-like neurones in man during voluntary contractions. *Journal of Physiology (London)* **449**, 673-687.

Cleland C, Rymer W & Edwards F (1982). Force-sensitive interneurones in the spinal cord of the cat. *Science* **217**, 652-655.

Darling WG & Hayes KC (1983). Human servo responses to load disturbances in fatigued muscle. *Brain Research* **267**, 345-351.

Duchateau J & Hainaut K (1993). Behaviour of short and long latency reflexes in fatigued human muscles. *Journal of Physiology (London)* **471**, 787-799.

Eldred E, Granit R & Merton PA (1953). Supraspinal control of the muscle spindles and its significance. *Journal of Physiology (London)* **122**, 498-523.

Enoka RM & Stuart DG (1992). Neurobiology of muscle fatigue. *Journal of Applied Physiology* **72**, 1631-1648.

Gandevia SC, Burke D, Macefield G & McKenzie DK (1992). Human motor output, muscle fatigue and muscle afferent feedback. *Proceedings of the Australian Physiological and Pharmacological Society* **23**, 59-67.

Gandevia SC, Macefield G, Burke D & McKenzie DK (1990). Voluntary activation of human motor axons in the absence of muscle afferent feedback. The control of the deafferented hand. *Brain* **113**, 1563-1581.

Gandevia SC, Macefield VG, Bigland-Ritchie B, Gorman R & Burke D (1993). Motoneuronal output and gradation of effort in attempts to contract acutely paralyzed leg muscles in man. *Journal of Physiology (London)* **474**, 411-427.

Garland SJ (1991). Role of small diameter afferents in reflex inhibition during human muscle fatigue. *Journal of Physiology (London)* **435**, 547-558.

Garland SJ, Garner SH & McComas AJ (1988). Reduced voluntary electromyographic activity after fatiguing stimulation of human muscle. *Journal of Physiology (London)* **401**, 547-556.

Garland SJ & McComas AJ (1990). Reflex inhibition of human soleus muscle during fatigue. *Journal of Physiology (London)* **429**, 17-27.

Gregory JE, Mark RF, Morgan DL, Patak A, Polus B & Proske U (1990). Effects of muscle history on the stretch reflex in cat and man. *Journal of Physiology (London)* **424**, 93-107.

Gregory JE, Morgan DL & Proske U (1987). Changes in size of the stretch reflex of cat and man attributed to aftereffects in muscle spindles. *Journal of Neurophysiology* **58**, 628-640.

Grimby L, Hannerz J & Hedman B (1981). The fatigue and voluntary discharge properties of single motor units in man. *Journal of Physiology (London)* **316**, 545-554.

Hagbarth KE (1993). Microneurography and applications to issues of motor control: Fifth Annual Stuart Reiner Memorial Lecture. *Muscle & Nerve* **16**, 693-705.

Hagbarth KE, Kunesch EJ, Nordin M, Schmidt R & Wallin EU (1986). Gamma loop contributing to maximal voluntary contractions in man. *Journal of Physiology (London)* **380**, 575-591.

Hagbarth KE, Nordin M & Bongiovanni LG (1995). After-effects on stiffness and stretch reflexes of human finger flexor muscles attributed to muscle thixotropy. *Journal of Physiology (London)* **482.1**, 215-223.

Hayward L, Wesselmann U & Rymer WZ (1991). Effects of muscle fatigue on mechanically sensitive afferents of slow conduction velocity in the cat triceps surae. *Journal of Neurophysiology* **65**, 360-370.

Hultborn H, Meunier S, Pierrot-Deseilligny E & Shindo M (1987). Changes in presynaptic inhibition of Ia fibres at the onset of voluntary contraction in man. *Journal of Physiology (London)* **389**, 757-772.

Hunter IW & Kearney RE (1983). Invariance of ankle dynamic sensitivity during fatiguing muscle contractions. *Journal of Biomechanics* **16**, 985-991.

Häkkinen K & Komi PV (1983). Electromyographic and mechanical characteristics of human skeletal muscle during fatigue under voluntary and reflex conditions. *Electroencephalography and Clinical Neurophysiology* **55**, 436-444.

Johansson H, Djupsjöbacka M & Sjölander P (1993). Influences on the gamma-muscle spindle system from muscle afferents stimulated by KCl and lactic acid. *Neuroscience Research* **16**, 49-57.

Jovanovic K, Anastasijevic R & Vuco J (1990). Reflex effects on gamma fusimotor neurones of chemically induced discharges in small-diameter muscle afferents in decerebrate cats. *Brain Research* **521**, 89-94.

Kaufman MP, Longhurst JC, Rybicki KJ, Wallach JH & Mitchell JH (1983). Effects of static muscular contraction on impulse activity in groups III and IV afferents in cats. *Journal of Applied Physiology* **55**, 105-112.

Kirsch RF & Rymer WZ (1987). Neural compensation for muscular fatigue: evidence for significant force regulation in man. *Journal of Neurophysiology* **57**, 1893-1910.

Kirsch RF & Rymer WZ (1992). Neural compensation for fatigue-induced changes in muscle stiffness during perturbations of elbow angle in human. *Journal of Neurophysiology* **68**, 449-470.

Kniffki KD, Mense S & Schmidt RF (1978). Responses of group IV afferent units from muscle to stretch, contraction and chemical stimulation. *Experimental Brain Research* **31**, 511-522.

Kniffki KD, Shomburg ED & Steffens, H (1981). Synaptic effects from chemically activated fine muscle afferents upon alpha motoneurones in decerebrate and spinal cats. *Brain Research* **206**, 361-370.

Kukulka CG, Moore MA & Russell AG (1986). Changes in human alpha-motoneuron excitability during sustained maximum isometric contractions. *Neuroscience Letters* **68**, 327-333.

Kumazawa T & Mizumura K (1977). Thin-fibre receptors responding to mechanical, chemical and thermal stimuli in the skeletal muscle of the dog. *Journal of Physiology (London)* **273**, 179-194.

Lagier-Tessonnier F, Balzamo E & Jammes Y (1993). Comparative effects of ischemia and acute hypoxemia on muscle afferents from tibialis anterior in cats. *Muscle & Nerve* **16**, 135-141.

Ljubisavljevic M, Jovanovic K & Anastasijevic R (1992a). Changes in discharge rate of cat hamstring fusimotor neurones during fatiguing contractions of triceps surae muscles. *Brain Research* **579**, 246-252.

Ljubisavljevic M, Jovanovic K & Anastasijevic R (1992b). Changes in discharge rate of fusimotor neurones provoked by fatiguing contractions of cat triceps surae muscles. *Journal of Physiology (London)* **445**, 499-513.

Ljubisavljevic M & Anastasijevic R (1995). Muscle spindle afferent discharges during fatiguing muscle contractions in decerebrate cats. In: Stuart DG, Gantchev GN, Gurfinkel VS, Wiesendanger M (eds.), *Motor Control VII,* pp. 00-00. Tucson AZ: Motor Control Press. In press

Loeb GE & Hoffer JA (1981). Muscle spindle function during normal and disturbed locomotion. In: Taylor A, Prochazka A (eds.), *Muscle Receptors and Movement*, pp. 219-228. London: Macmillan.

Macefield G, Hagbarth K-E, Gorman R, Gandevia SC & Burke D (1991). Decline in spindle support to alpha motoneurones during sustained voluntary contractions. *Journal of Physiology (London)* **440**, 497-512.

Macefield VG, Gandevia SC, Bigland-Ritchie B, Gorman R & Burke D (1993). The firing rates of human motoneurones voluntarily activated in the absence of muscle afferent feedback. *Journal of Physiology (London)* **474**, 429-443.

Marsden CD, Meadows JC & Merton PA (1983) "Muscular wisdom" that minimized fatigue during prolonged effort in man: peak rates of motoneuron discharge and slowing of discharge during fatigue. In: Desmedt JE (ed.), *Motor Control Mechanisms in Health and Disease*, pp. 169-211. New York: Raven Press.

Matthews PB (1986). Observations on the automatic compensation of reflex gain on varying the pre-existing level of motor discharge in man. *Journal of Physiology (London)* **374**, 73-90.

Matthews PBC (1981). Evolving views on the internal operation and functional role of the muscle spindle. *Journal of Physiology (London)* **320**, 1-30.

Mense S & Stahnke M (1983). Responses in muscle afferent fibres of slow conduction velocity to contractions and ischaemia in the cat. *Journal of Physiology (London)* **342**, 383-397.

Merton PA (1953). Speculations on the servo-control of movement. In: Wolstenholme, GEW (ed.), *The Spinal Cord*, pp. 247-307. London: Churchill.

Merton PA (1954). Voluntary strength and fatigue. *Journal of Physiology (London)* **123**, 553-564.

Meunier S & Pierrot-Deseilligny E (1989). Gating of the afferent volley of the monosynaptic stretch reflex during movement in man. *Journal of Physiology (London)* **419**, 753-763.

Nelson DL & Hutton RS (1985). Dynamic and static stretch responses in muscle spindle receptors in fatigued muscle. *Medicine and Science in Sports Exercise* **17**, 445-450.

Person RS (1974). Rhythmic activity in a group of human motoneurones during voluntary contraction of a muscle. *Electroencephalography and Clinical Neurophysiology* **36**, 585-595.

Polus BI, Patak A, Gregory JE & Proske U (1991). Effect of muscle length on phasic stretch reflexes in humans and cats. *Journal of Neurophysiology* **66**, 613-622.

Prochazka A & Wand P (1981). Independence of fusimotor and skeletomotor systems during voluntary movements. In: Taylor A, Prochazka A (eds.), *Muscle Receptors and Movement*, pp. 229-243. London: Macmillan.

Ribot-Ciscar E, Tardy-Gervet MF, Vedel JP & Roll JP (1991). Post-contraction changes in human muscle spindle resting discharge and stretch sensitivity. *Experimental Brain Research* **86**, 673-678.

Rotto DM & Kaufman MP (1988). Effect of metabolic products of muscular contraction on discharge of group III and IV afferents. *Journal of Applied Physiology* **64**, 2306-2313.

Sinoway LI, Hill JM, Pickar JG & Kaufman MP (1993). Effects of contraction and lactic acid on the discharge of group III muscle afferents in cats. *Journal of Neurophysiology* **69**, 1053-1059.

Vallbo ÅB, Hagbarth K-E, Torebjörk HE & Wallin BG (1979). Somatosensory, proprioceptive and sympathetic activity in human peripheral nerves. *Physiological Reviews* **59**, 919-957.

Windhorst U, Christakos CN, Koehler W, Hamm TM, Enoka RM & Stuart DG (1986). Amplitude reduction of motor unit twitches during repetitive activation is accompanied by relative increase of hyperpolarizing membrane potential trajectories in homonymous alpha-motoneurons. *Brain Research* **398**, 181-184.

Woods JJ, Furbush F & Bigland-Ritchie B (1987). Evidence for a fatigue-induced reflex inhibition of motoneuron firing rates. *Journal of Neurophysiology* **58**, 125-137.

ROLE OF MUSCLE AFFERENTS IN THE INHIBITION OF MOTONEURONS DURING FATIGUE

S. J. Garland[1] and M. P. Kaufman[2]

[1] Department of Physical Therapy
Elborn College, University of Western Ontario
Ontario, Canada N6G 1H1
[2] Division of Cardiovascular Medicine, Departments of Internal Medicine and Human Physiology
University of California
Davis, California 95616

ABSTRACT

In conscious humans, fatiguing muscular contractions are accompanied by a decrease in the discharge rate of alpha motoneurons. The association between alpha motoneuron discharge rate and the generation of force by skeletal muscle has been called "muscle wisdom" (Marsden et al., 1983). Its purpose is believed to ensure that central neural drive to skeletal muscle, which is fatigued, matches that needed to generate the required force. In addition, muscle wisdom may be one mechanism that functions either to decrease or to postpone central neural fatigue (Enoka & Stuart, 1992). Bigland-Ritchie and colleagues (1986) have suggested that a reflex arising from fatigued skeletal muscle is responsible, at least in part, for muscle wisdom. This chapter has two purposes. The first is to evaluate the evidence that a reflex arising from fatigued skeletal muscle causes muscle wisdom, and the second is to examine the discharge properties of muscle afferents to determine which ones are most likely to initiate reflexly this phenomenon.

INTRODUCTION

Before discussing the role played by muscle afferents in reflexly inhibiting motoneuron discharge during fatigue, we need to provide some basic information about the sensory innervation of limb skeletal muscles, the discharge properties of the afferents and in selected circumstances, the locations of their endings. Limb muscles are innervated by five types of sensory nerves. These have been classified as group I through IV, with the first group having two subtypes, Ia and Ib. The classification scheme is based, in part, on the

Fatigue, Edited by Simon C. Gandevia et al.
Plenum Press, New York, 1995

diameter of the afferent fibers. Specifically, the thicker the fiber (i.e., the more myelin it has surrounding it), the faster it conducts impulses (Hunt, 1954; Boyd & Davy, 1968).

Group Ia and Ib muscle afferents are thickly myelinated and conduct impulses between 72 and 120 m/s in cats and dogs. The receptors of group Ia afferents are primary muscle spindle endings and those of group Ib afferents are Golgi tendon organs. Primary spindle afferents (i.e., group Ia) are located *in parallel* with the skeletal muscle fibers that they innervate. Consequently, the discharge of primary spindle afferents is increased by stretch (i.e., lengthening) of the muscle and is decreased by contraction (i.e., shortening). Golgi tendon organs, in contrast, are situated in series with the muscle fibers they innervate; consequently, the discharge of these afferents is increased by both contraction and muscle stretch.

Group II afferents, also called secondary spindle endings, conduct impulses between 31 and 71 m/s in cats and dogs. The receptors of these afferents (i.e., the spindles), like their group Ia counterparts, are situated *in parallel* with the muscle fibers that they innervate. Although secondary spindle afferents (i.e., group II) are stimulated by muscle stretch, they are not capable of signaling the rate at which this occurs. In contrast, primary spindle afferents (i.e., group Ia) are capable of signaling the rate of stretch; consequently, they are *dynamically sensitive*. Muscular contraction decreases the discharge rates of group II spindle afferents. There are also group II afferent fibers, which are not innervated by muscle spindles. These group II fibers, unlike group II spindle afferents, are stimulated by contraction.

Group III afferents, also called Aδ fibers, are thinly myelinated and conduct impulses between 2.5 and 30 m/s in cats and dogs. Their receptors are free nerve endings. Group IV afferents, also called C-fibers, are unmyelinated and conduct impulses at less than 2.5 m/s. Like their group III counterparts, the receptors of group IV afferents are free nerve endings. Nevertheless, the endings innervating both group III and IV afferents are almost entirely surrounded by Schwann cells. A small portion of the ending is bare, and this area is thought to be the site of action for the stimuli that activate them. Moreover, the bare areas of the nerve ending contain mitochondria and other structures whose presence is consistent with the notion that these areas are indeed the receptive part of the sensory neuron (Andres et al., 1985).

Serial reconstruction of electron micrographs have unmasked the locations of the endings of group III and IV afferents in the calcaneal (i.e., Achilles) tendon of the cat (Andres et al., 1985). Five locations were found for the endings of group III afferents. Two of these locations were in vessels (venules and lymphatics), two were in the connective tissue of the peritenoneum externum and internum, and the other was in the endoneurium. The locations of the endings of group IV afferents showed a marked topographic relationship to the blood and lymphatic vessels of the tendon. Moreover, group IV endings, but not group III, contain granulated vesicles, the content of which is unknown. These granulated vesicles may contain neuropeptides, whose release via the *axon reflex* evokes vasodilation. Obviously, the locations of the group IV endings in blood vessels is ideal to cause this effect. In addition, preliminary description of the locations of the endings of group III and IV afferents in the triceps surae muscles of the cat appear to parallel closely the locations of these endings in the calcaneal tendon (von During & Andres, 1990). In both the tendon and the muscles, the diameters of the group III endings were found to be larger than those of the group IV endings. Also, group III endings had more mitochondria and had a more distinct receptor matrix than did group IV endings (Andres et al, 1985; von During & Andres, 1990).

HUMAN STUDIES

During a fatiguing maximal voluntary contraction (MVC), electromyographic (EMG) activity has been shown to decline roughly in parallel with the loss of force

(Bigland-Ritchie, 1981). The origin of the declining EMG is largely central to the neuromuscular junction (Woods et al., 1987; Garland et al., 1988) and has been attributed to reflex inhibition of the motoneuron pool (Bigland-Ritchie et al., 1986; Garland & McComas, 1990). The existence of a reflex whereby motoneurons are inhibited by sensory input from the fatigued muscle is not limited to sustained MVCs (Bigland-Ritchie et al., 1986) but also has been demonstrated in fatiguing intermittent submaximal contractions performed under ischemic conditions (Garland & McComas, 1990). This reflex is thought to match the motoneuron output to the functional status of the muscle fibers during fatigue.

The contribution of different sized afferents to reflex inhibition during fatigue remains unresolved. Two hypotheses have been forwarded: 1) input from small diameter afferents (i.e., group III and IV) responsive to chemical or mechanical stimuli during fatigue serves to decrease motoneuron activity (Bigland-Ritchie, 1986); and 2) a reduction in discharge of muscle spindle afferents (i.e., group Ia and II) results in disfacilitation (i.e., reduced excitation of motoneurons by a decrease in fusimotor-driven feedback from muscle spindles) of motoneurons during fatigue (Macefield et al., 1991). A third hypothesis, that is unrelated to muscle afferents, is that motoneuron discharge decreases as a result of intrinsic motoneuron adaptation. Support for this hypothesis has come from cat motoneurons subjected to constant intracellular or extracellular current injection (Kernell & Monster, 1982; Spielmann et al., 1993). This mechanism is less likely in voluntary contractions, especially in intermittent or submaximal contractions, because the motoneuron does not experience constant input.

Small Diameter Afferents

The role of small diameter afferents in reflex inhibition has been established in human muscle fatigue. When fatigue was induced following compressive blockade of large diameter afferents, the EMG during MVCs declined (Garland, 1991); this decline was comparable to that found in a previous study in which the sensory input was unaltered (Garland et al., 1988). The nature of the sensory input, i.e. mechanical or chemical, has not been demonstrated, although the latter may be more likely than the former.

Bigland-Ritchie and colleagues (1986) found that EMG associated with MVCs following fatigue remained depressed for 2 or 3 min while the limb was rendered ischemic. In that study, the EMG demonstrated near-full recovery in 3 min if the blood supply was intact; this suggested a chemical stimulus. Release of an ischemic cuff after fatigue is associated with a reduction in muscle pain (Garland et al., 1988), an effect presumably due to wash-out of metabolites. Moreover, many of the chemosensitive afferents are known to be nociceptive. If, however, fatigue is induced by low-force submaximal intermittent contractions performed while the muscles are freely perfused, motor units showed a constant or an increased discharge rate (Garland et al., 1988). In addition, Marsden and colleagues (1983) were unable to demonstrate any change in motor unit firing rates by slowing muscle contraction through cooling. Similarly, Bigland-Ritchie and colleagues (1992) found no difference in motoneuron discharge rates when the twitch contraction was slowed by changing muscle length. Hence, in the absence of chemical changes associated with fatigue, the motoneuron discharge was unaffected.

Large-Diameter Afferents

Feedback from large-diameter afferents has been demonstrated to influence motor unit discharge rates in non-fatigued muscle. Hagbarth and coworkers (1986) recorded motoneuron discharge rates following blockade of fusimotor efferent activity during brief nonfatiguing MVCs. Under these conditions, irregular and reduced motoneuron discharge was evident; this effect could be reversed by muscle vibration, which strongly activates Ia spindle afferents (e.g., Burke et al., 1976). Local anesthetic has been used to block all muscle afferents in subjects attempting sustained MVCs (Gandevia et al., 1990; Macefield et al., 1993). Under these conditions, subjects also demonstrated lower than normal motoneuron discharge rates yet the motoneuron discharge rates did not show the decline during sustained maximal voluntary effort evident in normally innervated muscle. This provides further support for the role of muscle afferents (but not specifically large-diameter afferents) in modulating motoneuron discharge during fatigue.

The only direct measure of spindle activity in humans has indicated disfacilitation from muscle spindles evoked by low-force submaximal contractions of less than one min duration (Macefield et al., 1991). However, the amount of fatigue in this situation would be minimal. Bongiovanni and Hagbarth (1990) found that vibration was able to increase the discharge rate of motoneurons during sustained MVCs for only a short time, after which the motoneuron discharge rate declined despite continued vibration. In that experiment, there was no verification (i.e., through superimposed electrical stimulation) that subjects were able to activate fully the musculature. Thus the vibration could have simply increased the motoneuron firing rates by supplementing declining descending central drive, rather than by removing any disfacilitation arising from the muscle spindles. Further, the proprioceptive ability of subjects remained unchanged following fatiguing MVCs (Sharpe & Miles, 1993). It is possible that any fatigue-induced reductions in muscle spindle discharge may have been compensated for by concurrent changes in central processing of the sensory input, thereby rendering the proprioceptive ability unchanged. Thus, although a decline in muscle spindle input would result in disfacilitation of the motoneuron pool, there remains some uncertainty as to its relative importance during fatigue.

Studies that demonstrated a decrease in the muscle's response to stretch following fatigue, evoked by repetitive stretch-shortening cycles, do not enable a definitive distinction between the type of afferent that may be involved (Balestra et al., 1992). The stretch reflex could have been reduced by decreased sensitivity of the muscle spindles or alternatively decreased alpha motoneuron excitability resulting from inhibition mediated by small diameter afferents. Others have demonstrated an increase in the stretch reflex evoked by mechanical tendon tap following fatigue induced by stretch-shortening exercise (Hartobagyi et al., 1991) or by submaximal sustained isometric contractions (Hakkinen & Komi, 1983). In these studies, the effect of muscle fiber membrane hyperpolarization, the phenomenon accounting for transient M-wave potentiation during fatigue (Hicks & McComas, 1989), could explain the increased amplitude of the response to stretch without any effect on the stretch reflex, per se. The aforementioned studies have not included the controls necessary to interpret the data in terms of spindle sensitivity during fatigue.

ANIMAL STUDIES

Effects of Fatigue on the Discharge of Muscle Afferents

Experimental paradigms examining the effects of fatigue on afferent discharge in anesthetized animals require some explanation. Obviously, the essence of the paradigm is to measure an afferent's response to some maneuver before and during muscle fatigue. The specific criterion used to define fatigue differs among investigators, but the one used by the Hayward and colleagues (1991) may prove informative. These investigators defined fatigue as a reduction in force output by the contracting gastrocnemius muscles to 30% of that developed by the initial contraction, which was induced by electrically stimulating the muscle nerve using 25 Hz at 1.3 times motor threshold. This stimulation protocol produced a submaximal contraction.

Muscle spindles, if responsible for causing fatigue-induced reflex inhibition of alpha motoneuron discharge, should display a reduction in their responses to a maneuver (such as tendon stretch or contraction) during fatigue as compared with their responses to the maneuver before fatigue. The animal literature provides little support for this hypothesis. For example, although spindle firing has been shown to decrease temporarily at the onset of fatigue (Windhorst & Kokkoroyiannis, 1991), this recovers quickly. As the contraction period progressed, fatigue enhanced rather than reduced the responses of muscle spindles to contraction in either chloralose or barbiturate anesthetized cats (Nelson & Hutton, 1985; Hayward et al., 1991). It is also possible that muscle spindle afferent feedback could combine with recurrent inhibition from Renshaw cells as proposed by Windhorst and Kokkoroyiannis (1991).

Golgi tendon organs, if responsible for causing fatigue-induced reflex inhibition of alpha motoneuron discharge, should display an increase in their responses to a maneuver such as tendon stretch during fatigue as compared with their responses to stretch before fatigue. This hypothesis has not been supported by the animal literature. Fatigue has been shown to decrease the responses of Golgi tendon organs to stretch (Hutton & Nelson, 1986; Thompson et al, 1990); in another study, fatigue had little effect on the responses of tendon organs to stretch (Hayward et al., 1991).

Fatigue affects the discharge properties of group III and non-spindle group II afferents in a manner consistent with the notion that they are responsible, in part, for reflexly decreasing motoneuron discharge rates either directly or by Renshaw cell inhibition. Muscular fatigue, induced in barbiturate anesthetized cats, has been shown to increase the spontaneous discharge rate of group II non-spindle and group III afferents as well as to increase their responses to tendon stretch, to surface pressure (i.e., probing their receptive fields) and in a few instances to contraction (Hayward et al., 1991). Unfortunately, the effect of fatigue on the discharge properties of group IV afferents has not been determined.

Discharge Properties Of Small Diameter Afferents

The stimulus to the group III and non-spindle group II afferents causing their fatigue-induced sensitization to mechanical stimuli, such as tendon stretch and surface pressure, is likely to be metabolic in nature. Many group III as well as group IV muscle afferents have been shown to be stimulated by intraarterial injection of metabolic products of muscular contraction. These products include bradykinin (Mense & Schmidt, 1974; Mense, 1977; Kaufman et al., 1983), arachidonic acid and prostaglandin E_2 (Mense, 1981; Rotto & Kaufman, 1988), potassium (Kniffki et al., 1978; Rybicki et al., 1985), as well as lactic acid (Thimm & Barm, 1987; Rotto & Kaufman, 1988; Sinoway et al., 1993). Moreover, intraarterial injection of these metabolic products has little, if any, direct effect on the discharge of spindle afferents and Golgi tendon organs (e.g., Prochazka & Somjen, 1986). Conceivably, these injections could stimulate indirectly these group Ia afferents by activating gamma motoneurons (Jovanovic et al., 1990; see Hagbarth & Macefield, Chapter 18).

The discharge pattern of group III and IV muscle afferents in response to a tetanic contraction of 1-2 min might also provide a clue about the role played by these small diameter afferents in reflexly inhibiting the discharge of motoneurons supplying fatigued muscles. Group III afferents often respond vigorously at the onset of a contraction, but then decrease their discharge rate as the tension developed by the tetanized muscle decreases; i.e., fatigue (Kaufman et al., 1983; Mense & Stahnke, 1983). Group IV afferents, on the other hand, respond weakly, or sometimes not at all, at the onset of a brief tetanic contraction. After this latent period, which usually lasts 10 to 30 s, group IV afferents respond strongly to contraction and often maintain their response to contraction as the tension developed by the tetanized muscle decreases (Kaufman et al., 1983; Mense & Stahnke, 1983). Ischemia has little effect on the responses of most group III afferents to tetanic contraction; in contrast, ischemia markedly increased the responses to contraction of a substantial number of group IV afferents (Kaufman et al., 1983; Mense & Stahnke, 1983).

The above information might lead one to conclude that group III afferents appear to be sensitive to mechanical events in the muscle, whereas group IV afferents appear to be sensitive to metabolic events in the muscle. While there is some basis for such a conclusion, it is a gross oversimplification. For example, the mechanical sensitivity of group III muscle afferents to contraction is increased by the prior administration of bradykinin (Mense & Mayer, 1988) and arachidonic acid (Rotto et al., 1990). Moreover, both substances are produced by hindlimb skeletal muscle when it contracts (Rotto et al., 1989; Stebbins et al., 1990). In addition, the mechanical sensitivity of group III afferents to contraction is decreased by administration of dichloroacetate, a substance which decreased the working muscle's ability to produce lactic acid (Sinoway et al., 1993). These findings lead to the speculation that fatigue-induced excitation of group III and IV afferents might be caused by the production of bradykinin, lactic acid, and arachidonic acid or its metabolites in the working muscles.

SUMMARY

In humans, the inhibitory reflex effect of both large and small diameter muscle afferents on motoneuron output during fatigue has been demonstrated. This reflex inhibition has been demonstrated with isometric contractions with occluded blood flow utilizing a sustained MVC (Bigland-Ritchie et al., 1986) or submaximal contractions under ischemic conditions (Garland & McComas et al., 1990). Reflex inhibition is less evident in non-ischemic conditions (Garland et al., 1988). Disfacilitation of motoneurons has been supported by the depressed spindle afferent discharge at the onset of sustained submaximal contractions (Macefield et al., 1991). Bongiovanni and Hagbarth (1990) found that the declining motoneuron discharge rates during MVCs could be alleviated, albeit temporarily, with vibration. The actions of large- and small-diameter muscle afferents may not be mutually exclusive. Rather, the two effects could occur together, with spindle disfacilitation having a larger impact at the onset of fatigue and inhibition from small-diameter afferents may dominate as fatigue progresses (Gandevia et al., 1990).

In animals, the inhibitory reflex effect by small diameter muscle afferents on motoneuron output during fatigue has been demonstrated (Hayward et al., 1988). In addition, there is substantial evidence that the stimulus to these small diameter afferents is metabolic in nature. In contrast to the findings in humans, the findings in animals offer no support for an inhibitory reflex effect by large diameter afferents on motoneuron output during fatigue. In a way, this is surprising because the inhibition from group III and IV afferents may be exerted on both alpha and gamma motoneurons. We can offer no explanation for the discrepant findings in humans and in animals, other than the different periods of recording in each paradigm may have prevented the mapping of the full fatigue-related response. Alternatively, the explanation may be found between differences in the methods used to induce fatigue in humans and those used to induce fatigue in animals. In any event, further work in this important area should focus on this discrepancy and should seek to confirm and extend previous findings.

ACKNOWLEDGMENTS

The authors' laboratories are supported by the Natural Sciences and Engineering Research Council (NSERC) of Canada (*S.J.G.*) and United States Public Health Service grant HL 30710 (*M.P.K.*). Their attendance at the 1994 Bigland-Ritchie conference was supported, in part, by NSERC (*S.J.G.*), the Muscular Dystrophy Association (USA; *M.P.K.*), and the University of Miami (*S.J.G.*).

REFERENCES

Andres KH, von During M & Schmidt RF (1985). Sensory innervation of the Achilles tendon by group III and IV afferent fibers. *Anatomy and Embryology* **172**, 145-156.

Balestra C, Cuchateau J & Hainaut K (1992). Effects of fatigue on the stretch reflex in a human muscle. *Electroencephalography and Clinical Neurophysiology* **85**, 46-52.

Bigland-Ritchie B (1981). EMG and fatigue of human voluntary and stimulated contractions. In: Porter R, Whelan J (eds.), *Human Muscle Fatigue, Physiological Mechanisms*, pp 130-156. London: Pitman Medical.

Bigland-Ritchie B, Dawson NJ, Johansson RS & Lippold OCJ (1986). Reflex origin for the slowing of motoneurone firing rates in fatigue of human voluntary contractions. *Journal of Physiology (London)* **379**, 451-459.

Bigland-Ritchie B, Furbush F, Gandevia SC & Thomas CK (1992). Voluntary discharge frequencies of human motoneurons at different muscle lengths. *Muscle & Nerve* **15**, 130-137.

Bongiovanni LG & Hagbarth K-E (1990). Tonic vibration reflexes elicited during fatigue from maximal voluntary contractions in man. *Journal of Physiology (London)* **423**, 1-14.

Boyd IA & Davy MR (1968). *Composition of Peripheral Nerves*. Edinbugh: Livingston.

Burke D, Hagbarth K-E, Logstedt L & Wallin BG (1976). The response of human muscle spindle endings to vibration of non-contracting muscles. *Journal of Physiology (London)* **261**, 695-711.

Enoka RM & Stuart DG (1992). Neurobiology of muscle fatigue. *Journal of Applied Physiology* **72**, 1631-1648.

Gandevia SC, Macefield G, Burke D & McKenzie DK (1990). Voluntary activation of human motor axons in the absence of muscle afferent feedback. *Brain* **113**, 1563-1581.

Garland SJ. (1991). Role of small diameter afferents in reflex inhibition during human muscle fatigue. *Journal of Physiology (London)* **435**, 547-558.

Garland SJ, Garner SH & McComas AJ (1988). Reduced voluntary electromyographic activity after fatiguing stimulation of human muscle. *Journal of Physiology (London)* **408**, 547-556.

Garland SJ & McComas AJ (1990). Reflex inhibition of human soleus muscle during fatigue. *Journal of Physiology (London)* **429**, 17-29.

Hagbarth K-E, Kunesch EJ, Nordin M, Schmidt R & Wallin EU (1986). Gamma loop contributing to maximal voluntary contraction in man. *Journal of Physiology (London)* **380**, 575-591.

Hakkinen K & Komi PV (1983). Electromyographic and mechanical characteristics of human skeletal muscle during fatigue under voluntary and reflex conditions. *Electroencephalography and Clinical Neurophysiology (Limerick)* **55**, 436-444.

Hayward L, Breitbach D & Rymer WZ (1988). Increased inhibitory effects on close synergists during muscle fatigue in the decerebrate cat. *Brain Research* **440**, 199-203.

Hayward L, Wesselmann U & Rymer WZ (1991). Effects of muscle fatigue on mechanically sensitive afferents of slow conduction velocity in the cat triceps surae. *Journal of Neurophysiology* **65**, 360-370.(Abstract)

Hicks A & McComas AJ (1989). Increased sodium pump activity following repetitive stimulation of rat soleus muscles. *Journal of Physiology (London)* **414**, 337-349.

Hortobagyi T, Lambert NJ & Kroll WP (1991). Voluntary and reflex responses to fatigue with stretch-shortening exercise. *Canadian Journal of Sport Sciences* **16**, 142-150.

Hunt CC (1954). Relation of function to diameter in afferent fibers of muscle nerves. *Journal of General Physiology* **38**, 117-131.

Hutton RS & Nelson DL (1986). Stretch sensitivity of Golgi tendon organs in fatigued gastrocnemius muscle. *Medicine and Science in Sports and Exercise* **1**, 69-74.

Jovanovic K, Anastasijevic R & Vuco J (1990). Reflex effects on gamma fusimotor neurones of chemically induced discharges in small diameter muscle afferents in decerebrate cats. *Brain Research* **521**, 89-94.

Kaufman MP, Longhurst JC, Rybicki KJ, Wallach JH & Mitchell JH (1983). Effects of static muscular contraction on impulse activity of groups III and IV afferents in cats. *Journal of Applied Physiology* **55**, 105-112.

Kaufman MP, Rybicki KJ, Waldrop TG & Ordway GA (1984). Effect of ischemia on responses of group III and IV afferents to contraction. *Journal of Applied Physiology* **57**, 644-650.

Kernell D & Monster AW (1982). Motoneurone properties and motor fatigue: an intracellular study of gastrocnemius motoneurones of the cat. *Experimental Brain Research* **46**, 197-204.

Kniffki K-D, Mense S & Schmidt RF (1978). Responses of group IV afferent units from skeletal muscle to stretch, contraction and chemical stimuli. *Experimental Brain Research* **31**, 511-522.

Macefield VG, Gandevia SC, Bigland-Ritchie B, Gorman RB & Burke D (1993). The firing rates on human motoneurones voluntarily activated in the absence of muscle afferent feedback. *Journal of Physiology (London)* **471**, 429-443.

Macefield G, Hagbarth K-E, Gorman R, Gandevia SC & Burke D (1991). Decline in spindle support to alpha-motoneurones during sustained voluntary contractions. *Journal of Physiology (London)* **440**, 497-512.

Marsden CD, Meadows JC & Merton PA (1983). "Muscular wisdom" that minimizes fatigue during prolonged effort in man: peak rates of motoneuron discharge and slowing of discharge during fatigue. In: Desmedt JE (ed.), *Motor Control Mechanisms in Health and Disease*, pp 169-211. New York: Raven Press.

Mense S (1977). Nervous outflow from skeletal muscle following chemical noxious stimulation. *Journal of Physiology (London)* **267**, 75-88.

Mense S (1981). Sensitization of group IV muscle receptors to bradykinin by 5-hydroxytryptamine and prostaglandin E-2. *Brain Research* **225**, 95-105.

Mense S & Meyer H (1988). Bradykinin-induced modulation of the response behaviour of different types of feline group III and IV muscle receptors. *Journal of Physiology (London)* **398**, 49-63.

Mense S & Schmidt RF (1974). Activation of group IV afferent units from muscle by algesic agents. *Brain Research* **72**, 305-310.

Mense S & Stahnke M (1983). Responses in muscle afferent fibers of slow conduction velocity to contractions and ischemia in the cat. *Journal of Physiology (London)* **342**, 383-397.

Nelson DL & Hutton RS (1985). Dynamic and static stretch responses in muscle spindle receptors in fatigued muscle. *Medicine and Science in Sports and Exercise* **17**, 45-450 (Abstract)

Prochazka A & Somjen GG (1986). Insensitivity of cat muscle spindles to hyperkalaemia in the physiological range. *Journal of Physiology (London)* **372**, 26P.

Rotto DM & Kaufman MP (1988). Effects of metabolic products of muscular contraction on the discharge of group III and IV afferents. *Journal of Applied Physiology* **64**, 2306-2313.

Rotto DM, Massey KD, Burton KP & Kaufman MP (1989). Static contraction increases arachidonic acid levels in gastrocnemius muscles of cats. *Journal of Applied Physiology* **66**, 2721-2724.

Rotto DM, Schultz HD, Longhurst JC & Kaufman MP (1990). Sensitization of group III muscle afferents to static contraction by products of arachidonic acid metabolism. *Journal of Applied Physiology* **68**, 861-867.

Rybicki KJ, Waldrop TG & Kaufman MP (1985). Increasing gracilis interstitial potassium concentrations stimulates group III and IV afferents. *Journal of Applied Physiology* **58**, 936-941.

Sharpe MH & Miles TS (1993). Position sense at the elbow after fatiguing contractions. *Experimental Brain Research* **94**, 179-182.

Sinoway LI, Hill JM, Pickar JG & Kaufman MP (1993). Effects of contraction and lactic acid on the discharge of group III muscle afferents in cats. *Journal of Neurophysiology* **69**, 1053-1059.

Spielmann JM, Laouris Y, Nordstrom MA, Robinson GA, Reinking RM & Stuart DG (1993). Adaptation of cat motoneurons to sustained and intermittent extracellular activation. *Journal of Physiology (London)* **464**, 75-120.

Stebbins CL, Carretero OA, Mindroiu T & Longhurst JC (1990). Bradykinin release from contracting skeletal muscle of the cat. *Journal of Applied Physiology* **69**, 1225-1230.

Thimm F & Baum K (1987). Response of chemosensitive nerve fibers of group III and IV to metabolic changes in rat muscles. *Pflügers Archiv* **410**, 143-152.

von During M & Andres KH (1990). Topography and ultrastructure of group III and IV nerve terminals of cat's gastrocnemius-soleus muscle. In: Zenker W, Neuhuber WL (eds.), *The Primary Afferent Neuron: A Survey of Recent Morpho-Functional Aspects*, New York: Plenum.

Windhorst U & Kokkoroyiannis T (1991). Interaction of recurrent inhibitory and muscle spindle afferent feedback during muscle fatigue. *Neuroscience* **43**, 249-259.

Woods J, Furbush F & Bigland-Ritchie B (1987). Evidence for a fatigue-induced reflex inhibition of motoneuron firing rates. *Journal of Neurophysiology* **58**, 125-137.

SECTION VI

The Case for Central Fatigue

This section explores the status of central fatigue, ranging from its definition and measurement to evidence that serotonergic pathways within the CNS are involved.

Chapter 20 (Gandevia and colleagues) introduces a definition of central fatigue as an exercise-induced decrease in muscle force that is attributable to a decline in motoneuronal output. In isometric voluntary contractions in human subjects, the technique of twitch interpolation has revealed that progressive central fatigue develops during a range of exercise protocols. Its contribution to the reduction in force may vary in different tasks, perhaps being less important during sustained maximal voluntary contractions than intermittent submaximal contractions.

It does not necessarily develop to an equal extent in all muscle groups either. The available evidence suggests that the diaphragm, the principal inspiratory muscle, shows less central fatigue than some limb muscles. Clearly this arrangement would have survival value. For limb muscles, transcranial motor cortical stimulation has revealed that corticospinal output is submaximal during the development of central fatigue.

To study the cognitive processes that may underlie or co-exist with central fatigue, **Chapter 21** (Popivanov and colleagues) reviews the evidence that *premovement* cerebral potentials extracted from the EEG are altered during fatigue. Also, they demonstrate that these processes can, in future, be evaluated by appropriate non-linear analysis of single trials. This possibility provides a potential new opening in the study of central fatigue although the simple effect of recruitment of more muscles during fatigue will need to be critically addressed.

Next, the way in which fatigue distorts the sensation of force is discussed (**Chapter 22**, L. Jones). The observation that a suitcase held for long enough *feels* heavy underlies the formal psychophysical finding that force is usually overestimated during muscle fatigue. It is unlikely that this error is due purely to altered signals from Golgi tendon organs. The sense of effort, which is related to the level of the motor command to drive motoneurons, is enhanced during fatigue and it is often difficult for subjects to dissociate this perception from the sensation of the absolute force being generated.

In **Chapter 23** (Newsholme & Blomstrand) changes in cerebral biochemistry that might contribute to central fatigue are considered. There are many central transmitter systems and it is unlikely that only one would be involved in all the cerebral manifestations of fatigue. Evidence is presented that the plasma ratio of tryptophan to branched-chain amino acids falls during exercise and it is speculated that this may alter the discharge of some 5-hydroxytryptamine neurons within the CNS. The biochemical conversion of tryptophan to 5-hydroxytryptamine is not likely to be easily saturated. Increased levels of 5-hydroxytryptamine have been found in the brainstem and hypothalamus of rats after exhaustive exercise. Studies in human subjects suggest that the manipulation of 5-hydroxytryptamine levels may alter the perceived levels of fatigue and mental effort.

During muscle fatigue there are not only changes in the perception of performance due to the altered feedback from the active muscles, but also altered humoral factors that affect CNS performance. In addition, both reflex feedback and humoral factors will change the performance of the segmental motor apparatus. While central fatigue has a specific definition and is measurable using interpolated nerve stimuli, the processes within the CNS likely to be directly involved in its production are still not well understood.

CENTRAL FATIGUE

Critical Issues, Quantification and Practical Implications

S. C. Gandevia,[1,3] G. M. Allen,[1] and D. K. McKenzie[1,2,3]

[1] Prince of Wales Medical Research Institute
[2] Department of Respiratory Medicine
Prince of Wales Hospital
[3] Division of Medicine
University of New South Wales
Sydney, Australia

ABSTRACT

Central fatigue during exercise is the decrease in muscle force attributable to a decline in motoneuronal output. Several methods have been used to assess central fatigue; however, some are limited or not sensitive enough to detect failure in central drive. Central fatigue develops during many forms of exercise. A number of mechanisms may contribute to its development including an increased inhibition mediated by group III and IV muscle afferents along with a decrease in muscle spindle facilitation. In some situations, motor cortical output is shown to be suboptimal. A specific terminology for central fatigue is included.

HISTORICAL PERSPECTIVE

Philosophers and psychologists have long recognized that volition underlies the limits of human motor performance. Late last century Mosso wrote perceptively about muscle performance and the role played by the central nervous system (CNS) in a volume entitled simply *Fatigue*. He measured fatigue in a simple repetitive task in which a weight was moved as far as possible by flexors of a finger at a set rate and he realized that the role played by the CNS could be deduced by electrical stimulation of the relevant muscles. In the twentieth century however, physiologists have either ignored the importance of volition by assuming that near maximal performance is achieved in an attempted maximal task, or considered volition too difficult to measure or not worthy of study. For example, A.V. Hill (1926) clearly recognized that for other than the elite athlete, sub-maximal voluntary drive could limit performance: "*If one took a patient from the hospital and made him work till he could barely move, one could never be sure that he had driven himself to his limit. . . . With young athletic people one may be sure that they really have gone 'all out'.*"

Fatigue, Edited by Simon C. Gandevia et al.
Plenum Press, New York, 1995

This chapter sets out some historical and phenomenological problems in the study of central fatigue, specifies appropriate terminology, and notes some, but not all, of the sites where central fatigue occurs.

TERMINOLOGY

Given that muscle fatigue represents any reduction in force generating capacity of a muscle (e.g., Gandevia, 1992), what should the best performance by a subject be termed? Traditionally, the best trial from a number of *maximal* efforts is called the *maximal voluntary contraction*, usually abbreviated to *MVC*. However, it is obvious that in some attempted *maximal* efforts, performance can be less than maximal. Even when a trained subject receives continuous feedback, and is able to eliminate any poor efforts, voluntary drive measured by twitch interpolation (see below) ranges from 90-100% (maximal) in repeated brief maximal isometric contractions of 2-3 s duration with the elbow flexor muscles (e.g., Allen et al., 1993a; their Fig. 1). In deference to this variability, Bigland-Ritchie and colleagues occasionally ask subjects to produce not *maximal* but *super* efforts (Bigland-Ritchie et al., 1978)! Therefore, we use the term *optimal force* for that which would be obtained when the motoneuron pool is activated maximally by volition, but retain MVC for the best contraction observed on that occasion in the laboratory. We use the term *maximal effort* for every contraction which the subject believes to be maximal, regardless of the force actually achieved. *Voluntary activation* denotes the level of motoneuronal drive achieved voluntarily under laboratory conditions with continuous feedback, loud exhortation and the ability to disallow poor efforts. We reserve the term *central fatigue* for the reduction in *voluntary activation* during a particular maneuver or exercise regimen (e.g., Bigland-Ritchie et al., 1978). Because a *maximal voluntary contraction* can usually be increased by motor nerve stimulation (see below), the term *maximal evocable force* would indicate that voluntary force which cannot be increased by interpolated peripheral nerve stimuli. Under this terminology, *optimal voluntary activation* is the level of drive that produces maximal evocable force. This terminology is given in Table 1.

Processes associated with central fatigue should theoretically be demonstrable not simply at the level of the final common pathway (i.e., the motoneuron pool), but also at other sites within the neuraxis. However, the more remote these sites are from the motoneurons

Table 1. Terminology for evaluation of human muscle fatigue

Term	Definition
Muscle fatigue	any exercise-induced reduction in the ability to exert muscle force or power, regardless of whether or not the task can be sustained
Maximal voluntary contraction	a voluntary contraction which a subject believes to be maximal that is performed with continuous feedback and encouragement
Maximal evocable force	the force in a maximal voluntary contraction which cannot be increased by interpolated supramaximal motor-nerve stimuli
Voluntary activation	the level of motoneuronal drive during a voluntary contraction
Optimal voluntary activation	the level of motoneuronal activation which produces maximal evocable force
Central fatigue	a progressive exercise-induced reduction in voluntary activation of a muscle (usually assessed in maximal voluntary contractions with twitch interpolation)

These definitions are adapted from those used to evaluate supraspinal influences on muscle fatigue (Gandevia et al., 1995).

the more difficult it is to establish a causal relation to central fatigue. Motor cortical neurons show a variety of patterns of discharge during movement but some decline during a sustained contraction (see Phillips & Porter, 1977; Porter & Lemon, 1993). Indirect evidence from transcranial cortical stimulation might indicate a reduced role of motor cortical input to motoneuronal discharge during a sustained effort (Brouwer et al., 1989). Muscle spindle afferent discharge diminishes progressively during strong isometric contractions although it is not technically feasible to measure this decline during maximal efforts (Macefield et al., 1991). Processes associated with central fatigue at the segmental level will include inhibitory feedback from Golgi tendon organs and group III and IV afferents.

MEASUREMENT OF MAXIMAL VOLUNTARY DRIVE: EXPERIMENTAL APPROACHES

There are few techniques to assess the degree to which voluntary isometric muscle performance approaches the optimum. The simplest is to measure the maximal muscle force and/or EMG activity during a voluntary task, but these are inadequate measures because the true maximal output is unknown. Optimal EMG and force for most muscle groups are likely to be slightly greater than actually observed, and this would lead to an unknown degree of overestimation of voluntary activation. When muscles are not fatigued this overestimation is probably small in normal subjects, but when measurements are made in patients during exercise, it may be much larger. Indeed, the wide normal range of maximal voluntary strength for all muscle groups means that weakness is difficult to establish with a measurement at one point during the evolution of a disease. Also, voluntary strength within the normal range may lead to the false conclusion that voluntary activation is adequate. An example of the latter error comes from estimation of maximal inspiratory muscle strength in patients with moderately severe asthma. Strength is within the *normal* range (e.g., McKenzie & Gandevia, 1986), but surprisingly, a significant proportion of asthmatics cannot achieve optimal performance of their diaphragm when assessed with twitch interpolation (Allen et al., 1993b). Another example of fallacious measurement of maximal neural drive to muscles comes from measurement of the efferent phrenic neurogram in animal experiments: even when the chemical stimulus to ventilation is massive, this estimate of *drive* is well below maximal, as judged by transdiaphragmatic pressures (Sieck & Fournier, 1989). Hence *untethered* estimates of force, EMG or even the neurogram are flawed for measurement of maximal voluntary output.

Mosso (1904), Merton (1954), and Bigland and Lippold (1954) correctly perceived that maximal voluntary force could only be judged by reference to what can be achieved by non-voluntary electrical stimulation of motor output. The first approach was to compare the force in tetanic stimulation with that in voluntary contraction. These forces are similar for thumb adduction achieved by voluntary effort and by tetanic stimulation of the ulnar nerve (Merton, 1954; see also Bigland & Lippold, 1954) although the biomechanics of thumb movements are such that it is not difficult to exceed the tetanic force voluntarily (Gandevia, unpublished observations). Stimulus rates above 100 Hz are said to be needed to produce the largest tetanic force for this muscle group (Marsden et al., 1983).

Interpolation of an electrical stimulus to the motor nerve during a maximal effort should produce an increment in force at the appropriate latency either if the stimulated axons are not *all* recruited voluntarily or if they are discharging at sub-tetanic rates. This technique has now been applied not only to intrinsic muscles of the hand (Merton, 1954; Gandevia & McKenzie, 1988) but also to proximal muscles such as quadriceps and elbow flexors (e.g., Bigland-Ritchie et al., 1978; Bigland-Ritchie et al., 1983; McKenzie & Gandevia, 1991;

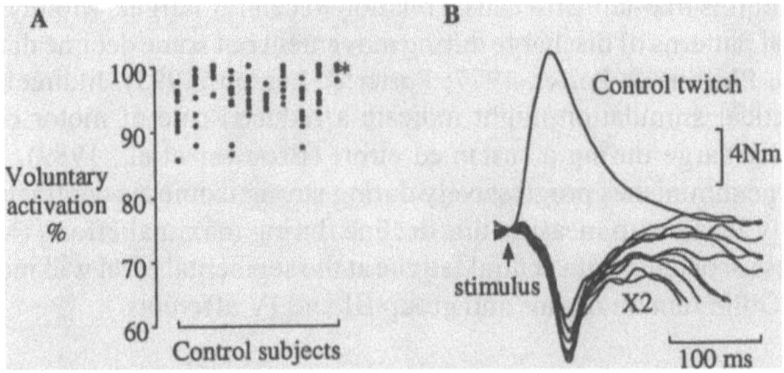

Figure 1. *A*, percentage of voluntary activation of biceps brachii (brachialis) during maximal elbow flexion measured in 10 normal subjects using twitch interpolation. Each vertical column of dots represents 10 scores for a single subject (i.e., one dot per score). *B*, raw traces for subject 10 (with asterisk) from *A*. One control twitch (large response) evoked by supramaximal stimulus delivered over relaxed elbow flexor muscles and 10 responses evoked by stimuli delivered at the peak force during attempted maximal elbow flexion. Amplitudes of the evoked response are expressed as a fraction of the control twitch amplitude and then subtracted from 1 and expressed as a percentage to give a voluntary activation score as shown in *A*. This data is from a subject with a very high level of voluntary activation.

Lloyd et al., 1991), tibialis anterior and/or soleus (Gandevia & McKenzie, 1988; Belanger & McComas, 1981) and even respiratory muscles (Bellemare & Bigland-Ritchie, 1984; Gandevia & McKenzie, 1985; Allen et al., 1993b). The usual context for its usage has been to show that the MVCs are rather close to optimal and any deficit, if detected (see below) is ignored.

The clinical utility of the twitch interpolation method for detection of *functional* components to muscular weakness has long been noted (e.g., McComas et al., 1983; Lloyd et al., 1991; Jacobsen et al., 1991). However, rarely has the sensitivity of the method been assessed. Unless separate amplification of DC-offset *twitches* and their superimposition on control twitches is used to show the electromechanical delay (e.g., Hales & Gandevia, 1988; their Fig. 1), the technique will not be sensitive enough to reveal whether more than 95% of the optimal force (from the stimulated axons) has been achieved voluntarily. This is clear from our measurements of Merton's original figure (Fig. 2). However, with appropriate techniques, superimposed twitches of <1% of the size of control potentiated twitches can be detected, and it is doubtful that further technical improvement would have much physiological significance. Pitfalls with the twitch interpolation technique abound (Gandevia, 1992). For example, stimulation must be constant and activate no axons producing antagonist torque. Therefore, common peroneal nerve stimulation is technically inadequate for twitch interpolation for ankle dorsiflexors, and ulnar stimulation is not accurate for abduction/adduction of the fingers. Sensitivity of the technique will be higher when a large twitch or brief tetanus is interpolated. In addition, the compliance of the myograph used to measure force must be minimal (Loring & Hershenson, 1992).

For twitch interpolation, the stimulus or stimuli may be applied to the nerve or intramuscular nerve fibers of the relevant muscle. Transcranial magnetic or electrical stimulation can also be used to recruit relevant motor axons (Marsden et al., 1980) or alternatively, to show that not all the potential synergists are driven maximally in complex muscle actions, depending on the task (Gandevia et al., 1990b). A new approach has been to examine the transcranially evoked compound muscle action potentials before and after exercise (Brasil-Neto et al., 1993; see also Gandevia et al., 1994).

Conventional twitch interpolation has revealed that although most subjects have the ability to activate motoneuron pools to a high level (voluntary activation scores ≈ 95%), this

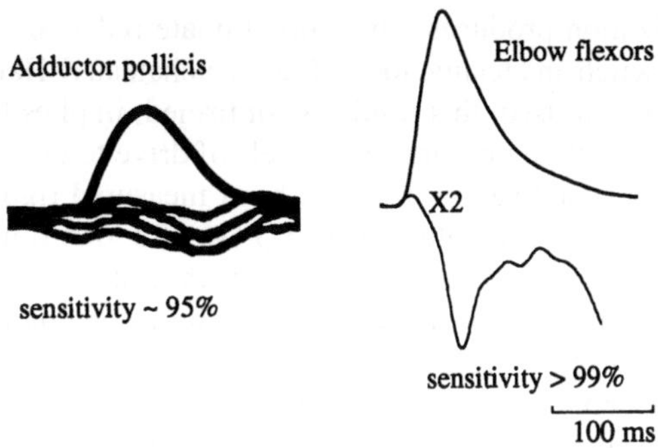

Figure 2. Example of the differing sensitivities of the twitch interpolation technique. On the left are shown the original traces from Merton (1954), on the right our more sensitive version of the technique. One large control twitch evoked by stimuli delivered to a relaxed muscle is shown with superimposed twitches (smaller responses) evoked at the peak force during maximal voluntary contractions. High resolution of superimposed twitches may reveal failure in voluntary drive previously undetected. As calculated from the original figure, Merton's system allows detection of a failure of 5% from optimal drive, with the more sensitive technique able to resolve failure of voluntary drive of less than 1%.

level is not consistently 100%. (This is not surprising given that the fluctuations in voluntary force during a MVC are absent in a fused tetanic contraction.) The elbow flexors are a muscle group for which high levels of maximal drive can be achieved with minimal training and hence they are appropriate for exercise testing (e.g., Lloyd et al., 1991; Allen et al., 1993a). Even so, on repeated testing under ideal conditions some normal subjects activate the muscle fully in only 2-5% of contractions. If 10 brief MVCs are performed with a sufficient interval to prevent peripheral fatigue, 95% of the population will have a median level of voluntary activation for biceps/brachialis at or above 92%. Data from 10 typical normal subjects are shown in Fig. 1. Furthermore, this level will not improve significantly with repeated testing (Allen et al., 1993a). It is as if this level is set at a high but characteristic level in each subject. Not surprisingly, the level of maximal voluntary drive varies between muscles. In direct comparisons it was lower for the diaphragm than for the elbow flexors (Allen et al., 1993b), and lower for the plantar-flexors than dorsiflexors of the ankle (Belanger & McComas, 1981).

Transcranial stimulation of the motor cortex has recently revealed that motor cortical output is close to optimal for the elbow flexors at the onset of maximal contractions when they are not fatigued, but that it is suboptimal as central fatigue develops (Gandevia et al., 1995). Ultimately, it will be necessary to focus on the factors effectively *upstream* of the motor cortex because they can clearly limit performance.

VARIATION IN *MAXIMAL* VOLUNTARY DRIVE

Voluntary activation will be suboptimal when the muscle, or nearby skin provides a nociceptive input either due to disease, recent surgery (Rutherford, et al., 1986) or deliberately induced painful restraint of the limb (Gandevia & McKenzie, 1988). Interestingly, when joint effusions were produced artificially in normal subjects, maximal voluntary knee torques were reduced (e.g., Wood et al., 1988). Many reflex inputs can reduce motoneuronal output, and while the pathways for this are not established, oligosynaptic spinal pathways are likely contributors.

Limb immobilization produces a disproportionate reduction in maximal voluntary forces compared with twitch and tetanic forces (e.g., Duchateau & Hainaut, 1987). Although twitch interpolation was not used, this result, if confirmed, implies that voluntary usage is necessary to retain the ability to produce high levels of drive to the motoneuron pool.

Levels of voluntary activation have not been measured rigorously under environmental extremes. However, at simulated altitude (PO_2, 282 torr) some subjects had poorer neural drive to ankle dorsiflexors, although the effect was not well quantified (Garner et al., 1990). In extreme heat, muscle performance diminishes, perhaps in part because twitch contraction times shorten requiring higher motoneuronal discharge frequencies, although other central factors probably contribute (cf. Bigland-Ritchie et al., 1992b).

Finally, if the results obtained for isometric contractions are to be generalized, we need to know the extent to which high levels of voluntary drive can be achieved (in the absence of fatigue) during tasks involving many muscle groups and during non-isometric tasks. When untrained subjects contract multiple muscle groups isometrically, lower levels of force and EMG may occur than for contractions of the individual muscles (e.g., Howard & Enoka, 1991) and this has been confirmed with twitch interpolation (Herbert & Gandevia, unpublished observations). While it is technically more demanding to apply twitch interpolation during concentric contractions, some results suggest that voluntary drive can be high (see Newham et al., 1991).

EXERCISE-INDUCED REDUCTIONS IN VOLUNTARY DRIVE: *CENTRAL FATIGUE*

As indicated above, during attempted *maximal* contractions of the unfatigued elbow flexors, the level of voluntary activation measured by twitch interpolation is about 95% for brief efforts of 2-3 s duration. How does this change when the muscle contracts for longer and develops peripheral fatigue?

Thomas and colleagues (1989) used prolonged isometric contractions of FDI and tibialis anterior. They found that while voluntary drive showed some diminution, it could be restored by an additional *extra* effort. This paralleled earlier findings (Bigland-Ritchie et al., 1978) in which central fatigue was observed with sustained quadriceps contractions in 4 of 9 subjects. Inspection of their records suggests that central fatigue probably occurred in them all. Maximal contractions of elbow flexors sustained for 3 min are accompanied by a progressive reduction in force. While voluntary drive is initially high (>95%), it usually declines and becomes more variable with time (Gandevia & McKenzie, 1993; text deleted Fig. 3). In an alternative approach subjects performed intermittent MVCs of 10 s duration with duty cycles less than 50%. These were also accompanied by concomitant peripheral

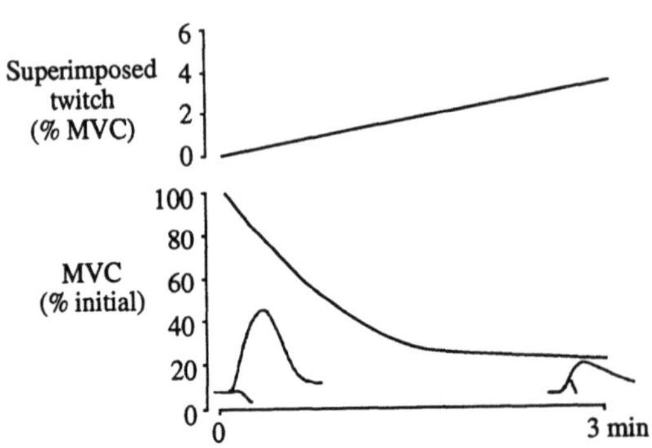

Figure 3. Diagrammatic representation of the force decline during a sustained maximal elbow flexion (3 min duration) which occurs with a simultaneous increase in the superimposed response evoked by supramaximal stimuli over biceps brachii from highly-trained subjects. An example of decline in peripheral twitch amplitude across the 3 min is shown at bottom and the accompanying increase in responses superimposed on the maximal effort.

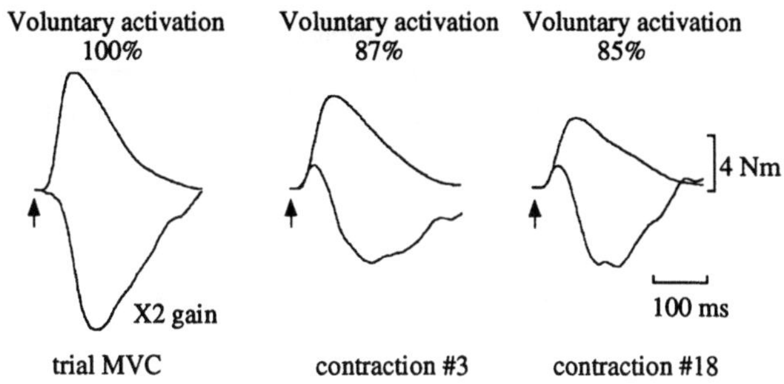

Figure 4. Data from one subject during a fatigue protocol (50% duty cycle). The larger response is that evoked by supramaximal paired (10 ms interval) stimuli in the relaxed muscle (control), the smaller (shown at increased gain X2) is that evoked by the stimuli delivered during a MVC. Stimuli are delivered at the arrow. Note the decline in control twitch amplitude which occurs as the response evoked by paired stimuli during a MVC to the elbow flexors increases. Voluntary activation, initially at 100%, declines as the contraction number increases. This illustrates the simultaneous development of both peripheral and central fatigue (data redrawn from McKenzie & Gandevia, 1991).

and central fatigue (McKenzie & Gandevia, 1991). Data from a typical experiment are given in Fig. 4: voluntary activation was 100% in the trial contraction but declined thereafter. Submaximal exercise at 30% MVC (60% duty cycle) produces a small but measurable degree of central fatigue (e.g., Lloyd et al., 1991). However, when this level of drive is plotted together with that for the maximal voluntary force, it is evident that perhaps half of the voluntary force decline after 45 min of exercise may be attributed to central fatigue. Isokinetic exercise is also accompanied by the development of central fatigue (e.g., Newham et al., 1991).

The development of central fatigue is accompanied by changes in the excitability of the human motor cortex (Gandevia et al., 1994). Evidence for simultaneous reductions in the thresholds for excitation and inhibition has been obtained. However, once central fatigue has developed, motor cortical output is less than optimal and, recent studies have shown that the changes in cortical excitability can be *dissociated* from the presence of central fatigue (Gandevia et al., 1995). Hence, it is unlikely that the changes in cortical excitability per se are necessarily the direct cause of central fatigue.

CONTRIBUTION OF MOTOR UNIT FIRING RATES

It is obvious that to obtain optimal output from a muscle all motor units must be driven at or above a rate sufficient to produce a fused tetanus. Indeed, motor unit firing rates are often below 20 Hz and rarely beyond 50 Hz in maximal isometric efforts, not rates that would necessarily imply full tetanic fusion of all motor units (even if they were all recruited) (Bellemare et al., 1983; Gandevia et al., 1990a; Thomas et al., 1991; Vander Linden et al., 1991; Bigland-Ritchie et al., 1992a). However to exceed the rate which can just achieve maximal force output risks failure of neuromuscular transmission.

One concept that has gained support is that motor unit firing rates are appropriately matched to the contractile machinery's twitch properties. This arose initially from the realization that *red* and *white* muscles subserved different tasks, and had different contractile speeds. There is also heterogeneity of contractile speed between units within muscle and between muscles. Moreover, contractile properties of muscle vary with temperature, contraction history and fatigue.

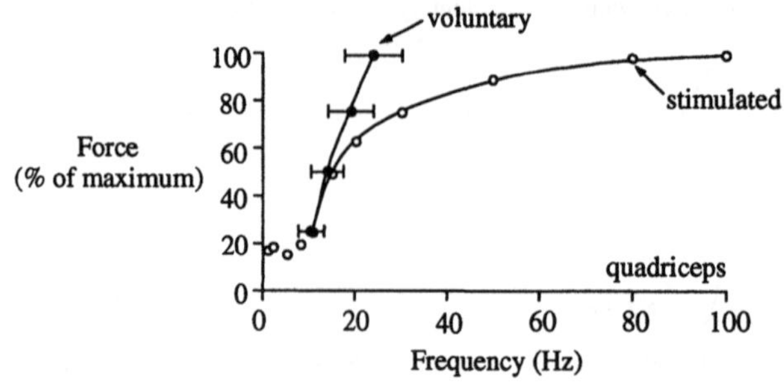

Figure 5. Force-frequency curves for stimulated (open circles) and voluntary contractions (closed circles). Up to 50% MVC, the stimulated and voluntary curves are similar, but above this the voluntary contractions generated higher forces at a particular frequency of excitation. Note the almost linear increase in force and motor unit discharge frequency for voluntary but not stimulated contractions (data redrawn from Bigland-Ritchie & Rice, 1994).

The properties of the force-frequency relationship require that motor unit firing frequencies remain in the ascending portion of the curve to allow frequency modulation of force (see Fig. 5). There is a natural tendency for a decline in firing rate of motor units during maximal voluntary contractions (e.g., Marsden et al., 1969; see also Grimby et al., 1981; Fig. 6). Evidence exists for the mechanisms that allow this alteration. First, motoneurons subjected to intracellular current injection discharge with progressively lower frequencies, a property linked appropriately to their type (Kernell & Monster, 1982). However, it is unlikely that this mechanism occurs during natural contractions (cf. Gandevia et al., 1990a; Macefield et al., 1993). Second, Bigland-Ritchie and colleagues found that motor unit firing rates diminish during single maximal isometric contractions which produce contractile slowing, and that this diminution remains so long as the muscle is held ischemic with a cuff inflated above systolic pressure (Woods et al., 1987). This suggests that while only minimal central fatigue developed, the reflex reduction in motor unit firing rate would match the

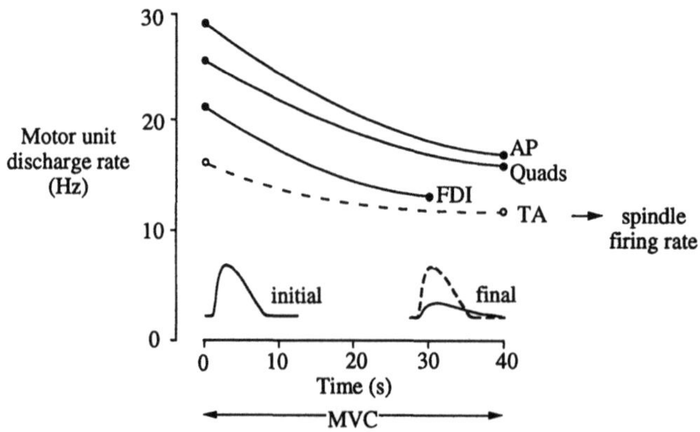

Figure 6. Motor unit firing rate declines in three human muscles with fatigue induced by a sustained maximal contraction. For adductor pollicis, mean firing rate declined by 42% over a 40 s contraction (Bigland-Ritchie et al., 1983), in quadriceps mean firing rate declined by 38% over 40 s (Woods et al., 1987) and for first dorsal interosseous, the mean motor unit firing rate declined by 39% over a 30 s contraction (Gandevia, et al., 1990). Muscle spindle firing rate (- - - -) also declines in isometric contractions at 20-60% MVC (Macefield et al., 1991). Decline in amplitude of response to stimulation over the relaxed muscle is shown diagrammatically at bottom of figure.

slowing of contractile speed. This would preserve flexibility in the operating point on the force-frequency relationship such that units did not continue to discharge at *supratetanic* rates.

In contrast, during intermittent isometric contractions at 30% MVC (60% duty cycle, Vøllestad et al., 1988; Lloyd et al., 1991) peripheral fatigue develops but motor unit firing rates are relatively well preserved. This occurs despite a tendency for contractile speed to increase and for some central fatigue to develop (Lloyd et al., 1991). Clearly, the relationship between muscle speed and motor unit firing rate during exercise will be complex in the face of potentiation, temperature changes, peripheral and central fatigue, and the variable behavior of individual motor units. It will be further complicated during dynamic exercise.

A new approach has involved recording the discharge of motor cortical cells in the monkey during phasic isotonic/isometric elbow flexion: reductions in discharge rates were detected for some cells (Maton, 1991). An alternative approach has been to study the discharge of motor units during sustained *contractions* in the absence of any peripheral feedback. This has been achieved for motor units supplying intrinsic hand muscles (Gandevia et al., 1990a) and tibialis anterior (Macefield et al., 1993) by recording with a microelectrode from a fascicle proximal to a complete nerve block produced by injection of local anesthetic. In these circumstances, subjects can maintain a constant output to the paralyzed muscle provided that ancillary auditory or visual feedback of the neurogram is provided. However, most importantly, the motor unit firing rates (across a range of efforts, including maximal ones) are lower during the blockade of peripheral afferents, implying that fusimotor-mediated spindle facilitation is necessary to achieve peak rates. Secondly, the motor unit discharge rates in sustained efforts do not show the progressive decline noted for intact axons. This implies that alterations in reflex input normally contribute to this decline (see also Gooch et al., 1990).

Both a decline in spindle facilitation, as directly observed in nerve recordings, and an increase in the inhibition mediated by group III or IV afferents could contribute to the reduction in motor unit firing rates with isometric contractions (Hayward et al., 1988). Ample evidence exists for activation of small-diameter group III and IV muscle afferents by muscle contraction, with group III afferents discharging more in response to static than dynamic contractions (Kaufman et al., 1984b) and group IV afferents discharging more under ischemic conditions (Paintal, 1960; Kaufman et al., 1984a). These inputs may provide late facilitation to fusimotor neurons when contractile force declines (Ljubisavljevic et al., 1992), but it is probably insufficient to affect spindle sensitivity such that the tendon jerk is restored to control levels following fatigue (Balestra et al., 1992; see also Kukulka et al., 1986). The observation that muscle vibration of a type likely to activate muscle spindles can increase force during a sustained MVC suggests that additional peripheral facilitation can replace the reduced spindle input (Bongiovanni & Hagbarth, 1990). However, this would be expected whether motoneuronal output had diminished due to a loss of reflex or descending excitation.

The neural substrate exists for potent modulation of motoneuronal output during sustained contractions. Future studies will no doubt tease out the relative reflex contributions of different afferent classes, the irradiation of relevant reflexes, and the temporal gating of excitation reaching the motoneuronal pool. It is all very well to know how the discharge of the various muscle afferents alters, but the effectiveness of these inputs at the motoneuron will need to be assessed under a range of conditions. For example, studies in the cat indicate that Ib homonymous inhibition may be gated out by presynaptic inhibition during electrically induced contractions (Lafleur et al., 1992) and studies in humans show that Ia excitation may be reduced through presynaptic inhibition (Meurnier & Pierrot-Deseilligny, 1989).

CENTRAL FATIGUE IN A BROADER CONTEXT

Our studies began with the use of modified twitch interpolation to show that the phrenic motoneuron pool could be driven by volition to produce the near optimal force for the diaphragm (Gandevia & McKenzie 1985). These studies complemented those of Bigland-Ritchie which provided similar conclusions for combined inspiratory and expulsive maneuvers (Bellemare & Bigland-Ritchie, 1984). Initially, we interpolated brief tetani to enhance the size of the force increments during maximal efforts and subsequently devised a special amplifier to capture the small superimposed twitches evoked by single stimuli (Hales & Gandevia, 1988). The more carefully we looked, the more clear it was that voluntary activation was less than optimal. In limited training sessions, this capacity remains unchanged (Allen et al., 1993a), although others have exploited the potential capacity to improve activation with merely imagined effort (Yue & Cole, 1992). As variable voluntary activation occurs in virtually every normal subject over a range of ages and athletic abilities, it is not an oddity but biological reality. If it was not recognized previously, this reflects in part the available methods.

Just because a biological phenomenon can be measured does not establish its importance. One implication from the studies on suboptimal drive to the motoneuron pool with an unexercised muscle is that some motoneurons during an MVC are not on the plateau of their force-frequency relationship. This assumes that all motoneurons have been recruited, a possibility which seems likely at least for distal muscles given that they tend to use predominantly frequency modulation to increase force (De Luca et al., 1982). Indeed, from 25 - 100% MVC, motor unit firing frequency increases linearly without evidence for a plateau (Bigland-Ritchie et al., 1992a; Bigland-Ritchie & Rice, 1994; their Fig. 5). Thus, motor unit frequencies and hence force must remain sensitive to any change in descending, interneuronal or reflex drive impinging on the motoneurons. Apart from preserving the flexibility of the force-frequency relationship, is there anything else that may explain or provide a functional correlate with the observation of suboptimal activation and central fatigue? When might the phenomenon have pathophysiological or evolutionary significance?

First, a neural limitation on the upper bounds of muscle output could reside at a motor cortical level or *upstream* of it (Gandevia et al., 1995). This would act to limit recruitment of many muscle groups simultaneously and limit the progress of peripheral fatigue. It may not be coincidental that athletes approach but rarely achieve their theoretical maximal breathing capacity during severe exercise. A second argument is suggested by observations in asthmatic subjects: some fail to achieve optimal output to their diaphragm during self-selected maximal efforts despite greater use of this muscle due to increased airway resistance (Allen et al., 1993b). Were this to occur during severe airway narrowing, those with the lowest levels of voluntary drive could be predisposed to acute ventilatory failure and death. Hence the capacity to drive some muscles voluntarily at a high level may have survival value. By analogy, central drive to major limb muscles may have survival necessity. The concept that central fatigue may develop for the diaphragm in a dramatic way during loading has been demonstrated during expulsive contractions when the diaphragm and abdominal muscles are coactivated to elevate abdominal pressure (Bellemare & Bigland-Ritchie, 1987; McKenzie et al., 1992). However, the diaphragm shows much less central fatigue when purely inspiratory contractions are required. Therefore, the development of central fatigue is dependent on the task (McKenzie et al., 1992). This is not especially surprising: the selection of motor cortical neurons and the setting of spinal and long-latency reflexes behave in a task-dependent way (Porter & Lemon, 1993).

Thirdly, discussion so far has focused on signs consistent with central fatigue during different forms of exercise. Why do we stop exercise: is it because we reach theoretical limits to cardiac, pulmonary or muscle performance? In a detailed statistical analysis of results from subjects undergoing cycle ergometry, Killian (1992) concluded that normal subjects in attempted *maximal* exercise, cycle at heart rates, levels of ventilation and even levels of symptom intensity (breathlessness and leg effort) which are submaximal. He likened this to giving up when a certain threshold for tolerance had been exceeded. To express the data in another way: central fatigue had terminated the exercise.

Finally, these arguments focus the spotlight on what is known about other supraspinal processes specific to muscle fatigue. It is not simple to establish whether a change in any aspect of CNS function reflects the contribution to the *central* fatigue or is caused by it. Nonetheless, we are at a stage where these questions need to be addressed. Already, changes in premovement potentials have been reported with isometric fatigue (Fruede & Ullsperger, 1987) although what caused them to change is unclear. However, it is tempting to ascribe them to the greater subjective effort and recruitment of synergists which occurs during repeated efforts at 80% MVC. Motivation alters with exercise and may directly or indirectly change exercise performance (Heyes et al., 1985; Chaouloff, 1991).

During exercise, changes occur at all steps in force production from upstream of the motor cortex to the motoneuron, and peripherally from the muscle mitochondria to the myofibril. It would be surprising and biologically unsound if force production always failed because of an impairment at one step in this *chain*. Evolutionary design dictates that multiple steps will tend to fail together; but ultimately, it is the central nervous system which may decide when enough is enough.

ACKNOWLEDGMENTS

The authors wish to thank Ms. Jane Butler and Dr. Janet Taylor for comments on the manuscript. Our laboratory has been supported by National Health & Medical Research Council of Australia, and the Asthma Foundation of New South Wales, Australia. Attendance of the authors at the 1994 Bigland-Ritchie conference was supported, in part, by the Prince of Wales Medical Research Institute, Australia (*G.M.A.*), the Muscular Dystrophy Association (USA; *S.C.G.* and *D.K.McK.*), and the University of Miami (*G.M.A.* and *S.C.G.*).

REFERENCES

Allen GM, Gandevia SC & McKenzie DK (1993a). Accurate measurements of maximal strength and maximal drive with twitch interpolation. *Electroencephalography and Clinical Neurophysiology* **86**, 52P.

Allen GM, McKenzie DK, Gandevia SC & Bass S (1993b). Reduced voluntary drive to breathe in asthmatic subjects. *Respiration Physiology* **93**, 29-40.

Balestra C, Duchateau J & Hainaut K (1992). Effects of fatigue on the stretch reflex in a human muscle. *Electroencephalography and Clinical Neurophysiology* **85**, 46-52.

Belanger AY & McComas AJ (1981). Extent of motor unit activation during effort. *Journal of Applied Physiology: Respiratory, Environmental & Exercise Physiology* **51**, 1131-1135.

Bellemare F & Bigland-Ritchie B (1984). Assessment of human diaphragm strength and activation using phrenic nerve stimulation. *Respiration Physiology* **58**, 263-277.

Bellemare F & Bigland-Ritchie B (1987). Central components of diaphragmatic fatigue assessed by phrenic nerve stimulation. *Journal of Applied Physiology* **62**, 1307-1316.

Bellemare F, Woods JJ, Johansson R & Bigland-Ritchie B (1983). Motor-unit discharge rates in maximal voluntary contractions of three human muscles. *Journal of Neurophysiology* **50**, 1380-1392.

Bigland B & Lippold OCJ (1954). Motor unit activity in the voluntary contraction of human muscles. *Journal of Physiology (London)* **125**, 322-335.

Bigland-Ritchie B, Dawson NJ, Johansson RS & Lippold OCJ (1986). Reflex origin for the slowing of motoneurone firing rates in fatigue of human voluntary contractions. *Journal of Physiology (London)* **379**, 451-459.

Bigland-Ritchie B, Furbush FH, Gandevia SC & Thomas CK (1992a). Voluntary discharge frequencies of human motoneurons at different muscle lengths. *Muscle & Nerve* **15**, 130-137.

Bigland-Ritchie B, Johansson R, Lippold OCJ & Woods JJ (1983). Contractile speed and EMG changes during fatigue of sustained maximal voluntary contractions. *Journal of Neurophysiology* **50**, 313-324.

Bigland-Ritchie B, Jones DA, Hosking GP & Edwards RHT (1978). Central and peripheral fatigue in sustained maximum voluntary contractions of human quadriceps muscle. *Clinical Science and Molecular Medicine* **54**, 609-614.

Bigland-Ritchie B & Rice CL (1994). Comparison of force-frequency relations in human voluntary and stimulated contractions. *Proceedings of the Physiological Society,* C103.

Bigland-Ritchie B, Thomas CK, Rice CL, Howarth JV & Woods JJ (1992b). Muscle temperature, contractile speed, and motoneuron firing rates during human voluntary contractions. *Journal of Applied Physiology* **73**, 2457-2461.

Bongiovanni LG & Hagbarth K-E (1990). Tonic vibration reflexes elicited during fatigue from maximal voluntary contractions in man. *Journal of Physiology (London)* **423**, 1-14.

Brasil-Neto JP, Pascual-Leone A, Valls-solé J, Cammarota A, Cohen LG & Hallet M (1993). Postexercise depression of motor evoked potentials: a measure of central nervous system fatigue. *Experimental Brain Research* **93**, 181-184.

Brouwer B, Ashby P & Midroni G (1989). Excitability of corticospinal neurons during tonic muscle contractions in man. *Experimental Brain Research* **74**, 649-652.

Chaouloff F (1991). Cerebral monoamines and fatigue. In Atlan G, Beliveau L, Bouissou P (eds.), *Muscle Fatigue: Biochemical and Physiological Aspects,* pp. 234-240. Paris: Masson.

De Luca CJ, Lefever RS, McCue MP & Xenakis AP (1982). Behaviour of human motor units in different muscles during linearly varying contractions. *Journal of Physiology (London)* **329**, 113-128.

Duchateau J & Hainaut K (1987). Electrical and mechanical changes in immobilized human muscle. *Journal of Applied Physiology* **62**, 2168-2173.

Fruede F & Ullsperger P (1987). Changes in Bereitschaftspotential during fatiguing and non-fatiguing hand movements. *European Journal of Applied Physiology and Occupational Physiology* **56**, 105-108.

Gandevia SC (1992). Some central and peripheral factors affecting human motoneuronal output in neuromuscular fatigue. *Sports Medicine* **13**, 93-98.

Gandevia SC, Allen GM, Butler JE & Taylor JL (1995). Supraspinal factors in human muscle fatigue: evidence for suboptimal output from the motor cortex. *Journal of Physiology (London)* In press.

Gandevia SC, Butler JE, Allen GM & Taylor JL (1994) Prolongation of the 'silent' period following transcranial magnetic stimulation. *Proceedings of the Physiological Society,* C138.

Gandevia SC, Macefield G, Burke D & McKenzie DK (1990a). Voluntary activation of human motor axons in the absence of muscle afferent feedback. The control of the deafferented hand. *Brain* **113**, 1563-1581.

Gandevia SC & McKenzie DK (1985). Activation of the human diaphragm during maximal static efforts. *Journal of Physiology (London)* **367**, 45-56.

Gandevia SC & McKenzie DK (1988). Activation of human muscles at short muscle lengths during maximal static efforts. *Journal of Physiology (London)* **407**, 599-613.

Gandevia SC & McKenzie DK (1993). Central factors in human muscle performance. *Proceedings of the International Union of Physiological Sciences* 122.6/O.

Gandevia SC, McKenzie DK & Plassman BL (1990b). Activation of human respiratory muscles during different voluntary manoeuvres. *Journal of Physiology (London)* **428**, 387-403.

Garner SC, Sutton JR, Burse RL, McComas AJ, Cymerman A & Houston CS (1990). Operation Everest II: neuromuscular performance under conditions of extreme simulated altitude. *Journal of Applied Physiology* **68**, 1667-1172.

Gooch JL, Newton BY & Petajan JH (1990). Motor unit spike counts before and after maximal voluntary contraction. *Muscle & Nerve* **13**, 1146-1151.

Grimby L, Hannerz J & Hedman B (1981). The fatigue and voluntary discharge properties of single motor units in man. *Journal of Physiology (London)* **316**, 545-554.

Hales JP & Gandevia SC (1988). Assessment of maximal voluntary contraction with twitch interpolation: an instrument to measure twitch responses. *Journal of Neuroscience Methods* **25**, 97-102.

Hayward L, Breitbach D & Rymer Z (1988). Increased inhibitory effects on close synergists during muscle fatigue in the decerebrate cat. *Brain Research* **440**, 199-203.

Heyes MP, Garnett ES & Coates G (1985). Central dopaminergic activity influences rats ability to exercise. *Life Sciences* **36**, 671-677.

Hill AV (1926). *Muscular Activity*. Baltimore: Williams & Wilkins.

Howard JD & Enoka RM (1991). Maximum bilateral contractions are modified by neurally mediated interlimb effects. *Journal of Applied Physiology* **70**, 306-316.

Jacobsen S, Wildschiodtz G & Danneskiold-Samsoe B (1991). Isokinetic and isometric muscle strength combined with transcutaneous electrical muscle stimulation in primary fibromyalgia syndrome. *Journal of Rheumatology* **18**, 1390-1393.

Kaufman MP, Rybicki KJ, Waldrop TG & Ordway GA (1984a). Effect of ischemia on responses of group III and IV afferents to contraction. *Journal of Applied Physiology* **57**, 644-650.

Kaufman MP, Waldrop TG, Rybicki KJ, Ordway GA & Mitchell JH (1984b). Effects of static and rhythmic twitch contractions on the discharge of group III and IV muscle afferents. *Cardiovascular Research* **18**, 663-668.

Kernell D & Monster AW (1982). Motoneurone properties and motor fatigue. *Experimental Brain Research* **46**, 197-204.

Killian K (1992). Symptoms limiting exercise. In: Jones NL, Killian KJ (eds.), *Breathlessness*, pp. 132-142. Hamilton, Canada: Boehringer Ingelheim.

Kukulka CG, Moore MA & Russell AG (1986). Changes in human alpha-motoneuron excitability during sustained maximum isometric contractions. *Neuroscience Letters* **68**, 327-333.

Lafleur J, Zytnicki D, Horcholle-Bossavit G & Jami L (1992). Depolarization of Ib afferent axons in the cat spinal cord during homonymous muscle contraction. *Journal of Physiology (London)* **445**, 345-354.

Ljubisavljevic M, Jovanovic K & Anastasijevic R (1992). Changes in discharge rate of fusimotor neurones provoked by fatiguing contractions of cat triceps surae muscles. *Journal of Physiology (London)* **445**, 499-513.

Lloyd AR, Gandevia SC & Hales JP (1991). Muscle performance, voluntary activation, twitch properties and perceived effort in normal subjects and patients with the chronic fatigue syndrome. *Brain* **114**, 85-98.

Loring SH & Hershenson MB (1992). Effects of series compliance on twitches superimposed on voluntary contractions. *Journal of Applied Physiology* **73**, 516-521.

Macefield G, Hagbarth K-E, Gorman R, Gandevia SC & Burke D (1991). Decline in spindle support to alpha motoneurones during sustained voluntary contractions. *Journal of Physiology (London)* **440**, 497-512.

Macefield VG, Gandevia SC, Bigland-Ritchie B, Gorman RB & Burke D (1993). The firing rates of human motoneurones voluntarily activated in the absence of muscle afferent feedback. *Journal of Physiology (London)* **471**, 429-443.

Marsden CD, Meadows JC & Merton PA (1969). Muscular wisdom. *Journal of Physiology (London)* **200**, 15P.

Marsden CD, Meadows JC & Merton PA (1983). "Muscular wisdom" that minimizes fatigue during prolonged effort in man: peak rates of motoneuron discharge and slowing of discharge during fatigue. *Advances in Neurology* **39**, 169-211.

Marsden CD, Merton PA & Morton HB (1980). Maximal twitches from stimulation of the motor cortex in man. *Journal of Physiology (London)* **312**, 5P.

Maton B (1991). Central nervous changes in fatigue induced by local work. In Atlan G, Beliveau L, Bouissou P (eds.), *Muscle Fatigue: Biochemical and Physiological Aspects*, pp. 207-221. Paris: Masson.

McComas AJ, Kereshi S & Quinlan J (1983). A method for detecting functional weakness. *Journal of Neurology, Neurosurgery & Psychiatry* **46**, 280-282.

McKenzie DK, Bigland-Ritchie B, Gorman RB & Gandevia SC (1992). Central and peripheral fatigue of human diaphragm and limb muscles assessed by twitch interpolation. *Journal of Physiology (London)* **454**, 643-656.

McKenzie DK & Gandevia SC (1986). Strength and endurance of inspiratory, expiratory, and limb muscles in asthma. *American Review of Respiratory Disease* **134**, 999-1004.

McKenzie DK & Gandevia SC (1991). Recovery from fatigue of human diaphragm and limb muscles. *Respiration Physiology* **84**, 49-60.

Merton PA (1954). Voluntary strength and fatigue. *Journal of Physiology (London)* **123**, 553-564.

Meurnier S & Pierrot-Deseilligny E (1989). Gating of the afferent volley of the monosynaptic stretch reflex during movement in man. *Journal of Physiology (London)* **419**, 753-763.

Mosso A (1904). *Fatigue*. London: Sonnenschein & Co.

Newham DJ, McCarthy T & Turner J (1991). Voluntary activation of human quadriceps during and after isokinetic exercise. *Journal of Applied Physiology* **71**, 2122-2126.

Paintal AS (1960). Functional analysis of group III afferent fibers of mammalian muscles. *Journal of Physiology (London)* **152**, 250-270.

Phillips CG & Porter R (1977). *Corticospinal Neurones, Their Role in Movement*. London: Academic Press.

Porter R & Lemon R (1993). *Corticospinal Function and Voluntary Movement*. Oxford: Clarendon Press.

Rutherford OM, Jones DA & Newham DJ (1986). Clinical and experimental application of the percutaneous twitch superimposition technique for the study of human muscle activation. *Journal of Neurology, Neurosurgery and Psychiatry* **49**, 1288-1294.

Sieck GC & Fournier M (1989). Diaphragm motor unit recruitment during ventilatory and nonventilatory behaviours. *Journal of Applied Physiology* **66**, 2539-2545.

Thomas CK, Bigland-Ritchie B & Johansson RS (1991). Force-frequency relationships of human thenar motor units. *Journal of Neurophysiology* **65**, 1509-1516.

Thomas CK, Woods JJ & Bigland-Ritchie B (1989). Impulse propagation and muscle activation in long maximal voluntary contractions. *Journal of Applied Physiology* **67**, 1835-1842.

Vander Linden DW, Kukulka CG & Soderberg GL (1991). The effect of muscle length on motor unit discharge characteristics in human tibialis anterior muscle. *Experimental Brain Research* **84**, 210-218.

Vøllestad NK, Sejersted OM, Bahr R, Woods JJ & Bigland-Ritchie B (1988). Motor drive and metabolic responses during repeated submaximal contractions in humans. *Journal of Applied Physiology* **64**, 1421-1427.

Wood L, Ferrell WR & Baxendale RH (1988). Pressures in normal and acutely distended human knee joints and effects on quadriceps maximal voluntary contractions. *Quarterly Journal of Experimental Physiology* **73**, 305-314.

Woods JJ, Furbush F & Bigland-Ritchie B (1987). Evidence for a fatigue-induced reflex inhibition of motoneurone firing rates. *Journal of Neurophysiology* **58**, 125-137.

Yue G & Cole KJ (1992). Strength increases from the motor program: comparison of training with maximal voluntary and imagined muscle contractions. *Journal of Neurophysiology* **67**, 1114-1123.

21

SINGLE-TRIAL READINESS POTENTIALS AND FATIGUE

D. Popivanov, A. Mineva, and J. Dushanova

Institute of Physiology
Bulgarian Academy of Sciences
Acad. G. Bonchev Str., Bl. 23, 1113 Sofia, Bulgaria

ABSTRACT

The authors propose that the cognitive processes related to internal motivation and volition (e.g., intention and preparation of a voluntary action), influenced by central fatigue, could be identified and characterized by cerebral readiness potentials (RP) using methods of chaotic dynamics. The boundaries of single trial RP and its successive phases can be detected by tracking the data dynamics, and are represented by chaotically behaved short EEG transitions.

INTRODUCTION

The performance of a voluntary movement is preceded by a foreperiod during which the programming process takes place. The segment of scalp recorded EEG occurring in the 1-2 s period before movement onset, is considered to reflect this process and is known as the readiness potential (RP). Since the time of their discovery by Kornhuber and Deecke (1964), the brain potentials preceding the performance of voluntary actions have been extensively investigated to determine to what extent they reflect the organization of the voluntary motor acts (e.g., Libet, 1985; Barrett et al., 1986; Tarkka & Hallett, 1990). Accordingly, the relations between parameters of the movement and the preceding RP have been studied. The results of Hashimoto and colleagues (1980) revealed that the amplitude and duration of the averaged cortical slow potential preceding voluntary movements in monkeys were positively correlated with increasing loads. Similar results have been published for humans by Kristeva and colleagues (1990), who showed a significant effect of the load on the late amplitudes of the RP during unilateral finger flexion.

The participation of the CNS during muscular fatigue has been investigated at different levels. There exists data supporting the role of monoaminergic systems in the perception of fatigue during exercise (Chaouloff, 1991; see Newsholme & Blomstrand, Chapter 23). Maton (1991) suggested that during the fatiguing task, central adaptive processes are required to compensate for the fatigue in order to keep the performance as

Fatigue, Edited by Simon C. Gandevia et al.
Plenum Press, New York, 1995

constant as possible. His results on monkeys indicated a positive correlation between the force and firing rate of precentral cortical cells, but this correlation did not always increase with muscular fatigue. Characterizing central fatigue by the increased perceived effort required to maintain desired workload, Nethery (1991) suggested that there were attentional influences on the effort sense during prolonged exercise. Thus the participation of cognitive processes in the sensation of the exertion was proposed. There are also cognitive processes related to internal motivation and volition (e.g., expectancy of an event, intention and preparation of a voluntary action) which are accompanied by the specific trend-like pattern of RP. To the extent that these processes can be measured by brain electrical activity, it seems reasonable to expect changes in the RP parameters during compensatory adjustment to fatigue. The results of Freude and colleagues (1987) showing an earlier onset and increased amplitude of averaged RP during fatigue, support this assumption. However, the changes of RP during fatigue have not been widely investigated. This could be due to two main reasons: 1) because of the classical *technology* to obtain RP, based on averaging, the dynamics of rapid changes are lost, which makes the method inappropriate for detecting the onset and time course of fatigue and its possible phases; and 2) the perception of fatigue is a gradual process and its onset should be detected by tracking the EEG dynamics. However, the detection of RP with single trials is difficult because of its low amplitude and low signal-to-noise ratio. Nevertheless, such an approach has been developed for single-trial RP in order to identify its onset and successive phases (Popivanov et al., 1995). It would be expected that using appropriate methods for detecting the onset of single-trial RP, the onset of fatigue might also be detected provided the central fatigue changes the RP parameters.

The general feature of RP is its ramp-like shape which may consist of successive phases. This *slow-transient*, supposedly governed by a simple mathematical law, could be the *signal* related to the preparation of voluntary movement. It means that the estimation of voluntary movement preparation in the sense of RP identification can be reduced to the determination of a few parameters. The statistical hypotheses that could be tested to identify RP are based on the assumption that EEG activity is a stochastic process. Using the stochastic dynamics approach the following hypothesis could be tested: RP is a sum of a smooth function and an autoregressive (AR) process having an order similar to the EEG background activity. Although this hypothesis was not rejected, neither for averaged RP (Popivanov, 1992) nor for single-trial RP (Popivanov et al., 1995) the dynamic behavior of EEG preceding voluntary movement is surely more complicated.

The promising results in the application of nonlinear dynamics to the EEG activity offer a new approach to the identification of changes in EEG activity preceding voluntary action. For the sake of simplicity we shall denote these EEG segments as the RP time series. The description of the dynamic behavior of the RP time series could be done not only in the time or frequency domain, but also in the phase (state) space (PS). In general, PS is identified with a topological manifold. Using Takens theorem (Takens, 1981), it is possible to reconstruct the values $x_i(t), x_i(t+T),..., x_i(t+(M-1)T)$ as a vectors in M-dimensional PS, where M is called an embedding dimension and T is a fixed time increment. These vectors form the so-called trajectory matrix X. Every instantaneous state of a process is represented by a set $(x_1,...,x_M)$ which defines a point in the PS. The sequence of such points (states) over the time defines a curve in a PS called a trajectory. As time increases, the trajectories either penetrate the entire PS or they converge to a lower-dimensional subset. In this case, the subset is called an attractor. If the dimension of the attractor is a non-integer, i.e., fractal, the attractor is strange and its properties characterize the so-called deterministic chaos. A comprehensive review of the modern algorithms of chaotic dynamics, applied in physiology, was published recently by Elbert and colleagues (1994).

In experimental practice, the data have unknown dynamics, they are noisy and usually are presented as short time series. Strictly speaking, the dimension is defined as a measure of the number of independent variables needed to specify the state of the dynamical system at a given instant. In order to distinguish deterministic (or multiply-periodic), chaotic and random processes, two other dimensions have been involved, namely the correlation dimension D2 and the statistical dimension S, satisfying the relations $D2 \leq S(M) \leq M$ (Broomhead & King, 1986; Vautard & Ghil, 1989) and $M \geq 2 * D2 + 1$ (Takens, 1981). Methods are proposed for estimating D2, S and M, which can help to analyze the dynamical behavior of RP data.

HYPOTHESES AND AIM

In the following sections we describe our experience in application of chaotic dynamics (methods and procedures) for detecting the onset of a RP (and its phases) in each single-trial record, assuming that each of these instances reflects the change in the state of EEG during the preparation of the voluntary motor act. Further we suppose that these instances considered as boundaries of more deterministic segments, could be changed by central fatigue. Our hypothesis is that RP time series consists of successive *steady state* segments and shorter transitions between them. While the former could be described by a simple mathematical law (e.g., by appropriate curve fitting), the latter are chaotic processes that pre-adjust the system to pass from one stable state to another.

METHODS AND PROCEDURES

The applied procedures comprise two main steps:

1. Checking whether chaotic dynamics exists within the RP time series without detection of the chaotic segments in time. This includes computing the correlation dimension D2, nonlinear prediction, and computing the statistical dimension S.
2. Tracking the dynamics of the RP time series in order to detect the instances when chaotic transients appear. This includes singular spectrum analysis (SSA), non-linear prediction, point D2 correlation dimension, and nonlinear filtering.

Some examples are shown in Fig. 1 using RP-data files from experiments in which subjects performed abrupt flexion of the right hand fingers, voluntarily, at irregular intervals (Popivanov et al., 1995). The data were collected using Ag/AgCl electrodes placed at Cz, Fz, C4, Pz and C3 according to the 10/20 system with linked earlobes as a reference.

The amplifiers (Nihon Kohden EEG-4314) were set to an upper frequency (-3 dB) of 120 Hz and a time constant of 5 s. A 12 bit ADC with sampling rate of 250 samples/s was used (CED 1401) for EEG, EOG, EMG and mechanogram channels, the latter used to synchronize the data record to the movement onset. Thus the typical file containing RP time series consists of 1024 samples, 90% of which were prior to the movement onset.

DETECTION OF A *STATIC* CHAOTIC BEHAVIOR

Correlation Dimension

One of the measures of deterministic chaos is the so-called correlation dimension D2. The idea is to compute how the number of pair of vectors x_i which are separated by distances less then r, changes as a function of r. This is the so-called correlation integral defined by Grassberger and Procaccia (1983) as $C(r,n) \sim (1/n^2) \bullet$ [number of pairs (i,j) for which $|x_i - x_j| \leq r$] for large number of pairs. D2 is estimated as a slope in the linear region of

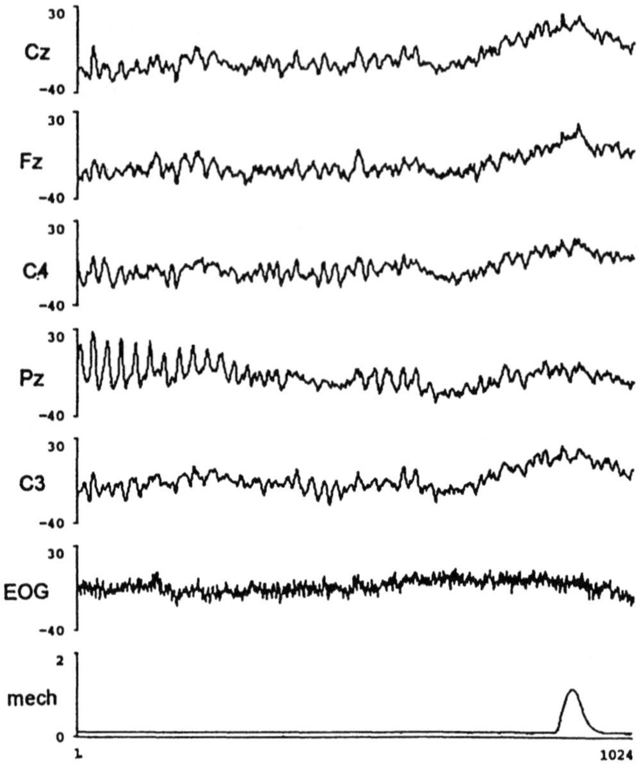

Figure 1. An example of a single-trial RP records at Cz, Fz, C4, Pz and C3, EOG and mechanogram (top to bottom). Amplitudes in μV, 1024 samples, sampling interval 4 ms (trace duration 4 s). The onset of the mechanogram is accepted as the onset of a finger flexion.

log C(r,n) *vs.* log(r) for incrementing embedding M, i.e., D2 = d[logC(r,n)]/d[log(r)] is valid for increasing embedding M. Thus D2 is estimated by a plot D2(M,r) *vs.* log(r). It is concluded that a chaotic attractor is present if D2(M,r) = f(log(r)) is converging to a plateau with increasing M and D2 is a non-integer. Thus D2 measures how much subspace of the PS is really occupied by the attractor, which may be less than Mmax. If D2 is an integer, the process is multiply-periodic, which is indicated with a torus in PS. For a random process it is believed that there would be no saturation of D2 with increasing M, (although this is not the case for some colored noises). Since the calculation of D2 for very small r is dominated by noise, particularly when dealing with experimental data, there is no convergence of the slope for very small r. In practice, D2 is defined in a scaling region (r_1, r_2), where the curves saturate to a flat plateau, which does not change with increasing M. It could be concluded that D2 is enough to distinguish the chaotic from periodic or random sets of data. However, the experimental data represent a complex process and its dynamics are unknown. Moreover, the reliability of D2 is less for short data sets (and/or mixed with colored noise) as it is in our case. In some cases the curves D2 *vs.* log(r) converged asymptotically to a plateau for RP time series data. However, the behavior of D2 should be interpreted with caution because of nonstationarity of the data and insufficient data samples. For example, Fig. 2*A*,*B* illustrates the behavior of C(r) *vs.* log (r) and D2(M,r) *vs.* log(r) for RP data at Cz, respectively. The computations were performed for n = 1024 samples, time delay T = 4 ms and embedding M = 2,4,...,20. The curves in the second plot do not converge to a plateau, which means that

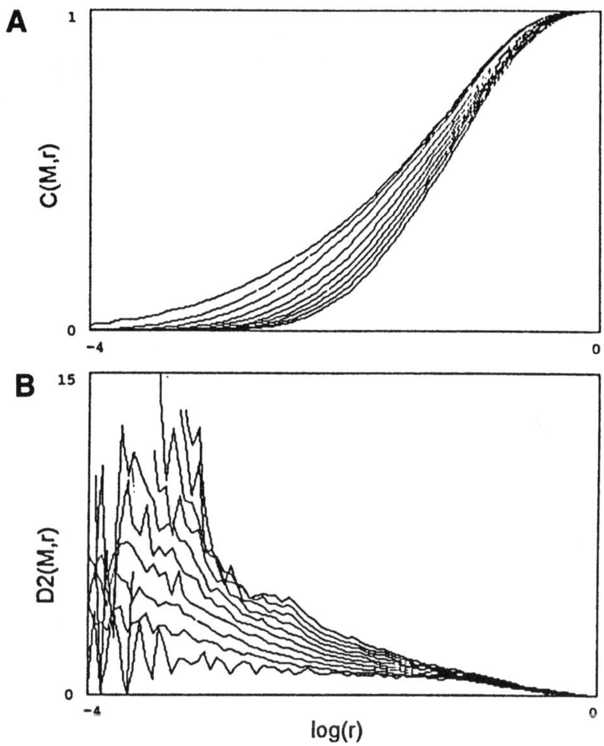

Figure 2. Plots of the correlation integral C(r) *vs.* log(r) (*A*) and correlation dimension D2 *vs.* log(r) (*B*) for increasing values of the embedding dimension M (M = 2,4,...,20; bottom to top). The computations are performed using 1024 samples, time delay T = 4 ms of RP time series at Cz. The slopes of C(r) are not parallel (*A*) and the curves D2(M,r) do not converge to a plateau (*B*).

D2 could not be determined exactly. Thus, for RP time series, D2 is not appropriate, moreover its computation is time consuming. Hence, it should be considered as a qualitative measure.

Nonlinear Prediction

We recommend the method proposed by Sugihara and May (1990), which can detect chaotic behavior in short data segments. The method is based on a library of past patterns in a time series, which is used to compute the prediction of the future pattern. If the time series is chaotic, the correlation coefficient Ro between predicted and actual values falls down as a function of a time step T_p; and the normalized error E_n as a function of T_p is an increasing curve like a mirror of Ro = f(T_p). If En = 0, the prediction is perfect; En = 1 indicates that the prediction is not better than a data mean value. Because the above results are sensitive to the chosen embedding M, the curve Ro = f(M) can show the *optimal* M, where Ro is max. This value can be used for further analysis. Fig. 3*A* shows Ro = f(T_p) and E_n = f(T_p) for RP time series at Cz. The same data values were taken as a library; that is, the library and forecasts span the same time period. This is recommended if some systematic changes (i.e., trends) exist (nonstationarity). The steep decay of the upper curve is an indication of chaotic behavior of the data. In Fig. 3*B* the two *optimal* values of M are indicated; that is, M = 4 and M = 7. These results correspond to the attractors with D2 ≤ 3, according to Takens formula $M \geq 2 * D2+1$.

Statistical Dimension

Statistical dimension S is defined as a number of eigenvalues of the matrix ($X^T \bullet X$), which are above the *noise level.* Here X is the trajectory matrix for a given embedding M. Thus the *significant* part could be distinguished from the *noise* part of a given data record.

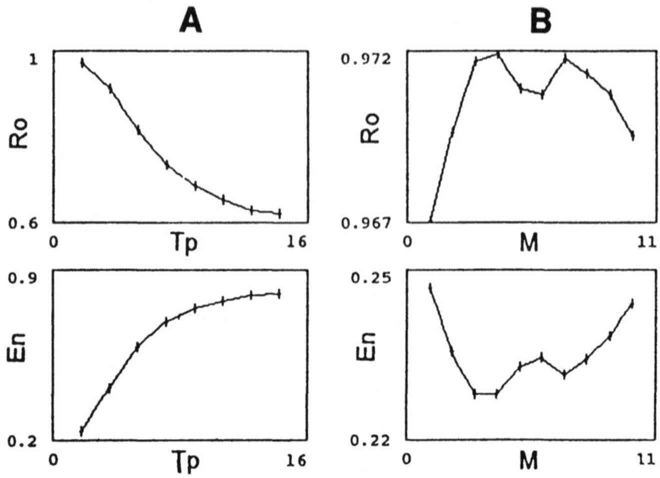

Figure 3. Top to bottom: *A*, plot of the correlation coefficient Ro between predicted and observed values as a function of prediction time for M = 4 and T = 1. Plot of the normalized error E_n as a function of prediction time. The prediction accuracy decreases with increasing time, which characterizes the chaotic dynamics generated by some low-dimensional attractor. RP- data are at Cz. The E_n was computed according to Farmer and Sidorowich (1987, p. 846). *B*, plot of the Ro as a function of M for Tp=1, T=1. Plot of the En as a function of M. The highest value of Ro and the lowest En give the *optimal* M (see text).

We have used the algorithms of Broomhead and King (1986) and Vautard and Ghil (1989), known as Singular Spectrum Analysis (SSA). SSA gives quantitative and qualitative information about the stochastic and dynamic components of the process even if the time series is short and noisy. Fig. 4 shows normalized eigenvalues spectra $\log(\sigma_k/\Sigma\sigma_k)$ *vs.* k, where k = 1,2,,...,M for single RP time series at C3, computed for several embedding dimensions M. The curves represent decreasing sequences of eigenvalues converging to the noise level. The end of steep decay determines S for each M. To verify the computed D2 (e.g., the value of D2 where the plateau is reached) the following relations should be satisfied: $D2 \leq S(M) \leq M$ for $2 * D2 + 1 \leq M \leq M_{sat}$, where M_{sat} is the value of M, when D2 reached its saturation value (see above, Correlation Dimension). If D2 > S, the time series could contain noise or its length is too short. In the examples shown above, the corresponding to M values of S are 4, 6, 7, 9 and 10. However, the left term of these relations could not be checked.

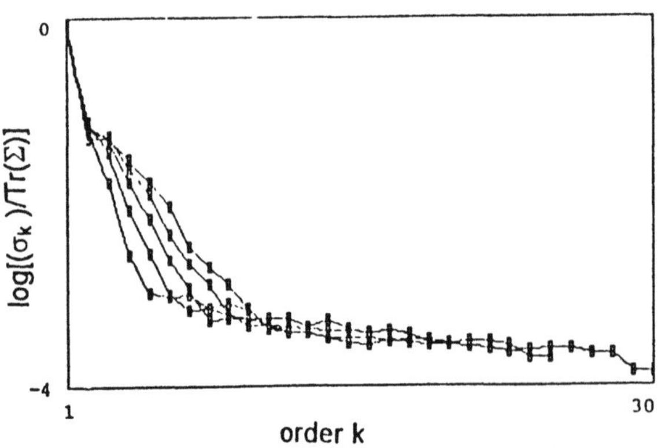

Figure 4. SSA eigenvalues spectra for RP time series at C3 computed for M = 10, 15, 20, 25 and 30 (bottom to top). The end of the steep decay determines statistical dimension S for each M (see text).

DETECTION OF THE CHAOTIC TRANSIENTS IN TIME DOMAIN

Analysis of the Principal Components

The principal components (PC) are defined as the orthogonal projection coefficients of the original series into their corresponding eigenvectors. Thus PC are the filtered versions of the original time series. It is accepted that PC corresponding to the eigenvalues above the *noise level*, reflect the intrinsic (significant) part of the process, while those corresponding to the eigenvalues below the noise level have low amplitude and reflect the noise. The disadvantage of PC is that their length is n - M + 1 (n = number of data samples); that is, they are shorter than the original data record. In the analysis of RP time series, we used the so-called reconstructed components (RC) instead of PC, described by Vautard and colleagues (1992, formulae 2.17a,b,c). RC represent series of length n corresponding to a given sets of eigenvalues (selected by the researcher). They are obtained from the PC and its eigenvectors. From the spectral point of view eigenvectors correspond to data-adaptive moving-average filters. Using RC, parts of the time series could be extracted, which are characterized according to some feature selected by the researcher (e.g., amplitude or frequency). Plots of three significant RCs from RP time series at C4 are shown in Fig. 5. The remaining RC made a small contribution and were considered as noise. However, this interpretation should be accepted with caution. It is possible, that the so-called noise contains hidden significant components, which may indicate chaotic transients between the bursts.

Point-D2 Correlation Dimension

The *point-D2* (PD2) estimate of the correlation dimension has been proposed by Skinner and associates (Molnar & Skinner, 1992, Elbert et al., 1994) in order for the correlation dimension to be less sensitive to nonstationarities. The PD2 does not accept every data point as a reference point but only those that pass two criteria: linear scaling in the log(C(r,n,nref)) *vs.* log(r) plots and convergence of slope *vs.* M (see above, Correlation

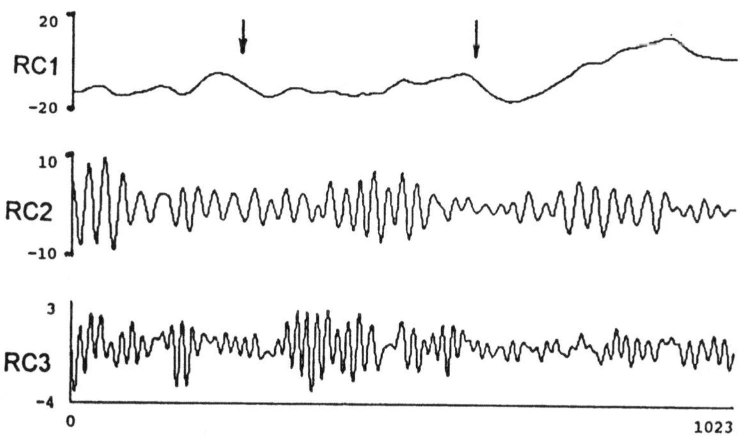

Figure 5. Plots of three RCs from RP time series at C4. The first plot (RC1) represents the trend, which is obtained from the first principal component. The second plot (RC2) shows the component reflecting the frequency band around 8 Hz (alpha rhythm). The diminishing of alpha spindle could be seen around 1 s before the movement onset, which could indicate the RP onset (Pfurtscheller, 1992). The third plot (RC3) shows a similar picture for the frequency band around 16 Hz (RC3). RC2 and RC3 could be interpreted as a *detrended* significant parts.

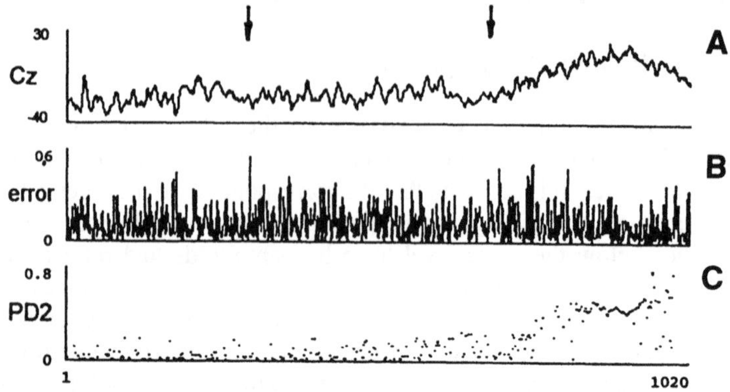

Figure 6. Plots of RP time series at Cz (*A*), absolute error between predicted and actual data values (*B*) and PD2 (*C*). The arrows mark the instances of highest error, which coincide with diminishing PD2. These instances indicate the possible onset of chaotic transients.

Dimension). Thus the method seeks stationary subepochs of the same type as that in which the reference vector is located, then tests and rejects those vectors for which linear scaling and convergence cannot be found. PD2 can be used to track the changes in both time and dimension, although there is, as yet, no mathematical proof that PD2 is a temporal characterization of the D2. Using mean PD2 for different subepochs containing event-related potentials (ERP), Molnar & Skinner demonstrated PD2 decreasing around ERP. We used PD2 to detect the onset of RP. Fig. 6*C* shows the plot of PD2 computed for RP time series at Cz.

Nonlinear Prediction Error

If the absolute error is computed between predicted and actual data values (see above, Nonlinear Prediction), it can indicate the instants in time that the chaotic transients are located. It can serve as an additional test for detecting the chaotic transients in time domain (Smith, 1992). Fig. 6*B* shows the plot of absolute error for RP time series at Cz. The same data values were taken as a library, i.e., the library and forecasts span the same time period.

Nonlinear Filtering

A nonlinear digital filtering approach is proposed for tracking the changes in chaotic features of a nonstationary time series (Saltzberg et al., 1988). By using the recurrence relations for logistic, cubic, and tent maps and the least-squares estimation procedure, the recurrence relations for the filter output were able to detect chaotic transients in short, noisy time series (Popivanov et al., 1993). Using this approach and the results for D2 and M obtained by the methods mentioned above, the logistic map proved to be appropriate for tracking the dynamics of RP time series. The filter output recovered the bifurcation parameter of the logistic map when short overlapping segments of data were fed into the filter. These segments served as a moving window shifted in steps along the record. The filter output (i.e., the bifurcation parameter) was displayed in each step until the end of the record. The output values over 3.57 indicated the chaotic transients. The details of the procedure are given elsewhere (Popivanov et al., 1993). For RP time series, M was selected as a length of the moving window and T was selected as the time decay corresponding to 1/e of the autocorrelation function initial value (e ~ 2.72).

DISCUSSION

The application of several methods to test the dynamics of RP time series show that our hypothesis with regard to single trial RP was not rejected. Chaotic transients were detected within data records near the same places by different methods. These instances are marked by arrows in Figs. 5 and 6. It is interesting that the instances of chaotic transients do not coincide always with the beginning of the local trend components. This means that the changes in the dynamical behavior of the process is not reflected in the trend components only. Instead, the key to understanding the intrinsic dynamics could be hidden in the faster components that are erroneously considered as a noise.

All the methods described above seem to be necessary for tracking the dynamics of RP data. At this stage of the study, we feel that the preprocessing procedures, including preliminary tests for existence of chaotic transients, are necessary before the application of any algorithm for tracking the dynamics. Unfortunately, the computations are very time consuming. The most important methods are those recommended for short, noisy time series. They could be used on data segments even when these overlap. However, more research is needed to select the proper segments and the time lag T (Elbert et al., 1994).

The results shown above suggest the following: 1) there exists chaotic transients somewhere within RP data records. However, because of the unknown dynamics of the data and unknown effect of central fatigue on the single RP, time consuming preprocessing is needed to obtain evidence of chaotic behavior; and 2) several tests are required to confirm chaotic behavior of RP data. This is because data is acquired in various experimental tasks, and chaotic behavior may be task specific.

ACKNOWLEDGMENTS

The authors wish to thank Prof. Douglas Stuart for his encouragement to contribute this work, and for his and Dr. Robert Lansing's helpful comments. The authors' laboratory has been supported, in part, by the National Fund for Scientific Research of Bulgaria (grant B-18/1994).

REFERENCES

Barrett G, Shibasaki H & Neshige R (1986). Cortical potentials preceding voluntary movement: evidence for three periods of preparation in man. *Electroencephalography and Clinical Neurophysiology* **83**, 327-339.

Broomhead DS & King GP (1986). Extracting qualitative dynamics from experimental data. *Physica* **20D**, 217-236.

Chaouloff F (1991). Cerebral monoamines and fatigue. In: Atlan G, Beliveau L, Bouissou P, (eds.), *Muscle Fatigue. Biochemical and Physiological Aspects*, pp. 234-240. Paris: Masson.

Elbert T, Ray WJ, Kowalik ZJ, Skinner JE, Graf KE & Birbaumer N (1994). Chaos and physiology: deterministic chaos in excitable cell assemblies. *Physiological Reviews*, **74**, 1-47.

Farmer JD & Sidorowich JJ (1987). Predicting chaotic time series. *Physical Review Letters* **59**, 845-848.

Freude G, Ullsperger P & Pietschmann M (1987). Are self-paced repetitive fatiguing hand contractions accompanied by changes in movement-related brain potentials? In: Gantchev GN, Dimitrov B, Gatev P (eds.), *Motor Control*, pp. 99-103. New York: Plenum.

Grassberger P & Procaccia I (1983). Measuring the strangeness of strange attractors. Physica **9D**, 189-208.

Hashimoto S, Gemba H & Sasaki K (1980). Premovement slow cortical potentials and required muscle force in self-paced hand movements in the monkey. *Brain Research* **117**, 415-423.

Kornhuber HH & Deecke L (1964). Hirnpotentialanderungen beim Menschen vor und nach Willkurbewegungen, dargestellt mit Magnetbundspeicherung und Ruckwartsanalyse. *Pflügers Archiv* **281**, 52.

Kristeva R, Cheyne D, Lang W, Lindinger G & Deecke L (1990). Movement-related potentials accompanying unilateral and bilateral finger movements with different inertial loads. *Electroencephalography and Clinical Neurophysiology* **75**, 410-418.

Libet B (1985). Unconscious cerebral initiative and the role of conscious will in voluntary action. *Behavioral and Brain Sciences* **8**, 529-566.

Maton B (1991). Central nervous changes in fatigue induced by local work. In: Atlan G, Beliveau L, Bouissou P (eds.), *Muscle Fatigue. Biochemical and Physiological Aspects*, pp. 207-221. Paris: Masson.

Molnar M & Skinner JJ (1992). Low-dimensional chaos in event-related brain potentials. *International Journal of Neuroscience* **66**, 263-276.

Nethery VM (1991). Central fatigue: the contribution of attentional focus. In: Atlan G, Beliveau L, Bouissou P (eds.), *Muscle Fatigue. Biochemical and Physiological Aspects*, pp. 243. Paris: Masson.

Pfurtscheller G (1992). Event-related synchronization (ERS): an electrophysiological correlate of cortical areas at rest. *Electroencephalography and Clinical Neurophysiology* **83**, 62-69.

Popivanov D. (1992). Time series analysis of brain potentials preceding voluntary movements. *Medical & Biological Engineering & Computing* **30**, 9-14.

Popivanov D, Dushanova J, Mineva A & Krekule I (1993). Identification of slow EEG-transients by using chaotic models. In: Rosenfalck A (ed.), *Proceedings of the IMIA-IFMBE Working Conference on "Biosignal Interpretation"*, pp. 198-201. Aalborg, Denmark.

Popivanov D, Dushanova J, Mineva A & Krekule I (1995). Single-trial readiness potentials: stochastic and chaotic dynamical behavior. In: Stuart DG, Gantchev GN, Gurfinkel VS, Wiesendanger M (eds.), *Motor Control VII*, pp. 00-00. Tucson, AZ: Motor Control Press.

Saltzberg B, Burton WD & Skinner JE (1988). Filtering for chaos.In: Harris G, Walker C (eds.), *Proceedings of the 10th Annual International Conference of the IEEE in Medicine & Biology Society*, pp. 1072-1073. New Orleans, LA

Smith LA (1992). Identification and prediction of low dimensional dynamics. *Physica* **D58**, 50-76.

Sugihara G & May RM (1990). Nonlinear forecasting as a way of distinguishing chaos from measurement error in time series. *Nature* **344**, 734-741.

Takens F (1981). Detecting strange attractors in turbulence. In: Rand DA, Young LS (eds.), *Lecture Notes in Mathematics*, vol. 898, pp. 366. Berlin: Springer.

Tarkka IM & Hallett M (1990). Cortical topography of premotor potentials preceding self-paced, voluntary movement of dominant and non-dominant hands. *Electroencephalography and Clinical Neurophysiology* **75**, 36-43.

Vautard R & Ghil M (1989). Singular spectrum analysis in nonlinear dynamics with applications to paleoclimatic time series. *Physica* **D35**, 395-424.

Vautard R, Yiou P & Ghil M (1992). Singular spectrum analysis: A toolkit for short, noisy chaotic signals. *Physica* **D58**, 95-126.

22

THE SENSES OF EFFORT AND FORCE DURING FATIGUING CONTRACTIONS

L. A. Jones

Department of Mechanical Engineering, Room 3-148
Massachusetts Institute of Technology
77 Massachusetts Ave, Cambridge, Massachusetts 02139

ABSTRACT

The sensory basis upon which judgments of force are made during fatiguing isometric contractions is reviewed. At issue is whether the perception of force is based on centrally generated sensations arising from the motor command, known as the sense of effort, or from peripheral sensations originating in the muscle, and termed a sense of force. The results from a number of studies indicate during sustained constant-force contractions the perceived magnitude of the force increases, which is consistent with the idea that judgments of force are based on centrally generated signals. It is, however, possible for some subjects to dissociate effort and force and to make accurate judgments of the magnitude of forces during fatigue.

INTRODUCTION

The basis on which judgments of force and weight are made has been subject to much debate and experimentation over the last century (for reviews see McCloskey, 1981; Jones, 1986). At issue, is whether the perception of force is primarily derived from centrally generated sensations arising from the innervation of the efferent pathways (i.e., sensations of innervation) as Helmholtz (1866/1925) proposed, or from peripheral sensations originating in the muscles, skin and joints as Bell (1826) hypothesized. Centrally mediated sensations are often termed a *sense of effort* (McCloskey et al., 1974), and are assumed to arise from internal neural correlates (corollary discharges) of the descending motor command. They presumably reflect the magnitude of the voluntary motor command generated (McCloskey, 1981). The second source of sensory information, described as a *sense of force or tension*, is derived from peripheral receptors in muscles, tendons and the skin and is presumed to signal intramuscular tension and, therefore, provides a measure of the actual force exerted by the muscle (Roland & Ladegaard-Pedersen, 1977). Changes in muscle force are signaled by the Golgi tendon organs (Crago et al., 1982) and these are the most likely source of a sense of force (Jami, 1992). The relation between the force generated by a muscle and the discharge rate of tendon organs does, however, depend on a number of factors (Proske &

Fatigue, Edited by Simon C. Gandevia et al.
Plenum Press, New York, 1995

Gregory, 1980; Gregory, 1990; Jami, 1992), but under natural conditions of muscle activation it has been reported to be linear (Crago et al., 1982). Although the senses of effort and force are often presented as dichotomous systems, they should be considered as complementary rather than competitive, in that both appear to be involved in mediating the perception of force.

In most situations, the sense of effort and force provide congruent information; that is, as the motor command increases, perhaps due to a change in the mass of an object being supported, so, too, would the discharge rate of muscle receptors signaling the force of contraction (Crago et al., 1982; Jami, 1992). Under some conditions, however, it is possible to dissociate the senses of force and effort by decoupling the relation between the motor input to the muscle and peripheral feedback, and in these situations one can evaluate whether subjects' judgments of force appear to be based on centrally or peripherally arising sensations. For example, during sustained submaximal constant-force contractions, there is an increase in the amplitude of the electromyogram (EMG) of the fatiguing muscle, reflecting the recruitment of additional motor units as contractile failure occurs in those units already active (Bigland-Ritchie, 1981). Under these conditions one can investigate whether judgments of the force generated by the fatiguing muscle covary with the excitatory input to the muscle (EMG), which is increasing, or remain essentially constant, consistent with the actual force being produced.

The smoothed rectified EMG has been used as an indirect measure of the excitatory input or neural drive sent to a muscle, as it reflects the number of active motor units and their discharge rates (Cafarelli, 1988). An indeterminate amount of this signal is, however, due to the activation of alpha motoneurons via the gamma loop and muscle spindle afferent fibers. The contribution of these segmental reflex pathways to the generation of muscle force is small, but not insignificant (Stein et al., 1995), and in the absence of this feedback there is a substantial decline (about 30%) in the firing rates of motoneurons (Gandevia et al., 1990). Support for using the surface EMG as an estimate of the descending motor command comes from a study by Cafarelli and Bigland-Ritchie (1979) in which they varied the force-generating capacity of a muscle by changing its length and asked subjects to match the forces in two corresponding muscles held at different lengths. Although the slope of the resulting force-matching function was proportional to the ratio of the maximal voluntary contractions (MVCs) of the two muscle groups, the relation between the EMGs recorded from the two muscles was not influenced by muscle length, which suggests that when the excitatory inputs to the muscles were the same, the sense of effort associated with each contraction was similar.

MEASUREMENT

A variety of methods have been used to study the perception of force ranging from ratio-scaling procedures (Gescheider, 1985) which require that subjects assign numbers to forces in proportion to the perceived magnitude of the force (Eisler, 1965), to the contralateral limb-matching method (McCloskey et al., 1974) in which the forces exerted by a muscle group in one limb (the reference limb) are matched in subjective magnitude by contractions of the corresponding muscle group on the contralateral side (the matching limb). The subject usually has feedback of the force produced by the reference limb and attempts to match the force sensations in both limbs during the matching contraction (Cafarelli, 1982). In addition to these methods which measure how subjects scale force, sensory thresholds have been estimated and these provide an index of the sensitivity of subjects to changes in force (Ross & Reschke, 1982; Pang et al., 1991). The relation between force discrimination and judgments of the magnitude of forces is not known, and there is no reason *a priori* to assume that the ability to estimate force accurately is related to the capacity to discriminate changes in force (Holmes, 1917; Jones, 1989). Ross and colleagues (1984) have shown that the relation is not monotonic in that force discrimination can deteriorate (i.e., thresholds increase) under conditions in which forces are perceived to be smaller than normal (e.g., in a 0-G environment). The results from a number of studies indicate that the differential threshold or just noticeable difference for force is approximately 6% (Ross & Reschke, 1982; Jones, 1989; Pang et al., 1991).

When subjects make numerical estimates of the perceived magnitude of forces, the perceived force grows as the 1.7 power of the force exerted. This means that when the force exerted increases by 50%, perceptually its magnitude doubles (Eisler, 1965; Cain & Stevens, 1971). The exponent of the power function does change with variations in experimental procedure (range: 0.8-2.0), which affect the way in which observers give judgments, but when the same procedure is used, consistent results are obtained even when the muscle group varies (Eisler, 1965; Stevens & Cain, 1970; Jones, 1986). In contrast to this nonlinear relation between force amplitude and perceived intensity, a linear relation between the reference and matching forces is obtained with the contralateral limb matching procedure. This result is consistent with those based on ratio-scaling methods in that they predict that the exponents of the force functions for the two limbs are identical, and hence the relation linear. The contralateral limb-matching method avoids some of the response biases associated with numerical estimation of stimulus magnitude (discussed in Poulton, 1979), and has the additional advantage that subjects are not able to remember previous responses with the same degree of accuracy with which numbers can be recalled. This is important in experiments in which the effects of varying the instructions given to subjects on judgments of sensory magnitude are studied (e.g., estimate force or effort during a constant force contraction).

FORCE PERCEPTION DURING MUSCLE FATIGUE

Changes in the perceived magnitude of forces occur as a function of the duration of the muscular contraction. The perceptual changes associated with maintaining a constant force over time have been studied by asking subjects to estimate numerically the perceived magnitude of a force sustained for varying periods of time. Stevens and Cain (1970) found that the perceived magnitude of a handgrip contraction grew as a power function of both the isometric force exerted (exponent of 1.7) and the duration of the contraction (exponent of 0.7). This means that when subjects are asked to judge the magnitude of forces exerted by fatigued muscles, there is a uniform ratio increase in the perceived amplitude of the forces, and so the exponent of the power function relating perceived to physical force does not change with fatigue (Cain & Stevens, 1971). This finding also indicates that from the point of view of endurance, it is better to bear a light load for a long period of time than a heavy load for a short time (Stevens & Cain, 1972).

Using a different psychophysical procedure, known as cross-modal matching, in which the loudness of tones was matched to the force of handgrip contractions, Teghtsoonian and colleagues (1977) reported that muscle fatigue was associated with a substantial change in the exponent of the force power function. In their experiment, fatigue was induced by requiring subjects to produce five consecutive MVCs prior to the matching trials. After the fatiguing contractions, tones of higher sound pressure levels were matched with forces that were smaller than those produced prior to fatigue, and softer tones were now matched with larger forces. This latter unexpected finding suggests that the absolute threshold for detecting force may increase with muscle fatigue (Teghtsoonian et al., 1977). Production of a brief, presumably non-fatiguing, MVC (i.e., 5 s) does influence the perception of force, but in this situation small amplitude forces (between 2 and 4% MVC) are over and not under estimated in magnitude (Hutton et al., 1987; Thompson et al., 1990). This error in estimating force has

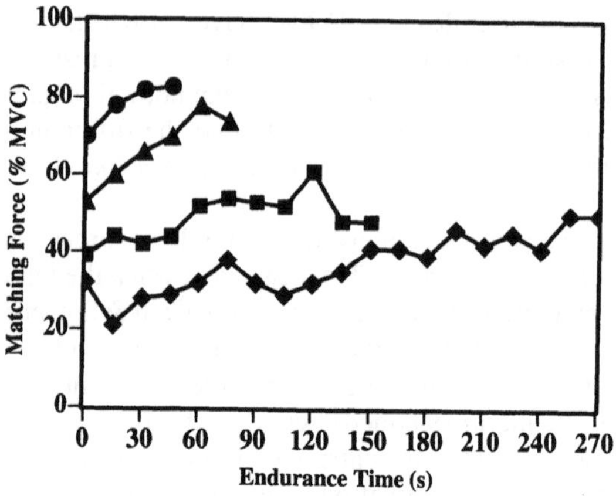

Figure 1. The matching forces exerted by the left arm as one of four constant forces (30%[◆], 45%[■], 60%[▲], and 75%[●] MVC) was maintained by the right arm until maximal endurance (data from one subject).

been attributed to post-contraction potentiation of stretch reflex pathways (Hutton et al., 1987) and tendon organ desensitization (Thompson et al., 1990).

The effect of fatigue on either absolute or differential thresholds for force does not appear to have been measured. Knowledge about changes in these sensory thresholds would contribute to our understanding of the performance decrements associated with fatigue. In the auditory modality, it has been known since the early 1940s that there is an increase in the detection threshold after prolonged exposure to a (fatiguing) sound (Botte & Scharf, 1980). Similar findings have been reported for visual (Geisler, 1979) and olfactory stimuli (Cain & Engen 1969). If fatigue does result in an elevation in the detection threshold for force, then this would affect the control of force in those tasks highly dependent on feedback (e.g., fine manual tasks such as threading a small nut on a bolt).

When subjects are asked to judge the magnitude of a sustained constant force by producing a brief matching contraction on the contralateral side, there is a progressive and linear increase in the perceived amplitude of the force being maintained, with the rate of increase dependent on the amplitude of the constant force being exerted, as shown in Fig. 1 (McCloskey et al., 1974; Gandevia & McCloskey, 1978; Jones & Hunter, 1983a, 1983b; Cafarelli & Layton-Wood, 1986). The overestimation in force as the muscle fatigues is considerable, with errors at the limit of endurance being in the order of 70% (McCloskey et al., 1974) to 130% (Jones & Hunter, 1983a).

In these studies, subjects were simply instructed to make both arms the *same* or make the forces produced by the two arms the *same*, and were unaware during the experiment of the extent to which they were overestimating the sustained forces (Jones & Hunter, 1983b). These results suggest that subjects base their judgments of force on a signal related to the motor command sent to the muscle, or some variable proportional to it, rather than on the force actually generated by the fatiguing muscle. The overestimation of force results from efferent signals of comparable magnitude being transmitted to fatigued and unfatigued muscles, the latter having greater force-generating capacity. This interpretation is supported by the relation observed by Jones and Hunter (1983b) between the EMG of the fatiguing muscle and the matching force produced on the contralateral side. As the EMG increased, due to either a larger force or progressive fatigue in the muscle, there was a corresponding increase in the magnitude of the matching force as shown in Fig. 2.

It seems unlikely that the overestimation of force during fatigue can be attributed to the activity of receptors in the muscle. Macefield and colleagues (1991) have shown that when subjects maintain a submaximal voluntary contraction, there is a decline in the discharge rate of most muscle spindle afferents and this occurs in the presence of a

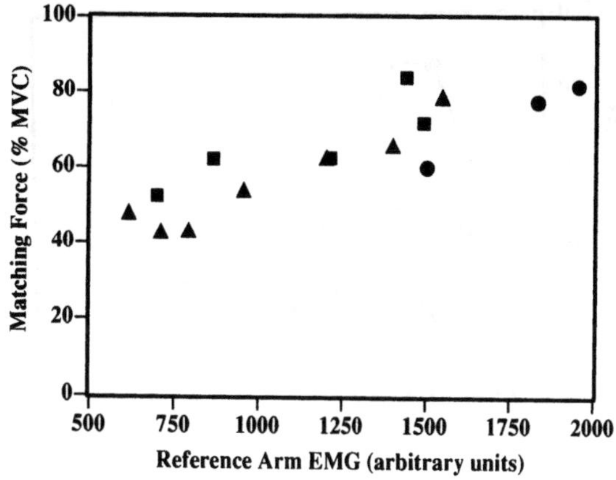

Figure 2. Relation between surface EMG recorded from the biceps muscle of the right arm during constant force contractions maintained at 35%[▲], 50%[■] and 65%[●] MVC and the matching forces produced by the left arm (data from one subject).

progressive increase in the EMG. The data on the responses of Golgi tendon organ receptors under these conditions are more limited, but they do suggest that there is a decay in the discharge rates of some receptors with time (Macefield et al., 1991). It also seems probable that their response rates saturate at high forces and that there is decrease in tendon organ sensitivity during fatigue (Hutton & Nelson, 1986; Gregory, 1990). The hypothesis that peripheral receptors are not the primary source of force information during fatigue is supported by Cafarelli and Layton-Wood's (1986) finding that vibrating a fatigued muscle has no effect on force sensation. They reported that whereas vibration increased the sensation of force in fresh (unfatigued) muscles, it had no effect on the perception of forces generated by fatigued muscles, which continued to be overestimated during vibratory stimulation.

Dissociation between Effort and Force during Fatigue

The results from the studies reviewed above suggest that subjects do not perceive the actual forces exerted by fatiguing muscles, and that their judgments of force are always based on the effort associated with generating the contraction. Under some conditions, however, it has been shown that subjects can distinguish between a sense of effort and a sense of force. Roland and Ladegaard-Pedersen (1977) demonstrated that during partial curarization, subjects could dissociate a sense of effort from force if they were instructed to disregard the increased effort that was required to produce the muscle contraction. Although the subjects' judgments of the forces generated by the paretic muscles were more variable than under control conditions, there was not a consistent tendency to overestimate the force as would be expected if subjects were basing their judgments on the descending motor command. This result was obtained during isometric force matching and when subjects were matching the force of compression of two hand-held springs. Despite disruption of the normal relation between motor signals and muscle output that resulted from the neuromuscular blockade, these subjects were able to produce equivalent forces in two limbs, a task that must have involved peripheral afferent feedback.

When the same instructions were given to subjects during sustained constant-force contractions, Jones (1983) did not find any consistent evidence of a dissociation between force and effort. On some trials (26%), however, subjects were able to match the force accurately during the fatiguing contraction, suggesting that it may be possible to dissociate effort from force during fatigue. Subjects did not perceive that their performance on these trials was any different from that on trials in which forces were overestimated, and were not able to describe the cues they used to match forces accurately. The results from one subject

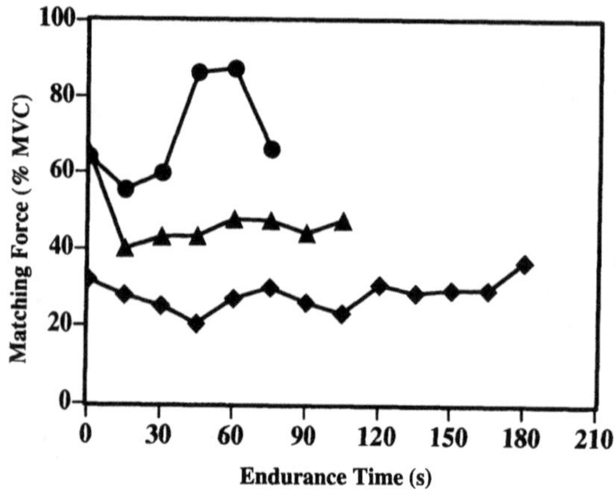

Figure 3. The matching forces exerted by the left arm as one of three constant forces (35%[◆], 50%[▲] and 65%[●] MVC) was maintained by the right arm until maximal endurance. This subject was the most accurate in judging the amplitude of the sustained force.

who was able to estimate the forces accurately are shown in Fig. 3. The typical performance of subjects in this study, as in the other experiments on fatigue, was for forces to be overestimated. There are data, however, that do suggest that during fatigue of the inspiratory muscles, most subjects (6/8) can judge accurately the inspiratory pressure developed (i.e., muscle tension) as well as the effort associated with producing the contraction (Gandevia et al., 1981).

It appears that the sense of effort dominates the perception of muscle force during fatigue and that it is extremely difficult for subjects to estimate the actual forces being produced by the muscle, although for some this is possible (Gandevia et al., 1981; Jones, 1983). With training and feedback of performance, it may be possible to facilitate this dissociation between effort and force. The preponderance of a sense of effort during fatiguing contractions may permit a subject to gage the endurance capacity of the muscle and so may function as a protective mechanism in that the subject can anticipate when muscle failure is imminent.

MUSCLE WEAKNESS AND PERCEIVED EFFORT AND FORCE

The changes in the perception of force during fatigue are also evident during other states or conditions that affect the force-generating capacity of the muscle. Among the clinical symptoms first described by Holmes (1917) in his studies of patients with unilateral cerebellar lesions who were without sensory loss were disturbances in the appreciation of weight (and force). Even though there was no difference between the patients' hands in their ability to discriminate weights, the majority of these people overestimated the heaviness of weights lifted on the affected side. The overestimation of weight was attributed to the asthenia of the limb and a loss of muscle tone. This presumably resulted from a reduction in the excitability of the motoneuronal pool which then required an increased neural input in order to achieve any specified level of force. Similar results have been reported in patients with paresis due to upper motoneuron disease who are without any clinical evidence of sensory loss. Gandevia and McCloskey (1977a) found that these subjects consistently overestimated the magnitude of forces generated on the hemiparetic side, and hypothesized that this was due to the increased neural output of unaffected corticofugal pathways required to compensate for the pathways no longer functional.

Paresis can be experimentally induced by partially blocking transmission at the myoneural junction using a neuromuscular blocking agent. This procedure has been used to study the perception of force and effort during paresis and to examine the time course of the perceptual changes. When a muscle is weakened by local infusion of a blocking agent, the forces generated by the muscle are perceived to be greater than those produced under control conditions, as reflected in the magnitude of the forces generated on the contralateral (unaffected) side (Gandevia & McCloskey, 1977b; Roland & Ladegaard-Pedersen, 1977). This overestimation of force during partial curarization is consistent with the idea that judgments of force are based on the magnitude of the centrally generated motor command producing the contraction, rather than on the sensory signals arising from the muscle. With paresis, there is an increase in the efferent signal required to produce a given level of muscle force (Tang & Rymer, 1981), but the peripheral afferent discharges from muscle spindle and tendon organ receptors continue to signal the actual force of contraction.

SUMMARY

The results from numerous experiments indicate that whenever there is a change in the voluntarily generated motor command producing a muscle contraction, there is a parallel change in the perceived amplitude of the force produced by the muscle. Although these changes occur in parallel, there is no directly proportional relation between the two. For example, during partial curarization, the maximal strength of the muscle may be reduced to 10% of its normal value, but the perceived heaviness of a weight is overestimated by only 40% (Gandevia & McCloskey, 1977b). Similarly, at the point of endurance of a sustained submaximal contraction, the force produced by the contralateral unfatigued muscle is not the MVC, but a submaximal force (Jones & Hunter, 1983a). An exception to these findings is the results of Roland (1978) who found that when the subjective effort of isometric contractions exerted by a non-paretic and paretic hand were matched, the matching forces produced could be predicted quite accurately from the maximal strength of the paretic hand.

The absence of a proportional relation between judgments of force and the descending efferent command is not unexpected given the numerous systems involved in generating and sensing muscle force. A peripheral signal from the muscle must be used to indicate the success of a muscle contraction (Gandevia & McCloskey, 1978), that is that the force generated is adequate for the task at hand. A signal of intramuscular force could also be used to scale a corollary discharge by modulating the sensory effects of these signals as McCloskey and colleagues (1983) have proposed. Although the evidence presented here strongly implicates a centrally mediated perception of force, the contribution of intramuscular receptors to force perception cannot be discounted. Unfortunately, it is not possible to devise an experiment in which tendon organs would be the sole source of force information. For many of the experiments described in this article, assumptions were made, on the basis of rather limited data, about the behavior of tendon organ receptors during these perturbations. In some of this work, it was also assumed that they were the only available signals of peripheral force. There is clearly a need for more detailed studies of muscle afferent discharge patterns during normal conditions of muscle activation and as the muscle fatigues.

ACKNOWLEDGMENTS

The author wishes to thank Dr. Simon Gandevia for his helpful comments on the manuscript. Her laboratory has been supported by the Medical Research Council of Canada and the Office of Naval Research (USA). Her attendance at the 1994 Bigland-Ritchie conference was supported, in part, by the University of Miami.

REFERENCES

Bell C (1826). On the nervous circle which connects the voluntary muscles with the brain. *Philosophical Transactions of the Royal Society of London* **116**, 163-173.

Bigland-Ritchie B (1981). EMG/force relations and fatigue of human voluntary contractions. *Exercise and Sport Sciences Reviews* **9**, 75-117.

Botte M-C & Scharf B (1980). La sonie - Effets simultanés de fatigue et de masque. *Acustica* **46**, 100-106.

Cafarelli E (1982). Peripheral contributions to the perception of effort. *Medicine and Science in Sports and Exercise* **14**, 382-389.

Cafarelli E (1988). Force sensation in fresh and fatigued human skeletal muscle. *Exercise and Sports Sciences Reviews* **16**, 139-168.

Cafarelli E & Bigland-Ritchie B (1979). Sensation of muscle force in muscles of different length. *Experimental Neurology* **65**, 511-525.

Cafarelli E & Layton-Wood J (1986). Effect of vibration on force sensation in fatigued muscle. *Medicine and Science in Sports and Exercise* **18**, 516-521.

Cain WS & Engen T (1969). Olfactory adaptation and the scaling of odor intensity. In: Pfaffman C (ed.), *Olfaction and Taste III*, pp. 127-141. New York: Rockefeller University Press.

Cain WS & Stevens JC (1971). Effort in sustained and phasic handgrip contractions. *American Journal of Psychology* **84**, 52-65.

Crago PE, Houk JC & Rymer WZ (1982). Sampling of total muscle force by tendon organs. *Journal of Neurophysiology* **47**, 1069-1083.

Eisler H (1965). The ceiling of psychophysical power functions. *American Journal of Psychology* **78**, 506-509.

Enoka RM & Stuart DG (1992). Neurobiology of fatigue. *Journal of Applied Physiology* **72**, 1631-1648.

Gandevia SC, Killian KJ & Campbell EJM (1981). The effect of respiratory muscle fatigue on respiratory sensations. *Clinical Science* **60**, 463-466.

Gandevia SC, Macefield G, Burke D & McKenzie DK (1990). Voluntary activation of human motor axons in the absence of muscle afferent feedback. *Brain* **113**, 1563-1581.

Gandevia SC & McCloskey DI (1977a). Sensations of heaviness. *Brain* **100**, 345-354.

Gandevia SC & McCloskey DI (1977b). Changes in motor commands as shown by changes in perceived heaviness, during partial curarization and peripheral anaesthesia in man. *Journal of Physiology (London)* **272**, 673-689.

Gandevia SC & McCloskey DI (1978). Interpretation of perceived motor commands by reference to afferent signals. *Journal of Physiology (London)* **283**, 493-499.

Geisler WS (1979). Evidence for the equivalent-background hypothesis in cones. *Vision Research* **19**, 799-805.

Gescheider GA (1985). *Psychophysics: method, theory, and application* (2nd Ed.). Hillsdale, NJ: Erlbaum.

Gregory JE (1990). Relations between identified tendon organs and motor units in the medial gastrocnemius muscle of the cat. *Experimental Brain Research* **81**, 602-608.

Helmholtz H von (1925). *Treatise on Physiological Optics* vol. 3, Southall JPC (ed., translator). Menasha WI: Optical Society of America (original work published 1866).

Holmes G (1917). The symptoms of acute cerebellar injuries due to gunshot injuries. *Brain* **40**, 461-535.

Hutton RS, Kaiya K, Suzuki S & Watanabe S (1987). Post-contraction errors in human force production are reduced by muscle stretch. *Journal of Physiology (London)* **393**, 247-259.

Hutton RS & Nelson DL (1986). Stretch sensitivity of Golgi tendon organs in fatigued gastrocnemius muscle. *Medicine and Science in Sports and Exercise* **18**, 69-74.

Jami L (1992). Golgi tendon organs in mammalian skeletal muscle: Functional properties and central actions. *Physiological Reviews* **72**, 623-666.

Jones LA (1983). Role of central and peripheral signals in force sensation during fatigue. *Experimental Neurology* **81**, 497-503.

Jones LA (1986). Perception of force and weight: Theory and research. *Psychological Bulletin* **100**, 29-42.

Jones LA (1989). Matching forces: Constant errors and differential thresholds. *Perception* **18**, 681-687.

Jones LA & Hunter IW (1983a). Perceived force in fatiguing isometric contractions. *Perception & Psychophysics* **33**, 369-374.

Jones LA & Hunter IW (1983b). Effect of fatigue on force sensation. *Experimental Neurology* **81**, 640-650.

Macefield G, Hagbarth K-E, Gorman R, Gandevia SC & Burke D (1991). Decline in spindle support to α-motoneurones during sustained voluntary contractions. *Journal of Physiology (London)* **440**, 497-512.

McCloskey DI (1981). Corollary discharges: Motor commands and perception. In: Brookhart JM, Mountcastle VB (sec. eds.), Brooks VB (vol. ed.), *Handbook of Physiology, sec. 1. vol. II, pt 2, The Nervous System: Motor Control.*, pp. 1415-1448. Bethesda, MD: American Physiological Society.

McCloskey DI, Ebeling P & Goodwin GM (1974). Estimation of weights and tensions and apparent involvement of a sense of effort. *Experimental Neurology* **42**, 220-232.

McCloskey DI, Gandevia SC, Potter EK & Colebatch JG (1983). Motor commands and judgements about muscular contractions. In: Desmedt JE (ed.), *Motor Control in Man*, pp. 151-167. New York: Raven Press.

Pang XD, Tan HZ & Durlach NI (1991). Manual discrimination of force using active finger motion. *Perception & Psychophysics* **49**, 531-540.

Poulton EC (1979). Models for biases in judging sensory magnitude. *Psychological Bulletin* **86**, 777-803.

Proske U & Gregory JE (1980). The discharge rate: tension relation of Golgi tendon organs. *Neuroscience Letters* **16**, 287-290.

Roland PE (1978). Sensory feedback to the cerebral cortex during voluntary movement in man. *Behavioral and Brain Sciences* **1**, 129-171.

Roland PE & Ladegaard-Pedersen H (1977). A quantitative analysis of sensations of tensions and of kinaesthesia in man. *Brain* **100**, 671-692.

Ross HE, Brodie E & Benson A (1984). Mass discrimination during prolonged weightlessness. *Science* **225**, 219-221.

Ross HE & Reschke MF (1982). Mass estimation and discrimination during brief periods of zero gravity. *Perception & Psychophysics* **31**, 429-436.

Stein RB, Hunter IW, Lafontaine SR & Jones LA (1995). Analysis of short latency reflexes in human elbow flexor muscles. *Journal of Neurophysiology* In press

Stevens JC & Cain WS (1970). Effort in isometric muscular contractions related to force level and duration. *Perception & Psychophysics* **8**, 240-244.

Stevens JC & Cain WS (1972). Effort in isometric contractions: Buildup and recovery. *17th International Congress, International Association of Applied Psychology* **1**, 399-407.

Tang A & Rymer WZ (1981). Abnormal force-EMG relations in paretic limbs of hemiparetic human subjects. *Journal of Neurology, Neurosurgery and Psychiatry* **44**, 690-698.

Teghtsoonian R, Teghtsoonian M & Karlsson JG (1977). The effects of fatigue on the perception of muscular effort. In: Borg G (ed.), *Physical Work and Effort*, pp. 157-180. Oxford: Pergamon Press.

Thompson S, Gregory JE & Proske U (1990). Errors in force estimation can be explained by tendon organ desensitization. *Experimental Brain Research* **79**, 365-372.

TRYPTOPHAN, 5-HYDROXYTRYPTAMINE AND A POSSIBLE EXPLANATION FOR CENTRAL FATIGUE

E. A. Newsholme[1] and E. Blomstrand[2]

[1] Cellular Nutrition Research Group, Department of Biochemistry
University of Oxford
South Parks Road, Oxford OX1 3QU, United Kingdom
[2] Pripps Research Laboratory
Pripps Bryggeri
Bromma, Sweden

ABSTRACT

In prolonged exercise the plasma level of branched-chain amino acids (BCAA) may fall and that of fatty acid increases: the latter increases the free tryptophan level, so that the plasma concentration ratio, free tryptophan/BCAA may increase leading to higher levels of tryptophan and therefore of 5-hydroxytryptamine (5-HT) in brain. The latter increases the activity of some 5-HT neurons in the brain which can cause sleep and which could, therefore, increase the mental effort necessary to maintain athletic activity. Drinks containing branched-chain amino acids should restore vigor to athletes whose performance is depressed by an excess of cerebral 5-HT. Recent work suggests that intake of branched-chain amino acids may improve performance in slower runners in the marathon and decrease perceived physical and mental exertion in laboratory experiments. This suggestion is supported by pharmacological manipulations that result in either increased or decreased physical performance.

INTRODUCTION

The mechanisms that underlie fatigue during physical exercise have attracted the attention of physiologists and biochemists for many years. Many studies have been published concerning muscle (peripheral) fatigue and several hypotheses have been put forward including accumulation of protons, depletion of muscle glycogen or failure of neuromuscular transmission. However, little is known about the biochemical mechanisms of central fatigue (i.e., fatigue resulting from changes within the central nervous system). Mechanisms that have been suggested to cause central fatigue, include changes in brain monoamine concentrations (Newsholme, 1986) or accumulation of ammonia in the brain during exercise (Okamura et al., 1987). Changes in plasma amino acid concentrations could play a role in

Fatigue, Edited by Simon C. Gandevia et al.
Plenum Press, New York, 1995

central fatigue by increasing the rate of synthesis and hence the level of the neurotransmitter 5-HT in parts of the brain (Newsholme, 1986). Thus, it has been considered that 5-HT is involved in sleep, aggression and mood, so that it was suggested that this neurotransmitter might also be involved in the mental fatigue that can occur during and after vigorous and sustained physical exercise.

PLASMA FREE TRYPTOPHAN CONCENTRATION AND FATIGUE

The following is a summary of several important cellular nutritional facts that form the basis for the hypothesis.

1. Tryptophan is converted in the brain to the neurotransmitter 5-HT.
2. Branched-chain amino acids (leucine, isoleucine and valine) are not taken up by liver but by muscle and their rate of uptake from the blood by muscle is increased during exercise.
3. Both branched-chain amino acids and tryptophan enter the brain upon the same amino acid carrier, so that competition between the two types of amino acids for entry into brain can occur (Fernstrom, 1990)
4. None of the enzymes involved in the conversion of tryptophan to 5-HT appears to approach saturation with substrate (i.e., there is no flux-generating step in this series of reactions) (Fernstrom, 1990; see Newsholme & Leech, 1983 for a discussion of flux-generating steps).
5. An increased level of tryptophan will, therefore, be expected to increase the rate of formation of 5-HT and hence increase the level of this neurotransmitter in the brain. This could result in increased firing of some 5-HT neurons. It is known that 5-HT is involved in sleep so that it may also result in tiredness and possibly fatigue; that is, it could increase the mental effort necessary to maintain the pace of running (Newsholme & Leech, 1983; Newsholme, 1986).
6. Tryptophan is unique amongst the amino acids in that it is bound to plasma albumin, so that it exists as a bound and a free form which are in equilibrium; this equilibrium is changed in favor of free tryptophan when the plasma fatty acid level is increased (Blomstrand et al., 1988).

AN HYPOTHESIS FOR CENTRAL FATIGUE

An increase in the plasma fatty acid level and/or a decrease in that of branched-chain amino acids could markedly influence the plasma concentration ratio of free tryptophan to branched-chain amino acids to favor entry of tryptophan into the brain and consequently increase the level of 5-HT in at least some parts of the brain.

In exercise, either intermittent or continuous, there is elevation in the blood catecholamine levels and a decrease in that of insulin which will result in fatty acid mobilization from adipose tissue. This increases the plasma fatty acid level. If there is precise control between the mobilization of fatty acids, the extent of vasodilation in muscle, and the stimulation of fatty acid oxidation in muscle, the increased rate of fatty acid oxidation by muscle may occur with little increase in the plasma level of fatty acids. Hence the free tryptophan level may not change much. However, if this coordination is poor, due to lack of training or due to hypoglycemia, the blood fatty acid concentration could be increased to sufficiently high levels to increase the plasma concentration of free tryptophan. Furthermore, in intermittent exercise in which there is usually a greater dependence upon *anaerobic* exercise and, therefore, less opportunity to oxidize these fatty acids, the plasma level of fatty acid could rise and hence increase the concentration of free tryptophan.

In prolonged exercise it is likely that muscle will use branched-chain amino acids for energy when the muscle glycogen level is depleted. Similarly, when the liver glycogen store is depleted, fatty acid mobilization may increase raising the blood level of fatty acids, so that the plasma level of free tryptophan will increase. Thus, an interesting possibility is fatigue that is caused by the depletion of glycogen in muscle could be due not to a direct effect of glycogen depletion on the muscle but to an increase in the level of 5-HT in a specific area of the brain. Failure of the motor centers in the brain to stimulate muscle to contract would mean that the power output would fall, i.e. fatigue due to a change in the balance of concentration of key amino acids in the blood but initiated by depletion of liver and muscle glycogen stores.

Testing of the Hypothesis

Recent experiments have provided evidence in support of the hypothesis. A summary of the experimental findings supporting the hypothesis is as follows: 1) the plasma concentration ratio, free tryptophan/branched-chain amino acids is increased in man and in the rat after prolonged and exhaustive exercise (Blomstrand et al., 1988; 1989); 2) in the rats, the levels of tryptophan were increased in all areas of the brain studied: in contrast the level of 5-HT was increased in only two areas, the brain stem and the hypothalamus: the level of 5-hydroxyindole acetic acid was also increased in these areas (Blomstrand et al., 1989); 3) such changes in neurotransmitter levels even in behaviorally significant areas of the brain such as the brain stem and hypothalamus, provides only indirect evidence in support. How is it possible to provide more direct evidence? Two approaches have been tried. First, athletes during various events have taken a drink containing branched-chain amino acids sufficient to maintain the resting free tryptophan/branched-chain amino acids concentration ratio. The effects on physical and mental fatigue have then been studied. Secondly, the effect of pharmacological manipulation of the 5-HT neurotransmitter system on physical performance have been studied in both rats and man.

A summary of the results of key experiments is as follows:

1. In a Stockholm marathon, 193 volunteers were randomly divided into two groups and the experiment was performed double-blind. In the experimental group, a drink containing a mixture of branched-chain amino acids was given four times during the race and the other group was given a placebo drink. Considering the whole group of subjects, the difference in performance between the experimental and placebo groups was small and did not reach significant levels. However, when the subjects were divided into two subgroups based on their marathon time, the performance was statistically significantly better for the slower runners (3.05-3.30 h to complete this marathon) in the experimental group compared with the placebo group. The difference in time at this pace would mean an improvement in performance of 5-6 min (Blomstrand et al. 1991a). The reason for the lack of effect on performance in the faster runners (<3.05 h to complete this marathon) is not known. It might be that the more well-trained (faster) runners are more resistant to fatigue, both central and peripheral fatigue, and, therefore, less sensitive to a supply of branched-chain amino acids. Alternatively, the slower runners may have depleted their glycogen stores at an earlier stage of the race so that they would increase the plasma level of fatty acids and decrease that of branched-chain amino acids at an earlier time in the race compared with the better runners. In this case, they would change their free tryptophan/branched-chain amino acid ratio earlier

in the race and this would cause fatigue according to the hypothesis. For this reason, the effect of the amino acid supplementation would be easier to detect in these slower runners.

2. A preliminary experiment was carried out in eight runners who completed two 24 km cross-country runs separated in time by one week. After each race the runners were asked to recall their perceived physical and mental effort. On the first occasion four subjects drank the placebo drink before the run and after 13 km, and four subjects drank the branched-chain amino acids at the same time periods. The placebo and the branched-chain amino acid were indistinguishable from one another and the experiment was performed *double-blind*. Drinks were then switched and the procedure repeated one week later. There was no difference in running time between the two occasions. However, both the perceived physical and mental efforts, as measured on a modified Borg scale, were lower when branched-chain amino acids were taken, but the difference was only statistically significant for the perceived mental effort (Newsholme et al., 1991).
3. Twenty-four participants were questioned after a marathon on their mental attitude towards the race. All of the 12 participants who drank the placebo drink experienced aversion to running the race during the last 10 km; in contrast, of the 12 participants who drank the branched-chain amino acids, only seven experienced aversion as indicated from a questionnaire. The difference was statistically significant (Newsholme et al., 1991).
4. The Stroop color and word test (CWT) was given to 16 subjects who participated in a 30 km cross-country race (Blomstrand et al., 1991b). Research on the CWT has established that the test provides a useful tool in the study of neuropsychological and cognitive processes. In the group who took the branched-chain amino acids during the race, performance in this test improved after the race in comparison with before the race, while no statistically significant difference was found for the subjects who took the placebo drink.

It is possible to change the effectiveness of neurotransmitters in a synapse in the brain by pharmacological manipulations: agents that can prevent the binding of the neurotransmitter to the post-synaptic receptor and elicit no response from the receptor are known as antagonists: those that bind and do elicit a response are known as agonists. In addition, some agents can prevent the uptake of the neurotransmitter from the synaptic cleft so reducing its effect. All these approaches have been used in testing the 5-HT and central fatigue hypothesis: 1) administration of a 5-HT agonist to rats impairs running performance in a dose-related manner (Bailey et al., 1992); 2) in contrast, administration of a 5-HT antagonist to the rats improved running performance (Bailey et al., 1993); 3) administration of a 5-HT re-uptake blocker to human subjects lowered physical performance; exercise time to exhaustion during standardized exercise was decreased in comparison to a control condition (Wilson & Maughan, 1992).

Although no signs of circulatory effects of the drug were observed, it cannot be ruled out that these drugs, especially at high does, might have effects on the peripheral metabolism or central circulation and such changes could influence performance.

EFFECT OF INGESTION OF BRANCHED-CHAIN AMINO ACIDS AND/OR CARBOHYDRATE ON FATIGUE

It has been suggested that an intake of BCAA might be detrimental to physical performance due to an increased production of ammonia originating from the metabolism of BCAA during exercise. However, when small amounts of BCAA are supplied but sufficient to maintain the plasma concentration ratio of free tryptophan/BCAA (6-8 g of BCAA during exercise) at resting levels, no increase in the plasma concentration of ammonia has been found as a consequence of BCAA supplementation, that is, exercise caused the same increase in plasma ammonia concentration when BCAA or water were supplied (Blomstrand & Newsholme unpublished observations).

When BCAA were supplied together with carbohydrates during exercise, no difference in the perceived mental and overall fatigue could be detected during 80 min of exercise at 75% of the maximal oxygen uptake when compared with provision of carbohydrate alone. Provision of carbohydrate before or during exercise may prevent or decrease the increase in plasma fatty acid concentration that can occur during exercise. This may be caused by a stimulation of insulin secretion, or, more interestingly, by increasing the rate of esterification of fatty acids in the liver. Whatever the mechanism, a decrease in the plasma concentration of fatty acids would cause a decrease in the free tryptophan level in the plasma. Hence, carbohydrate intake may be effective in delaying fatigue not only by preventing a fall in the blood glucose level but also by decreasing or preventing the increase in the free tryptophan concentrations in the bloodstream. In a recent study (Davis et al., 1992) it has been reported that provision of carbohydrate attenuates, in a concentration-related manner, the increase in the plasma free tryptophan concentration during sustained exercise to fatigue (2-4 h duration). However, when the exercise continues for longer than 2-3 h there may be an increase in the plasma concentration of fatty acids and hence that of free tryptophan even when carbohydrate is consumed (Wright et al., 1991; Davis et al., 1992). This might explain why it was possible to detect an effect of BCAA supplementation on the marathon performance even when carbohydrate was provided during the race (Blomstrand et al., 1991a), but not in shorter events. A competitive marathon is extremely heavy exercise and, in the experiment reported above, almost 200 volunteers were involved in the study which made it possible to detect a small difference in performance (3%).

Training and the 5-HT-System

Recently some evidence has been presented that endurance-training alters the sensitivity of the 5-HT system. The sensitivity to a neurotransmitter can be modified by changing the number of postsynapic receptors. Since the release of prolactin from the pituitary is controlled, in part, by the activity of the 5-HT-system in the hypothalamus, prolactin release can be used as test of 5-HT sensitivity (Cowen et al., 1990). This sensitivity is lower in endurance-trained athletes compared with untrained individuals (Jakeman et al., 1994). This provides further support for the theory that the 5-HT system is involved in central fatigue during endurance exercise and indicates, furthermore, that central fatigue, like peripheral fatigue, is affected by physical training. Thus, effects of BCAA supplementation on physical and mental fatigue may depend not only on the changes in the plasma free tryptophan concentrations but also on the sensitivity of the 5-HT system to changes in the level of 5-HT in neurons in specific parts of the brain.

ACKNOWLEDGMENTS

Support from Pripps Bryggeria (Stockholm, Sweden) is gratefully acknowledged.

REFERENCES

Bailey SP, Davis JM & Ahlborn EA (1992). Effect of increased brain serotonergic activity on endurance performance in the rat. *Acta Physiologica Scandinavica* **145**, 75-76.

Bailey SP, Davis JM & Ahlborn EA (1993). Neuroendocrine and substrate responses to altered brain 5-HT activity during prolonged exercise to fatigue. *Journal of Applied Physiology* **74**, 3006-3012.

Blomstrand E, Celsing F & Newsholme EA (1988). Changes in plasma concentrations of aromatic and branched-chain amino acids during sustained exercise in man and their possible role in fatigue. *Acta Physiologica Scandinavica* **133**, 115-121.

Blomstrand E, Hassmén P, Ekblom B & Newsholme EA (1991a). Administration of branched-chain amino acids during sustained exercise—effects on performance and on plasma concentration of some amino acids. *European Journal of Applied Physiology* **63**, 83-88.

Blomstrand E, Hassmén P & Newsholme EA (1991b). Effect of branched-chain amino acid supplementation on mental performance. *Acta Physiologica Scandinavica* **143**, 225-226.

Blomstrand E, Perrett D, Parry-Billings M & Newsholme EA (1989). Effect of sustained exercise on plasma amino acid concentrations and on 5-hydroxytryptamine metabolism in six different brain regions of the rat. *Acta Physiologica Scandinavica* **136**, 473-481.

Cowen PJ, Anderson IM & Grahame-Smith DG (1990). Neuroendocrine effects of azapirones. *Journal of Clinical Psychopharmacology* **10**, 215-255.

Davis JM, Bailey SP, Woods JA, Galiano FJ, Hamilton MT & Bartoli WP (1992). Effects of carbohydrate feedings on plasma free tryptophan and branched-chain amino acids during prolonged cycling. *European Journal of Applied Physiology* **65**, 513-519

Fernstrom JD (1990). Aromatic amino acids and monamine synthesis in the CNS: influence of diet. *Journal of Nutritional Biochemistry* **10**, 508-517.

Jakeman PM, Hawthorne JE, Maxwell SRJ, Kendall MJ & Holder G (1994). Evidence for down regulation of hypothalamic 5-hydroxytryptamine receptor function in endurance-trained athletes. *Experimental Physiology* **79**, 461-464.

Newsholme EA (1986). Application of knowledge of metabolic integration to the problem of metabolic limitations in middle distance and marathon running. *Acta Physiologica Scandinavica* **128** (suppl. 556), 93-97.

Newsholme EA, Blomstrand E, Hassmén P & Ekblom B (1991). Physical and mental fatigue: do changes in plasma amino acids play a role? *Biochemical Society Transactions* **19**, 358-362

Newsholme EA, & Leech AR (1983). *Biochemistry for the Medical Sciences*. Chichester, England: John Wiley & Sons.

Okamura K, Matsubara F, Yoshioka Y, Kikuchi N, Kikuchi Y & Kohri H (1987). Exercise-induced changes in branched chain amino acid/aromatic amino acid ratio in the rat brain and plasma. *Japanese Journal of Pharmacology* **45**, 243-248.

Wilson WM & Maughan RJ (1992). Evidence for a possible role of 5-hydroxytryptamine in the genesis of fatigue in man: administration of paroxetine, a 5-HT re-uptake inhibitor, reduces the capacity to perform prolonged exercise. *Experimental Physiology* **77**, 921-924.

Wright DA, Sherman WM & Dernbach AR (1991). Carbohydrate feedings before, during or in combination improve cycling endurance performance. *Journal of Applied Physiology* **71**, 1082-1088.

SECTION VII

Task Dependency of Fatigue Mechanisms

This section specifically addresses how fatigue varies during different tasks. Historically, the realization of the different tasks performed by *red* and *white* muscles was one of the forerunners of this discussion. The term *task dependency* reminds us that the pattern and recruitment of motor units, factors influencing blood flow, and variations in the state of the segmental and supraspinal apparatus will ultimately influence muscle force and the duration for which it can be sustained. Consequently, the duration that a task can be sustained will vary as the details of the task vary.

In **Chapter 24,** Sargeant and Jones review the relationship between fatigability and the speed of contraction. They assess how the known properties of human single muscle fibers would affect maximal power output. Analysis of human cycling is used to explore the factors that alter the relationship between velocity (pedaling rate) and maximal peak power. These factors include the proportions of the different fiber types, the extent to which they are recruited, and muscle temperature. Their modeling relies on results from experiments on single muscle fibers but, most valuably, exposes gaps in our knowledge about the neural control of muscle during concentric and eccentric exercise.

When a muscle contracts the pressure within it increases to levels 2-3 fold greater than the arterial blood pressure. Elevated intramuscular pressures have three important implications for muscle fatigue. First, the increased pressure will impede the blood flow and thus impose relatively ischemic conditions; second, it may interfere mechanically with the transmission of force into the tendon. Third, the elevated pressure will alter the forces that govern the transfer of fluids across the capillary wall. Sejersted and Hargens (**Chapter 25**) analyze the precise mechanism for the increase in intramuscular pressure and its implications for muscle fatigue under different circumstances, ranging from isometric contractions to walking. Interestingly, under some conditions the intramuscular pressure is highly correlated with intramuscular force and thus the effective neural drive to the muscle. Their analysis emphasizes the need to consider the biomechanical effects of intramuscular architecture on pressure and force generation: the so-called clinical *compartment syndromes* reinforce this view.

In **Chapter 26**, Botterman begins, like many others, with the assumption that the neuromuscular system has evolved to minimize fatigue. After reviewing the differences in fatigability among type-identified motor units in the cat, he then describes how the discharge rate of a motor unit must change if the desired outcome is to hold constant (or *clamp*) the force produced by a single motor unit. Such a paradigm is one that occurs regularly under natural conditions, of course. Not surprisingly, type S units can produce a constant submaximal force with a constant discharge frequency for extended periods of time. The concurrent expression of potentiation and fatigue (including contractile slowing), however, adds a complication to how the CNS might control all motor units to produce a constant force. This chapter also brings out the value of using a variety of fatigue-inducing stimulation paradigms to bring out different features of motor unit fatigue and recovery.

Bigland-Ritchie and colleagues attempt a global view of fatigue in **Chapter 27**. Fatigue is defined as "any reduction in a person's ability to exert force or power in response to voluntary effort, regardless of whether or not the task itself can be performed successfully". Rather than accepting that fatigue will occur because of a failure at one site in a chain from the motor cortex to the actomyosin crossbridge, they hypothesize that fatigue will depend on the amount of *stress* placed at each site. A corollary they suggest is that impairment at one site is balanced by an improvement at another. They indicate that there is no simple relation between muscle metabolism and muscle fatigue, and that impairment of excitation-contraction coupling is a major reason why the relation is not simple. Overall, because the neuromuscular system has evolved to operate in humans under a wide variety of conditions, the authors argue that fatigue, at least conceptually, is a *unitary* process.

The chapters in this section reinforce the view that the mechanisms responsible for fatigue depend on the exact task performed by the neuromuscular system.

THE SIGNIFICANCE OF MOTOR UNIT VARIABILITY IN SUSTAINING MECHANICAL OUTPUT OF MUSCLE

A. J. Sargeant[1] and D. A. Jones[2]

[1] Department of Muscle and Exercise Physiology, Faculty of Human Movement Sciences
Vrije University
Van der Boechorstraat 9, 1081 BT Amsterdam, The Netherlands
[2] School of Sport and Exercise Sciences
The University of Birmingham
Edgbaston, Birmingham B15 2TT, UK

ABSTRACT

Neuromuscular function and fatigue have been studied using a wide variety of preparations. These range from sections of single fibers from which the cell membrane has been removed to whole muscles or groups of muscles acting about a joint in the intact animal. Each type of preparation has its merits and limitations. There is no ideal preparation; rather the question to be answered will determine the most appropriate model in each case and sometimes a combination of approaches will be needed. In particular, it is important to understand how the mechanical output of whole muscle can be sustained to meet the demands of a task and to take into account the organized variability of the constituent motor units.

INTRODUCTION

In a mature muscle one motoneuron will, through its axonal branches, supply a number of muscle fibers throughout the muscle. In a healthy muscle the innervation is such that adjacent fibers will most probably be supplied by branches from different motoneurons. A motoneuron and all of the muscle fibers that it innervates forms a motor unit. As a consequence of a common pattern of activation, muscle fibers within a single motor unit will be relatively homogenous in terms of contractile and metabolic properties. There may, however, be considerable variability *between* motor units within a single muscle, although the degree of variability will depend on the range of mechanical output required from a muscle in a particular species. Not only will muscle fiber properties be influenced by the pattern of activation but also by the environment in which they operate. Thus some limited variability can be expected even within a single motor unit as a consequence of the

Fatigue, Edited by Simon C. Gandevia et al.
Plenum Press, New York, 1995

geographical scattering of the component fibers. This *scattering* may, for example, lead to marked differences in operating temperature and intra-muscular pressure depending on whether a fiber is located superficially or deep within the muscle (Sargeant, 1994a). In addition, changes in interstitial concentrations of potassium. for example, consequent upon activation of some fibers, may influence the excitability of other previously quiescent fibers. These changes may be important in influencing mechanical output during fatiguing submaximal contractions (for discussion, see McComas et al., 1993; see Sjøgaard & McComas, Chapter 4).

VARIABILITY OF MUSCLE FIBERS

As early as the seventeenth century, a number of authors had commented on the fact that muscles differ in their appearance. However, it was not until 1873 that Ranvier recognized that skeletal muscles not only differ in color, but that they also had different contractile properties. For example, in most mammals the soleus muscle is red in appearance and slow to contract while the extensor digitorum longus is white and contracts and relaxes rapidly. Subsequently, various combinations of histochemical stains have been used to identify different muscle fiber types. The most commonly used stains are: 1) myosin ATPase with preincubation at either alkaline or acid pH; 2) NADH tetrazolium reductase or succinic dehydrogenase, a marker of mitochondrial activity; and 3) myophosphorylase, a marker of glycolytic activity. Using this histochemical approach, three main fiber types can be distinguished. However, the histochemical techniques are, to a greater or lesser extent, manipulated to create apparently discrete fiber types, whereas in reality there is a continuum in most if not all contractile and metabolic properties (Figs. 1*A* and 2).

Although the histochemically based classifications may be adequate for many purposes, it is now possible to show that there are a complex set of genes controlling the expression of the contractile proteins which determine the intrinsic contractile properties. In human muscle fibers, for example, it has been shown that many type II muscle fibers co-express variable proportions of types A and B rather than expressing a single isoform of the myosin heavy chain. The continuum in the degree of co-expression is also reflected by: 1) the staining intensity for myosin ATPase with pre-incubation at pH 4.6 (Sant'Ana et al., 1995); 2) by the maximal rate of myofibrillar ATPase (Sant'Ana Pereira & van der Laarse, personal communication); and 3) the maximal velocity of shortening (Fig. 2; Larsson & Moss, 1993).

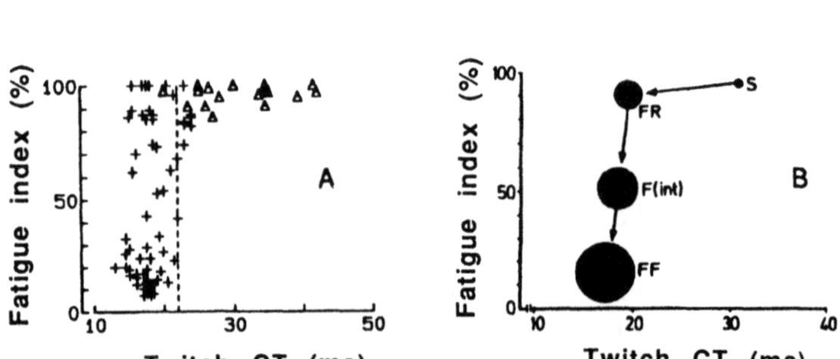

Figure 1. Co-variation between different contractile properties for muscle units of cat's peroneus longus muscle. Fatigue index (%) plotted *vs.* twitch time-to-peak (ms) for 80 units (*A*) and for average values of the same units after categorization into types S, FR, F(int) and FF (*B*). In *B*, the diameter of each symbol is proportional to the maximal tetanic force of the respective unit category (mean force for FF units: 27.9 g) (reproduced from Kernell et al., 1983 (*A*), by permission of Springer-Verlag; and Kernell, 1986 (*B*)).

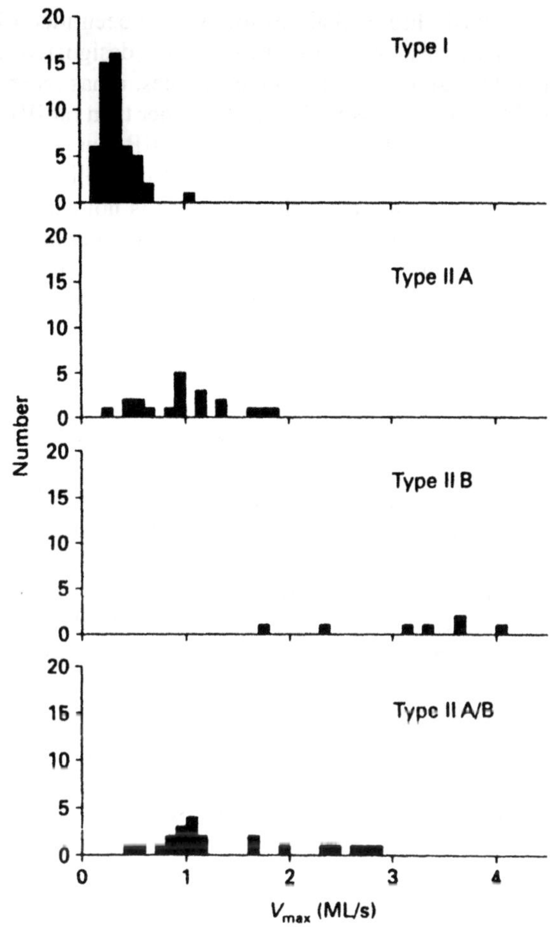

Figure 2. Distributions of V_{max} values from human quadriceps and soleus muscle fibers classified according to myosin heavy chain type, i.e., type I, IIA, IIB or IIA/B MHCs. Shortening velocities from two fibers co-expressing types I and IIA MHCs (0.3 and 1.0 ML/s) are not shown in the figure (reproduced with permission from Larsson & Moss, 1993).

The Relationship between Contractile Properties and Histochemistry

Experiments in which individual motor units are stimulated enables characterization of their contractile properties. Subsequent repetitive stimulation to deplete the constituent muscle fibers of glycogen then allows, in serial sections of the muscle, identification and histochemical characterization of the muscle fibers of that same motor unit (Burke et al., 1973; Kugelberg, 1973). On the basis of size, speed, and fatigability, motor units fall between two extremes: large, fast and fatigable or small, slow and fatigue resistant. Histochemically, the large units tend to be made up of type IIB fibers while the small units are predominantly composed of type I fibers. Type IIA motor units span a range of size and fatigue resistance which is reflected in the broad spectrum of their mitochondrial enzyme activities. While mammalian species show similar varieties of fibers when classified by histochemical techniques, there may be considerable differences in the contractile characteristics of an identified fiber type between species, between muscles in the same species, and possibly even within the same muscle. It should also be noted that compared with other species, there is a much less striking covariation of contractile properties and a smaller endurance range in at least the human thenar muscles (Thomas et al., 1991; see Thomas, Chapter 10). However, it is unclear whether this finding can be extrapolated to larger human limb muscles, to other exercise protocols or even during voluntary contractions. Furthermore, while three fast myosin heavy chain isoforms have been identified in rat and rabbit muscle (IIA, IIX, and IIB), only two major fast isoforms (designated IIA and IIB) have been found in human muscle. Using gene cloning techniques, it has recently been shown that human type IIB is highly homologous to the rat IIX rather than rat IIB (Ennion et al., 1995). This finding has functional significance since the type IIB muscle fibers in rodents have been shown to have the highest velocities of shortening (Bottinelli et al., 1991). It is possible therefore that the type IIB myosin heavy chain isoform is not expressed in humans because it is associated with too high an intrinsic velocity for larger animals (see the discussion of Hill, 1950, with respect to the constraints of animal size and the number of sarcomeres in series, on the intrinsic speed of shortening). Thus, although general principles may be similar across species there is a need for some caution in extrapolating from animal models to humans since there may not be an exact correspondence or covariance of properties.

The Control of Fiber Type Expression - Usage Dependent Plasticity

Contractile activity imposed on the muscle unit by the motoneuron is an important factor determining the contractile and metabolic properties of the constituent muscle fibers. This was first demonstrated in the classic cross-innervation studies of Buller and colleagues (1960) and in subsequent chronic stimulation studies (e.g., Salmons and Vrbova, 1969). These observations have led to the widely held belief that activity is the primary determinant of motor unit properties. Recent observations, on the effects of inactivity and re-innervation of adult muscle, indicate, however, that factors independent of activity may also play a significant role (e.g., Pierotti et al., 1991; Unguez et al., 1993).

The major changes seen after chronic stimulation include an increase in capillary density, proliferation of mitochondria, decrease in sarcoplasmic reticulum and the expression of different troponin and myosin isoforms. These changes also occur with different time courses (Pette & Vrbová, 1992). Increased capillary density and mitochondrial content, which are amongst the first changes to be seen in response to prolonged activity, will predominantly affect the fatigability of the muscle. Changes in the sarcoplasmic reticulum and contractile proteins, which require more activity, influence the speed of the muscle. Although there has been a great deal of work on stimulated animal muscle there are relatively few investigations of the effects of chronic stimulation on human muscle. The studies that have been undertaken show that changes in fatigability can be produced but probably without altering the myosin isoform expression (Rutherford & Jones, 1988). Similarly while training programs based on voluntary activation increase fatigue resistance of human muscle, it is difficult to demonstrate changes in myosin expression. Perhaps in both cases there was not a significant enough departure from basic postural and locomotory activities to reach the threshold needed to change the myosin isoform expression. Furthermore, the normal hierarchical pattern of recruitment used in training exercise will mitigate against changing the myosin of high threshold motor units. Nevertheless, if training is extremely intense, it may be possible to effect changes in the contractile properties (e.g., V_{max}, as shown by Fitts et al., 1989, using human skinned single fibers from the deltoid muscle of competitive swimmers).

FORCE MODULATION BY MOTOR UNIT RECRUITMENT AND RATE CODING

In order to meet the demand for steady and finely gradable force output, motor units are organized in a hierarchy according to size and contractile properties. There is in addition

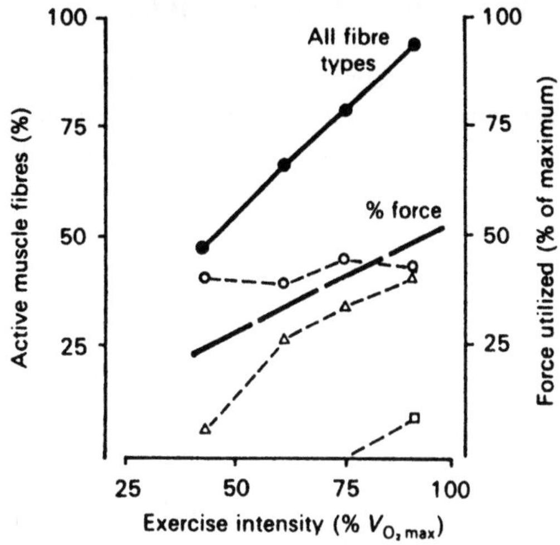

Figure 3. Proportion of maximal force utilized, and muscle fibers active, in relation to exercise intensity. Values are given for the total and component fiber type populations; type I, O; type IIA Δ; IIAB & IIB combined ❐ (reproduced with permission from Greig et al., 1985).

to this a task related recruitment strategy. Thus at low levels of required force, small (low force) slow motor units with thin axons and small, easily excitable motoneurons are recruited. As the requirement for force increases, there is systematic recruitment of larger (high force) motor units so that the force steps caused by successive recruitment remains rather constant relative to the prevailing force output. (Kernell, 1992, for review; see Kernell, Chapter 9). Not surprisingly, and as a consequence of the patterns of activation implicit in such a recruitment hierarchy there is, in general, an association with increasing size and increasing fatigue sensitivity and power output (Fig. 1*B*).

As well as varying the number of motor units recruited, muscle force can also be modulated by changing the frequency of stimulation, so called *rate-coding*. Modulation by frequency changes in the steep part of the force frequency curve provide an additional very sensitive force control for the neuromuscular system. The extent to which the demand for force in human muscle contractions is met by the recruitment of more motor units or by *rate-coding* is not clear and may vary somewhat between, for example, the small hand muscles compared to large limb muscles (Kukulka & Clamann, 1981). In the latter context, there is evidence that in human whole body dynamic exercise such as cycling, there is, in general, an orderly intensity dependent hierarchical recruitment of motor units such that type I fibers are recruited first followed by type IIA then type IIB (Gollnick et al., 1974; Vøllestad et al., 1984; Vøllestad & Blom, 1985). Nevertheless, there is also strong evidence for a large concurrent element of rate-coding as evidenced by metabolic activity in type II human fibers at low exercise intensities (Ivy et al., 1987). Similarly in cycling exercise at intensities close to VO_{2max}, which require only ~50% of the maximal muscle force (measured at the same velocity), there is evidence that virtually the whole muscle fiber population has been, to some degree, metabolically active (Fig. 3; Greig et al., 1985). This observation implies a significant element of *rate-coding* with presumably the more powerful, but fatigable, motor units at the top of the hierarchy being relatively less activated than the less powerful fatigue resistant motor units which are lower in the recruitment order. Such an approximate matching of fatigue resistance and degree of activation could be a useful strategy for sustaining power output at the highest possible level during prolonged activity. Nevertheless, it does seem that fatigue of whole muscle power output during high intensity submaximal exercise can be the result of a mismatch of activation level and fatigue sensitivity leading to a selective fatigue of fast fatigue sensitive motor units even though they may be activated at relatively low firing frequencies (Beelen & Sargeant, 1991; Beelen & Sargeant, 1993; Beelen et al., 1993).

TYPES OF CONTRACTION

The mechanical output required from skeletal muscle in-vivo may require one, or a combination of, three types of contraction. These include: 1) isometric contraction, 2) eccentric contraction; and/or 3) concentric contractions as described below.

Isometric Contraction

This term implies that muscle is activated but is held at a constant length. In fact in whole body human and other animal studies, it is usually defined operationally as an activation of muscle(s) when the joint(s) spanned by the muscle(s) are held in a fixed position. The important point here is that there will always be an element of shortening of the muscle due, amongst other things, to stretching of the tendinous connections and compression of joints. For example, in metabolic experiments involving short duration repetitive *isometric* contractions (that is, with a fixed joint position) the whole activation period may be taken up with shortening of the muscle fibers. In addition to this situation, there may be rather complex and simultaneous lengthening and shortening of sarcomeres in series along the length of the individual muscle fibers, especially when fatigue is occurring due to heterogeneity of metabolic and mechanical properties between sarcomeres. Notwithstanding these qualifications, the generation of *isometric* force is for largely technical reasons, the most commonly studied form of contraction. It is also of considerable importance, for stabilizing joint complexes during locomotion and maintaining posture.

Eccentric Contractions

In these contractions, the muscle is activated but instead of shortening it is lengthened due to an external force which exceeds that generated by the degree of activation. Eccentric contractions are a normal and common element of everyday human muscle function. The role of different fiber types in the generation of eccentric force is not well understood although a rather small number of fibers may be activated to generate eccentric as compared to concentric force (Bigland-Ritchie & Woods, 1976) and some investigators have suggested that there may be a selective recruitment of type II fibers during such contractions (Nardone et al., 1989). In relation to activity involving eccentric contractions, it is worth noting that single fiber studies indicate that levels of fatigue that reduce the maximal velocity of shortening and modify the (concentric) force velocity relationship may not affect eccentric force generation to the same extent (Curtin, 1990). This may be of significance in movement patterns involving a pre-stretch. Here the preservation of high force generation in the initial eccentric phase will maintain the amount of elastic energy stored. During the subsequent power generating concentric phase, the contribution of the recovered elastic energy could offset the effect of fatigue when compared to a pure concentric contraction (de Haan et al., 1991).

Concentric Contractions

These occur when the muscle is activated and shortens, power is generated and work is done by the muscle. Human muscles have been studied much less often under dynamic than isometric conditions, and often on the implicit assumption that isometric function will systematically reflect the functional status of the muscle during concentric contractions. This is not always the case as shown, for example, by the independent effects of fatigue on maximal isometric force and maximal velocity of shortening (de Haan et al., 1989; Beelen

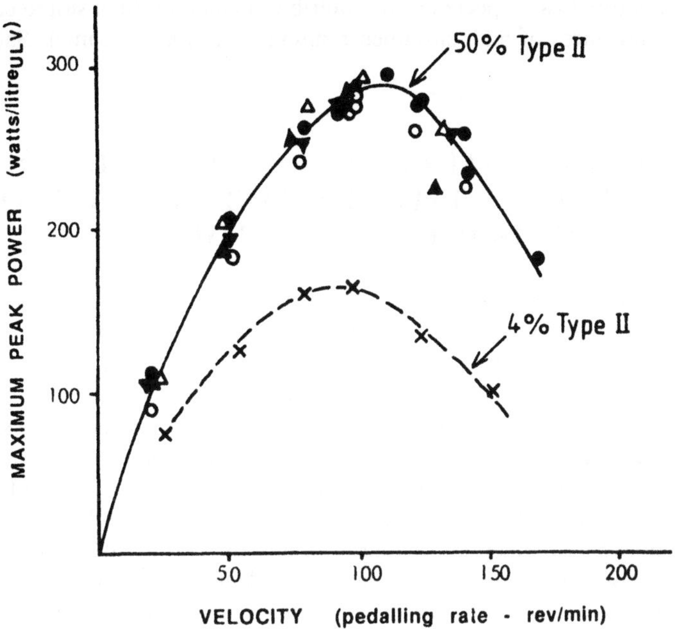

Figure 4. Relationship of maximal peak power to pedaling rate. Power is standardized for upper leg muscle volume. Data for 5 subjects (different symbols), with 50% type II fibers (—). For comparison, data are given for an ultra-marathon runner with only 4% type II fibers (- - -) (group data from Sargeant et al., 1981) (reprinted with permission from Sargeant, 1994b).

& Sargeant, 1991). Clearly, in human locomotion the capability of human muscles to generate power in concentric contractions is of fundamental importance. As pointed out previously, the measurement of true maximal human power is dependent upon ensuring that the muscles will be contracting at optimal velocity for maximal power production as defined by the force velocity relationships. The fact that in whole body exercise this velocity can only be determined globally for all the contributing muscles does not detract from the importance of this requirement (Sargeant & Beelen, 1993). In our own studies, we have addressed the problem by controlling the pedaling frequency in maximal cycling exercise using an isokinetic system (see e.g., Sargeant et al., 1981; Beelen et al., 1994). These studies have indicated that maximal power in this cyclic leg extension exercise is achieved at an optimal pedaling rate of about 120 crank rev/min (Fig. 4).

Stretch-Shortening Cycle

As well as these pure forms of contraction, there are many situations where a combination of contraction types occurs. The most important of these is the stretch-shortening cycle where an eccentric muscle contraction immediately precedes a concentric power generating phase. For example in running, the leg extensor muscle-tendon complex can act not simply as a *brake* or *shock absorber* on landing but also as an energy store. On landing, activation of the leg extensor muscles results in their series-elastic tendinous structures being stretched. Some of this stored elastic energy may then be recovered and used to add to the power generated in the subsequent push-off (see e.g., Cavagna et al., 1968; Alexander, 1984; Komi, 1984). The re-use of stored elastic energy in this way is common throughout the animal kingdom, but perhaps the most well known development of the principle in mammals is seen in the kangaroo. Less expected is the contribution that re-use of stored elastic energy makes to the achievement of high sustained running speeds in the camel (Saltin & Rose, 1994).

WHAT IS THE POTENTIAL CONTRIBUTION OF DIFFERENT HUMAN MUSCLE FIBER TYPES TO MAXIMAL POWER OUTPUT AT DIFFERENT CONTRACTION VELOCITIES?

This is an important but unanswered question in relation to dynamic muscle function in humans. Faulkner and colleagues (1980) measured the force-velocity characteristics of bundles of human muscle fibers obtained at surgery. Taking account of the proportion of type I and II fibers present, they suggested a ratio for the V_{max} of around 1:4. A similar ratio is indicated for the normal control subjects studied by Fitts and colleagues (1989) and this ratio seems theoretically reasonable on dimensional grounds relative to other species (see e.g., Hill, 1950, 1956; Rome et al., 1990). However, the recent investigation of Larsson and Moss (1993) reports a wider range, although the *mean* value for type II fibers appears to be of the same order. What is important to note from this last study is the wide range of V_{max} within the type II population and the large number of *hybrid* fibers co-expressing variable proportions of myosin heavy chain isoforms (see also Sant'Ana Pereira et al., 1995).

The implications for power output from human muscles that have fiber populations with different force-velocity characteristics can be shown by reference to a model. For illustrative purposes the following assumptions and simplifications are made:

1. There are two discrete populations of muscle fibers (type I and II) with a 1:4 ratio for their maximal velocities of shortening.
2. The specific force of these two fiber populations is such that each generates 50% of the whole muscle maximal isometric force (after account is taken of their respective cross-sectional areas).
3. The length-tension relationships, relative to the whole muscle, are the same for both fiber populations.
4. The a/P_0 constant which defines the curvature of the force-velocity relationship is the same for both populations.

Of these 1) is a simplification based on mean values to make the model manageable and comprehensible. In fact, there is wide range of V_{max} values, especially for type II fibers (Fig. 2; Larsson & Moss, 1993). Even among type II fibers expressing a single myosin heavy chain isoform, there may be variation in V_{max} related to the myosin light chain composition (Lowey et al., 1993; Bottinelli et al., 1994) (see also the apparent plasticity of contractile properties of human muscle fibers as indicated by the training study of Fitts et al., 1989; also training induced changes in myosin isoforms, Baumann et al., 1987; Schantz & Dhoot, 1987); 2) is arbitrary but not unreasonable for human muscle of mixed composition; 3) is a simplification, although on functional grounds one might expect differences related to the range over which task related recruitment of different fiber populations occurs.(see e.g., Herzog & ter Keurs, 1988; de Ruiter et al., 1995; Kernell, 1994); and 4) is a simplification made in the absence of systematic data (for a review of animal muscle see Woledge et al., 1985; and for human data see Fitts et al, 1989).

The relative force and power velocity relationships for types I and II fiber populations in our hypothetical human muscle are shown in Fig. 5. If we assume a summation of power production from the two populations in the maximally activated whole muscle, the combined power output will be as shown in Fig. 6. The great difficulty is to know how to relate the

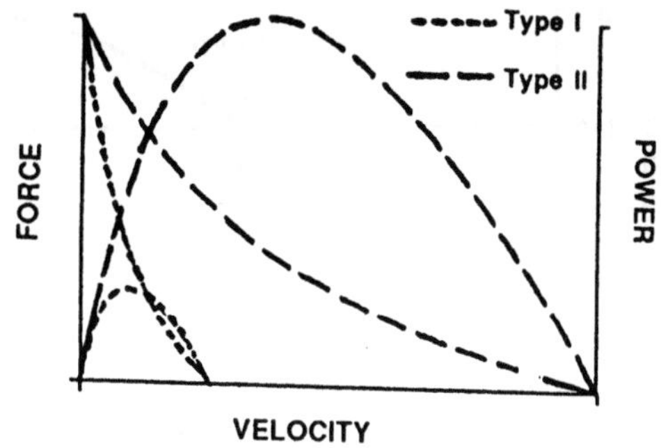

Figure 5. Force and power in relation to velocity for a slow (type I) and a fast (type II) population of fibers that have the same specific isometric force but V_{max} in the ratio 1 : 4 (note the very low power generated by the type I compared to type II fibers).

relative velocities of Figs. 5 and 6 to human locomotory movement. What we do know is that in cycling exercise the optimal pedaling rate (V_{opt}) for maximal power output is ~120 rev/min (Fig. 4). If it is assumed that this is the velocity at which the maximal power of the two populations combined is generated, then a pedaling rate of 60 rev/min approximates to the V_{opt} of the type I fibers. It also implies that the V_{max} of type I fibers will not be exceeded until a pedaling rate of ~ 165 rev/min is achieved and that the fastest fibers (type II) have an optimal velocity well in excess of that seen in normal locomotion. In general these implications from the model seem not unreasonable. The power/velocity data for the ultra-marathon runner shown in comparison to normal control subjects in Fig. 4 also seems to fit the general case.

One important point to note from the model is the much greater power available from the type II fibers within the locomotory range (see also the experimental human data in Fig. 4). As discussed by Faulkner and Brooks (1993), this implies that fast type II fibers could with relatively low levels of activation sustain a power output level equivalent to *maximally* activated slow type I fibers. This may be of particular significance for muscles containing a large proportion of type IIA fibers which have a high oxidative potential.

SUSTAINED SUBMAXIMAL POWER OUTPUT

In locomotory function, the ability to sustain submaximal levels of power is important. In Fig. 7, the x-axis from Fig. 6 is expanded and submaximal power output levels approximating 25 and 80% of the maximal power are indicated by dashed lines. At the 80% level of sustained power there is, by definition, a 20% reserve of the combined power generating capability when pedaling at 120 rev/min while at 60 rev/min there is no reserve. Similarly at the 25% level ,there is a reserve of 75% at 120 rev/min but only 55% at 60 rev/min. At first, the greater reserves at the fast pedaling rate might suggest that this rate is

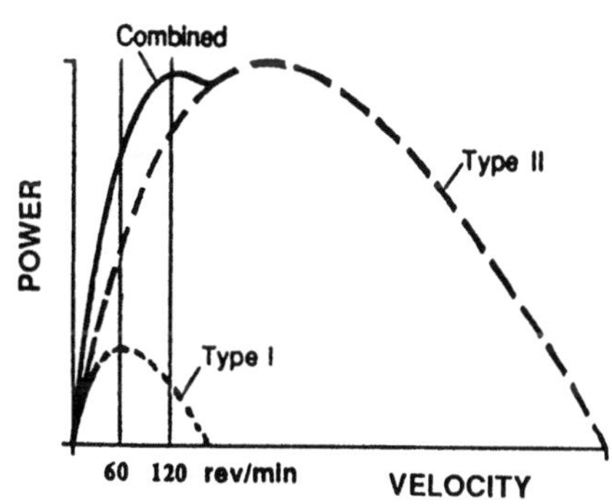

Figure 6. The component and combined power-velocity relationship for a whole muscle (modeled as if composed of two discrete populations of fiber types present in equal proportions, see text). The superimposed pedaling rates are derived by reference to the group data in Fig. 4.

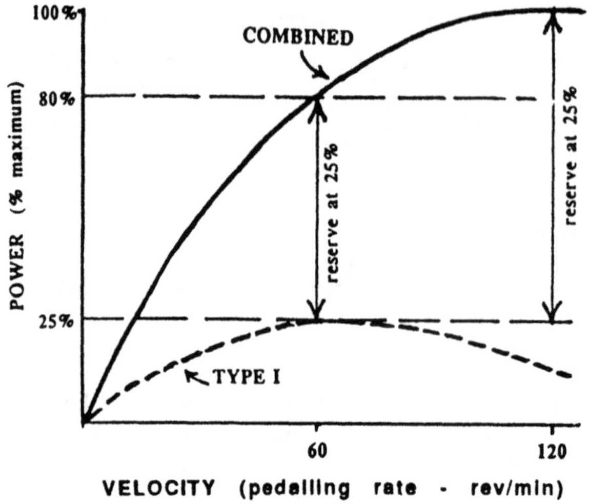

Figure 7. Submaximal levels of power (25% and 80% of maximum) in relation to the first part of the *combined* power velocity relationship expanded from Fig. 6. The maximal potential contribution from the type I fiber population is also shown (---).

preferable at *both* levels of power for resisting fatigue and hence sustaining power output. For the lower level of power output, however, this ignores the potential contribution to power output from the fatigue resistant type I fibers. In principle, a power output of 25% of maximum could be achieved at 60 rev/min by recruiting *only* the type I fibers. At a pedaling rate of 120 rev/min, however, the same power output can only be attained if there is a *minimal* contribution of ~15% power from the faster type II fiber population, which ultimately includes fatigue sensitive fibers. Calculating the *minimum possible* contribution from the type II fibers at different fractions of maximal power yields the data shown in Fig. 8. This suggests that at a low level of sustained power output (25% of maximum), the optimal pedaling rate to minimize the contribution required from type II fibers is 60 rev/min but at high levels of sustained power (80% of maximum) the optimum increases to 120 rev/min. However, as has already been pointed out, there is evidence for a significant element of rate coding superimposed on the hierarchical pattern of motor unit recruitment in this form of exercise. It should be emphasized therefore that this analysis is only describing the *minimum* contribution necessary from the *power generating capability* of the type II fibers - assuming an initial and full recruitment of the type I fiber population.

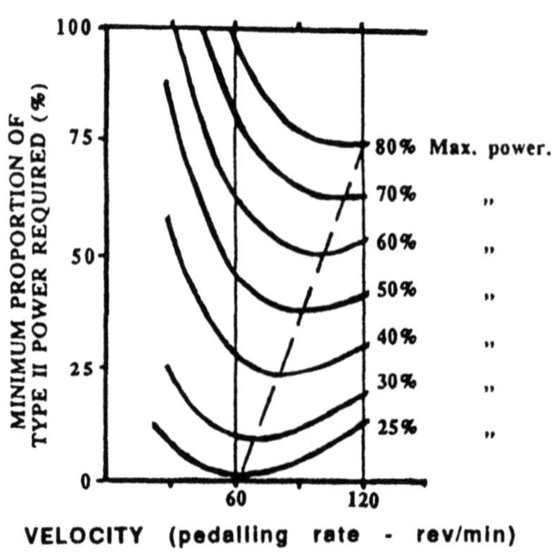

Figure 8. Relationship between velocity and the *minimum* proportion of the type II power required at different fractions of maximal power. The dashed line indicates the progressive increase in the velocity at which the minimum contribution from type II power occurs. This is derived from Fig. 7 and based on the assumption that the type I population is contributing maximally in all cases. As discussed in the text, this may not be true leading to underestimation, especially at the lower levels.

Interestingly, and despite all the assumptions and simplification, the model does seem to reflect muscle function. We recently had the opportunity to study a group of experienced competitive cyclists (Sargeant, 1994b). They were asked to cycle at different constant speeds on their own bicycles and they were allowed to choose freely the gear at each speed studied. As the exercise intensity increased so did their freely chosen pedal frequency; from ~60 rev/min at the lowest levels to ~110 at the highest (speed range, 20 - 47 km/hr; VO_2 range, 0.8 - 5.1 L/min). In addition, it is worth considering the data derived from the world cycling record for distance achieved in one hour. Knowing the gear ratios, the mean pedaling rate can be calculated. Over all the years analyzed, these world class athletes consistently chose a high pedaling rate of around 106 rev/min in order to maintain a high sustained level of power output (see Sargeant, 1994b; his Table 1).

MECHANICAL EFFICIENCY AND PEDALING RATE

As pointed out previously (Sargeant, 1988), the theoretical advantage described above of choosing fast pedaling rates at high levels of sustained submaximal power would be negated if there was a disproportionate increase in the energy cost for the same external power delivered; that is, mechanical efficiency decreased. Clearly, however, at least in well trained cyclists, this is either not the case, or the advantage, in terms of fatigue resistance, outweighs the extra energy cost.

Unfortunately, very little is known about the mechanical efficiency/velocity relationships of different human muscle fiber types. On the basis of animal muscle experiments, it is usually proposed that maximal efficiency occurs at a velocity that is close to, but slightly below, the optimum for maximal power (see e.g., Goldspink, 1978; Lodder et al., 1991; Rome, 1993). Thus it might be speculated that the efficiency/velocity relationships for type I and II fibers might be of the general form shown in Fig. 11. By reference to the earlier discussion regarding the V_{opt} for maximal power in cycling, it can be seen that the maximal efficiency for the type I fibers might occur at ~ 50 rev/min and for type II fibers at ~150 rev/min. In this model, the mechanical efficiency of the type I and II fiber populations is equal at about 90 rev/min. Also within the normal locomotory range there is a reciprocal change in the efficiencies of the two types. As a consequence, calculations from the model predict that mechanical efficiencies in high intensity cycling exercise should be largely independent of pedaling rate over a wide range from ~60 to 110 rev/min, a prediction confirmed by experimental data (Zoladz et al., 1995). It should be remembered, however, that this is again a greatly simplified two compartment model. The reality is that there should be a continuum of efficiency/velocity relationships reflecting the continuum of expression in the contractile proteins.

ACUTE PLASTICITY OF THE NEUROMUSCULAR SYSTEM

The contractile properties of muscle and potential for power output can change markedly and acutely as a consequence of exercise itself. In some circumstances, exercise can lead to potentiation of mechanical output (see e.g., Lodder et al., 1991). More commonly, the effect of exercise is studied in relation to the generation of fatigue and this is often seen as a *failure* of the system, but this approach is misleading and can limit comprehension of the phenomenon. Alternatively, fatigue might be seen as *protective*, a down-regulation of output to prevent an energy crisis (de Haan & Koudijs, 1994). Thus the muscle is temporarily *transformed* becoming slower and much less powerful as demonstrated in animal and human experiments (Figs. 9 and 10; de Haan et al., 1989; Beelen & Sargeant, 1991). Such a

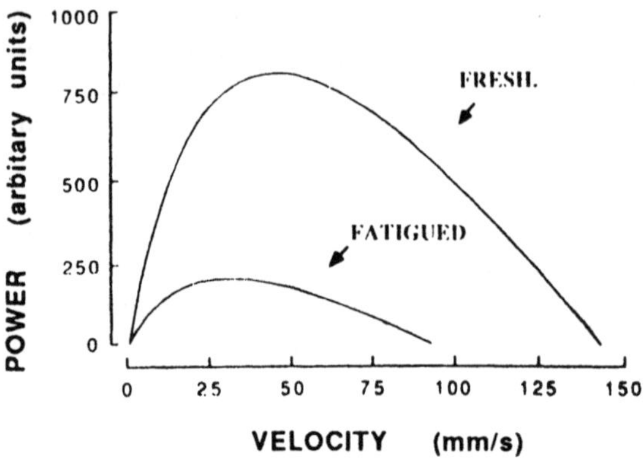

Figure 9. Power (arbitrary units) in relation to velocity for fresh and fatigued rat medial gastrocnemius muscle. Note: in the fatigued state isometric force was reduced to ~50% but power to ~25% of the value for fresh muscle due to the additive effect of the decrease in V_{max} (derived from de Haan et al., 1989).

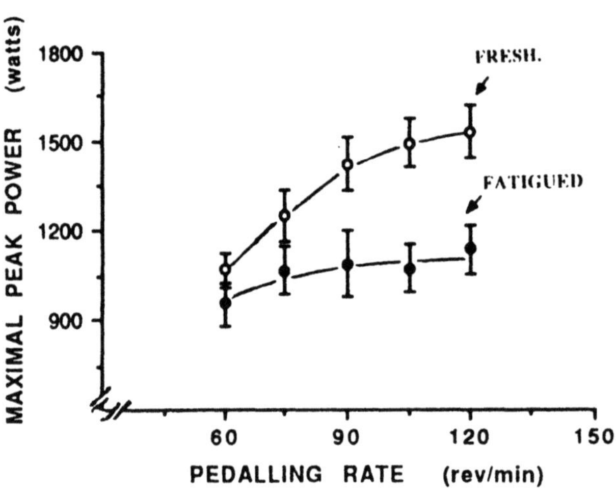

Figure 10. Human maximal peak power cycling at 5 different pedaling rates in fresh and fatigued states. Mean (± SE) data for 6 subjects (adapted from Beelen & Sargeant, 1991).

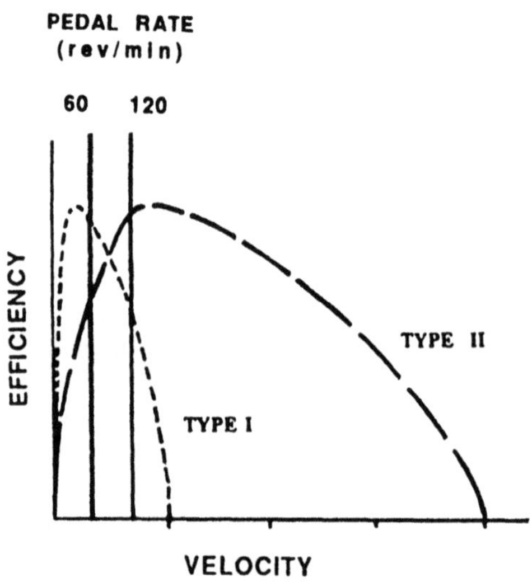

Figure 11. Schematic of the possible relationships between mechanical efficiency and velocity for the modeled fiber types. In the absence of systematic data, no relative difference between the maximal efficiencies is given; each type is normalized to the same maximum. The velocity range equivalent to pedaling rates of 60 to 120 rev/min is derived from Fig. 6.

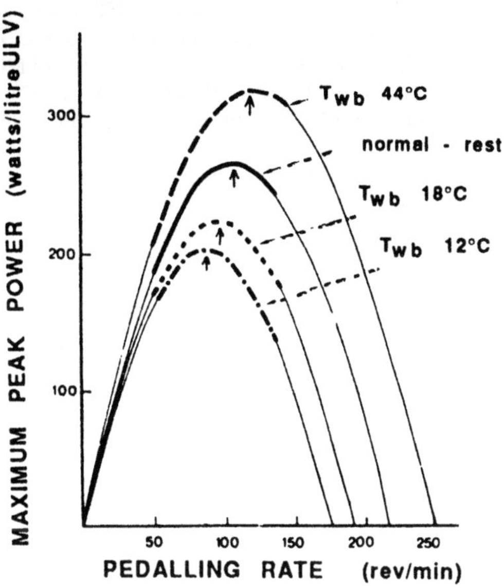

Figure 12. The relationship of maximal peak power during isokinetic cycling and pedaling rate. Power is expressed as watts per liter of upper leg muscle volume. Data is given for normal resting conditions at room temperature and following 45 min immersion in water baths at 44°C, 18°C and 12°C. Limits of the experimental data are given by thicker sections of the lines. Arrows indicate the optimal velocity which increases with temperature (adapted from Sargeant, 1987).

transformation might under some circumstances have the additional benefit of improving the economy of isometric contractions (Jones & Bigland-Ritchie, 1986) and possibly, depending on contraction velocity and recruitment patterns, even the efficiency of repetitive dynamic contractions (Sargeant, 1994b).

The contractile properties of muscle will also be acutely modified by temperature changes, whether as a consequence of metabolic heat production by the muscle itself or due to environmental conditions. In humans, as would be expected from isolated animal muscle experiments, there is a velocity-dependent effect of manipulating muscle temperature (Fig. 12, Sargeant, 1987). Thus, just as exercise-induced fatigue may *transform* muscle into a slower type, the exercise-induced increase in muscle temperature will make the muscle *faster* which may obscure the true magnitude of the fatigue (Beelen & Sargeant, 1991).

ACKNOWLEDGMENTS

Attendance of the authors at the 1994 Bigland-Ritchie conference was supported, in part, by the Muscular Dystrophy Association (USA), and the University of Miami (*A.J.S.*).

REFERENCES

Alexander RMcN (1984). Elastic energy stores in running vertebrates. *American Zoologist* **24**, 85-94.

Baumann H, Jaggi M, Soland F, Howald H & Schaub MC (1987). Exercise training induces transitions of myosin isoform subunits within histochemically typed human muscle fibers. *Pflügers Archiv* **409**, 349-360.

Beelen A & Sargeant AJ (1991). Effect of fatigue on maximal power output at different contraction velocities in humans. *Journal of Applied Physiology* **71**, 2332-2337.

Beelen A & Sargeant AJ (1993). Effect of prior exercise at different pedal frequencies on maximal power in humans. *European Journal of Applied Physiology* **66**, 102-107.

Beelen A, Sargeant AJ, Lind A, de Haan A, Kernell D & van Mechelen W (1993). Effect of contraction velocity on the pattern of glycogen depletion in human muscle fiber types. In: Sargeant AJ, Kernell D (eds.), *Neuromuscular Fatigue*, pp. 93-95. Amsterdam: Royal Netherlands Academy of Arts and Sciences.

Beelen A, Sargeant AJ & Wijkhuizen F (1994). Measurement of directional force and power during human submaximal and maximal isokinetic exercise. *European Journal of Applied Physiology* **69**, 1-5.

Bigland-Ritchie B & Woods JJ (1976). Integrated EMG and O_2 uptake during positive and negative work. *Journal of Physiology (London)* **260**, 267-277.

Bottinelli R, Betto R, Schiaffino S & Reggiani C (1994). Unloaded shortening velocity and myosin heavy chain and alkali light chain isoform composition in rat skeletal muscle fibers. *Journal of Physiology (London)* **478**, 3421-349.

Bottinelli R, Schiaffino S & Reggiani C (1991). Force-velocity relations and myosin heavy chain isoform compositions of skinned fibers from rat skeletal muscle. *Journal of Physiology (London)* **437**, 655-667.

Buller AJ, Eccles JC & Eccles RM (1960). Interactions between motoneurones and muscles in respect of the characteristic speeds of their responses. *Journal of Physiology (London)* **150**, 417-439.

Burke RE, Levine DN, Tsairis P & Zajac, FE (1973). Physiological types and histochemical profiles in motor units of the cat gastrocnemius. *Journal of Physiology (London)* **234**, 723-748

Cavagna GA, Dusman B & Margaria R (1968). Positive work done by a previously stretched muscle. *Journal of Applied Physiology* **24**, 21-32.

Curtin NA (1990). Force during stretch and shortening of frog sartorius muscle: effects of intracellular acidification due to increased carbon dioxide. *Journal of Muscle Research and Cell Motility* **11**, 251-257.

de Haan A, Jones DA & Sargeant AJ (1989). Changes in power output, velocity of shortening and relaxation rate during fatigue of rat medial gastrocnemius muscle. *Pflügers Archiv* **413**, 422-428.

de Haan A & Koudijs JCM (1994). A linear relationship between ATP degradation and fatigue during high-intensity dynamic exercise in rat skeletal muscle. *Experimental Physiology* **79**, 865-868.

de Haan A, Lodder MAN & Sargeant AJ (1991). Influence of an active pre-stretch on fatigue of skeletal muscle. *European Journal of Applied Physiology* **62**, 268-273.

de Ruiter CJ, de Haan A & Sargeant AJ (1995). Physiological characteristics of two extreme muscle compartments in gastrocnemius medialis of the anaesthetized rat. *Acta Physiologica Scandanavica* **153**, 313-324.

Ennion S, Sant'Ana Pereira JA, Sargeant AJ, Young A & Goldspink G (1995). Characterisation of human skeletal muscle fibers according to the myosin heavy chains they express. *Journal of Muscle Research and Cell Motility* **16**, 35-43

Faulkner JA & Brooks SV (1993). Fatigability of mouse muscles during constant length, shortening, and lengthening contractions: interactions between fiber types and duty cycles. In: Sargeant AJ, Kernell D (eds.) *Neuromuscular Fatigue*, pp. 116-123. Amsterdam: Royal Netherlands Academy of Arts and Sciences.

Faulkner JA, Jones DA, Round JM & Edwards RHT (1980). Dynamics of energetic processes in human muscle. In: Cerretelli P, Whipp BJ (eds.), *Exercise Bioenergetics and Gas Exchange*, pp. 81-90. Amsterdam: Elsevier/North-Holland Biomedical Press.

Fitts RH, Costill DL & Gardetto PR (1989). Effect of swim training on human muscle fiber function. *Journal of Applied Physiology* **66**, 465-475.

Goldspink G (1978). Energy turnover during contraction of different types of muscle. In: Asmussen E, Jørgensen K (eds.), *Biomechanics VI-A*, pp. 27-39. Baltimore: University Park Press.

Gollnick PD, Piehl K & Saltin B (1974). Selective glycogen depletion pattern in human muscle fibers after exercise of varying intensity and at varying pedaling rates. *Journal of Physiology (London)* **241**, 45-57.

Greig CA, Sargeant AJ & Vøllestad NK (1985). Muscle force and fiber recruitment during dynamic exercise in man. *Journal of Physiology (London)* **371**, 176P.

Herzog W & ter Keurs HEDJ (1988). Force-length relation of in-vivo human rectus femoris muscles. *Pflügers Archiv. European Journal of Physiology* **411**, 642-647.

Hill AV (1950). The dimensions of animals and their muscular dynamics. *Proceedings Royal Institute of Great Britain* **34**, 450-473.

Hill AV (1956). The design of muscles. *British Medical Bulletin* **12**, 165-166.

Ivy JL, Chi M-Y, Hintz CS, Sherman WM, Hellendall RP & Lowry OH (1987). Progressive metabolic changes in individual human muscle fibers with increasing work rates. *American Journal of Physiology* **252**, C630-C639.

Jones DA & Bigland-Ritchie B (1986). Electrical and contractile changes in muscle fatigue. In Saltin B (ed.), *Biochemistry of Exercise* VI, pp. 337-392. Champaign, Ill: Human Kinetics Publishers.

Kernell D (1983). Functional properties of spinal motoneurons and gradation of muscle force. In: Desmedt JE (ed.), *Motor Control Mechanisms in Health and Disease*, pp. 213-226. New York: Raven Press.

Kernell D (1986). Organization and properties of spinal motoneurones and motor units. *Progress in Brain Research* **64**, 21-30.

Kernell D (1992). Organized variability in the neuromuscular system: a survey of task-related adaptations. *Archives Italiennes de Biologie* **130**, 19-66.

Kernell D (1994). Motor tasks and functional organization of the neuromuscular system. Plenary lecture. *Proceedings for Joint Meeting of the Dutch Physiological Society and The Physiological Society*, 2P.

Kernell D, Eerbeek O & Verhey BA (1983). Motor unit categorization on basis of contractile properties: an experimental analysis of the composition of the cat's m. peroneus longus. *Experimental Brain Research* **50**, 211-219.

Komi PV (1984). Physiological and biomechanical correlates of muscle function. Effects of muscle structure and stretch shortening cycle on force and speed. In: Terjung R (ed.), *Exercise and Sports Science Reviews*, vol. 12, pp. 18 -121. Lexington: Collamore Press.

Kugelberg E (1973). Histochemical composition, contraction speed, and fatiguability of rat soleus motor units. *Journal of Neurological Sciences* **20**, 177-198.

Kukulka CG & Clamann HP (1981). Comparison of the recruitment and discharge properties of motor units in human biceps brachii and adductor pollicis during isometric contractions. *Brain Research* **219**, 45-55.

Larsson L & Moss RL (1993). Maximum velocity of shortening in relation to myosin isoform composition in single fibers from human skeletal muscles. *Journal of Physiology (London)* **472**, 595-614.

Lodder MAN, de Haan A & Sargeant AJ (1991). Effect of shortening velocity on work output and energy cost during repeated contractions of the rat EDL muscle. *European Journal of Applied Physiology* **62**, 430-435.

Lowey S, Waller GS & Trybus KN (1993). Skeletal muscle myosin light chains are essential for physiological speed of shortening. *Nature* **365**, 454-456.

McComas AJ, Galea V, Einhorn RW, Hicks AL & Kuiack S (1993). The role of the Na^+, K^+-pump in delaying muscle fatigue. In: Sargeant AJ, Kernell D (eds.), *Neuromuscular Fatigue*, pp. 35-43. Amsterdam: Royal Netherlands Academy of Arts and Sciences.

Nardone A, Romnano C & Schieppati M (1989). Selective recruitment of high-threshold human motor units during voluntary isotonic lengthening of active muscles. *Journal of Physiology (London)* **409**, 451-471.

Pette D & Vrbová G (1992). Adaptation of mammalian skeletal muscle fibers to chronic electrical stimulation. *Reviews in Physiology, Biochemistry and Pharmacology* **120**, 115-202.

Pierotti DJ, Roy RR, Bodine-Fowler SC, Hodgson JA & Edgerton VR (1991). Mechanical and morphological properties of chronically inactive cat tibialis anterior motor units. *Journal of Physiology (London)* **444**, 175-192.

Rome LC (1992). The design of the muscular system. In: Sargeant AJ, Kernell D (eds.), *Neuromuscular Fatigue*, pp. 129-136. Amsterdam: Royal Netherlands Academy of Arts and Sciences

Rome LC, Sosnicki AA & Goble DO (1990). Maximum velocity of shortening of three fiber types from horse soleus muscle: implications for scaling with body size. *Journal of Physiology (London)* **431**, 173-185.

Rutherford OM & Jones DA (1988). Contractile properties and fatiguability of the human adductor pollicis and first dorsal interosseous: a comparison of the effects of two chronic stimulation patterns. *Journal of Neurological Science* **85**, 319-331.

Salmons S & Vrbová G (1969). The influence of activity on some contractile characteristics of mammalian fast and slow muscles. *Journal of Physiology (London)* **201**, 535-549.

Saltin B & Rose RJ (eds.) (1994). The racing camel (*Camelus dromedarius*). Physiology, metabolic functions and adaptations. *Acta Physiologica Scandinavica Supplement* **617**, 9-18.

Sant'Ana Pereira JA, Wessels A, Nijtmans L, Moorman AFM & Sargeant AJ (1995). New method for the accurate characterisation of single human skeletal muscle fibers demonstrates a relation between mATPase and MyHC expression in pure and hybrid fibers types. *Journal of Muscle Research and Cell Motility* **16**, 21-34.

Sargeant AJ (1987). Effect of muscle temperature on leg extension force and short-term power output in humans. *European Journal of Applied Physiology* **56**, 693-698.

Sargeant AJ (1988). Optimum cycle frequencies in human movement. *Journal of Physiology (London)* **406**, 49P.

Sargeant AJ (1994a). 'Acute' and 'chronic' plasticity of human muscle and power output. Motor Control Symposium. Proceedings of the 2nd World Congress of Biomechanics. Stichting World Biomechanics, Nijmegen, The Netherlands, Volume II, 119.

Sargeant AJ (1994b). Human power output and muscle fatigue. *International Journal of Sports Medicine* **15**, 116-121.

Sargeant AJ & Beelen A (1993). Human muscle fatigue in dynamic exercise. In: Sargeant AJ, Kernell D (eds.), *Neuromuscular Fatigue*, pp. 81-92. Amsterdam: Royal Netherlands Academy of Arts and Sciences.

Sargeant AJ, Hoinville E & Young A (1981). Maximum leg force and power output during short-term dynamic exercise. *Journal of Applied Physiology* **51**, 1175-1182.

Schantz PG & Dhoot GK (1987). Coexistence of slow and fast isoforms of contractile and regulatory proteins in human skeletal muscle fibers induced by endurance training. *Acta Physiologica Scandinavica* **131**, 147-154.

Thomas CK, Johansson RS & Bigland-Ritchie B (1991). Attempts to physiologically classify human thenar motor units. *Journal of Neurophysiology* **65**, 1501-1508.

Unquez GA, Bodine-Fowler SC, Roy RR, Pierotti DJ & Edgerton VR (1993). Evidence of incomplete neural control of motor properties in cat anterior tibialis after self-reinnervation. *Journal of Physiology (London)* **472**, 103-125.

Vøllestad NK & Blom PCS (1985). Effect of varying exercise intensity on glycogen depletion in human muscle fibers. *Acta Physiologica Scandinavica* **125**, 395-405.

Vøllestad NK, Vaage O & Hermansen L (1984). Muscle glycogen depletion patterns in type I and subgroups of type II fibers during prolonged severe exercise in man. *Acta Physiologica Scandinavica* **122**, 433-441.

Woledge RC, Curtin NA & Homsher E (1985). Energetic aspects of muscle contraction. *Monographs of the Physiological Society* No. **41**. London: Academic Press.

Zoladz J, Rademacker A & Sargeant AJ (1995). Oxygen uptake does not increase linearly with power output at high intensities in humans. *Journal of Physiology (London)* In press.

25

INTRAMUSCULAR PRESSURES FOR MONITORING DIFFERENT TASKS AND MUSCLE CONDITIONS

O. M. Sejersted[1] and A. R. Hargens[2]

[1] Institute for Experimental Medical Research
University of Oslo
Ullevaal Hospital, Oslo, Norway
[2] Gravitational Research Branch
NASA
Ames Research Center, Moffett Field, California

ABSTRACT

Intramuscular fluid pressure (IMP) can easily be measured in man and animals. It follows the law of Laplace which means that it is determined by the tension of the muscle fibers, the recording depth and by fiber geometry (fiber curvature or pennation angle). Thick, bulging muscles create high IMPs (up to 1000 mmHg) and force transmission to tendons becomes inefficient. High resting or postexercise IMPs are indicative of a compartment syndrome due to muscle swelling within a low-compliance osseofascial boundary. IMP increases linearly with force (torque) independent of the mode or speed of contraction (isometric, eccentric, concentric). IMP is also a much better predictor of muscle force than the EMG signal. During prolonged low-force isometric contractions, cyclic variations in IMP are seen. Since IMP influences muscle blood flow through the muscle pump, autoregulating vascular elements, and compression of the intramuscular vasculature, alterations in IMP have important implications for muscle function.

INTRODUCTION

IMP, the hydrostatic fluid pressure inside muscles can be measured using various techniques. It is primarily thought to reflect the hydrostatic pressure within the interstital matrix gel and the osmotic pressure of the immobilized macromolecules, the sum of which is the pressure that equilibrates with the free fluid in the interstitium (Wiig, 1990). In resting muscles fiber tension is low, so pressure gradients inside the muscle are small. In most resting muscles compliance of the interstitial fluid compartment is high and hydrostatic and colloid osmotic pressures are of the same order of magnitude (Reed & Wiig, 1981). There is a small hydrostatic pressure gradient out of the capillaries (Aukland & Reed, 1993). However, during

Fatigue, Edited by Simon C. Gandevia et al.
Plenum Press, New York, 1995

contraction, the rigid muscle cells are able to withstand pressure just like the heart wall, and quite large pressure gradients can develop.

The physiological and clinical interest in IMP is related to several aspects of muscle function, of which three seem the most important. First, fluid balance inside muscles is highly dependent on hydrostatic pressures since these pressures constitute part of the Starling forces that govern fluid transfer across the capillary wall and therefore also the volume of the interstitial space. High IMP will impair fluid filtration, but promotes lymph formation (Levick, 1991). However, as IMP increases, blood flow through the muscle will decrease either due to an increased resistance at the level of venous outlet from the muscle or because the capillaries collapse. Reduced microcirculation will cause ischemia. Hence, the second important factor relates to metabolic control of the muscle cells. High pressures will promote anaerobic metabolism. The third aspect is purely biomechanical. High IMP will reduce transmission of force to muscle tendons as outlined below.

With this background, we will review the literature on IMP. First, we describe the factors that determine IMP and how it can be measured in various muscles. Second, we will discuss the task dependency of IMP and its relation to the EMG signal. Finally, the importance of IMP as an independent determinant of muscle function will be discussed.

FACTORS THAT DETERMINE INTRAMUSCULAR FLUID PRESSURES

Laplace's Law

In physical terms pressure within a compartment is determined by the properties and the geometry of the wall. This was also pointed out by Hill (1948) who stated that muscle fibers "that are curved . . . must necessarily exert a pressure inwards when they contract, depending on their tension and radius of curvature." Modified for skeletal muscle with one fiber direction, the law of Laplace states that

$$P = P_0 + n\Delta h\frac{s}{r}$$

where P is IMP at any point, P_0 is pressure beneath the fascia, n is the number of concentric muscle layers, Δh is thickness of each layer, s is tension or stress of the muscle cells and r is radius of fiber curvature (Sejersted et al., 1984). Since we know little of the actual geometry of muscle layers and individual fibers in skeletal muscle, the above version of the law might not predict pressures accurately in all muscles. However, the law of Laplace has been successfully applied to the heart (Huisman et al., 1980; Skalak, 1982). Two extreme situations will illustrate the consequences of the important principles of this law. First, imagine a muscle with straight, parallel fibers attached to a broad tendon sheet so that the tendon tension vector has exactly the same direction as the muscle fibers. During an isometric contraction there will be no curving of fibers and all of the force of contraction will be transmitted to the tendon. Since there is no force vector in other directions, no pressure will develop. This prediction has been corroborated by measurements in the diaphragm and trapezius muscle in which low IMP was found during contraction (Supinski et al., 1990; Decramer et al., 1990; Järvholm et al., 1991). Second, the other extreme would be a circular muscle fiber. During contraction no force will be transmitted to any tendon, and again if the fiber is prevented from shortening, all of the developed tension will be directed inwards towards the center of the ring. In the heart this is how pressure develops in the ventricular cavity. Most skeletal muscles will have fiber geometries in which the direction of muscle fiber force development does not align with the direction that force is transmitted through the tendons. Hence, fibers will tend to curve and force vectors will be present perpendicular to force in the tendon. These force vectors will elevate IMP. Thus, in spindle shaped muscles that bulge during contraction, and in pennate structured muscles, one would predict high contraction pressures (Sejersted & Hargens, 1986). In fact pressures close to 600 mmHg have been found in the human vastus medialis and supraspinatus muscles (Sejersted et al., 1984; Järvholm et al., 1991).

In addition to muscle fiber tension (s) and radius of curvature (r), muscle thickness ($n \cdot \Delta h$) determines IMP according to the law of Laplace. This has been extensively corroborated. IMP often increases linearly as a function of recording depth in many muscles (Sejersted et al., 1984; Järvholm et al., 1988). Large bulging muscles will not necessarily mean that a lot of force is transmitted to tendons, but rather that increased muscle fiber tension creates high IMPs.

Starling Forces

An important term in the equation above is the pressure beneath the fascia (P_0). If the muscle is surrounded by a thick fascia or bone so that compliance of the whole muscle compartment is small, P_0 will rise if the muscle swells. Provided there is no tension of the muscle($s = 0$, which is probably never quite true), the pressure will be homogenously elevated throughout the compartment. Clearly contraction pressures will also be elevated. Thus, elevated resting IMP is a cardinal sign of the chronic compartment syndrome that can develop in the tibialis anterior muscle, for example (Pedowitz et al., 1990). Supraspinatus muscle and various forearm muscles are also encapsulated in a compartment (Mubarak et al., 1978; Jensen, 1991). Most other muscles have elastic fasciae, and volume changes can occur with very little change of P_0.

Hence, muscle volume is important not only because it might affect fiber geometry, but because in certain muscles it will determine P_0. Muscle volume can vary by at least 10-15 % under normal conditions. During exercise, fluid is taken up by the active muscle. The reason for this fluid transfer is not quite clear (see McComas & Sjøgaard, Chapter 4). Vasodilation leads to higher capillary perfusion pressures, and probably also a larger capillary surface area which will promote ultrafiltration. However, muscle swelling is better related to exercise intensity than to muscle blood flow, since swelling is linearly related to power also at very high intensities whereas perfusion seems to reach a maximum at lower powers, at least during running or bicycling (Gullestad et al., 1993). Intracellular breakdown of creatine phosphate and accumulation of lactate probably creates an osmotic driving force for fluid uptake by the cells and interstitial colloid osmotic pressure will rise. Interestingly, this combination of hydrostatic and osmotic forces can cause the muscle to swell within seconds, whereas restoration of the muscle volume is a slow process (Sejersted et al., 1986). During recovery, the intracellular osmoles are removed quite rapidly so that fluid will leave the cells and accumulate in the interstitium. However, reabsorption of fluid into the capillaries and drainage into peripheral lymphatic vessels is slow since it is driven primarily by the colloid osmotic pressure gradient and small hydrostatic pressure gradients, respectively. Hence, elevated post-exercise IMP is quite characteristic of the chronic compartment syndrome (Pedowitz et al., 1990).

MEASUREMENTS OF INTRAMUSCULAR FLUID PRESSURES

Methods of Measurements

IMP has been measured using several techniques. Hill (1948) measured the temperature rise in an oil-filled syringe attached to a needle inserted into a frog muscle. Compression of oil will cause a rise in temperature. Subsequent investigators have used similar fluid filled systems, but with different pressure sensors. For example, pressure inside implanted porous or perforated capsules or cylinders have been used (Guyton, 1963; Wiig, 1990). The wick technique was introduced by Scholander and colleagues (1968) and later modified by Fadnes and colleagues (1977) to a wick-in-needle method. Pressure can also be measured by micropipette techniques. None of these methods are well suited for measurements in the contracting muscle. Wick catheters respond slowly to pressure changes, and micropipettes are difficult to keep in place and do break. The simplest technique and one which is similar to the one used by Hill (1948) is to insert a fluid-filled small-bore catheter into the muscle and record pressure with an ordinary pressure transducer (Sejersted et al., 1984). The tip of the catheter has been modified in various ways to ensure that it will remain open, and other investigators have infused saline at slow or intermittent rates (Styf & Körner, 1986a). Alternatively transducer-tipped catheters have been inserted into muscle (Sejersted et al., 1984; Crenshaw et al., 1992; Ballard et al., 1994b). These transducers are now available in very small sizes and have been used to record pressures in the small thyroarythenoid muscle of dogs and in human muscles (Cooper et al., 1993; Crenshaw et al., 1992). The virtue of the transducer-tipped catheters is of course that one avoids the signal dampening effect and oscillations of fluid filled systems. The frequency response is also very high. Potential problems comprise mechanical interference with the sensor area and lack of ways to zero-adjust during recording. There seems to be little difference between steady state pressures measured by these methods (Wiig, 1985; Wiig, 1990), but the open catheters or transducer-tipped catheters are the only means by which to measure rapid changes (Sejersted et al., 1984).

Intramuscular Fluid Pressures in Different Muscles

IMP has been measured in a variety of muscles in many species. Resting pressures range from -2 to +14 mmHg in normal muscle and are higher than those reported in subcutaneous tissue (Styf & Körner, 1987; Wiig, 1990; Pedowitz et al., 1990). This range of variation is primarily due to the position of the muscle relative to the rest of the body and the tightness of osseofascial boundaries. It is also difficult to compare IMP obtained from intact muscles with data from experiments in which the muscle surface has been exposed and compression forces due to skin and fasciae have been removed. In an upright or sitting position, IMP of resting thigh muscles in humans normally ranges between 0 to 10 mmHg (Sejersted et al., 1984).

During contraction, IMP will rise in most muscles. During maximal voluntary contraction IMP can vary from below 40 mmHg in the masseter muscle to above 1000 mmHg (Sylvest & Hvid, 1959; Sadamoto et al., 1983) in the vastus medialis muscle. Later studies have confirmed this variation range, but it is interesting that even in quite small muscles like the frog gastrocnemius, the dog arythenoid and parasternal intercostal muscles, maximum pressures can be 100 mmHg or more (Hill, 1948; Leenaerts & Decramer, 1990; Cooper et al., 1993). The variability is explained by Laplace's law and underscores the importance of muscle morphology.

Similarly, repeated or multiple simultaneous measurements within the same muscle will show large differences due to variable positioning of the pressure sensor. The depth of recording is especially important (Sejersted et al., 1984; Jensen, 1991; Nakhostine et al., 1993). Fig. 1 shows a typical example of force and pressure recordings in the vastus medialis muscle of one human subject. IMP was recorded by three different techniques and varied by almost three-fold between recording sites. In this experiment two open-tipped catheters and a transducer-tipped catheter were used. In another study IMPs in four shoulder muscles were compared. Maximum contraction pressures of 400-600 mmHg were observed in the supra-

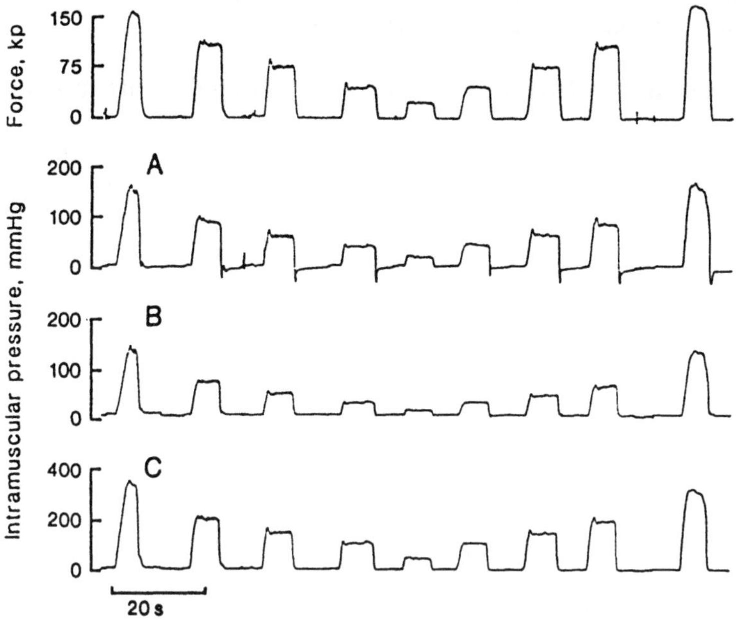

Figure 1. Force and three IMPs in the right human vastus lateralis muscle during isometric knee extension in the sitting position. Force was recorded at the ankle. First and last contractions are MVCs, and 3 to 5 min rest was allowed between contractions at 70, 50, 30 and 15% MVC. Pressure was recorded by a Millar Mikrotip pressure transducer (*A*), Bentley Trantec transducer (*B*), and AE840 transducer (*C*) (reproduced with permission from Sejersted et al., 1984)

and infraspinatus muscles, whereas pressure in the deltoid and trapezius muscles were just above 100 mmHg (Järvholm et al., 1991). Muscle fiber geometry and muscle thickness probably explain this variability.

TASK DEPENDENCY OF INTRAMUSCULAR FLUID PRESSURES

Relationship to Muscle Length and Mode of Contraction

Since fiber geometry and tension are important determinants of IMP, muscle length might affect both resting and contraction pressures. During passive stretch the radius of curvature of the fiber will increase (pennation angle will decrease). The actual outcome will depend on which factor dominates. For example, in dog parasternal intercostal muscles, stretch causes quite high pressures (Leenaerts & Decramer, 1990), whereas stretching in the tibialis anterior muscle of humans causes a pressure rise up to 35 mmHg (Baumann et al., 1979; Gershuni et al., 1984).

During contraction it has been shown first by Mazella (1954), and later by many investigators that IMP is linearly related to force or torque (Sadamoto et al., 1983; Sejersted et al., 1984; Sjøgaard et al., 1986; Järvholm et al., 1989; Leenaerts & Decramer, 1990; Järvholm et al., 1991; Aratow et al., 1993). The slope of this relationship, however, varies both within and between muscles (Sejersted et al., 1984; Jensen, 1991; Järvholm et al., 1991). As seen from Fig. 2, the linear relationship between IMP and torque is independent of contraction velocity, intensity and mode of contraction (isometric or isokinetic). Comparisons between eccentric and concentric exercise also indicate that IMP does not differ greatly in those contractions except at extreme muscle lengths (Ballard et al., 1994a). As expected, IMP was greater per unit torque at short muscle lengths.

Cyclic Variations of IMP at Low Force, Prolonged Contractions

There are no studies that have examined contraction pressure during intermittent contractions or dynamic exercise lasting more than a few minutes. It seems, however, that with continuous recording in one place, pressure measurements are reproducible between contractions, except when muscle compartment compliance decreases due to swelling (as discussed below), and the IMP force relationship is constant for a given recording. However, repeated measurements may give different slopes due to variability in pressure gauge position.

The tight linear relationship between overall force and IMP seems to be broken during low force prolonged isometric contractions. Two reports show that IMP at a given location may change considerably with contractions maintained for more than a couple of minutes. Sejersted and colleagues (1984) found oscillations of IMP amounting to 60% of the highest recorded pressure in the vastus medialis muscle when force was maintained at 15% of MVC. The oscillation frequency was about 1 min^{-1}. These authors also reported an experiment during which the subject maintained one pressure constant for one catheter at about 180 mmHg. Over a two minute period force declined by 30 % and the pressure in a nearby catheter fell initially and thereafter remained fairly constant. Similarly Sjøgaard and colleagues (1986) reported that during a maintained 5 % MVC contraction IMP in the rectus femoris muscle was quite variable. In one location it was maintained stable for one hour. In other locations cyclic variations were observed with a pressure variation range and periodicity similar to that seen by Sejersted and colleagues (1984). On one occasion a period of low pressure was closely associated with changes of the EMG signal. The importance of these observations has never been examined properly. The pressure oscillations must reflect variation in local muscle fiber tension. The absolute muscle volume that the pressure catheter can *see* is unknown. It is probably in the range of several cm^3 and is highly dependent on the depth of recording and interstitial compliance. Hence, it seems that quite large portions of the muscle can be recruited and turned off in a synchronous and cyclic manner, possibly to prevent fatigue.

In conclusion, task dependency of IMP is related primarily to the required force output, and under normal circumstances other task requirements will not affect pressure. The only known exception is prolonged isometric contractions during which cyclic changes in local IMP occur. These changes probably reflect the motor unit recruitment pattern and are observations that need further study.

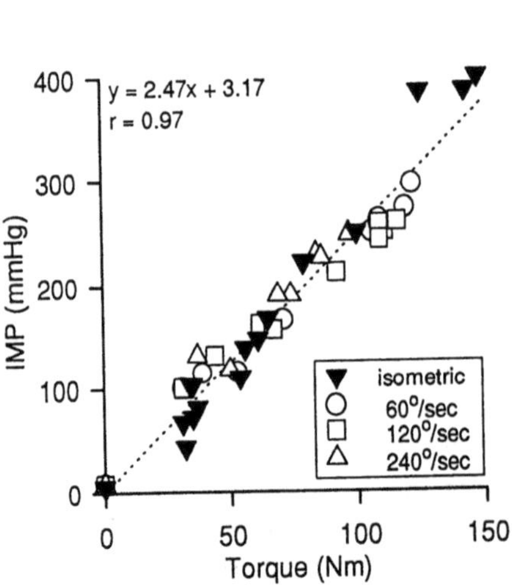

Figure 2. Soleus IMP and ankle joint torque during isometric and isokinetic (concentric) contractions in one subject. Contractions of various intensities were performed at velocities of 0 (isometric), 60, 120 and 240 degrees/s. Isometric contractions were performed at 5 different joint angles. IMP was measured using a 3F Milar transducer-tipped catheter. Knee and hip joints were flexed at 90°. The relationship between IMP and torque is independent of velocity, intensity, and mode of contraction.

THE RELATIONSHIP BETWEEN INTRAMUSCULAR FLUID PRESSURE AND THE ELECTROMYOGRAPHIC SIGNAL (EMG)

The linear relationship between force and IMP has prompted several investigators to suggest that IMP can be used as a force indicator during various tasks where force cannot be directly measured (Parker et al., 1984; Järvholm, 1990; Aratow et al., 1993). Especially with complex movements involving many muscles, the force contribution from one single muscle might be assessed from pressure measurements. The alternative and widely used non-invasive method is to measure EMG. However, there are various disadvantages with this technique, for example: non-linear calibration curves, low reproducibility, and fatigue related changes (Westgaard, 1988).Surface EMG can also pick up signals from several muscles. During brief isometric contractions there seems to be a fairly good relationship between the integrated or root mean square EMG signal and IMP (Järvholm et al., 1989; Järvholm et al., 1991; Aratow et al., 1993). Due to fatigue, however, IMP was superior as an estimator of muscle force (Parker et al., 1984). Interestingly, when movement is introduced, the relationship between EMG and force seems to change. In contrast, the relationship between IMP and force is preserved as outlined above (Aratow et al., 1993). Hence, during concentric contractions of either the soleus or tibialis anterior muscles in humans, the coefficients of determination (R^2) for the torque *vs.* IMP are significantly higher than those between torques and EMG. The discrepancy becomes much more pronounced during eccentric work. Fig. 3 shows that IMP is almost linearly related to torque, whereas EMG

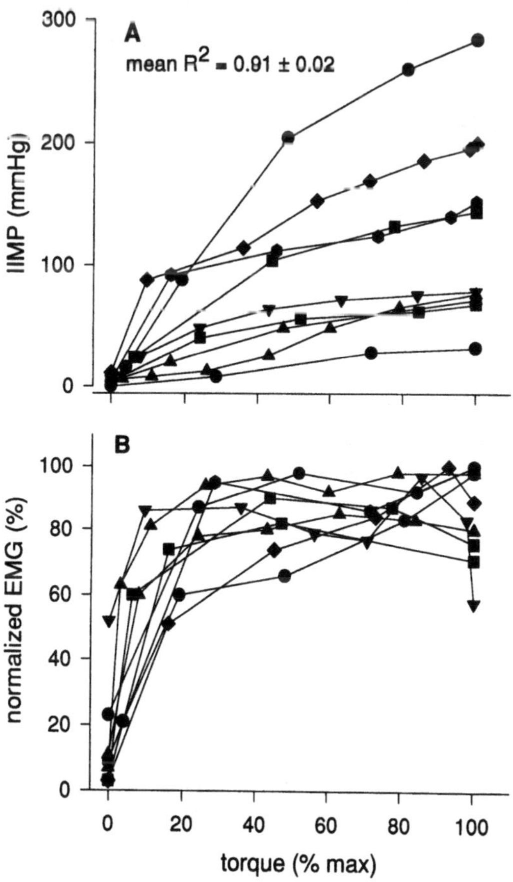

Figure 3. Soleus IMP (*A*; n-9) and EMG (*B*; n = 9) during eccentric exercise. Each line represents one MVC from a subject up to a maximal torque. Linear regression analysis of each line yielded an R^2 value. Mean ± SE of all R^2 values is shown (reproduced with permission from Aratow et al., 1993).

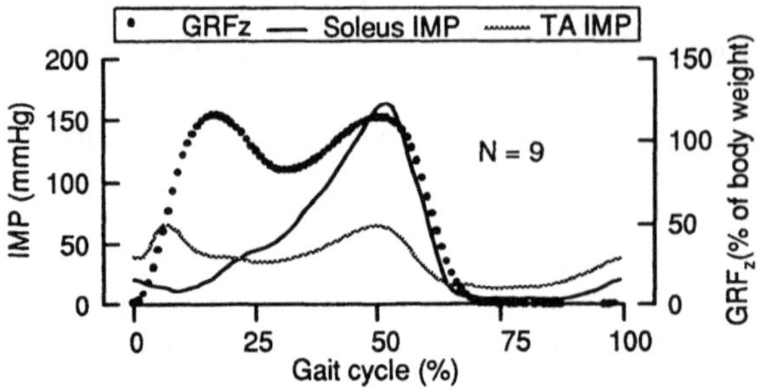

Figure 4. Soleus and tibialis anterior IMP during treadmill walking. The graph represents average data from nine subjects, four gait cycles per subject. IMPs were measured using 3 F Millar transducer-tipped catheters. GRF_z is the vertical ground reaction force measured by an insole force sensor. In all subjects, soleus IMP showed a single peak near the push-off phase of gait while tibialis anterior was biphasic.

shows a highly curved relationship. The reason is that during eccentric exercise there is a force peak late during the contraction that is not reflected in the EMG signal, only in IMP. Thus for normal movements comprising a mixture of eccentric, concentric as well as isometric contractions, IMP seems by far the best estimator of local muscle force.

IMP measurements have, therefore, been applied to analyze muscle function during walking. In an early study by Baumann and colleagues (1979), wick catheters were used, but responded too slowly. In a more recent study, transducer-tipped catheters were used (Ballard et al., 1994b). Fig. 4 shows a recording during one gait cycle in 9 subjects. At heel strike, there is a pressure peak in the tibialis anterior muscle. It then relaxes while IMP builds to a peak in soleus at push-off. There is a simultaneous second IMP peak in the tibialis anterior at this point. Because IMP correlates linearly with force, IMP magnitude provides information about the magnitude of contraction force for specific muscles during movement.

THE IMPORTANCE OF IMP AS A DETERMINANT OF MUSCLE FUNCTION

Muscle Pump

An increase in IMP will of course squeeze blood out of the muscle. This is the mechanism by which blood is propelled back to the heart, and the efficiency of this system depends on functional venous valves (Hargens et al., 1987).

During rhythmic exercise muscle blood flow increases rapidly, probably due to vasodilation Folkow and colleagues (1970) and more recently Sheriff and coworkers (1993) proposed that since the small venules and veins inside the muscle will be emptied during contraction, the elastic recoil of these vessels on relaxation creates a low intraluminal pressure. Hence, the pressure gradient across the capillary bed will be greatly enhanced. This would be sufficient to increase flow substantially. This is in keeping with the finding that the initial flow increment and its timing is independent of cardiac output, arterial blood pressure, power, or autonomic function. Hence, the muscle pump is not only a push-forward mechanism, but also exerts a sort of pull backwards.

Autoregulation of Arteriolar Resistance (Bayliss Reflex)

Another mechanism by which IMP might affect muscle blood flow is through autoregulation of the resistance vessels. Autoregulation is very strong in the vascular beds of organs like the kidneys, the brain and also the coronary circulation in the sense that blood flow is maintained constant over a wide perfusion pressure range. In skeletal muscle autoregulation is also present, but has a moderate gain (Mellander et al., 1987; Braakman et al., 1990). One important trigger of autoregulatory dilation during reduction in perfusion pressure is the Bayliss reflex. An decreased pressure gradient across the arteriolar wall will sometimes elicit dilation, and it can be created by raised IMP. However, it is well known that increased interstitial pressure in the kidneys will elicit preglomerular vasodilation. The role of autoregulatory vasodilation when IMP is elevated is moderate (Mellander & Albert, 1994).

Compression of Vessels of the Intramuscular Circulation

The third relation between IMP and nutritional flow relates to how IMP affects nutritional flow. Clearly, during contractions arterial inflow to the muscle is abruptly reduced, and may be transiently reversed due to expulsion of blood by compression of vessels (Vøllestad et al., 1990). Compression of capillaries and small vessels may occur at pressures above 30-50 mmHg (Hargens et al., 1989). During intermittent contractions (dynamic or isometric) perfusion may proceed during relaxation periods and the periods with high IMP limit average muscle perfusion very little (Styf et al., 1987; Vøllestad et al., 1990). Normally, resting pressures are rapidly regained when contraction ceases (Sejersted et al., 1984; our Fig. 1).

Previously, it has been believed that the intramuscular vascular bed behaves like a collapsible tube, a Starling resistor, which means that with increasing IMP a *waterfall* will appear on the venous side. Intravascular pressures will build up proximal to this waterfall since flow is reduced, and to some extent also because of vasodilation on the arteriolar side. Thus, total occlusion of the vascular bed only occurs when pressures exceed the arterial pressure head. It is also important that with rising IMP, but before occlusion of the capillaries, the driving force for flow through the muscle is no longer the arteriovenous pressure difference but the difference between arterial pressure and IMP. The concept of a Starling resistor is to a large extent correct, but Mellander and Albert (1994) have recently argued that the final short segment of the vein is compressed before it leaves the muscle creating a *venous outflow orifice resistance*. This differs from the behavior of a collapsible tube. Thus, during sustained isometric contractions or in instances where resting pressures are elevated, nutritional blood flow may be insufficient. Due to the large variability in the IMP-force relationship in these situations, IMP is a much better predictor of intramuscular blood flow than force. Again it is important to recognize that IMP during contraction is highly dependent on depth, whereas with compartment syndromes, the high pressure is usually more uniform throughout the compartment. Thus, with maintained isometric contractions only regional ischemia may occur, whereas with compartment syndromes the ischemia is usually more general.

The case of continuous isometric contraction is especially interesting, because fatigue may occur subsequent to ischemia. With repeated, prolonged episodes of low level isometric contractions, it has been claimed that the microcirculation is disturbed which can cause chronic myalgia (Henriksson & Bengtsson 1991). There is no evidence for this. In the study of Sjøgaard and colleagues (1988), flow was well maintained at contraction forces up to 10% of maximal voluntary force in the quadriceps muscles. Fatigue could not be related to lactate output either, because the levels were quite low. Furthermore after isometric contractions at this relative force, there is usually no relaxation hyperemia. Only at higher contraction forces and IMPs does ischemia occur.

Compartment Syndromes

Ischemia due to compression of the microvasculature is accompanied by lactate accumulation, which promotes further muscle swelling and elevation of pressure. As outlined above, muscle swelling during exercise may raise resting IMP and reduce compliance of the compartment further. Therefore, in chronic exertional compartment syndromes, elevated post-exercise or relaxation pressures are primary diagnostic features. Interestingly, delayed muscle soreness of the tibialis anterior muscle after eccentric exercise was also associated with elevation of both resting and contraction pressures probably secondary to increased muscle volume and strain-induced trauma (Fridén et al., 1986). In one study of patients with trapezius myalgia, elevation of both resting and contraction pressures was associated with extensive fibrosis (Hagert & Christenson, 1990). The authors suggest that this kind of myalgia is a functional compartment syndrome. Similarly, it seems that the supraspinatus muscle is surrounded by a tight compartment, and it has been suggested that shoulder pain could emanate from this muscle (Jensen, 1991).

Clearly, muscle swelling, elevated resting IMP and ischemia as well as ischemic pain seem to be related causally. Successful treatment therefore depends on increasing compliance of the muscle compartment by fasciotomy or some other means (Qvarfordt et al., 1983; Styf & Körner, 1986b; Styf et al., 1987).

CONCLUSIONS

IMP is a sensitive and direct measure of muscle force. It is easily monitored both in animals and humans. The large variability within and between muscles is explained by the law of Laplace, but at any one location, IMP is highly reproducible. IMP is related linearly to force independent of the mode or speed of contraction (isometric, concentric, eccentric). One exception is the cyclic variations in IMP during low-force prolonged isometric contractions. Furthermore, IMP is an important determinant of muscle blood flow and as such, has important influences on muscle function.

ACKNOWLEDGMENTS

This research was supported by the Anders Jahre's Fund for the Promotion of Science and by National Aeronautics and Space Administration grants 199-14-12-04 and 199-26-12-38. Attendance of *O.M.S.* at the 1994 Bigland-Ritchie conference was supported, in part, by the Muscular Dystrophy Association (USA).

REFERENCES

Aratow M, Ballard RE, Crenshaw AG, Styf J, Watenpaugh DE, Kahan NJ & Hargens AR (1993). Intramuscular pressure and electromyography as indexes of force during isokinetic exercise. *Journal of Applied Physiology* **74**, 2634-2640.

Aukland K & Reed RK (1993). Interstitial-lymphatic mechanisms in the control of extracellular fluid volume. *Physiological Reviews* **73**, 1-78.

Ballard RE, Styf J, Watenpaugh DE, Aratow M, Crenshaw AG & Hargens AR (1994a). Intramuscular pressure per unit torque varies with joint angle. *Transactions Orthopedic Research Society* **19**, 686.

Ballard RE, Watenpaugh DE, Breit GA, Murthy G, Whalen RT & Hargens AR (1994b). Intramuscular pressure measurement for assessing muscle function during locomotion. *Medicine and Science in Sports and Exercise* **26**, 5141

Baumann JU, Sutherland DH & Hänggi A (1979). Intramuscular pressure during walking: An experimental study using the wick catheter technique. *Clinical Orthopaedics and Related Research* **145**, 292-299.

Braakman R, Sipkema P & Westerhof N (1990). Two zero-flow pressure intercepts exist in autoregulating isolated skeletal muscle. *American Journal of Physiology* **258**, H1806-H1814.

Cooper DS, Pinczower E & Rice DH (1993). Thyroarytenoid intramuscular pressures. *Annals of Otology Rhinology and & Laryngology* **102**, 167-175.

Crenshaw AG, Styf JR & Hargens AR (1992). Intramuscular pressures during exercise - an evaluation of a fiber optic transducer-tipped catheter system. *European Journal of Applied Physiology and Occupational Physiology* **65**, 178-182.

Decramer M, Jiang TX & Reid MB (1990). Respiratory changes in diaphragmatic intramuscular pressure. *Journal of Applied Physiology* **68**, 35-43.

Fadnes HO, Reed R & Aukland K (1977). Interstitial fluid pressure in rats measured with a modified wick technique. *Microvascular Research* **14**, 27-36.

Folkow B, Gaskell P & Waaler BA (1970). Blood flow through limb muscles during heavy rhythmic exercise. *Acta Physiologica Scandinavica* **80**, 61-72.

Fridén J, Sfakianos PN & Hargens AR (1986). Muscle soreness and intramuscular fluid pressure - comparison between eccentric and concentric load. *Journal of Applied Physiology* **61**, 2175-2179.

Gershuni DH, Yaru NC, Hargens AR, Lieber RL, O'Hara RC & Akeson WH (1984). Ankle and knee position as a factor modifying intracompartmental pressure in the human leg. *Journal of Bone and Joint Surgery* **66A**, 1415-1420.

Gullestad L, Hallén J & Sejersted OM (1993). Variable effects of β-adrenoceptor blockade on muscle blood flow during exercise. *Acta Physiologica Scandinavica* **149**, 257-271.

Guyton AC (1963). A concept of negative interstitial pressure based on pressures in implanted perforated capsules. *Circulation Research* **12**, 399-414.

Hagert CG & Christenson JT (1990). Hyperpressure in the trapezius muscle associated with fibrosis. *Acta Orthopaedica Scandinavica* **61**, 263-265.

Hargens AR, Akeson WH, Mubarak SJ, Owen CA, Gershuni DH, Garfin SR, Lieber RL, Danzig LA, Botte MJ & Gelberman RH (1989). Tissue fluid pressures: from basic research tools to clinical applications. *Journal of Orthopaedic Research* **7**, 902-909.

Hargens AR, Millard RW, Pettersson K & Johansen K (1987). Gravitational haemodynamics and oedema prevention in the giraffe. *Nature* **329**, 59-60.

Henriksson KG & Bengtsson A (1991). Fibromyalgia - a clinical entity? *Canadian Journal of Physiology & Pharmacology* **69**, 672-677.

Hill AV (1948). The pressure developed in muscle during contraction. *Journal of Physiology (London)* **107**, 518-526.

Huisman RM, Sipkema P, Westerhof N & Elzinga G (1980). Comparison of models used to calculate left ventricular wall force. *Medical and Biological Engineering and Computing* **18**, 133-144.

Järvholm U (1990). *On Shoulder Muscle Load. An Experimental Study of Muscle Pressures, EMG and Blood Flow*, (PhD Thesis). Gothenburg: University of Gothenburg.

Järvholm U, Palmerud G, Herberts P, Högfors C & Kadefors R (1989). Intramuscular pressure and electromyography in the supraspinatus muscle at shoulder abduction. *Clinical Orthopaedics and Related Research* **245**, 102-109.

Järvholm U, Palmerud G, Karlsson D, Herberts P & Kadefors R (1991). Intramuscular pressure and electromyography in four shoulder muscles. *Journal of Orthopaedic Research* **9**, 609-619.

Järvholm U, Palmerud G, Styf J, Herberts P & Kadefors R (1988). Intramuscular pressure in the supraspinatus muscle. *Medical and Biological Engineering and Computing* **6**, 230-238.

Jensen BR (1991). *Isometric Contractions of Small Muscle Groups*, (Ph.D. Thesis). Copenhagen: Danish National Institute of Occupational Health.

Leenaerts P & Decramer M (1990). Respiratory changes in parasternal intercostal intramuscular pressure. *Journal of Applied Physiology* **68**, 868-875.

Levick JR (1991). Capillary filtration absorption balance reconsidered in light of dynamic extravascular factors. *Experimental Physiology* **76**, 825-857.

Mazella H (1954). On the pressure developed by the contraction of striated muscle and its influence on muscular circulation. *Archives Internationales de Physiologie* **62**, 334-347.

Mellander S & Albert U (1994). Effects of increased and decreased tissue pressure on haemodynamic and capillary events in cat skeletal muscle. *Journal of Physiology (London)* **481**, 163-175.

Mellander S, Maspers M, Björnberg J & Andersson LO (1987). Autoregulation of capillary pressure and filtration in cat skeletal muscle in states of normal and reduced vascular tone. *Acta Physiologica Scandinavica* **129**, 337-351.

Mubarak SJ, Owen CA, Hargens AR, Garetto LP & Akeson WH (1978). Acute compartment syndromes: diagnosis and treatment with the aid of the wick catheter. *Journal of Bone and Joint Surgery* **60A**, 1091-1095.

Nakhostine M, Styf JR, van Leuven S, Hargens AR & Gershuni DH (1993). Intramuscular pressure varies with depth. The tibialis anterior muscle studied in 12 volunteers. *Acta Orthopaedica Scandinavica* **64**, 377-381.

Parker PA, Körner LM & Kadefors R (1984). Estimation of muscle force from intramuscular pressure. *Medical and Biological Engineering and Computing* **22**, 453-457.

Pedowitz RA, Hargens AR, Mubarak SJ & Gershuni DH (1990). Modified criteria for the objective diagnosis of chronic compartment syndrome of the leg. *American Journal of Sports Medicine* **18**, 35-40.

Qvarfordt P, Christenson JT, Eklöf B, Ohlin P & Saltin B (1983). Intramuscular pressure, muscle blood flow, and skeletal muscle metabolism in chronic anterior tibial compartment syndrome. *Clinical Orthopaedics and Related Research* **179**, 284-291.

Reed RK & Wiig H (1981). Compliance of the interstitial space in rats. *Acta Physiologica Scandinavica* **113**, 297-305.

Sadamoto T, Bonde-Petersen F & Suzuki Y (1983). Skeletal muscle tension, flow, pressure and EMG during sustained isometric contractions in humans. *European Journal of Applied Physiology and Occupational Physiology* **51**, 395-408.

Scholander PF, Hargens AR & Miller SL (1968). Negative pressure in the interstitial fluid of animals. *Science* **161**, 321-328.

Sejersted OM & Hargens AR (1986). Regional pressure and nutrition of skeletal muscle during isometric contraction. In: Hargens AR (ed.), *Tissue Nutrition and Viability*, pp. 263-283. New York: Springer Verlag.

Sejersted OM, Hargens AR, Kardel KR, Blom P, Jensen Ø & Hermansen L (1984). Intramuscular fluid pressure during isometric contraction of human skeletal muscle. *Journal of Applied Physiology* **56**, 287-295.

Sejersted OM, Vøllestad NK & Medbø JI (1986). Muscle fluid and electrolyte balance during and following exercise. *Acta Physiologica Scandinavica Supplementum* **556**, 119-127.

Sheriff DD, Rowell LB & Scher AM (1993). Is rapid rise in vascular conductance at onset of dynamic exercise due to muscle pump. *American Journal of Physiology* **265**, H1227-H1234.

Sjøgaard G, Kiens B, Jørgensen K & Saltin B (1986). Intramuscular pressure, EMG and blood flow during low-level prolonged static contraction in man. *Acta Physiologica Scandinavica* **128**, 475-484.

Sjøgaard G, Savard G & Juel C (1988). Muscle blood flow during isometric activity and its relation to muscle fatigue. *European Journal of Applied Physiology and Occupational Physiology* **57**, 327-335.

Skalak R (1982). Approximate formulas for myocardial fiber stresses. *Journal of Biomechanical Engineering* **104**, 162-163.

Styf J & Körner LM (1986a). Microcapillary infusion technique for measurement of intramuscular pressure during exercise. *Clinical Orthopaedics and Related Research* **207**, 253-262.

Styf J & Körner LM (1986b). Chronic anterior-compartment syndrome of the leg - results of treatment by fasciotomy. *Journal of Bone and Joint Surgery* **68A**, 1338-1347.

Styf J & Körner LM (1987). Diagnosis of chronic anterior compartment syndrome in the lower leg. *Acta Orthopaedica Scandinavica* **58**, 139-144.

Styf J, Körner LM & Suurkula M (1987). Intramuscular pressure and muscle blood-flow during exercise in chronic compartment syndrome. *Journal of Bone and Joint Surgery* **69B**, 301-305.

Supinski GS, DiMarco AF & Altose MD (1990). Effect of diaphragmatic contraction on intramuscular pressure and vascular impedance. *Journal of Applied Physiology* **68**, 1486-1493.

Sylvest O & Hvid N (1959). Pressure measurements in human striated muscles during contraction. *Acta Rheumatologica Scandinavica* **5**, 216-222.

Vøllestad NK, Wesche J & Sejersted OM (1990). Gradual increase in leg oxygen uptake during repeated submaximal contractions in humans. *Journal of Applied Physiology* **68**, 1150-1156.

Westgaard RH (1988). Measurement and evaluation of postural load in occupational work situations. *European Journal of Applied Physiology and Occupational Physiology* **57**, 291-304.

Wiig H (1985). Comparison of methods for measurement of interstitial fluid pressure in cat skin subcutis and muscle. *American Journal of Physiology* **249**, H929-H944.

Wiig H (1990). Evaluation of methodologies for measurement of interstitial fluid pressure (P_i): physiological implications of recent P_i data. *Critical Reviews in Biomedical Engineering* **18**, 27-54.

26

TASK-DEPENDENT NATURE OF FATIGUE IN SINGLE MOTOR UNITS

B. R. Botterman

Department of Cell Biology and Neuroscience
University of Texas Southwestern Medical Center
Dallas, Texas 75235

ABSTRACT

The loss of force production during sustained activity presents the CNS a unique control problem. Different tasks stress the neuromuscular system at different sites and times, and involve different cellular mechanisms. The functional organization of muscles and their motor units has evolved to avoid fatigue processes that impair motor performance. The purpose of this brief review is to examine the fatigue properties of type-identified motor units and to speculate what these properties reveal about the organization and control of muscle.

INTRODUCTION

Most muscles contain three distinct types of motor units (slow twitch or type S; fast twitch fatigue resistant or type FR; and fast twitch, fatigable or type FF), each having different susceptibilities to fatigue. Within each motor unit type, the range of fatigabilities can also vary greatly, depending on the task that produces the fatigue. Prolonged activation of a motor unit, regardless of type, leads to a decline in force production. As Bigland-Ritchie and colleagues have stressed (Bigland-Ritchie & Woods, 1984; Bigland-Ritchie et al., 1986), the fatigue process begins immediately after the onset of activity. The site of fatigue might be due to failure anywhere along the path that ultimately results in force production, from the descending command that activates α-motoneurons of a motor pool, to interaction of the contractile proteins within single muscle fibers. During activity, whether it is brief or sustained, each component of the motor unit (motor axon, neuromuscular junction, muscle fibers) undergoes fatigue-related changes. These changes can be subtle or dramatic. The component which fails depends on the task that the motor unit has to perform. An important feature of motor units is how well their components are matched with respect to their fatigue resistance and their recovery from fatigue. Fatigue-resistant units show both electrical and mechanical endurance, while highly fatigable units show a rapid decline in electromyographic (EMG) activity and force after brief activation. This feature has important conse-

Fatigue, Edited by Simon C. Gandevia et al.
Plenum Press, New York, 1995

quences for muscle control. Different motor unit types fail for different reasons. Because of their different susceptibilities to fatigue, it may not be too surprising to find that the predominant site and mechanism for fatigue may differ between active motor units, depending on the task (for review, see Enoka & Stuart, 1992). On the other hand, given the appropriate stimulation parameters, it is likely that each type of motor unit can be made to fail at the same site(s).

What advantage is there to studying single, isolated motor units during fatigue? After all, a motor unit can be viewed as a very small muscle containing a fairly homogenous set of muscle fibers when compared to muscle fibers of other motor units within a muscle. In human studies, motor unit performance during fatiguing contractions has been largely inferred from the changing EMG pattern of the whole muscle or from the discharge patterns of single motor units. The advantage of the former approach is that the performance of the muscle can be evaluated for the task chosen by the investigator, but has the disadvantage that the contribution of individual motor units to the fatigue process cannot be evaluated. Using the latter approach, single motor unit studies in humans have provided important insights into how muscle is controlled during fatiguing contractions. For example, Bigland-Ritchie and colleagues (Bigland-Ritchie et al., 1983a; Bigland-Ritchie & Woods, 1984; Gandevia et al., 1990) have shown that the firing rates of single motor units decrease during maximal voluntary contractions. The decrease in firing rates is matched to the slowing of contractile speed associated with fatiguing contractions, thereby avoiding unnecessarily high rates that could hasten fatigue. However, these types of studies also have their limitations. It is not possible to control easily the activation history of individual units, and there is always the uncertainty of which types of units are active, the force that they produce, and the degree of fatigue that they show during the contraction. Single motor unit studies in animals overcome many of these limitations because motor units can be identified as to type and studied in isolation. By using intramuscular or intraneural microstimulation techniques, human motor units can also be isolated and their contractile properties recorded with an acceptable degree of certainty (Garnett et al., 1979; Westling et al., 1990). However, while isolated motor units can be subjected to a multitude of stimulation protocols, their environment differs from that in human studies, where many motor units are active simultaneously. Under these conditions, changes in the surrounding extracellular fluid may hasten fatigue of newly recruited motor units (Westerblad et al., 1991; see McComas & Sjøgaard, Chapter 4). The degree to which extracellular metabolites contribute to the development of fatigue in different motor unit types is unknown.

The purpose of this chapter is to examine briefly the fatigue properties of type-identified motor units and to speculate what these properties reveal about the organization and control of muscle. Moreover, the tasks used to delineate these fatigue properties will be considered, since the extent and site of motor unit fatigue depend on the task.

MOTOR UNIT TYPE AND FATIGUE

Muscle fatigue has been evaluated under a wide range of experimental protocols. In most human studies, the output level of the muscle is voluntarily held at maximum or some percentage of maximum for various lengths of time. By contrast, in most animal studies investigators have used constant or intermittent stimulation, usually of fixed frequency, to produce contractions of various duration. Fatigue is usually measured as a decline in force production relative to some predetermined rested state of the muscle. The muscle is usually stimulated under isometric conditions, and held at or near optimal length for force production during the fatiguing contraction. The type of stimulus protocol or task used to induce fatigue is often intended to produce significantly impaired function in only one or two processes involved in force production.

The most widely used fatigue test for single motor units is that of Burke and colleagues (Burke et al., 1973). Because motor unit type provided insight into the functional organization of the cat medial gastrocnemius (MG) muscle (Burke, 1981), the test has been used as a benchmark for comparison of motor units in other muscles and across species.

As applied to cat MG motor units, the test separated units into two groups. Motor units were studied under isometric conditions at optimal length for tension development and were stimulated intermittently (40 Hz for 330 ms of every second) for 2-4 min. The intermittent stimulation protocol had a minimal effect on the amplitude of the EMG, so fatigue was thought to be primarily restricted to the muscle fibers themselves. However, more recent work has shown that EMG waveform reductions in amplitude and area do occur in the more fatigable units (e.g., Sandercock et al., 1985; Gardiner & Olha, 1987; Enoka et al., 1992). The test essentially divided fast-twitch (type F) motor units into two, non-overlapping groups (type FR and FF units). As applied to other limb muscles in cat and other species (e.g., Botterman et al., 1985; Gardiner & Olha, 1987; cf. Kugelberg & Lindegren, 1979; Thomas et al., 1991a), the test produced a more continuous, albeit bimodal, distribution among fast-twitch units. While many FR units show only modest amounts of fatigue after 4 min of stimulation (Burke & Tsairis, 1974), they are much more fatigable as a group than S units. The test does not reveal the relative fatigabilities of S units (Enoka et al., 1992).

Another method of testing for the relative fatigability of motor units is by maintaining their output at some fixed percentage of their maximal force, similar to what is done when human subjects are asked to maintain force at a percentage of their maximal voluntary contraction. Motor unit force is maintained or *clamped* at a desired level through computer feedback control, by altering the stimulation rate of the motor axon (Fig. 1). When force can no longer be maintained by increasing stimulation rate, the unit's endurance time is reached. In the cat MG muscle, the endurance time of F units studied at an output level of 25% of maximal tension ranged between 20 and 2000 s. In contrast to the Burke fatigue test, where fatigue index had a bimodal distribution, endurance time was distributed continuously. If a Burke fatigue test is given 15 min later, the expected percentages of FR and FF units in MG are still found (cf. Clamann & Robinson, 1985). So although the absolute force is diminished following the first fatiguing contraction (force-clamp), units can still be separated into two groups, one representing FR units and the other group consisting of F units that have fatigue

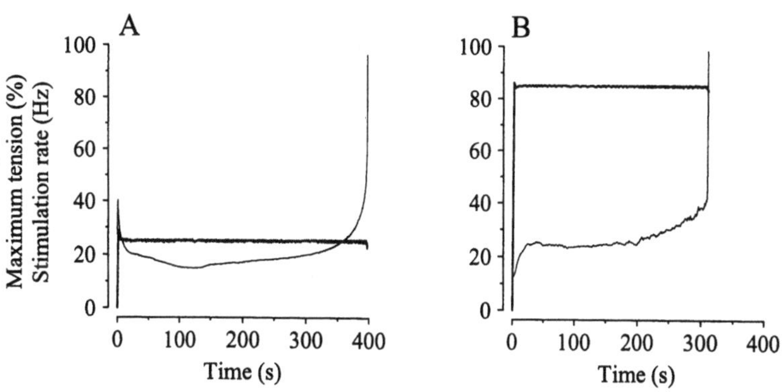

Figure 1. Typical force-clamp contractions of a FR unit in cat medial gastrocnemius (*A*) and an S unit in cat soleus (*B*). Tension was maintained by altering the stimulation rate applied to a unit's axon through computer feedback control. The endurance time of a unit was reached when target tension could no longer be maintained with a stimulation rate ≤ 100 Hz. Target levels were 25 and 85% of maximal tetanic tension of the FR and S units, respectively. In both *A* and *B*, the thicker traces represent the averaged tension maintained by feedback control, the thinner traces the instantaneous stimulation rate.

indices below 0.25. In this latter group, some FR and FI units are undoubtedly classified as FF because they were subjected to two fatigue tests.

Unlike F units, the endurance times of S units could not be established at an output level of 25% of maximum because of practical reasons; their endurance times exceeded 3000 s. When S units (MG and soleus muscles of the cat) were clamped at 85% of maximum (Fig. 1), endurance times were found between 14 and 800 s, roughly half the range found for F units at 25% of maximum. As with F units, the distribution of endurance times was continuous. The stimulation rates needed to maintain force at 85% of maximum was generally below 35 Hz for S units, in comparison to rates less than 30 Hz for F units held at 25% of maximum. At both output levels, only during the last 5% of the fatiguing contraction did stimulation rates increase dramatically in order to maintain force. Also during this period, EMG waveforms were found to change rapidly, substantially broadening in width and decreasing in peak amplitude (Botterman & Cope, 1988a; Cope et al., 1991). It appears that endurance time mainly reflects fatigue processes beyond sarcolemma excitability.

Although the exact cause for fatigue during either the Burke fatigue test or force-clamp contractions is not known, the electrical and mechanical failure observed during each test is clearly related to motor unit type. A number of studies have noted the relationship between motor unit and susceptibility to electrical failure (EMG decline) during various types of fatiguing contractions (e.g., Kugelberg & Lindegren, 1979; Clamann & Robinson, 1985; Sandercock et al., 1985; Gardiner & Olha, 1987; see also Enoka et al., 1992).

Recovery of tension after a fatiguing contraction, whether it is induced by high- or low- frequency stimulation and delivered continuously or intermittently, is also related to motor unit type. Fatigue produced by intense, high frequency (80 Hz) stimulation affects both S and F units, but more so type F units. This type of fatigue is thought to involve failure in electrical excitation during the contraction (Clamann & Robinson, 1985; Sandercock et al., 1985), and both tension and EMG recover quickly after the contraction. In contrast, less intense activity, such as continuous (10 Hz) or intermittent (40 Hz) stimulation lasting for several minutes, produces a delayed fatigue that develops 30-60 min after the initial fatiguing contraction (Jami et al., 1983; Sandercock et al., 1985). Even modest amounts of activity (several 10-s trains at 20 Hz) reduce tension significantly (Powers & Binder, 1991). This type of delayed fatigue or *low-frequency* fatigue (Edwards et al., 1977) is most evident when motor units are stimulated at lower rates (<40 Hz) and can be partly overcome with sustained activity at these rates (e.g., Jami et al., 1983). Low-frequency fatigue is most often attributed to disturbances in excitation-contraction coupling, since EMG waveform characteristics return to normal within several minutes (Edwards et al., 1977; Sandercock et al., 1985; Powers & Binder, 1991; cf. Metzger & Fitts, 1987). It occurs with more regularity in FF than FR units (Jami et al., 1983; Sandercock et al., 1985; Powers & Binder, 1991), while there is very little evidence for it in S motor units (Sandercock et al., 1985).

RATE MODULATION DURING FATIGUING CONTRACTIONS

A constant feature of sustained, maximal voluntary contractions in humans is the decrease in motor unit firing rate that occurs during the contraction. Even during sustained submaximal contractions, where the firing rates of active units may be expected to rise (e.g., Maton & Gamet, 1989), rates often either remain the same (Bigland-Ritchie et al., 1986) or decrease (Garland et al., 1994). To compensate for fatigue during these types of contractions, additional motor units are recruited to maintain target force. It has been proposed by Bigland-Ritchie and Woods (1984) that the CNS adjusts the firing rates of motor units so that the lowest possible rate produces the maximum required force. By optimizing force production in this way, premature fatigue of the muscle may be avoided. The reduction in motoneuron discharge rates under inhibitory reflex control from the exercising muscle (Woods et al., 1987) occurs in the presence of increased excitatory drive to the motor pool. An important issue is, to what extent do the contractile properties of motor units contribute to this seemingly paradoxical decrease in their firing rates, while other motor units are being recruited during the fatiguing contraction?

The relationship between the firing rate of a motor unit and its force output is strongly influenced by its prior activation history. Without knowledge of the pattern and duration of the preceding activity, predicting whether a given firing rate will produce an increase, a decrease or no change in force output is not possible (Binder-Macleod & Clamann, 1989). The force-firing rate relationship is influenced by two opposing processes; one that enhances force (potentiation), while the other one diminishes force (fatigue). The mechanisms underlying these two opposing processes are different (Krarup, 1981; Gordon et al., 1990b; Powers & Binder, 1991), and their influence on force production can last for several seconds to several hours. Another important influence on force production is the contractile slowing that accompanies fatiguing contractions (Burke et al., 1973; Bigland-Ritchie et al., 1983b; Dubose et al., 1987; Gordon et al., 1990a; Fitts, 1994).

The two mechanisms that promote decreased firing rates, potentiation and contractile slowing, often occur simultaneously and shift the force-frequency curve to the left (e.g., Powers & Binder, 1991; Thomas et al., 1991b). The presence of potentiation, a predominate factor early in sustained contractions, can still be found after fatigue processes have overwhelmed most of its effects (Kernell et al., 1975; Jami et al., 1983; Gordon et al., 1990b). Potentiation occurs chiefly in F units (FR > FF) and, depending on how its presence is measured, enhances force more in FR than FF units (Jami et al., 1983; Powers & Binder, 1991; cf. Gordon et al., 1990b). The susceptibility of motor units to contractile slowing, like potentiation, is also related to their fatigability. During the first 30 s of the Burke fatigue test, Clamann and colleagues (Dubose et al., 1987) have observed significant increases in contraction and half-relaxation times for FF units, whereas FR units responded with minor increases in half-relaxation times. In response to the initial 4-5 trains, S units actually sped up and then remained fairly constant for the remainder of the testing period. Also using the Burke fatigue test, Gordon and colleagues (1990a) found no change in the rate of tetanic force development and relaxation for S and FR units, while relaxation increased for FF units. However, during long, sustained force-clamp contractions (Fig. 1), FR units, and to a lesser extent S units, show evidence of contractile slowing (Botterman and Cope, unpublished observations).

Another mechanism that may contribute to lower firing rates is the marked hysteresis observed for the force-frequency relationship (Binder-Macleod & Clamann, 1989). When the force output of a motor unit is rapidly increased with a brief high-frequency stimulus train, higher force levels can subsequently be maintained with dramatically lower rates. This effect was seen for all unit types. As Binder-Macleod and Clamann (1989) have pointed out, this *high to low* strategy is an excellent way to generate maximal force at a given stimulation frequency. This strategy also enhances force during a fatigue test similar to that of Burke and colleagues (Bevan et al., 1992). Stimulus trains that contained two brief interpulse intervals (10 ms), followed by constant rate stimulation at a much lower rate (e.g., 25 Hz), produced more force than constant-rate trains at the same lower rate over the course of the test (360 s). Over a few seconds, *optimized* activation patterns are a means of delaying fatigue. However, once this *high to low* strategy is invoked and the optimal rate is achieved, the force produced by that rate can only be maintained by that same rate or higher rates as the motor unit becomes more fatigued (e.g., Fig. 1).

Many of the features seen in the discharge behavior of human motor units during sustained submaximal contractions (either a decrease or no change in firing rate) can also be seen in the stimulation patterns of cat motor units studied under force-clamp conditions. At an output level of 25% of maximum, F units showed an average reduction of 50% in stimulation rate during the first part of the contraction. In the more fatigue-resistant FR units, the continuous decline in stimulation rate can last in excess of 100 s (Fig. 1). This initial decline in stimulation rate was seen for all F units studied, and likely is in response to the combined processes of potentiation and contractile slowing during this phase of the contraction. On the other hand, S units clamped at either 25, 70 or 85% of maximum rarely showed decreases in stimulation rate and, for long periods, remained fairly constant (Fig. 1; Botterman & Cope, 1988a; Cope et al., 1991; Botterman et al., 1992). Therefore, for the units (S and FR) most likely to be recruited during submaximal contractions, their firing rates can remain the same (S) or *must* decrease (FR) if their force output is to remain constant. Because of the adaptive properties of motoneurons (Kernell & Monster, 1982), only small changes in the synaptic drive to S and FR motor units would be needed to maintain their force at a constant level. It is assumed, of course, that the rates used to maintain force were comparable to those seen during submaximal contractions in humans, which may be the case when differences in the contractile speed of cat and human motor units are taken into account (Kernell et al., 1983; Botterman et al., 1986; Thomas et al., 1991b).

The force-frequency relationship is strongly influenced by fatigue processes that reduce force production. Fatigued motor units require higher stimulation rates to produce the same force as they did before they were fatigued (Kernell et al., 1975; Thomas et al., 1991b). In an attempt to gain more information about the changes in the force-frequency relationship during fatiguing contractions, Botterman and colleagues (1992) varied the force of cat soleus (type S) units between three levels (50-70-90% of maximum), resulting in an averaged output of 70% of maximum (Fig. 2). Under computer feedback control, stimulation rate was altered to achieve the desired output level. During the contraction, the stimulation rate needed to maintain 50 and 70% of maximum remained fairly constant (~7 and 10 Hz, respectively) or modestly increased throughout the contraction, while the rate needed to maintain 90% of maximum steadily increased from ~30 Hz to 60 Hz, after which it increased rapidly to the experimentally set upper limit of 100 Hz. This form of *high-frequency* fatigue was found for all units. However, in 40% of the units the rate needed to maintain tension at 90% of maximum actually decreased from the initial 90% step of the contraction, between 4 and 49%, before increasing during the later stages of the contraction. Whether this phenomenon operating at such a high output level is due to potentiation or contractile slowing or another mechanism is not known. It does, however,

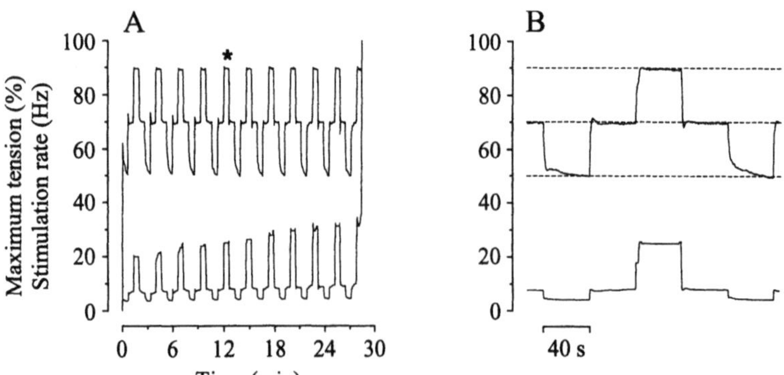

Figure 2. Typical step-clamp contraction of a S unit in cat soleus. Target levels were 50-70-90% of maximal tetanic tension of the unit, with a step duration of 40 s. Asterisk indicates one step cycle, which is shown in *B* on an expanded time scale. *B*, dashed lines indicate the three target levels. In both *A* and *B*, the upper traces represent the averaged tension maintained by feedback control, the lower traces the instantaneous stimulation rate.

probably provide another example of the need to reduce motoneuron firing rate to match the changing contractile properties of the muscle unit.

FORCE-FATIGABILITY RELATIONSHIP IN SINGLE MOTOR UNITS

Two important force-fatigability relationships can be identified in muscle, one that describes the generalized relationship for any force producing element, whether it is a muscle, motor unit or single muscle fiber, and the other that describes the relation of a motor unit's force capacity and resistance to fatigue relative to other motor units within a muscle. The first relationship can be intuitively appreciated: the greater the force exerted or work produced, the more rapid the fatigue. The second relationship, also an inverse one, reveals that the more force capacity a motor unit has, the more fatigable it is (see Thomas, Chapter 10).

That muscle endurance depends heavily on the absolute force produced by a muscle can be shown by comparing sustained isometric contractions at different output levels, or by varying the duty cycle (i.e., changing the amount of work) of intermittent isometric contractions (reviewed by Enoka & Stuart, 1992). Even when muscles are fully activated at different muscle lengths, endurance is found to be lower at the length that yields the most absolute force (e.g., Fitch & McComas, 1985). While different tasks can alter the form of the generalized force-fatigability relationship, there appears to be no structure or mechanism that when stressed invalidates its inverse nature. As Enoka and Stuart (1992) have stated, the underlying mechanisms responsible for fatigue interact in such a way as to scale with force.

Many of the important observations on the generalized force-fatigability relationship have been made using whole muscle contractions. Some of these observations have recently been confirmed for cat soleus motor units studied under force-clamp conditions. Isometric contractions of constant tension, clamped at 70 and 85% of maximum, revealed that units studied at the lower level had on average significantly longer endurance times than units at the higher level (1449 and 272 s, respectively). The mean force-time integral (proportional to work done) was also greater for units studied at the lower compared to the higher level by a factor of 4.38 to 1. Moreover, when contractions of constant force at 70% of maximum were compared with those that varied between 50-70-90% of maximum, constant-force endurance times were greater by nearly a factor of 2 (1449 *vs.* 741 s). Although both types of contractions had the same averaged output (70% of maximum), the energy cost appeared to be much higher to vary force than to maintain it at a fixed level, in agreement with earlier studies on single muscle fibers and whole muscles (Dawson et al., 1978; Loiselle & Walmsley, 1982; Duchateau & Hainaut, 1985; Bergström & Hultman, 1988).

While the relationship between the force capability (i.e., maximal tetanic tension) of a motor unit and its fatigability has been appreciated for many years (Henneman & Olson, 1965; Burke et al., 1973), the precision of the relationship has only recently been systematically studied (Botterman & Cope, 1988b; Cope et al., 1991; Botterman et al., 1992). For F motor units in individual limb muscles of the cat, there is a remarkably high correlation (Spearman rank correlation coefficent > -0.90) between tetanic tension and the endurance time of contractions clamped at 25% of maximum (Botterman & Cope, 1988b). Kugelberg and Lindegren (1979) also found a high correlation between unit tension and fatigability for F units of rat tibialis anterior muscle. In their study, fatigue was produced by stimulating units 20 times at 100 Hz twice a second for 4 min. In contrast, the strong tetanic tension-endurance time relationship seen for F units was not present for S units of the cat soleus when their force outputs were clamped at 70 or 85% of maximum (Botterman et al., 1992).

However, when force output was varied between three levels (Fig. 2), yielding an averaged output of 70% of maximum, a significant correlation between tetanic tension and endurance time was found. Thus, it appears that the tetanic tension-endurance time relationship has a task-dependent component to it. Whether this explanation is also valid for the observations of Nordstrom and Miles (1990), who found no correlation between twitch tension and fatigability in human masseter, is unclear. Fatigue resistance was evaluated by comparing a unit's twitch amplitude before and after 15 min of continuous activation at 10 Hz. The lack of a relationship may be due to the masseter's unusual physiological and histochemical features (Nordstrom & Miles, 1990; see also Miles & Nordstrom, Chapter 31). Further studies are clearly needed to clarify the effects that various tasks have on the force-fatigability relationship, as well as the influence of motor unit type, muscle, and species on the relationship.

CONCLUDING REMARKS

One of the more important tasks that the CNS performs is to engage the motor units of a muscle in a way that reduces or avoids fatigue of the muscle. From the pioneering work of Henneman and colleagues (Henneman et al., 1965a,b; Henneman & Olson, 1965) came the simplifying principle that motoneurons of a pool are recruited in a fixed order, and the speed, forcefulness and endurance of the muscle units that they supply are tightly coupled. Since most muscles contain motor units that have widely different susceptibilities to fatigue (Burke et al., 1973; Botterman & Cope, 1988a,b; Cope et al., 1991), the order in which motor units are recruited and the pattern of their activation probably strongly influences the time course of muscle fatigue (e.g., Botterman & Cope, 1988b; Bevan et al., 1992). While it is generally accepted that motor units of a muscle are recruited in order of increasing fatigability and forcefulness, few studies have actually tested for the relationship directly (e.g., Stephens & Usherwood, 1977; Zajac & Faden, 1985; Nordstrom & Miles, 1990; Yee et al., 1990; Cope & Clark, 1991). Moreover, the degree to which recruitment order and fatigability are related remains uncertain. In animal studies, investigators have used the Burke fatigue test to evaluate fatigue resistance, which is of limited utility in determining the relative fatigability of the more fatigue resistant units in a muscle. Human studies using traditional recording approaches have their limitations as well, where the activation history of recruited units is difficult to control and, by necessity, units are often restricted to the low-threshold range. Human studies that characterize the discharge properties of single motor units during voluntary contractions and subsequently record their contractile properties by stimulating their axons within the nerve fascicle may overcome many of these limitations (Westling et al., 1990; Thomas et al., 1991a,b). Because of the powerful effect that fatigue has on the threshold and discharge behavior of motoneurons (Bigland-Ritchie et al., 1986; Enoka et al., 1989), future animal and human studies are needed that examine the activity of well-characterized single motor units to various fatiguing tasks. By doing so, it may be possible to decipher the strategies that the CNS uses to manage muscle fatigue.

ACKNOWLEDGMENTS

I wish to thank my collaborators, Dr. Timothy Cope, Dr. Keith Tansey, Larry Graf and Andy Yee, for their invaluable contributions to the force-clamp studies. The work was supported by United States Public Health Service grant NS 17683. Attendance of B.R.B at the 1994 Bigland-Ritchie conference was supported, in part, by NS 17683.

REFERENCES

Bergström M & Hultman E (1988). Energy cost and fatigue during intermittent electrical stimulation of human skeletal muscle. *Journal of Applied Physiology* **65**, 1500-1505.

Bevan L, Laouris Y, Reinking RM & Stuart DG (1992). The effect of the stimulation pattern on the fatigue of single motor units in adult cats. *Journal of Physiology (London)* **449**, 85-108.

Bigland-Ritchie B, Cafarelli E & Vøllestad NK (1986). Fatigue of submaximal static contractions. *Acta Physiologica Scandinavica Supplementum* **556**, 137-148.

Bigland-Ritchie B, Johansson R, Lippold OCJ, Smith S & Woods JJ (1983a). Changes in motoneurone firing rates during sustained maximal voluntary contractions. *Journal of Physiology (London)* **340**, 335-346.

Bigland-Ritchie B, Johansson R, Lippold OCJ & Woods JJ (1983b). Contractile speed and EMG changes during fatigue of sustained maximal voluntary contractions. *Journal of Neurophysiology* **50**, 313-324.

Bigland-Ritchie B & Woods JJ (1984). Changes in muscle contractile properties and neural control during human muscular fatigue. *Muscle & Nerve* **7**, 691-699.

Binder-Macleod SA & Clamann HP (1989). Force output of cat motor units stimulated with trains of linearly varying frequency. *Journal of Neurophysiology* **61**, 208-217.

Botterman BR & Cope TC (1988a). Motor-unit stimulation patterns during fatiguing contractions of constant tension. *Journal of Neurophysiology* **60**, 1198-1214.

Botterman BR & Cope TC (1988b). Maximum tension predicts relative endurance of fast-twitch motor units in the cat. *Journal of Neurophysiology* **60**, 1215-1226.

Botterman BR, Graf LB & Tansey KE (1992). Fatigability of cat soleus motor units activated at varying force levels. *Society for Neuroscience Abstracts* **18**, 1556.

Botterman BR, Iwamoto GA & Gonyea WJ (1985). Classification of motor units in flexor carpi radialis muscle of the cat. *Journal of Neurophysiology* **54**, 676-690.

Botterman BR, Iwamoto GA & Gonyea WJ (1986). Gradation of isometric tension by different activation rates on motor units of cat flexor carpi radialis muscle. *Journal of Neurophysiology* **56**, 494-506.

Burke RE (1981). Motor units: anatomy, physiology, and functional organization. In: Brookhart JM, Mountcastle VB (eds.), Brooks VB (vol. ed.), *Handbook of Physiology, sec. 1, vol. II, pt. 1, The Nervous System: Motor Control*, pp. 345-422. Bethesda, MD: American Physiological Society.

Burke RE, Levine DN, Tsairis P & Zajac FE (1973). Physiological types and histochemical profiles of motor units of cat gastrocnemius. *Journal of Physiology (London)* **234**, 723-748.

Burke RE & Tsairis P (1974) The correlation of physiological properties with histochemical characteristics in single muscle units. *Annals New York Academy of Sciences* **228**, 145-158.

Clamann HP & Robinson AJ (1985). A comparison of electromyographic and mechanical fatigue properties in motor units of the cat hindlimb. *Brain Research* **327**, 203-219.

Cope TC & Clark BD (1991). Motor unit recruitment in the decerebrate cat: several unit properties are equally good predictors of order. *Journal of Neurophysiology* **66**, 1127-1138.

Cope TC, Webb CB, Yee AK & Botterman BR (1991). Nonuniform fatigue characteristics of slow-twitch motor units activated at a fixed percentage of their maximum tetanic tension. *Journal of Neurophysiology* **66**, 1483-1492.

Dawson MJ, Gadian DG & Wilkie DR (1978). Muscular fatigue investigated by phosphorus nuclear magnetic resonance. *Nature (London)* **274**, 861-866.

Dubose L, Schelhorn TB & Clamann HP (1987). Changes in contractile speed of cat motor units during activity. *Muscle & Nerve* **10**, 744-752.

Duchateau J & Hainaut K (1985). Electrical and mechanical failures during sustained and intermittent contractions in humans. *Journal of Applied Physiology* **58**, 942-947.

Edwards RHT, Hill DK, Jones DA & Merton PA (1977). Fatigue of long duration in human skeletal muscle after exercise. *Journal of Physiology (London)* **272**, 769-778.

Enoka RM, Robinson GA & Kossev AR (1989). Task and fatigue effects on low-threshold motor units in human hand muscle. *Journal of Neurophysiology* **62**, 1344-1359.

Enoka RM & Stuart DG (1992). Neurobiology of muscle fatigue. *Journal of Applied Physiology* **72**, 1631-1648.

Enoka RM, Trayanova N, Laouris Y, Bevan L, Reinking RM & Stuart DG (1992). Fatigue-related changes in motor unit action potentials of adult cats. *Muscle & Nerve* **14**, 138-150.

Fitch S & McComas A (1985). Influence of human muscle length on fatigue. *Journal of Physiology (London)* **362**, 205-213.

Fitts RH (1994). Cellular mechanisms of muscle fatigue. *Physiological Reviews* **74**, 49-94.

Gandevia SC, Macefield G, Burke D & McKenzie DK (1990). Voluntary activation of human motor axons in the absence of muscle afferent feedback. The control of the deafferented hand. *Brain* **113**, 1563-1581.

Gardiner PF & Olha AE (1987). Contractile and electromyographic characteristics of rat plantaris motor unit types during fatigue *in situ*. *Journal of Physiology (London)* **385**, 13-34.

Garland SJ, Enoka RM, Serrano LP & Robinson GA (1994). Behavior of motor units in human biceps brachii during a submaximal fatiguing contraction. *Journal of Applied Physiology* **76**, 2411-2419.

Garnett RAF, O'Donovan MJ, Stephens JA, Taylor A (1979). Motor unit organization of human medial gastrocnemius. *Journal of Physiology (London)* **287**, 33-43.

Gordon DA, Enoka RM, Karst GM & Stuart DG (1990a). Force development and relaxation in single motor units of adult cats during a standard fatigue test. *Journal of Physiology (London)* **421**, 583-594.

Gordon DA, Enoka RM & Stuart DG (1990b). Motor-unit force potentiation in adult cats during a standard fatigue test. *Journal of Physiology (London)* **421**, 569-582.

Henneman E & Olson CB (1965). Relations between structure and function in the design of skeletal muscles. *Journal of Neurophysiology* **28**, 581-598.

Henneman E, Somjen G & Carpenter DO (1965a). Functional significance of cell size in spinal motoneurons. *Journal of Neurophysiology* **28**, 560-580.

Henneman E, Somjen G & Carpenter DO (1965b). Excitability and inhibitability of motoneurons of different sizes. *Journal of Neurophysiology* **28**, 599-620.

Jami L, Murthy KSK, Petit J & Zytnicki D (1983). After-effects of repetitive stimulation at low frequency on fast-contracting motor units of cat muscle. *Journal of Physiology (London)* **340**, 129-143.

Kernell D, Ducati A & Sjöholm H (1975). Properties of motor units in the first deep lumbrical muscle of the cat's foot. *Brain Research* **98**, 37-55.

Kernell D, Eerbeek O & Verhey BA (1983). Relation between isometric force and stimulation rate in cat's hindlimb motor units of different twitch contraction time. *Experimental Brain Research* **50**, 220-237.

Kernell D & Monster AW (1982). Motoneurone properties and motor fatigue. An intracellular study of gastrocnemius motoneurones of the cat. *Experimental Brain Research* **46**, 197-204.

Krarup C (1981). Enhancement and diminution of mechanical tension evoked by staircase and by tetanus in rat muscle. *Journal of Physiology (London)* **311**, 355-372.

Kugelberg E & Lindegren B (1979). Transmission and contraction fatigue of rat motor units in relation to succinate dehydrogenase activity of motor unit fibres. *Journal of Physiology (London)* **288**, 285-300.

Loiselle DS & Walmsley B (1982). Cost of force development as a function of stimulus rate in rat soleus muscle. *American Journal of Physiology* **243**, C242-C246.

Maton B & Gamet D (1989). The fatigability of two agonistic muscles in human isometric voluntary submaximal contraction: and EMG study. II. Motor unit firing rate and recruitment. *European Journal of Applied Physiology & Occupational Physiology* **58**, 369-374.

Metzger JM & Fitts RH (1987). Fatigue from high- and low-frequency muscle stimulation: contractile and biochemical alterations. *Journal of Applied Physiology* **62**, 2075-2082.

Nordstrom MA & Miles TS (1990). Fatigue of single motor units in human masseter. *Journal of Applied Physiology* **68**, 26-34.

Powers RK & Binder MD (1991). Effects of low-frequency stimulation of the tension-frequency relations of fast-twitch motor units in the cat. *Journal of Neurophysiology* **66**, 905-918.

Sandercock TG, Faulkner JA, Albers JW & Abbrecht PH (1985). Single motor unit and fiber action potentials during fatigue. *Journal Applied Physiology* **58**, 1073-1079.

Stephens JA & Usherwood TP (1977). The mechanical properties of human motor units with special reference to their fatiguability and recruitment threshold. *Brain Research* **125**, 91-97.

Thomas CK, Bigland-Ritchie B & Johansson RS (1991b). Force-frequency relationships of human thenar motor units. *Journal of Neurophysiology* **65**, 1509-1516.

Thomas CK, Johansson RS & Bigland-Ritchie B (1991a). Attempts to physiologically classify human thenar motor units. *Journal of Neurophysiology* **65**, 1501-1508.

Westerblad H, Lee JA, Lännergren J & Allen DG (1991). Cellular mechanisms of fatigue in skeletal muscle. *American Journal of Physiology* **261**, C195-C209.

Westling G, Johansson RS, Thomas CK & Bigland-Ritchie B (1990). Measurement of contractile and electrical properties of single human thenar motor units in response to intraneural motor-axon stimulation. *Journal of Neurophysiology* **64**, 1331-1346.

Woods JJ, Furbush F and Bigland-Ritchie B (1987). Evidence for a fatigue-induced reflex inhibition of motoneuron firing rates. *Journal of Neurophysiology* **58**, 125-137.

Yee AK, Tansey KE & Botterman BR (1990). Relative endurance and recruitment order among pairs of fast-twitch motor units in the cat medial gastrocnemius muscle. *Society for Neuroscience Abstracts* **16**, 888.

Zajac FE & Faden JS (1985). Relationship among recruitment order, axonal conduction velocity, and muscle-unit properties of type-identified motor units in the cat plantaris muscle. *Journal of Neurophysiology* **53**, 1303-1322.

TASK-DEPENDENT FACTORS IN FATIGUE OF HUMAN VOLUNTARY CONTRACTIONS

B. Bigland-Ritchie,[1] C. L. Rice,[2] S. J. Garland,[3] and M. L. Walsh[1]

[1] Department of Pediatrics, Yale University School of Medicine
New Haven, Connecticut 06520
[2] School of Physical and Occupational Therapy and Department of Anatomy, McGill University
Montreal, Quebec, Canada
[3] Department of Physical Therapy, University of Western Ontario
London, Ontario, Canada

ABSTRACT

This chapter explores the hypothesis that fatigue is not caused uniquely by any common set of factors, but rather that the amount of stress placed on each site depends on the type of exercise from which fatigue develops. Evidence supporting this idea is presented by comparing results from various studies in which fatigue was caused by different exercise protocols. However, the way in which human endurance capacity changes with the type or intensity of the task performed suggest a unitary process. Thus, perhaps the neuromuscular system as a whole is so well adjusted that any task-related additional impairment at one site is compensated by corresponding functional improvements at others. We suggest that nature has had a long time in which to "get it right".

INTRODUCTION

Other chapters in this book provide detailed descriptions of functional changes at each of the various sites within the neuromuscular system that contribute to fatigue. Each process is treated individually. Our aim is to provide an overview of how each of these events may be influenced by the different types of exercise used to generate fatigue and how all these events combine. Our hypothesis is that the amount of stress placed on each site depends on the type of exercise, or task, from which fatigue develops. To address this question we compare results from different studies, deliberately selecting those that yield contradictory results. In fact, many of the inconsistent findings appear to be related to differences in the particular exercise protocols used to generate fatigue.

Fatigue is defined here as any reduction in a person's ability to exert force or power in response to voluntary effort, regardless of whether or not the task itself can still be performed successfully (Bigland-Ritchie & Woods, 1984; cf. Edwards, 1981; Enoka & Stuart

Fatigue, Edited by Simon C. Gandevia et al.
Plenum Press, New York, 1995

1992). Hence, this definition includes both central and peripheral factors. The term *task*, used to compare task-related changes with fatigue, includes not only the performance of different types of exercise by any given muscle group (isometric, dynamic, etc.), but also similar types of exercise performance by various muscles that have different contractile properties. We also compare muscle responses to stimulation with those in which fatigue is caused by voluntary contractions.

To establish that fatigue is always caused by a unique process or combination of processes requires that the time course of its changes match those of force or power output, both during the period of exercise and also throughout recovery. A second criterion, which is much less frequently addressed, is to establish that this match between force and the process being considered also exists under all exercise conditions. In this article, we point out some situations where these rules do not always apply. We have not tried to be comprehensive and many of the examples we use to illustrate each point were selected only because they occurred to us. In speculating as to how incompatible findings could perhaps be explained, we recognize that our views may not all be correct. But if so, we hope they may serve to stimulate new lines of research.

Different tasks are performed by using muscles in different ways (i.e., together in groups or more rarely in isolation) to generate either high or low forces that may either be sustained or repeated intermittently. In each contraction, the muscle may shorten or lengthen, or its length may remain unchanged. Energy requirements differ between isometric and dynamic contractions of comparable intensity. The rate at which muscular performance declines is determined mainly by the degree to which substrate delivery or removal is compromised by restriction of muscle blood flow. But performance must also decline, regardless of what metabolic changes take place within the muscle, if the central nervous system (CNS) fails to activate the muscle adequately. This can occur if there are changes in central motor drive, or in reflex input to spinal motoneurons, or if impulse conduction at peripheral sites becomes impaired. Many of these processes are extremely sensitive to the stimulation rate at which a muscle is excited. Thus, the relative contribution of each process to fatigue must depend on many factors, but particularly on the contraction intensity involved in each task. These factors must all be considered when trying to account for force loss during isometric contractions, but changes of muscle contractile speed must also be included when reduced power-output is considered. Hence there are good reasons to suspect that fatigue is caused by a combination of processes which occur at different sites, and the way these contribute to the final outcome is likely to depend on the task.

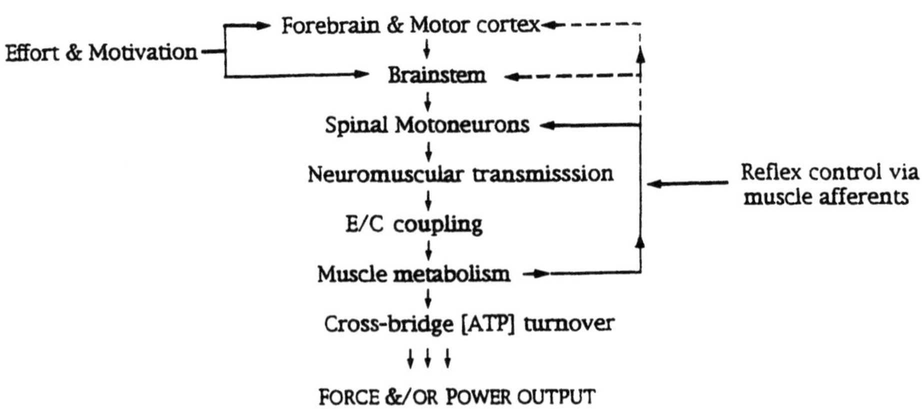

Figure 1. Sites of fatigue at which functional impairment may vary with the task.

The human body functions remarkably well under a wide variety of exercise conditions. In fact, the smooth transition of responses suggests that, when a task is changed, any increased stress on one site may be compensated by relief at others, so that the overall outcome appears to operate like a single process. This type of interaction between changes at each site could allow the overall system to make subtle adjustments that may optimize performance when different physical demands are placed upon it. At present, it is not clear to what extent the overall performance a person can sustain is limited by additive interactions between the various fatigue-related changes that occur simultaneously at different sites. Alternatively, a change at any one site may itself become rate-limiting, regardless of what occurs elsewhere.

The various sites or processes we consider in this context are shown in Fig. 1. starting with the biochemical events most immediately responsible for force generation.

CHANGES IN MUSCLE FUNCTION

Biochemical Events

Numerous studies have attempted to establish a close relation between changes in the metabolic profile of muscles and force loss during fatigue, but inconsistent findings are often evident when the data are compared. Important factors include: first, that force usually declines in a near-linear manner, but metabolic changes are usually most rapid at the onset of the exercise, their levels stabilizing thereafter. Second, the concentrations of many substances sometimes fail to change in proportion to the force exerted, particularly for forces above about 50% of the maximal voluntary contraction (MVC).

Figs. 2*A* and 2*B* show results adapted from Weiner and colleagues (1990) who measured PCr and Pi changes by NMR from tibialis anterior during (*A*) sustained MVCs each held for 4 min, and (*B*) intermittent MVCs continued for 20 min. The total force decline was similar in each case, as were the overall profiles of PCr and Pi changes for each task, despite the large differences in endurance times. Initially PCr fell rapidly, then leveled off with a tendency to return toward resting values in the second half of each contraction protocol. Thus, the relation between PCr and force loss is not a simple one. During sustained MVCs, force loss was well correlated with pH changes, but after the first 2 min of intermittent exercise pH remained stable while the force continued to fall.

These findings are compared in Fig. 2*C* with data from Vøllestad and colleagues (1988) in which fatigue was caused by intermittent isometric quadriceps contractions of low-intensity (30% MVC). Every 5 min, the subject made a brief 3 s MVC to test changes in the maximal force-generating capacity. This declined linearly throughout the exercise period by almost 50% after 30 min of exercise. However, biopsy samples taken after 5, 15, and 30 min showed that the force loss was not accompanied by significant changes in ATP or PCr. Muscle glycogen was minimally depleted, and lactate levels remained extremely low. In contrast, the metabolic changes (PCr, Pi, pH, lactate, etc.) reported during dynamic exercise are much larger than those seen during corresponding periods of isometric contractions (9.3 mM ATP/s and 4.7 mM ATP/s, respectively; Jones, 1993).

The force a muscle can generate depends ultimately on the number of actomyosin cross-bridges that form and on their turnover rates. Curtin and Edman (1994) found a progressive decline in ATP turnover rate during fatigue when frog muscle fibers were stimulated periodically for 1 s at 100 Hz. A similar decline in ATP turnover rate is also evident during fatigue of human voluntary contractions (Jones et al., 1993) in which the highest turnover rates are recorded during dynamic shortening contractions (running, cycling, etc.). While turnover rates can clearly be influenced by many factors (e.g., pH), any reduction in

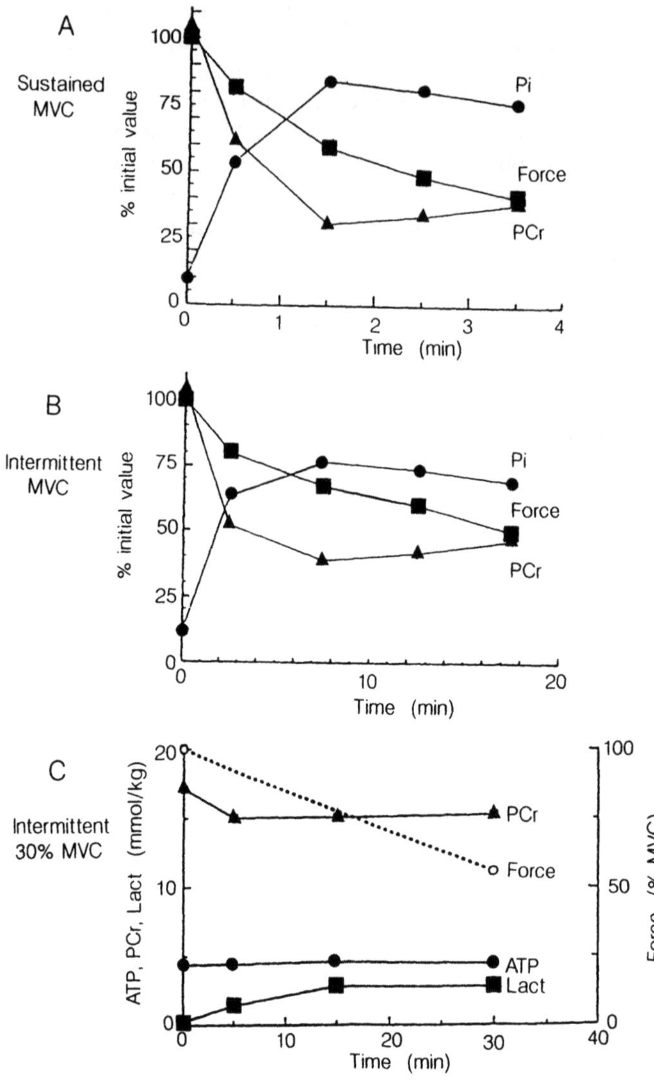

Figure 2. Metabolic changes in relation to force reduction during fatigue from (*A*) sustained and (*B*) intermittent MVCs of tibialis anterior (adapted from Weiner et al., 1990), and (*C*) intermittent 30% MVCs of quadriceps (adapted from Vøllestad et al., 1988).

ATP turnover reduces a muscle's ability to generate maximal force or power. However, since examples can be found that show no simple relation between metabolism and muscle performance, other possible causes of fatigue must also be considered.

Excitation/Contraction Coupling

In the study by Vøllestad and coworkers (1988), the MVC force decline could not be attributed to metabolic factors nor to other causes such as failure of neuromuscular transmission or of CNS motor drive. Thus, by exclusion, force loss was attributed to failure of excitation/contraction (EC) coupling. Indeed, recent evidence indicates that failure at this site may underlie and limit the rate of force or power loss in many laboratory conditions (e.g., Lännergren et al., 1993; Fitts, 1994). But since it is not yet possible to make direct measurements of functional changes at this site during human voluntary contractions, it is unwise to speculate as to how failure of EC coupling may be influenced by a change of task. Nevertheless, it is now clear that conduction in t-tubules, and hence muscle activation, becomes progressively impaired when excitation rates are high, but remains intact or can be restored if stimulus rates are low (see Lännergren et al., 1993; Fitts, 1994). Impaired conduction when excitation rates are high probably depends on the rates at which K^+, H^+, and other metabolites accumulate within the t-tubules. However, fatigue from low force contractions is more likely to reflect changes in the amount of Ca^{2+} released per impulse from the sarcoplasmic reticulum (SR) into the cytosol. Thus, in voluntary contractions, impaired conduction in t-tubules is more likely to occur when high forces are exerted.

Experiments on both intact human and isolated rat muscles show that a fatigued muscle can generate more force when it is stimulated at low rather than high excitation rates (Bigland-Ritchie et al., 1979; Jones et al., 1979). The stimulus protocol (50 Hz, turned down abruptly to 20 Hz) is similar to that used by Allen and coworkers (1989) on single frog muscle fibers to separate the influence of t-tubular conduction failure from that of reduced Ca^{2+} release per impulse.

Reduced Ca^{2+} release per impulse appears responsible for the faster rate of force decline of twitches, or response to low-frequency stimulation, compared to the corresponding changes in either maximal tetanic or voluntary contraction force (often termed *low-frequency fatigue*), a phenomenon seen readily in human muscle responses when fatigue is induced by most types of contraction protocols. Thus, the term *low frequency* refers to the stimulus rate used *to test* the response, rather than to that responsible *for inducing it*, as is sometimes assumed. However, as at many other sites, EC coupling seems to include a large safety factor, since the application of caffeine to either whole muscles (Jones & Bigland-Ritchie, 1986) or single fibers (Allen et al., 1989) causes a large release of additional Ca^{2+} with near full force recovery, even after the force from stimulation has been reduced to low values.

Role of Occluded Muscle Blood Flow

The degree to which the contractile properties of fatiguing muscle are influenced by depletion of energy substrates and accumulation also depends heavily on how muscle blood flow is affected by the exercise. In a sustained MVC, for example, muscle blood flow is totally occluded by the high intramuscular pressure generated by the contraction itself, and force usually declines by about 30-50% in 60 s. However, this rate of MVC force loss is substantially reduced if the same force-time integral is divided into segments with rest periods between, presumably because the event(s) that cause fatigue are partially reversed by reactive hyperemia in each rest period between contractions. If the same intermittent MVC protocol is repeated with the blood supply occluded by a cuff, the rate of force decline now matches that of the sustained MVC (Rice et al., 1991). If the intervals between contractions are sufficiently increased or if the contraction force is sufficiently reduced, a balance is reached between vascular supply and the increased metabolic demand of the exercise such that no fatigue ensues. The ratio between contraction time and repetition rate has been used to calculate a *fatigue threshold* (e.g., Bellemare & Grassino, 1982; McKenzie & Gandevia, 1988). Thus, this product is deemed to provide both a qualitative and quantitative estimate of the degree to which blood flow meets metabolic demands.

Blood flow becomes occluded whenever the intramuscular pressure exceeds systolic pressure. But systolic pressure has a relatively fixed value at any given time, while maximal forces of large and small muscles vary widely. Thus, it seems unlikely that this rule can be applied equally to all muscles. Estimates of the contraction intensity at which blood flow fails to meet metabolic demands are complicated by differences in muscle architecture. The site of greatest constriction is usually not in the blood vessels themselves, but due to their compression by adjacent muscle fibers or bone (Hudlicka, 1973). Intramuscular blood vessels lie mainly between and parallel to muscle fibers, so that blood flow restriction is likely to be greatest in muscles with a pennate fiber arrangement. When muscles contract their blood flow is probably unevenly distributed because intramuscular pressure is generally greatest at its center. It is interesting that the highest density of oxidative, fatigue-resistant fibers is often found in this central location, the first area to become ischemic, and that, under ischemic conditions the fatigue properties of all muscle fiber types can be characterized as fast-fatigable (Sahlin et al., 1987).

The presence or absence of ischemia has a major influence not only on changes in force generating capacity but also on muscle contractile speed, and probably also determines whether inhibitory reflexes are initiated (see below). An open question is the extent to which muscle blood flow is restricted in the types of contractions used in different types of voluntary exercise executed at various submaximal intensities and duty cycles, as well as how these estimates may vary between muscles. Early work (e.g., Edwards et al., 1972; Saltin et al., 1981) found that intramuscular pressure in quadriceps increases linearly up to 100% MVC. Arterial blood pressure also increases, but does so non-linearly and plateaus at about 50% MVC. This muscle is therefore considered to be ischemic at forces above 50% MVC (Sjøgaard et al., 1986). However, it is likely that blood flow becomes restricted and unevenly distributed when the force rises above about 10 or 20% MVC (Folkow & Halicka, 1968; Sjøgaard et al., 1986). Blood flow restriction is more severe in isometric contractions as compared with dynamic contractions. These studies confirm a general rule of thumb that blood flow is reduced at forces > 25% MVC (Barcroft & Millen, 1939).

While the importance of muscle blood flow is well recognized in principle, the profound differences in how muscles behave under aerobic and ischemic conditions, and how this behavior varies between muscles and with time, often tends to be overlooked.

Transmission at the Neuromuscular Junction and Muscle Surface Membrane

Failure of impulse transmission across the neuromuscular junction, detected as a reduction in the size of the muscle compound action potential (M-wave), has been recognized as a potential cause of fatigue for many decades. The rate at which M-waves decline depends largely on the frequency at which impulses are delivered (Krnjevic & Miladi, 1958; Bigland-Ritchie et al., 1979). Hence, if transmission failure contributes to fatigue of voluntary contractions, it would be most evident in tasks in which excitation rates are high.

M-wave size can be influenced independently of force by many factors such as changes in the dynamics of the Na^+/K^+ pump and muscle fiber conduction velocity (Hicks & McComas, 1989). Thus, there is either a wide safety margin between membrane depolarization and muscle activation, or the events that occur at the muscle fiber membrane are not always reflected accurately by M-waves recorded from the muscle surface. Indeed, Grabowski and colleagues (1972) found twitch forces of amphibian muscle fibers remained constant even when transmembrane action potentials were reduced 40% by lowering extracellular Na^+, a decline far larger than any reported during fatiguing voluntary contractions. An increased duration and amplitude reduction is also evident in M-waves from unfatigued fibers of both rat and amphibian muscles when extracellular K^+ is reduced to levels similar to those recorded during fatigue of human muscles (Hnik et al., 1972; Jones

& Bigland-Ritchie, 1986; Sjøgaard et al., 1986). Under these conditions, action potential amplitudes decrease while twitch forces increase (Lännergren & Westerblad, 1986).

One difficulty in comparing reports from different studies arises because no consensus has been reached as to how M-wave *size* should most properly be measured. Some measure peak-to-peak amplitudes and durations, others the area of either the whole M-wave or only its first negative phase with respect to the isoelectric baseline. Each method is vulnerable to some distortion due, for example, to interelectrode distance, changes in conduction velocity, etc., and other factors that become important during fatigue. When Thomas and colleagues (1989) measured the same M-wave data both ways they found peak-to-peak amplitudes and durations always changed more than did the area of the first half-waveform (see also Bigland-Ritchie et al., 1982). The importance of reporting changes in *both* M-wave amplitude and area is also emphasized by Milner-Brown and Miller (1986).

While it is easy to demonstrate that M-waves become reduced in size when muscles are stimulated at rates above 10-20 Hz, no overall consensus has yet been reached about changes in M-waves when measured during fatigue of human voluntary contractions. For example, despite repeated efforts, Bigland-Ritchie and colleagues (1979, 1982) and Thomas and colleagues (1989) found no decline in M-wave size during fatigue from sustained maximal contractions of adductor pollicis (AP), tibialis anterior (TA), or during submaximal contractions of the human diaphragm (Bellemare & Bigland-Ritchie, 1987; McKenzie et al., 1992). Similar findings have also been reported by Garland and coworkers (1988), Zijdewind and Kernell (1994) and others. However, these reports differ from others in the literature, including: Stephens and Taylor (1972), and Bellemare and Garzaniti (1988) in sustained MVCs of FDI and AP, respectively; Fuglevand and coworkers (1993) in sustained submaximal FDI contractions of various intensities; and Aldrich (1987) and others for diaphragm. Thomas and coworkers (1989) found near-linear rates of force loss in TA and FDI during sustained MVCs held for 5 min. While M-wave areas recorded from TA did not change throughout the contraction, those from FDI did decline, but only during the first 2 min of each contraction, remaining stable thereafter, while force continued to fall.

The extent to which impaired transmission may contribute to fatigue of voluntary contractions is likely to depend on the motoneuron discharge rates and hence the contraction intensity. Excitation rates recorded during voluntary activity are generally lower than those used for stimulation and may decline with time (Bigland-Ritchie et al., 1983a,b). Moreover, the maximal firing rates delivered simultaneously to different units vary widely. A few units may fire as fast as 50 Hz, while others never seem to exceed 10-15 Hz (DeLuca et al., 1982; Bellemare et al,. 1983). It is clearly beneficial if the CNS can keep excitation rates as low as possible, without compromising force, since this reduces the likelihood of t-tubular blockage, the rise of extracellular K^+, and depletion of metabolic substrates, as well as transmission failure at the neuromuscular junction. This goal is best achieved if the impulse rates delivered to each unit vary in relation to its contractile properties, which seems to be the case. This goal cannot be achieved by nerve stimulation whatever fixed frequency is chosen. However, it would be surprising if impaired transmission does not occur during fatigue from some type of exercise (e.g., Fuglevand et al., 1993) since safety margins seldom exceed those sometimes called upon in one task or another.

The comparison of M-wave behavior in different muscles (Thomas et al., 1989) suggests that perhaps there is a real difference between the EMG responses of FDI and other muscles. Fuglevand and coworkers (1993) found that M-wave amplitudes recorded from FDI also decline during fatigue from sustained submaximal contractions. Surprisingly, the rates at which they declined were faster for low compared to high forces. However, these low force contractions could be sustained much longer. This unexpected behavior may perhaps be explained, in part, by recent reports from Zijdewind and colleagues. (1995). They found that, during fatiguing submaximal contractions of FDI, the EMG responses may either increase or decrease depending on the precise position of the recording electrodes. In submaximal contractions, the amount of motor drive directed to each part of this muscle is unevenly distributed and is strongly task related. Thus, the CNS can achieve a common goal (e.g., a 50% MVC abduction of the index finger) by recruiting different unit subpopulations that can be used in a variety of task-specific ways.

Relation between Muscle Force and Speed Changes

It is now clear that the force a muscle can exert does not necessarily change in parallel with its contractile speed. The relation between force and contractile speed depends on the type of exercise from which fatigue develops as well as on the muscle fiber types recruited. Fig. 3*A* shows changes of contractile speed recorded during an MVC of AP sustained for 60 s, in which force typically declines by about 30-50%. In this case the maximal rate of relaxation (MRR) declines almost in parallel with the force, but maximal contraction rates (MCR) are much less affected (Bigland-Ritchie et al., 1983b). MRR changes are similar whether measured from voluntary or stimulated contractions. However, when the same MVC force-loss is caused by low-force quadriceps contractions repeating for 30 min (Fig. 3*B*), contraction rates get faster. Twitch contraction and half-relaxation times both get significantly shorter while their maximal rates increase (Bigland-Ritchie et al., unpublishsed data;

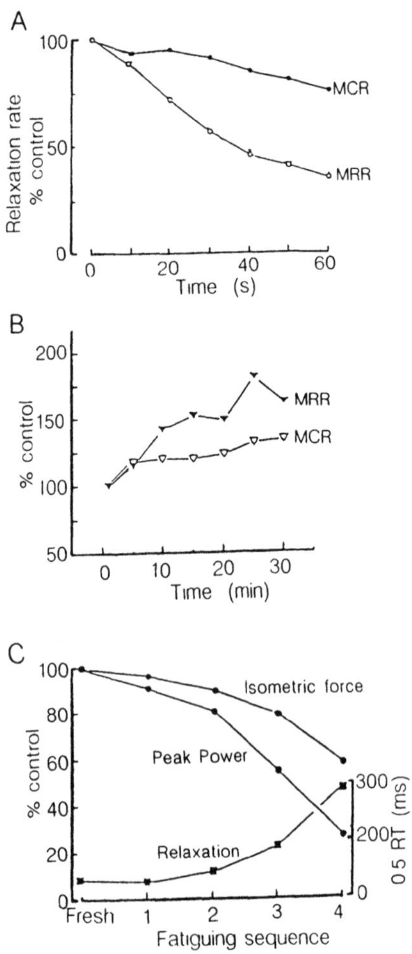

Figure 3. Whole muscle contractile speed changes during: *A*, isometric MVC of adductor pollicis (adapted from Bigland-Ritchie et al., 1983b); *B*, repeated 30% MVCs of quadriceps (Bigland-Ritchie unpublished data); *C*, dynamic leg extension (adapted from Jones, 1993).

Rice et al., 1992). In contrast, Fig. 3*C* illustrates the contribution of contractile slowing to the greater reduction in power output (force x velocity) compared to isometric force, which occurs during fatigue from dynamic exercise (Jones, 1993).

Changes of contractile speed are also influenced by the fiber type composition of the particular muscles or motor units recruited in each task. In experiments on cats, Botterman and Cope (1988), Dubose and colleagues (1987) and others found that fast-fatigable (FF) units became markedly slower when fatigued, while the speed of fatigue resistant (FR) and slow (S) units either remained unaltered or got faster (Botterman, Chapter 26). If different fiber types are recruited sequentially, the contractile speeds recorded from whole muscles, together with their rate of change during fatigue, must vary with the contraction intensity of the particular exercise involved.

Differences between Human and Animal Models

The contractile properties of human muscle cannot always be closely predicted from knowledge of their fiber-type composition determined histochemically. For example, the human biceps brachii muscle usually has about equal numbers of Type I (slow) and Type II (fast) units while both AP and soleus are reported to be about 70% Type I (Johnson et al., 1973). Thus, one would expect corresponding differences in their contractile speeds. However, Bellemare and coworkers (1983) and Round and colleagues (1984) both found little difference between either the contraction or half-relaxation times measured for biceps brachii and adductor pollicis; while those of the soleus were 50% longer. These findings raised the question of how accurately histochemical staining relates to muscle contractile performance measured physiologically (Rice et al., 1988) and also how far contractile properties of motor unit types in animals apply to human muscle. While large interspecies differences are known, it has not been possible to obtain comparable measurements from humans.

The contractile properties of human motor units can be examined by intraneural stimulation of single motor axons (Westling et al., 1990). This method has the advantage over spike-triggered averaging or stimulation of intramuscular nerve branches in that a wide range of stimulus rates can be delivered, including those sufficiently high to elicit maximal tetanic forces. Thus, force-frequency relations can be measured before and after fatigue induced by various excitation patterns, including the standard Burke fatigue test and those recorded during voluntary activity. Moreover, the conditions under which changes in motor unit force and speed are measured in these experiments are identical with those in most animal studies. However, the values measured in both animal and human stimulation studies neglect any influence on force that may result from the simultaneous activity of other surrounding units (Binder-Mcleod & Guerin, 1990). Nevertheless, Thomas and coworkers (1991b) were not able to classify human thenar motor units into three types using the criteria normally used in most animal studies. The most noticeable differences were that: 1) while the contractile speeds of different motor units varied widely, the individual values did not correlate with either the unit's force (size) or its fatigue index, measured after a standard 2 min Burke fatigue test; and 2) their fatigue indices showed no sign of a bimodal distribution, but were distributed continuously. In fact, human thenar motor units were remarkably fatigue-resistant, with no units responding like FF units in animal muscles (i.e., no unit lost more than 75% of its initial force). If FF units are present, we would expect to find them since motor axons are encountered randomly in microneurograghic procedures. If FF units are lacking in human intrinsic hand muscles, they may be rare in all human muscles. Alternatively, their absence may be due to an evolutionary adaptation peculiar to human intrinsic hand muscles. This idea seems reasonable since hands are used for precision grip and other tasks not performed by other animals.

Force-frequency relations measured for individual units before and after fatigue showed another puzzling feature; namely, that only about 1/3 of all units required an increase in excitation rates to elicit the same force after fatigue. Equal numbers of other units needed either no change in rates or lower rates compared to those required before fatigue (Thomas et al., 1991a). At first sight, this presents a major problem for the CNS if it is to regulate motoneuron firing rates so that a constant load, such as a heavy suitcase, can continue to be supported as fatigue develops.

Force-Velocity Relations in Dynamic Contractions

The term force and power are often used interchangably to describe fatigue, thus assuming these two parameters change in parallel. However, it is clear that force and power can follow quite different time courses as fatigue progresses (see Fig. 3*C*). Curtin and Edman (1994) found that isometric force of isolated frog muscle fibers was reduced by 26%, compared to only a 10% reduction in maximal shortening velocity when 1 s tetani were delivered every 2, 3, or 5 min. During muscle lengthening, force-velocity relations showed that the ability to resist stretch was well preserved. The fibers also became less stiff, indicating fewer attached cross-bridges in fatigue. These changes in V_{max} could also be induced by intracellular acidification with CO_2 although this caused smaller changes in isometric force than those seen during fatigue. They conclude that changes in isometric force may be due to an increase in both [Pi] and [H^+], while V_{max} is affected by [H^+] alone.

For human quadriceps or AP muscles, Jones (1993) found contractile slowing under ischemic conditions is greater when fatigue is induced by isokinetic shortening, compared to that from sustained or intermittent isometric contractions. For example, they found that when the isometric force of AP was reduced by 25% by repeated bouts of 50 Hz supramaximal nerve stimulation, relaxation times increased; thus, power was reduced by about 50%. Under other conditions, isometric force fell by only 10%, while V_{max} values were reduced by about 50% thus dissociating any parallel change in force and power during fatigue. Hence, tests for fatigue that depend on examining only changes in either MVC or stimulated forces are misleading if applied to changes in human performance capacity during dynamic exercise. These authors also provide evidence that voluntary effort can fully activate human muscles during shortening as well as isometric contractions (see also Gandevia et al., Chapter 20).

RELATION BETWEEN MUSCLE PROPERTIES AND MOTONEURON FIRING RATES

Role of Excitation Frequency

The rates of motor unit discharge vary widely from one unit to another, with maximal rates ranging from 10-50 Hz, distributed relatively normally between units. Mean rates recorded during brief MVCs of different muscles generally vary between about 25-30 Hz, sometimes showing consistent variations between subjects. For soleus, they did not exceed about 15 Hz. These values vary roughly in proportion to differences in each muscle's contractile speed (Bellemare et al., 1983; Bigland-Ritchie et al., 1983b, 1991). A similar relation between excitation rates and speed is also evident within a single motor unit population (Thomas et al., 1991b).

Since the force from maximal voluntary contraction usually matches that from supramaximal nerve stimulation, it is generally concluded that voluntary effort can recruit all units and drive them to generate maximal tetanic force, and also that the rates needed to elicit maximal force are higher for fast units. Conversely, if firing rates exceeded the minimum needed to generate maximal force, the slope of the EMG/force relation would increase sharply at high forces, something that is not seen (Woods & Bigland-Ritchie, 1983). Thus, it is reasonable to suppose that, even in the absence of fatigue, the CNS adjusts spinal motoneuron firing rates to match differences between the contractile properties of the individual motor units they supply.

When an MVC is prolonged, relaxation rates slow and this reduces the excitation rate needed to achieve maximal force. Thus, a decline in firing rates (or surface EMG) recorded during a prolonged MVC does not necessarily imply that the motor drive to a muscle now fails to activate it fully (i.e., it is not necessarily a sign that *central fatigue* is present). The balance between force and excitation rates is well illustrated in experiments that show that both the typical force and EMG decline, seen in a sustained MVC, can be mimicked if the motor nerve is stimulated initially at high rates and then these rates are progressively reduced (Bigland-Ritchie et al., 1979; Jones et al., 1979). The precise excitation rates are critical. If they are reduced either too fast or not fast enough, force falls below its optimal value. One benefit of keeping motoneuron firing rates as low as possible relates to the frequency-dependence of transmission failure and other peripheral causes of fatigue.

At present, it is not clear whether the firing rates of all motor units decline equally with time during a sustained MVC or whether the fall in mean rates is caused by a more precipitous decline for those units that fire initially at the highest rates (Fuglevand et al., 1994). We know that, in sustained submaximal isometric contractions, the change in motor unit discharge rate during the fatigue task was related to the initial discharge rates (Garland et al., 1994). That is, those units that increased discharge rate were among those motor units with low initial discharge rates (<10 Hz) and no motor units with high initial discharge rate (> 20 Hz) displayed an increase in discharge rate during the fatiguing contraction. Furthermore, in that experiment, the discharge rate of most motor units that were active from the beginning of the contraction declined, whereas the discharge rates of most newly recruited units were either constant or increased slightly. However, when Garland and coworkers utilized a fatigue task incorporating dynamic contractions, only a minority of motor units exhibited a significant decline in firing rate despite the similarity in duty cycle between the two fatigue paradigms. Thus the decline in discharge rate of motor units was not uniform across all units and diminished with a dynamic fatigue task.

Intermittent contractions, executed at either maximal or submaximal forces with blood supply intact, reduce a muscle's capacity to generate force and endurance times are inversely related to contraction intensity and duty cycle. When a subject makes repeated target force contractions of either 30% or 50% MVC (duty cycle, 0.6) the integrated EMG (IEMG) increases progressively until, at the limit of endurance, it matches that recorded in an MVC (Bigland-Ritchie et al., 1986a,b). Muscle speed did not get slower during the 30% MVC contraction protocol. While motor unit firing rates recorded during the target-force contractions increased gradually with fatigue, there was no decline in those recorded during brief test MVCs made every 5 min throughout the exercise period (Bigland-Ritchie & Vøllestad, unpublished data). Similarly, repeated MVCs, each held for only 10 s (duty cycle, 0.5), produced only small changes in IEMG, firing rate, and contractile speed. But when these were repeated with the blood supply occluded, all indices of fatigue changed rapidly, and were not significantly different from those of a sustained MVC held for an equal total integrated force x time period (Rice et al., 1991). Garland and McComas (1990) found that the IEMG of soleus also declined when fatigue was caused by periodic stimulation at 15 Hz under ischemic conditions. Thus, a decline in motoneuron firing rates with fatigue seems only to occur in the presence of contractile slowing, and when metabolic factors are involved. Hence, both contractile speed and firing rate changes depend on whether or not sufficient rest periods are allowed between contractions to avoid ischemia, and to replenish or remove the metabolic consequences of the previous contractions. However, again it makes good sense that a common factor should control both these parameters.

In human voluntary contractions, the match between changes in motoneuron firing rate and contractile speed (mainly relaxation rate), although close, is not always exact. Thus it is possible that the large differences between endurance times for different individuals and their rates of MVC force decline may be due, in part, to variations in the sensitivity with which their CNS adjusts motor unit firing rates to changes in contractile speeds. The difficulties presented to the CNS become even greater when contractile speed changes due to fatigue are compounded by speed changes imposed by other factors, such as changes in muscle length or its velocity of shortening.

Firing Rates and Muscle Afferent Feedback

It is not yet clear how changes in motoneuron firing rates are regulated during fatigue. If the excitatory input to spinal motoneurons remains constant, then firing rates are probably dictated by changes in the time course of their excitability (Kernell & Monster, 1982). Hagbarth and colleagues (1986) suggest that the decline in discharge rates with fatigue results mainly from withdrawal of Ia facilitation from muscle spindles; others favor an inhibitory reflex carried by groups III and IV muscle afferents initiated by metabolic changes within the fatigued muscle (Bigland-Ritchie et al., 1986c; Woods et al., 1987; Garland & McComas, 1990). Evidence for all three mechanisms is persuasive and it seems probable that all participate in different ways.

Stimulation *vs* Voluntary Contraction

For studies of fatigue in intact or isolated animal preparations, contractions are elicited by nerve or muscle stimulation. Human muscles are often stimulated in experiments in which the degree of muscle activation must be controlled, free from fluctuations imposed by spinal reflexes or changes in a subject's effort. They are also stimulated for training purposes and in studies designed to restore function to paralyzed muscles of paraplegics, etc. Clearly, the rates and patterns of impulse delivery have a major impact on the outcome.

Although electrical stimulation has been advocated to increase muscle strength, Hainaut and Duchateau (1989) found that individuals trained with submaximal 100 Hz electrical stimulation became more fatigable and showed less response to training (when tested with MVCs) than those subjects trained with submaximal voluntary contractions. This provides an example of the well-known phenomenon of task specificity of training (Duchateau & Hainaut, 1984) and illustrates the potential for lack of transfer of training from electrical stimulation to voluntary contractions. Binder-MacLeod and colleagues have demonstrated a reduction in the fatigability of quadriceps femoris muscle when stimulation parameters were adjusted to mimic voluntary motoneuron discharge (Binder-MacLeod & Guerin, 1990; Binder-MacLeod & Barker, 1991). Elucidating the best electrical stimulation protocol, however, has obvious implications for the field of functional electrical stimulation that may be used in the rehabilitation of individuals with spinal cord injury, for example.

Electrical stimulation may fail to address the task-dependent nature of muscle fatigue. For instance, the task-specific training response is affected by the motor command rather than just the type of muscle contraction (Behm & Sale, 1993); electrical stimulation would fail to address the effectiveness of the motor command. In addition, motor unit recruitment and rotation of motor units have been implicated in fatiguing contractions (for review, Enoka & Stuart, 1992). Electrical stimulation excites different motor units from voluntary contractions (Trimble & Enoka, 1990) and may not afford the possibility for rotation of motor unit activity within the motoneuron pool.

CNS STRATEGIES: CENTRAL FACTORS

Motivation, Force, and Effort Sensations

A subject's ability to judge the intensity of the force exerted is thought to be mediated by an efferent copy to the cortex of the motor command that activates the spinal motoneuron pool. The sense of force or effort increases with fatigue, for example, during a submaximal contraction held at constant force. This over-estimation probably results from the concurrent increase in neural drive (EMG) required to compensate for contractile failure in units already active (Jones & Hunter, 1983; Gantchev et al., 1989; see Jones, Chapter 22).

The sense of effort is a major factor influencing motivation. As fatigue ensues, the increased sense of effort tends to make subjects think they can no longer drive their muscles maximally or continue to exercise at all. The ability to activate a muscle during high force contractions or to prolong endurance time is clearly improved by practice. While this process is attributed mainly to peripheral events, the greater willingness of subjects to put up with discomfort and push themselves harder is important. This willingness is clearly enhanced if the event takes place in a competitive environment. Hence, sports records are more often beaten in competitions, in which participants make a greater effort to get *psyched up*, than in practice sessions. The problem of motivation is also vital in laboratory tests, particularly for protocols that continue until exhaustion or where fatigue is measured by endurance times. In the laboratory, this atmosphere can be mimicked if subjects are encouraged to compete against their own previous performance or work in groups, vying with each other as to who can do best.

Voluntary Muscle Activation

Methods for assessing central motor drive and for measuring the magnitude of central and peripheral fatigue are described elsewhere (see Gandevia, 1993; Gandevia et al., Chapter 20). Here we consider only whether the ability of the CNS to activate muscles during fatigue varies between tasks.

In 1954, Merton showed that the force of a maximal voluntary effort matches that elicited by supramaximal tetanic stimulation, and also that no additional force can be elicited when single shocks are applied at any time during a 3 min sustained MVC (*twitch occlusion*). Similar results have since been shown for many other muscles, including quadriceps, biceps brachii, TA, and FDI. Thus, the CNS is generally thought to be capable of recruiting all motor units and continuing to excite them at rates sufficient to elicit near-maximal force, despite the presence of fatigue (see McKenzie & Gandevia, 1991; Rice et al., 1991; cf. Gandevia et al., Chapter 20). It is not surprising that, in the early stages of prolonged exercise, muscles can be near-maximally activated, but later this ability often declines. However, most studies report that force loss due to reduced central motor drive is small (< 10%) compared to that due to other causes.

However, when either soleus or the diaphragm were fatigued by repeated submaximal contractions, central factors appeared to play a larger role (Bigland-Ritchie et al., 1986a; Bellemare & Bigland-Ritchie, 1987). Unlike quadriceps responses, twitches elicited from the diaphragm did not become occluded, and for both muscles the EMG associated with brief test MVCs showed a marked decline with time. It is unlikely that the difference between the behavior of quadriceps and that of either the diaphragm or soleus can be attributed to lack of motivation because most experiments were performed by the same subjects. However, later experiments showed that the amount of *central fatigue* evident in the diaphragm is task-dependent (see below). Central fatigue is clearly evident when a subject is asked to contract their diaphragm against a fatiguing inspiratory load (opposed by the weaker intercostal muscles), but almost absent when it is fatigued by expulsive contractions, in which maximal diaphragmatic force is less than that of the abdominal muscles which oppose it (McKenzie et al., 1992).

Role of Synergists

Experiments on limb muscles are often hampered by an unavoidable recruitment of unfatigued synergists, particularly when fatigue is elicited by intermittent voluntary contractions (cited in Bigland-Ritchie et al., 1986a, for contractions of AP). Similarly, in the respiratory system, when one muscle becomes fatigued the motor drive is always switched to elicit force from other less fatigued synergists whenever possible (Yan et. al., 1993). This strategy is evident in the *paradoxical breathing* often seen in patients in respiratory failure in which periods of contractions during inspiration alternate between the diaphragm and intercostal muscles. Similarly, Johnson and coworkers (1993) showed that the relative contribution of the diaphragm to total respiratory motor drive is progressively reduced during fatigue to exhaustion from dynamic exercise performed at 95% and 85% VO_2max. During the first 5-10 min of exercise, the decline in transdiaphragmatic pressure (Pdi) was proportional to that of esophageal pressure (Pe), after which ∫Pdi /min stayed constant, while ∫Pe/min, Ve, and inspiratory flow rates continued to rise. In marathon runners, Chevrolet and colleagues (1993) found that Pdi declined but with no change in Pe, maximal voluntary ventilation (MVV), or lung mechanics, indicating that other respiratory muscles play an the increasing role in breathing when the diaphragm becomes fatigued.

Possible Overall CNS Responses

The intensity and duration of exercise that can be maintained is determined mainly by the contractile properties and fatigue resistance of the motor units in the particular muscles involved. This can also be influenced by interactions between different muscles. Some subjects have difficulty in excluding co-contractions by muscle synergists during fatigue, even of those not closely related to the prime mover. Howard and Enoka (1991) provide evidence that full muscle activation is possible when an isometric MVC is performed by one leg, but not when both legs are involved, indicating that the amount of muscle mass may be important. While it is generally assumed that the CNS is organized to control respiratory and limb muscles independently, and that limb exercise is not limited by the capacity for respiration, recent evidence suggests that this may not be so under some exercise conditions (Boutellier & Piwko, 1992).

The concept that respiration does not limit exercise relies mainly on the fact that voluntary efforts can achieve maximal rates of ventilation that are significantly higher than the maximal rates elicited involuntarily by any kind of normal exercise. For example, when cycling to exhaustion at 80% of the maximal work capacity, ventilation was found to be only 50% of that recorded during 12 s periods of MVV. Thus, the chemical drive from exercise fails to activate respiratory muscles maximally; in fact, they always seem to retain a plentiful reserve (Mador et al., 1993). This design may be essential for preservation of life in that it protects respiratory muscles from excessive fatigue. For example, at the end of a race no harm is done if a runner's skeletal muscles are exhausted to the point where the runner can no longer stand, but their respiratory muscles must continue to function well above resting levels for some time against a markedly increased CO_2 load.

Hypoxia

Some data presented from the Everest II experiments showed interesting interactions between fatigue and chronic hypoxia. During chronic hypoxia 1) the CNS was usually capable of fully activating a small muscle mass (triceps surae) and of keeping it maximal activated when fatigued by intermittent MVCs for 180 s; and 2) in contrast, during dynamic exercise performed by both legs simultaneously (cycling), maximal power output was markedly depressed; and 3) under these hypoxic conditions, there was no similar depression of power output by respiratory muscles, since maximal exercise ventilation rates were not reduced at extreme altitudes. Why should hypoxia have such a profound effect on limb muscle performance, while the work capacity of respiratory muscles seemed unaffected, when both muscle groups were performing similar types of dynamic exercise? These observations seemed incompatible with accepted theory. To explain them, Bigland-Ritchie and Vøllestad (1988) postulated that it is essential for survival that somehow respiratory muscles must avoid the extremes of fatigue experienced by limb muscles; they suggested that this could be achieved if CNS strategy involves some kind of reciprocal inhibition between the motor drive to limb and respiratory muscles, with that from the respiratory system dominating. According to this scheme, the motor drive to limb muscles can achieve and retain maximal muscle activation, provided this does not increase the metabolic demand above that which the respiratory muscles can deliver. However, if either the muscle mass is too large or the exercise sufficiently demanding, such that metabolism exceeds the capacity of the oxygen delivery system, a balance between them is restored by an automatic reduction of motor drive to limb muscles. In this case, the limb muscles can no longer be fully activated and *central fatigue* ensues. They suggested that this may be brought about by reflex inhibition arising mainly from the fatigued diaphragm, but it could equally well be caused by reciprocal inhibition between the two systems acting centrally within the brainstem.

Direct evidence in support of this hypothesis comes from experiments by Kayser and coworkers (1993) who found that the rate of IEMG increase, measured during fatigue of arm muscles, was the same at sea level and at high altitude. A similar IEMG increase was recorded from leg muscles while cycling at sea level. Thus, in all three cases the IEMG behaved in the expected way, indicating that additional motor units were recruited and those already active were driven harder as fatigue developed. However, when cycling at high altitude, the IEMG responses of leg muscles did not change throughout the exercise. The usual increase in motor drive which compensates for contractile failure did not occur. These authors also confirmed that chronic hypoxia reduces the normal increase in muscle lactate and other fatigue-related compounds. Thus, fatigue developed under these conditions may have been caused more by a reduced motor drive than by peripheral factors. This concept also fits the observations that the IEMG of the diaphragm only increased marginally before stabilizing when this muscle was fatigued by repeated contractions at 50% MVC (Bellemare & Bigland-Ritchie, 1987). Furthermore, Walsh and Banister (1995) showed that, when cycling to exhaustion in acute normobaric hypoxia, both the maximal ventilation rate and respiratory delivery of oxygen per unit work rate were similar to the normoxic values, although acute hypoxia markedly reduced both VO_2max and the maximal work rate of leg muscles.

Taken together, these observations support the concept that the motor drive to limb muscles is reduced whenever the metabolic demand of skeletal muscle exceeds that which the respiratory muscles can supply. Consequently, leg muscles accumulate less lactate, $[H^+]$, and other signs of fatigue than those that accumulate in smaller muscles kept fully activated for a similar period. It is logical to conclude that this neural design has evolved because small muscles can never impose dangerously high metabolic demands on the respiratory system. Indeed, such a neural interaction of this kind between the skeletal and respiratory motor systems would provide a mechanism whereby respiratory muscles are protected from extremes of fatigue commonly experienced by leg muscles. If this operates under chronic hyperbaric hypoxic conditions, it seems likely that similar interactions are also available under other stressful conditions which are more commonly encountered.

ALTERNATIVE EXPLANATIONS

Many reasons have been presented to suggest why muscles are likely to respond differently when fatigued by different tasks, and why the responses of different muscles may vary from each other, even when they perform the same tasks. However, we must also recognize that the human body does, in fact, perform remarkably well when challenged by tasks that require a wide range of strength and speed. Indeed, the endurance time for which any task can be sustained varies hyperbolically in relation to either the force or power exerted, whether the task be a sustained (Romert, 1960) or intermittent isometric contraction (Monod & Scherrer, 1965), or power-output measure in dynamic exercise (Moritani et al., 1981; Poole et al., 1990).

At first sight, the smooth transitions of fatigue-related responses seen when the type of exercise is varied suggests that fatigue depends on only one single process. For example, Walsh and Bigland-Ritchie (1995) present a model of fatigue, based on a single process, that generates the well-documented hyperbolic relation between power and fatigue. This simple model accurately predicts exhaustion times and training influences for a variety of exercise conditions. It assumes that fatigue can either accumulate or partially recover during exercise, depending on the intensity. Initially, the hyperbolic relations calculated in this model were based on only the amount of K^+ release per impulse, measured in experiments reported by Sjøgaard (1991), but similar curves have also been constructed based on other factors, such as lactate accumulation, pH changes, etc. However, current data also suggest that the single limiting factor that dominates the rate at which fatigue develops in all types of exercise may prove to be failure of EC coupling, regardless of whatever functional changes may occur in parallel when measured individually at other sites during any particular type of exercise. Current knowledge about EC coupling failure and its influence on fatigue is accumulating rapidly, but is not yet sufficient to be able to judge how its magnitude may be affected by changes in the task (see Stephenson et al., Chapter 2).

It seems likely that the changes in performance during different tasks do depend on many factors, and that nature's experiment, extending over several million years, has got the balance between them just right. This concept would require that changing the type of exercise shifts the major stress from one site to another, but in such a way that, in so doing, the added stress on one site results in improved function at another. In that case, the net outcome would cause only small but progressive change in overall performance that simulate a single process, regardless of which type of exercise is undertaken.

ACKNOWLEDGMENTS

The authors' laboratories have been supported by the Natural Sciences and Engineering Research Council (NSERC) of Canada (*S.J.G.* and *C.L.R.*), and United States Public Health Service grants NS 14756 and HL 30026 (*B.B-R.* and *M.L.W.*). Attendance of the authors at the 1994 Bigland-Ritchie conference was supported by NSERC (*S.J.G.* and *C.L.R.*), McGill University (*C.L.R.*), NS 14756 (*B.B-R.* and *M.L.W.*), and the University of Miami (*B.B-R.*, *S.J.G.* and *C.L.R.*).

REFERENCES

Aldrich TK (1987). Transmission failure of the rabbit diaphragm. *Respiration Physiology* **69**, 307-319.

Allen DG, Lee JA & Westerblad H (1989). Intracellular calcium and tension during fatigue in isolated single muscle fibers from *Xenopus laevis*. *Journal of Physiology (London)* **415**, 433-458.

Barcroft H & Millen JLE (1939). The blood flow through muscle during sustained contraction. *Journal of Physiology (London)* **97**, 17-31.

Behm CG & Sale DG (1993). Intended rather than actual movement velocity determines velocity-specific training response. *Journal of Applied Physiology* **74**, 359-368.

Bellemare F & Bigland-Ritchie B (1987). Central components of diaphragmatic fatigue assessed from bilateral phrenic nerve stimulation. *Journal of Applied Physiology* **62**, 1307-1316.

Bellemare F, Bigland-Ritchie B, Johansson R, Smith S & Woods JJ (1983). Motor unit discharge rates in maximal voluntary contractions of three human muscles. *Journal of Neurophysiology* **50**, 1380-1392.

Bellemare F & Garzaniti N (1988). Failure of neuromuscular propagation during human maximal voluntary contraction. *Journal of Applied Physiology* **64**, 1084-1093.

Bellemare F & Grassino A (1982). Effect of pressure and timing of contraction on human diaphragm fatigue. *Journal of Applied Physiology* **53**, 1190-1195.

Bigland-Ritchie B, Cafarelli E & Vøllestad NK (1986b). Fatigue of submaximal static contractions. Lars Hermanson Memorial Symposium: Exercise in Human Physiology. *Acta Physiologica Scandinavica* **128** (suppl 556), 137-148.

Bigland-Ritchie B, Dawson NJ, Johansson RS & Lippold OCJ (1986c). Reflex origin for the slowing of motoneurone firing rates in fatigue of human voluntary contractions. *Journal of Physiology (London)* **379**, 451-459.

Bigland-Ritchie B, Furbush F, Gandevia SC & Thomas CK (1991). Voluntary discharge frequencies of human motoneurons at different muscle lengths. *Muscle & Nerve* **15**, 130-136.

Bigland-Ritchie B, Furbush F & Woods JJ (1986a). Fatigue of intermittent, submaximal voluntary contractions: central and peripheral factors. *Journal of Applied Physiology* **61**, 421-429.

Bigland-Ritchie B, Johansson R, Lippold OCJ, Smith S & Woods JJ (1983a). Changes in motoneurone firing rates during sustained maximal voluntary contractions. *Journal of Physiology (London)* **340**, 335-346.

Bigland-Ritchie B, Johansson R, Lippold OCJ & Woods JJ (1983b). Contractile speed and EMG changes during fatigue of sustained maximal voluntary contractions. *Journal of Neurophysiology* **50**, 313-324.

Bigland-Ritchie B, Jones DA & Woods JJ (1979). Excitation frequency and muscle fatigue: Electrical responses during human voluntary and stimulated contractions. *Experimental Neurology* **64**, 414-427.

Bigland-Ritchie B, Kukulka CG, Lippold OCJ & Woods JJ (1982). The absence of neuro-muscular transmission failure in sustained maximal voluntary contractions. *Journal of Physiology (London)* **330**, 265-278.

Bigland-Ritchie B & Vøllestad NK (1988). Fatigue and hypoxia: how are they related? In: Sutton JR, Houston CS, Costes G (eds.), *Hypoxia: The Tolerable Limits*, pp. 337-352. Indianapolis, IN: Benchmark Press.

Bigland-Ritchie B & Woods JJ (1984). Changes in muscle contractile properties and neural control during human muscular fatigue. *Muscle & Nerve* **7**, 691-699.

Binder-MacLeod SA & Barker CB (1991). Use of catchlike property of human skeletal muscle to reduce fatigue. *Muscle & Nerve* **14**, 850-875.

Binder-MacLeod SA & Guerin T (1990). Preservation of force output through progressive reduction of stimulation frequency in human quadriceps femoris muscle. *Physical Therapy* **70**, 619-625.

Botterman BR & Cope TC (1988). Motor-unit stimulation patterns during fatiguing contractions of constant tension. *Journal of Neurophysiology* **60**, 1198-1214.

Boutellier U & Piwko P (1992). The respiratory system as an exercise limiting factor in normal sedentary subjects. *European Journal of Applied Physiology* **64**, 145-152.

Chevrolet J-C, Tschopp J-M, Blanc Y, Rochat T & Junod AF (1993). Alterations in inspiratory and leg muscle force and recovery pattern after a marathon. *Medicine and Science in Sports and Exercise* **25**, 501-507.

Cooper RG, Edwards RHT, Gibson H & Stokes MJ (1988). Human muscle fatigue: frequency dependence of excitation and force generation. *Journal of Physiology (London)* **397**, 585-599.

Curtin NA & Edman KAP (1994). Force-velocity relation for frog muscle fibres - effects of moderate fatigue and of intracellular acidification. *Journal of Physiology (London)* **475**, 483-494.

DeLuca CJ, Lefever RS, McCue MP & Xenakis AP (1982). Control scheme governing concurrently active human motor units during voluntary contractions. *Journal of Physiology (London)* **29**, 129-142.

Dubose L, Scherlhorn TB & Clamann HP (1987). Changes in contractile speed of cat motor units during activity. *Muscle & Nerve* **10**, 744-752.

Duchateau J & Hainaut K (1984). Training effects on muscle fatigue in man. *European Journal of Applied Physiology* **53**, 248-253.

Edwards (1981). Human muscle function and fatigue. In: Porter R, Whelan J (eds.), *Human Muscle Fatigue: Physiological Mechanisms*, pp 1-18. London: Pitman Medical.

Edwards RHT, Hill DK, Jones DA & Merton PA (1977). Fatigue of long duration in human skeletal muscle after exercise. *Journal of Physiology (London)* **272**, 769-778.

Edwards RHT, Hill DK & McDonald M (1972). Myothermal and intramuscular pressure measurements during isometric contractions of the human quadriceps. *Journal of Physiology (London)* **224**, 58P-59P.

Enoka RM & Stuart DG (1992). Neurobiology of muscle fatigue. *Journal of Applied Physiology* **72**, 1631-1648.

Fitts R (1994). Cellular mechanisms of muscle fatigue. *Physiological Reviews* **74**, 49-94.

Folkow B & Halicka HD (1968). A comparison between red and white muscle with respect to blood supply, capillary surface area and oxygen uptake during rest and exercise. *Microvascular Research* **1**, 1-14.

Fuglevand AJ, Macefield VG, Howell JN & Bigland-Ritchie B (1994). Adaptation of motor unit discharge rate and variability during sustained maximal voluntary contractions. *Society for Neuroscience Abstracts.* **20**, 1759.

Fuglevand AJ, Zackowski KM, Huey KA & Enoka RM (1993). Impairment of neuromuscular propagation during human fatiguing contractions at submaximal forces. *Journal of Physiology (London)* **40**, 549-572.

Gandevia SC (1993). Central and peripheral components to human isometric muscle fatigue. In: Sargeant AJ, Kernell, D (eds.), *Neuromuscular Fatigue*, pp 156-164. Amsterdam: Royal Netherlands Academy of Arts and Sciences.

Gantchev GN, Gatev P & Ivanova T (1989). Muscle force sensation during sustained voluntary contraction. *Biomedica Biochimica Acta* **48**, S515-S520.

Garland SJ, Enoka RM, Serrano LP & Robinson GA (1994). Behavior of motor units in human biceps brachii during a submaximal fatiguing contraction. *Journal of Applied Physiology* **76**, 2411-2419.

Garland SJ, Garner SH & McComas AJ (1988). Reduced voluntary electromyographic activity after fatiguing stimulation of human muscle. *Journal of Physiology (London)* **401**, 547-556.

Garland SJ & McComas AJ (1990). Reflex inhibition of human soleus muscle during fatigue. *Journal of Physiology (London)* **429**, 17-29.

Grabowski W, Lobsiger EA & Luttgau, HC (1972). The effect of repetitive stimulation at low frequencies upon the electrical and mechanical properties of single muscle fibres. *Pflügers Archiv* **334**, 222-239.

Hagbarth K-E, Kunesch EJ, Nordin M, Schmidt R & Wallin EU (1986). Gamma loop contributions to maximal voluntary contractions in man. *Journal of Physiology (London)* **380**: 575-591.

Hainaut K & Duchateau J (1989). Muscle fatigue, effects of training and disuse. *Muscle & Nerve* **12**, 660-669.

Hicks A & McComas AJ (1989). Increased sodium pump activity following repetitive stimulation of rat soleus muscles. *Journal of Physiology (London)* **414**, 337-340.

Howard JD & Enoka RM (1991). Maximal bilateral contractions are modified by neurally mediated interlimb effects. *Journal of Applied Physiology* **70**, 306-316.

Hnik P, Vyskocil F, Kriz N & Holas M (1972). Work-induced increase in extracellular potassium concentration in muscle measured by ion selective electrodes. *Brain Research* **40**, 559-562.

Hudlicka O (1973). *Muscle Blood Flow: Its Relation to Muscle Metabolism and Function*. Amsterdam: Swets & Zeitlinger.

Johnson BD, Babcock MA, Suman OE & Dempsey JA (1993). Exercise-induced diaphragmatic fatigue in healthy humans. *Journal of Physiology (London)* **460**, 385-405.

Johnson MA, Polgar J, Weightman D & Appleton D (1973). Data on the distribution of fibre types in thirty-six human muscles: An autopsy study. *Journal of the Neurological Sciences* **18**, 111-129.

Jones DA (1993). How far can experiments in the laboratory explain the fatigue of athletes in the field? In: Sargeant AJ, Kernell D (eds.), *Neuromuscular Fatigue*, pp 100-108. Amsterdam: Royal Netherlands Academy of Arts and Sciences.

Jones DA & Bigland-Ritchie B (1986). Electrical and contractile changes in muscle fatigue. In: Saltin B (ed.), *Biochemistry of Exercise VI*, pp. 377-392. Champaign, IL: Human Kinetics Publishers.

Jones DA, Bigland-Ritchie B & Edwards RHT (1979). Excitation frequency and muscle fatigue: Mechanical response during voluntary and stimulated contractions. *Experimental Neurology* **64**, 401-413.

Jones L & Hunter I (1983). Effect of fatigue on force sensation. *Experimental Neurology* **81**, 640-650.

Kayser B, Narici M, Binzoni T, Grassi B & Cerretelli P (1993). Fatigue and exhaustion in chronic hypobaric hypoxia: influence of exercising muscle mass. *Journal of Applied Physiology* **76**, 634-640.

Kernell D & Monster AW (1982). Time course and properties of late adaptation in spinal motoneurones of cat. *Experimental Brain Research* **46**, 197-204.

Krnjevic K & Miledi R (1958). Failure of neuromuscular propagation in rats. *Journal of Physiology (London)* **140**, 440-461.

Lännergren J & Westerblad H (1986). Force and membrane potential during and after fatiguing, continuous high-frequency stimulation of single *Xenopus* muscle fibers. *Acta Physiologica Scandinavica* **128**, 359-368.

Lännergren J, Westerblad H & Allen DG (1993). Mechanisms of fatigue as studied in single muscle fibers. In: Sargeant AJ, Kernell D (eds.), *Neuromuscular Fatigue*, pp 3-9. Amsterdam: Royal Netherlands Academy of Arts and Science.

Mador MJ, Magalang UJ, Rodis A & Kufel TJ (1993). Diaphragmatic fatigue after exercise in healthy human subjects. *American Review of Respiratory Disease* **148**, 1571-1575.

McKenzie DK, Bigland-Ritchie B, Gorman RB & Gandevia SC (1992). Central and peripheral fatigue of human diaphragm and limb muscles assessed by twitch interpolation. *Journal of Physiology (London)* **454**, 643-656.

McKenzie DK & Gandevia SC (1988). Human diaphragmatic endurance during different maximal respiratory efforts. *Journal of Physiology (London)* **395**, 625-638.

McKenzie DK & Gandevia SC (1991). Recovery from fatigue of human diaphragm and limb muscles. *Respiration Physiology* **84**, 49-60.

Merton PA (1954). Voluntary strength and fatigue. *Journal of Physiology (London)* **123**, 553-556.

Miller RG, Giannini D, Milner-Brown HS, Layzer RB, Koretsky AP, Hooper D & Weiner MW (1987). Effects of fatiguing exercise on high-energy phosphates, force, and EMG: Evidence for three phases of recovery. *Muscle & Nerve* **10**, 810-821.

Milner-Brown HS & Miller RG (1986). Muscle membrane excitation and impulse propogation velocity are reduced during muscle fatigue. *Muscle & Nerve* **9**, 367-374.

Monod H & Scherrer J (1965). Work capacity of a synergistic muscle group. *Ergonomics*. **8**, 329-338.

Moritani T, Nagata A & DeVries HA (1981). Critical power as a measure of work capacity and anaerobic threshold. *Ergonomics* **24**, 339-350.

Newham DJ & Cady EB (1990). A ^{31}P study of fatigue and metabolism in human skeletal muscle with voluntary, intermittent contractions at different forces. *NMR in Biomedicine* 3, 211-219.

Poole DC, Ward S & Whipp BJ (1990). The effects of training on the metabolic and respiratory profile of high-intensity cycle ergometer exercise. *European Journal of Applied Physiology* **59**, 421-429.

Rice CL, Andrews MA & Bigland-Ritchie B (1991). Ischemic and non-ischemic fatigue from intermittent maximal voluntary contractions. *Medicine and Science in Sport and Exercise* (Suppl.) **23**, S126.

Rice CL, Cunningham DA, Taylor AW & Paterson DH (1988). Comparison of the histochemical and contractile properties of human triceps surae. *European Journal of Applied Physiology* **58**, 165-170.

Rice CL, Vøllestad NK & Bigland-Ritchie B (1992). Dissociation of fatigue-related neuromuscular events. *Medicine and Science in Sport and Exercise* (Suppl.) **24**, S56.

Romert W (1960). Ermittung von Erholungspausen für statische Arbeit des Menschen. *Internationale Zeitschrift fur Angewandte Physiologie* **18**, 123.

Round JM, Jones DA, Chapman SJ, Edwards RHT, Ward PS & Fodden DL (1984). The anatomy and fiber type composition of the human adductor pollicis in relation to its contractile properties. *Journal of the Neurological Sciences* **66**, 263-292.

Sahlin K, Edstrom L & Sjoholm H (1987). Force, relaxation and energy metabolism of rat soleus muscle during aerobic contractions. *American Physiological Society* **129**, 1-7.

Saltin B, Sjøgaard G, Gaffney FA & Rowell LB (1981). Potassium, lactate, and water fluxes in human quadriceps muscle during static contractions. *Circulation Research* **48** (Suppl. I), 118-124.

Sjøgaard G (1991). Role of exercise-induced potassium fluxes underlying muscle fatigue: a brief review. *Canadian Journal Physiology Pharmacology* **69**, 238-245.

Sjøgaard G, Kiens B, Jorgensen K & Saltin B (1986). Intramuscular pressure, EMG and blood flow during low level prolonged static contractions in man. *Acta Physiologica Scandinavica* **128**, 475-484.

Stephens JA & Taylor A (1972). Fatigue of maintained voluntary muscle contraction in man. *Journal of Physiology (London)* **220**, 1-18.

Thomas CK, Bigland-Ritchie B & Johansson RS (1991a). Force-frequency relationships of human thenar motor units. *Journal of Neurophysiology* **65**, 1509-1516.

Thomas CK, Johansson R & Bigland-Ritchie B (1991b). Attempts to classify human thenar motor units. *Journal of Neurophysiology* **65**, 1501-1508.

Thomas CK, Woods JJ & Bigland-Ritchie B (1989). Neuromuscular transmission and muscle activation in prolonged maximal voluntary contractions of human hand and limb muscles. *Journal of Applied Physiology* **67**, 1835-1842.

Trimble MH & Enoka RM (1990) Mechanisms underlying the training effects associated with neuromuscular electrical stimulation. *Physical Therapy* **71**, 273-282.

Vøllestad NK & Saugen E (1993). Gradual increase in metabolic rate during repeated isometric contractions. In: Sargeant AJ, Kernell D (eds.), *Neuromuscular Fatigue,* 96-97. Amsterdam: Royal Netherlands Academy of Arts and Sciences.

Vøllestad NK, Sejersted OM, Bahr R, Woods JJ & Bigland-Ritchie B (1988). Motor drive and metabolic responses during repeated submaximal voluntary contractions in man. *Journal of Applied Physiology* **64**, 1421-1427.

Walsh ML & Banister EW (1995). Acute ventilatory responses to ramp exercise while breathing hypoxic, normoxic, or hyperoxic air. In: Semple SJG, Adams L (eds.), *Advances in Experimental Medicine and Biology: Modelling and Control of Ventilation*, pp. 00-00. New York: Plenum Publishing, In press.

Walsh ML & Bigland-Ritchie B (1995). A unitary model of fatigue and exhaustion. *Abstracts, International Symposium on Neural and Neuromuscular Aspects of Muscle Fatigue* (Miami, FL, November 10-13, 1994, pp. 27. Miami, FL: Miami Project to Cure Paralysis, University of Miami.

Weiner MW, Moussavi RS, Baker AJ, Boska MD & Miller RG (1990). Constant relationships between force, phosphate concentration, and pH in muscles with differential fatigability. *Neurology* **40**, 1888-1893.

Westling G, Johansson RS, Thomas CK & Bigland-Ritchie B (1990). Measurement of contractile and electrical properties of single human motor units in response to intraneural motor axon stimulation. *Journal of Neurophysiology* **64**, 1331-1338.

Woods JJ & Bigland-Ritchie B (1983). Linear and non-linear surface EMG/Force relations in human muscles: An anatomical/functional argument for the existence of both. *American Journal of Physical Medicine* **62**, 287-299.

Woods JJ, Furbush F & Bigland-Ritchie B (1987). Evidence for a fatigue-induced reflex inhibition of motoneuron firing rates. *Journal of Neurophysiology* **58**, 125-136.

Yan S, Lichros I, Zakynthinos S & Macklem PT (1993). Effect of diaphragmatic fatigue on control of respiratory muscles and ventilation during CO2 breathing. *Journal of Applied Physiology* **75**, 1364-1370.

Zijdewind I & Kernell D (1994). Fatigue associated EMG behavior of the first dorsal interosseous and adductor pollicis muscles in different groups of subjects. *Muscle & Nerve* **17**, 1044-1054.

Zijdewind I, Kernell D & Kukulka CG (1995) Spacial differences in fatigue-associated EMG behaviour of the human first dorsal interosseous muscle. *Journal of Physiology (London)* **483,** 499-509.

SECTION VIII

Integrative Systems Issues

One aspect of fatigue that is sometimes ignored by neurobiologists in their rush to examine smaller and smaller units within the neuromuscular system (or even within a single muscle cell) is how the whole body uses oxygen and delivers it to the mitochondria. One potent method to examine this integrative issue is by assessment of how the problem has been solved in different parts of the evolutionary tree. In **Chapter 28** (Lindstedt & Hoppeler) oxygen delivery is quantified from the muscle back to the lung, and ultimately to the central respiratory controllers that generate the rate and depth of breathing. Design in oxygen delivery via the cardiorespiratory system seems to be driven by the necessity for consumption of oxygen by the muscle fiber. Some intriguing design problems have occasionally been encountered in evolution; one of them is how the humming bird uses so much oxygen without trading too much of its myofibrillar space for the necessary mitochondria.

Next (**Chapter 29**), Dempsey and Babcock apply some of the principles of the preceding chapter to consider the likelihood of a ventilatory limit to normal human exercise. Fortunately for us, but not the thoroughbred racehorse, our apparatus for oxygen delivery is *overbuilt* and a respiratory limit is achieved in only a few elite athletes. Studies reviewed in this chapter have revealed an important effect of whole-body exercise on the development of fatigue in the diaphragm, a muscle with much evolutionary and survival value. Whether this fatigue is mediated by suboptimal perfusion of the diaphragm, when the demands for blood flow are increased elsewhere, remains to be determined.

Does fatigue of respiratory muscles play a major role in patients with respiratory disease? This question has long been controversial despite a realization that the diaphragm has adaptations which suit it for endurance, such as a high capillary and mitochondrial supply. **Chapter 30** (McKenzie & Bellemare) reviews this area. Appropriate techniques, such as phrenic nerve stimulation, have been applied to assess the extent to which a failure to produce force by the diaphragm is mediated peripherally or centrally by the CNS. However, given what we know about fatigue of limb muscles during voluntary exercise, it would be naive to think that fatigue would occur at only one of these two sites; or, that there was no interaction between them.

Finally, morphometric and histochemical specializations exhibited by the muscles involved in chewing and speech are described (**Chapter 31**, Nordstrom & Miles). Human jaw-closing muscles exhibit extraordinary strength and their endurance in sustained voluntary contractions is probably high and limited as much by pain as by peripheral fatigue. Interestingly, the fibers in jaw-closing muscles are smaller than those in limb muscles, which would be advantageous for oxygen delivery.

Overall, the concept emerges that while individual muscles may be specialized for particular physiological tasks, their performance may sometimes be compromised by the need to use other muscles. In this way, the design features of some muscles are subservient to factors that must have influenced the evolution of the whole animal. Understanding the fatigue properties of every muscle studied in isolation will not necessarily reveal how fatigue will develop in a coordinated task that involves multiple muscle groups.

FATIGUE AND THE DESIGN OF THE RESPIRATORY SYSTEM

S. L. Lindstedt[1] and H. Hoppeler[2]

[1] Department of Biological Sciences
Northern Arizona University
Flagstaff, Arizona 86011-5640
[2] Department of Anatomy
University of Bern
CH-3000 Bern, Switzerland

ABSTRACT

One source of muscle fatigue may be the failure to provide the required oxygen by any step in the oxygen transport cascade or a lack of the necessary machinery to utilize that oxygen. We favor abandoning the concept of a single rate-limiting step for the concept of tuned resistors, each contributing to the overall resistance to oxygen flow. However, because some of these steps have considerably less phenotypic plasticity than others, these are the component parts of the respiratory system that must be built with adequate "reserve" to accommodate adaptive increases in the other steps (Lindstedt et al., 1988; Weibel et al., 1992; Lindstedt et al., 1994). These structures will usually appear to be over built except in those rare individual animals at the species-specific limit of $\dot{V}_{O_2}$ in which these less malleable structures may be limiting.

INTRODUCTION

When animals perform the vast majority of the tasks encountered during their lives, fine motor control is generally more critical than are extremes of force generation, power output or duration of muscular activity. Hence, it may be rare when an animal's performance is limited by inadequate force generation by its muscles. However, there will be times in all animals' lives in which one of these extremes of performance is necessary; in those instances, muscular fatigue defines the limit of performance and may be of survival value. We adopt the definition of fatigue as a failure to maintain the required or expected force.

Our focus is on those factors relating to the delivery and utilization of oxygen and hence we restrict our discussion to *aerobically* functioning muscles only. This restriction is not intended to imply that fatigue occurs only in these muscles, rather that when muscles are active over extended durations (i.e., durations beyond which the on-board fuel can supply the required ATP anaerobically), an additional potential source of fatigue is introduced that

Fatigue, Edited by Simon C. Gandevia et al.
Plenum Press, New York, 1995

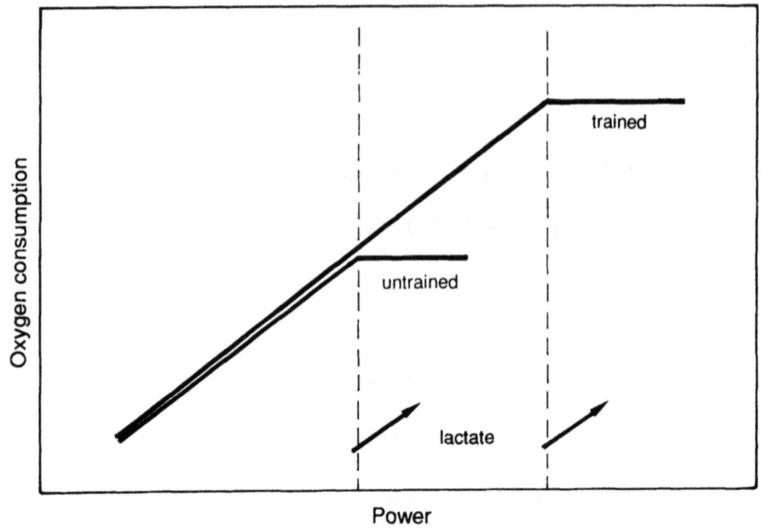

Figure 1. Maximum oxygen uptake ($\dot{V}_{O_2max}$) is defined as the plateau of $\dot{V}_{O_2}$ independent of increasing power output. Because additional power output above that point is not fueled aerobically, it must be accompanied by an abrupt increase in lactate production. Endurance training results in an increase in $\dot{V}_{O_2max}$ and a shift to the right of the power output at which lactate levels increase.

resides outside the neuromuscular machinery itself. When the supply of oxygen is inadequate to meet the demand for oxidative phosphorylation, the muscle will be compromised in its ability to produce force. How often does this occur and is there an identifiable weak link in the delivery and utilization of oxygen to fuel muscular activity? These questions form the focus of this review.

Aerobic fatigue may take two different forms. First, when an animal exercises, its *whole body* oxygen uptake increases greatly. In *average* mammals, this increase is equal to about ten times the resting or basal oxygen uptake and it may be much greater in some athletic species. However, in all animals there is a point at which an increase in power output (e.g., running, swimming or flying speed) is no longer accompanied by an increase in oxygen uptake. This plateau of oxygen consumption is defined as $\dot{V}_{O_2max}$ (Fig. 1). When exercise involves power output (or force generation) at or beyond this point it can be maintained for short durations only. However, this point of aerobic fatigue is adaptable; i.e., it is a phenotypically plastic trait that will respond to, for example, endurance training. As more oxygen is supplied and utilized by the musculature, the point of aerobic muscle fatigue is shifted upward or downward in detraining (Fig. 1; see also Gordon, Chapter 32).

Aerobic fatigue can take another form when high levels of muscular activity are required over extended periods of time. The sustainable level of O_2 supply and utilization decreases in whole body exercise as a function of the duration of exercise. Hence, $\dot{V}_{O_2max}$ may be maintained for a few minutes only while longer durations of exercise are coupled to ever decreasing fractions of $\dot{V}_{O_2max}$, as is demonstrated in the speed and duration of human running records (Fig. 2). We will briefly address this interesting form of aerobic fatigue, which likely is related to the supply of fuel rather than oxygen delivery *per se*.

To examine these aspects of fatigue in whole body exercise, we exploit two natural models and an experimental one (see Weibel et al., 1992). Across the range of mammalian body sizes, both basal and maximum oxygen uptake vary predictably with body size. Thus, *allometric variation* provides a range of about 20 fold in (weight specific) $\dot{V}_{O_2max}$ values

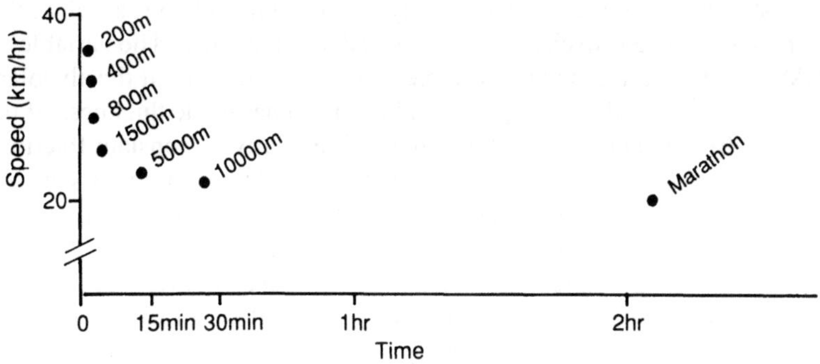

Figure 2. World record standings in running (men). Sustained power output decreases with the length of time the power is maintained, as is seen in current world record performances in men's track events. While the sprints are entirely anaerobic, all distance events require aerobic resynthesis of ATP.

comparing a 2 g shrew with a 5000 kg elephant. Comparing athletic and sedentary animals of the same body size, there is a 2.5 to 5 fold range of $\dot{V}_{O_2max}$, the consequence of *adaptive variation*, that can provide additional insights into possible design constraints of the respiratory system. Finally, we draw upon the 50% increase in $\dot{V}_{O_2max}$ that can be achieved within a given animal in response to specific training. This *induced variation* provides the final experimental model discussed below.

In this chapter, we define the respiratory system broadly, as all of the structures through which oxygen flows en route from atmospheric air to the mitochondria of exercising muscles. Because it is only at $\dot{V}_{O_2max}$ when these structures are used to their maximum capacities, we focus on this single physiological state to examine the structure-function links in the respiratory system. As a working hypothesis, we assume that at each step, the capacity to transport oxygen should be quantitatively tuned to the demand for oxygen of the working muscles (Fig. 3). We then ask when and under what conditions is the supply of oxygen inadequate; are there consistent or occasional *weak links* in the cascade of respiratory system structures?

O_2 UPTAKE IN THE LUNGS

The transfer of O_2 from alveoli to blood in the lungs is driven by the partial pressure difference and the conductance for O_2 (D_{LO2}) of the intervening tissue of the gas exchanger (line 1 in Fig. 3). D_{LO2} in turn can be decomposed into two sequential conductances, the lung membrane diffusing capacity, DM_{O2}, and the erythrocyte diffusing capacity, D_{eO2}. DM_{O2} is

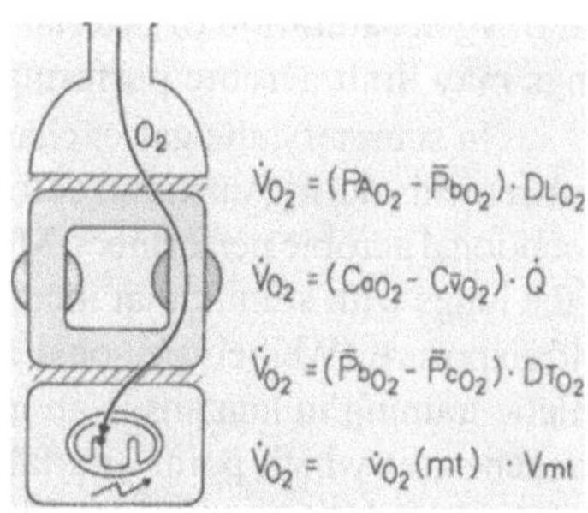

Figure 3. As oxygen flows through the respiratory system, the oxygen flow ($\dot{V}_{O_2}$) through each of the components must be equal. If any one or all of these structures is unable to supply oxygen at a rate equal to its demand in the working muscles, the muscle will be unable to maintain force. We consider each step independently below.

determined by structural parameters, specifically the total lung volume as well as the density and characteristics of the gas exchange units packed into the lungs. The variables that may modulate DM_{O2} are therefore, the lung volume, the packing density of the alveolar surface, the capillary loading of the alveolar surface and the mean harmonic thickness of the barrier separating the air from the blood. The *lung volume* is a relatively constant fraction of body size ranging from 35 $ml \cdot kg^{-1}$ in a 3 g shrew to 60 $ml \cdot kg^{-1}$ in large mammals such as humans or steers. With adaptive variation, athletic species such as horses and pronghorn antelopes may reach considerably larger weight-specific lung volumes (100 and 200 $ml \cdot kg^{-1}$, respectively). The packing of alveolar septa, *alveolar surface density*, is relatively constant (400-500 $cm^2 \cdot cm^{-3}$) in animals weighing between 1 and 100 kg. In larger animals, such as horses and steers, the alveolar surface density falls to 250 $cm^2 \cdot cm^{-3}$ while in shrews it is increased to 1500 $cm^2 \cdot cm^{-3}$). The packing density of the alveolar surface is similar in athletic and sedentary species of the same size and probably does not change with training (Weibel et al., 1992). There is a weak allometric variability of *capillary loading* of the alveolar surface as well as of the harmonic mean *barrier thickness* both of which increase with body mass ($M_b^{0.05}$). Capillary loading varies from 0.7 to 1.7 $ml \cdot m^{-2}$ from shrews to steers while barrier thickness varies from 0.27 to 0.65 μm in the same species. The erythrocyte diffusing capacity, characterized by the rate of oxygen uptake of the pulmonary capillary unit, is dependent on both body size and athletic capability. It ranges from 42 $mlO_2 min^{-1} \cdot ml^{-1}$ in animals weighing 500 g to 12 $ml\ O_2 \cdot min^{-1} ml^{-1}$ in animals weighing 250 kg, athletic animals having higher D_{eO2} values than their weight matched sedentary counterparts.

Intuitively one might expect total pulmonary diffusing capacity, D_{LO2}, to vary directly with $\dot{V}_{O_2max}$ in allometric and adaptive variation as well as possibly in response to induced variation. However, the experimental evidence is at odds with intuition as D_{LO2}/Mb varies with $Mb^{-0.11}$; a 500 kg animal has 3 times as much lung diffusing capacity than a 50 g animal. Likewise, athletic species have a larger D_{LO2}/Mb than sedentary species of the same body weight; however, the difference in D_{LO2} does not match the difference in $\dot{V}_{O_2max}$. As a consequence, the driving force, ΔPO_2, is larger both in small animals (Lindstedt, 1984) as well as in highly active species (Weibel et al., 1992). The partial pressure difference between the alveolar air and the lung capillary blood depends on the alveolar ventilation as well as on the perfusion of capillaries by the circulation. In this context, the capillary transit time (t_c) is an important variable indicating the time available for loading of O_2 onto red blood cells. Combining morphological estimates for lung capillary blood volume and measurements of cardiac output at $\dot{V}_{O_2max}$, the available transit time is sufficient for full saturation of blood during its passage through the lungs in all but the smallest mammals (Lindstedt, 1984; Weibel et al., 1992). However, athletic species such as dogs use up to 80% of the available transit time; whereas, sedentary species such as goats may reach full saturation after less than 50% of t_c. In some highly athletic species such as thoroughbred horses (Jones et al., 1989) and in top human endurance athletes (Dempsey et al., 1984), there is evidence for a drop of arterial PO_2 and oxygen saturation of arterial blood during maximal aerobic work indicating that the lungs may limit aerobic performance.

In summary, the gas exchanger of the lung is overbuilt with regard to the functional requirements during maximal aerobic whole body exercise in all mammals except for a few exceptional aerobic performers. Moreover, there seems to be a lack of epigenetic malleability of the lungs with stimuli that increase $\dot{V}_{O_2max}$ such as endurance exercise training or chronic cold exposure (Weibel, personal communication). We hypothesize that a consequence of athletic training in humans is an increase in $\dot{V}_{O_2max}$ until the diffusion capacity of the lungs is reached, at which point arterial oxygen pressure and saturation begin to drop, limiting oxygen availability to the periphery.

CONVECTIVE TRANSPORT OF O_2 BY THE HEART

The transport of oxygen to the periphery by the heart is described by the Fick equation (line 2, Fig. 3). The heart as a pump has conventionally been thought as the major determinant of aerobic performance capacity, limiting $\dot{V}_{O_2max}$ in most species and certainly in humans (Saltin & Strange, 1992). Cardiac output, $\dot{Q}$, is the product of the functional variable, heart rate (f_H), and the structural variable, stroke volume (V_s), the latter being directly proportional to heart volume or mass. The arteriovenous oxygen concentration difference depends on the partial pressure of O_2 in the arterial and mixed venous blood, the hematocrit and the O_2 capacitance of the blood (assumed to be identical in mammalian species). Heart size is a similar fraction of close to 0.6% of the body mass in all mammals (Stahl, 1967). As a consequence, the stroke volume at $\dot{V}_{O_2max}$ is found to be invariant with body size (Fig. 4). Likewise, hematocrit at $\dot{V}_{O_2max}$ is similar in mammals of all sizes. However, both stroke volume and hemoglobin concentration are consistently increased in athletic relative to sedentary species. Furthermore, endurance training increases stroke volume but not usually hemoglobin concentration. Because both heart size and hemoglobin concentration are invariant with body size, it is obvious that allometric differences in circulatory oxygen transport must entirely be brought about by varying heart rate. Resting heart rates vary with basal metabolic rate and relate to $Mb^{-0.25}$ (Stahl, 1967; Fig. 4). Heart rate at $\dot{V}_{O_2max}$ decreases at a somewhat more shallow slope ($Mb^{-0.17}$), because small animals can increase their heart rate less during heavy exercise than large animals. In adaptive pairs, the maximal heart rate is similar while athletic animals may have lower resting heart rates. The same is observed with endurance exercise training in humans which increases stroke volume and decreases resting heart rate.

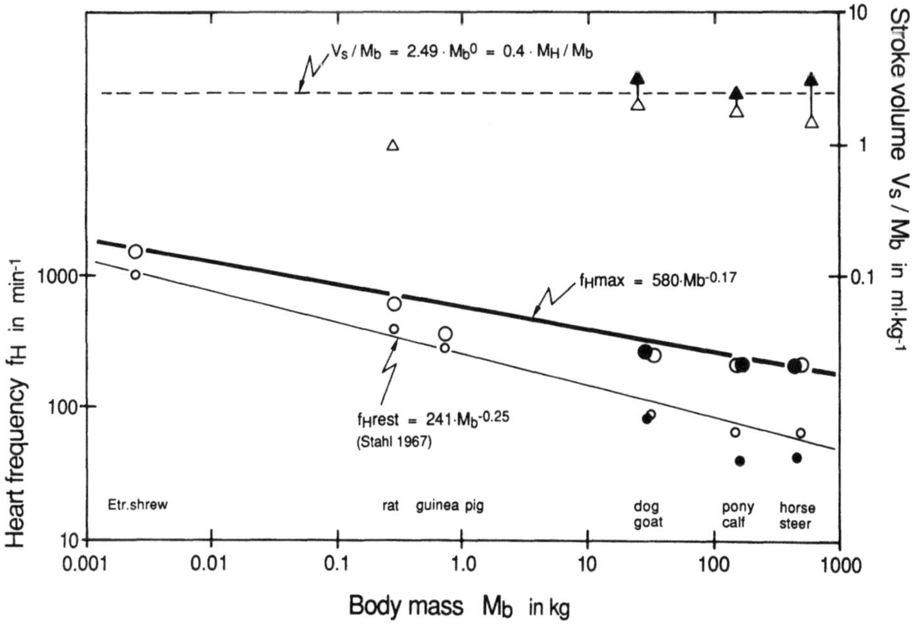

Figure 4. Stroke volume (V_s) is a nearly constant fraction of body mass in all mammals while heart rate (f_H) falls as a regular function of body mass. As a consequence, cardiac output, the product of V_s and f_H, is systematically higher in small than large animals relative to body size. Quantitatively, this relation is identical to that of the metabolic oxygen requirement.

The question as to whether cardiac output limits aerobic exercise performance can be answered negatively; however, it may be the wrong question to ask in the first place. One might rather ask how much of the resistance in the pathway of oxygen from lung to skeletal muscle mitochondria is related to convective oxygen transport by the heart under a given set of conditions (see di Prampero, 1985). Experiments using blood doping as an experimental tool to vary oxygen supply to the periphery acutely and independently from peripheral oxygen demand indicate that there is considerable resistance residing in the microcirculation, the transfer of oxygen to mitochondria and/or within the metabolic machinery in mitochondria (Turner et al., 1993).

MICROVASCULAR O_2 DELIVERY TO MITOCHONDRIA

O_2 flow from capillaries to mitochondria is functionally determined by the pressure head between average skeletal muscle capillary partial pressure and the O_2 partial pressure in the immediate vicinity of mitochondria believed to be as low as a few torr, and the diffusing capacity of the intervening tissue (Fig. 3, line 3; Gayeski & Honig, 1986). The tissue diffusing capacity, DT_{O2}, depends on a number of variables such as the capillary blood volume, the capillary hematocrit, the capillary and sarcolemmal diffusion surfaces, the average distance from the capillary to mitochondria, and the influence that the presence of myoglobin might have on the O_2 conductance of myofibers. There is, currently, no comprehensive model available that integrates these variables into a model for calculating skeletal muscle DT_{O2}. The design of the microvascular system will, therefore, be described by relating its main structural parameters to mitochondrial oxygen demand. Capillary diameter is between 4.1 and 4.6 μm with no systematic variation due to body size or athletic ability. The key structural parameter, therefore, is capillary length J(c). Capillary length can be calculated from counts of capillary profiles on tissue sections with appropriate morphometric methods. Once capillary length is known, capillary volume (relevant for mass flow of gases and solutes) and capillary surface area (relevant for transcapillary transport phenomena) can easily be obtained given the constant capillary diameter. Capillary length (and hence capillary surface and volume) per unit mass decreases with increasing body mass with a scaling exponent of -0.11. In adaptive variation, active species have a larger weight-specific capillary length than inactive species. This difference is increased by active species having a higher hematocrit at $\dot{V}_{O_2max}$ than inactive species as discussed previously. The higher hematocrit of active species is, therefore, critical for diffusive transport of O_2 both in the lungs and in muscles as well as for convective transport of O_2 in the circulation. When both capillary volume and capillary blood flow are known, mean capillary transit time can be calculated; skeletal muscle capillary time increases with increasing body size from 0.4 s in rats to 0.8 s in horses with no systematic difference between active and inactive species. Time for unloading in the periphery is about double that for loading in the lungs and is likely sufficient for unloading in all but the smallest mammals (Lindstedt & Thomas, 1994). Capillary density and hence J(c) increases with endurance exercise training (Hudlicka, 1985), such that there is a constant proportion relationship between capillary length, (characterizing oxygen supply) and mitochondrial volume (characterizing oxygen demand) in skeletal muscle tissue (Fig. 5).

The question as to how resistance to oxygen flow over the sequential steps between erythrocytes and mitochondria must be attributed cannot presently be answered. It is unclear, therefore, whether capillary, mitochondria, or the tissue between them is the more likely site of *tissue diffusion limitation*.

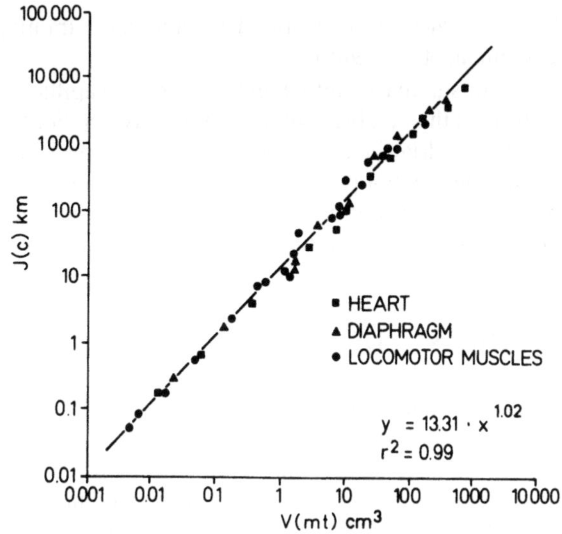

Figure 5. Total capillary length (J(c)) varies systematically and linearly with total mitochondrial volume (V(mt)) in heart, diaphragm and locomotor muscles among the mammals. This linkage suggests that capillaries and mitochondria form a single functional unit in skeletal muscle, mitochondria setting the demand and capillaries, the supply.

MITOCHONDRIAL OXYGEN SINK

The mitochondria are solely responsible for the resynthesis of ATP by oxidative phosphorylation. Therefore, if the amount of mitochondrial machinery in the muscle is inadequate to meet the ATP demand by the myosin and sarcoplasmic reticulum ATPases, the muscle will be unable to maintain its force production (line 4, Fig 3). The source of muscle fatigue in this instance can be directly and quantitatively linked to the total volume of mitochondria within the skeletal muscle. Among mammals, the density of inner mitochondrial membrane per unit volume of mitochondria is relatively invariant at 35 $\mu m^2 \cdot m^{3}$ (Schwerzmann et al., 1989). Thus, the total volume of mitochondria within any single muscle or muscle group is a reliable marker of the total quantity of respiratory chain enzymes within the muscle (Weibel et al., 1992). Because 90-95% of the oxygen consumed at the lungs at $\dot{V}_{O_2max}$ is destined for the skeletal muscle mitochondria, the muscle mitochondria are a single functional oxygen sink during exercise. We would expect the size of this sink to be not only directly proportional to $\dot{V}_{O_2max}$; but indeed, it is the very source of the oxygen demand. It is the mitochondria that ultimately set the demand for oxygen, the *upstream* structures discussed above must be subservient to that demand.

We have been interested in quantifying the size of the sink for the past decade (Hoppeler and Lindstedt, 1985). When $\dot{V}_{O_2max}$ is expressed as a function of the size of the oxygen sink, the total skeletal muscle mitochondrial volume, these two variables are linked via a consistent quantitative coupling, each ml of mitochondrial volume consumes roughly 4.5 ml of O_2 per minute. Recently, this relationship has been expanded to include other vertebrates in addition to mammals. If hummingbirds operated with the same mitochondrial oxygen uptake as mammals, their flight muscles would have to be composed of two-thirds mitochondria and only one-third myofibrils! To solve this problem, hummingbirds have managed a trick that is likely unique among the vertebrates; they have packed twice the density of mitochondrial membranes in their mitochondria. As a consequence, mitochondrial oxygen uptake in these animals is identical to that of other vertebrates when expressed per unit mitochondrial membrane normalized for differences in body temperature (Wells, 1990; Suarez, 1991; Schaeffer & Lindstedt, 1992) and the volume density of the mitochondria remains around one-third. Finally, there is evidence to suggest that mitochondria are among the first structures to respond to endurance training within an individual animal (e.g., Reichmann et al., 1985).

In summary, mitochondria (and the capillaries supplying them) respond adaptively to shifts in the aerobic functioning of muscles. For that reason, we propose that the density of mitochondria (i.e., the size of the oxygen sink) will reflect the current aerobic demand of the muscle. Such a proposal does not implicate this single step as the rate limiting step in the oxygen transport cascade; rather, that mitochondrial volume is quantitatively tuned to the ATP demand of aerobically functioning muscles.

ACKNOWLEDGMENTS

The authors thank Profs. C. Richard Taylor and Ewald Weibel for their numerous contributions. The authors' laboratories have been supported by United States Public Health Service grant HL 41986, and National Science Foundation grant IBN 9317527 (*S.L.L.*), and the National Science Foundation of Switzerland (*H.H.*). Attendance of the authors at the 1994 Bigland-Ritchie conference was supported, in part, by the Muscular Dystrophy Association (USA; *H.H.*) and the University of Arizona Regents' Professor funds of Douglas Stuart (*S.L.L.*).

REFERENCES

Dempsey JA, Hanson PG & Henderson KS (1984). Exercise-induced arterial hypoxia in healthy human subjects at sea level. *Journal of Physiology (London)* **355**, 161-175.

Dempsey JA, Johnson BD & Saupe KW (1990). Adaptations and limitations in the pulmonary system during exercise. *Chest Supplement* **97**, 81S-87S.

di Prampero PE (1985). Metabolic and circulatory limitations to $\dot{V}_{O_2max}$ at the whole animal level. *Journal of Experimental Biololgy* **115**, 319-331.

Gayeski TEJ & Honig CR (1986). O_2 gradients from sarcolemma to cell interior in red muscle at maximal $\dot{V}_{O_2max}$. *American Journal of Physiology* **251**, H789-H799.

Hoppeler H & Lindstedt SL (1985). Malleability of skeletal muscle tissue in overcoming limitations: Structural elements. *Journal of Experimental Biology* **115**, 355-364.

Hudlicka O (1985). Development and adaptability of microvasculature in skeletal muscle. *Journal of Experimental Biology* **115**, 215-228.

Jones JH, Taylor CR, Lindholm A, Straub R, Longworth KE & Karas RH (1989). Blood gas measurements during exercise: errors due to temperature correction. *Journal of Applied Physiology* **67**, 879-884.

Lindstedt SL (1984). Pulmonary transit time and diffusing capacity in mammals. *American Journal of Physiology* **246**, R384-R388.

Lindstedt SL & Thomas RG (1994). Exercise performance in mammals: An allometric perspective. In: Jones, JH (ed.) *Advances in Veterinary Science and Comparative Medicine* **38B**, pp 325-348. New York: Academic.

Lindstedt SL, Thomas RG & Leith DE (1994). Does peak inspiratory flow contribute to setting $\dot{V}_{O_2max}$? A test of symmorphosis. *Respiration Physiology* **95**, 109-118.

Lindstedt SL, Wells DJ, Jones JH, Hoppeler H & Thronson HA, Jr (1988). Limitations to aerobic performance in mammals: Interaction of structure and demand. *International Journal of Sports Medicine* **9**, 210-217.

Reichmann H, Hoppeler H, Mathieu-Costello O, von Bergen F & Pette D (1985). Biochemical and ultrastructural changes of skeletal muscle mitochondria after chronic electrical stimulation in rabbits. *Pflügers Archiv* **404**, 1-9.

Saltin B & Strange S (1992). Maximal oxygen uptake: "old" and "new" arguments for a cardiovascular limitation. *Medicine and Science in Sports and Exercise* **24**, 30-37.

Schaeffer P & Lindstedt S L (1992). Structure function coupling in the fastest contracting vertebrate muscle: the rattlesnake tail shaker muscle. *The Physiologist* **35**, 224 (abstract).

Schwerzmann K, Hoppeler H, Kayar SR & Weibel ER (1989). Oxidative capacity of muscle and mitochondria: Correlation of physiological, biochemical, and morphometric characteristics. *Proceedings of the National Academy of Sciences* **86**, 1583-1587.

Stahl WR (1967). Scaling of respiratory variables in mammals. *Journal of Applied Physiology* **22**, 453-460.

Suarez RK (1991). Oxidative Metabolism in Hummingbird Flight Muscles. In: Bicudo JEPW (ed.), *The Vertebrate Gas Transport Cascade,* pp. 279-285. Boca Raton: CBC Press.

Turner DL, Hoppeler H, Noti C, Gurtner HP, Gerber H, Schena F, Kayser B & Ferretti G (1993). Limitations to V_{O2}max in humans after blood retransfusion. *Respiration Physiology* **92**, 329-341.

Weibel ER, Taylor CR & Hoppeler H (1992). The concept of symmorphosis: A testable hypothesis of structure-function relationship. *Proceedings of the National Academy of Sciences, USA* **88**,10357-10361.

Wells DJ (1990). *Hummingbird Flight Physiology: Muscle Performance and Ecological Constraints*. Ph.D Dissertation, Univ. Microfilm Int. Cit. No. 9105412. Laramie: University of Wyoming.

AN INTEGRATIVE VIEW OF LIMITATIONS TO MUSCULAR PERFORMANCE

J. A. Dempsey and M. A. Babcock

The John Rankin Laboratory of Pulmonary Medicine
University of Wisconsin-Madison
Department of Preventive Medicine
Madison, Wisconsin

ABSTRACT

First we describe the changing site of limitation to maximal O_2 transport with increasing fitness in mammals. The capacity for diffusion and airway/parenchymal flow rate and volume are markedly overbuilt in the sedentary subject's lung, but undergo little change with increased training/fitness; accordingly, as demand for O_2 transport increases in the highly fit, the limits for maximal diffusion and ventilation are surpassed or met at maximal exercise. Secondly, low-frequency diaphragmatic fatigue occured with by heavy endurance exercise. This fatigue resulted from increased diaphragmatic work together with the major contribution from the secondary effects of increased locomotor muscle activity; namely, metabolic acidosis and increased requirement for blood flow.

INTRODUCTION

The physiologic mechanisms responsible for limiting muscular performance have long been debated. In exercise involving large muscle mass, oxygen delivery to locomotor muscle capillaries is the dominant key determinant of locomotor muscle performance and the heart's maximal stroke volume and cardiac output are the major limiting factors to sustain O_2 delivery (Andersen & Saltin, 1985; Wagner et al., 1991). So this is the general concept; but in an integrated physiological system, there are many exceptions even in health to these generalizations.

We make two points concerning limitations to exercise to underscore the complexity of the integrative system. First, we illustrate using the respiratory system that the site of the primary limitation to oxygen transport and to performance of healthy subjects is changeable because the capacities of the different organ systems adapt at different rates to physical training and to the aging. Secondly, we show that the incurred fatigue in respiratory skeletal muscles during whole-body exercise depends not only on the amount of mechanical work performed by the respiratory muscles themselves but also on the *milieu* created by the limb

Fatigue, Edited by Simon C. Gandevia et al.
Plenum Press, New York, 1995

locomotor muscles. The occurrence of clinical diaphragmatic fatigue is discussed by McKenzie and Bellemare, Chapter 30.

CHANGING SITES OF LIMITATION TO MAXIMAL O_2 TRANSPORT

As the external work rate is increased, total $\dot{V}_{O_2}$ increases linearly and then begins to plateau with further increases of work rate heralding the attainment of *maximal* $\dot{V}_{O_2}$ ($\dot{V}_{O_2max}$). The determinants of $\dot{V}_{O_2max}$ are classically described by the Fick equation:

$\dot{V}_{O_2max}$ = {maximal cardiac output (*stroke volume * heart rate*)}*{arterial O_2 content - mixed venous O_2 content}

The product of cardiac output and arterial O_2 content define the O_2 delivery and the magnitude of the arterial to mixed venous O_2 content difference measures the degree of O_2 extraction, primarily by the locomotor muscles. At $\dot{V}_{O_2max}$ the maximal muscle blood flow and maximal arteriovenous oxygen difference (a-$\bar{v}O_2$) are achieved simultaneously and plateau with the $\dot{V}_{O_2}$, so it is difficult to distinguish between the relative importance of these functions as limiting factors based purely on these correlative data. Additional data strongly implicate O_2 delivery to the muscles and in turn the maximal stroke volume as the critical limiting factors (Andersen & Saltin, 1985). In most mammals arterial O_2 content is maintained near resting levels throughout exercise. Accordingly, much has been made of the substantial *reserve* enjoyed by the healthy pulmonary gas exchanger and its *overbuilt* nature in terms of several characteristics:

1. Large diffusion surface area and minimal diffusion distances,
2. Large capacity of pulmonary capillary blood volume with respect to the maximal velocity of pulmonary blood flow,
3. The capacity of the lymphatic drainage system of the lung would far exceed the increased turnover of extravascular lung water during exercise,
4. The distribution of mechanical time constants (resistance x compliance) throughout the airways is sufficiently uniform so that $\dot{V}_A$ and $\dot{V}_A/\dot{Q}$ distributions remain relatively uniform even at the high rates of perfusion and shortened inspiratory times present during maximal exercise,
5. The high elastic recoil in healthy lungs allows the airways to maintain patency and avoids flow limitation during forced expirations at high levels of hyperpnea,
6. The capacity of the inspiratory muscles to generate negative intrapleural pressure far exceeds that demanded by the normal maximal ventilation.

This is a situation in the *normal human* in whom most data has been generated and from whom generalizations are made concerning performance limitations. However, the critical limiting factor is not fixed under all circumstances and as maximal work capacity and $\dot{V}_{O_2max}$ change, critical determinants of gas transport will also change.

We have examined physical training/fitness and healthy aging as adaptive processes during which the relative importance of the pulmonary system to maximal gas transport changes. Achieving greater than normal aerobic capacity *via* physical training and/or greater genetic endowment requires that appropriate increases in capacity occur at those links in O_2 transport which were previously limiting factors. Thus, maximal stroke volume and cardiac output are increased as are muscle capillarization and locomotor muscle oxidative capacity.

Whether the lung and chest wall continue to be *overbuilt* depends upon their relative adaptabilities as *demand* in the form of $\dot{V}_{O_2max}$ is altered.

Alveolar to Arterial O_2 Transport

In mammals, some aspects of lung and chest wall structure and function clearly lag far behind cardiovascular and skeletal muscle adaptations to increased energy demands while others appear to parallel the increasing O_2max. For example, in many highly fit humans, with $\dot{V}_{O_2max}$ 40 to 50% or more greater than normal, arterial PO_2 is no longer maintained at resting levels (Dempsey et al., 1984). The ultimate expression of this imbalance between demand *vs.* capacity in pulmonary gas exchange is the thoroughbred horse who because of centuries of genetic engineering is capable of exercising at maximal work rates and O_2 uptakes which are double that in the elite human athlete. These animals show substantial exercise-induced hypoxemia (Bayly et al., 1989). One key reason for these failures in pulmonary gas exchange in heavy exercise in the highly fit is the structural limits to the dimensions of the pulmonary vascular bed which are not changing with training even though the systemic vasculature is highly adaptable. This results in a pulmonary capillary blood volume which is not increased in the highly trained while pulmonary blood flow keeps increasing - thus red cell transit time in the lung's capillaries is shortened substantially. Furthermore, the high pulmonary blood flow also leads to high pulmonary arterial pressure and pulmonary vascular resistance. In extreme cases of increased capillary perfusion pressure it is speculated that the alveolar-capillary interface may actually undergo *stress failure* leading to the accumulation of extravascular lung water during heavy exercise causing diffusion limitation (West et al., 1991).

Air Flow Limitation

Intrathoracic airway structure and lung elastic recoil are also not affected by training/fitness in the human. The highly trained individual retains the same susceptibility as the untrained to airway closure during active, forced expiration and the area of their maximal flow:volume envelope is also normal. Accordingly, as maximal metabolic rate increases and $\dot{V}_E$ rises proportionately, signification limitation to expiratory flow begins to occur in the very highly trained human and ventilation reaches its maximal limit at maximal work rate (Johnson et al., 1992). Consequences of this increased flow limitation are to increase the work and the O_2 cost of breathing (Aaron et al., 1992) which is offset to some extent during submaximal exercise because the magnitude of the ventilatory response to maximal exercise is reduced in the highly fit. Nevertheless, while the magnitude of the hyperventilatory response to maximal exercise may be constrained slightly by mechanical limitations to flow and volume in the highly trained, frank CO_2 retention does not occur in the healthy, fit human, although it does in the thoroughbred horse (Bayly et al., 1989).

Respiratory Muscle Capacity

The capacity of inspiratory muscles to generate pressure at any given length or velocity of shortening is only slightly affected by physical training/fitness. Accordingly, because of the high ventilatory requirement at $\dot{V}_{O_2max}$, the pressure generating capacity of the inspiratory muscles is also achieved in the highly fit (Johnson et al., 1992). On the other hand, the endurance capacity of inspiratory muscles may well adapt with increased training/fitness, as evidenced by the finding that even though the highly trained endurance athlete does show clear evidence of diaphragm fatigue as a result of heavy endurance exercise, these trained subjects are capable of sustaining much higher levels of ventilation and diaphragmatic work during exercise for the same level of diaphragm fatigue as the untrained subject (Babcock et al., 1994).

These examples demonstrate how major determinants of maximal exercise capacity will change with increased training/fitness, specifically because of the relative lack of plasticity in the diffusion capacity of the lung, in the maximal airflow rates acceptable by the airways and in pressure generating capabilities of the inspiratory muscles. In other words, in the highly fit at $\dot{V}_{O_2max}$ demand on each of these sub-systems rises to reach capacity in the case of expiratory flow rate, maximal ventilation and inspiratory muscle pressure generation and to exceed capacity in the case of diffusion equilibrium at the level of the alveolar-capillary interface. In the latter case, the resulting exercise-induced arterial hypoxemia presents a significant limitation to $\dot{V}_{O_2max}$.

RESPIRATORY MUSCLE FATIGUE IN THE PHYSIOLOGICAL ENVIRONMENT OF WHOLE BODY EXERCISE

Whole body endurance exercise at very high intensity causes diaphragmatic fatigue, as shown by the 15-40% reduction in the trans-diaphragmatic pressure developed in response to supramaximal phrenic nerve stimulation (BPNS) immediately following exercise (Fig. 1). This reduction in diaphragmatic force 1) occurs at several different lung volumes (i.e., diaphragmatic lengths); 2) occurs at stimulation frequencies of 1-20 Hz; 3) persists for 60-90 min following exercise; 4) is potentiated by hypoxia; 5) occurs most consistently at endurance work loads that are greater than 85% of $\dot{V}_{O_2max}$ (Fig. 2) and lasting greater than 8-10 min; 6) is correlated with the relative intensity of the exercise; 7) occurs to a similar extent in human subjects with a wide variety of fitness levels from normal to twice normal $\dot{V}_{O_2max}$ (Johnson et al., 1993; Babcock et al., 1995a; Babcock et al., 1995b).

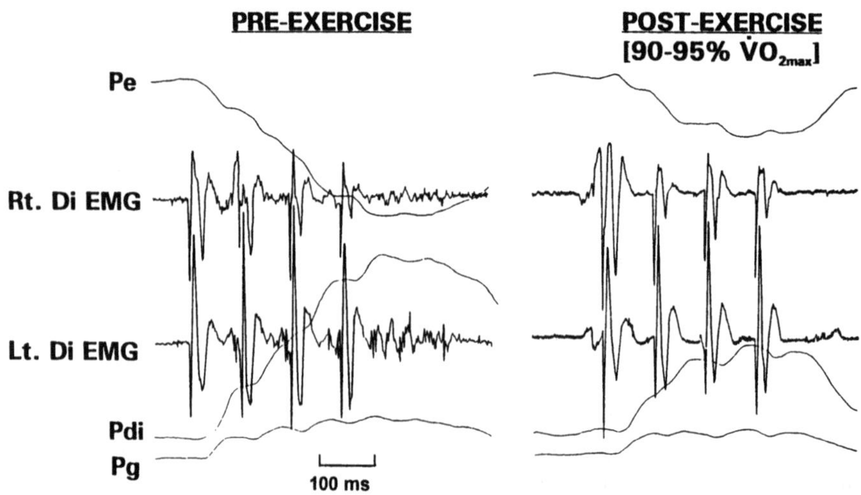

Figure 1. Typical responses of esophageal (P_e), gastric (P_g) and transdiaphragmatic (P_{di}) pressures and diaphragm EMG (i.e., M-wave) to 10 Hz bilateral phrenic nerve stimulation (BPNS) before and immediately after whole body exercise at 90% $\dot{V}_{O_2max}$. Transdiaphragmatic pressure (P_{di}) was significantly reduced following the whole-body exercise. There was no change in the M-wave response of the diaphragm to the supramaximal stimulation (modified from Johnson et al., 1993).

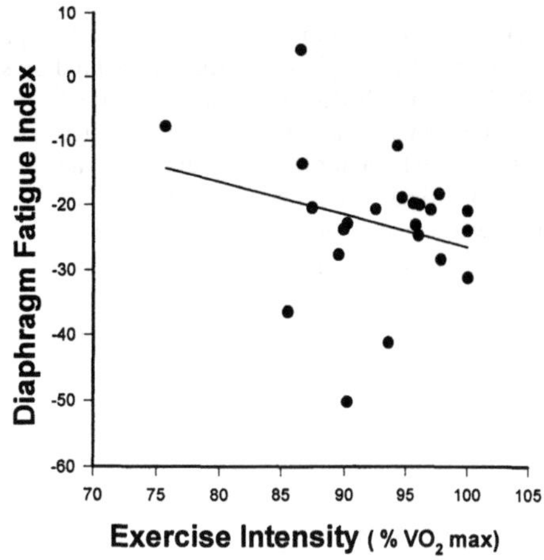

Figure 2. This figure illustrates that the level of exercise-induced diaphragm fatigue increased with increased intensity of the endurance exercise (measured as a percentage of $\dot{V}_{O_2max}$). Diaphragm fatigue index = % change in P_{di} during supramaximal BPNS at 1, 10 and 20 Hz, pre- *vs.* post-endurance exercise. The exercise time to exhaustion at 85% $\dot{V}_{O_2max}$ was 26.0 ± 2.6 min and at 95% $\dot{V}_{O_2max}$ was 13.0 ± 1.7 min. Fitness level varied widely among the subjects, $\dot{V}_{O_2max}$ was 58.7 ± 2.2 ml·kg^{-1}·min^{-1} (range 39.8 to 78.6 ml·kg^{-1}·min^{-1}, n = 24).

What causes this exercise-induced diaphragmatic fatigue? Is it dependent on the amount of work done by the diaphragm? . . . *or* do the *milieu* and/or demands created by the locomotor muscles exert significant independent influences?

To determine the effects of diaphragmatic work, per se, on diaphragmatic fatigue we had normal subjects voluntarily produce high levels of normocapnic ventilation and transdiaphragmatic pressure while at rest (Babcock et al., 1995b). Then we used the supramaximal BPNS (Bellemare & Bigland-Ritchie, 1984; see also Gandevia & McKenzie, 1985) to determine the change in maximal pressure generation by the diaphragm as affected by these sustained hyperpneic trials. Two levels of diaphragmatic work loads were voluntarily produced for 10 to 20 minutes, i.e., equal to the subject's whole-body exercise performance time. We used visual feedback for volume and pressure and audio feedback for breath timing. First, we mimicked the Vt and $\int P_{di} \cdot f_b$ · (diaphragmatic *work*) achieved throughout whole body endurance exercise. Secondly, the subjects voluntarily produced (for the same duration) a 50-70% increase in the $\int P_{di} \cdot f_b$ achieved during the exercise.

Two findings arose which pointed to a major effect of the whole body exercise, per se, on diaphragmatic fatigue. First, we found (using BPNS) that a fatigue threshold for diaphragmatic work (in the resting subject) was present during sustained inspiratory efforts requiring an increase in $\int P_{di} \cdot f_b$ of 4 to 5 times resting levels (Fig. 3). This fatigue threshold also coincided with the time:tension index of the diaphragm (TTdi) being less than 0.15 - a value previously associated with task failure during voluntary inspiratory efforts against high resistive loads (Bellemare & Grassino, 1982); or when the peak P_{di} during the sustained hyperpnea exceeded 60-70% of the dynamic capacity of the diaphragm for pressure generation. This fatigue threshold was only rarely achieved by the diaphragm during endurance exercise. Rather, exercise-induced diaphragmatic fatigue most commonly occurred when the diaphragmatic work was substantially *less* than the fatigue threshold as determined in the resting subjects (Fig. 3). We interpret these findings to mean that the whole-body endurance exercise itself must be contributing significantly to the exercise-induced diaphragmatic fatigue.

Several exercise-related mechanisms appear feasible to explain this finding. Firstly, extracellular metabolic acidosis created by locomotor muscles during whole body exercise may induce intracellular acidosis in the diaphragm and promote fatigue. This is indicated indirectly by the high levels of lactate accumulated in the rat diaphragm as a result of heavy endurance exercise (Fregosi & Dempsey, 1986). Secondly, the high levels of diaphragmatic work achieved during the hyperpnea do require high blood flows. Meeting this flow requirement is not a problem during the voluntary hyperpnea trials at rest when the

Figure 3. Relationship between the level of diaphragmatic pressure generating *work* as reflected by $\int P_{di} \cdot f_b$ and the level of diaphragm fatigue measured by supramaximal BPNS (mean % ΔP_{di} from 1, 10 and 20 Hz. BPNS). The hatched area represents the effects of sustained voluntary hyperpnea (at *rest*) generating a continuum of $\int P_{di} \cdot f_b$ levels, on diaphragm fatigue. Individual results (n = 24) from whole body endurance exercise are superimposed (■) and show that in most cases during exercise the $\int P_{di} \cdot f_b$ was below the fatigue threshold for the diaphragm (< 550 cm H_2O s·min^{-1}). The fact that significant diaphragmatic fatigue was found following whole body exercise means that some factor is contributing to the fatigue in addition to the diaphragm *work* per se. In exceptional cases, some individuals generate $\int P_{di} \cdot f_b$ levels greater than 550 cm H_2O s·min^{-1} throughout exercise and in these situations the diaphragmatic pressure generation *by itself* may be responsible for all or much of the exercise-induced diaphragm fatigue.

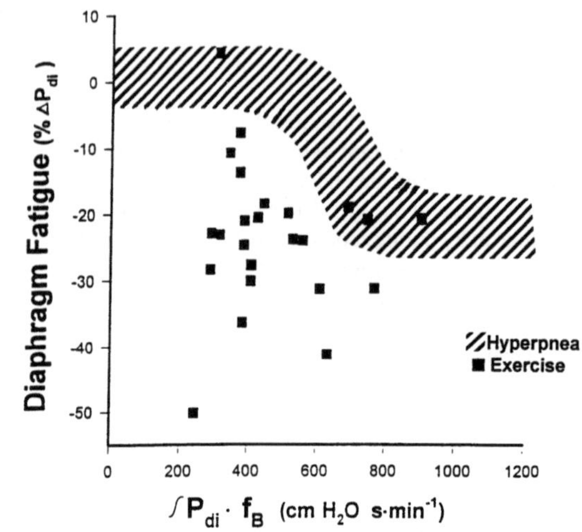

inspiratory muscles represent the major, unopposed requirement for increased perfusion. However, it might be a problem during whole body exercise when the diaphragm must compete with the locomotor muscles for the available cardiac output.

We cannot distinguish between these postulated mechanisms but we did determine that the amount of work that non-locomotor (including respiratory) muscles do during exercise is important in determining whether fatigue does occur. We used supramaximal ulnar nerve stimulation to show that whole body exercise did *not* cause fatigue of the first dorsal interosseous muscles. Thus, even though a resting skeletal muscle is *exposed* to the circulating metabolic acids and must compete for blood flow during heavy exercise, these factors alone will *not* cause fatigue. These data are consistent with recent findings in dogs which show glycogen depletion by the active muscles during leg exercise is critically dependent upon the intensity of contraction carried out by the muscle not primarily engaged in locomotion (Gladden et al., 1994). Furthermore, it has been shown that increasing contraction intensity enhances lactate uptake by human forearm muscle during leg exercise (Catchside & Scroop, 1993).

These data underscore the importance of the interaction occurring between the independent effects of locomotor muscle activity (which causes a competition for blood flow and increased circulating metabolic acidosis) and the work produced specifically by the diaphragm, as determinants of exercise-induced diaphragmatic fatigue. These data also indicate that fatigue during whole body exercise is a multifaceted process which involves responses beyond the primary muscles of interest. Furthermore, our first example concerning changing sites of limitation to maximal O_2 transport in the face of changing fitness levels, emphasizes that the various determinants of transport are *not* equally malleable in the face of changing performance capabilities. Accordingly, the major site of limitation may well change radically, and even come to include the organ systems and mechanisms which at one time were the most over-built.

In summary, our study epitomizes our gratitude to Brenda Bigland-Ritchie for her many contributions to our understanding of muscle fatigue. We are especially grateful to Brenda for her development of the BPNS technique - as a unique method for quantifying diaphragmatic force development, *in vivo*.

ACKNOWLEDGMENTS

We wish to thank Gundula Birong for preparation of the manuscript and David Pegelow for his invaluable technical assistance. The authors also wish to thank Dr. Simon Gandevia for his helpful comments on the manuscript. The author's laboratory has been supported by grants from United States Public Health Service, National Heart, Lung, and Blood Institute. *M.B.* is a Parker B. Francis fellow. Attendance of *J.A.D.* at the 1994 Bigland-Ritchie conference was supported, in part, by the University of Arizona Regents' Professor funds of Douglas Stuart.

REFERENCES

Aaron EA, Seow K, Johnson BD & Dempsey JA (1992). Oxygen cost of exercise hyperpnea: implications for performance. *Journal of Applied Physiology* **72**, 1818- 1825.

Andersen P & Saltin B (1985). Maximal perfusion of skeletal muscle in man. *Journal of Physiology (London)* **366**, 233-249.

Babcock MA, Johnson BD, Pegelow DF, Suman OE, Griffin D & Dempsey JA (1995a). Hypoxic effects on exercise-induced diaphragmatic fatigue in normal healthy humans. *Journal of Applied Physiology,* **78**(1), 82-92.

Babcock MA, Pegelow D & Dempsey JA (1994). Aerobic fitness effects on exercise-induced diaphragm fatigue. *American Journal of Respiratory and Critical Care Medicine* **149**, A799.

Babcock MA, Pegelow DF, Suman O, McClaran SR & Dempsey JA (1995b). Contribution of diaphragmatic work to exercise-induced diaphragm fatigue. *Journal of Applied Physiology* **78**, 1710-1719.

Bayly WM, Hodgon DR, Schulz DA, Dempsey JA & Gollnick PD (1989). Exercise-induced hypercapnia in the horse. *Journal of Applied Physiology* **67**, 1958-1966.

Bellemare F & Bigland-Ritchie B (1984). Assessment of human diaphragm strength and activation using phrenic nerve stimulation. *Respiration Physiology* **58**, 263-277.

Bellemare F & Grassino A (1982). Effect of pressure and timing of contraction on human diaphragm fatigue. *Journal of Applied Physiology* **53**, 1190-1195.

Catchside PG & Scroop GC (1993). Lactate kinetics in resting and exercising forearms during moderate in tensity supine leg exercise. *Journal of Applied Physiology* **74**, 435-443.

Dempsey JA, Hanson P G & Henderson KS (1984). Exercise-induced arterial hypoxemia in healthy human subjects at sea level. *Journal of Physiology (London)* **355**, 161-175.

Fregosi RF & Dempsey JA (1986). Effects of exercise in normoxia and acute hypoxia on respiratory muscle metabolites. *Journal of Applied Physiology* **60**, 1274-1283.

Gandevia SC & McKenzie DK (1985). Activation of the human diaphragm during maximal static efforts. *Journal of Physiology (London)* **367**, 45-56.

Gladden LB, Crawford R & Webster M (1994) Effect of lactate concentration and metabolic rate on net lactate uptake by canine skeletal muscle. *American Journal of Physiology: Regulatory, Integrative and Comparative Physiology* **35**, R1095- R1101.

Johnson BD, Babcock MA, Suman OE & Dempsey, JA (1993). Exercise-induced diaphragmatic fatigue in healthy humans. *Journal of Physiology (London)* **460**, 385-405.

Johnson BD, Saupe KW & Dempsey JA (1992). Mechanical constraints on exercise hyperpnea in endurance athletes. *Journal of Applied Physiology* **73**, 874-886.

Wagner PD, Hoppler H & Saltin B (1991). Determinants of maximal oxygen uptake. In: Crystal RG, West JB (eds.), *The Lung: Scientific Foundations*, vol. 2, pp. 1585-1593. New York: Raven Press.

West JB, Tsukimoto K, Mathieu-Costello O & Prediletto, R (1991). Stress failure in pulmonary capillaries. *Journal of Applied Physiology* **70**, 1731-1742.

ACKNOWLEDGMENTS

We wish to thank Gunilla Ewing for preparation of the manuscript and [illegible] for his invaluable technical assistance. The authors also wish to thank Dr. [illegible] Gandevia for his helpful comments on the manuscript. The author's laboratory has been supported by grants from United States Public Health Service, [illegible] [illegible] [illegible] and [illegible] [illegible] [illegible] by the [illegible] of [illegible] systems.

REFERENCES

[illegible] (1982). Oxygen and [illegible] hypopnea [illegible]. [illegible] 73, [illegible]

[illegible]

[illegible]

RESPIRATORY MUSCLE FATIGUE

D. K. McKenzie[1] and F. Bellemare[2]

[1] Department of Respiratory Medicine and Prince of Wales Medical Research Institute
Prince of Wales Hospital and Division of Medicine
University of New South Wales, Sydney, Australia
[2] Département d'Anesthésie
Hôtel-Dieu de Montréal
3840 St. Urbain, Montréal, Québec, Canada

ABSTRACT

Ventilatory failure may accompany a variety of pulmonary and neuromuscular diseases. There has been much controversy about whether this failure is due to respiratory muscle fatigue at peripheral sites or a failure of drive at sites within the central nervous system. The chapter reviews this topic.

INTRODUCTION

Is respiratory muscle fatigue an important cause of ventilatory failure? In 1919, the renowned physiologists Davies, Haldane and Priestley noted that "no very definite investigations have hitherto been made as to how the adaptation (of breathing to increased resistance) is brought about and at what point it begins to fail". They performed a series of experiments in which they subjected themselves to breathing through external resistances of varying severity. With extreme resistances they noted an inability to maintain ventilation with the onset of rapid shallow breathing and clinical features of hypoxaemia and hypercarbia (Fig. 1). They concluded that the *respiratory centre* had become *fatigued* and that the development of hypoxaemia "hastened this failure of the respiratory centre". About sixty years later Roussos, Macklem and colleagues (1977, 1979) performed similar experiments on themselves but imposed a set breathing pattern and a target respiratory pressure. When the target pressure exceeded a critical value the subject could not maintain the pressure and ventilation decreased. These authors concluded that this failure resulted largely from fatigue of respiratory muscles, especially the diaphragm, a conclusion which appeared to be supported by a shift in the EMG power spectrum to lower frequencies (Gross et al., 1979; Bellemare & Grassino, 1982a; their Fig. 2).

The concept of respiratory *muscle* fatigue as a major contributor to ventilatory failure subsequently gained wide acceptance and stimulated much research. Many pulmonary

Fatigue, Edited by Simon C. Gandevia et al.
Plenum Press, New York, 1995

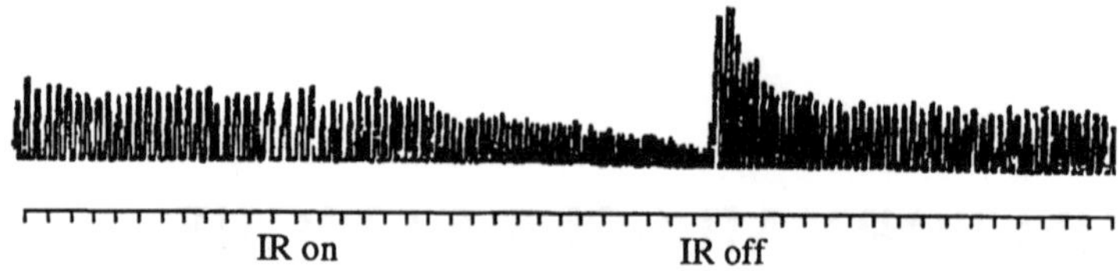

Effects of resistance

Figure 1. A resistive breathing experiment with J.S Haldane as the subject. With *excessive* resistance, respiratory rate fell initially from 27 to 12 per minute. Then tidal volume decreased and rate increased progressively until "the resistance had to be removed on account of threatening symptoms". The authors noted that the same resistance was tolerated by J.G. Priestley "without signs of fatigue. . . (or) distress". (adapted from Davies et al., 1919).

rehabilitation units implemented specific training programs with the object of increasing the strength and/or endurance of the respiratory muscles. However, in more recent years attention has again focused on respiratory control and the possibility that failure of central drive may be an inescapable consequence of the imposition of fatiguing loads to breathing. A number of researchers are examining the possible interaction and roles of volitional and involuntary central motor pathways to respiratory motoneuron pools and their relative importance in situations of critical loading. This chapter reviews the evidence for failure at central and peripheral sites both from human and animal studies (see also Dempsey & Babcock, Chapter 29). It highlights some recent studies which have employed new techniques in an attempt to resolve discrepancies. The first section outlines the clinical relevance of these studies and defines some of the terminology.

THE DEVELOPMENT OF VENTILATORY FAILURE

Ventilatory failure may occur as a result of any neuromuscular or musculoskeletal disorder which impairs the action of the ventilatory pump or as a result of insuperable loads to breathing imposed by disease of the lungs or chest wall.

Chronic airflow limitation due to diffuse airway narrowing is the commonest cause of respiratory failure. In these patients the work of breathing is increased substantially as a result of increased resistance to breathing, impaired gas exchange which necessitates an increase in minute ventilation, and the elastic loads imposed by hyperinflation of the lungs. The hyperinflation also reduces the operating length of inspiratory muscles and decreases the mechanical advantage of the diaphragm and rib-cage for producing volume expansion (for review see Rochester, 1991a,b).

The extent to which the inspiratory muscles of such patients adapt to these mechanical loads has not been studied exhaustively. It is likely that length adaptation occurs by shedding sarcomeres. It remains to be determined conclusively in humans whether the endurance capacity of these muscles is enhanced by the chronic loading of lung disease (McKenzie & Gandevia, 1986; Newell et al., 1989). Although the ability to develop inspiratory pressure is impaired at high lung volumes, the endurance capacity of the muscles is well preserved in relative terms (McKenzie & Gandevia, 1987; Clanton et al., 1993). It has also been proposed that adverse metabolic and nutritional consequences of advanced pulmonary disease may impair the function of the inspiratory muscles (see Rochester, 1991a,b). There is no information on the influence of the various consequences of airflow obstruction on blood flow to the diaphragm.

In view of all the factors which might compromise the function of the inspiratory muscles, it was logical to conclude that the muscles would develop peripheral fatigue. Fatigue is defined as any reduction in the force (or velocity) generating capacity of a muscle, regardless of whether the muscle is capable of sustaining a given task (NHLBI, 1990). Weakness refers to a reduced force generating capacity of rested muscle and the term should not be used when force is reduced by some change in the geometry of the muscle and/or its attachments. Failure of voluntary activation is used here to denote suboptimal force generation in a maximal effort with a *rested* muscle. Central fatigue denotes a decline in motoneuronal activation as force declines during voluntary muscle contractions (see Gandevia et al., Chapter 20).

ENDURANCE CAPACITY OF THE DIAPHRAGM

The diaphragm is the principal muscle of respiration during quiet breathing and moderate hyperpnoea. Under most circumstances of quiet breathing expiration is largely passive relying on the elastic recoil of the lungs. The diaphragm contracts intermittently from mid-gestation throughout life with no extended periods of rest. This pattern of phasic activity should endow the diaphragm with a high capacity for endurance and teleological argument would attach evolutionary importance to such properties.

As ventilation increases during exercise or as the loads to inspiration increase inspiratory synergists (primarily the inspiratory intercostals and scalenes) are progressively recruited. Eventually the expiratory muscles are also recruited. In terms of pressure generation during static effort these muscles can be considered as acting partly in series but this approach has led to conflicting results. At least in the mid-volume range, the diaphragm is the limiting muscle for pressure development - i.e., it approaches maximal activation while its synergists may not (Gandevia et al., 1990; cf. Hershenson et al., 1988). This finding could reflect either the greater muscle mass of the synergists or a mechanical advantage. The transmural pressure recorded across the diaphragm can exceed that across the rib cage during maximal inspiratory efforts due to recruitment of the abdominal muscles and elevation of abdominal pressure. This variation in transmural pressure, independent of diaphragmatic activation, can be explained largely by the length-tension and force-velocity properties of the diaphragm (Gandevia et al., 1992). If the diaphragm approaches maximal activation first its synergists are relatively protected from developing fatigue during inspiratory loading and any decline in output from the group of muscles is likely to reflect impaired diaphragmatic performance (Gandevia & McKenzie, 1988; cf. Hershenson et al., 1989).

Physiological, morphometric and biochemical studies indicate that the diaphragm is uniquely adapted for aerobic work. As early as 1916 Lee, Guenther and Melaney concluded from *in vitro* studies that the diaphragm was more resistant to fatigue than other skeletal muscles, a finding confirmed in humans by Gandevia, McKenzie and Neering (1983). In subsequent studies the latter group documented that the diaphragm recovered from fatigue induced by static inspiratory efforts at 10 times the rate of elbow flexors performing maximal static efforts (McKenzie & Gandevia, 1991a; their Fig. 3).

The upper limit of blood flow to the diaphragm is two to four times that available to limb muscles and there is a corresponding increase in capillary density around diaphragmatic muscle fibers compared with a range of limb muscles across a number of mammalian species (for review see McKenzie & Gandevia, 1991b). The volume density of mitochondria, the oxidative capacity of the muscle fibers and the maximal oxygen consumption of the diaphragm exceed those of other limb muscles by two to six times.

The diaphragm also appears to be situated advantageously such that its perfusion may be relatively preserved by the large negative intrathoracic pressures which are developed

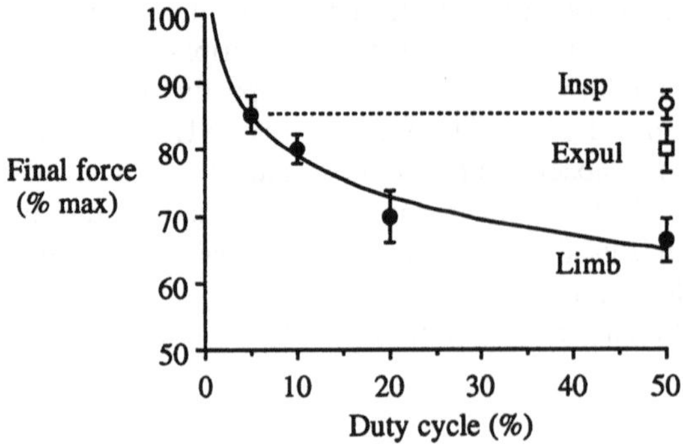

Figure 2. Relationship between final force of the elbow flexors after 18 maximal efforts of 10 s duration and duty cycle (filled circles). A logarithmic curve was fitted to the data (r =0.97). For comparison, data for the diaphragm during inspiratory efforts (open circles) and expulsive efforts (open squares) with a duty cycle of 50% have been added. The dashed line indicates the final force of elbow flexors with a duty cycle of 5%, a performance similar to that of the diaphragm with a 10-fold increase in duty cycle (from McKenzie & Gandevia, 1991a).

when the loads to breathing are high (Buchler et al., 1985), possibly contributing to its enhanced endurance capacity (Gandevia & McKenzie, 1988). This contrasts with the situation for most limb muscles and even the myocardium in which intramuscular perfusion is progressively impaired as tension increases. Diaphragmatic perfusion is reduced during expulsive contractions which increase abdominal pressure (Buchler et al., 1985) and, not surprisingly, endurance is impaired compared with that observed for inspiratory efforts performed with little change in abdominal pressure (Fig. 2).

CENTRAL AND PERIPHERAL COMPONENTS OF FATIGUE

Much of the earlier literature, both in man and mammalian species, lacked the methodological precision to enable the central and peripheral components of fatigue to be quantitated separately. Indeed, many studies did not provide any estimates of force producing capacity but relied on indirect estimates such as ventilation or a failure to sustain submaximal pressures. Studies of experimental animals, healthy subjects and patients are presented separately and the methodological limitations of each approach highlighted.

Animal Studies

Studies of respiratory muscle fatigue in animals have provided conflicting results, possibly explained by differences in level of anesthesia, the degree of loading and the physiological parameters measured. Species differences may also have contributed.

Although the inspiratory muscles are resistant to the development of fatigue, a number of studies have documented some failure of the contractile mechanism using supramaximal phrenic nerve stimulation (e.g., Mayock et al., 1987; De Vito & Roncoroni, 1993). When cardiac output was severely compromised by pericardial tamponade profound diaphragmatic fatigue was documented (Aubier et al., 1981).

By contrast Watchko and colleagues (1988) reported that awake infant monkeys exposed to high inspiratory resistances developed profound hypoventilation with no evidence of peripheral fatigue. A similar result has also been reported for adult dogs exposed to cardiogenic shock (Nava & Bellemare, 1989). Respiratory frequency declined and apnoea ensued with no evidence of peripheral diaphragmatic fatigue.

In awake adult sheep exposed to resistive loads hypoventilation appeared to be associated with evidence of both peripheral and central failure (Bazzy & Haddad, 1984; Sadoul et al., 1985) but in these studies contractile failure was not assessed by phrenic nerve stimulation and there was no direct estimate of central drive. In a subsequent study diaphragmatic action potentials were shown to decline late in extreme loading suggesting

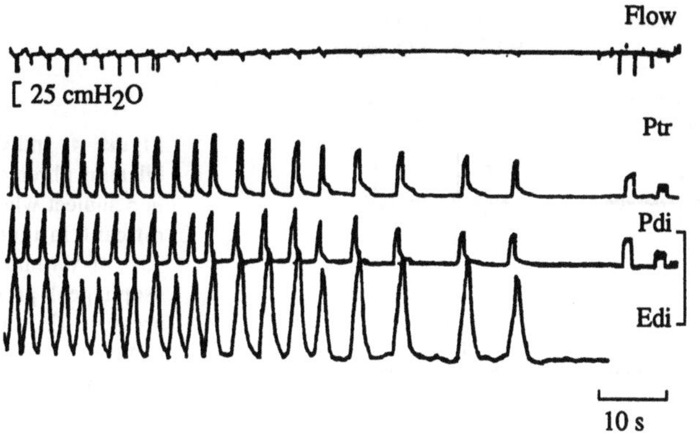

Figure 3. Typical record of air flow, tracheal pressure (P_{tr}), P_{di}, and E_{di} during resistive loading in an intact anesthetized dog just before respiratory arrest (RA). Frequency falls just before RA and is associated with preserved iEMGdi. P_{di} obtained by electrical stimulation at 60 and 10 Hz was 42.2 and 55.7%, respectively, of basal values immediately after RA (adapted from De Vito & Roncoroni, 1993).

that failure of neuromuscular transmission might also contribute to the development of ventilatory failure (Bazzy & Donnelly, 1993; see also Aldrich, 1987).

After allowing for differences in anesthesia and species a possible unifying explanation for these results is that the response of the respiratory system varies with severity of the obstruction. With critical airway narrowing respiratory arrest may occur early with little or no evidence of peripheral fatigue whereas ventilation may be sustained indefinitely with minimal peripheral fatigue in spite of moderately severe loads. There is presumably a intermediate range of loads which can be sustained for a prolonged period with a compromise of peripheral and central fatigue before respiratory arrest ultimately ensues. Fig. 3 shows the sudden development of respiratory arrest at a time when the stimulated diaphragm was still capable of generating 55% of maximal pressure.

The mechanisms underlying the profound central fatigue which can lead to respiratory arrest are unclear but several interesting observations have been made. Firstly, the administration of oxygen prevents secondary hypoxic depression of the respiratory centre during progressive hypercarbia (Chapman et al., 1980). Thus, more pronounced peripheral fatigue will occur with supplemental oxygen. Secondly, administration of naloxone to reverse the depressant effect of rising endorphin concentrations can restore ventilation (Scardella et al., 1986). Bilateral cervical vagotomy can also restore ventilation, suggesting an inhibitory role for pulmonary afferents (Adams et al., 1988). Road and colleagues (1987) have also suggested an inhibitory role for nociceptive phrenic afferents.

Studies in Human Volunteers

A number of early studies of respiratory muscle endurance employed isocapnic hyperpnoea and measured the maximal sustainable voluntary ventilation as a proportion of maximal voluntary ventilation (MVV). They mostly reported that very high levels of ventilation could be sustained. However, it has not been proven that MVV involves maximal activation of the inspiratory muscles. A decline in maximal inspiratory pressure (MIP) may occur in athletes after exertion but it is not known whether this is due to central or peripheral muscle fatigue. Dempsey and co-workers (Johnson et al., 1993; Dempsey et al., Chapter 29) demonstrated a small decline in diaphragmatic twitch and low frequency tetanic pressures

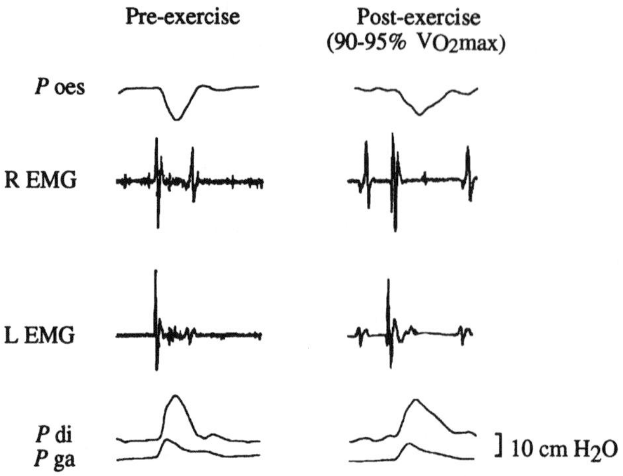

Figure 4. Individual examples of bilateral phrenic nerve stimulation in a subject who demonstrated significant fatigue post-exercise at 95% of V_{O2}max. The figure shows esophageal pressure (P_o), gastric pressure (P_g) and transdiaphragmatic pressure (P_{di}) in addition to the compound muscle action potentials from the left and right diaphragm (LD and RD). All stimuli were performed at FRC (adapted from Johnson et al., 1993).

after maximal intensity exercise but the subjects were still capable of developing their pre-exercise maximal P_{di} during Mueller maneuvers (Fig. 4).

Many studies of respiratory muscle endurance have employed either inspiratory resistive loading (e.g., Roussos & Macklem, 1977; Roussos et al., 1979) or threshold loading (Nickerson & Keens, 1981; see also Clanton et al., 1988, 1993). These studies involve the subject sustaining a given percentage of *maximal* pressure for as long as possible (cf. Clanton et al., 1993). Most investigators failed to document whether the *maximal* pressure was truly optimal for each subject and did not document the degree of failure of central drive when the target pressure could no longer be sustained. An important exception was the study of Bellemare and Bigland-Ritchie (1987) in which twitch interpolation was used to document the development of significant central fatigue.

Bellemare and Bigland-Ritchie (1984, 1987), Gandevia and McKenzie (1985, 1988), Bellemare and colleagues (1986), McKenzie and Gandevia (1986), and Gandevia and colleagues (1990) independently developed a method for quantitating the degree of voluntary activation of the diaphragm using twitch interpolation during maximal static efforts. They showed that the diaphragm could be activated maximally during inspiratory, expulsive and combined inspiratory/expulsive maneuvers. They also used twitch interpolation to investigate the development of central fatigue but reached contrasting conclusions. Whereas Gandevia and McKenzie (1988) reported little evidence of central fatigue during 6 min of repeated *maximal* inspiratory and expulsive efforts (duty cycle 50%), Bellemare and Bigland-Ritchie (1987) reported that, during submaximal expulsive efforts (30% maximal, duty cycle 67%) the ability to develop P_{di} declined by about 50% and half this loss of force could be attributed to central fatigue. In a collaborative study (McKenzie et al., 1992), the latter result was confirmed but the diaphragm appeared to be resistant to central fatigue during *inspiratory* efforts; i.e., it appeared to be task dependent (Fig. 5). That significant central fatigue occurred during inspiratory resistive breathing (Bellemare & Bigland-Ritchie, 1987) suggests a fundamental difference in neural control compared with the intermittent brief maximal static efforts used to test activation in the latter study (McKenzie et al., 1992). This study also showed that *low-frequency fatigue* was relatively small for the diaphragm compared with the elbow flexors performing a similar task and that there was no failure of neuromuscular transmission (Fig. 6). Other aspects of task dependence are considered by Bigland-Ritchie et al., Chapter 27.

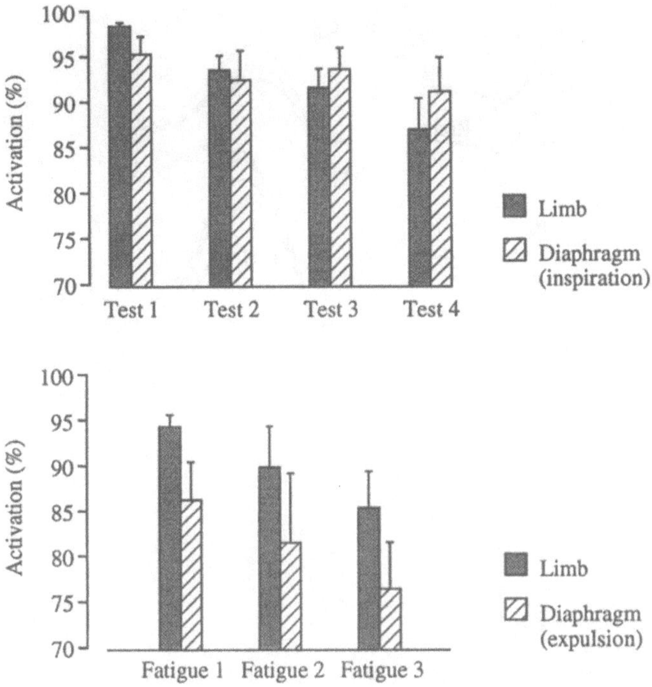

Figure 5. Index of voluntary activation (group data) for the diaphragm and the limb muscle during the four test sequences (above) and 3 fatigue sequences (below). The test sequences consisted of 3-5 brief MVCs with interpolation of twin stimuli (see also Fig. 6) while the intervening fatigue sequences consisted of 10 MVCs of 10 s duration (50% duty cycle). Means for all trials in each test and fatigue sequence shown. Diaphragm tested with *maximal inspiratory maneuvers* during *test* MVCs and *expulsive maneuvers* during *fatigue* MVCs. Stimuli were delivered without warning during fatigue sequences. Note the gradual reduction in voluntary drive with time and the reduced activation during the fatigue sequence, especially for diaphragm performing expulsive maneuvers (from McKenzie et al., 1992).

RESPIRATORY MUSCLE FATIGUE AND VENTILATORY FAILURE

In patients with chronic airflow limitation the balance between increased work of breathing in the face of resistive and elastic loads and the reserve capacity of the respiratory muscles to generate inspiratory pressures is shifted in an unfavorable direction. This balance can be expressed as the ratio of the pressure for a given breath (P_{breath}) and the maximal inspiratory pressure (MIP). As shown by Bellemare and Grassino (1982b, 1983) the likelihood of development of fatigue is predicted by the product of P_{breath}/MIP and the duty cycle for inspiratory work (i.e., the ratio of the duration of inspiration to total breath time: T_i/T_{tot}). They referred to this product as the pressure-time index (PTI) which can be increased by any combination of increased resistive or elastic work and reduced capacity to generate inspiratory pressure (hyperinflation, weakness, fatigue).

Hypercapnia results from a reduction in alveolar ventilation which is determined by the relative magnitudes of tidal volume and physiological dead space and the respiratory frequency. As the PTI increases, the ability to maintain tidal volume is impaired with the inescapable consequence that frequency must rise (Loverdige et al., 1986). Although this may reduce respiratory muscle efficiency and increase the relative proportion of dead space

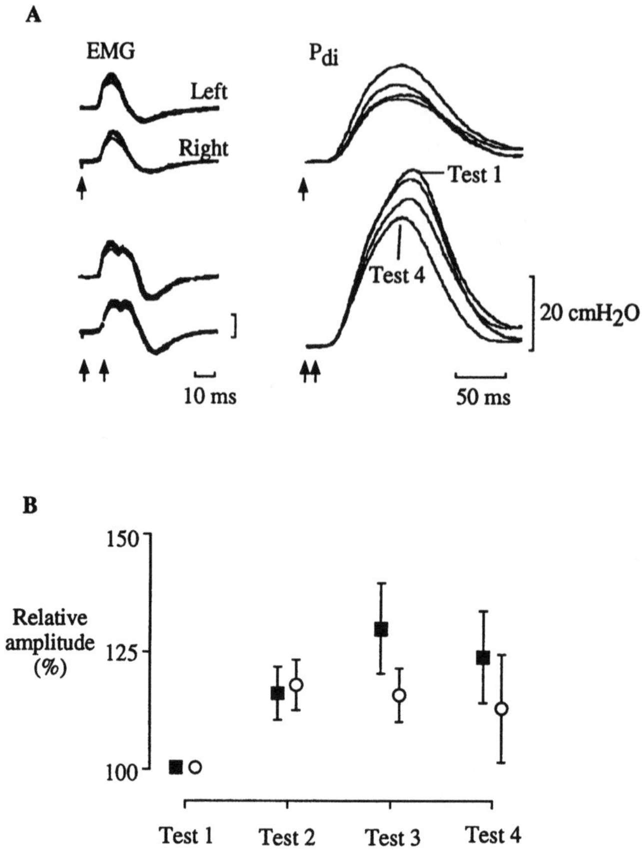

Figure 6. *A*, EMG (left traces) and twitch responses of relaxed diaphragm (P_{di}, right traces) to single stimuli (above) and twin stimuli (below) obtained during the four *test* sequences in a typical study (see Fig. 5). EMG responses of left and right costal diaphragm were obtained with surface electrodes. Note the decline in twitch amplitude following each fatigue sequence. The smallest EMG responses were obtained during the first test prior to development of fatigue. Vertical calibration for EMG 1mV. *B,* amplitude of compound muscle action potentials recorded from the hemidiaphragm during relaxation after each test sequence to a single stimulus. Pooled data from four experiments, normalized to value from first test (mean ± S.E.M.). The amplitude of the negative phase of the compound muscle action potential (closed symbols) and the peak-to-peak amplitude (open symbols) increased (from McKenzie et al., 1992).

ventilation it may reduce the load to breathing and minimize peripheral fatigue (Tobin et al., 1986). The perception of breathlessness and inspiratory effort are also predicted by the ratio P_{breath}/MIP (Killian & Jones, 1988). Thus the reduction of tidal volume may be as much a mechanism to reduce respiratory distress as to minimize inspiratory muscle fatigue.

There is no definitive evidence that inspiratory muscles are weakened as a result of chronic airflow limitation although their ability to develop inspiratory pressure is severely compromised by the length-tension and geometric consequences of hyperinflation. However, the metabolic consequences of advanced disease and acute ventilatory failure may include hypoxaemia, hypercapnia, acidosis and other electrolyte disturbances. Hypercapnic acidosis reduces twitch tension of adductor pollicis and reduces its endurance capacity (Vianna et al., 1990). The diaphragm may also be susceptible (Juan et al., 1984). Hydrogen ion accumulation may impair calcium release from the sarcoplasmic reticulum and also exert a negative influence on the contractile mechanism (e.g., Westerblad et al., 1991).

The diaphragm is resistant to utilizing anaerobic metabolism even under conditions of severe hypoxaemia and extreme loading, presumably because of its enhanced oxidative capacity and perfusion (see above). There is little data on this for the other respiratory muscles and lactate accumulation has been found in human subjects breathing through high inspiratory loads (Jardim et al., 1981; Freedman et al., 1983). Endurance times for resistive breathing are significantly decreased during hypoxaemia (Roussos & Macklem, 1977; Jardim et al., 1981) but it is not clear whether this finding reflects increased peripheral or central fatigue (see also Koulouris et al., 1984).

During prolonged submaximal work the development of fatigue is accompanied by glycogen depletion in the exercising muscle fibers. This has been documented in some animal studies of loaded breathing (Ferguson et al., 1990) but not all (Mayock et al., 1991). The discrepancy could reflect species and age differences or differences in loading protocols. One report has appeared suggesting a reduction in energy sources in biopsies of respiratory muscles from patients with respiratory failure (Gertz et al., 1977).

There is little evidence that respiratory failure in patients with chronic stable hypercapnia is associated with any degree of respiratory muscle fatigue. Bégin and Grassino (1991) studied 300 patients with chronic airflow limitation. Although the PTI for these patients was much higher than normal it did not lie in the fatiguing range (Bellemare & Grassino, 1982b), unless they were asked to breathe more slowly and deeply than they chose.

The study of Bégin and Grassino (1991) also demonstrated that the degree of hypercarbia was not tightly correlated with any single measure of airway narrowing or respiratory muscle compromise. Even when five variables were considered only half the variance of arterial carbon dioxide tension could be explained. This is not surprising in view of the many factors that can affect alveolar ventilation and CO_2 production and the influence of wake/sleep states. A major factor in determining the set point for arterial carbon dioxide is nocturnal hypoventilation due to repetitive central or obstructive apneas. The severity of respiratory disturbance during sleep is not well predicted by any daytime measure of function. Supporting this hypothesis, Bégin and Grassino reported that overweight patients were disproportionately represented among those with increasing levels of hypercarbia during wakefulness.

In summary, the major mechanism for chronic hypercarbia in these patients is a failure to increase central drive sufficiently to maintain alveolar ventilation. Central drive is *not reduced* compared with normal, except perhaps in the late stages of hypercarbia (Aubier et al., 1980; Gribben et al., 1983). This point has been recently confirmed directly by single motor unit recordings from the parasternal intercostals (De Troyer, Gandevia, Leeper and McKenzie, unpublished observations).

Although it seems that peripheral muscular fatigue is not prominent in stable hypercapnic patients the situation could be different in patients with acute exacerbations of critical airway narrowing. Useful data in this setting are scarce although there have been several studies of patients weaning from ventilators. The ratio P_{breath}/MIP is a useful guide and when the ratio is high weaning is likely to fail with development of rapid shallow breathing (Cohen et al., 1982; Tobin et al., 1986; Goldstone et al., 1994). Recently it has been shown that the maximal relaxation rate of the inspiratory pressure-time profile slows in patients who fail a trial of weaning while it remains unchanged in successful trials (Goldstone et al., 1994). While this finding points to a peripheral component of inspiratory muscle fatigue it does not rule out the possibility of a concomitant decline in central drive.

THE ROLE OF VOLUNTARY PATHWAYS IN LOADED BREATHING

For most of this century investigations of respiratory control mechanisms have centered on the groups of respiratory-related neurons located in the bulbo-pontine region of the brainstem. It has been widely accepted that this *respiratory centre* was the primary regulator of respiration in both sleep and wakefulness with only occasional intrusions from supra-pontine regions for specific tasks. For example it has long been assumed that a *voluntary* deep inhalation could be initiated by motor cortical neurons but direct confirmation has only recently been provided. Macefield and Gandevia (1991) demonstrated in recordings from scalp electrodes definite pre-motor D.C. potential shifts (Bereitschaftspotentials) with appropriate latencies prior to inhalation. Motor potentials were recorded 25 ms before the onset of EMG activity. PET scanning techniques have been used during *volitional* breathing to demonstrate increases in cerebral blood flow in the primary motor cortical area, the basal ganglia and the cerebellum (Colebatch et al., 1991).

It has also been accepted that respiratory center output can be inhibited during speech and breath-holding. In 1963 Agostoni documented cessation of diaphragmatic EMG for about 20 s during breath-holding and a potential neural substrate for this inhibition has been identified in cats trained to breath-hold (Orem & Netick, 1986).

Emotions have long been known to influence respiration but this is thought to involve subcortical pathways. For example, patients with lesions in the corticospinal tract which ablate volitional control of breathing still show alterations in respiration with emotions (Munschauer et al., 1991).

In recent years several groups have entertained the possibility that suprapontine pathways may predominate in the control of breathing during wakefulness while the brainstem respiratory neurons play the major role during sleep (and anesthesia). Evidence for this is circumstantial but mounting, especially for exercise hyperpnoea (see Eldridge & Waldrop, 1991). Until recently it has been assumed that the hyperpnoea of exercise was driven by afferent neural information and metabolites from exercising muscles but there have always been paradoxical results which were difficult to explain. Subjects with defective automatic control of respiration who hypoventilate or stop breathing during sleep have normal ventilatory responses to exercise (e.g., children with congenital hypoventilation: Shea et al., 1993; adults with medullary lesions: Heywood et al., 1993). There is even evidence that ventilation can be stimulated by anticipation or imagination of exercise (Tobin et al., 1986; Decety et al., 1991).

Voluntary control of breathing could be achieved either via direct corticospinal pathways or indirectly via putative cortico-bulbar projections (Bassal et al., 1981). The presence of powerful, rapidly-conducting corticospinal projections to a variety of neck and trunk muscles including the diaphragm has been documented by Gandevia and colleagues using high-voltage transcutaneous stimulation of the motor cortex and spinal outflow tracts (Gandevia & Rothwell, 1987; Gandevia & Plassman, 1988).

The partitioning between voluntary and automatic regulation of breathing may be of crucial importance in determining ventilatory responses to loaded breathing, especially in patients with critical airway narrowing (e.g., acute asthma or chronic airflow limitation). It is a common clinical observation that patients with critical airway narrowing cannot sleep. When such patients do *sleep* it means either that they have drifted into hypercapnic coma or their airflow obstruction has been relieved by the emergency therapy. As yet, however, there is no direct evidence to support the contention that volitional breathing is the predominant driving force in such patients. On the other hand there is ample evidence that such patients are at risk of developing hypoventilation during sleep (see Saunders & Sullivan, 1994).

Moreover, there is evidence from animal studies that the respiratory centre is not capable of maximal activation of the phrenic motoneuron pool, even under conditions of extreme hypercarbia and/or hypoxemia (Sieck & Fournier, 1989).

In conclusion, there has been a rapid growth in knowledge about the function of respiratory muscles in response to loading over the past fifteen years. Much of this progress has been stimulated by the development and application of sound neurophysiological methodology in the investigation of a complex set of muscles which act in series and in parallel. In contrast to the situation in the organ bath and for many limb muscles, it is difficult to measure the force, length and velocity of most respiratory muscles non-invasively compounding the difficulties in understanding the mechanics of contraction (cf. Gandevia et al., 1992; McKenzie et al., 1994). The challenge of the future will be to document the precise roles and interactions of the volitional and involuntary control systems in everyday activities and in response to increased loads.

ACKNOWLEDGMENTS

The authors wish to thank Dr. Simon Gandevia and Ms. Gabrielle Allen for comments on the manuscript, and G.A. and Ms. Jane Butler for its preparation. Our laboratory has been supported by the National Health & Medical Research Council of Australia and the Asthma Foundation of New South Wales, Australia. Attendance of the authors at the 1994 Bigland-Ritchie conference was supported, in part, by the Muscular Dystrophy Association (USA; *D.K.McK.*) and the University of Miami (*F.B.*).

REFERENCES

Adams JM, Farkas GA, Rochester DF (1988). Vagal afferents, diaphragm fatigue and inspiratory resistance in the anesthetized dog. *Journal of Applied Physiology* **64**, 2279-2286.

Agostoni (1963). Diaphragm activity during breath-holding: factors related to its onset. *Journal of Applied Physiology* **18**, 30-36.

Aldrich TK (1987). Transmission failure of the rabbit diaphragm. *Respiration Physiology* **69**, 307-319.

Aubier M, Murciano D, Fournier M, Milic-Emili J, Pariente R & Derenne JPh (1980). Central respiratory drive in patients with chronic obstructive lung disease in acute respiratory failure. *American Review of Respiratory Disease* **122**, 191-200.

Aubier M, Trippenbach T & Roussos Ch (1981). Respiratory muscle fatigue during cardiogenic shock. *Journal of Applied Physiology* **51**, 499-508.

Bassal M, Bianchi AL, Dussardier M (1981). Efféts de la stimulation des structures nerveuses centrâles sur l'activité des neurones réspiratoires chez le chat. *Journal of Physiology (Paris)* **77**, 779-795.

Bazzy AR & Donnelly DF (1993). Diaphragmatic failure during loaded breathing: role of neuromuscular transmission. *Journal of Applied Physiology* **74**, 1679-1683.

Bazzy AR & Haddad GG (1984). Diaphragmatic fatigue in unanesthetized adult sheep. *Journal of Applied Physiology* **57**, 182-190.

Bégin P & Grassino A (1991). Inspiratory muscle dysfunction and chronic hypercapnia in chronic obstructive pulmonary disease. *American Review of Respiratory Diseases* **143**, 905-912.

Bellemare F & Bigland-Ritchie B (1984). Assessment of human diaphragm strength and activation using phrenic nerve stimulation. *Respiration Physiology* **58**, 263-277.

Bellemare F & Bigland-Ritchie B (1987). Central components of diaphragmatic fatigue assessed by phrenic nerve stimulation. *Journal of Applied Physiology* **62**, 1307-1316.

Bellemare F, Bigland-Ritchie B & Woods JJ (1986). Contractile properties of the human diaphragm *in vivo*. *Journal of Applied Physiology* **61**, 1153-1161.

Bellemare F & Grassino A (1982a). Evaluation of the human diaphragm fatigue. *Journal of Applied Physiology* **53**, 1196-1206.

Bellemare F & Grassino A (1982b). Effect of pressure and timing of contraction on human diaphragm fatigue. *Journal of Applied Physiology* **53**, 1190-1195.

Bellemare F & Grassino A (1983). Force reserve of the diaphragm in patients with chronic obstructive pulmonary disease. *Journal of Applied Physiology* **55**, 8-15.

Buchler B, Magder, S, Katsardis H, Jammes Y & Roussos C (1985). Effects of pleural pressure and abdominal pressure on diaphragmatic blood flow. *Journal of Applied Physiology* **58**, 691-697.

Chapman RW, Santiago TV & Edelman NH (1980). Brain hypoxia and the control of breathing: neuromechanical control. *Journal of Applied Physiology* **49**, 497-505.

Clanton TL & Ameredes BT (1988). Fatigue of the inspiratory muscle pump in humans: an isoflow approach. *Journal of Applied Physiology* **64**, 1693-1699.

Clanton TL, Hartman E & Julian MW (1993). Preservation of sustainable inspiratory muscle pressure at increased end-expiratory lung volume. *American Review of Respiratory Disease* **147**, 385-391.

Cohen C, Zagelbaum G, Gross D, Roussos Ch & Macklem PT (1982). Clinical manifestations of inspiratory muscle fatigue. *American Journal of Medicine* **73**, 308-316.

Colebatch JG, Adams L, Murphy K, Martin AJ, Lammertsma AA, Tochon-Danguy HJ, Clark JC, Friston KJ & Guz A (1991). Regional cerebral blood flow during volitional breathing in man. *Journal of Physiology (London)* **443**, 91-103.

Davies HW, Haldane JS & Priestly JG (1919). The response to respiratory resistance. *Journal of Physiology (London)* **53**, 60-69.

Decety J, Jeannerod M, Germaine M & Pastene J (1991). Vegetative response during imagined movement is proportional to mental effort. *Behavioural Brain Research* **42**, 1-5.

De Vito EL & Roncoroni AJ (1993). Central and peripheral diaphragmatic fatigue in loaded normal and vagotomized dogs. *Journal of Applied Physiology* **74**, 2820-2827.

Eldridge FL &Waldrop TG (1991). Neural control of breathing during exercise. In: Whipp BJ, Wasserman K (eds.), *Exercise Pulmonary Physiology and Pathology*, pp. 309-370. New York: Marcel Dekker.

Ferguson GT, Irvin CG & Cherniack RM (1990). Effect of corticosteroids on diaphragm function and biochemistry in the rabbit. *American Review of Respiratory Disease* **141**, 156-163.

Freedman S, Cooke NT & Moxham J (1983). Production of lactic acid by respiratory muscles. *Thorax* **38**, 50-54.

Gandevia SC, Gorman RB, McKenzie DK & Southon FCG (1992). Dynamic changes in human diaphragm length: maximal inspiratory and expulsive efforts studied with sequential radiography. *Journal of Physiology (London)* **457**, 167-176.

Gandevia SC & McKenzie DK (1985). Activation of the human diaphragm during maximal static efforts. *Journal of Physiology (London)* **367**, 45-56.

Gandevia SC & McKenzie, DK (1988). Human diaphragmatic endurance during different maximal respiratory efforts. *Journal of Physiology (London)* **395**, 625-638.

Gandevia SC, McKenzie DK & Neering IR (1983). Endurance properties of respiratory and limb muscles. *Respiration Physiology* **53**, 47-61.

Gandevia SC, McKenzie DK & Plassman BL (1990). Activation of human respiratory muscles during different voluntary manoeuvres. *Journal of Physiology (London)* **428**, 387-403.

Gandevia SC & Plassman BL (1988). Responses in human intercostal and truncal muscles to motor cortical and spinal stimulation. *Respiration Physiology* **73**, 325-338.

Gandevia SC & Rothwell JC (1987). Activation of the human diaphragm from the motor cortex. *Journal of Physiology (London)* **384**, 109-118.

Gertz I, Hedenstierna G, Hellers G & Wahren J (1977). Muscle metabolism in patients with chronic obstructive lung disease and acute respiratory failure. *Clinical Science and Molecular Medicine* **52**, 395-403.

Goldstone JC, Green M & Moxham J (1994). Maximum relaxation rate of the diaphragm during weaning from mechanical ventilation. *Thorax* **49**, 54-60.

Gribben HR, Gardiner IT, Heinz CJ, Gibson TJ & Pride NB (1983). The role of impaired inspiratory muscle function in limiting ventilatory response to CO_2 in chronic airflow obstruction. *Clinical Science* **64**, 487-495.

Gross D, Grassino A, Ross WRD & Macklem PT (1979). Electromyogram pattern of diaphragmatic fatigue. *Journal of Applied Physiology* **46**, 1-7.

Hershenson MB, Kikuchi Y & Loring SH (1988). Relative strengths of the chest wall muscles. *Journal of Applied Physiology* **65**, 852-862.

Hershenson MB, Kikuchi Y, Tzelepis GE & McCool FD (1989). Preferential fatigue of the rib cage muscles during inspiratory resistive loaded ventilation. *Journal of Applied Physiology* **66**, 750-754.

Heywood P, Moosavi S, Morrell MJ & Guz A (1993). CO_2 sensitivity and breathing at rest and during exercise in human subjects with unilateral medullary lesions. *Journal of Physiology (London)* **459**, 353P.

Jardim J, Farkas G, Prefaut C, Thomas D, Macklem PT & Roussos Ch (1981). The failing inspiratory muscles under normoxic and hypoxic conditions. *American Review of Respiratory Disease* **124**, 274-279.

Johnson BD, Babcock MA, Suman OE & Dempsey JA (1993). Exercise-induced diaphragmatic fatigue in healthy humans. *Journal of Physiology (London)* **460**, 385-405.

Juan G, Calverly P, Talamo C, Schnader J & Roussos C (1984). Effect of carbon dioxide on diaphragmatic function in the human being. *New England Journal of Medicine* **310**, 874-879.

Killian KJ & Jones NL (1988). Respiratory muscles and dyspnea. *Clinical Chest Medicine* **9**, 237-248.

Koulouris N, Moxham J, Barnes N, Gray B, Heaton R & Green M (1984). Effect of hypoxia on respiratory and limb muscle endurance. *Thorax* **39**,714.

Lee FS, Guenther AE & Melaney HE (1916). Some of the general physiological properties of diaphragm muscle as compared with certain other mammalian muscles. *American Journal of Physiology* **40**, 446-473.

Loverdige B, West P, Kryger MH & Anthonisen NR (1986). Alteration in breathing pattern with progression of chronic obstructive pulmonary disease. *American Review of Respiratory Diseases* **134**, 930-934.

Macefield G & Gandevia SC (1991). The cortical drive to human respiratory muscles in the awake state assessed by premotor cerebral potentials. *Journal of Physiology (London)* **439**, 545-558.

Mayock DE, Badura RJ, Watchko JF, Standaert TA & Woodrum DE (1987). Response to resistive loading in the newborn piglet. *Pediatric Research* **21**, 121-125.

Mayock DE, Standdaert TA, Murphy TD & Woodrum DA (1991). Diaphragmatic force and substrate response to resistive loaded breathing in the piglet. *Journal of Applied Physiology* **70**, 70-76.

McKenzie DK, Bigland-Ritchie B, Gorman RB & Gandevia SC (1992). Central and peripheral fatigue of human diaphragm and limb muscles assessed by twitch interpolation. *Journal of Physiology (London)* **454**, 643-656.

McKenzie DK & Gandevia SC (1986). Strength and endurance of inspiratory, expiratory, and limb muscles in asthma. *American Review of Respiratory Diseases* **134**, 999-1004.

McKenzie DK & Gandevia SC (1987). Influence of muscle length on human inspiratory and limb muscle endurance. *Respiration Physiology* **67**, 171-182.

McKenzie DK & Gandevia SC (1991a). Recovery from fatigue of human diaphragm and limb muscles. *Respiration Physiology* **84**, 49-60.

McKenzie DK & Gandevia SC (1991b). Skeletal muscle properties: Diaphragm and chest wall. In: Crystal RG, West JB (eds.), *The Lung*, pp. 649-659. New York: Raven Press Ltd.

McKenzie DK, Gandevia SC, Gorman RB & Southon FCG (1994). Dynamic changes in the zone of apposition and diaphragm length during maximal respiratory efforts. *Thorax*, **49**, 634-638.

Munschauer FE, Mador MJ, Ahuja A and Jacobs L (1991). Selective paralysis of voluntary but not limbically influenced automatic respiration. *Archives of Neurology* **48**, 1190-1192.

Nava S & Bellemare F (1989). Cardiovascular failure and apnea in shock. *Journal of Applied Physiology* **66**, 184-189.

Newell SZ, McKenzie DK & Gandevia SC (1989). Inspiratory and skeletal muscle strength, endurance and diaphragmatic activation in patients with chronic airflow limitation. *Thorax* **44**, 903-912.

NHLBI Workshop on Respiratory Muscle Fatigue: Report of the Respiratory Muscle Fatigue Workshop Group (1990). *American Review of Respiratory Disease* **142**, 474-480.

Nickerson BG & Keens TG (1981). Measuring ventilatory muscle endurance in humans as sustainable inspiratory pressure. *Journal of Applied Physiology* **52**, 768-772.

Orem J & Netick A (1986). Behavioural control of breathing in the cat. *Brain Research* **366**, 238-253.

Road J, West NH & Van Vliet BN (1987). Ventilatory effects of stimulation of phrenic afferents. *Journal of Applied Physiology* **63**, 1063-1069.

Rochester DF (1991a). Effects of COPD on the respiratory muscles. In: Cherniack NS (ed.), *Chronic Obstructive Pulmonary Disease*, pp. 134-157, Philadelphia: WB Saunders Co.

Rochester DF (1991b). Respiratory muscle weakness, pattern of breathing, and CO2 retention in chronic obstructive pulmonary disease. *American Review of Respiratory Disease* **143**, 901-903.

Roussos Ch, Fixley M, Gross D & Macklem PT (1979). Fatigue of inspiratory muscles and their synergistic behaviour. *Journal of Applied Physiology* **46**, 897-904.

Roussos Ch & Macklem PT (1977). Diaphragmatic fatigue in man. *Journal of Applied Physiology* **43**, 189-197.

Sadoul N, Bazzy AR, Akabas SR & Haddad GG. (1985) Ventilatory response to fatiguing and nonfatiguing resistive loads in awake sheep. *Journal of Applied Physiology* **59**, 969-978.

Saunders NA & Sullivan CE (eds.) (1994). *Sleep and Breathing.* 2nd Ed. New York: Marcel Dekker.

Scardella AT, Parisi RA, Phair DK, Santiago TV & Edelman NH (1986). The role of endogenous opioids in the ventilatory response to acute flow-resistive loads. *American Review of Respiratory Disease* **133**, 26-31.

Shea SA, Andres LP, Shannon DC & Banzett RB (1993). Ventilatory responses to exercise in humans lacking ventilatory chemosensitivity. *Journal of Physiology (London)* **468**, 623-640.

Sieck GC & Fournier M (1989). Diaphragm motor unit recruitment during ventilatory and nonventilatory behaviours. *Journal of Applied Physiology* **66**, 2539-2545.

Tobin MJ, Press V & Guenther SM (1986). The pattern of breathing during successful and unsuccessful trials of weaning from mechanical ventilation. *American Review of Respiratory Disease* **134**, 1111-1118.

Vianna LG, Koulouris N, Lanigan C & Moxham J (1990). Effect of acute hypercapnia on limb muscle contractility in humans. *Journal of Applied Physiology* **69**, 1486-1493.

Watchko JF, Mayock DE, Standaert TA & Woodrum DE (1988). Ventilatory failure during loaded breathing: the role of central neural drive. *Journal of Applied Physiology* **65**, 249-255.

Westerblad H, Lee JA, Lannergren J & Allen DG (1991). Cellular mechanisms of fatigue in skeletal muscle. *American Journal of Physiology (Cell Physiology)* **261,** C195-C209.

FATIGUE OF JAW MUSCLES AND SPEECH MECHANISMS

T. S. Miles and M. A. Nordstrom

Department of Physiology
University of Adelaide
Adelaide SA 5005, Australia

ABSTRACT

Histochemical studies show that the distribution of fiber types in human jaw muscles is different from that in various limb muscles, no doubt representing different functional demands as well as a different embryological derivation. Jaw-closing muscles appear more resistant to fatigue than limb muscles with intermittent maximal contractions. Endurance of continuous isometric biting is limited by pain. Masseter motor unit fatigability in sub-maximal contractions is similar to the limb muscles. There are few physiological data for the jaw-opening muscles. The distribution of fiber types in human speech muscles is consistent with the high speeds of contraction that must be used in phonation. Although clinical syndromes of fatigue of speech muscles are recognized, there is little direct information on the fatigability of the muscle fibers themselves.

INTRODUCTION

The current world record for weightlifting in the *snatch* event, in which a barbell is lifted above the shoulders in one movement, is 216 kg (about 2120 N). However, it is not necessary to be an elite athlete to exert forces of this magnitude. In 1681, Borelli attached weights to a cord which passed over the mandibular teeth and reported that his subjects were able to lift about 200 kg (Rowlett, 1932-33). More recently, Gibbs and colleagues (1986) reported that the highest isometric, bilateral bite force recorded with an improved transducer was 4400 N, although the average value was 725 N. Another recent study has reported maximal values for unilateral bites of 847 N and 597 N for young adult males and females respectively (Waltimo & Könönen, 1993).

These observations serve to illustrate the power of the jaw-closing muscles. The muscles themselves are well adapted to produce the high forces that are required for breaking down tough foods (although these are lamentably rare in modern diets). The six muscles that act to elevate the jaw have a higher mechanical advantage than most limb muscles because they are short and act directly across the joint without the interposition of long, flexible tendons between the muscle and the bones. These muscles are phenotypically different from

Fatigue, Edited by Simon C. Gandevia et al.
Plenum Press, New York, 1995

those in the limbs. These differences may reflect not only their unique functions, but also their different embryological origin. In comparison to the limb and trunk muscles which are derived from myotones of the somites, the masticatory muscles arise from somitomeres in the first branchial arch (Noden, 1983). The masseter in particular expresses myosin isoforms that are observed only during ontogeny in trunk and limb muscles (Soussi-Yanicostas et al., 1990), including the *cardiac-specific* heavy chain that is found only in muscles (including the heart) that arise from the cranial part of the embryo (Bredman et al., 1990).

In marked contrast to the powerful jaw-closing muscles, the jaw-opening muscles are long and thin, and do not exert much force. They usually operate against minimal external load when opening the jaw, and are rarely tonically active.

The concept of fatigue of the masticatory muscles may seem strange to those of us accustomed to a modern diet in which the food is chewed a few times, then swallowed. Indeed, it is likely that jaw muscle fatigue is rare in these circumstances. However, the histochemical appearance of the jaw-closing muscles suggests that they evolved to deal with diets that required considerable force. Waugh (1937) reported that Alaskan Eskimos could exert bite forces as high as 158 kg (about 1550 N). Their masticatory prowess may reflect their diet at that time, in which the practice of chewing whale skins to remove the blubber gave their jaw-closing muscles intensive training. Masticatory fatigue in those eating a modern diet is probably limited mainly to individuals who grind their teeth for protracted periods, usually while sleeping. (The syndrome of night grinding, known as bruxism, is important clinically because it is associated with severe pain in the head and neck).

This review will also touch briefly on fatigue in the muscles involved in speech. This will be limited to the little that is known about the laryngeal muscles as the respiratory muscles are dealt with elsewhere (Dempsey & Babcock, Chapter 29; McKenzie & Bellamare, Chapter 30.

FATIGUE OF THE JAW MUSCLES

Histochemical Profile of Human Masticatory Muscles

The muscle fiber composition of the jaw muscles provides useful information on the likely fatigue resistance of the jaw muscles (Table 1). The fiber types of the human jaw muscles have been classified histochemically using biopsy (Ringquist, 1971; Ringquist, 1973; Serratrice et al., 1976; Ringquist et al., 1982) and autopsy techniques (Serratrice et al., 1976; Vignon et al., 1980; Eriksson et al., 1981; Eriksson et al., 1982; Eriksson & Thornell, 1983).

The main fiber type in the jaw closers and lateral pterygoid is type I, with type IIB the next most numerous. Very few type IIA fibers are found in the jaw muscles. The jaw-opening digastric muscle has a more even distribution of the three main fiber types, and its fiber type composition and appearance is more like the limb muscles. The histological appearance of the jaw-closing muscles (masseter, medial pterygoid, temporalis) and lateral pterygoid differs from that normally seen in the limb muscles. These differences are relevant for fatigue resistance of the muscles, and are summarized below:

1. The diameter of type I fibers is larger than that of type II fibers in nearly all sites in the muscle (Ringquist, 1973; Serratrice et al., 1976; Vignon et al., 1980; Eriksson & Thornell, 1983). This is in contrast to the normal situation in the limb muscles where type II fibers are either larger, or of similar diameter to type I fibers (Dubowitz & Brooke, 1973). The total cross-sectional area of the jaw-closing

Table 1. Relative proportion of histochemical fiber types in masticatory muscles

	Histochemical fiber type proportion				
Muscle	I	IM	IIA	IIB	IIC
Masseter*	63	6	2	27	3
Medial Pterygoid*	54	8	0	36	2
Temporalis*	53	5	0	39	2
Lateral pterygoid†	70	14	0	11	5
Digastric§	34	0	27	38	0

Values represent percent of fibers of each histochemical type in the jaw muscles.
*Eriksson and Thornell (1983).
†Eriksson and colleagues (1981).
§Eriksson and colleagues (1982).

muscles and lateral pterygoid occupied by type I fibers (up to 90%; Eriksson & Thornell, 1983) is therefore larger than suggested by the fiber type proportions.

2. Diameters of both the type I and II fibers are smaller than in other skeletal muscles (Ringquist, 1971; Eriksson & Thornell, 1983; Vignon et al., 1980). Small fiber diameters may enhance fatigue resistance by facilitating diffusion of metabolites and energy substrates in and out of muscle cells (Edstrom & Grimby, 1986).
3. The jaw-closing muscles contain a large proportion of fibers with intermediate staining for myosin-ATPase (MATPase) (Vignon et al., 1980; Ringquist et al., 1982; Eriksson & Thornell, 1983). Fibers of this type (IM and IIC) are rare or non-existent in normal adult limb muscles (Dubowitz & Brooke, 1973), but have been seen with extreme endurance training of muscles (Edstrom & Grimby, 1986).
4. It appears that the jaw-closing muscles and lateral pterygoid in general do not contain type IIA fibers (Eriksson et al., 1981; Ringquist et al. 1982, Eriksson & Thornell, 1983), although these may be found in significant proportions in some individuals. Type IIA fibers correspond to motor units of the FR type.
5. The masticatory muscles do not have the mosaic pattern of fiber types normally found in limb muscles, but rather have large groups of densely packed fibers of the same histochemical type (Eriksson & Thornell, 1983). Such an appearance in a limb muscle would be considered pathological (Dubowitz & Brooke, 1973).

In summary, the histological appearance of the masticatory muscles is sufficiently different from the normal appearance of limb muscles that care should be taken in inferring the physiological properties of the muscles from this criterion alone. The high proportion of type I fibers, and the small diameters of the fibers suggest that masticatory muscles should be reasonably resistant to fatigue.

Physiological Data from Animal Experiments

In contrast to the limb muscles, the physiological properties of jaw muscles and their motor units have not been well characterized in animal experiments. For example, there have been no studies in which electrical stimulation has been used to determine the mechanical properties of jaw-muscle motor units in animals. This is presumably because of the difficulty in gaining access to the motor nerves in these muscles without damaging the muscles themselves. Motor roots are not accessible for fiber-splitting as they are for limb muscles. It is feasible to activate motoneurons in the trigeminal motor nucleus by intracellular current injection through a microelectrode, but this technique has not been used to investigate the mechanical properties of motor units in the masticatory system. Consequently, the physiological properties of motor units in the jaw muscles of animals have mostly been inferred from whole-muscle studies. Even here, the literature is sparse. The consensus of the few animal studies in the literature is that the jaw muscles are fast-twitch muscles. Direct stimulation of the muscle has been used in most animal studies because of the difficulty of access to the motor nerves. Tamari and colleagues (1973) reported that the twitch time-to-peak tension (TTP) of the cat masseter was faster than temporalis (18 *vs.* 34 ms). Very fast twitch TTPs have also been reported in the masseter of the rat (14 ms; Nordstrom & Yemm, 1974) and jaw-closers in the possum (16-18 ms; Thexton & Hiiemae 1975). The most extensive study is that of Taylor and colleagues. (1973), who combined an investigation of the mechanical properties of masseter and temporalis muscle strips and a histochemical analysis. The muscle strips were very fast-contracting, with a twitch TTP of 11-13 ms. A fatigue test was applied to the muscle strips (train of pulses at 55-90 Hz, of 330 ms duration, repeated once every second). With this stimulation pattern there was a rapid loss of tension in the first minute to about 25% of the initial level, and then a gradual decline over the next 2-3 min. Although this test paradigm was chosen to be compatible with a widely used fatigue test (see Burke, 1981) it is possible that with these high rates of stimulation fatigue resulted primarily from failure of activation of the muscle fibers. It is not possible to assess this as EMG was not recorded. The fatigue profile was deemed to be consistent with the relative proportions of fiber types determined histochemically. The predominant fiber type was large in size and stained strongly for MATPase, and weakly for the oxidative enzyme succinate dehydrogenase (SDH). The second fiber type was of intermediate size, stained strongly for MATPase and SDH. These fiber types were assumed to correspond to motor unit types FF and FR, respectively, and in the cat masseter they comprised 90% of the muscle cross-sectional area. The remaining fibers had low MATPase activity, and stained strongly for SDH, corresponding to type S motor units.

Even if extensive data on jaw muscle fatigue were available from animals, its relevance to fatigue of human jaw muscles is questionable. There are significant differences in the fiber type composition of the human and animal jaw muscles. Human jaw-closing muscles are composed predominantly of type I fibers, whereas in the cat and rat the jaw-closers are almost exclusively type II fibers. Human jaw muscles appear to lack the type IIA fibers (corresponding to the physiological type FR units). The masseter in the rhesus monkey does appear to have a similar histochemical profile to humans (Maxwell et al., 1979) and might therefore be a valuable animal model for physiological data.

Physiological Data from Whole-Muscle Studies in Humans

Fatigue can be defined as a reduction of force-producing capacity of the neuromuscular system with prolonged activity (Asmussen, 1979). It is often useful to distinguish between central and peripheral fatigue, particularly when attempting to quantify fatigue. Central fatigue is characterized by a reduction of electrical activation of the muscle, accompanied by (but not necessarily causing) force loss. Peripheral (or contractile) fatigue is characterized by a reduction in force for a given level of muscle excitation. When measuring fatigue it is desirable to assess, and if possible control the level of excitation of the muscle in order to differentiate central from peripheral causes of force loss. For most practical purposes, it is desirable to test fatigue under conditions of constant excitation (i.e., in the absence of central fatigue), so that fatigue may be objectively measured in terms of loss of force. This ideal experimental design is possible in animal experiments, but is very difficult to achieve in human experiments. Unfortunately, in many studies of human jaw muscle fatigue, objective quantification of fatigue (deficit in force producing capacity) has been hampered by loose operational definitions of *fatigue*. The approaches to fatigue testing in the human jaw muscles can be grouped into three broad categories, which are considered below. Of these, only the use of the maximal voluntary contractile force (MVC) approaches the ideal of an objective measure of fatigue.

1. The endurance time, or the time elapsed until the subject is unwilling to continue a required task, has been used as an index of fatigue in both maximal (Christensen, 1979; Palla & Ash, 1981; Christensen & Mohamed, 1983; Christensen et al., 1985) and submaximal biting (Naeije, 1984; Clark et al., 1984; Clark & Carter, 1985; Clark & Adler, 1987; Maton et al., 1992). Endurance has been confused with resistance to fatigue in many of these studies but while they may be related, they are not necessarily the same. Endurance is the ability to withstand prolonged strain, and is limited by pain and motivational factors as well as contractile fatigue. Endurance times are shorter in the jaws than limb muscles for isometric contractions at comparable relative forces (Maton et al., 1992), but this may indicate that isometric contractions of jaw muscles are more painful rather than more fatigable than the limbs. Clark and colleagues have examined this issue in the jaw muscles in well-designed experiments. During sustained biting at force levels varying from 25-100% of maximal they demonstrated that the subject's endurance was limited by pain (Clark & Carter, 1985). The ability to perform intermittent MVCs at intervals during sustained submaximal biting was unaffected, indicating that fatigue or inability to produce the target force was not the limiting factor. Pain in a contracting muscle is believed to be related to an increased concentration of metabolic by-products of contraction such as H^+ and K^+ as a consequence of contraction-induced occlusion of blood flow through the muscle (Mense, 1977).
2. The shift in the power spectrum of the surface EMG signal towards greater low-frequency energy content with muscle fatigue (Lindstrom et al., 1970; Mills, 1982) has been used in numerous attempts to quantify jaw muscle fatigue in humans (Naeije & Zorn, 1981; Palla & Ash, 1981; Lindstrom & Hellsing, 1983; van Boxtel et al., 1983; Naeije, 1984; Kroon et al., 1986; Maton et al., 1992). All of these studies show the familiar increase in low frequency content in the EMG signal during a variety of fatiguing tasks yet, to be meaningful, any index of fatigue must consider force, and studies in which force has not been measured (Palla & Ash, 1981; Naeije & Zorn, 1981; van Boxtel et al., 1983; Naeije, 1984; Kroon et al., 1986) are difficult to interpret in terms of force-producing capacity. Clark and colleagues (1988) showed that increased low-frequency power in the masseter and temporalis EMG power spectrum during prolonged isometric biting was not accompanied by a reduction in MVC force. The change in power spectrum of the surface EMG signal appears to be closely related to metabolic changes in the active muscle and concomitant reductions in muscle fiber conduction velocity (Lindstrom et al., 1970; Eberstein & Beattie, 1985), but the relationship between the EMG signal and force-producing capacity during fatigue is complex (Mills, 1982; Sandercock et al., 1985).
3. The ability to produce maximal force during repeated contractions appears to give a reasonably objective measure of fatigue. In a comparison of the ability to maintain repeated maximal contractions of short duration, van Steenberghe and colleagues (1978) found that the jaw-closing muscles were not affected under conditions in which the muscles producing hand-grip, and arm-flexion forces were significantly weakened. The ability to preserve maximal force-producing capacity during and following prolonged submaximal contractions at various proportions of maximal force also appears to be greater in the jaw-closing muscles than other muscles studied (Clark et al., 1984; Clark & Carter, 1985).

Fatigue of the jaw-opening muscles does not seem to have been studied in humans. Jaw-opening muscles normally do not contract against a load, are very weak in relation to jaw-closers, and are used intermittently. Fatigue in jaw-opening muscles is not likely to be a problem in everyday life. Endurance tests of jaw-retruders (which include anterior digastric) show a similar pattern as the jaw-closers (Clark & Adler, 1987).

In summary, quantitative information on the fatigability of the human jaw muscles from whole-muscle studies of voluntary contractions is limited. There is evidence from maximal biting force experiments that the jaw-closing muscles are more resistant to fatigue of maximal contractions than non-masticatory muscles (van Steenberghe et al., 1978; Clark et al., 1984; Clark & Carter, 1985). The paucity of relevant data from animal studies, and the unusual histochemical appearance of the human jaw muscles, further complicate the assessment of fatigability of the human jaw muscles in submaximal contractions. For these reasons, it is useful to examine single motor units in the jaw muscles to estimate fatigability during submaximal efforts.

Physiological Data from Single Motor Units in Humans

In contrast to the limb muscles, there are no animal data on the physiological properties of individual motor units in the jaw muscles. The only information on this topic comes from human studies employing the spike-triggered averaging (STA) technique (Goldberg & Derfler, 1977; Yemm, 1977; Nordstrom & Miles, 1990; McMillan et al., 1990). With this technique, a needle or fine-wire electrode is inserted into the muscle to record the activity of a single motor unit. Subjects maintain a low discharge rate of the motor unit with the aid of feedback. The partially fused twitch force of a single motor unit is extracted from the total force signal using ensemble averaging (see Thomas, Chapter 10). Yemm (1977) reported a continuous range of contraction times for masseter units from 24-91 ms, with no evidence for separate populations of fast and slow motor units. Goldberg and Derfler (1977) also found a continuous distribution of contractile speeds in the masseter motor unit twitches, but reported that they were distributed over a narrower range (38-69 ms). A similar range (25-67 ms) was reported by McMillan and colleagues (1990).

Motor unit fatigability in human jaw muscles has been assessed only in the masseter muscle (Nordstrom & Miles, 1990). Most (95%) were fast-twitch (mean twitch TTP ± SD 34 ± 10 ms) and showed a broad range of fatigability. Although most units were recruited at less than 20% MVC, only 2 of 37 (5%) were classified as type S units on the basis of their twitch properties. Fig. 1 shows examples of twitch fatigue in 3 masseter motor units during a 15-min isometric bite. The averaged motor unit twitches in the first and fifteenth minute of the contraction are shown on the left. The time-course of the change in twitch amplitude is shown immediately to the right for each motor unit. The units display a range of fatigability, with unit *A* being fatigue resistant, *B* fatiguing gradually over the fifteen minutes, and *C* fatiguing rapidly in the first few minutes without subsequent recovery. Three indices of fatigue were calculated. 1) FI_6 = 100 x (the ratio of twitch amplitude in the 6th and 1st min). This index facilitated comparison with other human motor unit fatigue studies (Table 2); 2) FI_{15a} = 100 x (the ratio of twitch amplitude in the 15th and 1st min); and 3) FI_{15b} = 100 x (the ratio of twitch amplitude in the 5th 3-min epoch and the first 3-min epoch).

There were non-significant correlations of fatigability with both twitch amplitude and contractile speed in the predominantly fast-twitch masseter motor units (Nordstrom & Miles, 1990). There are several anomalies between the physiological data and the available histochemical data for the human masseter. Despite the histochemical evidence that a substantial proportion of the masseter is comprised of type I fibers (Eriksson & Thornell, 1983), we found very few physiological type S units in the masseter. This is not unprecedented for physiological fast-twitch muscle (Goslow et al., 1977). The human masseter

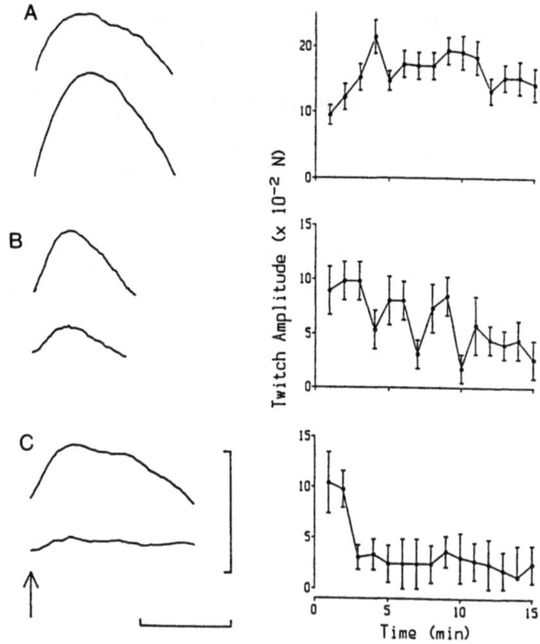

Figure 1. Twitch fatigue in 3 human masseter motor units. Averaged motor unit twitches calculated during the 1st and 15th min of contraction for 3 different motor units are shown on *left* in *A-C*. Uppermost trace in each pair, twitch calculated during the 1st min; arrow, time of occurrence of trigger. Vertical calibration bar, 200 mN, horizontal bar, 50 ms. Immediately to *right* of each pair of twitches, mean twitch amplitude measured in each 1-min epoch for corresponding unit is plotted over 15 min. Note different vertical scale in each. Error bars, SE (adapted from Nordstrom & Miles, 1990).

generally lacks the type IIA fibers which are believed to correspond to the physiological type FR motor units (Eriksson & Thornell, 1983). However, we found a substantial population of fast-twitch, fatigue resistant units (type FR) in masseter using STA. The most likely explanation for the conflicting physiological and histological evidence is that at least some type FR units in human masseter have muscle fibers which stain histochemically as type I. This suggests that the correlation between the physiological properties of motor units and the staining of their fibers for MATPase is not as strong as is presently believed. Unique isoforms of myosin are present in the jaw-closing muscles of cats and other carnivores (Hoh et al., 1993). The cat jaw-closers contain *superfast* fibers, which contract isometrically at about twice the speed as limb fast fibers, apparently because of their superfast myosin. Although these fibers are present in most primates, they are believed to be absent in man (reviewed in Hoh et al., 1993). Although a relationship between histochemical fiber type and motor unit fatigue resistance has been demonstrated in the cat hindlimb (Burke, 1981), there is evidence that fatigue resistance correlates with the activity of oxidative enzymes, not MATPase activity *per se* (Kugelberg & Lindegren, 1979; Hamm et al. 1988). It seems necessary to assess metabolic enzyme activities in the human jaw muscles to make reliable estimates of fatigue tolerance of the tissues.

How does the fatigability of human masseter motor units compare with motor units in other human muscles? A summary is presented in Table 2. The first 3 columns show the data for the 3 fatigue indices used for masseter units by Nordstrom and Miles (1990). In a study of human FDI motor units, Young and Mayer (1981) used intramuscular microstimulation to activate 41 motor units. The fatiguing paradigm was 30 Hz pulse trains for 330 ms, repeated at 1 s^{-1} for 2 min, i.e., about 1200 stimuli. This test was designed to be similar to the standard fatigue test which has been used to categorize motor unit populations of a number of animal muscles (see Burke, 1981); however this test does not seem appropriate for human muscle. Stimulation for 2 min is insufficient to induce contractile fatigue in human FDI, as 83% of the units were potentiated by the test (FI > 100). Similar results were found in human thenar muscles using intermittent 40 Hz stimulation of motor axons for 2 min

Table 2. Distribution of fatigability among motor units in several human muscles

	Human masseter			Human FDI			
	FI_6	FI_{15a}	FI_{15b}	Stephens and Usherwood*	Young and Mayer†	Human MG Garnet et al.§	Human thenar Thomas et al.§§
FI > 75	56	34	19	62	90	55	72
FI 25-75	28	38	59	24	7	39	28
FI < 25	16	28	22	14	3	6	0

Values represent percent of units studied with fatigue indices (FI) falling within a particular range. See text for explanation of FI_6, FI_{15a}, FI_{15b}. FDI, first dorsal interosseus; MG, medial gastrocnemius.
*Stephens and Usherwood (1977).
†Young and Mayer (1981).
§Garnett and colleagues (1978).
§§Thomas and coworkers (1991).

(Thomas et al., 1991). Two minutes of activity is also insufficient for fatigue in the masseter, as many units were potentiated during the first 2-4 minutes of continuous 10 Hz activation. It seems that human motor units are less fatigable than motor units studied in animals such as cat and rat.

Stephens and Usherwood (1977) used STA to assess twitch fatigue of 22 motor units in human FDI after 5 minutes of continuous activity at 10 Hz. A pressure cuff was used to prevent recovery. This fatigue index is directly comparable with our FI_6 in masseter, particularly as in both studies units were tested at comparable initial total force levels. In the masseter study, 75% of tested units had an activation threshold below 10% of maximal force, while in FDI approximately 63% of the tested units had an activation threshold below 10% of maximal. The distribution of FIs after 5 minutes of activity was similar in the ischaemic FDI and masseter units.

The only other fatigue data for human motor units is in the MG muscle (Garnett et al., 1978). Intramuscular microstimulation was used to fatigue 18 units by activation at 10-20 Hz for 0.5 s, repeated every 1 or 2 s. The FI was the ratio of the initial twitch or tetanic tension to that after 3000 stimuli, which is comparable with the FI_6 in the masseter study. Although the fatiguing paradigm activated a single unit in isolation (in contrast to the situation with voluntary activation), the distribution of motor unit fatigability in this muscle was also similar to the masseter.

This comparison suggests that the pattern of contractile fatigue in masseter motor units is similar to that seen in FDI and MG after about 3000 activations. Although different factors undoubtedly influence fatigability in maximal and low-force isometric contractions, the present findings suggest that the superiority of the masseter in repeated high-force contractions (van Steenberghe et al., 1978; Clark et al., 1984; Clark & Carter, 1985) is not due to an inherent advantage in aerobic fatigue-resistance of its low- and moderate-threshold motor units. Therefore, other explanations, such as superior oxygen delivery become more likely (van Steenberghe et al., 1978).

The motor unit organization in the masseter should be considered from a functional viewpoint. Unlike the limb muscles, the masseter has minimal tonic postural activity. At rest, the jaw is supported against gravity mostly by passive tension, with perhaps some activity in temporalis but not in masseter (Yemm, 1976). The paucity of physiological type S units in the masseter is therefore not surprising. The masseter is involved in a wide variety of activities which include mastication, speech, swallowing, and facial expression. These diverse tasks, and the rapid, intermittent activation of motor units necessary for many of these functions may produce quite different functional demands on the muscles than the more stereotyped functions of the limb muscles.

FATIGUE OF THE MUSCLES OF PHONATION

The primary role of the laryngeal muscles is to protect the airways against the invasion of foreign objects. In some species, notably humans, they have evolved a secondary role in sound production. Although any anesthetist will relate experiences of laryngospasm, in which the laryngeal muscles continue to contract until death is imminent, there is comparatively little direct information about fatigue properties of the laryngeal muscles.

Human voice muscles contain a higher proportion of oxidative fibers than sub-primate species. Histochemically, the posterior cricoaretynoid muscle in humans consists mostly of type I fibers, while the thyroaretynoid has about 42% type I fibers and 58% type II fibers (see Cooper & Rice, 1990). Claassen and Werner (1992) reported that type I, IIA and IIB fibers were present in the human thyroarytenoid and posterior cricoarytenoid muscles. The percentage of type IIB fibers was low in the thyroarytenoid and posterior cricoarytenoid muscles and in the other laryngeal muscles. The type IIA fiber content of the arytenoid muscle was significantly higher than in the posterior cricoarytenoid muscle.

Cooper and Rice (1990) stimulated canine vocal fold muscles with trains of electrical stimuli and observed that they are resistant to fatigue, but it is not clear how this would relate to the work done by these muscles in phonation when the contractions occur in brief, high-speed bursts. Robin and coworkers (1992) examined the strength and endurance of human tongue muscles in subjects who had acquired high skill levels with their tongues. The maximal strength of the tongue muscles was found to be similar in skilled and control subjects but, when pressing with their tongues on air-filled bulb, the skilled group could maintain 50% of their maximal pressure for longer than the controls, indicating an increased endurance time.

For obvious technical reasons, there is little direct information on the fatigability of speech muscles in humans. However, one interesting clinical syndrome which is believed to represent a discrete vocal fatigue syndrome has been described in professional voice users (Koufman & Blalock, 1988). This syndrome is characterized by muscle tension in the neck, poor control of the breath stream and an abnormally low-pitched speaking voice in both men and women. The evocative expression *Bogart-Bacall syndrome* has been coined to describe the characteristically husky voice in this disorder.

In summary, it is apparent that motor units in the human masticatory muscles possess combinations of motor unit physiological, histochemical, and anatomical properties which differ from the relationships widely accepted for animal limb muscles. One can only speculate that their unique features reflect their complex functional requirements. The fatigue properties of the speech muscles await further elucidation.

ACKNOWLEDGMENTS

The authors' laboratories are supported by the National Health & Medical Research Council of Australia. Attendance of *T.S.M.* at the 1994 Bigland-Ritchie conference was supported, in part, by a grant from the Muscular Dystrophy Association (USA) and the University of Arizona Regents' Professor funds of Douglas Stuart.

REFERENCES

Asmussen E (1979). Muscle fatigue. *Medicine Science and Sports* **11**, 313-321.

Bredman JJ, Weijs WA, Moorman AMF (1990). Expression of "cardiac specific" myosin heavy chain in rabbit cranial muscles. In Marechal G, Carraro U (eds.), *Muscles and Motility*, pp. 329-335. Andover, Hampshire, Intercept.

Burke RE (1981). Motor units: anatomy, physiology, and functional organization. In: Brookhart JM, Mountcastle VB (sec. eds.), Brooks VB (vol. ed.), *Handbook of Physiology, sec. 1, vol. II, pt 1, The Nervous System: Motor Control.*, pp. 345-422. Bethesda, MD: American Physiological Society.

Christensen LV (1979). Some subjective-experimental parameters in experimental tooth clenching in man. *Journal of Oral Rehabilitation* **6**, 119-136.

Christensen LV & Mohamed SE (1983). The possible activity of large and small jaw muscle units in experimental tooth clenching in man. *Journal of Oral Rehabilitation* **10**, 519-525.

Christensen LV, Mohamed SE & Rugh JD (1985). Isometric endurance of the human masseter muscle during consecutive bouts of tooth clenching. *Journal of Oral Rehabilitation* **12**, 509-514.

Claassen H & Werner JA (1992). Fiber differentiation of the human laryngeal muscles using the inhibition reactivation myofibrillar ATPase technique. *Anatomy and Embryology* **186**, 341-346.

Clark GT & Adler RC (1987). Retrusive endurance, fatigue and recovery of human jaw muscles at various isometric force levels. *Archives of oral Biology* **32**, 61-65.

Clark GT, Beemsterboer PL & Jacobsen R (1984). The effect of sustained submaximal clenching on maximum bite force in myofascial pain dysfunction patients. *Journal of Oral Rehabilitation* **11**, 387-391.

Clark GT & Carter MC (1985). Electromyographic study of human jaw-closing muscle endurance, fatigue and recovery at various isometric force levels. *Archives of oral Biology* **30**, 563-569.

Clark GT, Carter, MC & Beemsterboer PL (1988). Analysis of myoelectric signals in human jaw closing muscles at various isometric force levels. *Archives of oral Biology* **33**, 833-837.

Cooper DS & Rice DH (1990). Fatigue resistance of canine vocal fold muscle. *Annals of Otolaryngology Rhinology and Laryngology* **99**, 228-233.

Dubowitz V & Brooke MH (1973). *Muscle Biopsy- A Modern Approach.* London: Saunders.

Eberstein A & Beattie B (1985). Simultaneous measurement of muscle conduction velocity and EMG power spectrum changes during fatigue. *Muscle & Nerve* **8**, 768-773.

Edstrom L & Grimby L (1986). Effect of exercise on the motor unit. *Muscle & Nerve* **9**, 104-126.

Eriksson P-O, Eriksson A, Ringquist M & Thornell L-E (1981). Special histochemical muscle-fibre characteristics of the human lateral pterygoid muscle. *Archives of oral Biology* **26**, 495-507.

Eriksson P-O, Eriksson A, Ringquist M & Thornell L-E (1982). Histochemical fibre composition of the human digastric muscle. *Archives of oral Biology* **27**, 207-215.

Eriksson P-O & Thornell L-E (1983). Histochemical and morphological muscle-fibre characteristics of the human masseter, the medial pterygoid and temporal muscles. *Archives of oral Biology* **28**, 781-795.

Garnett RAF, O'Donovan MJ, Stephens JA & Taylor A (1978). Motor unit organisation of human medial gastrocnemius. *Journal of Physiology (London)* **287**, 33-43.

Gibbs CH, Mahan PE, Mauderli A, Lundeen HC & Walsh EK (1986). Limits of human bite strength. *Journal of Prosthetic Dentistry* **56**, 226-229.

Goldberg LJ & Derfler B (1977). Relationship among recruitment order, spike amplitude, and twitch tensions of single motor units in human masseter muscle. *Journal of Neurophysiology* **40**, 879-890.

Goslow GE, Cameron WE & Stuart DG (1977). The fast twitch motor units of cat ankle flexors. 1. Tripartite classification on basis of fatigability. *Brain Research* **134**, 35-46.

Hamm TS, Nemeth PM, Solanki L, Gordon DA, Reinking RM & Stuart DG (1988). Association between biochemical and physiological properties in single motor units. *Muscle & Nerve* **11**, 245-254.

Hoh JFY, Hughes S, Kang LDH, Rughani A & Qin H (1993). The biology of cat jaw-closing muscle cells. *Journal of Computer-Assisted Microscopy* **5**, 65-70.

Koufman JA & Blalock PD (1988). Vocal fatigue and dysphonia in the professional voice user. *Laryngoscope* **98**, 493-498.

Kroon GW Naeije M & Hansson TL (1986). Electromyographic power-spectrum changes during repeated fatiguing contractions of the human masseter muscle. *Archives of oral Biology* **31**, 603-608.

Kugelberg E & Lindergren B (1979). Transmission and contraction fatigue of rat motor units in relation to succinate dehydrogenase activity of motor unit fibres. *Journal of Physiology (London)* **288**, 285-300.

Lindstrom L & Hellsing G (1983). Masseter muscle fatigue objectively quantified by analysis of myoelectric signals. *Archives of oral Biology* **28,** 297-301.

Lindstrom L, Magnusson R & Petersen I (1970). Muscular fatigue and action potential conduction velocity changes studied with frequency analysis of EMG signals. *Electromyography* **4**, 341-353.

Maton B, Rendell J, Gentil M & Gay T (1992). Masticatory muscle fatigue: endurance times and spectral changes in the electromyogram during the production of sustained bite forces. *Archives of oral Biology* **37**, 521-529.

Maxwell LC, Carlson DS, McNamara JA & Faulkner JA (1979). Histochemical characteristics of the masseter and temporalis muscles of the rhesus monkey (*Macaca mulatta*). *Anatomical Record* **193**, 389-403.

McMillan A, Sasaki K & Hannam AG (1990). The estimation of motor unit twitch tensions in the human masseter muscle by spike-triggered averaging. *Muscle & Nerve* **13**, 697-703.

Mense S (1977). Muscular nociceptors. *Journal of Physiology (Paris)* **73**, 233-240.

Mills KR (1982). Power spectral analysis of electromyogram and compound muscle action potential during muscle fatigue and recovery. *Journal of Physiology (London)* **326**, 401-409.

Naeije M (1984). Correlation between surface electromyograms and the susceptibility to fatigue of the human masseter muscle. *Archives of oral Biology* **29**, 865-870.

Naeije M & Zorn H (1981). Changes in the power spectrum of the surface electromyogram of the human masseter muscle due to local muscle fatigue. *Archives of oral Biology* **26**, 409-412.

Noden DM (1983). The embryonic origins of avian cephalic and cervical muscles and associated connective tissues. *American Journal of Anatomy* **168**, 257-276.

Nordstrom MA & Miles TS (1990). Fatigue of single motor units in human masseter. *Journal of Applied Physiology* **68**, 26-34.

Nordstrom SH & Yemm R (1974). The relationship between jaw position and isometric active tension produced by direct stimulation of the rat masseter muscle. *Archives of oral Biology* **19**, 353-359.

Palla S & Ash Jr MM (1981). Power spectral analysis of the surface electromyogram of human jaw muscles during fatigue. *Archives of oral Biology* **26**, 547-553.

Ringquist M (1971). Histochemical fibre types and fibre sizes in human masticatory muscles. *Scandinavian Journal of Dental Research* **79**, 366-368.

Ringquist M (1973). Fibre sizes of human masseter muscle in relation to bite force. *Journal of Neurological Sciences* **19**, 297-305.

Ringquist M, Ringquist I, Eriksson P-O & Thornell L-E (1982). Histochemical fibre type profile in the human masseter muscle. *Journal of Neurological Sciences* **53**, 273-282.

Robin DA, Goel A, Somodi LB & Luschei ES (1992). Tongue strength and endurance: relation to highly skilled movements. *Journal of Speech and Hearing Research* **35**, 1239-1245.

Rowlett AE (1932-33). The gnathodynamometer and its use in dentistry. *Proceedings of the Royal Society for Medicine* **26**, 463-471.

Sandercock TG, Faulkner JA, Albers JW & Abbrecht PH (1985). Single motor unit and fiber action potentials during fatigue. *Journal of Applied Physiology* **58**, 1073-1079

Serratrice G, Pellisier JF, Vignon C & Baret J (1976). The histochemical profile of the human masseter. An autopsy and biopsy study. *Journal of Neurological Sciences* **30**, 189-200.

Soussi-Yanicostas N, Barbet JP, Laurent-Winter C, Barton P & Butler-Browne GS (1990). Transition of myosin isozymes during development of human masseter muscle. Persistence of developmental isoforms during postnatal stage. *Development* **108**, 239-249.

Stephens JA & Usherwood TP (1977). The mechanical properties of human motor units with special reference to their fatiguability and recruitment threshold. *Brain Research* **125**, 91-97.

Tamari JW, Tomey GT, Ibrahim MZM, Baraka A, Jabbur SJ & Bahuth N (1973). Correlative study of the physiologic and morphologic characteristics of the temporal and masseter muscles of the cat. *Journal of Dental Research* **52**, 538-543.

Taylor A, Cody FWJ & Bosley MA (1973). Histochemical and mechanical properties of the jaw muscles of the cat. *Experimental Neurology* **38**, 99-109.

Thexton AJ & Hiiemae KM (1975). The twitch tension characteristics of opossum jaw musculature. *Archives of oral Biology* **20**, 743-748

Thomas CK, Johansson RS, Bigland-Ritchie B (1991). Attempts to physiologically classify human thenar motor units. *Journal of Neurophysiology* **65**, 1501-1508.

Van Boxtel A, Goudswaard P, van der Molen GM & van den Bosch WEJ (1983). Changes in electromyogram power spectra of facial and jaw-elevator muscles during fatigue. *Journal of Applied Physiology* **54**, 51-58.

Van Steenberghe D, De Vries JH & Hollander AP (1978). Resistance of jaw-closing muscles to fatigue during repetitive maximal voluntry voluntary clenching efforts in man. *Archives of oral Biology* **23**, 697-701.

Vignon C, Pellisier JF & Serratrice G (1980). Further histochemical studies on the masticatory muscles. *Journal of Neurological Sciences* **45**, 157-176.

Waltimo A & Könönen M (1993). A novel bite force recorder and maximal isometric bite force values for healthy young adults. *Scandinavian Journal of Dental Research* **101**, 171-175.

Waugh LM (1937). Dental observations among Eskimo. VII. Survey of mouth conditions, nutritional study and gnathodynamometer data, in most primitive and populous native villagers in Alaska. *Journal of Dental Research* **16**, 355-356.

Yemm R (1976). The role of tissue elasticity in the control of mandibular resting posture. In: Anderson DJ, Matthews B (eds.), *Mastication*, pp. 81-89. Bristol: John Wright and Sons.

Yemm R (1977). The orderly recruitment of motor units of the masseter and temporal muscles during voluntary isometric contraction in man. *Journal of Physiology (London)* **265**, 163-174.

Young JL & Mayer RF (1981). Physiological properties and classification of single motor units activated by intramuscular microstimulation in the first dorsal interosseous muscle in man. In: Desmedt JE (ed.), *Progress in Clinical Neurophysiology, vol 9, Motor Unit Types, Recruitment and Plasticity in Health and Disease*, pp. 17-25. Basel: Karger.

SECTION IX

Fatigue of Adapted Systems: Overuse, Underuse and Pathophysiology

The first chapters in this section deal with the adaptability of the neuromuscular system to variations in usage and to the unavoidable effects of occasional damage, repair and aging. The final two chapters consider how pathological changes in muscle function should be approached in individual patients.

Muscles appear well adapted to the tasks they perform, in terms of endurance, strength and speed. However, the mechanisms that underlie this apparent adaptation and the limits to which it can be extended are less well understood. In **Chapter 32**, Gordon reviews the extensive evidence that the neuromuscular system exhibits remarkable plasticity such that it can respond dramatically to altered muscle use. Some limits to the various models of altered use are discussed. Overall, the forces exerted by muscles seem to preferentially affect strength, while the total daily amount of muscle activity sets the endurance capability. Systemic effects produced by the altered use are also likely to contribute to the performance capabilities of individual muscles. Review of this field emphasizes that motoneurons do not exert total trophic control over their muscle fibers. Rather the motoneuron *modulates* the properties of its muscle fibers within a limited (adaptive) range, which is partly a function of muscle architecture and mechanical factors, and partly a function of intrinsic factors preset during development (i.e., muscle genotype). Material in this chapter is relevant to a variety of pathological circumstances, as well as the development of functional electrical stimulation in rehabilitation, and understanding the effects of long-term space flight.

Eccentric exercise (i.e., lengthening contractions) produces familiar deleterious short-term effects that are manifested as delayed muscle soreness and increased fatigue. This syndrome is reviewed in **Chapter 33** (Clarkson & Newham). Eccentric exercise is metabolically attractive (as it uses less oxygen) and biomechanically useful (as it stores elastic energy, which can be used in the *stretch-shortening* cycle), but it is especially likely to damage muscle. This effect may partly be mediated by an uneven distribution of sarcomere strengths, such that the weakest sarcomeres *pop* during muscle lengthening, although other factors are likely to be involved. The exact biochemical triggering mechanisms within the active muscle cell that are responsible for the damage are not yet clear. However, the damaged cells release a cocktail of substances that induce reflex effects through the activation of small-diameter muscle afferents.

The decline in muscle strength and power accompanying aging is highlighted in **Chapter 34** (Faulkner & Brooks). These decreased capabilities are due to a reduced number and size of muscle fibers pursuant to a loss of motoneurons. Although inevitable, the physiological consequences can be partly offset by strength and endurance training. Intracellular processes are likely to mediate the decline in performance because the maximal force achieved by high calcium concentrations in permeabilized fibers of old animals is the same as that in younger animals. Large, fatigable motor units, with their extensive territories, are likely to be preferentially affected by aging. Contraction-induced damage may underlie the denervation of fast fibers with consequent reinnervation by slow fibers.

In the final two chapters of this section, some principles for understanding the patient with fatigue are presented. Edwards and colleagues (**Chapter 35**) describe the major advances in understanding human muscle fatigue from an historical perspective. These range from the technical advances which made new measurements possible (e.g., myographs, needle muscle biopsy, ^{31}P NMR), new conceptual advances, and insights that resulted from understanding pathophysiology, such as the deficits attributable to single biochemical mechanisms. The focus is extended beyond the 1980 symposium in London to then propose a new way of categorizing the muscle fatigue found in patients with neuromuscular and other disorders. In **Chapter 36** (McComas and colleagues), the clinical approach to the patient with fatigue is highlighted with some upper and lower motoneuron diseases discussed, including multiple sclerosis and poliomyelitis. There is a need to make comparatively simple assessments of muscle performance more often in the clinic.

FATIGUE IN ADAPTED SYSTEMS

Overuse and Underuse Paradigms

T. Gordon

Department of Pharmacology
Division of Neurosciences
University of Alberta
Edmonton, Canada, T6G 2S2

ABSTRACT

Alterations in structural and biochemical properties of muscles that underlie physiological parameters of contractile force, speed and fatigability are described under conditions of 1) overuse: imposed electrical stimulation, natural exercise and functional overload; 2) reinnervation of denervated muscles; and 3) underusage: conditions of restricted use after spinal cord injury, weightlessness, immobilization and drug-induced neuromuscular blockade. These conditions demonstrate the remarkable plasticity of muscle fibers with obvious implications in health and disease. They also identify that the amount of neuromuscular activity and loading of muscle contractions are major factors determining susceptibility to fatigue and muscle strength, respectively.

INTRODUCTION

The capacity of some skeletal muscles but not others to contract with little fatigue has often been related to their function. In the male frog, for example, the slow tonic muscles in the forearm develop slow contractures that can be maintained without fatigue for many hours (Kuffler & Vaughn-Williams, 1953). These muscles are clearly adapted for their function of clasping the female during mating. In the bird, the slow-tonic anterior latissimus dorsi muscle can maintain sufficient force to successfully hold the wings close to the body when the bird is not in flight. In mammals, the slow-twitch soleus muscle is ideally suited to maintain tension at the low frequencies of motoneuron discharge during posture. In contrast, fast-twitch muscles which develop much larger forces at higher frequencies of discharge are far more susceptible to fatigue but they are well suited to the development of high forces during movement. In all species, recruitment of the fast-twitch muscle fibers in the hindlimb can generate large forces for short periods required to accelerate the body during walking, running and jumping.

Fatigue, Edited by Simon C. Gandevia et al.
Plenum Press, New York, 1995

Differences in the susceptibility of muscle to fatigue were initially associated with obvious differences in the color of red and white skeletal muscles (Kuhne, 1863; Ranvier, 1874). These in turn, were recognized to indicate differences in capillary density and myoglobin content (Ranvier, 1874; reviewed by Burke & Edgerton, 1975; Burke, 1981; Saltin & Gollnick, 1983; Vrbova et al., 1994). Red and white muscles also differ in their rates of contraction. Red muscles were, therefore, denoted as red slow muscles and distinguished from white fast-twitch muscles (Denny-Brown, 1929). Slow contraction speeds combined with low fatigability are recognized as properties ideally suited for the continuous low force, high endurance requirements of postural slow-twitch muscles as opposed to the intermittent high force, low endurance characteristics of the fast-twitch muscles (Vrbova et al., 1994).

The matching of muscle properties to their characteristic activation patterns begged the question of whether the patterns of muscle activation are actually responsible for the expression of the slow or fast muscle phenotype. The natural extension of this question is how far muscle properties can adapt to altered usage in health and disease. These questions have been addressed directly or indirectly in many hundreds of studies of mature muscle phenotype, development, capacity for change during adult life, and for recovery after injury. Recent reviews of these studies include those of Roy and colleagues (1991), Henriksson (1992), Pette and Vrbova (1992); Gordon and Pattullo (1993) and Vrbova and colleagues (1994). Dramatic alterations in structural and biochemical properties that underlie the capacity of the muscle to develop force (muscle strength) over time (contractile speed) and to sustain muscle force (muscle endurance) have been demonstrated after the nerve supply (self- and cross-reinnervations) or the amount of neuromuscular activity (overuse and underuse paradigms) is altered. These experiments point to the remarkable plasticity of muscle fibers. Because motor control depends on muscle power, speed and endurance, the potential for manipulating muscle properties for required function was immediately recognized. Many applications have been extensively explored including sports training, rehabilitation of muscles after trauma, injury and disease, paralysis after lower and upper motoneuron lesions, limb immobilization, and space flight. To provide insight into this topic, the extent of muscle plasticity is considered in this chapter with emphasis on muscle endurance. The matching properties of muscles and their motoneurons are briefly reviewed prior to considering adaptability of muscle endurance under conditions of 1) overuse: imposed electrical stimulation of muscle, natural exercise and functional overload of muscles; 2) reinnervation of denervated muscles; and 3) under-usage: conditions of restricted use after spinal cord injury, weightlessness, immobilization and drug-induced neuromuscular blockade.

METABOLIC PROFILE OF SLOW AND FAST SKELETAL MUSCLE IN NORMAL FUNCTION

High resistance to fatigue of slow-twitch muscles is associated with dense capillarization, sustained blood flow through active muscles and generation of ATP via carbohydrate and fat oxidation. Anaerobic glycolysis which generates only 3 moles of ATP per mole of glycogen plays a minor role in contrast to aerobic metabolism via the mitochondrial pathway which generates 39 moles of ATP per mole of glycogen (Fox et al., 1988). In fast-twitch muscles, relatively poor blood flow during muscle contraction is associated with more reliance on anaerobic pathways and lower resistance to fatigue. Thus, reciprocal distribution of oxidative and glycolytic enzymes appears to be well adapted to the high and low endurance capabilities of slow- and fast-twitch muscles respectively (Vrbova et al., 1994).

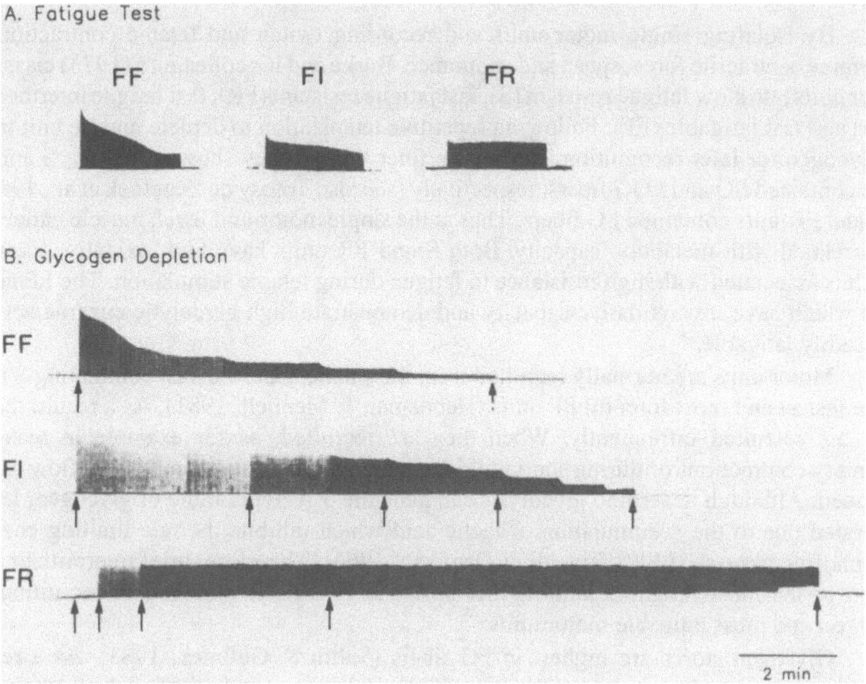

Figure 1. Fatigability of fast motor units in rat tibialis anterior muscle. *A*, decline in isometric force during a 2-min period of 300 ms duration tetanic stimulation at 40 Hz repeated every min. Force declines to less than 25% of initial force in FF units, between 25% and 75% in FI units and to more than 75% in FR units. *B*, repetitive tetanization with short bursts of 100 Hz stimulation (50 ms tetanus) repeated at progressively faster rates leads to fatigue in fast units. The pattern of fatigue was different for the 3 fast unit types with force declining most rapidly in the FF units. The repetition rate was increased at each arrow at the time point when force had declined to a steady level. The arrows from left to right are 1 Hz, 1.25 Hz, 1.7 Hz, 2.5 Hz and 0.5 Hz. The latter frequency was used to permit recovery of force. This tetanization protocol was repeated 6 or 7 times for each unit and is associated with complete glycogen depletion of the stimulated muscle fibers.

Presence of oxidative and glycolytic enzymes and their specific activities also reflect differences in substrate availability. Thus slow-twitch muscle fibers have large quantities of stored triglycerides which provide energy via β-oxidation and ATP generation via the Krebs cycle in contrast to the large glycogen stores in fast-twitch muscle fibers which rely on anaerobic glycolysis (Pette & Staron, 1990; for additional discussion see Kushmerick, Chapter 4).

Yet, oxidative and glycolytic enzymes are not always reciprocally related as both oxidative and glycolytic enzyme activities are high in some muscle fibers. These fibers were identified histochemically and designated as fast oxidative glycolytic fibers (FOG) on the basis of their acid labile myosin ATPase and high enzyme activities for oxidative and glycolytic enzymes (Peter et al., 1972). These were distinguished from 1) the fast glycolytic (FG) fibers which contained acid labile myosin ATPase, associated with fast isoforms of myosin heavy chains, and have high glycolytic but not oxidative enzyme activity (FG); and 2) the slow oxidative (SO) fibers which contained acid stable myosin ATPase and are highly reactive for oxidative but not glycolytic enzyme activity. Although an earlier classification of fibers into acid stable Type I and acid labile type IIA and IIB fibers is based only on pH sensitivity of mATPase (Brooks & Kaiser, 1970), there is generally good correspondence between the two systems (type I: SO; type IIA: FOG; type IIB: FG; Gordon & Pattullo, 1993).

By isolating single motor units and recording twitch and tetanic contractions to determine contractile force, speed and endurance, Burke and his colleagues (1973) classified motor units into slow fatigue resistant (S), fast fatigue resistant (FR), fast fatigue intermediate (Fint) and fast fatigable (FF). Following repetitive tetanization to deplete muscle unit fibers of glycogen for later recognition and muscle fiber typing, they showed that the S and FR units contained SO and FOG fibers, respectively (see also Totosy de Zepetnek et al., 1992b). Fint and FF units contained FG fibers. Thus at the single motor unit level, muscle endurance is correlated with metabolic capacity. Both S and FR units have high oxidative capacity which is associated with high resistance to fatigue during tetanic stimulation. The FF motor units which have low oxidative capacity and demonstrate high glycolytic enzyme activity are readily fatigable.

Motor units are normally recruited from the smallest and slowest contracting S units to the fastest and most forceful FF units (Henneman & Mendell, 1981). As a result, the FF units are recruited infrequently. When they are recruited, as for example in maximal voluntary contractions or during short rapid movements such as jumping, blood flow can be occluded. Although anaerobic glycolysis can generate 3 ATP per mole of glycogen, less is generated due to the accumulation of lactic acid which inhibits the rate limiting enzyme phosphofructokinase (PFK; Triveldi & Danforth, 1966). Therefore, brief intermittent activation of the most fatigable units is the most energy efficient method of recruiting the strongest and most fatigable motor units.

Glycogen stores are highest in FG fibers (Saltin & Gollnick, 1983). As a result, intermittent contractions in which there is a high work-to-rest ratio but in which lactic acid accumulation is minimized is the most effective method of utilizing glycogen stores. This principle is readily tested for single motor units in which stimulation regimes designed to elicit brief intermittent, high force tetani at progressively higher stimulation rates of 1 to 2.5 Hz were very effective in depleting all unit types of glycogen (Totosy de Zepetnek et al., 1992b; their Fig. 1). The more fatigue resistant units, however, required considerably longer periods of stimulation for fatigue and corresponding depletion of glycogen.

Skeletal muscles differ considerably in the proportion of the different types of motor units and their constituent muscle fiber types (Ariano et al., 1973; reviewed by Saltin & Gollnick, 1983). Generally, most muscles are mixed but contain either a majority of slow or fast muscle fibers depending on their function. Many muscles contain motor units that demonstrate a continuous distribution of fatigability and corresponding oxidative and glycolytic potential, for example the rat tibialis anterior (TA; Kugelberg & Lindegren, 1979; Totosy de Zepetnek et al., 1992b).

METABOLIC AND FUNCTIONAL ADAPTATION TO INCREASED USAGE DURING CHRONIC ELECTRICAL STIMULATION

Fast-to-Slow Conversion

In fast-twitch muscles, synchronous low-frequency stimulation of all muscle fibers for 3-24 hrs per day leads to a reproducible change in the ratio of oxidative to glycolytic enzyme activities and an associated increase in their resistance to fatigue (reviewed by Pette & Vrbova, 1992). Within days of daily stimulation, mRNA levels and the activities of enzymes of the citric acid cycle, fatty acid β-oxidation and the respiratory chain increase significantly (Pette et al., 1973; Henriksson et al., 1986; Williams et al., 1987; Hood et al.,

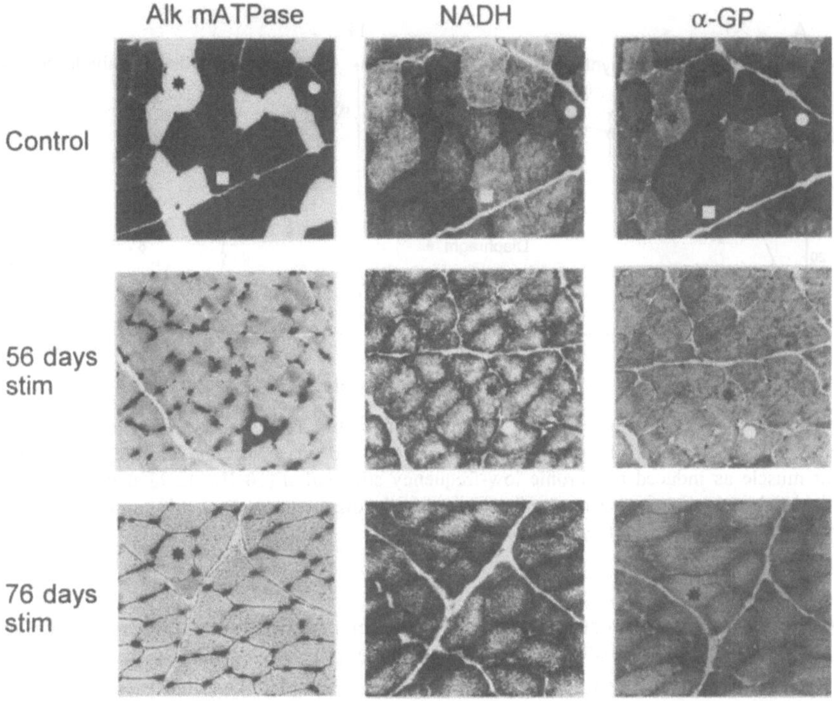

Figure 2. Fast-to-slow conversion of chronically stimulated cat medial gastrocnemius muscle. Daily low-frequency (20 Hz) stimulation in a 50% duty cycle (2.5 s on, 2.5 s off) led to increased expression of NADH and downregulation of α-glycerophosphate. The characteristic mosaic of muscle fibers in the normal muscle was replaced by a homogeneous fiber population in which oxidative enzyme (NADH) activity was high and glycolytic enzyme (α-GP) activity is low. This change was associated with a conversion of myosin heavy chain composition of the muscle fibers to the slow form, which accounts for the loss of all alkali-stable mATPase.

1989). These enzymes include succinic dehydrogenase and NADH, which increase concurrently with a decline in glycolytic enzymes such as α-glycerophosphate (Fig. 2). As shown in Fig. 3, the increase in oxidative- and decline in glycolytic enzyme activities follow an exponential time course. The time constants depend on the total amount of neuromuscular activity, being significantly longer when muscles were stimulated for 12 hrs than for 24 hrs (Pette & Vrbova, 1992). Consistent with the normal pattern in which enzyme activities are jointly regulated (Pette & Luh, 1962; Pette & Hofer, 1980), the conversion of glycolytic to oxidative potential is accompanied by an increase in the number of mitochondria per muscle fiber (Williams et al., 1986). The most rapid change is an increased activity of hexokinase, a glycolytic enzyme associated with uptake and oxidation of blood glucose (Bass et al., 1969). This is the only enzyme of the glycolytic pathway to be elevated by chronic stimulation (Pette et al., 1973). The transient increase presumably reflects the early response to high activity with generation of ATP via glycolysis and metabolism of pyruvate via the citric acid cycle rather than to lactate.

Associated with the metabolic adaptation, there is an early and dramatic increase in capillary density surrounding the stimulated muscle fibers (Brown et al., 1976; Hudlicka et al., 1977; Hudlicka, 1984) and an increase in myoglobin content (Pette et al., 1973). Thus, the high metabolic demand of the active muscles results in an increased oxygen uptake and metabolism to generate ATP for contraction. A significant decline in muscle fiber diameter

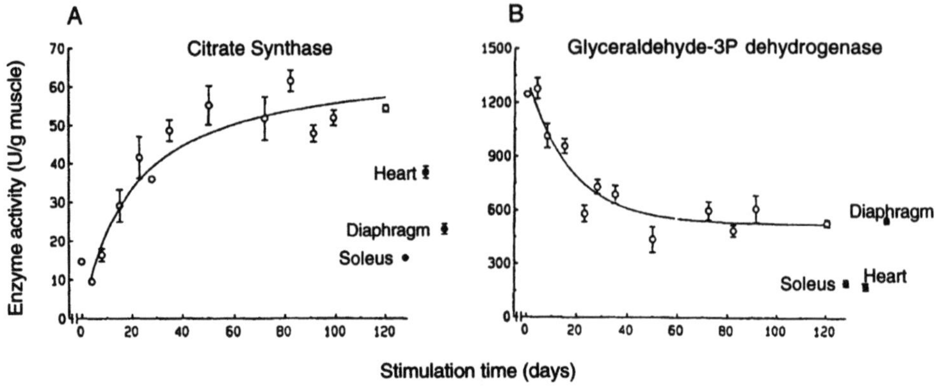

Figure 3. Time course of changes in the activity levels of two representative enzymes of aerobic (citrate synthase) and anaerobic (glyceraldehyde phosphate dehydrogenase) energy metabolism in rabbit tibialis anterior muscle as induced by chronic low-frequency stimulation (10 Hz, 12 hr daily). For comparison, activities of the two enzymes were also determined in the diaphragm, soleus, and cardiac muscles (means ± SE, n= 3-5; modified from Pette & Vrbova, 1992).

is also consistent with more efficient oxygen delivery to the mitochondria for oxidative metabolism (Pette et al., 1976; Donselaar et al., 1987).

Concurrent with reduced muscle fiber size, muscle force declines reciprocally with muscle endurance within the first month of stimulation (Fig. 4). As reviewed elsewhere, contractile speed and underlying changes in regulatory and contractile proteins occur with a slower time course. The extent of fast-to-slow conversion is species-dependent, possibly involving differences in thyroid status and/or adaptive ranges of fast muscles in different species (Pette & Vrbova, 1992; Gordon & Pattullo, 1993; Vrbova et al., 1994). Except in the rat where conversion of phenotype is more limited, slowing of contraction is associated with slower release of Ca^{2+} from the sarcoplasmic reticulum via ryanodine/Ca^{2+} release channels (Ohlendieck et al., 1991), expression of slow isoforms of the regulatory proteins troponin and tropomyosin (Roy et al., 1979; Hartner et al., 1989), and reduced rate of cross-bridge cycling as slow forms of myosin heavy and light chain isoforms are incorporated into the thick filaments (Sreter et al., 1973; Pette et al., 1976; Sweeney et al., 1988). In addition, the active state is prolonged as a result of reduced sarcoplasmic reticulum Ca^{2+}-ATPase activity. This, in turn, is associated with expression of the slow isoform of the enzyme and the regulatory protein, phospholamban and the downregulation of Ca^{2+} binding proteins, calsequestrin in the sarcoplasmic reticulum and parvalbumin in the cytoplasm (Leberer et al., 1986; 1989). Reduced t-tubular volume and enhanced Na^+/K^+ pumping (Eisenberg & Salmons, 1981) may be responsible for preventing K^+ accumulation and thereby ensuring transmission of muscle action potentials (Kernell et al., 1987b). Incorporation of increased numbers of myonuclei in stimulated muscles is consistent with the higher numbers in slow as compared to fast muscle fibers (Russell et al., 1992).

Thus a comprehensive documentation of changes in metabolism, protein expression and function has shown that adult skeletal muscle adapts to the mechanical signals imposed by electrical stimulation. This involves switching specific genes for conversion of muscle fibers to slow phenotype. Different genes are up-regulated or downregulated resulting in the metabolic adaptation with the result that muscles become more fatigue-resistant.

Despite the many studies that document the phenotypic changes in stimulated muscles, the question of how the changes come about has been largely neglected. The trigger for change may be linked to the muscle fatigue that occurs during repetitive low-frequency

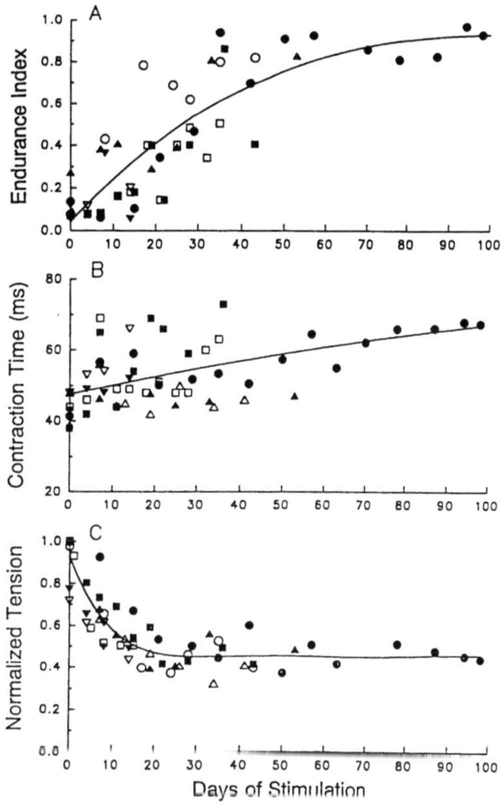

Figure 4. Fast-to-slow conversion of chronically stimulated cat medial gastrocnemius muscle. Daily low-frequency (20 Hz) stimulation in a 50% duty cycle (2.5 s on, 2.5 s off) led to: *A*, increased resistance to fatigue; endurance index: tetanic force recorded 2 min after tetanic stimulation at 13 pulses at 25 Hz every s for 2 min relative to tetanic force at 0 min; *B*, slowing of the contraction time; time to peak twitch force; and *C*, fall in contractile tension; isometric tension developed in response to 100 Hz stimulation normalized to tension developed by the muscles prior to onset of daily electrical stimulation. The data for each of 8 muscles are plotted with separate symbols (from Gordon & Mao, 1994; reprinted from *Physical Therapy* with the permission of the American Physical Therapy Association).

stimulation (Fig. 5*A*). The low levels of force generated by the unfused tetanic contractions do not occlude the blood supply (Sjøgaard, 1987). However, the relatively sparse capillary network does not deliver sufficient oxygen to the working muscle fibers even though cardiac output and blood pressure are reflexly elevated to deliver more blood to the muscle per unit time (Hudlicka, 1984). The oxygen deficit and poor oxidative capacity of the fast-twitch muscle fibers cannot sustain ATP production, and the relatively little ATP generated via anaerobic glycolysis is associated with glycogen depletion and rapid accumulation of lactate, free ADP and AMP (Green et al., 1992). The reliance on creatine phosphate as the immediate source of ATP in fast-twitch muscles results in a rapid depletion (Westerblad et al., 1991; Nagesser et al., 1992). In addition, the acidic pH inhibits PFK which, in turn, reduces the generation of ATP via anaerobic glycolysis (Trivedi & Danforth, 1966; Sahlin, 1978). The resulting metabolite accumulation of glucose-1,6, phosphate and fructose-2,6,biphosphate within the first day of stimulation has an inhibitory effect on glycolysis and later may stimulate glycogenolysis for replenishment of glycogen.

As reviewed earlier, generation of inorganic phosphate and decline in pH following lactate accumulation in the working muscle may account for the immediate decline in force (Edman, Chapter 1; Stephenson et al., Chapter 2; Allen et al., Chapter 3). Phosphate and hydrogen ions reduce the Ca^{2+} sensitivity of the contractile filaments and the force generated by the cross-bridges. As fatigue progresses, a fall in ATP may reduce Ca^{2+} release via ryanodine release channels (because ATP is required for opening of the channels, Smith et al., 1985), and increase levels of cytoplasmic Mg^{2+} (Lamb & Stephenson, 1991) and extracellular K^+. The latter arises partly as a result of increased opening of ATP-sensitive K^+ channels at low $[ATP]_i$ levels (Castle & Haylett, 1987) and despite increased Na^+/K^+

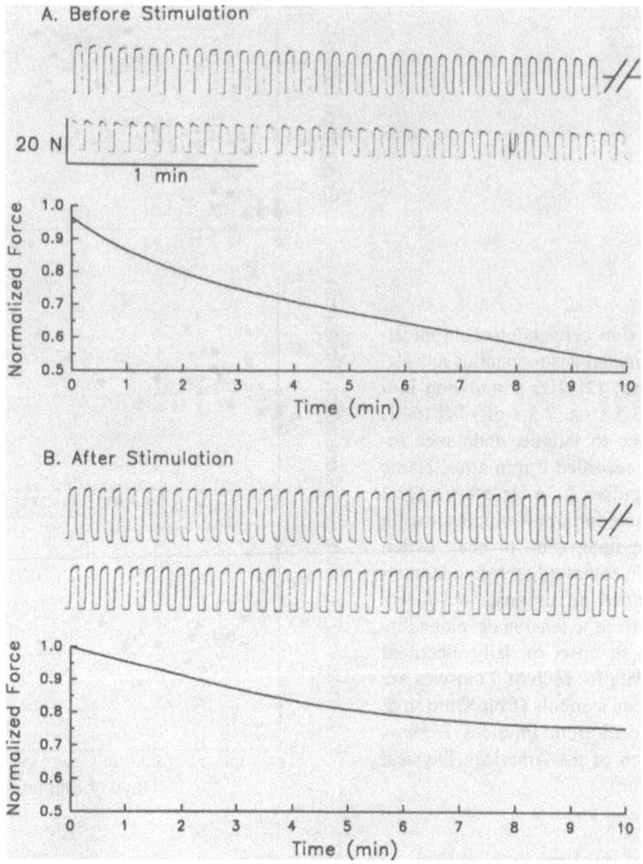

Figure 5. Fatigue of the cat medial gastrocnemius muscle during repetitive trains of low-frequency stimulation and long-term consequences. *A*, 2.5 s trains of 20 Hz repeated every 2.5 s in a 50% duty cycle was associated with a 50% fall in tetanic force. This fall was presumably due to the fatigue of the most fatigable FF units, which normally contribute 60% of the total force of the maximally stimulated muscle (Gordon & Mao, 1994). *B*, after subjecting the muscle to this stimulation pattern for 7 weeks, performance improved and the chronically stimulated muscle experienced less than a 20% decline in tetanic force in the same 10 min period. The majority of the motor units were classified as fatigue resistant, consistent with the presence of highly oxidative muscle fibers as shown in Fig. 2.

pumping (Hicks & McComas, 1989). Progressive failure of the Ca^{2+} ATPase to sequester Ca^{2+} leads to a progressive increase in free intracellular Ca^{2+} which in turn may further inhibit Ca^{2+} release (Allen et al., 1992). The high levels of free intracellular Ca^{2+} tend to continue during the first weeks of stimulation possibly as a result of the reduced Ca^{2+}-ATPase activity (Leberer et al., 1987).

Both the reduced energy supply and the increase in intracellular Ca^{2+} have been suggested to trigger change in gene expression (Saltin & Gollnick, 1983; Pette & Vrbova, 1992). It has been speculated that the [ATP]/[ADP][P_I] ratio is sensed by the genome of the nucleus and mitochondria to upregulate the expression of enzymes of the citric acid cycle and oxidative phosphorylation. Within the first 15 min of continuous 24 hr daily stimulation, the ratio declined almost to zero and did not recover in the 50 days of study (Pette & Vrbova, 1992). Thus, chronic ischemia, appears to be a potent stimulus to increase oxidative potential.

In addition, elevated substrate concentrations may be important for inducing gene expression for oxidative enzymes (Saltin & Gollnick, 1983).

A five-fold increase in free [Ca^{2+}] in the resting muscle within the first 2 weeks of stimulation has been postulated to trigger conversion of sarcoplasmic reticular and myofibrillar proteins to the slow type (Sreter et al., 1987). The mechanical loading of actively contracting muscles may also influence gene expression (Goldspink et al., 1992). The signal transduction mechanisms that mediate the changes in gene expression, however, are not yet understood.

Ischemia and increased K^+, H^+, and Ca^{2+} ionic concentrations could also be involved in protein catabolism and, in turn, reduced muscle fiber volume. Oxidative stress in stimulated muscles and associated damage to muscle fibers and connective tissue may lead to invasion of blood borne cells and release of growth factors. These, in turn, are likely to help increase the number of myonuclei and the growth of capillaries (angiogenesis; Thompson et al., 1988; Grounds, 1991; Russell et al., 1992). The satellite cells, which normally lie dormant under the muscle fiber basement membrane, are the most likely source of the increased number of nuclei found in the stimulated muscle fibers (Russell et al., 1992). Proliferation of endothelial cells is essential for the marked growth of capillaries in the stimulated muscles (Hudlicka, 1984).

The growth factors include TGF-β1 and platelet derived growth factor (PDGF) from platelets and basic fibroblast growth factor (bFGF) from macrophages. bFGF may also be released from endothelial cells and from binding sites on muscle extracellular matrix proteins in response to plasmin and heparinases released from lymphocytes and macrophages and plasmin from the endothelium (Klagsbrun, 1989; Grounds, 1991). TGF-β and PDGF, much like helper cells of the immune system, recruit cells such as macrophages into the site of injury and stimulate their release of bFGF, other angiogenic factors, and interleukin-1. Importantly, macrophages grown under conditions of low oxygen tension have been shown to secrete angiogenic factors (Knighton et al., 1983; Gimbrone, 1984). Oxidative stress may also trigger important changes in the active nerve terminals. Even low amounts of stimulation of fast motor units in the crayfish increase the mitochondrial content and synaptic efficiency of the terminals (Nguyen & Atwood, 1994).

Slow-to-Fast Adaptation

The reverse conversion from slow-to-fast muscle phenotype has been more difficult to demonstrate. Salmons & Vrbova argued that low-frequency stimulation promoted slow phenotype and high intermittent stimulation more typically seen in fast flexor muscles promoted fast phenotype (Vrbova, 1963; Salmons & Vrbova, 1969; Salmons & Sreter, 1976). In experiments in which reflexly elicited contractions of the soleus muscle were reduced by tenotomy and central drive eliminated by spinal cord transection, Vrbova (1963) demonstrated that slow contractions were maintained only if the muscles were stimulated with 10-20 Hz stimulation. If the muscles were stimulated intermittently with 40-100 Hz, the muscle contractions were fast. Even though tenotomy is associated with considerable muscle necrosis (McMinn & Vrbova, 1964), the results were convincing in their demonstration of the effects of low-frequency stimulation in promoting slow phenotype.

The effects of high-frequency intermittent stimulation are controversial. Intermittent stimulation of denervated slow- and fast-twitch muscles with high-frequency tetanic bursts increased the rate of contraction in the muscles (Gorza et al., 1988; Gundersen et al., 1988). However, slow-twitch soleus muscles retained their high fatigue resistance and did not express the type IIB fast myosin heavy chain isoform typical of the FF motor units (Gundersen et al., 1988; Ausoni et al., 1990). Thus, although conversion did occur, conversion was limited to fiber type I -> type IIA.

In innervated muscles, the effects of intermittent stimulation were quite different. Three laboratories observed, to their surprise, that high-frequency stimulation of innervated fast muscles resulted in a fast-to-slow conversion rather similar to that observed after low-frequency stimulation (Hudlicka et al., 1982; Sreter et al., 1982; Eerbeek et al., 1984). It appeared that the important factor controlling muscle speed and endurance was the amount of daily activity (total number of impulses per day) rather than the pattern of activity as originally suggested. A progressive increase in daily activity in paralysed flexor muscles in the cat from as little as 0.5% to 50% was associated with a progressive conversion of muscle contractile properties from FF -> FI -> FR -> S (Kernell et al., 1987a,b; Kernell & Eerbeek, 1989). In contrast, maximal force appeared to depend on both pattern and amount of chronic stimulation: the fall in muscle force that accompanies an increase in muscle endurance could be reversed by interposing occasional tetanic bursts at high stimulation frequencies (Kernell et al., 1987b). These results suggested that occasional high force contractions are required to maintain muscle strength and that this effect is independent of the effects of low-frequency stimulation on muscle endurance. Alternatively, the decline in muscle force during low-frequency stimulation can be reversed by preventing muscle shortening during contraction (Fig. 6).

In summary, oxidative capacity and muscle endurance increase with the duration of neuromuscular activity. Muscle strength and endurance are reciprocally regulated with high neuromuscular activities favoring high endurance but low force in small muscle fibers that have high mitochondrial content and high oxidative enzyme capacity. Muscle strength is promoted under conditions of relatively little neuromuscular activity but depends on the intermittent development of high forces or resisted contractions against a load.

EXERCISE TRAINING AND MUSCLE ENDURANCE, STRENGTH AND SPEED

In their excellent review of biochemical adaptations to endurance exercise, Holloszy and Booth (1976) stated that the nature of the exercise stimulus determines which of two distinct adaptive responses are induced in skeletal muscle. Increase in muscle strength but not endurance is exemplified by muscle adaptation in weight lifters and body builders. Increase in muscle endurance but not strength is found in the muscles of competitive long-distance runners and cross-country skiers, cyclists and swimmers. Because the training in the former primarily involves isometric or loaded muscle contractions as opposed to dynamic contractions in the latter, these are sometimes referred to as static *vs.* dynamic forms of exercise (Longhurst & Stebbins, 1992). However, since many strengthening activities also involve dynamic exercise, the terms must be used with caution.

Endurance or Dynamic Training

There are many parallels in the extent and time course of muscle adaptation to imposed low-frequency electrical stimulation and the adaptive changes during endurance exercise (reviewed by Henrikkson, 1992). Typical endurance training regimes are 2-3 months with bouts of 30-60 min of exercise corresponding to 70-80% of maximal oxygen uptake, 3-5 times per week in humans or strenuous running programs in rats (up to 2 hrs per day). These lead to a progressive increase in muscle endurance, which is associated with increased capacity of the trained muscles to generate ATP from the oxidative metabolism of carbohydrates and long chain fatty acid substrates (Holloszy & Booth, 1976; Saltin & Gollnick, 1983). These adaptations occur in all types of muscle fibers (Terjung, 1976). Increased

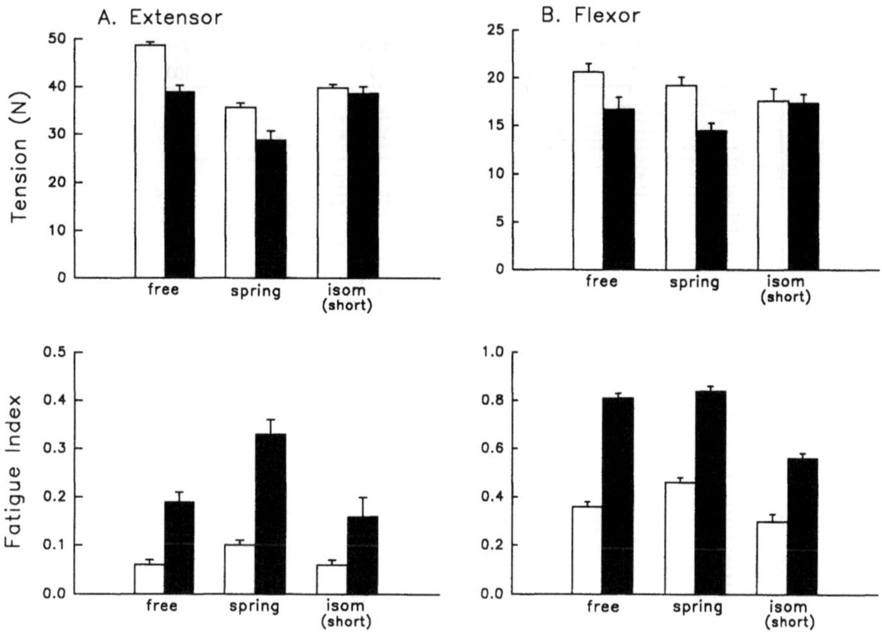

Figure 6. Effects of muscle loading on reduced tetanic tension and fatigability of paralysed cat (*A*) medial gastrocnemius extensor muscles and (*B*) common peroneal innervated flexor muscles in response to low-frequency (20 Hz) daily electrical stimulation in a 50% duty cycle for 3 hrs/day (5% total daily activity). Paralysis was induced by unilateral deafferentation and hemisection at T12. Isometric tension was recorded at regular intervals at optimal length by non-invasive coupling of the cat foot to an external-held force transducer, under Halothane anesthesia (chronic recording methods described by Davis et al., 1978). Fatigue index was calculated as the ratio of the tetanic tension at the end and the beginning of a 2-min period of repetitive trains of stimuli (13 pulses at 40 Hz every s). Since recordings were made in the paralysed muscles before and after daily stimulation, each animal served as its own control. The open bars are the mean (± S.E.) of 5-15 recorded values before the onset of stimulation. The closed bars are the average of 10-15 values recorded more than 60 days after stimulation. Under conditions in which muscle contractions were unopposed, tetanic tension declined by 20% (*A*) associated with an increase in endurance (a 3-fold increase above prestimulation values). Spring loading of the contractions had little effect on muscle strength but was associated with increased fatigue resistance, particularly in the extensor muscles. Preventing muscle shortening by fixing the ankle in the same boot used to spring load the contractions largely eliminated the fall in muscle force without severely compromising the endurance response of the muscles (Tyreman, Mao & Gordon, unpublished observations).

oxidative capacity arises as a result of increased oxygen uptake, myoglobin content, hexokinase activity, mitochondria number, oxidative enzyme content, and increased capillarization with little change in glycolytic enzyme activities (Lamb et al., 1969; Baldwin et al., 1973; Saltin & Gollnick, 1983). There is, in addition, an increased capacity for glycogen and triglyceride synthesis in trained muscles (Morgan et al., 1969; Taylor et al., 1972). The small changes in glycolytic enzyme activities in athletes (e.g., Baldwin et al., 1973) were attributed to the normally low levels in these individuals who tend to have high percentages of type I slow fibers in contrast to the high type II fiber content in sprinters (Gollnick et al., 1972; Mero et al., 1981; 1992). Another explanation is that exercise-induced decreases of glycolytic enzyme activity in type IIA fibers in muscles containing all three muscle fiber types were obscured by the increases in type I fibers and the small changes in glycolytic enzyme levels in the type IIB fibers (Holloszy & Booth, 1976). In rats, the question of whether the treadmill running and swimming programs were sufficiently severe to stress the

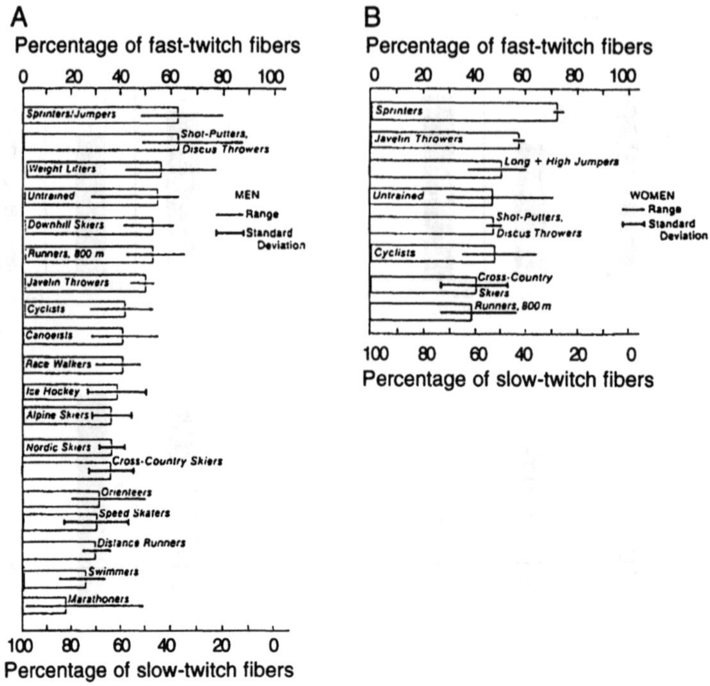

Figure 7. Distribution of fast-twitch (FT) and slow-twitch (ST) fibers in (*A*) vastus lateralis muscles of different groups of male athletes and (*B*) in medial gastrocnemius of female athletes. Although there is some degree of variation, on the average, endurance athletes tend to have greater percentages of ST fibers, whereas non-endurance athletes have greater percentages of FT fibers than their nonathletic counterparts (modified from Fox et al., 1993).

glycolytic potential (Saltin & Gollnick, 1983). The exception is the increase in hexokinase which, as described above, is the only glycolytic enzyme to be significantly elevated with increased activity (Barnard & Peter, 1969).

Increased oxidative capacity is associated with conversion of type IIB to IIA muscle fibers in human muscles but relatively little conversion from type II to type I (Anderson & Henriksson, 1977; Jansson & Kaijser, 1977; Schantz et al., 1983). The trend for endurance athletes to have a high percentage of type I muscle fibers in contrast to weight-lifters and sprinters (Fig. 7) was attributed to genetic factors (Komi et al., 1977; Edstrom & Grimby, 1986; Salmons & Henriksson, 1986; Bouchard et al., 1992; Mero et al., 1992), based very much on the early findings of Gollnick and colleagues (1973) that there was no interconversion of fiber types in trained muscles. More recently, however, the observation of increasing numbers of fast muscle fibers during detraining suggests that there may be some interconversion (Henriksson, 1992), consistent with histochemical findings of more intermediate fiber types (Schantz et al., 1983). The absence of adaptation in fast-twitch *white* muscles in rats after endurance training was explained by their lack of recruitment during exercise regimes including marathon running (Holloszy & Booth, 1976; Costill et al., 1977). More extreme and prolonged training, however, led to significant fiber conversions in the same direction as found with chronic electrical stimulation; namely type IIB-> IIA-> I (Green et al., 1984). Thus, failure to detect more type I fibers following endurance training may be attributed to the failure of exercise programs to recruit the larger fast motor units for sufficient time and intensity to result in extensive muscle fiber conversion.

The earliest adaptation to muscle exercise is the reflex increase in blood flow to the working muscle that is mediated by increased cardiac output. During endurance or dynamic exercise, muscle blood flow increases linearly with exercise intensity with little change in local vascular resistance (Hudlicka, 1984; Sjøgaard, 1987). The limiting factor to oxygen supply to the muscle is the transit time for the blood through the capillary network. As the capillary density increases with training (reviewed by Henriksson, 1992), this time is prolonged. As a result, oxygen, energy substrates, and metabolites are better exchanged across the capillary walls. However, an upper limit in cardiac output ultimately is a major factor that limits muscle performance particularly for whole muscle exercise (see Lindstedt & Hopeller, Chapter 28). Consequently, cardiac and respiratory training are a major focus in endurance training (e.g., Anholm et al., 1989; Boutellier et al., 1992). The vasodilatation in the working muscles constitutes a volume load to the heart. This leads to a significant ventricular hypertrophy in trained endurance athletes, which is an essential component of their cardiovascular training (Longhurst et al., 1980b). The volume load is considerably greater than the afterload due to increased blood pressure during static or isometric exercise (Longhurst & Stebbins, 1992).

The competitive athlete is, therefore, well adapted to use oxidative substrates to generate ATP during exercise. A resting bradycardia and lower absolute heart rate and trained capacity to increase cardiac output and ventilation ensures delivery of oxygen and substrates to the working muscles (Longhurst et al., 1980a; Brooks & Fahey, 1985). During the exercise, carbohydrate and fat are metabolized for energy with exhaustion being experienced only when glycogen is significantly depleted (Hagerman, 1992). During competitive marathon running which requires amongst the highest total energy expenditure, muscle glycogen was depleted to extremely low levels (Costill, 1986). Short, high-intensity exercise such as sprinting produces smaller augmentation of the oxidative pathways (Saubert et al., 1973).

Static or Isometric Exercise

Static or high resistance exercise training is the typical form of exercise for weight lifters and wrestlers. The duration of exercise is short but, most importantly, the muscles perform relatively little external work in comparison with the dynamic exercise of long-distance runners, for example. The muscles consume energy and liberate it as heat (Fenn, 1923) and there is little increase in oxygen consumption during exercise (Longhurst & Mitchell, 1983; Reid et al., 1987). During the exercise, blood flow to the active muscle is progressively impeded as intramuscular pressure exceeds arterial and capillary blood pressure. With forces that exceed 30% maximal voluntary contraction, intramuscular blood vessels may be completely occluded (Humphreys & Lind, 1963) and the muscles are highly adapted for anaerobic glycolysis, containing high glycolytic enzyme activities (Saltin & Gollnick, 1983). There is a tendency for these athletes to have a higher type II/type I fiber ratio but there is relatively little change in endurance with strictly isometric exercise training (Salmons & Henriksson, 1986; their Fig. 7). There is also little metabolic adaptation to this type of exercise; there is a small but significant increase in resting muscle creatine, CP, ATP, and glycogen (MacDougall et al., 1977). However, a disproportionate increase in myofibrillar proteins tends to dilute the metabolic changes (Gollnick et al., 1972). In rats, high intensity but brief exercise similarly results in little metabolic adaptation but significant fiber hypertrophy (Hoppeler, 1988).

The primary effect of training and the most important goal of resistance training is to increase the force generated by the muscle. Some of the increase in muscle strength and corresponding muscle bulk is undoubtedly due to increased muscle fiber cross-sectional area (hypertrophy) and/or increased force per unit area (specific force; Gollnick et al., 1972; MacDougall et al., 1977; 1982; Saltin & Gollnick 1983). Evidence for an increase in the number of muscle fibers (hyperplasia; reviewed by Antonio & Gonyea, 1993) has been criticized on methodological considerations of the difficulties in obtaining representative and accurate muscle fiber counts (see Saltin & Gollnick, 1983; Edstrom & Grimby, 1986; Antonio & Gonyea, 1993). Hypertrophy, however, cannot fully account for increased muscle bulk or increased force. Firstly, a dramatic increase in the hydrophillic connective tissue, which increases water content, accounts for a substantial component of the increased muscle bulk (Jablecki et al., 1973; Saltin & Gollnick, 1983). Secondly, comparison of evoked and voluntary muscle contractions has shown that the increase in volitional muscle strength is due, in part, to an enhanced capacity of the trained individual to recruit additional motor units and increase the firing rates of recruited units (Gollnick et al., 1974; Jones et al., 1989; Narici et al., 1989; reviewed by Edstrom & Grimby, 1986; Enoka, 1988). Thus, neural as well as muscle plasticity contribute to increased strength after weight training.

ADAPTATION TO EXPERIMENTAL MUSCLE OVERLOAD

The adaptive changes in the functionally overloaded muscles following experimental removal of muscle synergists have been compared to those subjected to endurance training (reviewed by Gardiner, 1991). In an early study, Schiaffino and Bormioli (1973) noted a dramatic shift in type IIB to IIA fiber type shift in remaining ankle flexor muscles when the TA muscle was extirpated in the neonatal rat. Clearly the overuse had induced an adaptive metabolic response. Yet, in the adult animal the adaptation is usually small by comparison, especially in physiological flexors (Olha et al., 1988; Frischknecht & Vrbova, 1991; Rosenblatt & Parry, 1993), unless muscle activity is increased by additional activation (Riedy et al., 1985; Roy et al., 1992) or the functional overload is maximized by removal of all synergists but the experimental muscle (Chalmers et al., 1992). In adult extensor muscles, there is a small but significant increase in the proportion of S motor units and corresponding SO fibers, particularly in the deep portion of the muscles (Walsh et al., 1978; Tsika et al., 1987; Olha et al., 1988; Crerar et al., 1989). Interestingly, all motor units show compensatory hypertrophy but conversion of fast-to-slow motor units is limited. The limited conversion was attributed to the minimal recruitment of the larger FI and FF units in accordance with the good correlation of motor unit endurance with total amount of neuromuscular activity (Kernell & Eerbeek, 1989). This is the same argument used for the limited fast-to-slow adaptation after endurance training, as discussed above. In contrast, when all but one muscle is left in a muscle compartment, the adaptation is more marked. Nonetheless, a disproportionate increase in fiber volume as compared with oxidative enzyme activity dilutes the influence of overload on the total enzyme activity (Chalmers et al., 1992), as described for static exercise below.

The more generalized muscle fiber hypertrophy, however, is consistent with the chronic stimulation experiments which showed that it is the type rather than the amount of muscular activity that is important in determining muscle fiber size and motor unit force. Increasing mechanical loading of contracting muscle is a powerful stimulus for muscle fiber enlargement (Riedy et al., 1985). Even passive muscle responds to loading since compensatory hypertrophy following synergist ablation even occurs in denervated muscle (Gutmann et al., 1971). With few exceptions, increased muscle mass and strength are due to hypertrophy rather than hyperplasia (James, 1976; Gollnick et al., 1981; but see Antonio & Gonyea, 1993) and requires incorporation of satellite cells (Rosenblatt & Parry, 1993). Recruitment of satellite cells as the source of new nuclei maintains the normal ratio of myonuclei to cell volume (Schmalbruch & Hellhamer, 1977; Russell et al., 1992). Studies of muscle regeneration after muscle trauma indicate that many of the growth factors implicated in the revascularization of damaged muscles are also important for the proliferation, differentiation and fusion of muscle cells from satellite cells (Grounds, 1991). Many of the same muscle regulatory factors are expressed as during the commitment and maturation of developing muscle fibers. For example, the expression of a muscle-specific regulatory factor qmf1 is increased several fold when avian muscles are stretched (Russell et al., 1992). Expression of skeletal muscle specific genes such as MyD1 and myogenin have been used to identify muscle precursor cells in regenerating adult muscle after direct muscle damage (Grounds, 1991). Using gamma irradiation to prevent satellite cell incorporation and compensatory hypertrophy, recent experiments show that the functional overload can affect a fiber-type conversion in the absence of compensatory hypertrophy (Rosenblatt & Parry, 1993). Thus, the progressive conversion in the direction of FF-> FI -> FR -> S with increased neuromuscular activity in the overload model provides additional evidence for the direct association between neuromuscular activity, metabolic enzyme activities, and metabolic adaptation.

MUSCLE ADAPTATION TO SELF- AND CROSS-REINNERVATION OF DENERVATED MUSCLES

The dramatic results of the cross-reinnervation experiments of Buller and colleagues (1960) that the fast flexor digitorum longus muscles became slowly contracting and the slow soleus muscles contracted more rapidly when their nerves were cross-united provided irrefutable evidence for muscle plasticity in the adult. Findings that these changes could be reversed if the original patterns of stimulation were superimposed provided strong evidence that the neural effects were mediated by neuromuscular activity (Salmons & Sreter, 1976).

Changes in contractile speed are associated with corresponding changes in myosin light and heavy chain isoforms, although cross-reinnervated motor units may continue to express their original myosin isoform in some of their muscle fibers (Sreter et al., 1976; Gauthier et al., 1983; Fu et al., 1992). Also extensively documented are changes in sarcoplasmic reticular and regulatory proteins that support the conversion of fiber types in cross-reinnervated muscles (e.g., Margreth et al., 1973; Amphlett et al., 1975; Sreter et al., 1976).

Increase in the number of S motor units and SO fibers in cross-reinnervated fast-twitch muscles is associated with increased oxidative and decreased glycolytic enzyme activity, accounting for the increased resistance of the muscles to fatigue (e.g., Dum et al., 1985; Gillespie et al., 1986; 1987; Gordon et al., 1988). The glycolytic enzyme activity increases in many fibers in cross-reinnervated soleus muscles in association with expression of fast contractile and regulatory proteins. However, the muscle fibers retain their high oxidative capacity with the result that all motor units classified as F units are FR units and the muscles retain their high resistance to fatigue (Chan et al., 1982; Gillespie et al., 1986; 1987).

Even after self-reinnervation of skeletal muscles, there is extensive mismatch of nerves and muscle fibers because reinnervating nerves do not selectively reinnervate their original muscle fibers (Kugelberg et al., 1970; Totosy de Zepetnek et al., 1992a). Newly reinnervated motor units, therefore, include muscle fibers of different types which formerly belonged to different motor units. Interestingly, many of the self-reinnervated motor units are FI (Gordon, 1987: Gordon et al., 1988) and variance in fiber size within reinnervated motor units is high (Totosy de Zepetnek et al., 1992b). These findings suggest that reinnervated muscle fibers retain some of their original phenotype. Normally the variance in oxidative and glycolytic enzyme activities between unit muscle fibers is very low (Nemeth et al., 1986; their Fig. 8). After self- or cross-reinnervation, however, the variance for representative enzymes is increased several fold (Sesodia et al., 1993; their Fig. 8). This heterogeneity could explain the trend for motor units to be fatigue intermediate. It is unlikely that the heterogeneity can be explained by incomplete activation of all motor unit fibers

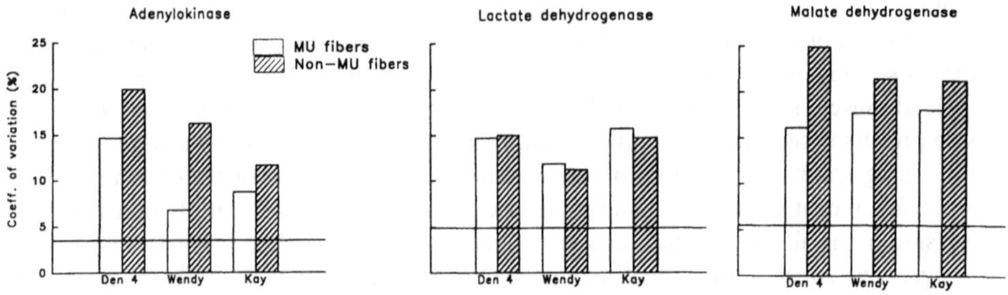

Figure 8. Coefficient of variation of the enzyme activity of oxidative and glycolytic enzymes in muscle fibers in reinnervated tibialis anterior muscles of the rat after self-reinnervation by common peroneal nerve (Den 4) or cross-reinnervation by the posterior tibial nerve (Wendy, Kay). At least 4 months after nerve repair, a single motor unit was isolated for glycogen depletion (methods described by Totosy de Zepetnek et al., 1992a) and the enzyme activities of oxidative (malate dehydrogenase) and glycolytic enzymes (lactate dehydrogenase, adenylokinase) analyzed using microdissection and microanalysis of 8-20 muscle fibers (methods described by Nemeth et al., 1986). The coefficient of variation expressed as a percentage of the normal variation in age-matched control muscles was obviously higher in muscle fibers of the same type randomly sampled outside of the depleted unit. Within motor units, the variability between fibers in enzyme activities were greatly increased above normal and sometimes as high as between fibers of the same type outside of the motor unit (unpublished data of Sesodia, Fu, Nemeth and Gordon).

because the heterogeneity was seen in long-term reinnervated motor units in which neuromuscular transmission was stable.

One explanation for the apparent retention of former phenotype is that adult muscle fibers can adapt within a limited range set by their developmental origins (reviewed by Gordon & Pattullo, 1993). Another consideration is that the architecture of reinnervated muscles is not altered, irrespective of the source of innervation. As a result, the length and load on the muscles are not necessarily altered after reinnervation and may contribute to the retention of original muscle phenotype. Experimental support for the second suggestion is provided by the findings that almost complete slow-to-fast conversion occurred in cross-reinnervated soleus muscles when the common peroneal nerve (CP) was the source of *fast* nerves but conversion was not complete when the *fast* nerves were synergistic flexor hallucis (FHL) or digitorum longus (FDL) muscle nerves (Dhoot & Vrbova, 1991). In the former case (CP), complete denervation of antagonist muscles removed the normal stretch on the soleus muscle. As a result, the cross-reinnervated soleus muscles experienced an unloaded contraction which possibly accounted for the transition to fast phenotype. In the latter case (FHL or FDL), denervation of one physiological extensor did not change the loading conditions of the soleus substantially, and the loaded soleus muscle retained a slow phenotype, despite cross-reinnervation.

Thus, metabolic and phenotypic adaptations in reinnervated muscles reveal a combined effect of neuromuscular activity and muscle loading which together determine the conversion of muscle properties by the innervating nerves. The important contribution of muscle loading to the expression of slow phenotype and metabolic adaptation is further revealed in underuse paradigms as described below.

METABOLIC ADAPTATION TO MUSCLE DISUSE

The effects of underusage paradigms are quite variable in natural and laboratory conditions, between muscles, species, procedures and different laboratories. Nevertheless, there is a general trend for muscles to express fast phenotype with respect to contractile and regulatory proteins and to upregulate enzymes of anaerobic glycolysis (reviewed by Gordon & Pattullo, 1993). However, in no case has complete conversion taken place. Many muscle properties are unchanged, in particular, susceptibility to fatigue in postural muscles which normally cross a single joint. These muscles do not fatigue readily and maintain their high oxidative capacity even though glycolytic enzyme activity increases. The adaptations have been recently reviewed (Roy et al., 1991; Gordon & Pattullo, 1993) and are, therefore, only briefly described here.

Spinal Cord Transection

Fast-twitch but not slow-twitch postural muscles become more fatigable after hemi- and complete spinal cord transections. There is a coordinated increase in glycolytic and mATPase enzyme activity associated with a limited fiber type I to II conversion (Gordon & Pattullo, 1993). In the cat fast-twitch MG muscle, for example, in which there are normally only 25% FF units, there is a significant increase in susceptibility to fatigue (Fig. 9). This in turn coincides with increased numbers of FF units, a corresponding increase in type IIB fibers (Gordon & Mao, 1994; see also Mayer et al., 1984; Munson et al., 1986) and increased glycolytic enzyme activity relative to oxidative enzyme activity (Hoffman et al., 1990; Jiang et al., 1990b).

In slow-twitch and postural muscles, retention of the high oxidative potential despite increased mATPase and glycolytic enzyme activity explains the high fatigue resistance of the paralyzed muscles (Hoffman et al., 1990; Jiang et al., 1990a,b) and the limited conversion of S to FR motor units and associated type I to IIA fiber type conversion (Cope et al., 1986). This adaptive pattern resembles that of chronically stimulated denervated muscles in the rat where intermittent electrical stimulation caused a type I to IIA conversion of fiber types. The same pattern is also seen in other disuse models, such as hindlimb suspension, spaceflight and immobilization.

Reduced Load Bearing: Hindlimb Suspension and Space Flight

In hindlimbs that are chronically unloaded by tethering the tail to a swivel support or by supporting the body weight with a sling, fast-twitch muscles become slightly more fatigable after an initial delay (Winiarski et al., 1987; Herbert et al., 1988; Templeton et al., 1988; Pierotti et al., 1990). The small increase was associated with a modest type I to type II fiber conversion, an increase in the ratio of glycolytic to oxidative enzyme activities (Roy et al., 1987; Jiang et al., 1992) and expression of the fast isoform of the sarcoplasmic reticulum Ca^{2+}-pump (Schulte et al., 1993). As above, soleus muscles retained their high fatigue resistance, high oxidative capacity (Ohira et al., 1992) and showed a modest increase in contractile speed (Corley et al., 1984; Templeton et al., 1984; Fitts et al., 1986).

Although muscle endurance has not been measured under the non-weight bearing conditions of space-flight, the direction of fiber-type conversion is the same (Roy et al., 1991). However, the metabolic changes are considerably less after short duration space flight (14 days) than hindlimb suspension (Jiang et al., 1992) particularly in the monkey (Bodine-Fowler et al., 1992). In the latter experiments, the little or no impact of spaceflight on muscle metabolism or muscle fiber size was attributed to the work load of the muscles in the constrained sitting position of the animal (Bodine-Fowler et al., 1992).

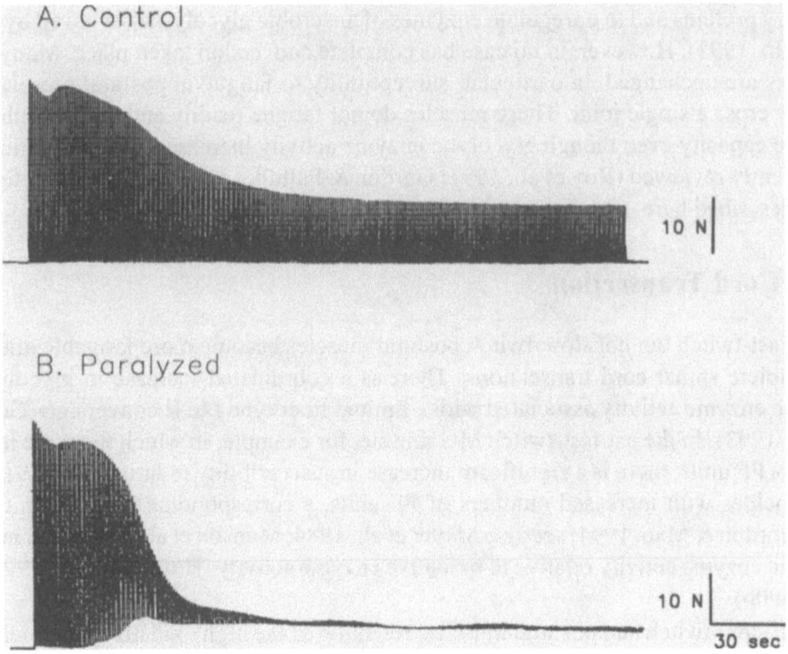

Figure 9. Fatigue of cat medial gastrocnemius muscle before and 93 days after paralysis by hemisection at T12 and unilateral deafferentation from L2 to S2. The MG nerve was stimulated maximally via an implanted cuff electrode. 13 pulses delivered at a rate of 40 Hz were repeated every second for 2 min and the ankle extensor tension was recorded by non-invasive coupling the foot to an external force transducer, under Halothane anesthesia (methods described in detail by Davis et al., 1978).

Limb Immobilization

There is partial type I to type II conversion in slow- and fast-twitch hindlimb muscles after limb immobilization. However, there is little or no change in susceptibility to fatigue despite a decline in the number of mitochondria and oxidative enzyme activity (Rifenberick et al., 1973; Krieger et al., 1980; Mayer et al., 1981: Witzmann et al., 1983). Many studies have noted a small increase in contraction speed in slow-twitch as well as fast-twitch muscles (reviewed by Gordon & Pattullo, 1993). The most dramatic and most well studied effects of limb immobilization relate to the atrophy of the immobilized muscles, particularly in a shortened position.

Pharmacological Paralysis

Application of tetrodotoxin (TTX) via cuff electrodes on sciatic or tibial nerves provides an effective block of evoked neuromuscular activity and has been explored over periods of 2-6 weeks as a model of hindlimb disuse. Increase in contractile speed associated with type I to II fiber conversion and expression of fast myosin heavy chains was not accompanied by any significant change in muscle fatigability (Spector, 1985a,b; St-Pierre & Gardiner, 1985, Gardiner et al., 1992). Decline in oxidative potential was noted but the change was small (Gardiner et al., 1992).

How Much Can Adaptive Changes Be Attributed to Loss of Neuromuscular Activity in Disuse Models

The direction of motor unit and fiber type conversion (from S to FF and type I to IIB, respectively) is consistent with the scheme of Kernell and coworkers (1987a,b; Kernell & Eerbeek, 1989) in which progressively less daily activity shifts properties progressively from S -> FR-> FI -> FF. Summed EMG activity after complete spinal transection in cats and kittens (Alaimo et al., 1984; Lovely et al., 1990) and humans (Stein et al., 1992) may be reduced to less than 10% of normal but depends on the extent of spasticity. Perhaps the remaining neuromuscular activity is sufficient to maintain slow phenotype in some fibers and limit progressive conversion to FF units. This does not appear to be so since many type I fibers remain in soleus, gastrocnemius and TA muscles that have been completely silenced by spinal isolation (Graham et al., 1992; Pierotti et al., 1991). Furthermore, the adaptive response of different muscles to the same loss of neuromuscular activity varies considerably (e.g., Roy & Acosta, 1986) and the adaptive response of any one muscle to different degrees of paralysis may be surprisingly similar (Jiang et al., 1990a,b; Jiang et al., 1991; Graham et al., 1992).

In most disuse models, with the exception of complete pharmacological blockade, remaining neuromuscular activity may be considerable. After hindlimb suspension for example, neuromuscular activity is reduced only temporarily, returning to normal within a week (Alford et al., 1989). Daily integrated EMG activity in immobilized soleus muscles may be as low as 5-15% but as high as 50-80% of normal (Fournier et al., 1983). In immobilized fast-twitch extensor and flexor muscles, neuromuscular activity approaches normal levels (Mayer et al., 1981; Hnik et al., 1985).

Similar to the adaptive responses in overuse paradigms and reinnervated muscles, muscle loading is likely to make important contributions to the adaptive responses of muscles in underuse paradigms. Muscle paralysis is usually associated with muscle unloading. This is particularly striking in postural muscles that are normally weight bearing and which show the most pronounced fiber type conversion and atrophic changes after paralysis (e.g., Lieber et al., 1988; reviewed by Gordon & Pattullo, 1993). Spastic contractions in spinal-cord-injured patients may exacerbate the atrophy because unloaded muscles, particularly single joint postural muscles are known to be damaged when muscle shortening is unopposed (Vrbova, 1963; McMinn & Vrbova, 1967; Tabary et al., 1972; Baker & Matsumoto, 1988). Loss of weight bearing is the major consequence of hindlimb suspension, which is the basis for this experimental model being used to simulate space-flight conditions with unresisted muscle contractions.

Expression of fast or slow phenotype appears to strongly influenced by muscle length, whether the muscle is active or not. Immobilization of soleus muscle in a shortened position, for example, initiates expression of fast myosin heavy chains (Goldspink et al., 1992), consistent with fiber type I to II conversion seen in immobilized muscles. Immobilization of the fast-twitch TA in a lengthened position repressed the normal expression of fast phenotype (Goldspink et al., 1992) suggesting that muscle stretch is an important determinant of slow phenotype and that unloading is an important contributing factor to type I to II conversion of immobilized muscles. Stretch and chronic stimulation enhances the expression of slow myosin heavy chains in these muscles indicating that loading of chronically stimulated muscles is likely to be an important contributing factor in their fast-to-slow conversion.

Changes in blood flow and bone density associated with muscle paralysis are also factors that may be involved in the adaptive changes of muscle and may also be linked to the loss of weight-bearing (see Gordon & Mao, 1994). Finally, generalized effects of stress must be considered in the adaptive responses of disused muscles (reviewed by Gordon & Pattullo, 1993). In animal models, considerable stress may be provoked by the experimental manipulations of spinal cord surgery, pinning of joints and the animal handling. In human patients, the immobilization and muscle paralysis after spinal cord and peripheral nerve, bone, tendon or ligament injuries must be considered in the context of the pain and stress associated with the injuries. Responses to pain and stress, which include raised levels of glucocorticoids, adrenal hypertrophy and gastric ulcers, have been recorded in animal models in association with muscle atrophy (Thomason & Booth, 1990). Glucocorticoids released in response to stress and binding to receptors induced in immobilized muscles may contribute to the increase in contractile speed (Dubois & Almon, 1980; Gardiner et al., 1980).

In summary, changes in muscle phenotype and disuse atrophy have often been attributed to reduced muscle use but review of the many studies using disuse models indicate that changes in muscle length and/or weight-bearing are likely to be important contributing factors. Fiber type I to type II conversions with remaining heterogeneity of muscle fibers and motor units occur in all these models which share in common the unloading of the muscle contractions even when they differ considerably in their total amount of daily activity. Hormonal factors and changes in blood flow associated with the traumatic manipulations may also contribute to the phenotypic expression of muscles, including their susceptibility to fatigue.

CONCLUSIONS

Adaptive responses of skeletal muscles to overuse, reinnervation and underuse paradigms may be striking with conversion of muscle speed, endurance and strength. Although dramatic conversion of muscle properties can be demonstrated under conditions of chronic electrical stimulation, many studies identify the limits of the muscle adaptability and the role of cardiovascular and central plasticity. Careful analysis of the adaptive responses of muscles to novel innervation and to underuse paradigms within the context of alterations in muscle activity, length and loading reveal the importance of the load on the passive and active muscles in expression of fast and slow muscle phenotype. Finally, systemic responses to altered neuromuscular activity must be recognized as significant contributing factors in the adaptive responses of skeletal muscles in over- and underuse paradigms. Overall, there is strong evidence to support the ideas that 1) the amount of neuromuscular activity strongly determines muscle endurance and the underlying metabolic adaptations; and 2) the loading against which a muscle contracts determines muscle strength.

ACKNOWLEDGMENTS

The support of the Medical Research Council and Muscular Dystrophy Association of Canada is gratefully acknowledged. The author would also like to thank many of her colleagues who participated in the studies referred to in the chapter, particularly Drs. Jian Mao and Susan Fu whose unpublished data is discussed, and Neil Tyreman for his tireless work in the laboratory, and assistance in preparation of the manuscript. The author's attendance at the 1994 Bigland-Ritchie conference was supported, in part, by the University of Miami.

REFERENCES

Alaimo MA, Smith AJL, Roy RR & Edgerton VR (1984). EMG activity of slow and fast ankle extensors following spinal cord transection. *Journal of Applied Physiology* **56**, 1608-1613.

Alford EK, Roy RR, Hodgson JA & Edgerton VR (1989). Electromyography of rat soleus, medial gastrocnemius, and tibialis anterior during hind-limb suspension. *Experimental Neurology* **96**, 635-649.

Allen DG, Westerblad H, Lee JA & Lannergren J (1992). Role of excitation-contraction coupling in muscle fatigue. *Sports Medicine* **13**, 116-126.

Amphlett GW, Perry SV, Syska H, Brown M & Vrbová G (1975). Cross innervation and the regulatory protein system of rabbit soleus muscle. *Nature* **257**, 602-604.

Andersen P & Henriksson J (1977). Training induced changes in the subgroups of human type II skeletal muscle fibers. *Acta Physiology Scandinavica* **99**, 123-125.

Anholm JD, Stray-Gundersen J, Ramanathan M & Johnson RL Jr (1989). Sustained maximal ventilation after endurance exercise in athletes. *Journal of Applied Physiology* **67**, 1759-1763.

Antonio J & Gonyea WJ (1993). Skeletal muscle fiber hyperplasia. *Medicine and Science in Sports and Exercise* **25**, 1333-1345.

Ariano MA, Armstrong RB & Edgerton VR (1973). Hindlimb muscle fiber populations of five mammals. *Journal of Histochemistry and Cytochemistry* **21**,51-55, 1973.

Ausoni S, Gorza L, Schiaffino S, Gundersen K & Lomo T (1990). expression of myosin heavy chain isoforms in stimulated fast and slow rat muscles. *Journal of Neuroscience* **10**, 153-160.

Baker JH & Matsumoto DE (1988). Adaptation of skeletal muscle to immobilization in a shortened position. *Muscle & Nerve* **11**,231-244.

Baldwin KM, Winder WW, Terjung RL & Holloszy J (1973). Glycolytic enzymes in different types of skeletal muscle: adaptation to exercise. *American Journal of Physiology* **225**, 962-966.

Barnard RJ & Peter JB (1969). Effect of training and exhaustion on hexokinase activity of skeletal muscle. *Journal of Applied Physiology* **27**, 691-695.

Bass A, Brdiczka D, Eyer P, Hofer S & Pette D (1969). Metabolic differentiation of distinct muscle types at the level of enzymatic organization. *European Journal of Biochemistry* **10**, 198-206.

Bodine-Fowler SC, Roy RR Rudolph W, Haque N, Kozlovskaya IB & Edgerton VR (1992). Spaceflight and growth effects on muscle fibers in the rhesus monkey. *Journal of Applied Physiology* **73**, 82S-89S.

Bouchard C, Dionne FT, Simoneau J-A & Boulay MR (1992). Genetics of aerobic and anaerobic performances. *Exercise and Sport Sciences Review* **20**, 27-58.

Boutellier U, Buchel R, Kundert A & Spengler C (1992). The respiratory system as an exercise limiting factor in normal trained subjects. *European Journal of Applied Physiology and Occupational Physiology* **65**, 347-353.

Brooks GA & Fahey TA (1985). *Exercise Physiology: Human Bioenergetics and its Applications*. New York: Macmillian.

Brooks MH & Kaiser KK (1970). Three "myosin adenosine trophosphatase" systems: the nature of their pH lability and sulfhydryl dependence. *Journal of Histochemistry and Cytochemistry* **18**, 670-672.

Brown MD, Cotter MA, Hudlická O & Vrbová G (1976). The effect of different patterns of muscle activity on capillary density, mechanical properties and structure of slow and fast rabbit muscles. *Pflügers Archiv* **361**, 241-250.

Buller AJ, Eccles JC & Eccles RM (1960). Interactions between motoneurones and muscles in respect of the characteristic speeds of their responses. *Journal of Physiology (London)* **150**, 399-416.

Burke RE (1981). Motor units: anatomy, physiology, and functional organization. In: Brookhart JM, Mountcastle VB (sec. eds.), Brooks VB (vol. ed.), *Handbook of Physiology, sec. 1, vol. II, pt 1, The Nervous System: Motor Control,* pp. 345-422. Bethesda, MD: American Physiological Society.

Burke RE & Edgerton VR (1975). Motor unit properties and selective involvement in movement. *Exercise and Sport Sciences Reviews* **3**, 31-81.

Burke RE, Levine DN, Tsairis P & Zajac FE (1973). Physiological types and histochemical profiles in motor units of the cat gastrocnemius. *Journal of Physiology (London)* **234**, 723-748.

Castle NA & Haylett DG (1987). Effect of channel blockers on potassium efflux from metabolically exhausted frog skeletal muscle. *Journal of Physiology (London)* **383**, 31-43.

Chalmers GR, Roy RR & Edgerton VR (1992). Variation and limitations in fiber enzymatic and size responses in hypertrophied muscle. *Journal of Applied Physiology* **73**, 631-641.

Chan M, Edgerton VR, Goslow GE Jr, Kurata H, Rasmussen S & Spector SA (1982). Histochemical and physiological properties of cat motor units after self- and cross-reinnervation. *Journal of Physiology (London)* **332**, 343-361.

Cope TC, Bodine SC, Fournier M & Edgerton VR (1986). Soleus motor units in chronic spinal transected cats: physiological and morphological alterations. *Journal of Neurophysiology* **55**, 1202-1220.

Corley K, Kowalchuk B & McComas AJ (1984). Contrasting effects of suspension on the hindlimb muscles in the hamster. *Experimental Neurology* **85**, 30-40.

Costill DL (1986). *Inside Running: Basics of Sports Physiology.* Indianapolis, IN: Bechmark Press, Incorporated.

Costill DL, Coyle EF, Dalsky G, Evans W, Fink W & Hoopes D (1977). Effects of elevated plasma FFA and insulin on muscle glycogen usage during exercise. *Journal of Applied Physiology* **43**, 695-699.

Crerar MM, Hamilton NC, Blank S, Urdea MS & Ianuzzo CD (1989). The genes for β-myosin heavy chain and glycogen phosphorylase are discoordinately regulated during compensatory growth of the plantaris muscle in the adult rat. *Molecular Cellular Biochemistry* **86**, 115-123.

Davis LA, Gordon T, Hoffer JA, Jhamandas J & Stein RB (1978). Compound action potentials recorded from mammalian peripheral nerves following ligation and resuturing. *Journal of Physiology (London)* **285**, 543-559.

Denny-Brown, D (1929). On the nature of postural reflexes. *Proceedings of Royal Society (Biology)* **104**, 252-301.

Dhoot GK & Vrbova G (1991). Conversion of cat and rabbit soleus muscle fibres by alien nerves. *Muscle & Motility* **2**, 125-130.

Donselaar Y, Eerbeek O, Kernell D & Verhey BA (1987). Fibre sizes and histochemical staining characteristics in normal and chronically stimulated fast muscle of cat. *Journal of Physiology (London)* **382**, 237-254.

Dubois DC & Almon RR (1980). Disuse atrophy of skeletal muscle is associated with an increase in number of glucocorticoid receptors. *Endocrinology* **107**: 1649-1651.

Dum RP, O'Donovan MJ, Toop J, Tsairis P, Pinter MJ & Burke RE (1985). Cross-reinnervated motor units in cat muscle. II. Soleus muscle reinnervated by flexor digitorum longus motoneurons. *Journal of Neurophysiology* **54**, 837-851.

Edstrom L & Grimby L (1986). Effect of exercise on the motor unit. *Muscle & Nerve* **9**, 104-126.

Eerbeek O, Kernell D & Verhey BA (1984). Effects of fast and slow patterns of tonic long-term stimulation on contractile properties of fast muscles in cat. *Journal of Physiology (London)* **352**, 73-90.

Eisenberg BR & Salmons S (1981). The reorganization of subcellular structure in muscle undergoing fast-to-slow type transformation. A streological study. *Cell Tissue Research* **220**, 449-471.

Enoka RM (1988). Muscle strength and its development: new perspectives. *Sports Medicine* **6**, 148-168.

Fenn WO (1923). A quantitative comparison between the energy liberated and the work performed by the isolated sartorius muscle of the frog. *Journal of Physiology (London)* **58**, 175-203.

Fitts RH, Metzger DA, Riley DA & Unsworth BR (1986). Models of disuse: a comparison of hindlimb suspension and immobilization. *Journal of Applied Physiology* **60**, 1946-1953.

Fournier M, Roy RR, Perham H, Simard CP & Edgerton VR (1983). Is immobilisation a model of disuse? *Experimental Neurology* **80**, 147-156.

Fox EL, Bowers RW & Foss ML (1988). *The Physiological Basis of Physical Education and Athletics*. New York: WB Saunders.

Fox EL, Bowers RW & Foss ML (1993). *The Physiological Basis for Exercise and Sport*, 5th Ed. Dubuque, IA: Wm. C. Brown Communications, Inc.

Frischknecht R & Vrbova G (1991). Adaptation of rat extensor digitorum longus to overload and increased activity. *Pflügers Archiv* **419**, 319-326.

Fu S, Gordon T, Parry DJ & Tyreman N (1992). Immunohistochemical analysis of fiber types within physiologically typed motor units of rat tibialis anterior muscle after long-term cross-reinnervation. *Society for Neuroscience Abstracts* **18**, 1557.

Gardiner PG (1991). Effects of exercise training on components of the motor unit. *Canadian Journal of Sports Sciences* **16**, 271-288.

Gardiner PG, Favron M & Corriveau P (1992). Histochemical and contractile responses of rat medial gastrocnemius to two weeks of complete disuse. *Canadian Journal of Physiology and Pharmacology* **70**, 1075-81.

Gardiner PG, Montanaro DS & Edgerton VR (1980). Effects of glucocortocoid treatment and food restriction on rat hindlimb muscles. *Pflügers Archiv* **385**, 147-153.

Gauthier GF, Burke RE, Lowey S & Hobbs AW (1983). Myosin isozymes in normal and cross-reinnervated cat skeletal muscle fibers. *Journal of Cell Biology* **97**, 756-771.

Gillespie MJ, Gordon T & Murphy PA (1986). Reinnervation of the lateral gastrocnemius and soleus muscles in the rat by their common nerve. *Journal of Physiology (London)* **372**, 485-500.

Gillespie MJ, Gordon T & Murphy PA (1987). Motor units and histochemistry in the rat lateral gastrocnemius and soleus muscles: evidence for dissociation of physiological and histochemical properties after reinnervation. *Journal of Neurophysiology* **57**, 921-937.

Gimbrone MA (1984). Macrophages, neovascularisation and the growth of vascular cells. In: Jaffe EA (ed), *Biology of Endothelial Cells*, pp. 97-107. Boston: Martinus Nihoff Publications.

Goldspink G, Scutt A, Loughna PT, Wells DJ, Jaenicke T & Gerlach GF (1992). Gene expression in skeletal muscle in response to stretch and force generation. *American Journal of Physiology* **262**, R356-R363.

Gollnick PD, Armstrong RB, Saltin B, Saubert CW, Sembrowich WL & Shepherd RE (1973). Effect of training on enzyme activity and fiber composition of human skeletal muscle. *Journal of Applied Physiology* **34**, 107-111.

Gollnick PD, Armstrong RB, Saubert IV CW, Piehl K & Saltin B (1972). Enzyme activity and fiber composition in skeletal muscle of untrained and trained men. *Journal of Applied Physiology* **33**, 312-319.

Gollnick PD, Piehl K & Saltin B (1974). Selective glycogen depletion pattern in human skeletal muscle fibres after exercise of varying intensity and at varying pedaling rates. *Journal of Physiology (London)* **216**, 1502-1509.

Gollnick PD, Timson BF, Moore RL & Riedy M (1981). Muscular enlargement and the number of fibers in the skeletal muscles of rats. *Journal of Applied Physiology: Respiratory Environment Exercise Physiology* **50**, 936-943.

Gordon T (1987). Muscle plasticity as demonstrated during sprouting and reinnervation. *American Zoology* **27**, 1055-1066.

Gordon T & Mao J (1994). Muscle atrophy and procedures for training for spinal cord injury. *Physical Therapy* **74**, 50-60.

Gordon T & Pattullo MC (1993). Plasticity of muscle fiber and motor unit types. *Exercise and Sports Sciences Reviews* **21**, 331-362.

Gordon T, Thomas CK, Stein RB & Erdebil S (1988). Comparison of physiological and histochemical properties of motor units after cross-reinnervation of antagonistic muscles in the cat hindlimb. *Journal of Neurophysiology* **60**, 365-378.

Gorza L, Gundersen K, Lomo T, Schiaffino S & Westgaard, RH (1988). Slow-to-fast transformation of denervated soleus muscles by chronic high-frequency stimulation in the rat. *Journal of Physiology (London)* **402**, 627-749.

Green HJ, Dusterhoft S, Dux L & Pette D (1992). Metabolite patterns related to exhaustion, recovery, and transformation of chronically stimulated rabbit fast-twitch muscle. *Pflügers Archiv* **420**, 359-366.

Green HJ, Klug GA, Reichman H, Seedorf U, Wiehrer W & Pette D (1984). Exercise-induced fibre type transitions with regard to myosin, parvalbumin, and sarcoplasmic reticulum in muscles of the rat. *Pflügers Archiv* **400**, 432-438.

Graham SC, Roy RR, Navarro B, Jaing B, Pierotti D, Bodine-Fowler S & Edgerton VR (1992). Enzyme and size profiles n chronically inactive cat soleus muscle fibers. *Muscle & Nerve* **15**, 27-36.

Grounds MD (1991). Towards understanding skeletal muscle regeneration. *Pathology Research Practice* **187**, 1-22.

Gundersen K, Leberer E, Lömo T, Pette D & Staron RS (1988). Fibre types, calcium-sequestering proteins and metabolic enzymes in denervated and chronically stimulated muscles of the rat. *Journal of Physiology (London)* **398**, 177-189.

Gutmann E, Schiaffino S & Hanzlikova V (1971). Mechanism of compensatory hypertrophy in skeletal muscle of the rat. *Experimental Neurology* **31**, 451-464.

Hagerman FC (1992). Energy metabolism and fuel utilization. *Medical Sciences Sports Exercise* **24**, S309-S314.

Hartner K-T, Kirschbaum BJ & Pette D (1989). The multiplicity of troponin T isoforms. Normal rabbit muscles and effects of chronic stimulation. *European Journal of Biochemistry* **179**, 31-38.

Henneman E & Mendell LM (1981). Functional organization of the motoneurone pool and its inputs. In: Brookhart JM, Mountcastle VB (sec. eds.), Brooks VB (vol. ed.), *Handbook of Physiology, sec. 1. vol. II, pt. 1, The Nervous System: Motor Control*, pp. 423-508, Bethesda: Ameriocan Physiological Society.

Henriksson J (1992). Effects of physical training on the metabolism of skeletal muscle. *Diabetes Care* **15**, 1701-1711.

Henriksson J, Chi MM-Y, Hintz CS, Young DA, Salmons S & Lowry OH (1986). Chronic stimulation of mammalian muscle: changes in enzymes of six metabolic pathways. *American Journal of Physiology* **251**, C614-C632.

Herbert ME, Roy RR & Edgerton VR (1988). Influence of one-week hindlimb suspension and intermittent high load exercise on rat muscles. *Experimental Neurology* **102**, 190-198.

Hicks A & McComas AJ (1989). Increased sodium pump activity following repetitive stimulation of rat soleus muscles. *Journal of Physiology (London)* **414**, 337-349.

Hnik P, Vejsada R, Goldspink DF, Kasicki S & Krekule I (1985). Quantitative evaluation of electromyogram activity in rat extensor and flexor muscles immobilized at different lengths. *Experimental Neurology* **88**, 515-528.

Hoffmann SJ, Roy RR, Bianco CE & Edgerton VR (1990). Enzyme profiles of single muscle fibers in the absence of normal neuromuscular activity. *Journal of Applied Physiology* **69**, 1150-1158.

Holloszy JO & Booth FW (1976). Biochemical adaptations to endurance exercise in muscle. *Annual Reviews of Physiology* **38**, 273-291.

Hood DA, Zak R & Pette D (1989). Chronic stimulation of rat skeletal muscle induces coordinate increases in mitochondrial and nuclear mRNAs of cytochrome c oxidase subunits. *European Journal of Biochemistry* **179**, 275-280.

Hoppeler H (1988). Exercise-induced structural changes of skeletal muscle. *ISI Atlas Science Biochemistry* **1**, 247-255.

Hudlicka O (1984). Development of microcirculation: capillary growth and adaptation. In: Renkin EM, Michel CC, Geiger SR (eds), *Handbook of Physiology, sec. 2, The Cardiovascular System*, pp 165-216. Baltimore, MD: Williams and Wilkins.

Hudlicka O, Brown M, Cotter M, Smith M & Vrbova G (1977). The effect of long-term stimulation of fast muscles on their blood flow, metabolism and ability to withstand fatigue. *Pflügers Archiv* **369**, 141-149.

Hudlicka O, Tyler KR, Srihari T, Heilig A & Pette D (1982). The effect of different patterns of long-term stimulation on contractile properties and myosin light chains in rabbit fast muscles. *Pflügers Archiv* **393**, 164-170.

Humphreys PW & Lind AR (1963). The blood flow through active and inactive muscles of the forearm during sustained hand-grip contractions. *Journal of Physiology (London)* **166**, 12-131.

Jablecki CK, Heuser JE & Kaufman S (1973). Auto-radiographic localisation of new RNA synthesis in hypertrophying skeletal muscle. *Journal of Cellular Biology* **57**, 743-759.

James NT (1976). Compensatory hypertrophy in the extensor digitorum longus muscle of the mouse. *Journal of Anatomy* **122**, 121-131.

Jansson E & Kaijser L (1977). Muscle adaptation to extreme endurance training in man. *Acta Physiology Scandinavica* **100**, 315-325.

Jiang B, Ohira Y, Roy RR, Nguyen Q, Ilyina-Kakueva EI, Oganov V & Edgerton VR (1992). Adaptation of fibers in fast-twitch muscles of rats to spaceflight and hindlimb suspension. *Journal of Applied Physiology Supplement* **73**, 58S-65S.

Jiang B, Roy RR & Edgerton VR (1990a). Expression of a fast fiber enzyme profile in the cat soleus after spinalization. *Muscle & Nerve* **13**, 1037-1049.

Jiang B, Roy RR & Edgerton VR (1990b). Enzymatic plasticity of medial gastrocnemius fibers in the adult chronic spinal cat. *American Journal of Physiology* **259**, C507-C514.

Jiang B, Roy RR, Navarro C, Nguyen Q, Pierotti D & Edgerton VR (1991). Enzymatic responses of cat medial gastrocnemius fibers to chronic inactivity. *Journal of Applied Physiology* **70**, 231-239.

Jones DA, Rutherford OM & Parker DF (1989). Physiological changes in skeletal muscle as a result of strength training. *Quarterly Journal Experimental Physiology* **74**, 233-256.

Kernell D, Donselaar Y & Eerbeek O (1987b). Effects of physiological amounts of high- and low-rate chronic stimulation on fast-twitch muscle of the cat hindlimb. 11. Endurance-related properties. *Journal of Neurophysiology* **58**, 614-627.

Kernell D & Eerbeek O (1989). Physiological effects of different patterns of chronic stimulation on muscle properties. In: Rose FC, Jones R (eds), *Neuromuscular Stimulation*, pp 193-200. New York: Demos.

Kernell D, Eerbeek O, Verhey BA & Donselaar Y (1987a). Effects of physiological amounts of high- and low-rate chronic stimulation on fast-twitch muscle of the cat hindlimb. I. Speed and force-related properties. *Journal of Neurophysiology* **58**, 598-613.

Klagsbrun M (1989). The fibroblast growth factor family: structural and biological properties. *Progress in Growth Factor Research* **1**, 207-235.

Komi PV, Viitasalo JHT, Havu M, Thorstensson A, Sjodin B & Karsson J (1977). Skeletal muscle fibres and muscle enzyme activities in monozygous and dizygous twins of both sexes. *Acta Physiology Scandinavica* **100**, 385-392.

Knighton DR, Hunt TK, Scheunesnstuhl H, Halliday BJ, Werb Z & Banda MJ (1983). Oxygen tension regulates the expression of angiogenic factor by macrophages. *Science* **221**, 1283-1285.

Krieger DA, Tate CA, McMillin-Wood J & Booth FW (1980). Populations of rat skeletal muscle mitochondria after exercise and immobilization. *Journal of Applied Physiology* **48**, 23-28.

Kuffler SW & Vaughn-Williams EM (1953). Properties of the "slow" skeletal muscle fibres of the frog. *Journal of Physiology (London)* **121**, 318-340.

Kugelberg E, Edstrom L & Abruzzese M (1970). Mapping of motor units in experimentally reinnervated rat muscle. *Journal of Neurology, Neurosurgery and Psychiatry* **33**, 310-329.

Kugelberg E & Lindegren B (1979). Transmission and contraction fatigue of rat motor units in relation to succinate dehydrogenase activity of motor unit fibres. *Journal of Physiology (London)* **288**, 285-300.

Kuhne W (1863). Uber die Endignung der Nerven in den Muskeln. *Vischow Archives* **27**, 508-533.

Lamb DR, Peter JB, Jeffress RN & Wallace HA (1969). Glycogen, hexokinase, and glycogen synthetase adaptations to exercise. *American Journal of Physiology* **217**, 1628-1632.

Lamb GD & Stephenson DG (1991). Effect of Mg^{2+} on the control of Ca^{2+} release in skeletal muscle fibres of the toad. *Journal of Physiology (London)* **434**, 507-528.

Leberer E, Hartner K-T, Brandl CJ, Fujii J, Tada M, MacLennan DH & Pette D (1989). Slow\cardiac sarcoplasmic reticulum Ca-ATPase and phospholamban mRNAs are expressed in chronically stimulated rabbit fast-twitch muscle. *European Journal of Biochemistry* **185**, 51-54.

Leberer E, Hartner K-T & Pette D (1987). Reversible inhibition of sarcoplasmic reticulum Ca-ATPase by altered neuromuscular activity in rabbit fast-twitch muscle. *European Journal of Biochemistry* **62**, 255-561.

Leberer E, Seedorf U & Pette D (1986). Neural control of gene expression in skeletal muscle Ca-sequestering proteins in developing and chronically stimulated rabbit skeletal muscles. *Biochemical Journal* **239**, 295-300.

Lieber RL, Friden JO, Hargens, AR, Danzig LA & Gershuni DH (1988). Differential response of the dog quadriceps muscle to external skeletal fixation of the knee. *Muscle & Nerve* **11**, 193-201.

Longhurst JC, Kelly AR, Gonyea WJ & Mitchell JH (1980a). Echocardiographic left ventricular mass in distance runners and weight lifters. *Journal of Applied Physiology* **48**, 154-162.

Longhurst JC, Kelly AR, Gonyea WJ & Mitchell JH (1980b). Cardiovascular responses to static exercise in distance runners and weight lifters. *Journal of Applied Physiology* **49**, 676-683.

Longhurst JC & Mitchell JH (1983). Does endurance training benefit the cardiovascular system? *Journal of Cardiovascular Medicine* **8**, 227-236.

Longhurst JC & Stebbins CL (1992). The isometric athlete. *Cardiology Clinics* **10**, 281-294.

Lovely RG, Gregor RJ, Roy RR & Edgerton VR (1990). Weight-bearing hindlimb stepping in treadmill exercised adult spinal cats. *Brain Research* **514**, 206-218.

MacDougall JD, Sale DG, Elder DG, Elder GB & Sutton JR (1982). Muscle ultrastructural characteristics of elite power lifters and bodybuilders. *European Journal of Applied Physiology* **48**, 117-126.

MacDougall JD, Ward GR, Sale DG & Sutton JR (1977). Biochemical adaptation of human skeletal muscle to heavy resistance training and immobilization. *Journal of Applied Physiology: Respiratory Environment Exercise Physiology* **43**, 700-703.

Margreth A, Salviati G & Carraro U (1973). Neural control of the activity of the calcium transport system in sarcoplasmic reticulum of rat skeletal muscle. *Nature* **241**, 285-286.

Mayer RF, Burke RE, Toop J, Hodgson JA, Kanda K & Walmsley B (1981). The effect of long-term immobilization on the motor unit population of the cat medial gastrocnemius muscle. *Neuroscience* **6**, 725-739.

Mayer RF, Burke RE, Toop J, Walmsley & Hodgson JA (1984). The effect of spinal cord transection on motor units in cat medial gastrocnemius muscles. *Muscle & Nerve* **7**, 23-31.

McMinn RMH & Vrbova G (1964). The effect of tenotomy on the structure of fast and slow muscle in the rabbit. *Quarterly Journal of Experimental Physiology* **49**, 424-429.

McMinn RMH & Vrbova G (1967). Motoneurone activity as a cause of degeneration in the soleus muscle of the rabbit. *Quarterly Journal of Experimental Physiology* **52**, 411-415, 1967.

Mero A, Komi PV & Gregor RJ (1992). Biomechanics of sprint running. A review. *Sports Medicine* **13**, 376-392.

Mero A, Luhtanen P, Viitasalo JT & Komi PV (1981). Relationships between the maximal running velocity, muscle fiber characteristics, force production and force relaxation of sprinters. *Scandinavian Journal of Sports Sciences* **3**, 16-22.

Morgan TE, Short FA & Cobb LA (1969). Effect of long-term exercise on skeletal muscle lipid composition. *American Journal of Physiology* **216**, 82-86.

Munson JB, Foerhing RC, Loften SA, Zengel JE & Sypert GW (1986) Plasticity of medial gastrocnemius motor units following cordotomy in the cat. *Journal of Neurophysiology* **55**, 619-634.

Nagesser AS, van der Laarse WJ & Elzinga G (1992). Metabolic changes with fatigue in different types of single muscle fibres of Xenopus Laevis. *Journal of Physiology (London)* **448**, 511-523.

Narici MV, Roi GS, Landoni L, Minetti AE & Cerretelli P (1989). Changes in force, cross-sectional area and neural activation during strength training and detraining of the human quadriceps. *European Journal of Applied Physiology* **59**, 310-319.

Nemeth PM, Solanki L, Gordon DA, Hamm TM, Reinking RM & Stuart DG (1986). Uniformity of metabolic enzymes within individual motor units. *Journal of Neuroscience* **6**, 892-898.

Nguyen PV & Atwood HL (1994). Altered impulse activity modifies synaptic physiology and mitochondrial rhodamine-123 florescence in crayfish phasic motor neurons. *Journal of Neurophysiology* **72**, 2944-2955.

Ohira Y, Jiang B, Roy RR, Oganov V, Ilyina-Kakueva E, Marini JF & Edgerton VR (1992). Rat soleus muscle fiber responses to 14 days of spaceflight and hindlimb suspension. *Journal of Applied Physiology Supplement* **73**, 51S-57S.

Ohlendieck K, Briggs KN, Lee KF, Wechsler AW & Campbell KP (1991). Analysis of excitation-contraction-coupling components in chronically stimulated canine skeletal muscle. *European Journal of Biochemistry* **02**, 739-747.

Olha AE, Jasmin BJ, Michel RN & Gardiner PF (1988). Physiological responses of rat plantaris motor units to overload induced by surgical removal of its synergists. *Journal of Neurophysiology* **60**, 2138-2151.

Peter JB, Barnard RJ, Edgerton VR, Gillespie CA & Stempel KE (1972). Metabolic profiles of three fibre types of skeletal muscle in guinea pigs and rabbits. *Biochemistry* **11**, 2627-2633.

Pette D & Hofer HW (1980). The constant proportion enzyme group concept in the selection of reference enzymes in metabolism. *Trends in Enzyme Histochemistry and Cytochemistry: Excerpta Medica* (Ciba Foundation Symposium) **73**, 231-244.

Pette D & Luh W (1962). Constant-proportion groups of multilocated enzymes. *Biochemistry Biophysics Research Communications* **8**, 283-287.

Pette D, Muller W, Leisner E & Vrbova G (1976). Time dependent effects on contractile properties, fibre population, myosin light chains and enzymes of energy metabolism in intermittently and continuously stimulated fast twitch muscle of the rabbit. *Pflügers Archiv* **364**, 103-112.

Pette D, Smith ME, Staudte HW & Vrbova G (1973). Effects of long-term electrical stimulation on some contractile and metabolic characteristics of fast rabbit muscles. *Pflügers Archiv* **338**, 257-272.

Pette D & Staron RS (1990). Cellular and molecular diversities of mammalian skeletal muscle fibres. *Reviews of Physiology, Biochemistry and Pharmacology* **116**, 1-76.

Pette D & Vrbová G (1992). Adaptation of mammalian skeletal muscle to chronic electrical stimulation. *Reviews of Physiology, Biochemistry and Pharmacology* **120**, 115-202.

Pierotti DJ, Roy RR, Bodine-Fowler SC, Hodgson JA & Edgerton VR (1991). Mechanical and morphological properties of chronically inactive cat tibialis anterior motor units. *Journal of Physiology (London)* **444**, 175-192.

Pierotti DJ, Roy RR, Flores V & Edgerton VR (1990). Influence of 7 days of hindlimb suspension and intermittent weight support on rat muscle mechanical properties. *Aviation Space Environmental Medicine* **61**, 205-210.

Ranvier L (1874). De quelques faits relatifs á l'histologie et á la physiologie des muscles stries. *Archives Physiology and Normal Pathology* **6**, 1-15.

Reid CM, Yeater RA & Ullrich H (1987). Weight training and strength, cardiorespiratory functioning and body composition of men. *British Journal of Sports Medicine* **21**, 40-44.

Riedy MR, Moore RL & Gollnick PD (1985). Adaptive response of hypertrophied skeletal muscle to endurance training. *Journal of Applied Physiology* **59**, 127-131.

Rifenberick DH, Gamble J & Max SR (1973). Response of mitochondrial enzymes to decreased muscular activity. *American Journal of Physiology* **225**, 1295-1299.

Rosenblatt JD & Parry DJ (1993). Adaptation of rat extensor digitorum longus muscle to gamma irradiation and overload. *Pflügers Archiv* **423**, 255-264.

Roy RR & Acosta L Jr (1986). Fiber type and fiber size changes in selected thigh muscles six months after low thoracic spinal cord transection in adult cats: Exercise effects. *Experimental Neurology* **92**, 675-685.

Roy RR, Baldwin KM & Edgerton VR (1991). The plasticity of skeletal muscle: effects of neuromuscular activity. *Exercise Sports Sciences Reviews* **19**, 269-312.

Roy RR, Bello MA, Bouissou P & Edgerton VR (1987). Size and metabolic properties of fibers in rat fast-twitch muscles after hindlimb suspension. *Journal of Applied Physiology* **62**, 2348-2357.

Roy RR, Hodgson JA, Chalmers GR, Buxton W & Edgerton VR (1992). Responsiveness of the cat plantaris to functional overload. In: Sato Y, Portmans J, Hashimoto I, Oshida Y (eds), *Integration of Medical and Sports Sciences*, *Medical Sport Sciences* **37**, pp. 43-51. Basel: Karger.

Roy RK, Mabuchi K, Sarkar S, Mis C & Sreter FA (1979). Changes in tropomyosin subunit pattern in chronic electrically stimulated rabbit fast muscles. *Biochemistry Biophysics Research Communications* **89**, 181-187.

Russell B, Dix DJ, Haller DL & Jacobs-El J (1992). Repair of injured skeletal muscle: A molecular approach. *Medicine and Science in Sports and Exercise* **24**, 189-196.

Sahlin K (1978). Intracellular pH and energy metabolism in skeletal muscle in man. *Acta Physiology Scandinavica Supplement* **455** 1-56.

Salmons S & Henriksson J (1986). The adaptive response of skeletal muscle to increased use. *Muscle & Nerve* **4**, 94-105.

Salmons S & Sreter FA (1976). Significance of impulse activity in the transformation of skeletal muscle type. *Nature* **263**, 30-34.

Salmons S & Vrbova G (1969). The influence of activity on some contractile characteristics of mammalian fast and slow muscles. *Journal of Physiology (London)* **201**, 535-549.

Saltin B & Gollnick PD (1983). Skeletal muscle adaptability: significance for metabolism and performance. In: Geiger SR, Adrian RH (eds.), Peachey LD (sec. ed.), *Handbook of Physiology, sec. 10, Skeletal Muscle. Specialization, Adaptation, and Disease*, pp. 555-631. Bethesda, MD: American Physiological Society.

Saubert CW, Armstrong IV RB, Shepherd RE & Gollnick PD (1973). Anaerobic enzyme adaptations to sprint training in rats. *Pflügers Archiv* **341**, 305-312.

Schantz PG, Henriksson J & Jansson E (1983). Adaptation of human skeletal muscle to endurance training of long duration. *Clinical Physiology* **3**, 141-151.

Schiaffino S & Bormioli SP ((1973). Adaptive changes in developing rat skeletal muscle in response to functional overload. *Experimental Neurology* **40**, 126-137.

Schmalbruch H & Hellhamer U (1977). The nuclei in adult rat muscles with special reference to satellite cells. *Anatomical Records* **189**, 169-176.

Schulte LM, Navarro J & Kandarian SC (1993). Regulation of sarcoplasmic reticulum calcium pump gene expression by hindlimb unweighting. *American Journal of Physiology* **264** *(Cellular Physiology* **33***)*, C1308-C1315.

Sesodia S, Gordon T & Nemeth P (1993). Increased metabolic enzyme heterogeneity of muscle unit fibers after reinnervation. *Society for Neuroscience Abstracts* **19**, 152.

Sjøgaard G (1987). Muscle fatigue. *Medical Sport Sciences* **26**, 98-109.

Smith JS, Coronado R & Meissner G (1985). Sarcoplasmic reticulum contains adenine nucleotide-activated calcium channels. *Nature* **316**, 736-738.

Spector SA (1985a). Effects of elimination of activity on contractile and histochemical properties of rat soleus muscle. *Journal of Neuroscience* **5**, 2177-2188.

Spector SA (1985b). Trophic effects on the contractile and histochemical properties of rat soleus muscle. *Journal of Neuroscience* **5**, 2189-2196.

Sreter FA, Gergely J, Salmons S & Romanul F (1973). Synthesis by fast muscle of myosin light chains characteristic of slow muscle in response to long term stimulation. *Nature New Biology* **241**, 17-19.

Sreter FA, Lopez JR, Alamo L, Mabuchi K & Gergely J (1987). Changes in intracellular ionized Ca concentration associated with muscle fiber type transformation. *American Journal of Physiology* **253**, C296-300.

Sreter FA, Luff AR & Gergely J (1976). Effect of cross-reinnervation on physiological parameters and on properties of myosin and sarcoplasmic reticulum of fast and slow muscles of the rabbit. *Journal of General Physiology* **66**, 811-821.

Sreter FA, Pinter K, Jolesz F & Mabuchi K (1982). Fast to slow transformation of fast muscles in response to long-term phasic stimulation. *Experimental Neurology* **75**, 95-102.

Stein RB, Gordon T, Jefferson J, Sharfenberger A, Yang JE, Totosy de Zepetnek & Belanger M (1992). Optimum stimulation of paralysed muscle in spinal cord patients. *Journal of Applied Physiology* **72**, 1393-1400.

St-Pierre D & Gardiner P (1985). Effect of "disuse" on mammalian fast-twitch muscle: joint fixation compared to neurally applied tetrodotoxin. *Experimental Neurology* **90**, 635-651.

Sweeney HL, Kushmerick MJ, Mabuchi K, Sreter FA & Gergely J (1988). Myosin light chain and heavy chain variations correlate with altered shortening velocity of isolated skeletal muscle fibers. *Journal of Biological Chemistry* **263**, 9034-9039.

Tabary JC, Tabary C, Tardieu C, Tardieu G & Goldspink G (1972). Physiological and structural changes in the cat's soleus muscle due to immobilization at different lengths by plaster casts. *Journal of Physiology (London)* **224**, 231-244.

Taylor AW, Thayer R & Rao S (1972). Human skeletal muscle glycogen synthase activities with exercise and training. *Canadian Journal of Physiology and Pharmacology* **52**, 119-122.

Templeton G, Padalino M, Manton J, Leconey T, Hagler H & Glasberg M (1984). The influence of rat suspension-hypokinesia on the gastrocnemius muscle. *Aviation Space Environmental Medicine* **55**, 381-386.

Templeton GH, Sweeney HL, Timson BF, Padalino M & Dudenhoeffer GA (1988). Changes in fiber composition of soleus muscle during hindlimb suspension. *Journal of Applied Physiology* **65**, 1191-1195.

Terjung RL (1976). Muscle fiber involvement during training of different intensities and durations. *American Journal of Physiology* **230**, 946-950.

Thomason DB & Booth FW (1990). Atrophy of the soleus muscle by hindlimb unweighting. *Journal of Applied Physiology* **68**, 1-12.

Thompson JA, Anderson KD, Di Pietro JM, Swiebel JA, Zametta M, Anderson WF & Maciag T (1988.) Site-directed neovessel formation in vivo. *Science* **241**, 1349-1352.

Totosy de Zepetnek J, Zung HV, Erdebil S & Gordon T (1992a). Innervation ratio is an important determinant of force in normal and reinnervated rat tibialis anterior muscle. *Journal of Neurophysiology* **67**, 1385-1403.

Totosy de Zepetnek JE, Zung HV, Erdebil S & Gordon T (1992b). Motor unit categorization on the basis of contractile and histochemical properties: a glycogen depletion analysis of normal and reinnervated rat tibialis anterior muscle. *Journal of Neurophysiology* **67**, 1404-1415.

Trivedi B & Danforth WH (1966). effect of pH on the kinetics of frog muscle phosphofructokinase. *Journal of Biological Chemistry* **241**, 4110-4112.

Tsika RW, Herrick RE & Baldwin KM (1987). Time course of adaptation in rat skeletal muscle isomyosin during compensatory growth and regression. *Journal of Applied Physiology* **63**, 2111-2121.

Vrbova G (1963). The effect of motoneurone activity on the speed of contraction of striated muscle. *Journal of Physiology (London)* **169**, 513-526.

Vrbova G, Gordon T & Jones R (1994). *Nerve-Muscle Interaction.* London: Chapman & Hall.

Walsh JV, Burke RE, Rymer WZ & Tsairis P (1978). Effect of compensatory hypertrophy studied in individual motor units in medial gastrocnemius muscle of the cat. *Journal of Neurophysiology* **41**, 496-508.

Westerblad H, Lee JA, Lannergren J & Allen DG (1991). Cellular mechanisms of fatigue in skeletal muscle. *American Journal of Physiology* **261**, C195-C209.

Williams RS, Garcia-Moll M, Mellor J, Salmons S & Harlan W (1987). Adaptation of skeletal muscle to increased contractile activity. Expression of nuclear genes encoding mitochondrial proteins. *Journal of Biological Chemistry* **262**, 2764-2767.

Williams RS, Salmons S, Newsholme EA, Kaufman RE & Mellor J (1986). Regulation of nuclear and mitochondrial gene expression by contractile activity in skeletal muscle. *Journal of Biological Chemistry* **261**, 376-380.

Winiarski AM, Roy RR, Alford EK, Chiang PC & Edgerton VR (1987). Mechanical properties of rat skeletal muscle after hind limb suspension. *Experimental Neurology* **96**, 650-660.

Witzmann FA, Kim DH & Fitts RH (1983). Effect of hindlimb immobilization on the fatigability of skeletal muscle. *Journal of Applied Physiology* **54**, 1242-1248.

ASSOCIATIONS BETWEEN MUSCLE SORENESS, DAMAGE, AND FATIGUE

P. M. Clarkson and D. J. Newham

[1] Department of Exercise Science
University of Massachusetts
Amherst, Massachusetts 01003
[2] Physiotherapy Group, Biomedical Sciences Division
King's College
London, England SE5 9RS

ABSTRACT

Eccentric exercise results in muscle soreness, structural damage, prolonged losses in strength and range of motion, and neuromuscular dysfunction. Greater and longer lasting fatigue occurs after eccentric compared with concentric and isometric exercise. Higher forces are achieved during eccentric contractions with less ATP usage and greater increases in temperature. Although mechanisms involved in the damage and repair process are not well understood, active strain during eccentric contractions is suggested to cause the initial damage which increases over 2-3 days, followed by regeneration.

INTRODUCTION

Both recreational and serious athletes have experienced delayed onset muscle pain, soreness, and stiffness following unaccustomed exercise or increased training workload. Although research on exercise-induced muscle soreness dates back to 1902 (Hough), the exact mechanism of soreness and pain, as well as the accompanying responses remain unclear. These responses include a prolonged loss in range of motion and strength, increases in muscle enzymes in the blood, swelling, and structural damage.

Several studies have shown that eccentric muscle contractions result in greater soreness, fatigue, and damage than isometric or concentric contractions. This explains why some activities such as hiking that incorporate a large degree of eccentric contractions result in considerable soreness while others, like cycling, that are not biased toward eccentric contractions produce little soreness. Exercise models developed to study muscle damage and soreness are those where eccentric contractions predominate such as downhill running and high-force exercise like lowering weights. The latter exercise results in the greater muscle damage, fatigue, and soreness.

Fatigue, Edited by Simon C. Gandevia et al.
Plenum Press, New York, 1995

CHARACTERISTICS OF ECCENTRIC MUSCLE ACTIVITY

During active muscle lengthening, the mechanical and energetic behavior is very different than during isometric or shortening contractions (see review by Woledge et al., 1985). Active muscles produce more force during stretch than when acting isometrically at the same length, but the net energy produced and the chemical changes are smaller. This suggests that the crossbridge itself is affected by stretch, which changes the way the muscle produces force.

Force-Velocity Relationship

The maximal forces generated during eccentric activity always exceed isometric forces. As velocity increases, the force initially increases rapidly and then reaches a plateau of nearly twice isometric force which is relatively independent of velocity. According to Huxley's sliding filament theory (1957), this occurs because the extensible parts of the myosin molecules forming the cross-bridges are stretched further than during isometric activity. As the velocity of stretch increases, the number of attached cross-bridges at a given muscle length is less, but those attached generate more force unless ripped apart by the extent of the stretch. However, the energetic consequences of lengthening are not correctly predicted by this theory in which the rate of crossbridge turnover, and thus ATP splitting, should increase but this does not happen. While the stretch of the myosin molecules accounts for some of the additional force generation during stretch, there is still an unexpectedly large force because, for an unknown reason, there are more crossbridges attached. The mechanism of ATP use during shortening contractions suggests that ATP is split to recharge the myosin with energy after it has produced work during shortening. During stretch, however, the myosin breaks from the actin in the backward position without doing work and does not need recharging with energy. Thus, cyclic interactions between actin and myosin can occur without ATP splitting (Woledge et al., 1985).

Whether this process fully accounts for the lower ATP splitting during stretch is not known. Furthermore, the energetics are different in slow and rapid stretches, where ATP turnover decreases as the velocity of the stretch increases. Katz (1939) was the first to make detailed studies of the force-velocity relationship during stretch and described a three phase response using isotonic stretches of tetanized muscles. These were 1) an instantaneous lengthening of elasticity due to the series elastic component; 2) a relatively rapid stretch followed by 3) slow lengthening of the fibers. During isovelocity stretches, the increased force also varies with stretch velocity. At lower lengthening velocities it continues to increase, becomes relatively constant at intermediate velocities, but peaks then declines at higher velocities. While forces greater than isometric are generated during stretch, and are affected by velocity, the force may also depend on mechanisms other than instantaneous lengthening. The nature of these mechanisms and their behavior is uncertain. One factor may be a disparity in sarcomere length along a stretched fiber, with simultaneous lengthening and shortening in different segments.

The force-velocity relationship in intact humans differs from that of isolated muscle, with relatively lower forces being generated during eccentric contractions (see Stauber, 1989). Biomechanical factors, such as joint angles, changes in the angle of pennation, and the length of individual fibers, need to be taken into account in human studies. Also, antagonist muscle activity appears to be greater in eccentric activity and this will affect the force-velocity relationship (Gülch, 1994). There is evidence of activation failure (reduced EMG activity) in some instances which is hypothesized to protect the musculoskeletal system from additional damage (e.g., Westing et al., 1991). However, other workers have recorded similar EMG activity during maximal concentric and eccentric activity (e.g., Komi, 1973).

A logical deduction is that the high force generated by stretch in each active muscle fiber causes mechanical trauma resulting in pain and damage. Warren and colleagues (1993a), in isolated rat soleus muscle, showed that damage was initiated by mechanical factors, with muscle tension being most important. However, it seems that damage is also a function of active length changes (Lieber & Fridén, 1993) and that, in humans, eccentric activity performed at long muscles lengths is the most damaging (Jones et al., 1989; Newham et al., 1988). These data indicate damage to the elastic cytoskeletal structures in series to the contractile proteins (Jones et al., 1989).

Energy Cost

During shortening contractions, work is done by the muscle, but during eccentric contractions, work is done on the muscle by the external lengthening forces. This is equivalent to the negative work done by the muscle. ATP splitting is reduced during eccentric contractions. It is less than an isometric tetanus of the same duration, and is strongly velocity dependent, being least at low velocities. ATP splitting increases at high stretch velocities, but internal energy, that is the sum of heat production and work done on the muscle by external forces (equivalent to the negative work done by it), does not. This supports the hypothesis that energy is being stored in another form, possibly an endothermic process. Bigland, Abbott and Ritchie (1952) studied oxygen consumption (assumed to be determined by the number of active motor units and their firing frequency) during submaximal concentric and eccentric activity of the same force. They demonstrated that oxygen intake was substantially lower during eccentric contractions.

Elegant studies on the effects of force and speed on oxygen intake during eccentric activity were performed by Abbott and Bigland (1953). When work rate was increased by requiring greater force and lengthening velocity remained constant, the oxygen intake increased rapidly. However, when work rate was increased by requiring greater speed, oxygen intake remained constant. These results were explained by the fact that greater forces can be generated at slower velocities, and at slower velocities more fibers are used. Abbott and colleagues (1952), and others (Knuttgen et al., 1971) have shown that oxygen intake during maximal eccentric activity is much less than during concentric. Furthermore, during eccentric contractions EMG was about half, and oxygen intake was one sixth, of that during equivalent concentric (submaximal cycling) activity (Bigland-Ritchie & Woods, 1973). Therefore, not only were fewer motor units recruited, but there was a 60% reduction in oxygen and thus ATP consumed by each active fiber.

Muscle Temperature

Heat production is greater during lengthening, except at very low velocities, than isometric contractions. However, the change in total internal energy is less during eccentric contractions. The extra heat seems to be less than expected by the work done on the muscle. The most likely explanation for this is that an extra endothermic process occurs during stretch, such as the storage of energy in a stressed component of the muscle.

Intramuscular temperatures of up to 42°C have been recorded in human muscles (Nadel et al., 1972; Sargeant & Dolan, 1987). Heat liberation during eccentric contractions is that produced by the metabolic processes, plus induced heat supplied by external energy. It appears that muscle blood flow is proportional to metabolic rate and therefore inadequate to dissipate the additional heat produced during eccentric exercise. The effects of these high temperatures are not known, but have been speculated to: 1) increase the metabolic cost by increasing the rate of crossbridge cycling; and 2) change mechanical properties to make the muscle more susceptible to damage. Neither of these possibilities seems likely because studies have shown that a progressive increase in oxygen intake during constant eccentric exercise only occurs in individuals unfamiliar with that type of exercise (e.g., Bonde-Petersen et al., 1973). Furthermore, similarly high muscle temperatures were found in trained individuals (Knuttgen et al, 1982) in whom pain and damage are substantially reduced or eliminated (Newham et al., 1987).

From these data, it is possible to deduce that the damage caused by eccentric activity is not likely to be initiated by chemical processes in the active muscle. The high mechanical tensions and strain, and perhaps temperature, are undoubtedly involved. However, their exact role and the mechanisms to explain changes in contractile and non-contractile tissue are unclear. In volitional exercise, the possibility for variability in all these factors is immense. For example, motor unit recruitment, the length and changes in length probably alter from one second to the next, and may be different in individual muscles and even within single fibers (Faulkner et al., 1993).

MANIFESTATIONS OF MUSCLE DAMAGE

Histological and Ultrastructural Changes

Muscle tissue can be structurally damaged by intense or long duration exercise, particularly those where eccentric contractions predominate (see Fridén & Lieber, 1992; Faulkner et al., 1993). Damage is present in the sarcolemma, T-tubules, myofibrils, and cytoskeletal system.

After eccentric exercise, the damage ranges from mild, which is only detectable with electron microscopy (EM), to more severe, which also can be seen with light microscopy (LM). In the case of mild damage, immediately after exercise only single sarcomeres throughout the muscle are affected and show evidence of Z-line streaming or bulging with A-band and myofibrillar disarray. Over the next two to three days, the disruption becomes more profound and extensive, with normal muscle architecture being lost from areas adjacent to affected sarcomeres and fibers (Newham et al., 1983b). This damage is repaired over approximately ten days in humans.

The more severe form of damage has been reported repeatedly in human studies after maximal eccentric exercise and is the type most often produced in animal studies. In human studies the changes are minimal for the first couple of days, but then cytoskeletal and myofibrillar damage, edema, invading mononuclear cells, muscle fiber atrophy and necrosis, and finally regeneration are seen (Jones et al., 1986). Full recovery and regeneration of fiber size take about three weeks.

The progressive nature of structural damage after eccentric exercise is poorly understood. The mechanism is thought to initially involve a mechanical insult followed by an accumulation of intracellular calcium which triggers calcium mediated processes. Type II fibers are most affected in humans and animals (Jones et al., 1986; Fridén & Lieber, 1992; Faulkner et al., 1993). Fridén and Lieber proposed that fast twitch fibers fatigue early in the exercise period and due to an inability to regenerate ATP enter a state of high stiffness. Subsequent stretch mechanically disrupts the fibers, causing the cytoskeletal and myofibrillar damage. Calcium chelators have been reported to reduce structural changes (Duan et al., 1990), but the buffering capacity may be inadequate to prevent increases in cytosolic calcium (Lowe et al., 1994).

Serum Levels of Muscle Proteins

When skeletal muscle is damaged, several intramuscular proteins leak out into the blood. Although the most commonly measured is creatine kinase (CK), serum levels of myoglobin, lactate dehydrogenase, aspartate aminotransferase, alanine aminotransferase and myosin heavy chain fragments are also elevated following eccentric exercise (Nosaka et al., 1992; Mair et al., 1992). Fatiguing concentric and isometric exercise result in relatively little or no change in serum proteins, unless the exercise is prolonged.

After eccentric exercise there is a delay in the appearance of these proteins in the blood (Clarkson et al., 1992). Elevated levels may not be evident until 24-72 hours and reach peak values 1 to 7 days later. Mair and colleagues (1992) reported that the time course of increases in myosin heavy chain fragments and CK paralleled that of increases in MRI signal intensity. It was suggested that the release of muscle specific proteins and their slow removal from the extracellular area via the lymphatic system provide the colloid osmotic pressure to produce edema.

The magnitude of increase in serum CK after eccentric exercise can be dramatic. For example, peak values generally range from 1,000 to 10,000 IU/L after high force eccentric exercise of the elbow flexors (Clarkson et al., 1992). However, it is not uncommon to see levels of 25,000 IU/L, and in two cases levels of 100,000 IU/L were found. Such high levels occasionally are found after exercises with a larger muscle mass, like those involved in downhill walking and bench stepping (Newham et al., 1983b; Jones et al., 1986). There are some subjects who show very little or no CK response. The explanation for this subject variability is not known.

Although several studies have attempted to quantify the amount of muscle damage based on serum CK levels, these studies failed to account for clearance rates. One recent study assessed clearance at rest by injecting a CK solution into horses (Volfinger et al., 1994). However, it was assumed that clearance rates would be unaltered by exercise or by high levels of CK in the blood. Nosaka and Clarkson (1993) found that eccentric exercise performed when serum CK was elevated by prior eccentric exercise of another muscle group resulted in almost no further increase in CK. It was suggested that elevated CK levels from the first exercise accelerated clearance. Further studies of muscle protein release and clearance are necessary to fully understand the serum protein response to eccentric exercise.

Soreness

Exercise-induced muscle soreness is different from other muscle pains such a cramps, trauma, or ischemic pain (Miles & Clarkson, 1994). Usually a pain response is immediate, while after eccentric exercise, pain and soreness do not appear for several hours and peak 24-48 hours post-exercise. Acute pain from a cramp or trauma is commonly described as a sharp intense pain, while after eccentric exercise it is described as dull and aching. Muscle cramps and pain due to trauma are characterized by ongoing pain, however the sensation of muscle soreness is more commonly experienced when the muscle is active or being palpated.

Although the exact cause of the soreness is not known, inflammation and swelling are considered prime factors (Smith, 1991). Initial studies found that the time course of swelling and the development of soreness were not similar. Increases in limb circumference peak about 5 days after exercise, approximately 3 days after peak soreness (Clarkson et al., 1992). However, circumference measures cannot differentiate between changes in different compartments. More recent studies using MRI techniques show that fluid accumulates in the muscle for 5 days after exercise and then moves into the subcutaneous area (Nosaka & Clarkson, unpublished observation). Swelling within the muscle compartment could produce pain and increase intramuscular pressures in low compliance compartments, as well as sensitize pain receptors to other noxious stimuli.

Damaged cells release substances such as bradykinin, histamines, prostaglandins, and potassium ions that may activate and sensitize pain receptors (Miles & Clarkson, 1994).

Slow release of these substances from the developing muscle damage may explain the delayed sensation of soreness and pain.

Muscle soreness appears more intense at the musculotendinous junction. More damage may occur at this site, and several studies of animals have documented that mechanical stress can alter the integrity of the junction (Tidball, 1991). An increased number of pain receptors in this region has also been suggested to account for the pain. However, there are no data to substantiate this suggestion, rather pain receptors (type IV afferents) appear to be located in the connective tissue throughout the muscle and not found in greater concentration in any region (Miles & Clarkson, 1994).

Magnetic Resonance Imaging (MRI) and Magnetic Resonance Spectroscopy (MRS)

These non-invasive techniques enable repeated studies on whole muscle, or a part of it. Increases in MRI signal intensity and T2 relaxation times are thought to indicate edema by detecting changes in intracellular water (Fried et al., 1988). Shellock and colleagues (1991) found that the T2 relaxation times increased for several days after eccentric but not concentric exercise, and subclinical abnormalities persisted for 60-80 days in some subjects. Muscle biopsies taken from abnormal areas observed on MRIs showed ultrastructural damage (Nurenberg et al., 1992).

Greater changes have been reported in the region of the myotendinous junction (Shellock et al., 1991), which is usually identified as the most painful area. This study and others (Fleckenstein et al., 1988; Rodenburg et al., 1994) found a large inter-subject variability in the intensity of the response and in changes within individual muscles of a functional group.

^{31}P MRS studies have shown changes in the inorganic phosphate:phosphocreatine ratio (Pi:PCr) with an unchanged muscle pH. This ratio appears normal immediately after eccentric exercise (Aldridge et al., 1986) but is elevated about an hour afterwards (McCully et al., 1988). The increase is maximal at about 24 hours after exercise and persists for up to 10 days. These changes clearly follow a different time course to pain. Moreover, the changes in force loss after eccentric exercise appear unrelated to the altered metabolic state (Rodenburg et al., 1994).

An elevated Pi:PCr has also been reported in patients with various neuromuscular diseases, many of which are not associated with pain or soreness. It has also been found after high force concentric contractions (Newham & Cady, 1990) and so is not unique to eccentric activity. McCully and colleagues (1988) suggested that it is caused by an increased resting muscle metabolism, possibly reflecting repair processes such as protein synthesis. Combined studies of MRI and MRS indicate that metabolic changes occur earlier than edema and are more evenly distributed throughout the muscle (Rodenburg et al., 1994).

FUNCTIONAL CONSEQUENCES OF MUSCLE DAMAGE

Changes in Force Generation

Eccentric exercise causes greater changes, including force fatigue, than comparable contractions of either isometric or concentric activity (Jones et al., 1989; Warren et al., 1993a). The resting membrane potential of isolated muscles fatigued by eccentric and isometric contractions is reported to be similar, despite the greater force loss after the former (Warren et al., 1993c). Therefore, there is no evidence of excitation failure, and the ability to conduct action potentials should be unimpaired. Human EMG studies have shown that eccentric exercise causes the greatest and longest lasting changes in the total electrical activity and mean power frequency (Kroon & Naejie, 1991).

Relatively few maximal eccentric contractions can reduce the isometric force of humans by approximately 50% (e.g., Newham et al., 1987). The time course of recovery of isometric force is particularly slow, and is incomplete after two weeks in untrained subjects. After exercise with a higher energy cost (either isometric or concentric), force recovery starts almost simultaneously with the cessation of exercise and restoration of the circulation and is usually recovered within minutes or hours.

The majority of studies have examined the effect of eccentric exercise on isometric force generation. It is assumed these changes are comparable to those in eccentric and concentric force, but there is no direct evidence that this is so. Most human studies have measured maximal voluntary isometric forces and it is possible that force generation is affected by the existence, or fear, of pain through either reflex or motivational mechanisms. This does not seem to be the case, since superimposed electrical stimulation has shown full voluntary activation during isometric contractions of painful human muscle (Rutherford et al., 1986; Newham et al., 1987; cf. Gandevia & McKenzie, 1988). Furthermore, the changes in human muscle are compatible with those of studies on electrically stimulated animal muscle (Warren et al., 1993c; Faulkner et al., 1993).

A number of mechanical influences such as muscle force, length and velocity, apparently play a part in the consequences of eccentric exercise. However their exact role, and the relationship between them is not clear. Eccentric exercise performed at a long muscle length causes greater pain in humans and fatigue in human and animal muscle than when carried out a shorter lengths (Jones et al., 1989; Warren et al., 1993a). Lieber and Fridén (1993) reported that the active strain during lengthening was the critical factor, rather than force. It also appears that a critical number of high force contractions are required to cause marked changes in force (Warren et al., 1993b). These data strongly support the hypothesis that non-contractile force-transmitting structures are damaged by eccentric exercise, in addition to the contractile elements.

In addition to the reduction in maximal force generation, eccentric exercise also affects contractile properties. This is demonstrated by a change in the force-frequency relationship so that relatively lower forces are generated at low (<20 Hz) frequencies (Newham et al., 1983a; Sargeant & Dolan, 1987). This is termed low-frequency fatigue (LFF) and has been suggested to be the consequence of decreased calcium release by each action potential or changes in the SR. Changes in both of these factors have been reported after eccentric contractions as mentioned elsewhere in this chapter. However, LFF is not unique to eccentric contractions and has been widely reported after both brief, high intensity isometric contractions and also after prolonged, low intensity isometric activity (for reviews see Kukulka, 1992; Enoka & Stuart, 1992). After eccentric contractions, the magnitude of LFF appears approximately the same as the decrease in maximal isometric force generation, that is it is greater than after isometric or concentric contractions. The recovery is rather faster than that of maximal force generation, but still takes longer after eccentric contractions and can take more than one week (Jones et al., 1989). The functional significance of LFF is unknown, but force generation is impaired in the physiological range of motor unit discharge rates for isometric and, presumably, eccentric activity as demonstrated by Bigland-Ritchie and others (see Kukulka, 1992; Enoka & Stuart, 1992). Therefore, LFF would be expected to decrease functional performance, particularly as the firing rate is decrease by fatigue, at least in isometric fatigue.

While repeated eccentric exercise decreases and eventually eliminates pain and also cellular structural changes, it appears to have less impact on the amount of force fatigue caused by the exercise. However, the recovery rate is increased (Newham et al., 1987;

Clarkson & Tremblay, 1988). It may be that repeated exercise has different effects on the contractile and non- contractile components of the muscle and that these have different time courses. However, both could be expected to affect force generation.

Work on the human elbow flexors show that the activation of individual muscles in a functional group varies according to whether concentric or eccentric exercise is being performed and also joint angle (Nakazawa et al., 1993). Thus the interaction of synergistic muscles is complex and variable. It cannot be assumed that changes in any one muscle are representative of all those in the same functional group.

Power Output

In functional terms, the ability to generate power and do work is more important than isometric force generation, but this has been studied infrequently. McCully and Faulkner (1985) reported a reduction in the maximal power output of mice after eccentric activity. In human studies, peak power was reduced immediately after eccentric exercise (Beelan & Sargeant, 1991), but only at pedaling speeds > 90 rpm. It is not known if the time course of recovery for power generation is the same as for isometric force, nor are the exact effects of velocity known. This area warrants further investigation.

The consequences of eccentric activity are not always to be viewed in terms of pain and damage. Work output can be increased by approximately 40% by a brief stretch immediately prior to shortening with no apparent increase in fatigue during short periods of activity (de Haan et al., 1990). This stretch-shortening cycle is frequently found during normal activities and is a way in which the body utilizes the additional force that can be produced by eccentric contractions.

Prolonged Loss in Range of Motion and Stiffness

When the elbow flexors are damaged by high force eccentric exercise, there is a loss in the ability to fully flex the joint. The peak change occurs immediately after eccentric exercise and gradually recovers in the next 5 - 7 days (Clarkson et al., 1992). Full flexion can be easily achieved passively.

The time course of change in the ability to flex the forearm coincides with the loss in muscle strength. Certainly damage to the contractile units and cytoskeleton as described previously would impact on the ability to generate tension. However, greatest damage occurs in the days after exercise while strength is improving. Also, it is less clear how damage affects the ability of muscles to fully shorten. Using modeling techniques, Morgan (1990) proposed that during eccentric exercise some sarcomeres become lengthened to the point that they are stretched until there is no overlap between the thick and thin filaments. Furthermore, upon relaxation the overstretched sarcomeres cannot fully return to the original position, there would be less overlap of the filaments, and less crossbridges will be able to form.

After eccentric exercise of the elbow flexors, there is an inability to fully extend the joint, producing a sensation of stiffness. When the arm is hanging in relaxed position, the elbow angle becomes approximately 10-20 degrees more acute immediately after exercise and for the next 48-72 hours. Thereafter there is a gradual improvement with full recovery about 10 days after exercise.

Jones and colleagues (1987) examined the force necessary to extend the elbow in subjects with sore muscles after eccentric exercise of the elbow flexors. This force increased in a similar time course to the development of muscle soreness. Similarly, Howell and colleagues (1993) extended the elbow in a stepwise progression from 90 to 180 degrees. The biceps and triceps muscles were electrically silent. The slope of the relationship between elbow angle and the force required for extension was used as a measure of stiffness. The slope doubled immediately after eccentric exercise and had not fully recovered 10 days later.

The mechanism to explain increased muscle stiffness is unclear. Either spontaneous contraction of the elbow flexors or a shortening of the non-contractile tissue has been proposed. Jones and colleagues (1987) found no surface EMG activity in the shortened elbow flexor muscles. MRI data show that greater changes in signal intensity generally occur in the brachialis, and surface EMG electrodes may not pick up electrical signals of the brachialis (Shellock et al., 1991).

Howell and colleagues (1993) assessed the influence of swelling on stiffness and found that swelling of the upper arm (from circumference measures) and flexor muscles (from ultrasound image analysis) was greatest 3-7 days after exercise. This time course differed from the peak change in stiffness at 1-4 days after exercise. However, MRI indicated the greatest increase in swelling of the forearm flexors at 1-5 days (Nosaka & Clarkson, unpublished observation). This coincides better, but not perfectly, with the changes in muscle shortening. Although swelling may be one contributing factor it is not the only factor. It has also been suggested that the spontaneous shortening could be due to an abnormal accumulation of calcium inside the muscle cell (Ebbeling & Clarkson, 1989), due either to a loss of sarcolemmal integrity or a dysfunction of the sarcoplasmic reticulum (SR). Damaged SR has been shown in biopsies taken from horses subjected to high-intensity exercise, along with a depression in calcium uptake ability, diminished release of calcium, and increased intracellular levels of free calcium (Byrd, 1992). Biopsies taken from a patient suffering from spontaneous contractures induced by exercise showed a dysfunction in calcium uptake by the SR (Brody, 1969). The role of calcium in exercise damage is still unclear.

Neuromuscular Dysfunction

Most studies of neuromuscular function after eccentric exercise focused on gross changes such as force generation and range of motion. Recently Saxton and colleagues (1994) examined changes in muscle tremor and proprioception after performance of high force eccentric exercise. Tremor was assessed using an accelerometer attached to the forearm while subjects maintained the elbow joint angle at 90 degrees. Tremor amplitude increased immediately after and gradually returned to baseline by 72 hours. The power frequency spectrum showed two dominant frequencies at approximately 3-5 and 8-12 Hz, which remained largely unchanged by exercise.

In another experiment, a force-matching task was used with the control (non-exercised) arm acting as the reference (Saxton et al., 1994). The target force was 35% of the pre-exercise maximal isometric contraction. Prior to exercise, subjects were well able to match this force. However, in the days afterwards they consistently undershot the target; they were generating less force than they thought. One subject could only produce 30% of her baseline strength and thus was unable to achieve the target of 35%, yet claimed that she was exerting sufficient force to match the control arm. Instead of matching the target force, subjects were producing 35% of the maximum that could be achieved with the weakened exercised arm. This persisted throughout recovery.

How muscles sense force is not understood, although muscle sensations probably arise from some combination of peripheral feedback and central feedforward mechanisms (see Cafarelli, 1988; see Jones, Chapter 22). The feedforward mechanisms are important in making relative judgments. For example, when two muscle groups have differing force ability, subjects can produce the same relative force in a matching task. Apparently this holds true when a muscle is damaged. From a performance standpoint, the CNS believed that the forces generated by the contralateral muscle groups were similar when they were not. When individuals have sore muscles from performance of strenuous exercise in the preceding days, they may be unaware of the accompanying strength loss and may overestimate how much force they are capable of achieving.

Another indication of neuromuscular dysfunction after eccentric exercise results from studies of joint position sense (Saxton et al., 1994). Testing was done using the exercised limb to 1) match specific joint angles of the control limb; and 2) to replicate specific angles set on the same limb. When the control arm acted as a reference, the exercised arm consistently overshot the target joint angle. However, when the exercised limb acted as its own reference, the joint positions were more accurately reproduced. The CNS appears to have difficulty integrating information between the damaged and control limbs. Further studies of neuromuscular control of exercise damaged limbs are warranted.

SUMMARY

Eccentric exercise results in muscle soreness and muscle damage. During eccentric contractions more force and greater heat is produced compared with isometric and concentric contractions, but the net energy generated and chemical changes are smaller. The high mechanical tensions, muscle strain, and raised muscle temperatures appear to be factors contributing to damage. Muscle biopsies taken from muscles that performed eccentric contractions show histological and ultrastructural evidence of damage. This damage is progressive until several days after exercise then regeneration ensues. A consequence of damage is the development of muscle soreness that peaks 24-48 hours after eccentric exercise. The exact cause of soreness is not known but inflammation and swelling are considered factors as increased bradykinin, histamines, prostaglandins, and potassium ions that are found in damaged muscle.

Eccentric exercise causes more fatigue, both maximal force generation and low-frequency fatigue, than other forms of activity. There is a slow recovery of force that may take longer than 3 weeks. Power output is also impaired, although this has been infrequently studied. There is a prolonged loss in range of motion and stiffness such that the ability to fully contract or lengthen the muscle is compromised. Evidence of neuromuscular dysfunction in the control of exercise-damaged muscle includes an increase in tremor amplitude and a decrease in proprioception.

ACKNOWLEDGMENTS

The authors' work has been supported by the Wellcome Trust (*D.J.N.*), the Ono Sports Science Foundation and Yokohama City University, Japan (*P.M.C.*, with K. Nosaka), the University of Walverhampton, England, and the University of Massachusetts (*P.M.C.*, with J.M. Saxton). Attendance of *D.J.N.* at the 1994 Bigland-Ritchie conference was supported, in part, by the Physiological Society, the Royal Society, the Wellcome Trust, the United States Public Health Service (National Center for Medical Rehabilitation Research, National Institute of Child Health and Human Development), and the Muscular Dystrophy Association (USA).

REFERENCES

Abbott BC & Bigland B (1953). The effects of force and speed changes on the rate of oxygen consumption during negative work. *Journal of Physiology (London)* **120**, 319-325.

Abbott BC, Bigland B & Ritchie JM (1952). The physiological cost of negative work. *Journal of Physiology (London)* **117**, 380-390.

Aldridge R, Cady EB, Jones DA & Obletter G (1986). Muscle pain after exercise is linked with an inorganic phosphate increase as shown by 31P NMR. *Bioscience Reports* **6**, 663-667.

Beelan A & Sargeant AJ (1991). Effect of fatigue on maximal power output at different contraction velocities in humans. *Journal of Applied Physiology* **71**, 2332-2337.

Bigland Ritchie B & Woods J (1973). Oxygen consumption and integrated electrical activity of muscle during positive and negative work. *Journal of Physiology (London)* **234**, 40P.

Bonde-Petersen F, Henriksson J & Knuttgen HG (1973). Effect of training with eccentric muscle contractions on skeletal muscle metabolites. *Acta Physiologica Scandinavica* **88**, 564-570.

Brody IA (1969). Muscle contracture induced by exercise. *New England Journal of Medicine* **281**, 187-192.

Byrd SK (1992). Alterations in the sarcoplasmic reticulum: a possible link to exercise-induced muscle damage. *Medicine and Science in Sports and Exercise* **24**, 531-536.

Cafarelli E (1988). Force sensation in fresh and fatigued human skeletal muscle. *Exercise and Sport Sciences Reviews* **16**, 139-168.

Clarkson PM, Nosaka K & Braun B (1992). Muscle function after exercise- induced muscle damage and rapid adaptation. *Medicine and Science in Sports and Exercise* **24**, 512-520.

Clarkson PM & Tremblay I (1988). Exercise induced muscle damage, repair and adaptations in humans. *Journal of Applied Physiology* **65**, 1-6.

de Haan A, Lodder MAN & Sargeant AJ (1990). Influence of active pre stretch on fatigue of skeletal muscle. *Journal of Applied Physiology* **62**, 268- 273.

Duan C, Delp MD, Hayes PD, Delp PD & Armstrong RB (1990). Rat skeletal muscle mitochondria [$Ca2^+$] and injury from downhill running. *Journal of Applied Physiology* **68**, 1241-1251.

Ebbeling CB & Clarkson PM (1989). Exercise-induced muscle damage and adaptation. *Sports Medicine* **7**, 207-234.

Enoka RM & Stuart DG (1992). Neurobiology of muscle fatigue. *Journal of Applied Physiology* **72**, 1631-1648.

Faulkner JA, Brooks SV & Opiteck JA (1993). Injury to skeletal muscle fibers during contractions. Conditions of occurrence and prevention. *Physical Therapy* **73**, 911-921.

Fleckenstein JL, Canby RC, Parkey RW, Peshock RM (1988). Acute effects of exercise on MR images of skeletal muscle in normal volunteers. *American Journal of Radiology* **151**, 480-485.

Fridén J & Lieber RL (1992). Structural and mechanical basis of exercise-induced muscle injury. *Medicine and Science in Sports and Exercise* **24**, 521- 530.

Fried R, Jolesz FA, Lorenzo AV, Francis H & Adams DF (1988). Developmental changes in proton magnetic resonance relaxation times of cardiac and skeletal muscle. *Investigative Radiology* **23**, 289-293.

Gandevia SC & McKenzie DK (1988). Activation of human muscles at short muscle lengths during maximal static efforts. *Journal of Physiology (London)* **407**, 599-613.

Gülch RW (1994). Force-velocity relations in human skeletal muscle. *International Journal of Sports Medicine* **15**, S2-S10.

Hough T (1902). Ergographic studies in muscular soreness. *American Journal of Physiology* **7**, 76-92.

Howell JN, Chlebourn G & Conatser R (1993). Muscle stiffness, strength loss, swelling and soreness following exercise-induced injury in humans. *Journal of Physiology (London)* **464**, 183-196.

Huxley AF (1957). Muscle structure and theories of contraction. *Progress in Biophysics and Biophysical Chemistry* **7**, 255-318.

Jones DA, Newham DJ & Clarkson PM (1987). Skeletal muscle stiffness and pain following eccentric exercise of the elbow flexors. *Pain* **30**, 233-242.

Jones DA, Newham DJ, Round JM & Tolfree SEJ (1986) Experimental human muscle damage: morphological changes in relation to other indices of damage. *Journal of Physiology (London)* **375**, 435-448.

Jones DA, Newham DJ & Torgan A (1989). Mechanical influences on long lasting human muscle fatigue and delayed onset pain. *Journal of Physiology (London)* **412**, 415-427.

Katz B (1939). The relation between force and muscular contraction. *Journal of Physiology (London)* **96**, 45-64.

Knuttgen HG, Bonde-Peterson F & Klausen K (1971). Oxygen uptake and heart rate responses to exercise performed with concentric and eccentric muscle contractions. *Medicine and Science in Sports* **3**, 1-5.

Knuttgen HG, Nadel ER, Pandolf KB & Patton JF (1982). Effects of training with eccentric muscle contractions on exercise performance, energy expenditure and body temperature. *International Journal of Sports Medicine* **3**, 13-17.

Komi PV (1973). Relationship between muscle tension, EMG and velocity of contraction under concentric and eccentric work. In: Desmedt JE (ed), *New Developments in Electromyography and Clinical Neurophysiology*, vol 1, pp 596-606. Basel: Karger.

Kroon GW & Naejie M (1991). Recovery of the human biceps electromyogram after heavy eccentric, concentric or isometric exercise. *European Journal of Applied Physiology* **63**, 444-448.

Kukulka CG (1992). Human skeletal muscle fatigue. In: Currier DP, Nelson RM (eds), *Dynamics of Human Biological Tissues*, pp 164-181. Philadelphia: FA Davis.

Lieber RL & Fridén J (1993). Muscle damage is not a function of muscle force but active muscle strain. *Journal of Applied Physiology* **74**, 520-526.

Lowe DA, Warren GL, Hayes DA, Farmer MA & Armstrong RB (1994). Eccentric contraction-induced injury of mouse soleus muscle: effect of varying $[Ca^{2+}]o$. *Journal of Applied Physiology* **76**, 1445-1453.

Mair J, Koller A, Artner-Dworzak E, Haid C, Wicke K, Judmaier W & Puschendorf, B (1992). Effects of exercise on plasma myosin heavy chain fragments and MRI of skeletal muscle. *Journal of Applied Physiology* **72**, 656-663.

McCully KK, Argov Z, Boden BP, Brown RL, Bank WJ & Chance B (1988). Detection of muscle injury with 31-P magnetic resonance spectroscopy. *Muscle & Nerve* **11**, 212-216.

McCully KK & Faulkner JA (1985). Injury to skeletal muscle fibers of mice following lengthening contractions. *Journal of Applied Physiology* **59**, 119- 126.

Miles MP & Clarkson PM (1994). Exercise-induced muscle pain, soreness, and cramps. *Journal of Sports Medicine and Physical Fitness* **34**, 203-216.

Miles MP, Ives J & Vincent KR (1993). Neuromuscular control following maximal eccentric exercise. *Medicine and Science in Sports and Exercise* **25**, S176.

Morgan DL (1990). New insights into the behavior of muscle during active lengthening. *Biophysics Journal* **57**, 209-221.

Nadel ER, Bergh U & Saltin B (1972). Body temperatures during negative work exercise. *Journal of Applied Physiology* **33**, 553-558.

Nakazawa K, Kawakami Y, Fukunaga T, Yano H & Miyashita M (1993). Differences in activation patterns in elbow flexor muscles during isometric, concentric and eccentric contractions. *European Journal of Applied Physiology* **66**, 214-220.

Newham DJ & Cady E (1990). A 31P study of fatigue and metabolism in human skeletal muscle with voluntary, intermittent contractions at different forces. *NMR in Biomedicine* **3**, 211-219.

Newham DJ, Jones DA, Ghosh G & Aurora P (1988). Muscle fatigue and pain after eccentric contractions at long and short length. *Clinical Science* **74**, 553-557.

Newham DJ, Jones DA & Clarkson PM (1987). Repeated high force eccentric exercise; effects on muscle pain and damage. *Journal of Applied Physiology* **63**, 1381-1386.

Newham DJ, Mills KR, Quigley BM & Edwards RHT (1983a). Pain and fatigue following concentric and eccentric muscle contractions. *Clinical Science* **64**, 55-62.

Newham DJ, McPhail G, Mills KR & Edwards RHT (1983b). Ultrastructural changes after concentric and eccentric contractions of human muscle. *Journal of the Neurological Sciences* **61**, 109-122.

Nosaka K & Clarkson PM (1993). Plasma creatine kinase response to a subsequent bout of eccentric exercise with the contralateral limb. *Medicine and Science in Sports and Exercise* **25**, S33.

Nosaka K, Clarkson PM & Apple FS (1992). Time course of serum protein changes after strenuous exercise of the forearm flexors. *Journal of Laboratory and Clinical Medicine* **119**, 183-188.

Nurenberg P, Giddings C, Stray-Gundersen J, Fleckenstein JL, Gonyea WJ & Peshock RM (1992). MR imaging-guided muscle biopsy for correlation of increased signal intensity with ultrastructural change and delayed-onset muscle soreness after exercise. *Radiology* **184**, 865-869.

Rodenburg JB, deBoer RW, Schiereck P, van Echeld CJA & Bar PR (1994). Changes in phosphorus compounds and water content in skeletal muscle due to eccentric exercise. *European Journal of Applied Physiology* **68**, 205-213.

Rutherford OM, Jones DA & Newham DJ (1986). Clinical and experimental application of the percutaneous twitch superimposition technique for the study of human muscle activation. *Journal of Neurology, Neurosurgery and Psychiatry* **49**, 1288-1291.

Sargeant AJ & Dolan P (1987). Human muscle function following prolonged eccentric exercise. *European Journal of Applied Physiology* **56**, 704-711.

Saxton JM, Clarkson PM, James R, Miles M, Westerfer M, Clark S & A.E. Donnelly (1994). Neuromuscular function following maximum voluntary eccentric exercise. *Medicine and Science in Sports and Exercise* **26**, S115.

Shellock FG, Fukunaga T, Mink JH & Edgerton VR (1991). Exertional muscle injury: evaluation of concentric versus eccentric actions with serial MR imaging. *Radiology* **179**, 659-664.

Smith LL (1991). Acute inflammation: the underlying mechanism in delayed onset muscle soreness? *Medicine and Science in Sports and Exercise* **23**, 542-551.

Stauber WT (1989). Eccentric action of muscles, physiology, injury and adaptation. *Exercise Sport Sciences Reviews* **17**, 157-185.

Tidball JG (1991). Myotendinous junction injury in relation to junction structure and molecular composition. *Exercise and Sport Sciences Reviews* **19**, 419-445.

Volfinger L, Lassourd V, Michaux JM, Braun JP & Toutain PL (1994). Kinetic evaluation of muscle damage during exercise by calculation of amount of creatine kinase released. *American Journal of Physiology* **266**, R434-441.

Warren GI, Hayes DA, Lowe DA & Armstrong RB (1993a). Mechanical factors in the initiation of eccentric contraction-induced injury in rat soleus muscle. *Journal of Physiology (London)* **464**, 457-475.

Warren GI, Hayes DA, Lowe DA, Prior BM & Armstrong RB (1993b). Materials fatigue initiates eccentric contraction-induced injury in rat soleus muscle. *Journal of Physiology (London)* **464**, 477-489.

Warren GI, Lowe DA, Hayes DA, Karowski CJ, Prior BM & Armstrong RB (1993c). Excitation failure in eccentric contraction-induced injury of mouse soleus muscle. *Journal of Physiology (London)* **468**, 487-499.

Westing SH, Cresswell AG & Thorstensson A (1991). Muscle activation during maximal voluntary eccentric and concentric knee extension. *European Journal of Applied Physiology* **62**, 104-108.

Woledge RC, Curtin NA & Homsher E (1985). Energetic aspects of muscle contraction. *Monographs of the Physiological Society No 41*. London: Academic Press.

[illegible]

MUSCLE FATIGUE IN OLD ANIMALS

Unique Aspects of Fatigue in Elderly Humans

J. A. Faulkner and S. V. Brooks

Institute of Gerontology and Bioengineering Program
University of Michigan
Ann Arbor, Michigan 48109-2007

ABSTRACT

Muscle atrophy, weakness, injury, and fatigue are inevitable and immutable concomitants of old age. Atrophy results from a gradual process of fiber denervation with loss of some fibers and atrophy of others. Fast fibers show more denervation and atrophy than slow fibers. Some fast fibers are reinnervated by axonal sprouting from slow fibers resulting in remodeling of motor units. With aging, the decreases in strength and power are greater than expected from the loss in muscle mass. Contraction-induced injury is proposed as a mechanism of the fast fiber denervation. With atrophy and weakness, human beings show a dramatic decrease in endurance and increase in fatigability with aging, but strength and endurance training slows the process.

INTRODUCTION

In old age, skeletal muscle structure and function are impaired significantly in all animals, including human beings (Brooks & Faulkner, 1994a). Through a gradual process of fiber denervation, fiber loss, and motor unit remodeling, muscle mass declines by 20% to 40% between 30 and 80 years of age in human beings and between 12 and 28 months of age in rodents (Brooks & Faulkner, 1994a). The muscle atrophy is caused by a combination of a reduction in the number of muscle fibers and in the mean single fiber cross-sectional area (CSA). Each factor appears to make an equal contribution to muscle atrophy in untrained elderly human beings (Lexell et al., 1988). Early studies of human beings concluded that muscle atrophy explained the totality of the decrease in strength with aging (Grimby & Saltin, 1983). A subsequent study on EDL and soleus muscles of male mice indicated that compared with the maximal force normalized for total fiber CSA of muscles in young (3 months) and adult (12 months) mice, the maximal specific force of those in old (28 months) mice was 20% lower (Brooks & Faulkner, 1988). More precise assessments of the total fiber CSA of muscles in human beings have resulted in new data which are in agreement with the data on

Fatigue, Edited by Simon C. Gandevia et al.
Plenum Press, New York, 1995

mice and rats and conclude that muscles in elderly human beings are about 20% weaker than those in adults (Bruce et al., 1989).

In all species, the decreases in muscle mass and strength appear similar for males and females (Brooks & Faulkner, 1994a). For female subjects, a sharp drop in maximal strength is associated in time with the onset of menopause. The loss in strength at menopause is not observed for women who receive hormone replacement (Phillips et al., 1992). The mechanism of the weakness in muscles of old animals, or in the absence of specific hormones is not known. When maximally activated by a high calcium concentration, single permeabilized skeletal muscle fibers from muscles of old mice develop the same maximal specific forces as those from adult mice (Brooks & Faulkner, 1994b). This suggests that either some substance in the cytosol of fibers in the muscles of old animals inhibits the force development of cross-bridges, or that cross-bridges in intact fibers are not fully activated by calcium (Brooks & Faulkner, 1994b). Following isometric contractions, a prolonged half relaxation time has been attributed to a delayed removal of calcium from the cytosol of fibers in the muscles of old animals (Larsson & Salviati, 1989). Compared with the deficits in specific force associated with increasing age, even greater decreases are reported for maximal normalized power (Kadhiresan, 1993). For muscles in old compared with adult animals, the decrease in maximal normalized power results in part from an equivalent loss of 20% for force during shortening compared with the loss observed for isometric force. In addition, muscles in old animals have a lower optimal velocity for the generation of power than those in adult animals. Consequently, for old animals the product of force and velocity, the power, is impaired more than force alone.

MECHANISM OF FIBER DENERVATION AND MOTOR UNIT REMODELING

Morphological, histochemical and electrophysiological data support the premise that throughout the life span of animals, terminals of nerve degeneration occurs at the motor end-plates of muscle fibers and the dennervation of the muscle fibers stimulates nerve terminal and nodal sprouting (Brown et al., 1981; Rosenheimer, 1990). The large fast motoneurons appear to be more likely to degenerate with aging than the small slow ones (Kanda & Hashizume, 1989). The increased vulnerability to degeneration of the large fast motoneurons is attributed to their greater surface area, higher metabolic enzyme activity, and more complex synaptic organization (Kanda & Hashizume, 1989). Coupled with the greater probability of denervation of fast fibers, slow motoneurons are more effective at reinnervating denervated muscle fibers than fast motoneurons, due in part to the more rapid growth of the slow axons (Desypris & Parry, 1990). In muscles of old animals, the incidence of denervated muscle fibers is increased and concurrently the ability of nerves to sprout and regenerate is decreased (Pestronk et al., 1980). Rosenheimer (1990) has proposed that the limited response of EDL muscles in old animals to partial denervation may indicate that the muscle has already achieved its maximal capacity for sprouting. The withdrawal of trophic factors has been proposed as the mechanism for both the denervation of muscle fibers and the impairment in reinnervation, but no compelling evidence supports the premise (Brown et al., 1981; Kanda & Hashizume, 1989).

As a result of the age-related remodeling, large fast motor units decrease in number and in their innervation ratios; whereas, slow motor units maintain motor unit number, but increase their innervation ratio dramatically (Kadhiresan, 1993; Kanda & Hashizume, 1989). The potential for motor unit remodeling is a function of the number of fast fibers available for denervation and the degree of competition between fast and slow fibers in the reinner-

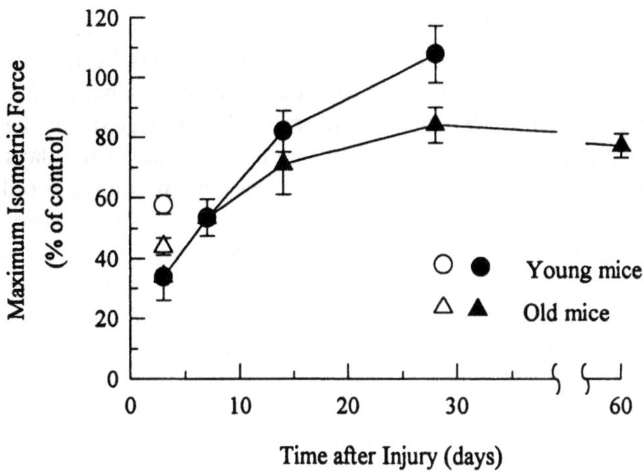

Figure 1. Maximal isometric force of extensor digitorum longus muscles (EDL) from young (*circles*) and old (*triangles*) mice at selected times after a protocol of repeated stretches during contractions. Values are expressed as a percentage of the maximal forces of the uninjured control muscles. The data were collected on an *in situ* EDL muscle preparation with the nerves and blood flow intact. Stimulation of the muscle was by electrical stimulation of the peroneal nerve. *Open* symbols represent data from a protocol of 75 contractions which resulted in significantly greater injury to the muscles in old compared with young mice. *Filled* symbols represent data from a 225 contraction protocol which resulted in prolonged deficits in the muscles of old mice (data from Brooks & Faulkner, 1990; Zerba et al., 1990).

vation of denervated fast fibers (Desypris & Parry, 1990). Consequently, the potential for remodeling is greatest in muscles with 50% fast and 50% slow fibers which is the characteristic distribution of fiber types for untrained muscles of human beings (Faulkner et al., 1986). Such an interpretation is supported by the greater remodeling observed in the 70% fast and 30% slow medial gastrocnemius muscle (Kanda & Hashizume, 1989; Kadhiresan, 1993) compared with the 97% fast and 3% slow EDL muscle (Edström & Larsson, 1987). The heterogeneous fiber population of most of the 400 muscles in human beings amplifies the remodeling of motor units that occurs in all mammalian muscles with aging and contributes to the greater atrophy and weakness observed in the muscles of human beings compared with those of most other mammalian species.

The mechanisms responsible for the transient state of nerve-muscle contacts throughout life and the tenuous nature of the contact in muscles of old animals are not known. Similar to the disruption of nerve-muscle contacts, contraction-induced injury to skeletal muscle fibers occurs intermittently throughout life and increases in frequency and severity in muscles of old animals (Fig. 1). In addition, recovery from contraction-induced injury, although complete in the muscles of young and possibly adult animals, is incomplete in muscles of old animals (Fig. 1). Furthermore, damaged and dennervated muscles in old animals reinnervate less well than those in young animals (Carlson & Faulkner, 1995). Based on these observations, we hypothesize that contraction-induced injury to skeletal muscle fibers contributes to the transient nature of nerve-muscle contacts throughout life and initiates the denervation of fibers in old animals.

CONTRACTION-INDUCED INJURY

Contraction-induced injury is defined as morphological damage to small focal groups of sarcomeres as a result of mechanical disruption of the interdigitation of the thick and thin filament arrays or of the Z-lines of single sarcomeres (Faulkner et al., 1993). The mechanical injury initiates a cascade of events that produce a more severe secondary injury after two or three days. The secondary injury involves an inflammatory response, free radical damage, appearance of cytosolic enzymes in the serum, and phagacytosis of elements within the cytosol of the damaged sarcomeres. Human beings report muscle pain associated with the secondary injury which, as a consequence, has been termed *late onset muscle soreness* (Faulkner et al., 1993; see Clarkson & Newham, Chaper 33). When muscles in old and adult mice are injured by comparable protocols of repeated stretches during maximal contractions, the muscles in old mice show a more severe injury at 3 days than the muscles in the young and adult mice. In addition, when muscles in adult and old mice are injured to the same degree, the recovery of the muscles in the old mice is slower and not complete (Fig. 1). For muscles in old compared with young animals, the greater susceptibility to contraction-induced injury (Zerba et al., 1990) combined with the impaired ability of muscle fibers to regenerate following injury (Brooks & Faulkner, 1990) contributes to muscle atrophy and has the potential to induce motor unit remodeling. Damage to muscle fibers may make fibers more susceptible to denervation. Since denervated muscle fibers in muscles of old animals reinnervate less successfully than those in young animals (Carlson & Faulkner, 1995), such a mechanism could explain the lack of complete recovery from contraction-induced injury for muscles in old animals (Brooks & Faulkner, 1990).

FATIGABILITY OF MUSCLES IN OLD ANIMALS

Fatigue may occur in a fiber, motor unit, whole muscle, or a group of muscles. Fatigue is highly specific to different types of fibers to such an extent that the characteristics of fatigue are an intrinsic part of fiber classifications (Burke et al., 1973). During repeated isometric contractions, slow (S) fibers or motor units are highly resistant to fatigue, fast-fatigable (FF) fibers or motor units fatigue rapidly to almost zero force within a few minutes, and fast fatigue resistant (FR) fibers or motor units are intermediate, fatiguing to low forces over a period of hours. The differences among the three fiber types are due to biochemical (Peter et al., 1972), metabolic (Peter et al., 1972) and mechanical (Brooks & Faulkner, 1991) properties of the fibers that determine cycling rates and efficiency of force or power production (Barclay et al., 1993) and consequently the capability of the fiber to generate and sustain force and power. Although challenged regarding the discrete nature of the three fiber types, the basic utility of these three functional types of fibers has withstood twenty years of rigorous investigation (Faulkner et al., 1994). Dependent on the circumstances, each element of the motor system in young, adult, and old animals may be involved in the development of fatigue, from the drive of suprasegmental centers to the interaction of the contractile proteins (Enoka & Stuart, 1992). In terms of task dependency, the force-fatigability relationship, the hyperbolic force or power-endurance relationships, and the sense of effort (Enoka & Stuart, 1992), the difference between the fatigue response of muscles in old animals, compared with those in young or adult animals, appears to be quantitative rather than qualitative. Furthermore, the multicausitive nature of fatigue and the mechanisms responsible for fatigue described for muscles in young animals appear to be equally appropriate to the phenomenon of fatigue in muscles of old animals.

Atrophy and weakness of muscles in animals of any age increase the susceptibility of the muscles to fatigue when performing a task of absolute force or power (Faulkner et al., 1994). Under circumstances of muscle atrophy and weakness, the task represents a greater proportion of maximal strength or power and based on the force-fatigue relationship this translates into a shorter endurance time (Enoka & Stuart, 1992). Consequently, when presented with the same task, fatigability is increased in the elderly (Makrides et al., 1985; Overend et al., 1992). Surprisingly, compared with control muscles atrophied muscles often show a greater resistance to fatigue when tests of fatigability are designed relative to maximal force or power (Larsson & Karlsson, 1978). The mechanism responsible for the resistance to fatigue of atrophied muscles is not clear, but the atrophy of fibers increases capillary density and may also increase the concentration of mitochondria and metabolic enzymes. In the muscles of old animals, the shift in fiber types from fast to slow (Kanda & Hashizume, 1989) has implications for the maintenance of force development because slow fibers are more resistant to fatigue during isometric contractions than are fast fibers (Burke et al., 1973). Consequently, the increased proportion of slow fibers even coupled with muscle and mean single fiber atrophy may increase endurance and decrease fatigability during low force tasks or tests of endurance at relative values of maximal voluntary contraction strength (Larsson & Karlsson, 1978). These factors explain the scattered reports of older subjects performing better than young subjects in relative tests of strength and endurance (Vandervoort et al., 1986).

Fatigue is most often observed during physical activities that require high sustained force, high power short term repetitive contractions, or low power long term repetitive contractions (Faulkner et al., 1994). The development of fatigue when muscles shorten during contractions is more complex than fatigue during isometric contractions (Brooks & Faulkner, 1991). The complexity arises from a fatigue due to decreases in both shortening force and the velocity of shortening (Faulkner et al., 1994). Despite the usefulness of the concept of fatigue in understanding the underlying mechanisms of impaired development of force and power, in general animals avoid inducing fatigue. This is evident in diaphragm function wherein both patients with chronic obstructive pulmonary disease and athletes performing maximal exhaustive exercise avoid inducing fatigue of the diaphragm muscle except under the most demanding conditions (Johnson et al., 1993). Exceptions wherein fatigue is rapid and profound are the dramatic chases of carnivores, the equally dramatic flight of the prey and the performance of human beings during sprint and endurance sports events. Even under these circumstances, the most direct correlate of the success of the hunt, the flight, or the athletic performance is on the maintenance of a high power output, rather than on the induction of fatigue (Faulkner et al., 1994).

MAXIMAL SUSTAINED NORMALIZED POWER OF MUSCLES IN OLD ANIMALS

We have proposed that the *maximal sustained normalized power* (*sustained power*) is the most valid measurement of the functional capability of a muscle, a limb, or a total organism (Brooks & Faulkner, 1991; Faulkner & Brooks, 1993). *Sustained power* output was selected as the most appropriate measure because movement, in addition to being critical for all animals, is the essence of the human experience. Although postural stability and the stretching of agonists are each a part of most, if not all, movements, the essential element is the driving stroke of cross-bridges during shortening (Faulkner et al. 1994). The *sustained power* can be measured by beginning at low levels of power output and increasing the output systematically with regard to both duration and intensity (Fig. 2). Beginning at a low power is necessary to allow the appropriate responses in muscle respiration, circulation and metabolism. A transition from rest to a high power output results in a premature onset of fatigue without attaining maximal sustainable power output (Brooks & Faulkner, 1991). The increments in the power required and the duration at each power must be designed to allow power output to plateau at each level. Consequently, the test must be designed for the capacities of the tissue or organism being tested.

For human beings with muscle temperatures of ~35°C, the estimates for the values for maximal normalized power during a single contraction are 225 W/kg for fast fibers and 65 W/kg for slow fibers (Faulkner et al., 1994). The difference results from the three-fold greater maximal velocity of unloaded shortening of fast compared with slow fibers and the

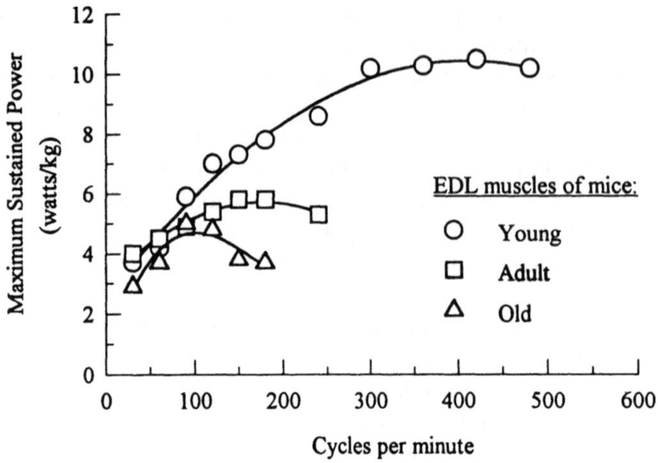

Figure 2. The relationship of the cycle rate of repeated contractions (cycles per minute) and the maximal sustained normalized power for extensor digitorum longus muscles in young (*circles*), adult (*squares*), and old (*triangles*) mice. As cycle rate is increased from very low values, the maximal power that can be sustained increases until a maximum is reached. Further increases in cycle rate result in rapid decline in power output (data from Brooks & Faulkner, 1991).

less curved force-velocity relationship of fast fibers. The optimal velocity for power of fast and slow fibers occurs at ~30% of the maximal velocity of unloaded shortening and ~30% of the maximal force (Faulkner et al., 1986). With aging, the decrease in the optimal velocity for power in muscles in old animals results, at least in part, from motor unit remodeling which reduces the fast to slow fiber ratio (Kanda & Hashizume, 1989; Larsson & Karlsson, 1978; Lexell et al., 1988). Other factors that upregulate the expression of slower and downregulate the expression of faster isoforms of myosin, or both may also contribute to a decrease in maximal normalized power.

The *sustained power* of a muscle is a function of the fiber composition and the oxidative capacity of the muscle. The capability of a muscle to sustain power output depends on the balance between energy output and energy supply (Brooks & Faulkner, 1991). The ability of the blood flow to supply oxygen to the mitochondria in muscle fibers and the capacity of the muscle fibers for oxidative metabolism will determine the percentage of maximal power output attained during a single contraction that can be sustained over time (Faulkner et al., 1994). After a warm-up at a moderate intensity of ~10% of maximal power during a single contraction, highly trained persons can maintain 15% of maximal power, or ~30 W/kg, for several hours. The dramatic decrease in power from the value during a single maximal contraction to a *sustained power* is a function of the introduction rest periods between each of a series of repeated contractions, a duty cycle, and the decrease from maximal force development to a metabolically determined equilibrium force development determined by the duty cycle and the metabolic capabilities of the muscle (Faulkner & Brooks, 1993).

The decreased contractile (Brooks & Faulkner, 1991) and metabolic (McCully et al., 1993) capabilities of muscles in old mice combine to produce a 50% decrease in the *sustained power* compared with the value for muscles in young mice and 25% of the value for muscles in adult mice (Fig. 2). These decrements in maximal sustained normalized power are in excellent agreement with the loss in endurance performance capacity of untrained human beings between the ages of 30 and 80 years (Fig. 3) The decrement in *sustained power* with age is due to decreases in specific force, optimal velocity for power, and energy delivery and conversion. A second phenomenon evident in the test of *sustained power* is the decrease in the cycles per minute at which maximal sustained power was attained (Fig. 2). The cycle rate at which *sustained power* is maximal decreased from 500 cycles per minute (cpm) for young mice to 350 cpm for adult mice and finally to 150 cpm for old mice. In addition, as cycle rates are further increased, the ability of muscles in mice of all ages to generate power decreases rapidly. For human beings, this decrease in the ability of muscles to perform a contraction and recover and be ready to perform another identical contraction is reflected in a decrease in strides per minute walking, stair-climbing, running, or skiing; strokes per minute in swimming; or revolutions per minute cycling. The loss in the *sustained power* with aging is a result of both the decreased cycling rate and the lower power generated with each cycle. Regardless of the measure made of the functional capability of skeletal muscles, the muscles of old animals demonstrate declines of up to 50% in endurance or an equivalent increase in fatigability.

ATROPHY, WEAKNESS, INJURY AND FATIGUE INTRINSIC FUNCTIONS OF AGING

At any age throughout the life span of animals, a decrease in physical activity initiates decreases in muscle mass and in the capacity to develop force and power (Faulkner et al., 1994). Consequently, a key issue is whether the muscle atrophy and weakness observed in old animals is simply a function of the decrease in physical activity or whether these impairments in skeletal muscle structure and function are immutable and inevitable consequences of basic biological changes that occur in skeletal muscles with aging. Three widely disparate sets of data strongly support the premise that some degree of atrophy and weakness is immutable and an inevitable consequence of aging. First, at any given age the muscle atrophy and weakness observed with immobilization of limbs, or a decrease in physical activity, is reversed completely by appropriate conditioning of the atrophied and weak skeletal muscles (Faulkner et al., 1994). Although some reversibility is evident even in very old people (Frontera et al., 1988; Meredith et al., 1989), as evidenced by gains in strength, power, and endurance, the older person never attains the status possible at an earlier age (Fig. 3).

Second, athletes competing in masters' track and field events are highly motivated to train as hard as younger athletes. Despite this high level of motivation and rigorous training, the records for performance in all events that involve strength, power, or endurance show an inevitable decline with increasing age (Fig. 3). The decline appears to be more rapid for events requiring high power than for the endurance events (Schulz & Curnow, 1988). The slower decline in endurance than in maximal power has given rise to the premise that aerobic power deteriorates less rapidly than muscle power. With regard to data on track and field events, this premise is suspect because of the varying roles of relative and absolute loads in the performance of track compared with field events, respectively. In field events, the muscle power involves tossing or throwing an absolute mass, the shot or the discus,

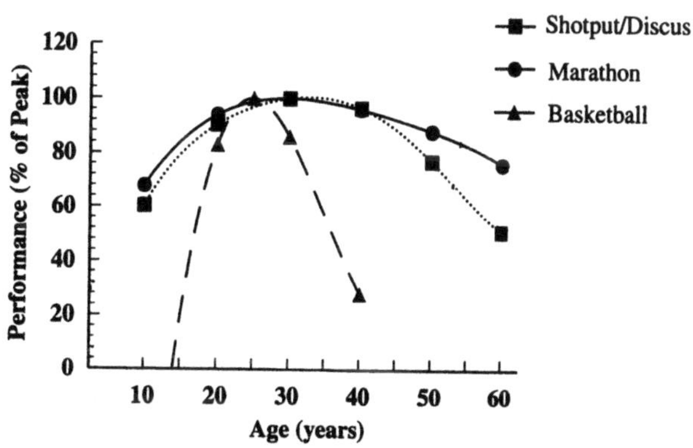

Figure 3. Physical performance data plotted against the age of the performer for the marathon run, the shot put/discus throw, and the number of points scored per game by a basketball player. The data for the marathon run and the shot put and discus throw are from Moore (1975) and the data for the basketball player are from the NBA Register 1992-1993 edition.

which constitutes an increasing percentage of maximal muscle power with aging. In contrast, the endurance events in track require movement of the total organism which reflects, at least in part, the smaller muscle mass in older runners. Clearly, the relative rates of decline are highly complex and subject to multiple influences. An even more precipitous decline is observed for the performance of professional athletes in the team sports of baseball, basketball, and football (Fig. 3). In the team sports, the decline in performance might be heightened by injuries, direct competition with younger athletes, or both.

Finally, during the 100 years of the modern Olympiad, the Olympic records for different power and endurance events have improved by 20% to 90% (Schulz & Curnow, 1988). The improvements in olympic records are attributable to improvements in training and equipment, intensity of training, and the skill of the competitors. Despite the improvements in the record-setting performances of both power and endurance events, in all cases, the ages of the record-setting athletes have remained between 16 and 31 years (Schulz & Curnow, 1988). Thus, declines in strength, power and endurance are due to immutable and inevitable changes in the basic properties of skeletal muscles which are not reversible with training.

With aging, lower absolute values of strength, power, and endurance for an untrained compared with a trained person will bring the untrained person to the threshold of impaired mobility, inability to perform the activities of daily living, and the increased probability of experiencing accidents and falls at an earlier age than trained subjects. The impairments in muscle function arise from muscle *atrophy, weakness, fatigue*, and *injury* (Brooks & Faulkner, 1994a; Faulkner et al., 1994). Consequently, the lower values of strength, power and endurance of untrained people initiate a cascade of events that lead to an earlier onset of disabilities that affect the quality of life. Even in very old untrained subjects, programs in strength (Frontera et al., 1988) and endurance (Overend et al., 1992) training result in significant improvements. Furthermore, careful progressive training in activities that involve stretches of maximally activated muscles enable old persons to perform such activities without injury. The conclusion is that strength and endurance training for old people provide the opportunity of more years free of immobility and disability. The solution for the individual and for health professionals administering programs for elderly people is the maintenance of sufficient strength, power, and endurance to remain above the threshold of an impaired ability to perform the activities of daily living throughout one's life. Such a life style will not prevent the inevitable decline in the functional capabilities of skeletal muscle, but it will defer the impingement on the quality of life to fewest possible years.

ACKNOWLEDGMENTS

This research was supported by United States Public Health Service grant AG 06157 (to *J.A.F.*), and a postdoctoral fellowship award, AG 00114 (to *S.V.B.*). Attendance of *J.A.F.* at the 1994 Bigland-Ritchie conference was supported, in part, by the University of Arizona Regents' Professor funds of Douglas Stuart.

REFERENCES

Barclay CJ, Constable JK & Gibbs CL (1993). Energetics of fast-and slow-twitch muscles of the mouse. *Journal of Physiology (London)* **472**, 61-80.

Brooks SV & Faulkner JA (1988). Contractile properties of skeletal muscles from young, adult, and aged mice. *Journal of Physiology (London)* **404**, 71-82.

Brooks SV & Faulkner JA (1990). Contraction-induced injury: recovery of skeletal muscles in young and old mice. *American Journal of Physiology* ***258*** *(Cell Physiology* **27**), C436-C442.

Brooks SV & Faulkner JA (1991). Maximum and sustained power of extensor digitorum longus muscles from young, adult and old mice. *Journal of Gerontology: Biological Science* **46**, B28-33.

Brooks SV & Faulkner JA (1994a). Skeletal muscle weakness in old age: underlying mechanisms. *Medical Science and Sports Exercises* **26**, 432-439.

Brooks SV & Faulkner JA (1994b). Isometric, shortening and lengthening contractions of muscle fiber segments from adult and old mice. *American Journal of Physiology* ***267*** *(Cell Physiology* **36**), C507-C513.

Brown MC, Holland RL & Hopkins WG (1981). Motor nerve sprouting. *Annual Reviews of Neuroscience* **4**, 17-42.

Bruce SA, Newton D & Woledge RC (1989). Effect of age on voluntary force and cross-sectional area of human adductor pollicis muscle. *Quarterly Journal of Experimental Physiology* **74**, 359-362.

Burke RE, Levine DN, Zajak FE, Tsairis P & Engel WK (1973). Physiological types and histochemical profiles in motor units of the cat gastrocnemius. *Journal of Physiology (London)* **234**, 723-748.

Carlson BM & Faulkner JA (1995). Skeletal muscle regeneration and aging: the influence of innervation. *Journal of Gerontology* In press.

Desypris G & Parry DJ (1990). Relative efficacy of slow and fast-motoneurons to reinnervate mouse soleus muscle. *American Journal of Physiology* **258** (*Cell Physiology 27*), C62-C70.

Edström L & Larsson L (1987). Effects of age on contractile and enzyme-histochemical properties of fast-and slow-twitch single motor units in the rat. *Journal of Physiology (London)* **392**, 129-145.

Enoka RM & Stuart DG (1992). Neurobiology of muscle fatigue. *Journal of Applied Physiology* **72**, 1631-1648.

Faulkner JA & Brooks SV (1993). Fatigability of mouse muscles during constant length, shortening, and lengthening contractions: interactions between fiber types and duty cycles. In: Sargeant T, Kernell D (eds.) *Neuromuscular Fatigue*, pp 116-123. Amsterdam: Royal Netherlands Academy of Arts and Sciences.

Faulkner JA, Brooks SV & Opiteck JA (1993). Injury to skeletal muscle fibers during contractions: conditions of occurrence and prevention. In: Binder-Macleod SA (ed.), *Physical Therapy, Special Edition*, pp 911-921. Alexandria, VA: American Physical Therapy Association.

Faulkner JA, Claflin DR & McCully KK (1986). Power output of fast and slow fibers from human skeletal muscles. In: Jones NL, McCartney N, McComas J, (eds.), *Human Power Output,* pp. 81-91. Champaign, IL: Human Kinetics Publishers Inc.

Faulkner JA, Green HJ & White TP (1994). Skeletal muscle responses to acute and adaptations to chronic physical activity. In: Bouchard C, Shephard RJ, Stephens T (eds.), *Physical Activity, Fitness & Health,* pp. 343-357. Champaign, IL: Human Kinetics Publishers, Inc.

Frontera WR, Meredith CN, O'Reilly KP, Knuttgen HG & Evans WJ (1988). Strength conditioning in older men: skeletal muscle hypertrophy and impaired function. *Journal of Applied Physiology* **64**, 1038-1044.

Grimby G & Saltin B (1983). The aging muscle. *Clinical Physiology* **3**, 209-218.

Johnson BD, Babcock MA, Suman OE & Dempsey JA (1993). Exercise-induced diaphragmatic fatigue in healthy humans. *Journal of Physiology (London)* **460**, 385-405.

Kadhiresan VA (1993). *Functional Properties of Motor Units in Medial Gastrocnemius Muscles of Rats: Remodelling in Old Age.* Ph.D. Dissertation, Univ. Microfilm Int.Cit. No. 9409723. Ann Arbor, MI: University of Michigan.

Kanda K & Hashizume K (1989). Changes in properties of the medial gastrocnemius motor units in aging. *Journal of Neurophysiology* **61**, 737-746.

Larsson L & Karlsson J (1978). Isometric and dynamic endurance as a function of age and skeletal muscle characteristics. *Acta Physiologica Scandanavica* **104**, 129-136.

Larsson L & Salviati G (1989). Effects of age on calcium transport activity of sarcoplasmic reticulum in fast- and slow-twitch rat muscle fibres. *Journal of Physiology (London)* **419**, 253-264.

Lexell J, Taylor CC & Sjostrom M (1988). What is the cause of the ageing atrophy? Total number, size and proportion of different fiber types studied in whole astus lateralis muscle from 15- to 83-year-old men. *Journal of Neurological Science* **84**, 275-294.

Makrides L, Heigenhauser GJ, McCartney N & Jones NL (1985). Maximal short term exercise capacity in healthy subjects aged 15-70 years. *Clinical Science* **69**, 197-205.

McCully KK, Fielding RA, Evans WJ, Leigh Jr, JS & Posner JD (1993). Relationships between in vivo and in vitro measurements of metabolism in young and old human calf muscles. *Journal of Applied Physiology* **75**, 813-819.

Meredith CN, Frontera WR, Fisher EC, Hughes VA, Herland JC, Edwards J & Evans J (1989). Peripheral effects of endurance training in young and old subjects. *Journal of Applied Physiology* **66**, 2844-2849.

Moore DH (1975). A study of age group track and field records to relate age and running speed. *Nature* **253**, 264-265.

Overend TJ, Cunningham DA, Paterson DH & Smith WD (1992). Physiological responses of young and elderly men to prolonged exercise at critical power. *European Journal of Applied Physiology* **64**, 187-193.

Pestronk A, Drachman DB & Griffin J (1980). Effects of aging on nerve sprouting and regeneration. *Experimental Neurology* **70**, 64-82.

Peter JB, Barnard RJ, Edgerton VR, Gillespie CA & Stemple KE (1972). Metabolic profiles of three fiber types of skeletal muscle in guinea pigs and rabbits. *Biochemistry* **14**, 2627-2633.

Phillips SK, Rook KM, Siddle NC & Woledge RC (1992). Muscle weakness in women occurs at an earlier age than in men, but strength is preserved by hormone replacement therapy. *Clinical Science* **84**, 95-98.

Rosenheimer JL (1990). Ultraterminal sprouting in innervated and partially denervated adult and aged rat muscle. *Neuroscience* **38**, 763-770.

Schulz R & Curnow C (1988). Peak performance and age among superathletes: track and field, swimming, baseball, tennis, and golf. *Journal of Gerontology: Psychological Sciences* **43**, P113-120.

Vandervoort AA, Hayes KC & Belanger AY (1986). Strength and endurance of skeletal muscle in the elderly. *Physiotherapy Canada* **38**, 167-173.

Zerba E, Komorowski TE & Faulkner J A (1990). Free radical injury to skeletal muscles of young, adult, and old mice. *American Journal of Physiology* **258** (*Cell Physiology* **27**), C429-C435.

HISTORICAL PERSPECTIVE: A FRAMEWORK FOR INTERPRETING PATHOBIOLOGICAL IDEAS ON HUMAN MUSCLE FATIGUE

R. H. T. Edwards, V. Toescu, and H. Gibson

Muscle Research Centre
Department of Medicine
University of Liverpool
P.O. Box 147
Liverpool L69 3BX
United Kingdom

ABSTRACT

The flow of ideas on the causes of human muscle fatigue appear to have been established in the literature during the last century. Critical analysis had to await innovative experimental designs or techniques. Progress has come particularly from the recognition of inconsistencies, particularly in clinical conditions in which there are alterations in the supply of energy or contractile function. While there is a continuing search for a unique cause of fatigue, much evidence points to there being different causes according to the type of muscular activity or clinical condition. "Nature's experiments" (patients exhibiting isolated defects of function or metabolism) offer unique opportunities for understanding the relative importance of particular levels of metabolic organization or physiological control. Theories of limiting biochemical processes and the Ca- kinetic basis of electromechanical coupling defects are both essentially "single-cell" models of fatigue. Functional requirements appear to determine the diversity of structure and organization of motor units. This would suggest that a "muscle cell population" approach would take into account the consequences of disease altering the number, type of functioning fibers or intrinsic strength of individual fibers. A graphical model is offered to allow a possible interpretation of the cause of fatigue in different forms of exercise and clinical conditions.

SETTING THE SCENE

This chapter begins by identifying a few of the most evident *milestones* in development of ideas on muscle fatigue.

Fatigue, Edited by Simon C. Gandevia et al.
Plenum Press, New York, 1995

ORIGINS OF BIOLOGICAL ELECTRICITY AND THE ROLE THAT STIMULATION HAD IN UNDERSTANDING FATIGUE

In 1786, Luigi Galvani studied the effects of atmospheric electricity upon dissected frog muscles. His 1791 *Commentary on the Effects of Electricity on Muscular Motion* (cited in Rasch & Burke, 1967) is probably the earliest explicit statement of the presence of electrical potentials in nerve and muscle, although it was not until 1838 that the word *tetanize* was introduced (Matteucci, 1838; cited in Rasch & Burke, 1967). The study of animal electricity at once became the absorbing interest of the physiological world. A prominent name among the early students of the subject was Emil DuBois-Reymond (1818-1896) to whom Rosenthal (1883) dedicated his book, *Muscles and Nerves*, because it was DuBois-Reymond who laid the foundations of modern electrophysiology.

Following the discoveries of Luigi Galvani (1737-1798) and those of Michael Faraday (1791-1867) in establishing the principles of electricity and magnetism, electrical stimulation became well established as a tool for studying muscle function and fatigue. Angelo Mosso, in the English translation of his book, *Fatigue* (Drummond & Drummond, 1915), describes how his interest was kindled in the subject after he met Professor Hugo Kronecker in Leipzig in 1873. He observed Kronecker's latest experiments on isolated muscles and these formed a model for Mosso's subsequent experiments in man. In muscles from frogs, Kronecker ". . . succeeded in obtaining 1,000 and even 1,500 contractions, one after the other, with the greatest regularity. As the contractions follow one another, their height diminishes in proportion as the fatigue increases, and goes on diminishing with regularity until it disappears altogether."

This is know as the first of Kronecker's *Laws of Fatigue* i.e., "The curve of fatigue of a muscle which contracts at regular intervals, and with equally strong induction shocks is represented by a straight line." While more recent observations have not confirmed the straight line relationship between force and time in all forms of fatigue, this is, nevertheless, a useful starting point. There are forms of muscular activity in which the loss of force with time in human muscle is fairly precipitate as in some individuals doing voluntary muscular activity (as illustrated in Mosso's book), and with some forms of exercise involving ischemia (e.g., Merton, 1954). The probability is that the isolated muscles in which Kronecker demonstrated this *Law* were somewhat anoxic as would be suggested today by the study of muscles treated with a metabolic inhibitor.

The second is: "The difference in the height of the contractions is less when the intervals of time are greater. In other words, the height of the contractions diminishes the more rapidly, the more rapid is the rhythm in which they are produced and vice versa" The second *Law* clearly establishes the dependence of fatigue on the frequency of electrical stimulation and thus on the frequency of contractions in voluntary muscular activity such as isometric contractions in which the tendency to fatigue is related to the frequency (and duration) of contractions.

In Julius Althaus' book, *A Treatise on Medical Electricity* (1873), which summarizes techniques for examining neuromuscular function in patients, early controlled experiments on fatigue in man are cited. A more analytical approach had already been developed by Duchenne and published in 1872 in what was probably his life's work (*De Electrisation Localisee et de son application a la pathologie et la therapeutique par courants induits et par courants galvaniques interompus et continus*). Duchenne, who is well known for his clinical accounts of the muscular dystrophy which bears his name, is less well known for his important physiological experiments in which he systematically stimulated virtually all the muscle groups in the human body in order to establish their actions. He devised cloth-covered electrodes for percutaneous stimulation and was the first to use alternating current, suggesting the name *faradic* for this form of stimulation. His work had earlier been published in 1867 under the title *Physiologie des Mouvements* (translated by Kaplan, 1959).

Jolly (1895), using a Marey capsule placed on the skin with a smoked drum kymograph for recording mechanical responses and a double coil stimulator for faradic stimulation (30-60 Hz) demonstrated for the first time that the fatigue and paresis of Erb-Goldflam disease depended on a failure in the neuromuscular system. He applied the faradic current at repeated short intervals or maintained it for about 1 min. This resulted in an unusually rapid deterioration of the size of the muscle responses which recovered after rest. These observations led him to propose the name *myasthenia gravis* (myasthenia gravis pseudoparalytica) for the disease.

These examples illustrate the early forging of ideas for the use of electricity in analyzing the function of muscle, now commonplace in physiology and clinical laboratories.

CONCEPTS OF SITES OF MUSCLE FATIGUE

Michael Foster, one of the founders of the Physiological Society, expressed the contemporary views on the neural contribution to fatigue and opens consideration of the metabolic factors responsible for the development of human muscle fatigue in his *Textbook of Physiology* (1883).

> The sense of fatigue of which, after prolonged or unusual exertion, we are conscious in our own bodies, is probably of complex origin, and its nature, like that of the normal muscular sense of which we shall have to speak hereafter, is at present not thoroughly understood. It seems to be in the first place the result of changes in the muscles themselves, but is possibly also caused by changes in nervous apparatus concerned in muscular action, and especially in those parts of the central nervous system which are concerned in the production of voluntary impulses. In any case it cannot be taken as an adequate measure of the actual fatigue of the muscles; for a man who says he is absolutely exhausted may under excitement perform a very large amount of work with his already weary muscles. The will in fact rarely if ever calls forth the greatest contractions of which the muscles are capable. . . . A certain number of single induction-shocks repeated rapidly, say every second or oftener, bring about exhaustive loss of irritability more rapidly than the same number of shocks repeated less rapidly, for instance every 5 or 10 seconds. Hence tetanus is a ready means of producing exhaustion. . . . In exhausted muscles, the elasticity is much diminished, the tired muscle returns less readily to its natural length than does the fresh one. . . . The exhaustion due to contraction may be the result: either of the consumption of the store of really contractile material present in the muscle. Or of the accumulation in the tissue of the products of the act of contraction. Or of both of these causes.

The sense of effort was approached in an analytical if rather discursive manner by Augustus Waller (1891), another early member of the Physiological Society, who argued as follows:

> The question is: does normal voluntary fatigue depend upon central expenditure of energy, or upon peripheral expenditure of energy, or upon both factors conjointly; and if upon both, in what proportion upon each? The form of this question is justified as follows; a maximum voluntary effort may decline: 1) by decline of cerebral motility; 2) by sub-central or spinal block; *Central.*, 3) by motor end-plate block; 4) by decline of muscular energy; *Peripheral.* On the human subject it is impossible to separate factors 1 and 2, we must therefore embrace them under the term central. The third factor can be more easily demonstrated upon isolated parts of lower animals; its existence can also - if less assuredly - be demonstrated upon man; but for the cleared appreciation of the central versus the peripheral factors I do not dwell upon this sub-distinction, and include the effects of factors 3 and 4 under the common term peripheral.

Following careful experimentation Waller concludes:

> That in voluntary fatigue the degree of change is on decreasing ratio from centre to periphery. In other words that the cell of higher function is, relatively to the amount of effect which it can produce, more exhaustible than the cell which is subordinate to it in the cerebro-muscular chain. So that central fatigue is protective from peripheral fatigue; and although normally in the body the dynamic effect of the physiological stimulus emitted by a "motor" centre far exceeds the maximal effect which can be elicited by direct experimental excitation, yet that physiological stimulus, by virtue of the rapid exhaustibility of the organ from which it proceeds, cannot normally "overdrive" and exhaust the subordinate elements of the motor apparatus.

Mosso devised the *ergograph* which he used on a colleague to repeat Kronecker's work and found that "We have a great change from the straight line found by Kronecker as the expression of fatigue in frog's and dog's muscles after separation from the body. This shows that in man the phenomenon is considerably more complex." However he goes on:

> To eliminate the mental element which might alter the fatigue curve of the muscle, I thought of stimulating the nerve of the arm, or rather the flexor muscles of the fingers. . . . Fatigue was produced with the same curve as when the muscles contracted voluntarily. . . one must conclude that the mental factor does not exercise a preponderating influence, and that fatigue may even be a peripheral phenomenon.

In Bainbridge's *Physiology of Muscular Exercise* (Bainbridge, 1931), the following summary of his views of the site of fatigue in different forms of muscular activity is to be found.

> Fatigue after exertion is characterised by diminished capacity for performing work, this being usually accompanied by certain subjective sensations. The sensation of fatigue does not necessarily correspond with the actual fatigue of an individual when this is measured by his capacity for work. There are two types of fatigue, one originating entirely within the central nervous system, the other arising partly in the nervous system and partly within the active muscles. The former is of common occurrence, whereas the latter occurs comparatively infrequently. Industrial fatigue is usually of the first type. There is no clear evidence that the products of muscular activity take any part in bringing about fatigue of the central nervous system.

Perhaps the most convincing argument for a peripheral cause of fatigue this century came from a study by Merton on a single individual. ". . . anyone who possesses a sphygmomanometer and an open mind can readily convince himself that the site of fatigue is in the muscles themselves " (Merton, 1956). "If it were central, recovery should occur when effort ceases, for there is no interference with the circulation to central structures." (Merton, 1954). This study has proved to have been a particularly important milestone in the understanding of human muscle fatigue. Not only did it show beyond any doubt (in isometric contractions under ischemic conditions) that fatigue was peripheral, but it also showed that force need not follow alterations in action potential amplitude (see below), and that recovery from the state of fatigue required restoration of the circulatory oxygen supply (see also Gandevia et al., Chapter 20).

The debate as to whether the site of fatigue was *central* or *peripheral* was thus well established by the early part of the twentieth century. Established too was the use of electrical stimulation as a tool for analyzing fatigue in particular forms of activity. Thereafter, debate has been ongoing as to the specific causes of fatigue; those looking for peripheral causes and other workers identifying the role of central mechanisms. While major milestones in the understanding of fatigue mechanisms have come about through the advance in the tools for the study of muscle, others have come about through the study of muscle disease, the pathobiological processes that result in the premature and delayed development of fatigue. The following paragraphs examine the evolution of concepts, the hypotheses that lay behind fatigue development in normal muscle through analysis of muscle disease.

CENTRAL MECHANISMS OF FATIGUE

The work of Waller, *The sense of effort: an objective study* (1891), alluded to above illustrates the extensive efforts put in by early workers in unraveling the complexities of the central processes contributing to fatigue. While much work has gone into identifying and quantifying central contributions, the manifestation of generalized fatigue and myalgia has been well documented over the last 200 years. This is well illustrated by an example from Lewis' *Soldier's Heart and Effort Syndrome* (1918).

> Fatigue is an almost universal complaint of our patients. In the "effort syndrome" group it varies in degree and, when it can be gauged, is an excellent index of the severity of the affection. The mildest cases suffer such fatigue on exercise as would a healthy man who is out of training, in the severer cases it is experienced after very brief and simple exercises. Lassitude is especially prominent in the early morning and late afternoon. More rarely, fatigue on exercise proceeds to exhaustion; actual collapse from exhaustion on exercise is not seen because it is guarded against. Fatigue and the early signs of exhaustion appear objectively in the expression of the face and in the droop of the body; a material feeling of weakness is usually accompanied by uncontrollable tremor of the hands, or shaking of the legs. The symptoms are those which are found in healthy subjects submitted to strenuous exercise. That they have the same origin is rendered probable by their association with a general feeling of malaise and in some patients with a rapidly developed and severe "stiffness" of the muscles.

Now known as the *chronic fatigue syndrome*, similar symptoms have been described over the years under many various names including irritable heart (Da Costa, 1871), Soldier's heart (Lewis, 1918), neurocirculatory asthenia (Oppenheimer et al, 1918), effort syndrome (Lewis, 1918; Wood, 1941) and autonomic imbalance (Kessel & Hyman, 1923). Shorter in his book, *From Paralysis to Fatigue* (1992), takes an historical perspective of psychosomatic illness that may contribute to the perception of this form of fatigue. Stokes and colleagues (1988) and Gibson and coworkers (1993a) have shown that this type of fatigue is centrally mediated and is not due to peripheral failure of force generation. However, as yet, no agreed explanation can be offered for the fatigue described in the *chronic fatigue syndrome*.

The role of peripheral inhibitory influences in the form of afferent reflexes altering central activity have long been debated. This century has seen a plethora of information on the alterations in motor control, particularly from the work of Marsden and colleagues (1971; 1983) who demonstrated the decline in motor unit discharge rates during contraction in relation to slowing of relaxation (*muscle wisdom*), and Bigland-Ritchie and colleagues (1986) who examined the origins of the reflex mechanisms involved. It is interesting to note that the methods enjoyed now to determine the central contribution of fatigue such as interpolation of stimuli on voluntary efforts (Merton, 1954; Belanger & McComas, 1981; Bigland-Ritchie et al., 1986) were, in fact, well established in the last century as reported by workers such as Waller (1891) in his studies of central fatigue.

While metabolic factors involved in the reflex hypotheses of central fatigue appear to receive much attention, a central metabolic cause has generally been excluded owing to the unlikely situation of changes in brain metabolism such as acidosis except in pathological conditions and that exercise lactacidosis has a negligible effect on brain function (Siesjo, 1982; see Newsholme & Blomstrand, Chapter 23). Nevertheless, the importance of the role of the reticular formation in sleep has received attention more latterly with Newsholme and colleagues (1987) suggesting a pivotal role for altered levels of the neurotransmitter 5-hydroxy-tryptamine and its close association with the plasma tryptophan:branched chain amino acid ratio. It is possible that such a mechanism may be involved in exercise of long duration, but whether such an explanation can be offered for fatigue in short held contractions is doubtful.

FAILURE OF MUSCLE EXCITATION AS A CAUSE OF FATIGUE

The credit for determining the motor nerve firing frequency is generally given to Adrian and Bronk (1929). Historical researches would suggest that the phenomenon of recording electrical potentials may have its origin far earlier with the existence of muscle action potentials, *negative variations*, being first proven in 1843 and action currents recorded from the arm of a man who contracted his muscle by using jars of liquid as electrodes by (DuBois-Reymond in 1851, Fig. 1).

The discovery of muscle action potentials led to the exclusion of failure of the nerve fiber in the genesis of fatigue as several workers in the latter part of the 19th century used novel methods to detect action potentials and to demonstrate that nerve was indefinitely excitable: Wedenski (1884, cited in Waller 1891) "... used a telephone to detect the negative variation of faradised nerve", while Waller (1891) employed a galvanometer to record *true negative variation and electrotonic currents*.

The identification of transmission failure of excitation in myasthenia gravis perhaps gave important clues as to the possible involvement of the neuromuscular junction in fatigue. With developments in electromyograpy (Adrian & Bronk, 1929) applied to muscular disorders together with an understanding of the chemical nature of the neuromuscular junction, Lindsley (1935) concluded that the early fatigue seen in myasthenia gravis was indeed due to "... a blocking of the transmission of excitation at the 'myoneural junction'". Techniques of supramaximal electrical stimulation with recording of the action potential

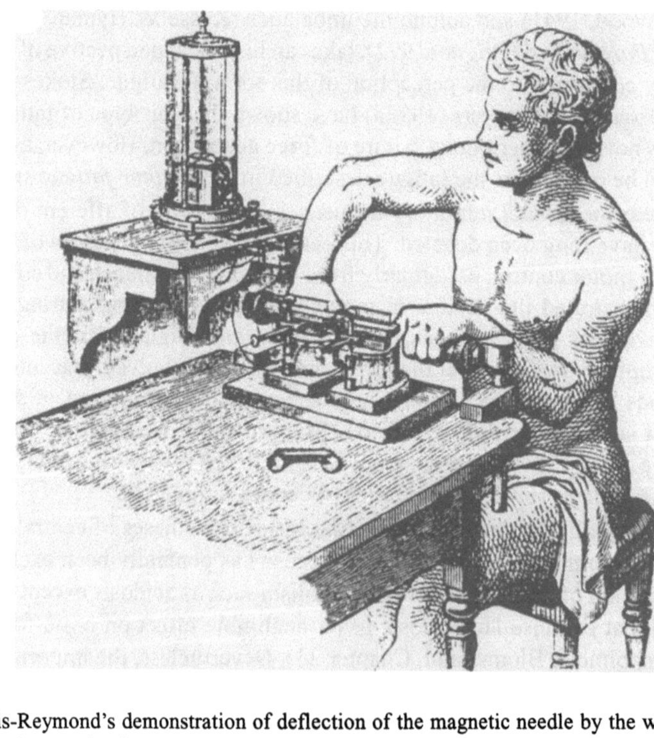

Figure 1. DuBois-Reymond's demonstration of deflection of the magnetic needle by the will. "... the man strongly contracts the muscle of one arm, the result is an immediate deflection of the multiplier, which indicates the presence of a current ascending in the contracted arm from the hand to the shoulder. If the muscles of the other arm are contracted, a deflection occurs in the opposite direction. We are, therefore, able by the mere power of the will to generate an electric current and to set the magnetic needle in motion." (reprinted with permission from Rosenthal, 1883).

were developed further by Harvey and Masland (1941) as a diagnostic tool. Various programs of repetitive nerve stimulation to permitted the identification of disorders of neuromuscular activation forming the basis of the electrophysiological diagnosis of myasthenia gravis, and later were refined with the introduction of single unit EMG methods.

There remained the question as to whether there was evidence of impaired neuromuscular transmission in any form of muscular activity in normal subjects. The finding that there was close correlation between the force and smoothed rectified EMG suggested to Edwards and Lippold (1956) and later Stephens and Taylor (1972) that in sustained isometric contractions of the first dorsal interosseous muscle, the site of fatigue was indeed at the neuromuscular junction. But earlier work by Merton (1954) which was supported later by Bigland-Ritchie and colleagues (1983) systematically examined the question of neuromuscular junction failure. By interpolating stimuli on a maximum voluntary contraction, these investigations were unable to demonstrate any evidence of failure during at least 60 s of sustained maximal isometric activity, clearly highlighting the involvement of processes that lie beyond the neuromuscular junction.

BIOCHEMICAL BASIS FOR FATIGUE

The growth of ideas about the biochemical basis of muscular contraction is covered in Dorothy Needham's classic *Machina Carnis* (Needham, 1971). Various clinical studies can be perceived in retrospect to strongly support the hypothesis that muscle pain and fatigue are closely related to failure of energy supply or undue accumulation of lactate. Direct evidence to support these hypotheses dates back to 1904 when Weichardt suggested an accumulation of a substance, which he called *kenotoxin*, was responsible for fatigue. Muller (1935) also demonstrated a faster onset of fatigue with circulatory occlusion. The finding that the disappearance of lactic acid showed a consistent correlation with recovery from fatigue led Hill and coworkers (1924) to propose that lactic acid was the fatigue substance. This had important implications in the future work on fatigue with many researchers seeking to show the importance of lactic acid and H^+ in the genesis of fatigue.

The description of the experimental procedure by Lewis (1942, p. 96) emphasizes these ideas:

> In the standard test, almost maximal voluntary gripping movements, which develop a tension of 20 to 28 lbs. weight, are made they are carried out rhythmically at the rate of usually one a second. Such rhythmic movements can be undertaken painlessly for very many minutes, provided that the circulation to the limb is free; but, if the circulation is stopped before exercise begins, pain is soon felt. It begins after 24 to 45 sec. (or the same number of contractions) and quickly grows in intensity until it renders the exercise so disagreeable that the exercise is stopped at a point between 60 and 80 sec. Using uniform conditions and adequate periods of intervening rest between tests, it is surprising how constant is the time taken to reach the intolerable point in repeated tests on a given individual. The pain is rather diffuse, once it has come, and it gathers steadily in intensity as the test proceeds. It is easy to ascertain that the pain, though diffuse, appears largely in the region of the muscles used. Thus, in the standard test, the pain is felt mainly over the flexor surface of the forearm; some may be felt over the thenar eminence. The most convincing examples are those in which small muscles and simple movements are employed, such as opposition of the thumb or abduction of the little finger. . . . It is concluded that the stimulus responsible for pain when muscular exercise is taken in the absence of blood supply arises directly or indirectly out of the contraction process.

The practical application of this clinical physiological approach of Sir Thomas Lewis came with the publication in 1951 of the description of the ischemic forearm lactate test. This test was applied in a patient who complained of muscle fatigue and cramps and who was unable to produce lactate during this type of exercise (McArdle, 1951). The implication of this finding for the study of energy metabolism was to provide a paradigm disorder consistent with isolated poisoned muscle models. Similarly, Lunsgaard (1930) demonstrated that contraction occurred in frog muscle despite inhibiting glycolysis with iodoacetate. The same observations are made in these patients in whom some exercise is possible, but the premature fatigue seen cannot be a consequence of the accumulation of lactate since none is produced.

Another group of individuals with a defect in metabolism are those with mitochondrial disorders (De Jesus, 1974, Morgan-Hughes et al., 1977). Profound exercise intolerance and marked fatigability with a tendency to develop lactic acidosis illustrate the primary defect is in oxidative phosphorylation. Such individuals again belong to *Nature's experiments*. Their counterpart, isolated cyanide poisoned preparations (Lüttgau, 1965), where there is a reliance on anaerobic metabolism for supply of energy. Another group of patients which have given important clues as to the metabolic dependence of fatigue are those with hyper- and hypothyroidism. Hypothyroid patients are able to sustain force for longer at a lower ATP cost than normal individuals or patients with hyperthyroidism (Wiles et al., 1979). Though attributed largely to an increase in prevalence of type I fatigue resistant fibers, the resistance to fatigue cannot be explained by this fact alone in view of changes in relaxation characteristics before changes in fiber composition with treatment.

Simultaneously and mostly in Scandinavia, there had progressed a major drafting of the principal changes in the energy metabolism associated with fatigue in muscular exercise. This had been possible only because of the introduction of the needle biopsy technique (Bergström, 1962). Such studies identified the importance of high energy phosphates in brief exercise (Hultman et al., 1967) as well as muscle glycogen as a determinant of endurance time in heavy dynamic exercise. The needle biopsy technique again allowed direct observation of the extent of recruitment of motor units in particular forms of dynamic exercise, from determination of the glycogen depletion patterns (Gollnick et al., 1974), thus confirming the motor unit recruitment patterns according to the *size principle* (Henneman, 1957), and fatigability of fiber types as demonstrated in animal muscle (Burke et al., 1973).

Recently, there has been much interest as well as rapid progress in our understanding of muscle chemistry due to the introduction of magnetic resonance spectroscopy. The use of this tool to study muscle energy metabolism in health and disease (Edwards et al., 1982; Kemp & Radda, 1994), and its applications to fatigue are discussed in Section 4, this volume. This major advance in technology spurred once again an interest in analyzing fatigue in patients with mitochondrial and glycolytic defects, elegantly demonstrating specific blocks in metabolism, and is now finding applications in other muscle diseases. The true value of magnetic resonance in the study of fatigue is only likely to be realized, however, when combined with objective measurement of muscle function (Gibson & Edwards, 1993; Vestergaard-Poulsen et al., 1994).

EXCITATION-CONTRACTION COUPLING FAILURE CAUSING FATIGUE

The demonstration in human muscle that the equivalent of the impairment of excitation-contraction coupling could occur with ischemic isometric exercise (Edwards et al., 1977) raised yet another candidate for the site of fatigue. More dramatically, this impairment can be induced by eccentric activity suggesting the involvement of mechanical

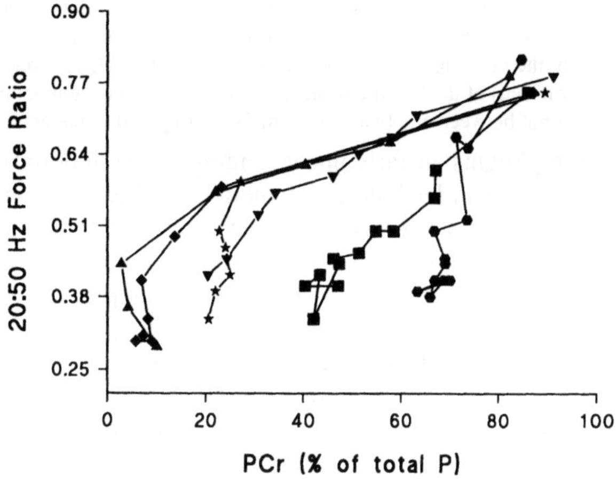

Figure 2. Voluntary, submaximal exercise involving repeated isometric contraction at 40% of maximum force generation (6 s with 4 s rest) was performed by six subjects Electrically evoked frequency-force relations are plotted against energy status determined by magnetic resonance spectroscopy (data from each subject are shown with a different symbol). Note that while there is a considerable variation among the subjects in their use of muscle PCr, there is consistent reduction in the ratio of forces achieved with 20 and 50 Hz electrical stimulation. These data indicate that impairment in electromechanical coupling was more important than energy depletion as a cause of fatigue in this form of exercise. (Gibson et al., 1993b)

damage (Newham et al., 1983). However, fatigue during isometric activity also occurs more rapidly in patients with mitochondrial disorders indicating a metabolic cause (Wiles, 1981).

The interrelations between *energy utilization* and *excitation/excitation-contraction failure* were first explored in an intuitive manner at the conclusion of the *milestone* Ciba Foundation Symposium on *Human Muscle Fatigue: Physiological Mechanisms.* Merton proposed "that muscle fatigue has everything to do with electricity and nothing to do with chemistry". That debate continues today helped by rapid progress in magnetic resonance spectroscopy (e.g., Miller et al., 1987; see Miller, Chapter 14; Bertocci, Chapter 15). There was further debate in the Chairman's concluding section. These discussions provided the foundation for the subsequent Catastrophe Theory of fatigue (Edwards, 1983). At the Amsterdam meeting on muscle fatigue (1992), more evidence was forthcoming as to the greater importance of the impairment of electromechanical coupling compared with energy limitations as determining the onset of fatigue. Fig. 2 illustrates in normal quadriceps muscle for six subjects the greater relative importance of these processes compared to energy lack. In this study, there were electrophysiological assessments and spectroscopic measures of metabolism in a whole body MR system during muscle fatigue resulting from prolonged submaximal activity. These results contrast with those of Vøllestad and colleagues (1988) who, using serial biopsy measures of energy metabolism, showed little change in metabolism until exhaustion.

IMPORTANCE OF RECOVERY PROCESSES IN THE ANALYSIS OF FATIGUE

There has been much interest (and there still is today) in the time course of recovery processes which are apparently associated with fatigue. In 1883, Foster wrote

> Absolute (temporary) exhaustion of the muscles, so that the strongest stimuli produce no contraction, may be produced even within the body by artificial stimulation; recovery takes place on rest.

> Out of the body absolute exhaustion takes place readily. Here also recovery may take place. Whether in any given case it does occur or not, is determined by the amount of contraction causing the exhaustion, and by the previous condition of the muscle. In all cases recovery is hastened by renewal (natural or artificial) of the blood-stream. The more rapidly the contractions follow each other, the less the interval between any two contractions, the more rapid the exhaustion.

This quotation highlights an early understanding of the importance of blood borne processes in recovery, eventually leading to a plethora of studies which examine the role of metabolism in fatigue. The major advances, however, have come about not through identifying those factors that correlate with fatigue, but rather from the divergence of processes during recovery from fatigue. Hence an early paradoxical finding by Davis and Davis (1932) on cat soleus was the observation that fatigue during high frequency stimulation could be reversed on acutely reducing the frequency, where force is seen to increase dramatically. This has subsequently been confirmed for human muscle by Bigland-Ritchie and colleagues (1979) who were also able to demonstrate a rapid concomitant recovery in the action potential. Clearly metabolic factors are not involved owing to the marked differences in the time course of recovery of excitation and metabolism, the reduction in frequency permitting time for the rapid redistribution of ionic fluxes.

Further recent insight into the processes of failure of excitation in fatigue have come from ischemic studies in McArdle's patients. A decline in the action potential amplitude is observed not to recover following activity when ischemia is maintained (Gibson & Edwards, 1988). The converse is seen in normal subjects. Moreover, the action potential does not appear to increase in duration, a phenomenon associated with a slowing of muscle conduction velocity, often attributed to accumulation of potassium. Such observations suggest the involvement of an energy-dependent system, and supports the notion that *dropping out* of fibers may occur during fatigue. These results give an opportunity to examine alternative concepts of fatigue. Selective dropping out of fibers may also explain the phenomenon described above where type II fibers may have become fatigued more readily. Such a scenario is illustrated in Fig. 3 where a *fiber population* paradigm of fatigue is presented for an admixture of type I and II fibers with differing fatigue characteristics alluded to earlier. High frequency stimulation results in a decline in force, but as stimulation frequency is reduced, force recovers. The insets illustrate the relative changes in two populations of fibers where one population fatigues more rapidly than the second (fibers *drop out*). Reduction of stimulation frequency allows rapid recovery of excitatory processes with subsequent recovery of both fiber populations. This is, of course, an oversimplification; the gradual decline

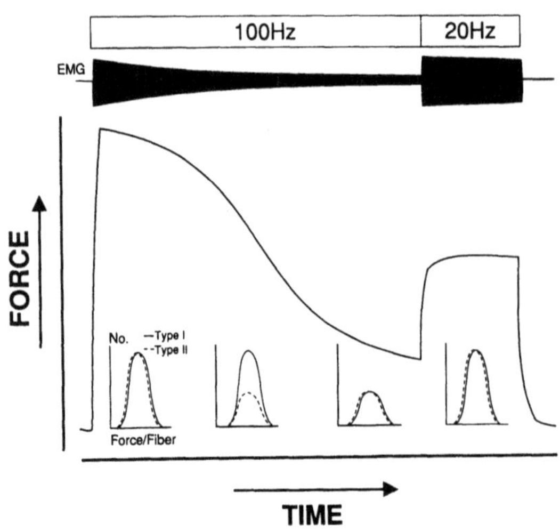

Figure 3. Hypothetical representation of differential fatigability of muscle fiber populations (insets) as a possible explanation of the paradoxical increase in force as stimulation frequency is reduced (experimental example based on Jones, 1981).

in force is likely to represent a statistical statement of events for many fibers. Thus fatigue in a whole muscle (often treated as a *single cell* model) in which fibers pull in parallel could be considered as 1) a shift in active fiber population (derecruitment); or 2) a reduction in the fiber force-generating performance.

A COMBINED CONCEPTUAL MODEL FOR CLASSIFYING SPECIFIC FORMS OF FATIGUE IN DIFFERENT MUSCLE OR ACTIVITY CIRCUMSTANCES

Fig. 4 summarizes the principal mechanisms resulting from the *milestone* advances that have contributed to our understanding of fatigue which may be applied to all forms of activity in health and disease and illustrates the contribution of the many elements of fatigue often studied in isolation.

The model incorporates the principal components:

- *Parallel Fatigue* i.e., loss of overall force generation due to loss of function of contractile elements (i.e., individual fibers) pulling in parallel
- *Contractile Fatigue* comprises interrelations between energy depletion and electromechanical coupling processes.

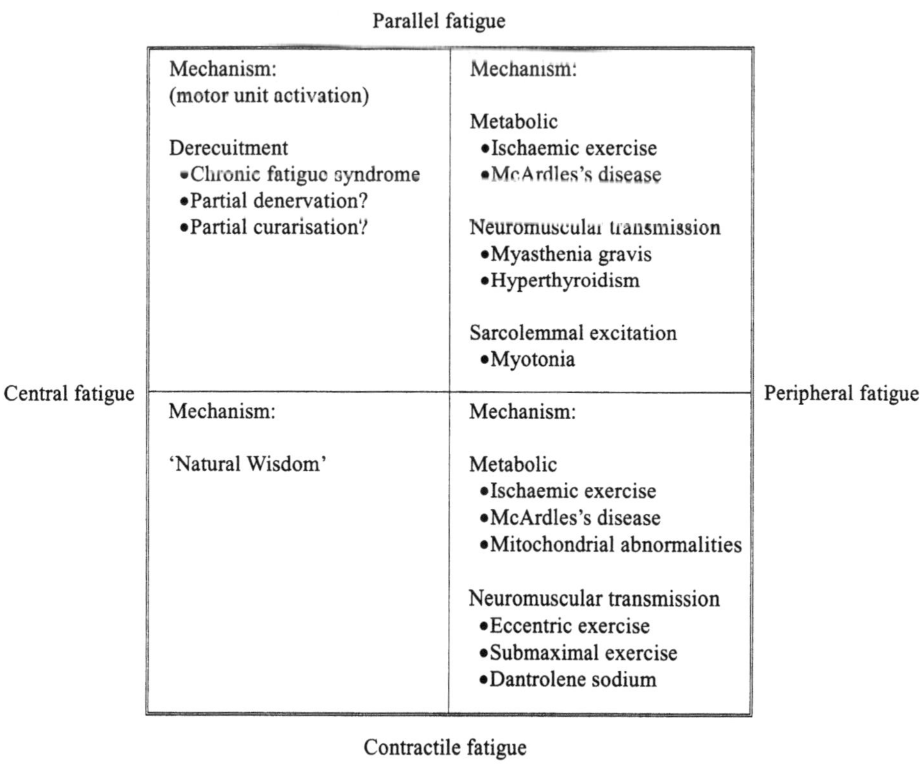

Figure 4. A four quadrant diagram to help classify type of fatigue in different types of exercise or clinical condition.

These have been combined to form a four quadrant diagram (Fig. 4) to illustrate how these components may contribute to:

- *Central Fatigue* resulting in parallel fatigue, due to derecruitment of motor units - (e.g., reduced motivation for continuing voluntary muscular activity)
- *Central Fatigue* with impaired force generation of population of muscle fibers (this represents the *natural wisdom* in which there is apparent matching of the firing frequency to match alterations in the contractile properties of fatiguing muscle fibers.
- *Peripheral Fatigue* (Parallel) due to selective fatigue of particular fiber populations either because of specific metabolic characteristics or because of preexisting alteration in the distribution of fiber types e.g., Type II fiber dominance and,
- *Peripheral Fatigue* (Contractile) due to impaired intrinsic force generating capacity of individual muscle fibers owing to impaired excitation-contraction coupling defect or energy depletion affecting all fibers.

CONCLUSIONS

This cursory review of the history of pathophysiological studies in fatigue shows that several of the fatigue phenomena that are studied in isolated experimental preparations can be studied in human muscle. All these studies have given clues as to the sites and mechanisms of fatigue in diseased muscle. Concepts clearly thought out by the latter part of the last century required new tools to verify the ideas. Findings in neuromyopathies such as myasthenia gravis, *Nature's experiments* (single enzyme defects or specific abnormalities of function) or altered central drive give the opportunity to understand the factors controlling neuronal firing frequency and to discover ways to optimize force generation of fatiguing muscle fibers to improve overall performance.

ACKNOWLEDGMENT

The authors gratefully acknowledge laboratory support from the Muscular Dystrophy Group of Great Britain and Northern Ireland, Association Française contre les Myopathies, and ICI Pharmaceuticals. Attendance of *R.H.T.E.* and *H.G.* at the 1994 Bigland-Ritchie conference was supported, in part, by the Wellcome Trust, the American Physical Therapy Association (Research and Analysis Division for *R.H.T.E.*; Section on Research for *H.G.*), and the Muscular Dystrophy Association (USA),

REFERENCES

Adrian ED & Bronk DW (1929). The discharge of impulses in motor nerve fibres. *Journal of Physiology (London)* **67**, 119-151.

Althaus J (1873). Theoretical and practical paralysis, neuralgia and other diseases In: *A Treatise on Medical Electricity*, 3rd Ed. London: Longmans, Green, and Co.

Bainbridge FA (1931). *The Physiology of Muscular Exercise*, p. 235. London: Longmans, Green & Co.

Belanger AY & McComas J (1981). Extent of motor unit activation during effort. *Journal of Applied Physiology* **51**, 1131-1135.

Bergström J (1962). Muscle electrolytes in man. *Scandinavian Journal of Clinical Laboratory Investigation* **14** (suppl 68), 9-88.

Bigland-Ritchie B, Dawson NH, Johansson RS & Lippold OCJ (1986). Reflex origin for the slowing of motoneurone firing rates in fatigue in human voluntary contractions. *Journal of Physiology (London)* **379**, 451-459.

Bigland-Ritchie B, Johansson R, Lippold OCJ & Woods JJ (1983). Contractile Speed and EMG changes during fatigue of sustained maximal voluntary contractions. *Journal of Neurophysiology* **50**, 313-324.

Bigland-Ritchie B, Jones DA & Woods JJ (1979). Excitation frequency and muscle fatigue: electrical responses during human voluntary and stimulated contractions. *Experimental Neurology* **64**, 414-427.

Burke RE, Levine DN, Tsairis P & Zajac FE (1973). Physiological types and histochemical profiles in motor units of the cat gastrocnemius. *Journal of Physiology (London)* **234**, 723-748.

Da Costa JKM (1871). On irritable heart: a clinical study of a form of functional cardiac disorder and its consequences. *American Journal of Medical Science* **61**, 17-52.

Davis H & Davis PA (1932). Fatigue in skeletal muscle in relation to the frequency of stimulation. *American Journal of Physiology (London)* **101**, 339-356.

De Jesus PV (1974). Neuromuscular physiology in Luft's syndrome. *Electromyography and Clinical Neurophysiology* **14**, 17-27.

Drummond M & Drummond WB (1915) (English translation of Mosso A book). *Fatigue*. pp. 81-102. London: George Allen & Unwin Ltd.

Duchenne L (1872). L'electrisation localisée et de son application a la pathologie et a la thérapeutique par courants induits et par courants galvaniques interrompus et continus. Paris: Bailliere, JB et fils.

Edwards RHT (1983). Biochemical basis of fatigue in exercise: Catastrophy theory of muscular fatigue In: Knuttgen HG, Vogel JA, Poortmans J (eds.), *Biochemistry of Exercise*, pp. 3-28. Champaign: Human Kinetics.

Edwards RHT, Dawson MJ, Wilkie DR & Gordon RE (1982). Clinical use of nuclear magnetic resonance in the investigation of myopathy. *Lancet* **1(8274)**, 725-731.

Edwards RHT, Hill DK, Jones DA & Merton PA (1977). Fatigue of long duration in human skeletal muscle after exercise. *Journal of Physiology (London)* **272**, 769-778.

Edwards RG & Lippold OCJ (1956). The relation between force an integrated electrical activity in fatigued muscle. *Journal of Physiology (London)* **132**, 677-681.

Foster M (1883). *A Textbook of Physiology*. 4th Ed, pp. 96-97. London: Macmillan and Co.

Gibson H & Edwards RHT (1988). No ischaemic recovery of the evoked compound muscle action potential in McArdle's disease. *Clinical Science* **75**, 40P.

Gibson H & Edwards RHT (1993). Inter-relation of electro-mechanical coupling and chemistry in the study of fatigue by ^{31}P-NMR spectroscopy In: Sargeant AJ, Kernell D (eds.), *Neuromuscular Fatigue*, pp. 30-31. Amsterdam: Royal Netherlands Academy of Arts and Sciences.

Gibson H, Carroll N, Clague JE & Edwards RHT (1993a). Exercise performance and fatiguability in patients with the chronic fatigue syndrome. *Journal of Neurology, Neurosurgery, and Psychiatry* **56**, 993-998.

Gibson H, Saugen E, Martin PA, Vollestad NK & Edwards RHT (1993b). Individual pathways for eletromechanical-coupling and energy utilization in submaximal exercise. *International Union of Physiological Sciences*, Glasgow, August 1-6, 216.

Gollnick PD, Karlsson J, Piehl K & Saltin B (1974). Selective glycogen depletion in skeletal muscle fibres in man following sustained contractions. *Journal of Physiology (London)* **241**, 59-67.

Harvey AM & Masland RL (1941). A method for the study of neuromuscular transmission in human subjects. *Bulletin of the Hopkins Hospital* **68**, 81-93.

Henneman E (1957). Relation between size of neurons and their susceptibility to discharge. *Science* **126**, 1345-1347.

Hill AV, Long CNH & Lupton H (1924). Muscular exercise lactic acid, and the supply and utilization of oxygen, parts I-III. *Proceedings of the Royal Society of London* **96**, 438-479.

Hultman E, Bergstrom J & McLennan AN (1967). Breakdown and resynthesis of adenosine triphosphate in connection with muscular work in man. *Scandinavian Journal of Clinical Laboratory Investigation* **19**, 56-66.

Jolly F (1895). Über myasthenia gravis pseudoparalytica. *Berliner Kliniscshe Wochenschrift* **32**, 1-7.

Jones DA (1981). Muscle fatigue due to changes beyond the neuromuscular junction In: Porter R, Whelan J (eds.), *Human Muscle Fatigue: Physiological Mechanisms.* Ciba Foundation Symposium 82, pp. 178-212. London: Pitman Medical.

Kaplan EB (1959) (English translation of Duchenne GB book). *Physiology of Motion Demonstrated by Means of Electrical Stimulation and Clinical Observation and Applied to the Study of Paralysis and Deformities*. Philadelphia: WB Saunders.

Kemp GJ & Radda GK (1994). Quantitative interpretation of bioenergetic data from ^{31}P and ^{1}H magnetic resonance spectroscopic studies of skeletal muscle: an analytical review. *Magnetic Resonance Quarterly* **10**, 43-63.

Kessel L & Hyman HT (1923). The clinical manifestations of disturbances of the involuntary nervous system (autonomic imbalance). *American Journal of Medical Sciences* **165**, 513.

Lewis T (1918). *The Soldier's Heart and The Effort Syndrome*. London: Shaw & Sons.

Lewis T (1942). *Pain*. New York: The Macmillan Company.

Lindsley DB (1935). Myographic and electromyographic studies of myasthenia gravis. *Brain* **58**, 470-482.

Lüttgau HC (1965). The effect of metabolic inhibitors on the fatigue of the action potential in single muscle fibres. *Journal of Physiology (London)* **178**, 45-67.

Lundsgaard E (1930). Über die Einwirkung der Monoiodoessigsäure aud den spaltungs - und Oxydationsstoffweschel. *Biochemishe Zeitschrift* **220**, 8-12.

Marsden CD, Meadows JC & Merton PA (1971). Isolated single motor units in human muscle and their rate of discharge during maximal voluntary effort. *Journal of Physiology (London)* **217**, 12-13P.

Marsden CD, Meadows JC & Merton PA (1983). "Muscular wisdom" that minimizes fatigue during prolonged effort in man: peak rates of motorneuron discharge and slowing of discharge during fatigue In: Desmedt JE (ed.), *Motor Control Mechanisms in Health and Disease*, pp. 169-211. New York: Raven Press.

McArdle B (1951). Myopathy due to a defect in muscle glycogen breakdown. *Clinical Science* **10**, 13-33.

Merton PA (1954). Voluntary strength and fatigue. *Journal of Physiology (London)* **123**, 553-564.

Merton PA (1956). Problems of muscular fatigue. *British Medical Bulletin* **12**, 219-239.

Miller RG, Giannini D, Milner-Brown HS, Layzer RB, Koretsky AP, Hooper D & Weiner MW (1987). Effects of fatiguing exercise on high-energy phosphates, force, and EMG: evidence for three phases of recovery. *Muscle & Nerve* **10**, 810-821.

Morgan-Hughes JA, Darveniza P, Kahn SN, Landon DN, Sherratt RM, Land JM & Clark JP (1977). A mitochondrial myopathy characterised by a deficiency in reducible cytochrome b. *Brain* **100**, 617-640.

Muller EA (1935). Die Erholung nach statischer Haltearbeit. *Arlbeitsphysiologie* **8**, 72-75.

Needham DM (ed) (1971). *Machina Carnis*. Cambridge: Cambridge University Press.

Newham DJ, Mills KR, Quigley BM & Edwards RHT (1983). Pain and fatigue after concentric and eccentric muscle contractions. *Clinical Science* **64**, 55-62.

Newsholme EA, Acworth IN & Blomstrand E (1987). Amino acids, brain neurotransmitters and a functional link between muscle and brain that is important in sustained exercise In: Benzi G (ed.), *Advances in Myochemistry*, pp. 127-133. London: John Libbey, Ltd.

Oppenheimer B, Levine SA, Moneson RA, Rothechild MA, St. Lawrence W & Wilson FN (1918). Appendix of illustrative cases of neurocirculatory asthenia. *Military Surgery* **42**, 409-420.

Rasch PJ & Burke RK (eds.) (1967). The history of kinesiology In: *Kinesiology and Applied Anatomy. The Science of Human Movement*, 3rd Ed., pp. 1-17. Philadelphia: Lea & Febiger.

Rosenthal I (ed.) (1883). *General Physiology of Muscles and Nerves*. London: Kegan Paul, Trench & Co.

Shorter E (1992). *From Paralysis to Fatigue*. New York: Free Press.

Siesjo BK (1982). Lactic acidosis in the brain: occurrence, triggering mechanisms and pathophysiological importance In: Porter R, Lawrenson G (eds.), *Metabolic Acidosis*, Ciba Foundation Symposium 87, pp. 77-88. Edinburgh: Churchill Livingstone.

Stephens JA & Taylor A (1972). Fatigue of maintained voluntary muscle contraction in man. *Journal of Physiology (London)* **220**, 1-18.

Stokes M, Cooper RG & Edwards RHT (1988). Normal muscle strength and fatigability in patients with 'effort syndromes'. *British Medical Journal* **297**, 1014-1017.

Vestergaard-Poulsen P, Thomsen C, Sinkjaer T & Henriksen O (1994). Simultaneous ^{31}P NMR spectroscopy and EMG in exercising and recovering human skeletal muscle: technical aspects. *Magnetic Resonance in Medicine* **31**, 93-102.

Vøllestad NK, Sejersted OM, Bahr R & Bigland-Ritchie B (1988). Motor drive and metabolic responses during repeated sub-maximal contractions in man. *Journal of Applied Physiology* **63**, 1-7.

Waller AD (1891). The sense of effort: an objective study. *Brain* **14**, 179-249.

Weichardt W (1904). Uber das ermudungtoxin und-antitoxin. *Muenchener Medizinsche Wochenschrift* **51**, 2121-2126.

Wiles CM, Jones DA & Edwards RHT (1981). Fatigue in human metabolic myopathy In: Porter R, Whelan J (eds.), *Human Muscle Fatigue: Physiological Mechanisms*, Ciba Foundation symposium 82, pp. 264-282. London: Pitman Medical.

Wiles CM, Young A, Jones DA & Edwards RHT (1979). Muscle relaxation rate, fibre-type composition and energy turnover in hyper- and hypo-thyroid patients. *Clinical Science* **57**, 375-384.

Wood P (1941). Da Costa's syndrome (or effort syndrome). *British Medical Journal* **i**, 767- 770.

FATIGUE BROUGHT ON BY MALFUNCTION OF THE CENTRAL AND PERIPHERAL NERVOUS SYSTEMS

A. J. McComas[1], R. G. Miller,[2] and S. C. Gandevia[3]

[1] Department of Biomedical Sciences
McMaster University, Hamilton, Canada
[2] Department of Neurology, California Pacific Medical Center
San Francisco, California
[3] Prince of Wales Medical Research Institute
Sydney, Australia

ABSTRACT

Increased fatigability necessarily occurs in every patient with muscle weakness, regardless of whether the latter is due to a central or peripheral neurological disorder. The tendency for disuse to increase fatigability, as a secondary phenomenon, must also be considered; disuse affects both motoneuron recruitment and the biochemical and physiological properties of the muscle fibers. In recent studies impaired recruitment has been observed in postpolio patients, while patients with multiple sclerosis or spinal cord injury have shown, in addition, altered neuromuscular function. Findings are also presented in ALS and the chronic fatigue syndrome. In general, the most dramatic increases in fatigability take place in disorders of the peripheral nervous system and almost any cell component can be incriminated. There is a need to study fatigability systematically in neurology and rehabilitation.

INTRODUCTION

Symptomatology

Patients with neurological disorders often have symptoms which suggest that they suffer from undue fatigability. Commonly heard complaints include the following: "My legs get tired", "I cannot walk as far as I used to", "My arms start to feel heavy", "My eyelids droop as the day goes on", "I start to see double when I am tired", and others. The choice of verb in these statements is significant, in that it leaves open whether or not the affected muscles are initially weak; instead, there is an insistence that some change takes place in the course of an activity which would normally be accomplished with relative ease. Indeed, the ability to distinguish between absolute weakness and fatigability may be crucial in leading

Fatigue, Edited by Simon C. Gandevia et al.
Plenum Press, New York, 1995

towards the correct diagnosis. An example is the fatigability of the neuromuscular disorder, myasthenia gravis (see below). Mildly affected patients will be quite clear that they have no problems in the mornings, but that the drooping of the eyelids, the tendency to see double and the *tiring* of the limbs declare themselves later in the day and then progress. Such information strongly suggests that the available muscle mass is normal.

If muscles *are* weak to begin with, however, fatigability will be increased if the same demanding tasks are attempted as before the illness. This conclusion depends on the fact that there is normally an inverse relationship between the percentage of maximal force attempted and the time to a set fatigue level (e.g., Bigland-Ritchie et al., 1986; Dolmage & Cafarelli, 1991). Suppose, then, that a patient with a neuromuscular disorder can only develop half the initial force that he/she was previously capable of. Clearly the time to fatigue will be less for a MVC (maximal voluntary contraction) in these circumstances than it would have been in the premorbid state, or in a control subject (Fig. 1*A*). On the other hand, if the diminished MVC is taken as the reference value, there is no *a priori* reason for fatigue to be faster for a MVC than in the premorbid state. Since the force that can be developed in a MVC varies markedly from one subject to another, depending on muscle bulk, training and effective lever arm, performance in a patient should be assessed against previous abilities, rather than in absolute terms. Thus a steelworker may have genuine fatigability and still be stronger than the examining physician.

Furthermore, regardless of whether or not a set physical task can be accomplished normally, patients with neuromuscular disorders commonly have more protracted recoveries with greater discomfort than before their illnesses.

Finally, although this review is concerned mainly with the changes in muscle physiology which accompany fatigue, the problem of *mental fatigue* must not be overlooked. Some patients appear to have low mental *energy* and would previously have been labeled as neurasthenic. Alternatively, or in addition, patients may find that the mental cost of a physical activity becomes disproportionately large, such that they feel *drained* afterwards. Such considerations apply especially to the chronic fatigue syndrome (see below).

Central and Peripheral Fatigue in Neurological Disorders

It is customary to divide fatigue into *central* and *peripheral* components (see elsewhere in this volume); in neurology such a division is analogous to the simple classification of motor disorders into those of the *upper motoneuron* (UMN) and *lower motoneuron* (LMN) types. A UMN disorder is one in which a CNS lesion may cause diminished effort and impaired organization or transmission of the motor command to motoneurons in the brainstem and spinal cord. A lower motoneuron lesion, in contrast, is one which involves the motoneuron, its axon, the neuromuscular junction or the muscle fiber. The correlation between UMN and LMN lesions, on the one hand, and central and peripheral fatigue on the other, is not exact, however, for dysfunction of the motoneuron soma would be seen as part of a LMN lesion but also as a cause of central fatigue. In contrast, dysfunction of the motor axon, for example at the branch points or at the nerve terminal, would be regarded as contributing to peripheral fatigue. This distinction is largely artificial, however, since dying-back changes in the nerve endings may be the first manifestation of a cell body disorder, presumably because of impaired axoplasmic transport and inadequate trophic support.

On general grounds, it would be expected that the LMN lesion would cause peripheral fatigue and that the UMN lesion would result in central fatigue. Closer inspection, however, suggests that such correlations may be too simple (Fig. 1*B*). Thus, weakness from a LMN lesion will require greater than normal effort to accomplish a set task and this would be expected to induce central fatigue. Again, weakness from either UMN or LMN lesions may be complicated by the effects of disuse, as the patient tries to conserve strength by resting

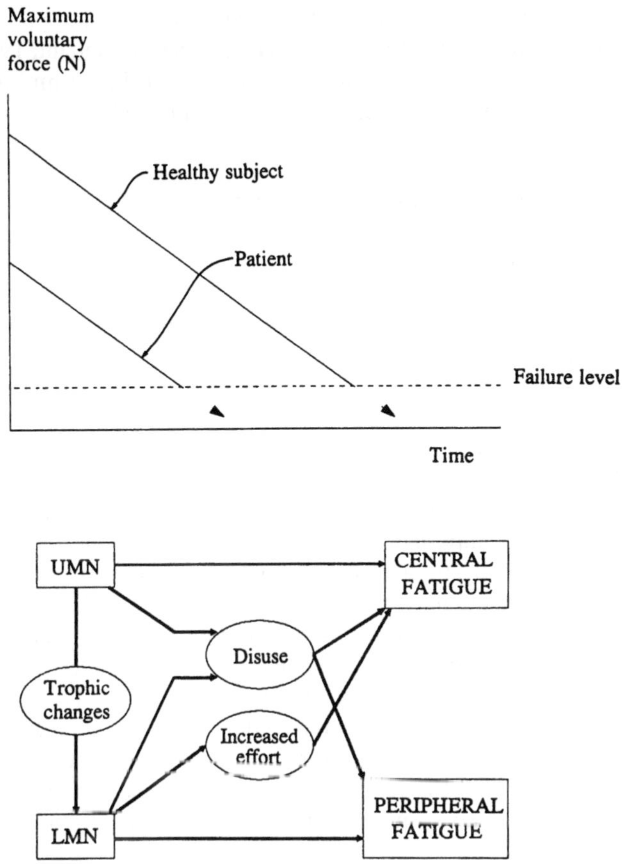

Figure 1. *A*, if a patient has become weak and attempts the same task as a healthy subject or as previously undertaken, fatigue must occur sooner. *B*, relationships between central and peripheral fatigue, and upper and lower motoneuron lesions.

or no longer attempts tasks that have become particularly tiring; the most extreme disuse effects would involve those patients who are largely confined to bed or to wheelchairs. On the one hand, disuse will cause or accentuate fatigue by reducing the ability of the motor *centers* to recruit motoneurons maximally. Perhaps the most unequivocal demonstration of this effect, unaccompanied by disease, was the finding of a reduction in the density of the EMG interference pattern when patients with previous bone fractures attempted to make MVCs in limbs that had been recently freed from immobilizing casts (Fuglsang-Frederiksen & Scheel, 1978; Duchateau & Hainaut, 1987). The transient nature of the EMG changes, with resumption of normality in a few days, indicated that impaired recruitment was unlikely to be associated with overt degeneration of the descending fibers, but the possibility of microscopic changes in the nerve terminals remains. On the other hand, disuse of UMN or LMN origin would also be expected to produce excessive fatigability in muscle fibers on the basis of animal experiments in which surgically denervated muscles were subjected to different regimens of direct electrical stimulation (Westgaard & Lømo, 1988). Although increased fatigability after immobilization has not been observed in some animal studies (Maier et al., 1976; Robinson et al., 1991; see, however, Petit & Gioux, 1993), the findings in humans are consistent with the results of Westgaard and Lømo. Thus, tetanic stimulation at 30 Hz produced faster fatigue in the adductor pollicis muscles of previously immobilized arms than in the contralateral control muscles (Duchateau & Hainaut, 1987). Similar results have recently been noted by Miller and coworkers (1990) in leg muscles of patients with spastic paraparesis of varying etiologies (Gordon, Chapter 32).

INVESTIGATION OF FATIGABILITY IN PATIENTS

Force Measurements

It is true to say that the rate of fatigue, and the possible peripheral and central components, are almost *never* investigated in the neuromuscular clinic. The reason for this unfortunate state of affairs is that very few clinics, even in university centers, have incorporated quantitative strength measurements as part of the diagnostic armamentarium, despite the commercial availability of hand-held myometers with chart recorders. Consequently the study of muscle fatigue has been left to neurological research laboratories; alternatively gross performance may be tested on treadmills or cycle ergometers in exercise laboratories. Whenever possible, it is important that stimulated force also be recorded. Although prolonged repetitive maximal stimulation of a muscle through its peripheral nerve is too uncomfortable for most patients, much useful information can be obtained from twitches. Not only is the twitch tension itself a useful index of muscle function in the resting state, but the interpolation of one or more stimuli can be used to detect the presence of central fatigue during a MVC (see Gandevia et al., Chapter 20). The presence of a substantial interpolated twitch, early in the MVC, is not conclusive evidence of muscle fatigue, since such results also occur in patients who, for one reason or another, choose not to make a full effort (McComas et al., 1983). The muscles in which contractions have been evoked by stimulation include the ankle dorsiflexors and plantarflexors, adductor pollicis, quadriceps, biceps brachii and diaphragm.

Subsequent Strategies

Further investigation will depend on whether or not a central or peripheral etiology is suspected. In the case of a *central* disorder, MR imaging should reveal a structural lesion, with electrophysiology playing a useful adjunct role. For example, cortical evoked potentials tend to be delayed and prolonged in multiple sclerosis, while transcranial magnetic stimulation of each motor cortex, in turn, will test the integrity of the corticospinal pathways. In a peripheral disorder, further examination may require some or all of the following procedures, together with muscle enzymes, and possibly blood lactate levels after ischemic forearm exercise.

Electromyography

Any patients suspected of having increased fatigue should undergo needle EMG and nerve conduction studies in an attempt to reveal an underlying neuromuscular disorder. If impaired neuromuscular transmission is suspected, repetitive nerve stimulation and single fiber electromyography (SFEMG) should be performed (see Trontelj & Stålberg, Chapter 7).

Muscle Biopsy

Muscle biopsy, preferably by needle, should be undertaken for diagnosis in any patient suspected of harboring a neuromuscular disorder. Leaving aside the usefulness of assisting in the diagnosis of denervation or myopathy, a biopsy may suggest a more specific fatigue-inducing pathology as in myophosphorylase deficiency, carnitine deficiency or a mitochondrial disorder (see below). Detailed testing for an enzyme disorder is best done by one of the laboratories offering this type of service.

NMR Spectroscopy

This powerful non-invasive technique enables changes in muscle ATP, creatine phosphate and H+ to be followed during voluntary or stimulated contractions, and has also been used to explore the cellular nature of fatigue in healthy subjects (Miller et al., 1987; see Miller, Chapter 14; Bertocci, Chapter 15).

FATIGUE STUDIES IN SELECTED CENTRAL NERVOUS SYSTEM DISORDERS

Postpolio Patients

A proportion of patients who have had poliomyelitis many years previously appear to make full recoveries and, indeed, have normal strength values and motor unit estimates when tested quantitatively. Others, more severely affected by the initial illness, are left with weakness and undue fatigability. A substantial fraction of the latter patients will complain that their muscle deficits are becoming worse, and are associated with further muscle wasting and with soreness after exercise. It is only this last group who should be considered to have the *postpolio syndrome*. Two of the authors have made detailed studies of muscle performance in postpolio patients with varying degrees of clinical involvement. In both instances endurance was tested using voluntary contractions which were 30% of maximal and which lasted for 6 s out of every 10 s; the exercise was maintained for 45 min in the Sydney patients (Allen et al., 1994) and for 15 min, with an additional 10 min at 50% of maximal force, in the San Francisco patients (Sharma et al., 1994). The Sydney study was performed on the elbow flexors, while the ankle dorsiflexors (especially tibialis anterior) were investigated in the San Francisco patients.

In both studies MVCs showed greater rates of decline in the postpolio patients than in controls. In the Sydney patients an important contributing factor was impaired voluntary activation of motoneurons, as tested by a double-pulse technique. Thus, after 45 min of fatiguing exercise, activation was 90.4 ± 9.5% (mean ± SD) in controls and 80.0 ± 17.7% in patients (Fig. 2). In an extension of the Sydney series (now greater than 100 patients), approximately one quarter of patients have been found to have impaired activation of unfatigued muscles. In keeping with earlier remarks the deficit in activation may be ascribed to the effects of disuse. Although no sign of impaired voluntary activation was found in the San Francisco study, this may have been due to the smaller sample size and the different technique employed for assessment (tetanic force instead of interpolated twitch).

Both the San Francisco and Sydney studies have been important in demonstrating significantly greater relative decreases in stimulated force in postpolio patients than in controls. In addition, the tetanic responses of the San Francisco patients were found to have significantly slower rates of tension development and relaxation. These findings, presumably due to differences in intracellular Ca^{2+} handling, raise the possibility that there may have been increased incidences of type I muscle fibers in the postpolio patients. However, had this been the case, fatigability to electrical stimulation should have been smaller, rather than greater.

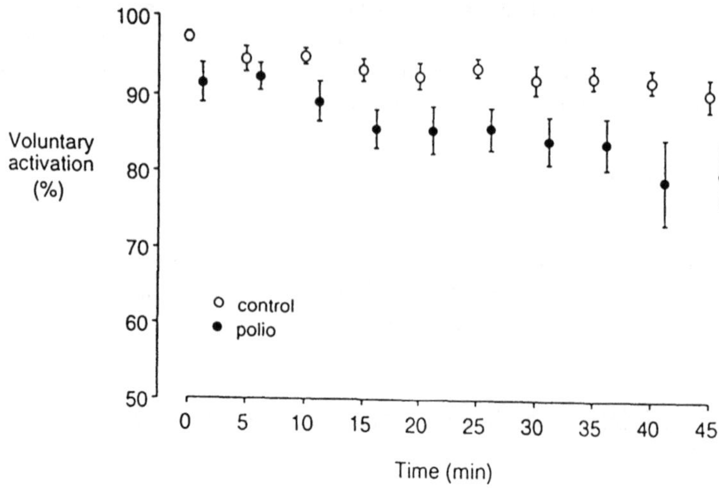

Figure 2. Progressive failure of voluntary activation in postpolio patients attempting repetitive elbow flexion against resistance, compared with controls. Values expressed as means ± SDs (reproduced from Allen et al., 1994, by permission of Oxford University Press).

Although there was no evidence of abnormal M-wave decrement to repetitive stimulation in the San Francisco patients, signs of synaptic dysfunction have been noted by others (Trojan et al., 1993). A feature of the San Francisco study was the application of NMR spectroscopy in some of the patients during and after the fatiguing exercise (Miller et al., 1988). No significant differences between postpolio patients and controls were detected in terms of phosphocreatine (PCr), inorganic phosphate (P_i) and intracellular pH. Interestingly, in both subject groups there was an almost complete recovery of metabolites within 5 min of stopping the exercise, whereas tetanic force was still considerably depressed after 15 min, especially in the patients. Such observations are, of course, crucial for an understanding of the cellular mechanisms responsible for prolonged low-frequency fatigue (Moussavi et al., 1989).

Amyotrophic Lateral Sclerosis Patients

Patients with amyotrophic lateral sclerosis (ALS) also have symptoms of increased fatigue, although the functional limitations imposed by the progressive weakness often dominate the clinical picture. In ALS, there are lesions involving both the upper motoneuron (corticospinal tract pathways) and the lower motoneuron (anterior horn cells). Thus, fatigue in these patients is even more complicated than postpolio syndrome patients in whom lower motoneuron disturbances predominate (see, however, Bruno et al., 1991). Fatigue in patients with ALS should involve both central and peripheral components because of the nature of the disease process. Few studies have examined the mechanisms of fatigue in patients with ALS. ^{31}P nuclear magnetic resonance spectroscopic studies of resting muscles in patients with ALS or peripheral neuropathy demonstrated a higher inorganic phosphate/phosphate ratio and higher intracellular pH compared to controls (Zochodne et al., 1988; Argov & Bank, 1991). The only metabolic study that has been performed during exercise in patients with ALS utilized bicycle ergometry and demonstrated a reduced maximum oxygen consumption and reduced work capacity (Sanjak et al., 1987). Moreover, increased muscular fatigability was found in the shoulder abductor muscles of ALS patients compared to controls, but no measures of metabolism or force evoked by electrical stimulation were made (Nicklin et al., 1987).

Ten San Francisco patients with ALS have now been examined with a protocol similar to that employed for postpolio patients (see above); all the patients had moderate weakness of ankle dorsiflexors, with force values 25-30% of control. As with the postpolio patients, the rate of fatigue was higher in ALS patients than in controls, and this was true both for MVCs and for tetanic stimulation. The ALS results resembled the postpolio ones in another respect, in that the rates of rise and relaxation of tetanic force were reduced, compared with those of control subjects. Again, no significant differences from controls were found in ^{31}P metabolites or in pH, using NMR spectroscopy; this was true before, during and after fatiguing exercise.

Surprisingly, in view of the pathophysiology of ALS, little evidence of impaired motoneuron activation in fatigue was found when MVCs and tetanic forces were compared. However, the NMR spectroscopy results suggested that there may have been some impairment, since an exercise-induced increase in muscle water could be demonstrated in controls but not in patients. Although the M-wave was maintained during tetanic stimulation, other investigators have noted decrements in a proportion of ALS patients (e.g., Denys & Norris, 1979); further declines in amplitude can sometimes be seen in the potentials of single motor units during voluntary contractions (McComas, unpublished observations).

Multiple Sclerosis Patients

Multiple sclerosis (MS) is a chronic neurologic disorder with demyelination often involving the long motor and sensory pathways of the central nervous system. Patients with MS who have motor control disturbances have both reduced firing rate modulation and impaired recruitment of motor units (Rice et al., 1992). In a recent San Francisco study it was hypothesized that fatigue, which is the most common complaint in patients with MS, was primarily central, partly as a result of the effort involved in activating and recruiting motor units via dysfunctional corticospinal pathways and partly because of the *constitutional impact* of the disease. The experimental approach was therefore to bypass the central nervous system by employing tetanic, rather than voluntary, contractions. The ankle dorsiflexors were the muscles chosen and the stimulation protocol was 50 Hz for 240 ms every 3 s, continued for 9 min; 29 patients with MS were examined.

From Fig. 3 it can be seen that fatigue was considerably greater in the MS patients than in controls, and that recovery of force was less complete after 15 min of rest. An example of the rapid reduction in force is given in Fig. 5 (*right, middle*), taken from another investigation in which intermittent tetanic stimulation of the ankle dorsiflexor muscles was employed to test fatigability in MS (Lenman et al., 1989). These results were not associated with excitation failure, as reflected in the M-waves of the San Francisco patients, and must therefore have been due to changes in the muscle fibers. This conclusion is supported by the results of NMR spectroscopy in a subgroup of patients, who showed greater depression of PCr, a larger rise in P_i and a more marked fall in pH, compared to control subjects (Fig. 4). However, less obvious changes were observed in a subsequent study, in which a delay was found in the resynthesis of PCr during the recovery period after fatiguing stimulation (Kent-Braun et al., 1994).

The most likely explanation for the altered physiological and metabolic properties of the muscle fibers is that they stem from the effects of disuse or altered patterns of use, a conclusion which is supported by the finding of similar fiber abnormalaties in patients with spastic paraparesis from causes other than MS (Miller et al., 1990). In this context, the results of the study by Lenman and colleagues (1989) are pertinent; they found very significant

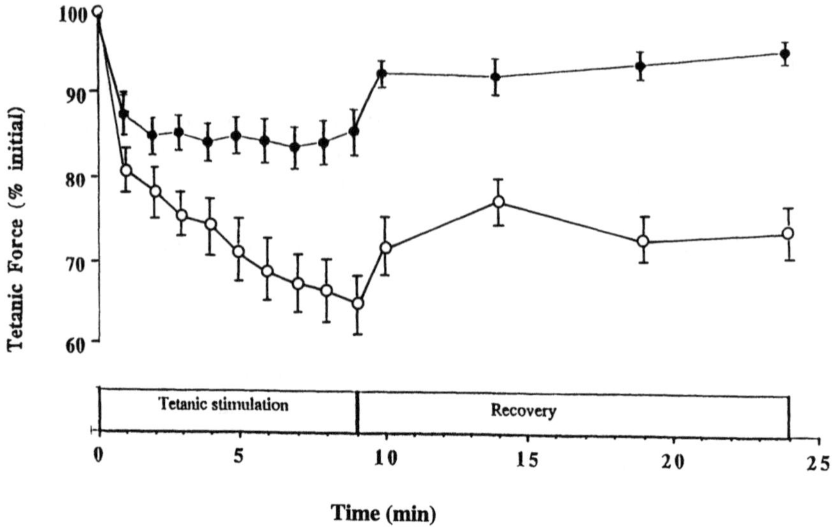

Figure 3. Greater fatigability of 29 MS patients (open circles) compared with control subjects (filled circles) during intermittent tetanic stimulation of ankle dorsiflexors. Values expressed as means ± SEMs. Note also the delayed and incomplete recovery in the patients.

reductions in force during tetanic stimulation in 16 patients with paraplegia resulting from spinal cord injury. An example from the study is included in Fig. 5.

Chronic Fatigue Syndromes

Numerous reports exist of syndromes which include chronic fatigue as a prominent symptom. It may follow a viral or other illness. However, a strict etiological sequence which has stood the test of time and multidisciplinary investigation has yet to be devised for the fatigue or the subjectively reduced endurance. Clinically, the patients may describe disabling subjective fatigue and compared with those patients with definite reasons for reduced performance (e.g., polio, or polymyositis), their symptoms often seem disproportionately severe relative to their usual daily activities. In particular, subjective recovery from simple daily activities is markedly prolonged. Not surprisingly, some have been recently exposed to viral infections, some have mild depressive symptoms, some have jobs with repetitive manual tasks and others have generalized pain disorders (such as fibromyalgia).

Patients with presumed post-viral fatigue syndromes have no diagnostic features on routine clinical EMG and nerve conduction studies. No specific findings occur in muscle biopsies. However, the highly sensitive technique of SFEMG reveals abnormal jitter in a proportion of patients (e.g., Jamal & Hansen, 1985; 1989; Connolly et al., 1993). The limitation of this finding is that it is non-specific, only involves the lowest threshold motor units, is not usually measured after objective fatigue, and is not accompanied by significant blocking. Blocking is mandatory if any peripheral force reduction is to be ascribed to defective neuromuscular transmission. Slightly increased fiber density has been reported in a subgroup of those presenting with unexplained fatigue in whom myalgia and muscle tenderness are major features (Connolly et al., 1993). An initial study suggesting an abnormality of muscle metabolism based on NMR spectroscopy (Arnold et al., 1984) has not been confirmed (Kent-Braun et al., 1993; Barnes et al., 1993). Immunological abnormalities (including reduced T cell subsets and impaired delayed hypersensitivity), together

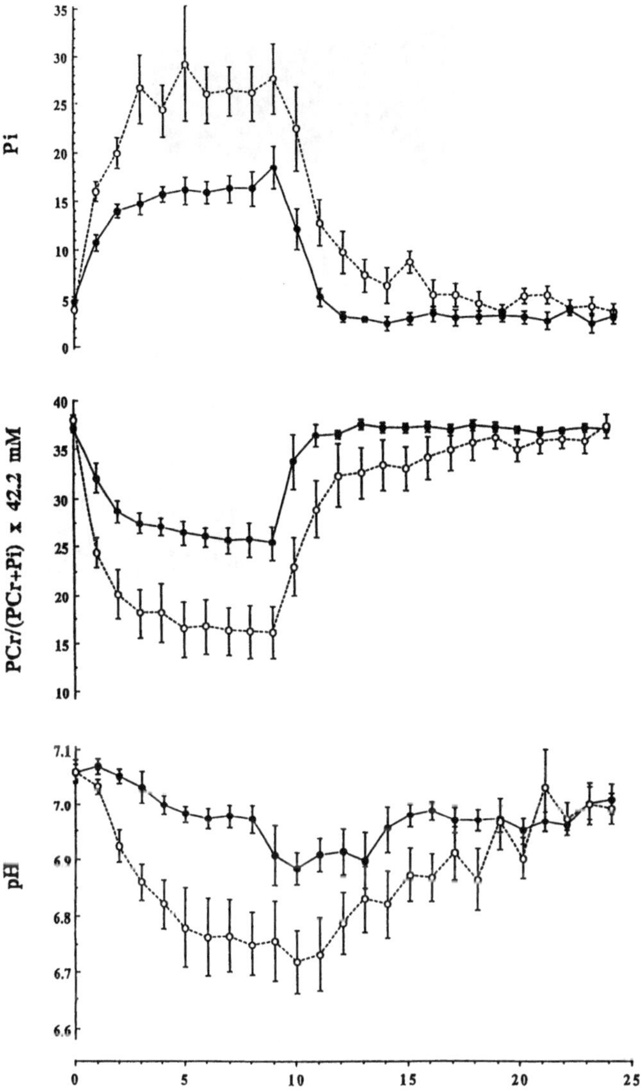

Figure 4. Results of ^{31}P NMR spectroscopy in MS patients during intermittent tetanic stimulation of ankle dorsiflexors. Values are means ± SEMs. Note that all three metabolites studied (P_i, PCr, and H^+) show significantly greater changes in patients (open circles) than in controls (filled circles).

with persistence of enteroviral RNA have been reported (Yousef et al., 1988), but not crucially linked to impaired peripheral muscle performance.

Several studies of small sample size have assessed muscle performance in these patients and relatively normal results have been obtained. Lloyd and colleagues (1988) found normal strength, maximal isometric endurance and recovery for the elbow flexors in 20 patients. In an extended test using interpolated stimulation and 45 min of submaximal exercise, voluntary activation to more than 95% occurred in all patients and the decline in maximal voluntary strength, voluntary drive, and the twitch force was not excessive.

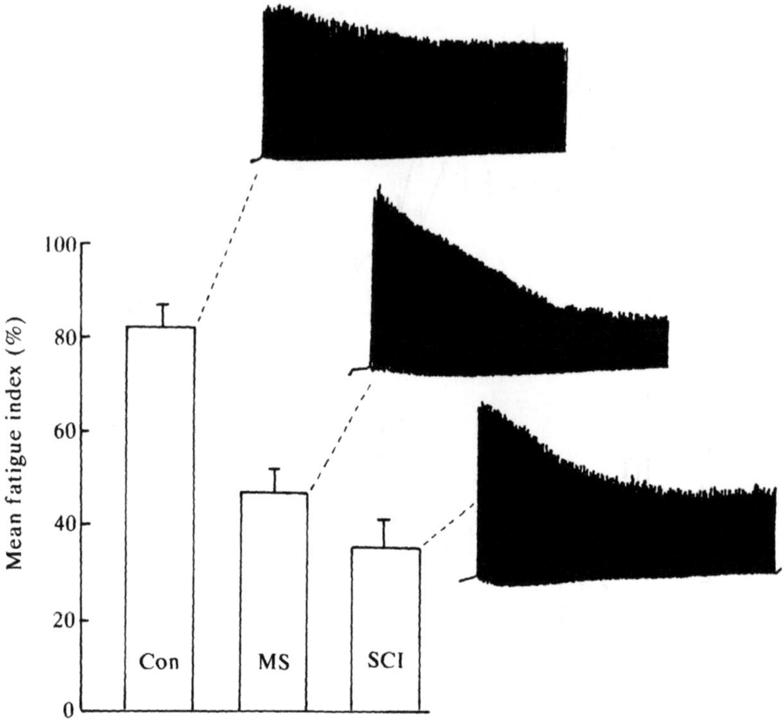

Figure 5. *Left*, mean fatigue indices (+ SEMs) of healthy subjects (*Con*), and of patients with MS (*MS*) or spinal cord injury (*SCI*). Values derived from intermittent tetanic stimulation of ankle dorsiflexors. *Right*, examples of typical responses from the three groups of subjects; the tetanic trains lasted a total of 3 min and different amplifications were employed, since the initial force was considerably greater for the control subject than for the two patients (reproduce from Lenman et al., 1989, by permission of John Wiley & Sons, Inc.).

Interestingly, the subjective effort measured on a Borg scale during the test did not grow abnormally during exercise in the patients (Lloyd et al., 1991; see also Gibson et al., 1993). Almost all patients in the two studies described above had documented immunological abnormalities. Quadriceps strength was normal in 11 patients with post-viral fatigue (mostly following infectious mononucleosis; Rutherford & White, 1991; see also Stokes et al., 1988) and endurance over 20 min was not impaired. Two patients had reduced ability to drive the muscles at rest. Performance in cycle ergometry was not impaired (Gibson et al., 1993).

Kent-Braun and colleagues have studied the ankle dorsiflexors in seven patients with chronic fatigue. The patients had lower maximal voluntary and twitch forces than controls (by about 30%), although these changes were not statistically significant, and nor were the patients' changes in pH and PCr with 25 min of submaximal isometric exercise excessive (see also Barnes et al., 1993). However, in this study, voluntary activation failure was present in the patients prior to exercise and had increased abnormally at the end of the exercise. This study highlights the difficulty in interpretation of voluntary endurance studies when voluntary activation is initially low. The level of exercise is overestimated for the patients (as %MVC) because their true MVC was underestimated. Unfortunately, it is not simple to correct for this, but one implication is that the sensitivity for detection of a peripheral abnormality is reduced. The discrepancy between this result and that of Lloyd and colleagues is likely to reflect patient selection because the twitch interpolation technique is likely to be sensitive in both studies (see Gandevia et al., Chapter 20). Another recent study using NMR and dynamic plantarflexion contractions found the usual peak and pattern of changes in PCr, P_i and pH in the patient group, but the changes occurred much more rapidly, suggesting accelerated glycolysis, relative to oxidative metabolism, in the patients (Wong et al., 1993).

In summary, biochemical, functional and electrophysiological indices of muscle performance are abnormal in comparatively few individuals with prominent subjective fatigue but no supporting pathology elsewhere. The *fatigue* in the so-called chronic fatigue syndrome is more a result of a misperception of what is a normal level of fatigue than a manifestation of abnormal muscle physiology at the level of the muscle or motor cortex and motoneurons responsible for voluntary muscle activation. Measurements of actual performance incorporating some type of electrical stimulation to assess the adequacy of voluntary drive are essential for accurate patient evaluation. Future studies must provide an adequate description of patient groups so that abnormal findings can be investigated by others. Second, given the difficulty in definition of anything resembling a homogeneous group, study groups must be sufficiently large so that some reliance can be placed on either positive or negative results. Third, the control groups must be appropriate for the patient group. This is easier to recommend than to achieve, but is critical when small changes consistent with decreased muscle conditioning are found.

FATIGUE STUDIES IN PERIPHERAL NERVOUS SYSTEM DISORDERS

Although, as already observed, fatigue has been rather poorly investigated, even in the research laboratory, there is a clinical impression that it may be a feature of almost any peripheral nervous system disorder. Rather than survey each disorder in turn, this review will divide the disorders into four classes and will look for general pathophysiological principles within each (Fig. 6).

Denervating Disorders

If there is a loss of excitable muscle mass as a result of denervation, fatigue will result for the reasons given in the introduction. However, there are four additional considerations which are applicable to all denervating disorders regardless of etiology.

First, in disorders of the motoneuron soma (polio, ALS, spinal muscular atrophy) or of its axon (peripheral neuropathy) there will be a strong tendency on the part of the surviving neurons to develop axonal sprouts and to innervate the muscle fibers which have been relinquished by the degenerating motoneurons. This process of collateral reinnervation is remarkably efficient and may be adequate to sustain muscle bulk and twitch tension until 80% of the motoneuronal population is lost (McComas et al., 1971). In the early stages of reinnervation, however, impaired impulse conduction in the new axon sprouts and dysfunction in their endings may be responsible for increased neuromuscular *jitter*, as revealed by SFEMG, and the latter phenomenon is probably also present in the early stages of a dying-back neuropathy. Impaired synaptic function in neuropathies can also be determined by monitoring M-wave responses to repetitive nerve stimulation, or even by assessing individual motor unit potentials during voluntary contraction. (e.g., McComas et al., 1971).

A second consideration, applicable to all chronic neuropathies, concerns the proportions of fast-twitch and slow-twitch motor units which will remain in the affected muscle, and which will have increased their sizes by collateral reinnervation. According to Rafuse and colleagues (1992), all motor units increase proportionately in size so that those which were previously the largest will adopt the greatest numbers of denervated muscle fibers. If

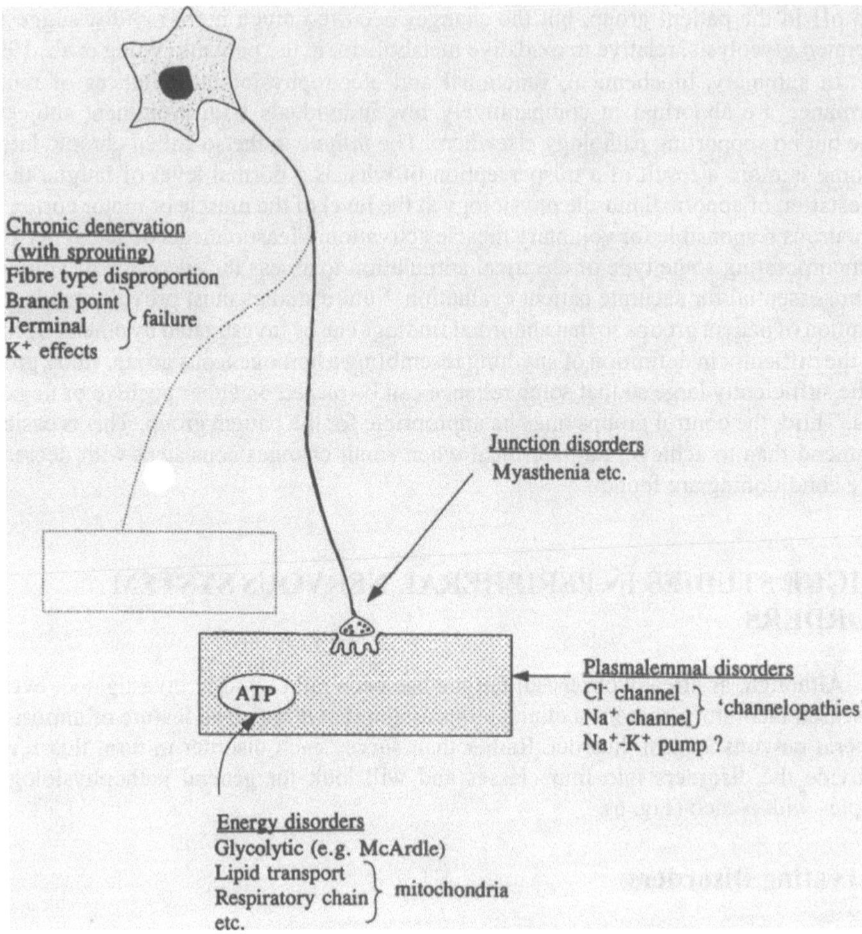

Figure 6. Summary of peripheral neuromuscular disorders capable of inducing fatigue, arranged in four categories.

the loss of fast-twitch and slow-twitch motor units is random, the expansionist proclivity of the fast-twitch units may result in normal proportions of fast-twitch and slow-twitch muscle fibers, albeit with fiber type grouping. In some cases, by chance, the proportion of fast-twitch fatigable muscle fibers may be distinctly higher or lower, and hence patients would be expected to demonstrate greater or lesser susceptibility to peripheral fatigue respectively. The same considerations might be expected to apply to those muscle disorders in which there was preferential involvement of one or other main fiber types. However, in congenital type I fiber predominance any difference in fatigability is over-ridden by significant weakness (Linssen et al., 1991).

A third consideration arises out of recent awareness of the importance of Na^+, K^+-pumping in offsetting K^+ effects during exercise (see Sjøgaard & McComas, Chapter 4). In a normal muscle the fibers of one motor unit are intermingled with those of 20 or so others and, hence, during submaximal contractions, K^+, released by the contracting fibers, can diffuse into the interstitial spaces surrounding the quiescent fibers. Further, the latter fibers can reduce the interstitial K^+ concentration by increasing Na^+, K^+ exchange through enhanced Na^+, K^+-pumping (Kuiack & McComas, 1992). In chronic denervation, however, the juxtaposition of active and inactive fibers is disrupted due to the presence of fiber type grouping and it seems likely that decreased excitability of muscle fibers, due to K^+-induced depolarization, may result.

Finally, if the denervating disorder is one that affects sensory neurons and fibers, as well as motor ones, the diminished impulse traffic from type Ia and II endings would be expected to reduce oliogosynaptic excitation of motoneurons during voluntary contractions, contributing to central fatigue. Similarly, in a severe neuropathy, the lack of afferent feedback would be anticipated to interfere with the sense of effort and with the planning, amplification and transmission of the motor command.

Neuromuscular Junction Disorders

The archetypal neuromuscular disorder is myasthenia gravis, in which fatigability, especially of the external ocular muscles, is the cardinal symptom. It has been known for some time that the disorder is due to an immunological attack on the acetylcholine (ACh) receptors. However, it is still not clear why fatigability should become worse as the day progresses, especially since there is no evidence for an associated deficiency of ACh in the motor nerve terminals (Ito et al., 1976). The diagnosis of myasthenia gravis may be confirmed in a number of ways, of which the most sensitive is SFEMG (Ekstedt & Stålberg, 1967), particularly as applied to muscles of the face. Estimation of ACh receptor antibodies in the serum is slightly less sensitive, followed by repetitive nerve stimulation. The ability of anticholinesterase medication to relieve weakness and fatigability may be dramatic, especially in the Tensilon test. The improvement in muscle function is due to the ability of the diminished ACh receptors to have multiple *hits* from ACh molecules which have escaped hydrolysis; thus the endplate potential is prolonged and may better reach threshold or produce multiple firing. In the Lambert-Eaton myasthenic syndrome, although there is severe weakness, the problem of fatigue is reversed inasmuch as performance improves at the task progresses. This condition resembles myasthenia gravis in being autoimmune in origin, with antibodies are directed against the Ca^{2+} channels of the motor nerve terminal (Vincent et al., 1989).

Disorders of the Muscle Plasmalemma

In the hereditary myotonic disorders, myotonia congenita and myotonic dystrophy, patients may have an exceptionally fast form of fatigue which appears within seconds of voluntary or stimulated contractions and can reduce force to less than half the initial value. As the contractions continue, however, the paresis gradually wears off (*warm-up* phenomenon). In both conditions, recordings of EMG activity or of M-waves evoked by nerve stimulation, show that the problem is one of impaired plasmalemmal excitability, due to K^+-mediated depolarization of the contracting fibers. In myotonia congenita the increased sensitivity is known to result from the low Cl^- conductance of the muscle plasmalemma as a consequence of the expression of abnormal Cl^- channel genotypes (Koch et al., 1992). The nature of the problem in myotonic dystrophy is more obscure, since no consistent channel abnormalities have been found so far. Fenton and colleagues (1991) have suggested, in keeping with various lines of investigation, that defective Na^+, K^+-pumping may be responsible.

In hyperkalemic familial periodic paralysis and the closely related paramyotonia congenita, fatigue is also due to diminished plasmalemmal excitability but it is rather less rapid than in myotonia, running a time-course measured in minutes rather than seconds. Further, the fatigue tends to become worse after the voluntary contractions cease, and the

M-waves may decline to as little as one quarter of the initial value within 30 min (McManis et al., 1986). In these conditions the genetic abnormality is one that affects the Na^+ channels (Fontaine et al., 1990).

Energy Disorders

In the contracting muscle fiber a renewed supply of ATP is essential, not only for the contractile process itself but also for the vigorous ion pumping which takes place across the membranes of the muscle fiber surface, T-tubules and sarcoplasmic reticulum.

The first report of an hereditary metabolic disorder characterized by rapid fatigue was that of McArdle (1951). The absence of a rise in blood lactate concentration during exercise led McArdle to postulate that muscle glycogen could not be broken down, a prediction confirmed 8 years later, when a deficiency of myophosphorylase was discovered (Mommaerts et al., 1959). A feature of McArdle's disease is severe muscle cramping on exercise. The absence of associated EMG activity suggests that the muscle fibers enter a rigor state, as would be expected if ATP was unavailable to detach the myosin cross-bridges from the actin filaments. There is also a marked decline in the amplitude of the M-wave (Dyken et al., 1967; their Fig. 7; also Cooper et al., 1989), a finding which could be attributed to diminished Na^+, K^+-pumping secondary to ATP deficiency. However, both biopsy and NMR spectroscopy studies have shown, surprisingly, that ATP concentration is not diminished in the muscle fibers (Wiles et al., 1981; Ross et al., 1981). Since there is no obvious reason for any inhibition of ATP hydrolysis, the biochemical basis for the physiological abnormalities remains enigmatic. In this context, the suggestion that there may be a compartmental reduction in ATP within the fiber, which does not reveal itself in whole muscle studies, seems rather unlikely. Clearly, more can be learned about the biochemical basis of muscle fatigue from further studies of this disorder.

After McArdle's seminal contribution, reports of other types of hereditary glycolytic disorders followed. Like myophosphorylase deficiency, these conditions are rare and most are associated with fatigue. A different type of metabolic disorder, also characterized by

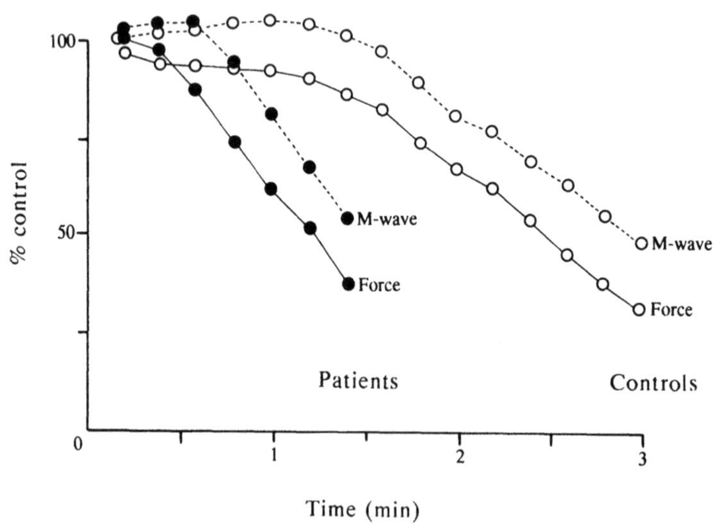

Figure 7. Fatigue studies in patients with McArdle's disease (filled circles) and control subjects (open circles). Mean values are shown without error bars (for clarity). Note that there is not only a dramatic loss of force, but that muscle excitability is also affected (adapted from Cooper et al., 1989).

excessive fatigability, is carnitine deficiency. As carnitine is required for the transport of fatty acids into mitochondria, there is an inability to use lipid as a fuel for the later stages of exercise, after the muscle glycogen has been consumed.

A third class of metabolic disorders comprises those in which there are defects in the mitochondrial respiratory chain or in which oxidative metabolism is dissociated from ATP production, as in Luft's disease (Luft et al., 1962). In the latter condition, the energy from oxidative metabolism is released as heat. Within this class of disorders mention should be made of cytochrome b deficiency, in view of the fatigue studies which were undertaken when this condition was first recognized (Morgan-Hughes et al., 1977). Mitochondrial dysfunction appears to be more frequent in old age, as the organelle population declines within the muscle fibers; it is probable that multiple enzyme defects result. A mitochondrial etiology should, however, be suspected in any patient with undue fatigability and in whom no other cause can be demonstrated (cf. Van Ekeren et al., 1991).

At the present time, it appears that different genetic abnormalities may affect almost all the enzymes involved in glycogenolysis, lipid metabolism, the citric acid cycle and the cytochrome chain, and that peripheral fatigue, with or without muscle pathology, is the inevitable consequence.

CONCLUSIONS

It is evident from this review that, regardless of the nature of the primary pathophysiological deficit, fatigue may be a complex process, with important contributions from one or more secondary factors. In this respect, the findings of the various studies have verified and emphasized themes which had been anticipated in the introduction. One example of this complexity is the impaired voluntary activation of motoneurons which occurs in those postpolio patients with moderate disability, and which must be attributed to disuse. The effects of disuse and/or altered patterns of use are also evident in the muscle fibers of patients with central lesions, as in those patients with MS or spinal cord injury. In the MS patients, not only the force recordings but the results of NMR spectroscopy also, show striking abnormalities. So far as the primary defects are concerned, the most dramatic fatigue effects have been seen in those patients with peripheral neuromuscular disorders, such as McArdle's disease, periodic paralysis and myotonic dystrophy.

Sufficient is now known about the cellular and body system mechanisms in fatigue, to justify more frequent application of fatigue testing to patients. Such studies will certainly tell us more about physiological perturbations in disease but, of equal importance, they will provide a firm reference point for assessing the results of therapeutic interventions. In this context, it is clear that considerable functional improvement would be expected from appropriate training, to overcome the central and peripheral effects of disuse. Indeed, we consider that there is a need for carefully designed training studies in neurological patients with fatigue, in whom the different elements in the physiological chain of muscle contraction can be dissected out for analysis. It is to be hoped that the results of such trials will become regular features of future fatigue symposia.

ACKNOWLEDGMENTS

The authors wish to thank Dr. Christine Thomas for her helpful comments on the manuscript. The authors also express their gratitude to their respective technical and secretarial staffs. The authors' laboratories have been supported by: the Muscular Dystrophy Association and the Natural Sciences and Engineering Research Council (NSERC) of

Canada, and the De Groote Foundation, Canada (*A.J.McC.*); the Multiple Sclerosis Society and the Muscular Dystrophy Association (USA; *R.G.M.*); and the National Health & Medical Research Council of Australia, and the Asthma Foundation of New South Wales, Australia (*S.C.G.*). Attendance of two of the authors at the 1994 Bigland-Ritchie conference was supported by NSERC (*A.J.McC.*), the University of Miami (*S.C.G.* and *A.J.McC.*), and by the Muscular Dystrophy Association (USA; *S.C.G.*).

REFERENCES

Allen GM, Gandevia SC, Neering IR, Hickie I, Jones R & Middleton J (1994) Muscle performance, voluntary activation and perceived effort in normal subjects and patients with prior poliomyelitis. *Brain* **117**, 661-670.

Argov Z & Bank WI (1991). Phosphorus magnetic resonance spectroscopy (^{31}P MRS) in neuromuscular disorders. *Annals of Neurology* **30**, 90-97.

Arnold DL, Bore PJ, Radda GK, Styles P & Taylor D (1984). Excessive intracellular acidosis of skeletal muscles in exercise in a patient with a post viral exhaustion fatigue syndrome. *Lancet* **1(8391), 1367-1369.**

Barnes PRJ, Taylor DJ, Kemp GJ & Radda GK (1993). Skeletal muscle bioenergetics in the chronic fatigue syndrome. *Journal of Neurology Neurosurgery and Psychiatry* **56**, 679-683.

Bigland-Ritchie B, Cafarelli E & Vøllestad NK (1986). Fatigue of submaximal contractions. *Acta Physiologica Scandinavica* **128** (suppl 556), 137-148.

Bruno RL, Frick NM & Cohen J (1991). Polioencephalitis, stress, and the etiology of post-polio sequelae. *Orthopedics* **14**, 1269-1275.

Connolly S, Smith DG, Doyle D & Rowler CJ (1993). Chronic fatigue: electromyographic and neuropathological evaluation. *Journal of Neurology* **240**, 435-438.

Cooper RG, Stokes MJ & Edwards RHT (1989). Myofibrillar activation failure in McArdle's disease. *Journal of Neurological Sciences*. **93**, 1-10.

Denys EH & Norris FH (1979). Amyotrophic lateral sclerosis: impairment of neurologic transmission. *Archives of Neurology* **36**, 202-205.

Dolmage T & Cafarelli E (1991). Rate of fatigue during repeated submaximal contractions of human quadriceps muscle. *Canadian Journal of Physiology and Pharmacology* **69**, 1410-1415.

Duchateau J & Hainaut K (1987). Electrical and mechanical changes in immobilized human muscle. *Journal of Applied Physiology* **62**, 2168-2173.

Dyken ML, Smith DM & Peake R (1967). An electromyographic diagnostic screening test in McArdle's disease and a case report. *Neurology* **17**, 45-50.

Ekstedt J & Stålberg E (1967). Myasthenia gravis. Diagnostic aspects by a new electrophysiological method. *Opuscula Medica* **12**, 73-76.

Fenton J, Garner S & McComas AJ (1991). Abnormal M-wave responses during exercise in myotonic muscular dystrophy. A Na+, K+-pump defect? *Muscle & Nerve* **14**, 79-84.

Fontaine B, Khurana TS, Hoffman EP, Bruns GAP, Hains JL, Trofatter JA, Hanson MP, Rich J, McFarlane H, McKenna Yasek D, Romano D, Gusella JF & Brown RH (1990). Hyperkalemic periodic paralysis and the adult muscle sodium channel a-subunit gene. *Science* **250**, 1000-1002.

Fuglsang-Frederiksen A & Scheel U (1978). Transient decrease in number of motor units after immobilization in man. *Journal of Neurology Neurosurgery and Psychiatry* **41**, 924-929.

Gibson H, Carroll M, Clague JE & Edwards RH (1993). Exercise performance and fatiguability in patients with chronic fatigue syndrome. *Journal of Neurology Neurosurgery and Psychiatry* **56**, 993-998.

Gordon T & Stein RB (1982). Reorganization of motor unit properties in reinnervated muscles of the cat. *Journal of Neurophysiology* **48**, 1175-1190.

Ito Y, Miledi R, Molenaar PC, Vincent A, Polak RL, Van Gelder M & Davis JN (1976). Acetylcholine in human muscle. *Proceedings of the Royal Society of London B Biological Series* **192**, 475-480.

Jamal GA & Hansen S (1985). Electrophysiological studies in the postviral fatigue syndrome. *Journal of Neurology Neurosurgery and Psychiatry* **48**, 691-694.

Jamal GA & Hansen S (1989). Postviral fatigue syndrome: evidence for underlying organic disturbance in the muscle fibre. *European Neurology* **29**, 273-276.

Kent-Braun JA, Sharma KR, Miller RG & Weiner MW (1994). Post-exercise phosphocreatine resynthesis is slowed in multiple sclerosis. *Muscle & Nerve* **17**, 835-841.

Kent-Braun JA, Sharma KR, Weiner MW, Massie B & Miller RG (1993). Central basis of muscle fatigue in chronic fatigue syndrome. *Neurology* **43**, 125-131.

Koch MC, Steinmeyer K, Lorenz C, Ricker K, Wolf F, Otto M, Zoll B, Lehmann-Horn F, Grzechik KH & Jentsch TJ (1992). The skeletal muscle chloride channel in dominant and recessive human myotonia. *Science* **257**, 797-800.

Kuiack S & McComas AJ (1992). Transient hyperpolarization of non-contracting muscle fibres in anaesthetized rats. *Journal of Physiology (London)* **454**, 609-618.

Lenman AJR, Tulley F, Vrbová G, Dimitrijevic M & Towle JA (1989). Muscle fatigue in some neurological disorders. *Muscle & Nerve* **12**, 938-942.

Linssen WH, Stegeman DF, Joosten EM, Merks HJ, ter Laak HJ, Binkhorst RA, & Notermans SL (1991). Force and fatigue in human type I muscle fibres. A surface EMG study in patients with congenital myopathy and type I fibre predominance. *Brain* **114**, 2123-2132.

Lloyd AR, Gandevia SC & Hales JP (1991). Muscle performance, voluntary activation, twitch properties and perceived effort in normal subjects and patients with the chronic fatigue syndrome. *Brain* **114**, 85-98.

Lloyd AR, Hales JP & Gandevia SC (1988). Muscle strength, endurance and recovery in the post-infection fatigue syndrome. *Journal of Neurology Neurosurgery and Psychiatry* **51**, 1316-1322.

Luft R, Ikkos D, Palmieri G, Ernster L & Afzelius B (1962). A case of severe hypermetabolism of nonthyroid origin with a defect in the maintenance of mitochondrial control - a correlated clinical, biochemical and morphological study. *Journal of Clinical Investigation* **41**, 1776-1804.

Maier A, Crocket JL, Simpson DR, Saubert CW & Edgerton VR (1976). Properties of immobilized guinea pig hindlimb muscles. *American Journal of Physiology* **231**, 1520-1526.

McArdle B (1951). Myopathy due to a defect in muscle glycogen breakdown. *Clinical Science* **10**, 13-33.

McComas AJ, Kereshi S & Quinlan J (1983). A method for detecting functional weakness. *Journal of Neurology Neurosurgery and Psychiatry* **46**, 280-282.

McComas AJ, Sica REP, Campbell MJ & Upton ARM (1971). Functional compensation in partially denervated muscles. *Journal of Neurology Neurosurgery and Psychiatry* **34**, 453-460.

McManis PG, Lambert EH & Daube JR (1986). The exercise test in periodic paralysis. *Muscle & Nerve* **9**, 704-710.

Miller RG, Boska MD, Moussavi RS, Carson PJ & Weiner MW (1988). ^{31}P nuclear magnetic resonance studies of high energy phosphates and pH in human muscle fatigue. *Journal of Clinical Investigation* **81**, 1190-1196.

Miller RG, Giannini D, Milner-Brown HS, Layzer RB, Koretsky AP, Hooper D, & Weiner MW (1987). Effects of fatiguing exercise on high-energy phosphates, force and EMG: evidence for three phases of recovery. *Muscle & Nerve* **10**, 810-821.

Miller RG, Green AT, Moussavi RS, Carson PJ & Weiner MW (1990). Excessive muscle fatigue in patients with spastic paraparesis. *Neurology* **40**, 1271-1274.

Mommaerts WFHM, Illingworth B, Pearson CM, Guillorg RJ & Seraydarian K (1959). A functional disorder of muscle associated with the absence of phorsphorylase. *Proceedings of the National Academy of Sciences of the United States of America* **45**, 791-797.

Morgan-Hughes JA, Darveniza P & Kahn SA (1977). A mitochondrial myopathy characterized by a deficiency in reducible cytochrome b. *Brain* **100**, 617-640.

Moussavi RS, Carson PJ, Boska MD, Weiner MW & Miller RG (1989). Nonmetabolic fatigue in exercising human muscle. *Neurology* **39**, 1222-1226.

Nicklin J, Karni Y & Wiles CM (1987). Shoulder abduction fatigability. *Journal of Neurology Neurosurgery and Psychiatry* **50**, 423-427.

Petit J & Gioux M (1993). Properties of motor units after immobilization of cat peroneus longus muscles. *Journal of Applied Physiology* **74**, 1131-1139.

Rafuse VF, Gordon T & Oroczo R (1992). Proportional enlargement of motor units after partial denervation of cat triceps surae muscles. *Journal of Neurophysiology.* **68**, 1261-1276

Rice CL, Volmer TL & Bigland-Ritchie B (1992). Neuromuscular responses of patients with multiple sclerosis. *Muscle & Nerve* **15**, 1123-1132.

Robinson GA, Enoka RM & Stuart DG (1991). Immobilization-induced changes in motor unit force and fatigability in the cat. *Muscle & Nerve* **14**, 560-573.

Ross BD, Radda GK, Gadian DG, Rocker G, Esiri M & Falconer-Smith J (1981). Examination of a case of suspected McArdle's syndrome by ^{31}P nuclear magnetic resonance. *New England Journal of Medicine* **304**, 1338-1343.

Rutherford OM & White PD (1991). Human quadriceps strength and fatiguability in patients with post-viral fatigue. *Journal of Neurology Neurosurgery and Psychiatry* **54**, 961-964.

Sanjak M, Paulson D, Sufit R, Reddan W, Beaulieu D, Erickson L, Shug A & Brooks BR (1987). Physiologic and metabolic response to progressive and prolonged exercise in amyotrophic lateral sclerosis. *Neurology* **37**, 1217-1220.

Sharma KR, Kent-Braun J, Mynhier MA, Weiner MW & Miller RG (1994). Excessive muscular fatigue in the postpoliomyelitis syndrome. *Neurology* **44**, 642-646.

Stokes MJ, Cooper RG & Edwards RHT (1988). Normal muscle strength and fatiguability in patients with effort syndromes. *British Medical Journal* **297**, 1014-1017.

Trojan DA, Gendron D & Cashman NR (1993). Anticholinesterase-responsive neuromuscular junction transmission defects in post-poliomyelitis. *Journal of Neurological Sciences* **114**, 170-177.

Van Ekeren GJ, Cornelissen EAM, Stadhouders AM & Sengers RCA. (1991). Increased volume density of peripheral mitochondrion in skeletal muscle of children with exercise intolerance. *European Journal of Pediatrics* **150**, 744-750.

Vincent A, Lang B & Newsom-Davis J (1989). Autoimmunity to the voltage-gated calcium channel underlies the Lambert-Eaton myasthenic syndrome, a paraneoplastic disorder. *Trends in Neuroscience* **12**, 496-502.

Westgaard R & Lømo T (1988). Control of contractile properties within adaptive ranges by patterns of impulse activity in the rat. *Journal of Neuroscience* **8**, 4415-4426.

Wiles CM, Jones DA & Edwards RHT (1981). Fatigue in human metabolic myopathy. *Ciba Foundation Symposium* **82**, 264-282.

Wong R, Lopaschuk F, Zhu G, Walker D, Catellier D, Burton D, Teo K, Collins-Nakai R & Montague T (1993). Skeletal muscle metabolism in the chronic fatigue syndrome. *Chest* **102**, 1716-1722.

Yousef GE, Bell EJ, Mann GF, Murugesan V, Smith DG & McCartney RA (1988). Chronic enterovirus infection in patients with post viral fatigue syndrome. *Lancet* **1(8578)**, 146-150.

Zochodne DW, Thompson RT, Driedger AA, Strong MJ, Gravelle D & Bolton CF (1988). Metabolic changes in human muscle denervation: topical ^{31}P NMR spectroscopy studies. *Magnetic Resonance in Medicine* **7**, 373-383.

EPILOGUE

NEUROBIOLOGY OF MUSCLE FATIGUE

Advances and Issues

S. C. Gandevia,[1] R. M. Enoka,[2] A. J. McComas,[3] D. G. Stuart,[4] and C. K. Thomas[5]

[1] Prince of Wales Medical Research Institute
Sydney, Australia
[2] Department of Biomedical Engineering
Cleveland Clinic Foundation
Cleveland, Ohio 44195
[3] Department of Biomedical Sciences, McMaster University
Hamilton, Ontario, L8N 3Z5, Canada
[4] Department of Physiology
University of Arizona College of Medicine
Tucson, Arizona 85724
[5] The Miami Project to Cure Paralysis
University of Miami School of Medicine
Miami, Florida 33136

When you cannot measure it,
When you cannot express it in numbers -
You have scarcely, in your thoughts,
Advanced to the stage of Science, whatever the matter may be.
—Lord Kelvin

In reading this volume one should ask what is new and true, and what will lead to new approaches and insight. Clearly, as the topic *muscle fatigue* is increasingly studied by muscle physiologists, neuroscientists and clinicians, it becomes more difficult to summarize the state of knowledge, even in a volume with 36 chapters. Since a major symposium on the physiology of muscle fatigue in London in 1980 (Porter & Whelan, 1981), much progress has been made in understanding both the intracellular, muscle-fiber mechanisms and, to some extent, mechanisms of CNS control as they relate to muscle fatigue. Some of this progress has been summarized at recent symposia in Paris in 1990 (Atlan et al., 1991), and Amsterdam in 1992 (Sargeant & Kernell, 1993). Despite this, we are far from having an equation that will predict the degree of fatigue under a variety of physiological circumstances, although some of the boundary conditions are becoming well established.

In this epilogue, we have addressed some points that became evident at the 1994 Bigland-Ritchie Symposium in Miami at which meeting the majority of the chapters were

Fatigue, Edited by Simon C. Gandevia et al.
Plenum Press, New York, 1995

partially presented, and other issues that arose when editing the volume and comparing it with the 1980, 1990, and 1992 Symposia. These issues are examined in the conventional framework, moving from the periphery to the central nervous system (CNS) while comments on the motoneuron are considered in a separate section termed, "*at the motoneuron interface*". Additionally, there is some consideration of fatigue in adapted states, and pathophysiological conditions. From this perspective, it is now widely accepted that sustained physical activity causes a reduction in the maximal force a muscle can exert due to the concurrent impairments of several (more or less) simultaneous processes which begin upstream of the motor cortex and go down to the myofibril.

As it is now possible to measure many factors that change as fatigue develops, the challenge is to determine if, when, and how much they are contributing to the reduction in force. This is a difficult task. For example, some biochemical changes, such as an increase in ADP concentration, affect force and the maximal velocity of shortening differentially; ADP increases maximal force but depresses maximal shortening velocity. The reduced speed of shortening combined with slowed relaxation diminish performance, and this effect can be just as disabling as the loss of force. Doubtless it is theoretically possible to construct a situation in which one, say, intracellular process is the crucial limit to force, but in typical exercise it is inevitable that more than one process is involved. One of the major challenges for the field of muscle fatigue will be to quantitate the mounting qualitative evidence (Section VII; see also Enoka & Stuart, 1992) that the mechanisms of fatigue and recovery, from the molecular to the whole organism level, are task dependent. This theme is prominent throughout the volume, from the introductory consideration of single muscle fibers to the concluding chapters on neuromuscular adaptations. In parallel with this emphasis on task-dependency, there is need to continue work on the individual molecular processes that generate force. Without such information, the fundamental mechanisms will remain unknown.

IN THE PERIPHERY

Muscle produces force and movement through the interaction of myosin crossbridges with actin filaments, and it is through these crossbridges that fatigue must be manifest. Expressed most simply, fatigue is due to a reduction in the number of interactions, or in the forces developed by individual crossbridges. In this context, it is remarkable that, with an *in vitro* laser trap system, it is now possible to measure the movement in a single crossbridge cycle, or the tension generated by a single crossbridge when movement is prevented (Finer et al., 1994). To further our understanding of fatigue mechanisms, it should be a relatively easy step to study the effects on the isolated crossbridge of alterations in the concentrations of ATP, ADP, P_i, Ca^{2+} and H^+, all of which alter in fatigue. These measurements would include not only the amplitudes of movement or force, but also the speeds with which these are generated. Equally relevant to an understanding of fatigue are three-dimensional models of the myosin head, as deduced by X-ray diffraction and computer analysis (Rayment et al., 1993).

Although future fatigue symposia will include observations at the level of the individual myosin molecule or crossbridge, for the present summary we must begin with the single fiber preparation. In Chapter 1, Edman has shown, by comparisons of electrical stimulation and K^+ contracture, that *moderate* fatigue appears to result from the reduced forces developed by individual crossbridges and that acidification of the fiber interior is likely to be one of the factors involved. Another of the intracellular candidates for force reduction during fatigue is the excitation-contraction coupling mechanism, a topic discussed in detail at the molecular level by Stephenson and colleagues in Chapter 2. Critical observations here are the measurements of Ca^{2+} flux within single muscle fibers by means of fluorescent dyes. First achieved by Ashley and Ridgway (1970) in the large fibers of the barnacle, the methodology was subsequently extended to amphibian muscle by Blinks and colleagues (1978) and, finally, to the smaller and more vulnerable fibers of the mouse by Allen, Westerblad and Lannergren (Chapter 3). The findings from these last studies agree with those of Edman in showing that early fatigue is not due to any failure of excitation-contraction coupling. As fatigue progresses, however, there is diminished Ca^{2+} release from the SR, as well as reduced sensitivity of the myofibrils to Ca^{2+}.

The term excitation-contraction coupling, while remaining conceptually simple, is being exposed as a cascade of events encompassing many processes, each modifiable in a multitude of interacting ways (Chapter 2). For example, the sarcoplasmic reticulum Ca^{2+} release channel seems differentially controlled by ATP (increasing channel opening) and Mg^{2+} (tonically inhibiting channel opening). Other biochemical factors, such as pH and inorganic phosphate are also able to modify the Ca^{2+} release channel.

Given the large number of possibilities, as described by Stephenson in Chapter 2, what is the cause of the decreased Ca^{2+} release? Here, there are major differences in viewpoint. On the one hand, Allen and colleagues find no evidence for failure of the inward spread of excitation in the T-tubules. On the other hand, Edman, using a wavy myofibrillar preparation originally developed by Gonzalez-Serratos (1971), reaches the opposite conclusion. If inward spread is indeed impaired, it seems likely that K^+-induced depolarization of the T-tubules is an important contributory factor, given the striking increases in K^+ concentration which have been observed, or deduced, in the interstial fluid of contracting muscle (Chapter 4). It is perhaps surprising, in view of the size of the impulse-mediated K^+ efflux and the narrowness of the interfiber spaces and T-tubules, that K^+ effects are not more prominent in fatigue. However, as reviewed by Sjøgaard and McComas, there are a number of cellular mechanisms, including enhanced electrogenic Na^+-K^+ pumping, which stave off K^+-induced paralysis. In addition to failure of inward excitation in fatigue, there are other factors including intracellular acidosis that will reduce Ca^{2+} release from the SR through the ryanodine channels.

Observations of the biochemical changes in muscle fibers are critical to an understanding of fatigue, and complement the findings of the various types of physiological experiments. The greatest advance in this field is NMR spectroscopy which enables the concentrations of high-energy phosphates, P_i and H^+ to be measured *in situ* in contracting human muscle. The technique of ^{31}P NMR has become an important part of the armamentarium of those examining the metabolic performance of whole human muscles (Chapters 12-14). Such studies have been important in confirming that only slight reductions in ATP concentrations occur in fatigue, at a time when phosphocreatine has all but disappeared and the H^+ concentration has markedly increased. Nevertheless, values obtained with NMR are averages, and ignore the fact that the different types of fiber will have been employed to greater or lesser extents in the fatiguing contractions and, because of their differing biochemistry, are likely to have considerable differences in metabolite concentration. Even in the same fiber, a value for ATP says little about the conditions likely to exist at such critical sites as the crossbridges, and the SR and surface membrane pumps.

Not only does NMR spectroscopy reveal the status of the high-energy phosphates, but it promises more with measurements of water and ionic fluxes across the cell membrane. Results with NMR appear to agree well with those obtained with muscle biopsies, although neither method reveals the metabolite concentrations at the level of the actomyosin crossbridges. This is unlikely to be relevant for ATP concentrations which probably do not drop below critical levels at the crossbridge under natural conditions, but may be important for other factors that modify crossbridge action. Furthermore, with NMR and usually with biopsy studies, the concentrations across the sampled tissue are averages so that the level of neural drive to the particular muscle or motor unit needs to be determined. At a molecular level, the way in which the intracellular changes in fatigue interact at the level of the crossbridge and actomyosin ATPases must be known. Of particular interest is what factors differentially impair crossbridge force as opposed to crossbridge cycling and how these change in eccentric exercise when the crossbridges sustain more force and do it more efficiently. Eccentric exercise produces excessive muscle damage perhaps because of inhomogeneity in the strength of individual sarcomeres and thus local sarcomere dynamics may generate fatigue, pain and altered muscle performance which take days to recover. In addition, unexpected evidence has emerged that the energy cost of contractions (assessed by oxygen consumption or heat output) remains constant during dynamic contractions but increases with isometric contractions.

At the risk of being overwhelmed by the complexity of molecular controls on muscle bioenergetics, Kushmerick (Chapter 14) attempts a simplification. He points out that human muscles are relatively homogeneous from a biochemical point of view compared with those of common experimental animals, and thus they can be regarded as consisting of a unimodal metabolic continuum. This view, of course, bolsters the global approach offered by NMR and reveals the likely way in which the products of the ATPases in muscle (ADP, Pi, H^+ and creatine) must ultimately regulate energy balance. Applicability of this view must be tempered by evidence that *activation* of crossbridges may sometimes be a critical impairment (Chapter 13). Furthermore, the concept of a unimodal metabolic continuum is at odds with the literature on muscle-fiber and motor-unit types. In this scheme, typing is accomplished by classifying muscle fibers and motor units on the basis of biochemical, morphological, and physiological properties. These classifications are not rigid distinctions based on absolute criteria but rather characterize the populations on the basis of relative values. These schemes have proven to be robust (Chapters 8-10, 16, 17, 24, 26, 27; Stuart et al., 1984; Rome, 1993; Stuart & Callister, 1993) and much remains to be done to assess the validity of the Kushmerick model as it applies to less constrained systems.

In addition to the widespread use of NMR spectroscopy in studies on muscle fatigue, there are a number of developments that promise new avenues for exploring human performance. Bertocci (Chapter 15) reviews the most promising of these new opportunities; these include ^{13}C magnetic resonance spectroscopy (MRS) to examine the regulation of substrate into the citric acid cycle, ^{23}Na and ^{39}K MRS to determine the distribution of sodium and potassium ions across the membrane, and ^{1}H protons combined with the mechanisms of nuclear spin relaxation to provide spatial information on the location and movement of water. There seems little doubt that the results obtained with these methods will advance our understanding of the mechanisms underlying muscle fatigue. For example, the measurement of longitudinal (T1) and transverse (T2) relaxation times with magnetic resonance imaging can provide spatial information, both static and dynamic spatial information on regions within a muscle or group of muscles that have been activated in the performance of a particular task. The information that can be obtained with these techniques may ultimately be as significant as that obtained with the development of the EMG.

In studies of human muscles *in vivo*, isolation of some of the force failure to excitation-contraction coupling is helpful but it says little about which processes are critically involved. An accepted hallmark of coupling of contraction to neural excitation is the selective impairment of force production with low compared with high rates of stimulation. This frequency-dependent alteration in force means that the extent of fatigue must depend on the exact range of frequencies used by motor units under natural conditions. Hence it will remain necessary to obtain information about the naturally-occurring discharge rates of motor units. A second reason for this approach is that factors favoring both fatigue and force potentiation may coexist (Chapter 9), so that the liability of the force-frequency relationship needs to be emphasized.

The potential for neuromuscular transmission failure to reduce force (Chapters 5-7) is undenied. However, the role of axonal factors, such as branch-point failure proximal to the presynaptic terminal in muscle fatigue during normal voluntary activities is not established. Although transmission across the neuromuscular junction is typically regarded to have a high safety factor, Fuglevand (Chapter 6) suggests that experiments have not adequately distinguished between a reduction in action potential amplitude and a depolarization-induced contraction (due to the extracellular accumulation of potassium) to determine if impairment of neuromuscular transmission contributes to the decline in force. Certainly, fast-twitch, fatigable motor units exhibit greater reductions in action potential amplitude with sustained activity. Neuromuscular propagation might become impaired in those neuromuscular diseases in which there is chronic continuous reinnervation of muscle fibers and the number of muscle fibers innervated by one motoneuron increases. Under these circumstances, there is both an increase in the number of branch points due to the enlarged motor unit territory and a reduction in the local *sink* provided by non-active fibers for a rising interstitial potassium concentration due to *grouping* of fibers from the same motor unit (see also Chapters 4-5).

One of the factors that seems to exacerbate impairment of neuromuscular propagation is ischemia (Chapter 7). Blood flow will be restricted if the intramuscular pressure is sufficiently high and hence the architecture of the muscle and its vasculature will be relevant. Muscles vary in the extent to which their blood supply is restricted as muscle force increases (Chapter 25). Here, thin, parallel-fibered muscles such as the diaphragm and some intrinsic muscles have an advantage. Intramuscular pressure is being used increasingly to estimate force in muscles during complex tasks such as walking, when EMG records may be hard to interpret. However, its role in assessment of the acute changes during fatigue is not yet established.

AT THE MOTONEURON INTERFACE

In the Sherringtonian sense, the motoneuron is the *final common* path and the last site at which any CNS (supraspinal, segmental, sensory feedback) commands to motoneurons can be modified. Conceding the powerful role of segmental interneurons in the control of motoneuron discharge (Chapter 17), the extent of the neuromodulation of discharge by factors other than synaptic input is remarkable (Hultborn & Kiehn, 1992; Binder et al., 1995). Furthermore, the motoneuron transduces the net input command or current into a discharge rate that is then translated into the force produced by a single motor unit (Chapters 8-9). These transformations are not linear, although when considered solely in terms of intrinsic properties and steady-state isometric contractions, the overall relation between input current to motoneurons in the brainstem/spinal cord and force output by the motor units is more linear than the component relationships (Binder et al., 1995). A challenge in the immediate future is to determine the extent to which these relationships are modified by synaptic input and other forms of neuromodulation. For example, motoneurons exhibit a bistable state in which discharge can be superimposed on a membrane potential that is relatively near or far from (depolarized) the resting level (Hultborn & Kiehn, 1992; Binder et al., 1995). The discharge rate is significantly different under these two conditions. If such bistability can occur during fatiguing contractions, it would confound interpretation of the relationship between stimulus current-firing frequency-force. Interestingly, such bistability has been inferred from the EMG firing patterns of single motor units in conscious experimental animals (Kiehn, 1991) but not yet from human data.

It is also necessary to emphasize that the transformation of motoneuron discharge frequency into motor unit force will also depend on the history of recent stimulation, the nature of the contraction (shortening *vs.* isometric *vs.* lengthening), the range of muscle lengths, the temperature within the muscle and the relative effects of potentiating and fatigue-inducing processes within the muscle fibers (Chapter 9). Despite the impressive progress over the last few years on the interactions between motoneurons and muscle fibers (Chapters 8 and 9), much remains to be learned about these factors during fatiguing contractions.

Since its intriguing discussion at the 1980 symposium (Porter & Whelan, 1981), several chapters in the present volume have addressed the substantial interest in the so-called *muscle wisdom* phenomenon. According to this scheme, motoneurons discharge at progressively lower rates during maximal voluntary contractions (Chapters 8, 9, 16-19, 26, 27). This produces a convenient matching of *speed* because the contraction and relaxation times of the motor units, particularly the fast-twitch ones, slows during maximal voluntary contractions. The term, muscle wisdom, is obviously a misnomer because if there is any wisdom in the appropriateness of the motoneuron firing rate, it certainly does not reside *in* the muscle. The decline in discharge rate during fatiguing contractions appears to be task dependent. For example, it is not apparent during submaximal contractions and may not be evident during eccentric exercise (Chapter 16). Also, the slowing of muscle relaxation and contraction, arguably a requirement for *muscle wisdom*, do not occur with all forms of muscle fatigue under even isometric conditions (Chapter 16).

Binder-Macleod (Chapter 16) emphasizes the need to distinguish between activation by voluntary command and the artificial wisdom of contractions elicited by imposed electrical stimulation. With imposed contractions, it is possible to capitalize on the catch-like property of striated muscle which involves an enhancement of muscle force due to the inclusion of a brief interpulse interval in the stimulus train. However, the extent to which catch-like effects are observed under natural circumstances remains an open issue. Nonetheless, muscle fatigue remains a major limitation of motor prostheses that use electrical stimulation to generate force. Further work on variable-frequency stimulus trains is necessary to advance the field of functional electrical stimulation.

The role of the fusimotor system in muscle fatigue is enigmatic. Originally, the system was conceived as automatically supplying excitation through the discharge of muscle spindle afferents during fatigue (Merton, 1953; Chapter 18). However, the limited recordings in conscious humans make this unlikely. In fact, the fusimotor-muscle spindle system appears to produce less reflex support to motoneuron output during fatigue. Ultimately, the effects of presynaptic inhibition and other segmental reflexes will need to be characterized before the full implications of the observations on muscle spindle behavior in human fatigue can be determined.

Finally, at the *interface* between the muscle fiber and the CNS, evolution appears to have introduced a substantial safety margin under most circumstances, particularly for axonal conduction and, to a lesser degree, for synaptic transmission across the neuromuscular junction. Consequently, the major adaptations to fatigue, disuse, and disease must be focused either in the muscle cell or in the segmental and supraspinal machinery.

CNS FACTORS

To what extent is the supraspinal and spinal circuitry concerned with the discharge of *single* motoneurons? Apart from insufficient neural circuitry for individual control of motoneurons, variable force-frequency relationships within and across motoneurons confound a rigid strategy that could be imposed by higher centers. These are far from trivial issues for the CNS, and ones which may well have been solved by the evolution of some basic strategies to organize a motoneuron pool, such as Henneman's size principle (Binder

& Mendell, 1991). It is clear that independent control of the discharge rate of each motoneuron is not a realistic strategy and this is probably why so many properties are organized systematically from the low- to high-threshold motor units (Henneman, 1991). Windhorst and Boorman (Chapter 17), describe the progress of the last four decades on CNS control based on intracellular recordings from largely passive (non-firing) motoneurons in deeply anesthetized, reduced *in vivo* (largely cat) preparations (see also Jankowska, 1992; McCrea, 1992). Within the past two decades, progress has also been made in understanding CNS control strategies by: 1) recording the reflex forces of several muscles simultaneously in less reduced, unanesthetized preparations (Nicholls, 1994); 2) measuring unitary firing patterns of single motor and sensory axons in freely moving animals (Prochazka, 1995); and 3) resorting to modeling and simulation (e.g., Loeb & Levine, 1989; Fuglevand et al., 1993; Binder et al., 1995).

In view of these developments, it is significant that Chapter 17 emphasizes how *little* is known about the segmental and suprasegmental factors that may modify motoneuron discharge rate during fatigue. This situation appears to be the consequence of several factors. *First*, there is a paucity of experiments on segmental mechanisms of fatigue but, also, the technical difficulty of studying the firing patterns of motoneurons and interneurons with in *in vivo* preparations during fatiguing contractions in which force is also measured. While there have been substantial advances on active firing properties using *in vitro* slice preparations (Binder et al., 1995) these beg the question of force development and its associated control. *Second*, there is the failure of investigators in this field to appreciate the full role of proprioceptive input in the control of muscle contractions and movements (Hasan & Stuart, 1988; Gandevia & Burke, 1992; Prochazka, 1995). *Third*, although the control of motoneuron discharge is influenced by many species of afferent feedback (Chapters 17-19), there is insufficient specificity in individual afferent interneuronal pathways to effect something like a precise decline in discharge rate during a fatiguing contraction.

Studies on CNS factors have underscored that *fatigue* encompasses more than simply the failure to maintain a particular level of force. When attempting to lift a weight that is half of one's maximum, fatigue can begin at the start of the task (not only when it is impossible to sustain the contraction) and the sensorium is soon alerted to the need to increase the motor command to drive the fatiguing muscles. As documented in Chapter 22, this increased command is manifest as the increased apparent heaviness of the load and the increased recruitment and discharge frequencies of motoneurons. Between the start of the lift and the detection of the need to increase the motor command, the altered sensory input will have modified the state of spinal and supraspinal circuits controlling the muscles. Such events will occur well before the task, which initially involved a submaximal contraction, cannot be sustained. For tasks involving submaximal forces, the increase in the motor command will occur before the force is reduced below the requisite amount. Most subjects appear to base their judgments of force during fatiguing isometric contractions, in part, on centrally generated signals (Chapter 22; Gandevia, 1995). More detailed examination of the underlying mechanisms may be possible with recently developed imaging technology, such as functional magnetic resonance imaging. This approach might well complement the information that can be gleaned from studies on the readiness potential (Chapter 21). These two techniques have the advantage of being able to examine the behavior of supraspinal centers in conscious human volunteers.

To what extent does the CNS act to prevent catastrophic changes in the muscle fibers? This question has attracted interest for over a century, although techniques required to examine it quantitatively were not devised until Merton's development of twitch interpolation (Merton, 1954; Gandevia et al., Chapter 20). Careful application of the technique has revealed first, that voluntary activation of muscles is usually less than maximal even in highly motivated subjects when a muscle is not fatigued, and second, that voluntary activation diminishes during exercise, i.e. central fatigue develops. A specific terminology to cover these findings is presented in Chapter 20. Major differences between muscles and tasks have been shown for the development of central fatigue (Chapter 20). Central fatigue, unlike many steps involved in peripheral fatigue, need not scale with the level and duration of force production. Indeed, they may be inversely related, with peripheral fatigue dominant in brief high-force tasks and central fatigue more obvious during more prolonged low-intensity exercise.

Transcranial magnetic stimulation of the motor cortex that is interpolated during maximal voluntary contractions produces twitch-like increases in force, particularly when central fatigue develops (Thomas, 1993; Gandevia et al., 1995). Hence, the output of the motor cortex and distal sites at the time of the stimulation is not maximal and thus limits force. This observation suggests that some of the mechanisms underlying the *reflex inhibition* of motoneuronal discharge during fatigue may involve supraspinal sites. Given that motor cortical output to the fatiguing muscle probably increases, these observations focus some attention on the causes for central fatigue that may reside *upstream* of the motor cortex. Here, the interaction with motivation becomes crucial. Much remains to be done to check the generality of these observations and investigate the supraspinal mechanisms. Already, some have speculated which central transmitters are involved (Chapter 23, Newsholme & Blomstrand). Of course, the further one moves from those areas within the CNS that directly modulate the firing of motoneurons, the more difficult it is to establish the cause rather than the effect of central fatigue.

NEUROMUSCULAR ADAPTATIONS

Fatigue is an acute response of the neuromuscular system to exercise. From a mechanistic perspective, it is instructive to determine the interactions between chronic adaptations and fatigue. Such an approach is adopted in Chapters 32-34 with an examination of some paradigms that increase use, others that reduce use, and some that involve remodeling of the neuromuscular system. Studies of the chronic application of electrical stimulation have demonstrated that the fatigability and strength of muscle appear to be controlled reciprocally (Chapter 32). Frequent stimulation improves fatigue resistance by increasing the mitochondrial content and oxidative-enzyme capacity of the activated muscle fibers. However, this regimen decreases muscle-fiber diameter and hence the maximal force that a muscle can exert; the retention of force capabilities depends on intermittent activation with high-frequency trains of stimuli. These contrasting effects of the amount and pattern of activity have also been observed with exercise training and with the models of compensatory hypertrophy (removal of synergist muscles). However, the adaptive capabilities are limited. Reinnervation studies (self- and cross-reinnervation) have indicated that reinnervated muscle fibers retain some of their original phenotype and that the motor unit properties, including fatigability, are determined by the combined effects of neuromuscular activity and muscle loading. Little is known about the factors that determine the adaptive limits of muscle fibers, motor units, and muscles.

The function of the neuromuscular system is impaired with increasing age, the ultimate chronic adaptation. Many of these changes are attributable to the loss of neurons and the subsequent reorganization of motor units (Chapter 34). For example, the decline in muscle mass and strength in elderly persons is largely due to the denervation of muscle fibers that occurs as motoneurons die. As in other paradigms that induce muscle atrophy and weakness (Chapter 32), however, the muscles of elderly subjects either exhibit an increased resistance to fatigue or, at the least, no effect of the chronic adaptation on fatigability (Chapter 34). This effect might be caused by three types of mechanisms: 1) an increased capillary density, concentration of mitochondria, or metabolic-enzyme activity due to muscle atrophy; 2) a reduced occlusion of blood flow because the absolute force exerted by the muscle, and hence the intramuscular pressure, is less for a task where force is normalized as a percentage of maximum; and 3) an increased proportion of slow-twitch motor units. However, absolute measures of fatigability, such as the maximal sustainable power, decrease with age, apparently due to reductions in specific force, optimum velocity for power, and energy delivery and conversion (Chapter 34).

Faulkner and Brooks (Chapter 34) hypothesize that contraction-induced injuries mediate such adaptations as muscle-fiber denervation with aging. Physical activity, especially when it involves eccentric contractions, can damage muscle and impair performance (Chapter 33). The strain experienced by active muscle fibers appears to cause damage that is perceived as soreness over the subsequent 2-3 days, followed by regeneration of the damaged fibers. After a bout of eccentric exercise, the magnitude of low-frequency fatigue is greater than after isometric or concentric contractions. Because motor unit discharge rates during voluntary contractions occur in this domain, eccentric contractions may also have a greater effect on the control of movements, as indicated by increased tremor and alterations of force perception after such activity (Chapter 33).

PATHOPHYSIOLOGY

As highlighted in Chapters 35 and 36, a range of diseases will produce excessive fatigue (and even no disease at all, merely a sedentary lifestyle) and the clinician's job is to determine the critical site or sites at which performance is impaired. Objective measurement of the peripheral and central components to fatigue is possible and has already been adopted by several laboratories using similar, but not identical, procedures. Given that fatigue has many task-dependent elements, the use of comparable testing procedures is important. Results are fairly reproducible and, although their specificity for a particular disease is low, this is outweighed by the practical information provided to the patient and the physician. Because of the variability in strength and endurance between individuals, longitudinal studies of peripheral and central fatigue are often helpful. Commonly, the patient presents with a story suggesting both peripheral and central components to their fatigue and weakness. As an example, patients with myasthenia gravis, with a frequency-dependent abnormality at the neuromuscular junction, may not drive their muscles fully, because, to do so, would result in excessive force loss. They, and others with disorders in which high motor unit firing rates lead to excessive fatigue, have developed a longer-term central strategy which has *taught* them to adopt low firing rates. In contrast, patients with muscular dystrophy show greater levels of voluntary activation, and less central fatigue than expected. This, too, is an appropriate CNS adaptation.

Clinically, absolute muscle strength should also be considered. After all, in many instances of disease, disability and disuse, small increments in strength may make large differences to the activities of daily life, such as the capacity to rise from a chair and to walk unaided. Thus, from a therapeutic viewpoint, much interest surrounds the attempts to unravel the underlying molecular signaling leading to adaptations within muscle fibers and their motoneurons.

As Bigland-Ritchie and her colleagues emphasize in Chapter 27, evolution has had a long time to sort out muscle fatigue. In a particular species, all aspects of muscle performance have adaptability within some boundaries. The critical limits differ between species, individuals and between muscles and even their constituent fibers. Ultimately, what drives the muscles to start and to stop contraction is the CNS and it also changes the behavior of reflexes and proprioceptors during fatigue - hence the importance of the phrase in the title of the chapter "the neurobiology of muscle fatigue".

SUMMARY

Throughout this epilogue, we have emphasized that rapid advances in understanding of neural and muscular aspects of fatigue have occurred since the 1980 London Symposium. However, in each instance of progress, from the single muscle fiber to the forebrain, the application of more precise techniques have raised important new questions. Neuroscientists and muscle physiologists have expanded opportunities for rigorous study of a topic of major scientific and social importance.

ACKNOWLEDGMENTS

Research by the authors is currently supported by: the National Health & Medical Research Council of Australia, and the Asthma Foundation of New South Wales, Australia (*S.C.G.*); United States Public Health Service (USPHS) grants AG 09000 and NS 20544 (*R.M.E.*); The Medical Research Council and Natural Sciences and Engineering Research Council (NSERC) of Canada (*A.J.McC.*); USPHS grants GM 08400, NS 20577, NS 01686 and NS 07309 (*D.G.S.*); and USPHS grant NS 30226 and the Miami Project to Cure Paralysis (*C.K.T.*). The authors' attendance at the 1994 Bigland-Ritchie conference was supported, in part, by NSERC (*A.J.McC.*), NS 30226 (*C.K.T.*), the Cleveland Clinic Foundation (*R.M.E.*), University of Arizona Regents' Professor funds (*D.G.S.*), the Muscular Dystrophy Association (USA; *S.C.G.*), and the University of Miami (*S.C.G.* and *A.J.McC.*).

REFERENCES

Ashley CC & Ridgway EB (1970). On the relationships between membrane potential, calcium transient and tension in single barnacle muscle fiber. *Journal of Physiology (London)* **209**, 105-130.

Atlan G, Beliveau L & Bouissou P (eds). (1991). *Muscle Fatigue: Biochemical and Physiological Aspects.* Paris: Masson.

Binder MD, Heckman CJ & Powers RK (1995). The physiological control of motoneuron activity. In: Rowell LB, Shepherd JT (eds.), *Handbook of Physiology, Integration of Motor, Respiratory and Metabolic Control During Exercise*, Sect. A, J. Smith (Sec. Ed.), *Neural Control of Movement,* pp. 00-00. Bethesda: American Physiological Society.

Binder MD & Mendell LM (eds.) (1990). *The Segmental Motor System*. New York: Oxford University Press.

Blinks JR, Rüdel R & Taylor SR (1978). Calcium transients in isolated amphibian skeletal muscle fibers: detection with aqueorin. *Journal of Physiology (London)* **277**, 291-323.

Enoka R & Stuart DG (1992). Neurobiology of muscle fatigue. *Journal of Applied Physiology* **72**, 1631-1648.

Finer JT, Simmons RM & Spudich JA (1994). Single myosin molecule mechanics: piconewton forces and nanometre steps. *Nature* **368**, 113-119.

Fuglevand AJ, Winter DA & Patla AE (1993). Models of recruitment and rate coding organization in motor-unit pools. *Journal of Neurophysiology* **70**, 2470-2488.

Gandevia SC (1995). Kinesthesia: roles for afferent signals and motor commands. In: Rowell LB, Shepherd JT (eds.), *Handbook on Integration of Motor, Circulatory, Respiratory and Metabolic Control during Exercise Sect. A,* J. Smith (Sec. Ed.), *Neural Control of Movement,* pp. 00-00. Bethesda: American Physiological Society.

Gandevia SC, Allen GM, Butler JE & Taylor JL (1995). Supraspinal factors in human muscle fatigue: evidence for suboptimal output from the motor cortex. *Journal of Physiology (London)* In press.

Gandevia SC & Burke D. (1992). Does the nervous system depend on kinesthetic information to control natural limb movements? *Behavioral and Brain Sciences* **15**, 614-632.

Gonzalez-Serratos H, Garcia M, Somlyo A, Somlyo AP & McClellan G (1981). Differential shortening of myofibrils during development of fatigue. *Biophysical Journal* **33**, 224a.

Hasan Z & Stuart DG (1988). Animal solutions to problems of movement control: the role of proprioceptors. *Annual Review of Neuroscience* **11**, 199-223.

Henneman E (1990). Henneman's contributions in historical perspective. In: Binder MD, Mendell, LM (eds.), *The Segmental Motor System*, pp. 3-19. New York: Oxford University Press.

Hultborn H & Kiehn O (1992). Neuromodulation of vertebrate motor neuron membrane properties. *Current Opinion in Neurobiology* **2**, 770-775.

Jankowska E (1992). Interneuronal relay in spinal pathways from proprioceptors. *Progress in Neurobiology* **38**, 335-378.

Kiehn O (1991). Plateau potentials and active integration in the 'final common pathway' for motoneuron behaviour. *Trends in Neurosciences* **14**, 68-73.

Loeb GE, He J & Levine WS (1989). Spinal cord circuits: are they mirrors of musculoskeletal mechanics? *Journal of Motor Behavior* **21**, 473-491.

McComas AJ (1995). Fatigue. Chapter 15. In: *Skeletal Muscle: Form and Function*, pp. 00-00. Champaign (IL): Human Kinetics.

McCrea DA (1992). Can sense be made of spinal interneuron circuits? *Behavioral and Brain Sciences* **15**, 633-643.

Merton PA (1953). Speculations on the servo-control of movement. In: Wolstenholme GEW (ed.), *The Spinal Cord*, pp. 247-260. London: Churchill.

Merton PA (1954). Voluntary strength and fatigue. *Journal of Physiology (London)* **123**, 553-564.

Nicholls TR (1994). A biomechanical perspective on spinal mechanisms of co-ordinated muscle action: an architecture principle. *Acta Anatomica* **151**, 1-13.

Porter R & Whelan J (1981). *Human Muscle Fatigue: Physiological Mechanisms* (CIBA Foundation Symposium 82). London: Pitman Medical.

Prochazka A (1995). The physiological control of motoneuron activity. In: Rowell LB, Shepherd JT (eds.), *Handbook Of Physiology, Integration of Motor, Respiratory and Metabolic Control during Exercise, Sect. A*, J. Smith (Sec. Ed.), *Neural Control of Movement,* pp. 00-00. Bethesda: American Physiological Society.

Rayment I, Holden HM, Whittaker M, Yohn CB, Lorenz M, Holmes KC & Mulligan RA (1992). Structure of the actin-myosin complex and its implications for muscle contraction. *Science* **261**, 58-65.

Rome LC (1993). The design of the neuromuscular system. In: Sargeant AJ, Kernell D (eds.), *Neuromuscular Fatigue,* pp. 129-136. Amsterdam: Royal Netherlands Academy of Arts and Sciences.

Sargeant AJ & Kernell D (1993). *Neuromuscular Fatigue.* Amsterdam: Royal Netherlands Academy of Arts and Sciences.

Stuart DG, Binder MD & Enoka RM (1984). Motor unit organization: Application of the quadripartite classification scheme to human muscles. In: Dyck PJ, Thomas PK, Lambert EH, Bunge R (eds.), *Peripheral Neuropathy,* pp. 1067-1090. Philadelphia: W.B. Saunders Co.

Stuart DG & Callister RJ (1993). Afferent and spinal reflex aspects of muscle fatigue: issues and speculations. In: Sargeant AJ, Kernell D (eds.), *Neuromuscular Fatigue,* pp. 169-180. Amsterdam: Royal Netherlands Academy of Arts and Sciences.

Thomas CK (1993). Muscle fatigue after incomplete human cervical spinal cord injury. *Journal of the American Paraplegia Society* **16**, 87.

CONTRIBUTORS

D.G. Allen
Department of Physiology
University of Sydney
NSW 2006
Australia

G.M. Allen
Prince of Wales Medical Research Institute
High Street, Randwick
Sydney NSW 2031
Australia

M.A. Babcock
The John Rankin Laboratory of Pulmonary Medicine
University of Wisconsin-Madison
Department of Preventive Medicine
Madison, Wisconsin 53705-2368

F. Bellemare
Département d'Anesthésie
Hôtel-Dieu de Montréal
3840 St. Urbain
Montréal, Québec
Canada

L. Bertocci
Institute for Exercise and Environmental Medicine
Presbyterian Hospital of Dallas
7232 Greenville Avenue
Dallas, Texas 75231-5129

B. Bigland-Ritchie
Department of Pediatrics
Yale University
P.O. Box 208-064
New Haven, Connecticut 06520

M.D. Binder
Department of Physiology & Biophysics, SJ-40
University of Washington
School of Medicine
Seattle, Washington 98195

S.A. Binder-Macleod
Department of Physical Therapy
University of Delaware
315 McKinly Laboratory
Newark, Delaware 19716

E. Blomstrand
Division of Technology and Development
Pripps Brewery
Bryggerivägen 10
S-161 86 Bromma
Sweden

G. Boorman
Departments of Clinical Neurosciences and Medical Physiology
The University of Calgary
Faculty of Medicine
3330 Hospital Drive N.W.
Calgary, Alberta T2N 4N1
Canada

B.R. Botterman
Department of Cell Biology and Neuroscience
University of Texas Southwestern Medical Center
Dallas, Texas 75235

S.V. Brooks
Institute of Gerontology and Bioengineering Program
University of Michigan
300 N. Ingalls
Ann Arbor, Michigan 48109-2007

P.M. Clarkson
Department of Exercise Science
Boyden Building, Box 31020
University of Massachusetts
Amherst, Massachusetts 01003

J.A. Dempsey
The John Rankin Laboratory of Pulmonary Medicine
University of Wisconsin-Madison
Department of Preventive Medicine
Madison, Wisconsin 53705-2368

R. Dengler
Department of Neurology and Clinical Neurophysiology
Medical School of Hannover
D-30623 Hannover
Germany

J. Dushanova
Institute of Physiology
Bulgarian Academy of Sciences
Acad. G. Bonchev Str., Bld. 23
1113 Sofia
Bulgaria

K.A.P. Edman
Department of Pharmacology
University of Lund
Sölvegatan 10
S-223 62 Lund
Sweden

R.H.T. Edwards
Muscle Research Centre
Department of Medicine
University of Liverpool
P.O. Box 147
Liverpool L69 3BX
United Kingdom

J.M. Elek
Department of Neurology and Clinical Neurophysiology
Medical School of Hannover
D-30623 Hannover
Germany

R.M. Enoka
Department of Biomedical Engineering
The Cleveland Clinic Foundation
9500 Euclid Avenue
Cleveland, Ohio 44195-5254

J.A. Faulkner
Institute of Gerontology and Bioengineering Program
University of Michigan
300 N. Ingalls
Ann Arbor, Michigan 48109-2007

M.W. Fryer
School of Physiology and Pharmacology
University of New South Wales
Kensington, NSW 2033
Australia

A.J. Fuglevand
The John B. Pierce Laboratory
290 Congress Ave.
New Haven, Connecticut 06519-1403

S.C. Gandevia
Prince of Wales Medical Research Institute
High Street, Randwick
Sydney NSW 2031
Australia

S.J. Garland
Department of Physical Therapy
The University of Western Ontario
Elborn College
London Ontario N6C 1H1
Canada

H. Gibson
Muscle Research Centre
Department of Medicine
University of Liverpool
P.O. Box 147
Liverpool L69 3BX
United Kingdom

T. Gordon
Department of Pharmacology
Division of Neurosciences
525 Heritage Medical Research Centre
University of Alberta
Edmonton T6G 2S2
Canada

K-E. Hagbarth
Department of Clinical Neurophysiology
University Hospital
S-751 85 Uppsala
Sweden

A.R. Hargens
Life Science Division
NASA Ames Research Center
Moffett Field, California 94035-1000

H. Hoppeler
Department of Anatomy
University of Bern
CH-3000 Bern
Switzerland

D.A. Jones
School of Sport and Exercise Sciences
The University of Birmingham
Edgbaston, Birmingham B15 2TT
United Kingdom

L.A. Jones
Department of Mechanical Engineering
77 Massachusetts Ave
Massachusetts Institute of Technology
Cambridge, Massachusetts 02139-4307

M.P. Kaufman
Division of Cardiovascular Medicine
Departments of Internal Medicine and Human Physiology
University of California
Davis, California 95616-8636

J.A. Kent-Braun
UCSF Medical Center
Veterans Administration
3150 Clement Street (11M)
San Francisco, California 94121-1676

D. Kernell
Department of Medical Physiology
University of Groningen
Bloemsingel 10
9712 KZ Groningen
The Netherlands

M.J. Kushmerick
NMR Research Laboratory
Department of Radiology, SB-05
University of Washington
Seattle, Washington 98195

G.D. Lamb
School of Zoology
La Trobe University
Bundoora
Victoria 3083
Australia

J. Lännergren
Department of Physiology and Pharmacology
Karolinska Institutet
S-171 77 Stockholm
Sweden

S.L. Lindstedt
Department of Biological Sciences
Northern Arizona University
Flagstaff, Arizona 86011-5640

V.G. Macefield
Prince of Wales Medical Research Institute
High Street, Randwick
Sydney 2031
Australia

A.J. McComas
Department of Biomedical Sciences
McMaster Health Science Centre
1200 Main Street West
Hamilton
Ontario L8N 3Z5
Canada

D.K. McKenzie
Department of Respiratory Medicine
Prince of Wales Hospital
High Street, Randwick
Sydney NSW 2031
Australia

T.S. Miles
Department of Physiology
University of Adelaide
Adelaide SA 5005
Australia

R.G. Miller
Department of Neurology
California Pacific Medical Center
3698 California St., Rm 545
San Francisco, California 94118-1702

A. Mineva
Institute of Physiology
Bulgarian Academy of Sciences
Acad. G. Bonchev Str., Bld. 23
1113 Sofia
Bulgaria

D.J. Newham
Physiotherapy Group
Biomedical Sciences Division
King's College
London, England SE5 9RS
United Kingdom

E. Newsholme
Department of Biochemistry
University of Oxford
South Parks Rd
Oxford, OX1 3QV
United Kingdom

M.A. Nordstrom
Department of Physiology
University of Adelaide
Adelaide SA 5005
Australia

D. Popivanov
Institute of Physiology
Bulgarian Academy of Sciences
Acad. G. Bonchev Str., Bld. 23
1113 Sofia
Bulgaria

R.K. Powers
Department of Physiology & Biophysics, SJ-40
University of Washington
School of Medicine
Seattle, Washington 98195

Y.S. Prakash
Departments of Anesthesiology & Physiology and Biophysics
Mayo Clinic and Foundation
Rochester, Minnesota 55905

C.L. Rice
School of Physical & Occupational Therapy and Dept. of Anatomy
McGill University
Montreal, Quebec
Canada

A.J. Sargeant
Department of Muscle and Exercise Physiology
Faculty of Human Movement Sciences
Vrije University
Van der Boechorstraat 9
1081 BT Amsterdam
The Netherlands

A. Sawczuk
Department of Oral Biology
University of Washington
Seattle, Washington 98195

O.M. Sejersted
Institute for Experimental Medical Research
The University of Oslo
Ullevaal Hospital
N-0407 Oslo
Norway

K.R. Sharma
Department of Neurology
University of Miami
Miami, Florida 33136

G.C. Sieck
Departments of Anesthesiology & Physiology and Biophysics
Mayo Clinic and Foundation
Rochester, Minnesota 55905

G. Sjøgaard
National Institute of Occupational Health
Lersø Parkalle 105
DK-2100 Copenhagen Ø
Denmark

E. Stålberg
Department of Clinical Neurophysiology
University Hospital
Uppsala
Sweden

D.G. Stephenson
School of Zoology
La Trobe University
Bundoora
Victoria 3083
Australia

G.M.M. Stephenson
Department of Chemistry and Biology
Victoria University of Technology
Footscray, Victoria 3011
Australia

D.G. Stuart
Department of Physiology
University of Arizona
College of Medicine
Tucson, Arizona 85724-5051

C.K. Thomas
The Miami Project to Cure Paralysis
University of Miami School of Medicine
1600 NW 10th Avenue, R-48
Miami, Florida 33136-1015

V. Toescu
Muscle Research Centre
Department of Medicine
University of Liverpool
P.O. Box 147
Liverpool L69 3BX
United Kingdom

J. V. Trontelj
University Institute of Clinical Neurophysiology
University Medical Center of Ljubljana
61105 Ljubljana, Zaloska 7
Slovenia

N.K. Vøllestad
Department of Physiology
National Institute of Occupational Health
Box 8149 Dep
N-0033 Oslo
Norway

M.L. Walsh
Department of Pediatrics
Yale University
P.O. Box 208-064
New Haven, Connecticut 06520

M.W. Weiner
Magnetic Resonance Unit
VA Medical Center
Departments of Medicine, Radiology and Psychiatry
University of California
San Francisco, California 94121

H. Westerblad
Department of Physiology and Pharmacology
Karolinska Institutet
S-171 77 Stockholm
Sweden

U. Windhorst
Departments of Clinical Neurosciences and Medical Physiology
The University of Calgary
Faculty of Medicine
3330 Hospital Drive N.W.
Calgary, Alberta T2N 4N1
Canada

INDEX

MIX
Papier aus verantwortungsvollen Quellen
Paper from responsible sources
FSC® C105338

If you have any concerns about our products,
you can contact us on
ProductSafety@springernature.com

In case Publisher is established outside the EU,
the EU authorized representative is:
Springer Nature Customer Service Center GmbH
Europaplatz 3, 69115 Heidelberg, Germany

Printed by Libri Plureos GmbH
in Hamburg, Germany